T0309655

Data Mining and Statistics for Decision Making

Wiley Series in Computational Statistics

Wiley Series in Computational Statistics is comprised of practical guides and cutting edge research books on new developments in computational statistics. It features quality authors with a strong applications focus. The texts in the series provide detailed coverage of statistical concepts, methods and case studies in areas at the interface of statistics, computing, and numerics.

With sound motivation and a wealth of practical examples, the books show in concrete terms how to select and to use appropriate ranges of statistical computing techniques in particular fields of study. Readers are assumed to have a basic understanding of introductory terminology.

The series concentrates on applications of computational methods in statistics to fields of bioinformatics, genomics, epidemiology, business, engineering, finance and applied statistics.

Titles in the Series

Biegler, Biros, Ghattas, Heinkenschloss, Keyes, Mallick, Marzouk, Tenorio, Waanders, Willcox – Large-Scale Inverse Problems and Quantification of Uncertainty
Billard and Diday – Symbolic Data Analysis: Conceptual Statistics and Data Mining
Bolstad – Understanding Computational Bayesian Statistics
Borgelt, Steinbrecher and Kruse – Graphical Models, 2e
Dunne – A Statistical Approach to Neutral Networks for Pattern Recognition
Liang, Liu and Carroll – Advanced Markov Chain Monte Carlo Methods
Ntzoufras – Bayesian Modeling Using WinBUGS

Data Mining and Statistics for Decision Making

Stéphane Tufféry

University of Rennes, France

Translated by Rod Riesco

A John Wiley & Sons, Ltd., Publication

First published under the title 'Data Mining et Statistique Decisionnelle' by Editions Technip

Authorised translation from French language edition published by Editions Technip, 2008

This edition first published 2011

Registered office
John Wiley & Sons Ltd, The Atrium, Southern Gate, Chichester, West Sussex, PO19 8SQ, United Kingdom

For details of our global editorial offices, for customer services and for information about how to apply for permission to reuse the copyright material in this book please see our website at www.wiley.com.

Library of Congress Cataloging-in-Publication Data

Tufféry, Stéphane
Data mining and statistics for decision making / Stéphane Tufféry.
p. cm. – (Wiley series in computational statistics)
Includes bibliographical references and index.
ISBN 978-0-470-68829-8 (hardback)
1. Data mining. 2. Statistical decision. I. Title.
QA76.9.D343T84 2011
006.3'12–dc22 2010039789

A catalogue record for this book is available from the British Library.

Print ISBN: 978-0-470-68829-8
ePDF ISBN: 978-0-470-97916-7
oBook ISBN: 978-0-470-97917-4
ePub ISBN: 978-0-470-97928-0

Typeset in by 10/12pt Times Roman by Thomson Digital, Noida, India

to Paul and Nicole Tufféry,
with gratitude and affection

Contents

Preface

All models are wrong but some are useful.
George E. P. Box[1]

[Data analysis] is a tool for extracting the jewel of truth from the slurry of data.
Jean-Paul Benzécri[2]

This book is concerned with data mining, which is the application of the methods of statistics, data analysis and machine learning to the exploration and analysis of large data sets, with the aim of extracting new and useful information for the benefit of the owner of these data.

An essential component of decision assistance systems in many economic, industrial, scientific and medical fields, data mining is being applied in an increasing variety of areas. The most familiar applications include market basket analysis in the retail and distribution industry (to find out which products are bought at the same time, enabling shelf arrangements and promotions to be planned accordingly), scoring in financial establishments (to predict the risk of default by an applicant for credit), consumer propensity studies (to target mailshots and telephone calls at customers most likely to respond favourably), prediction of attrition (loss of a customer to a competing supplier) in the mobile telephone industry, automatic fraud detection, the search for the causes of manufacturing defects, analysis of road accidents, assistance to medical prognosis, decoding of the genome, sensory analysis in the food industry, and others.

The present expansion of data mining in industry and also in the academic sphere, where research into this subject is rapidly developing, is ample justification for providing an accessible general introduction to this technology, which promises to be a rich source of future employment and which was presented by the Massachusetts Institute of Technology in 2001 as one of the ten emerging technologies expected to 'change the world' in the twenty-first century.[3]

This book aims to provide an introduction to data mining and its contribution to organizations and businesses, supplementing the description with a variety of examples. It details the methods and algorithms, together with the procedures and principles, for implementing data mining. I will demonstrate how the methods of data mining incorporate and extend the conventional methods of statistics and data analysis, which will be described reasonably thoroughly. I will therefore cover conventional methods (clustering, factor analysis, linear regression, ridge regression, partial least squares regression, discriminant

[1] Box, G.E.P. (1979) Robustness in the strategy of scientific model building. In R.L. Launer and G.N. Wilkinson (eds), *Robustness in Statistics*. New York: Academic Press.

[2] Benzécri, J.-P. (1976) *Histoire et Préhistoire de l'Analyse des Données*. Paris: Dunod.

[3] In addition to data mining, the other nine major technologies of the twenty-first century according to MIT are: biometrics, voice recognition, brain interfaces, digital copyright management, aspect-oriented programming, microfluidics, optoelectronics, flexible electronics and robotics.

analysis, logistic regression, the generalized linear model) as well as the latest techniques (decision trees, neural networks, support vector machines and genetic algorithms). We will take a look at recent and increasingly sophisticated methods such as model aggregation by bagging and boosting, the lasso and the 'elastic net'. The methods will be compared with each other, revealing their advantages, their drawbacks, the constraints on their use and the best areas for their application. Particular attention will be paid to scoring, which is still the most widespread application of predictive data mining methods in the service sector (banking, insurance, telecommunications), and fifty pages of the book are concerned with a comprehensive credit scoring case study. Of course, I also discuss other predictive techniques, as well as descriptive techniques, ranging from market basket analysis, in other words the detection of association rules, to the automatic clustering method known in marketing as 'customer segmentation'. The theoretical descriptions will be illustrated by numerous examples using SAS, IBM SPSS and R software, while the statistical basics required are set out in an appendix at the end of the book.

The methodological part of the book sets out all the stages of a project, from target setting to the use of models and evaluation of the results. I will indicate the requirements for the success of a project, the expected return on investment in a business setting, and the errors to be avoided.

This survey of new data analysis methods is completed by an introduction to text mining and web mining.

The criteria for choosing a statistical or data mining program and the leading programs available will be mentioned, and I will then introduce and provide a detailed comparison of the three major products, namely the free R software and the two market leaders, SAS and SPSS.

Finally, the book is rounded off with suggestions for further reading and an index.

This is intended to be both a reference book and a practical manual, containing more technical explanations and a greater degree of theoretical underpinning than works oriented towards 'business intelligence' or 'database marketing', and including more examples and advice on implementation than a volume dealing purely with statistical methods.

The book has been written with the following facts in mind. Pure statisticians may be reluctant to use data mining techniques in a context extending beyond that of conventional statistics because of its methods and philosophy and the nature of its data, which are frequently voluminous and imperfect (see Section A.1.2 in Appendix A). For their part, database specialists and analysts do not always make the best use of the data mining tools available to them, because they are unaware of their principles and operation. This book is aimed at these two groups of readers, approaching technical matters in a sufficiently accessible way to be usable with a minimum of mathematical baggage, while being sufficiently precise and rigorous to enable the user of these methods to master them and exploit them fully, without disregarding the problems encountered in the daily use of statistics. Thus, being based on both theoretical and practical knowledge, this book is aimed at a wide range of readers, including:

- statisticians working in private and public businesses, who will use it as a reference work alongside their statistical or data mining software manuals;
- students and teachers of statistics, econometrics or engineering, who can use it as a source of real applications of their statistical learning;

- analysts and researchers in the relevant departments of companies, who will discover what data mining can do for them and what they can expect from data miners and other statisticians;
- chief executive and IT managers which may use it a source of ideas for productive investment in the analysis of their databases, together with the conditions for success in data mining projects;
- any interested reader, who will be able to look behind the scenes of the computerized world in which we live, and discover how our personal data are used.

It is the aim of this book to be useful to the expert and yet accessible to the newcomer.

My thanks are due, in the first place, to David Hand, who found the time to carefully read my manuscript, give me his precious advice on several points and write a very interesting and kind foreword for the English edition, and to Gilbert Saporta, who has done me the honour of writing the foreword of the original French edition, for his support and the enlightening discussions I have had with him. I sincerely thank Jean-Pierre Nakache for his many kind suggestions and constant encouragement. I also wish to thank Olivier Decourt for his useful comments on statistics in general and SAS in particular. I am grateful to Hervé Abdi for his advice on some points of the manuscript. I must thank Hervé Mignot and Grégoire de Lassence, who reviewed the manuscript and made many useful detailed comments. Thanks are due to Julien Fournel for his kind and always relevant contributions. I have not forgotten my friends in the field of statistics and my students, although there are too many of them to be listed in the space available. Finally, a special thought for my wife and children, for their invaluable patience and support during the writing of this book.

This book includes on accompanying website. Please visit www.wiley.com/go/decision_making for more information.

Foreword

It is a real pleasure to be invited to write the foreword to the English translation of Stéphane Tufféry's book *Data Mining and Statistics for Decision Making*.

Data mining represents the merger of a number of other disciplines, most notably statistics and machine learning, applied to the problem of squeezing illumination from large databases. Although also widely used in scientific applications – for example bioinformatics, astrophysics, and particle physics – perhaps the major driver behind its development has been the commercial potential. This is simply because commercial organisations have recognised the competitive edge that expertise in this area can give – that is, the business intelligence it provides - enabling such organisation to make better-informed and superior decisions.

Data mining, as a unique discipline, is relatively young, and as with other youngsters, it is developing rapidly. Although originally it was secondary analysis, focusing solely on large databases which had been collated for some other purpose, nowadays we find more such databases being collected with the specific aim of subjecting them to a data mining exercise. Moreover, we also see formal experimental design being used to decide what data to collect (for example, as with supermarket loyalty cards or bank credit card operations, where different customers receive different cards or coupons).

This book presents a comprehensive view of the modern discipline, and how it can be used by businesses and other organizations. It describes the special characteristics of commercial data from a range of application areas, serving to illustrate the extraordinary breadth of potential applications. Of course, different application domains are characterised by data with different properties, and the author's extensive practical experience is evident in his detailed and revealing discussion of a range of data, including transactional data, lifetime data, sociodemographic data, contract data, and other kinds.

As with any area of data analysis, the initial steps of cleaning, transforming, and generally preparing the data for analysis are vital to a successful outcome, and yet many books gloss over this fundamental step. I hate to think how many mistaken conclusions have been drawn simply because analysts ignored the fact that the data had missing values! This book gives details of these necessary first steps, examining incomplete data, aberrant values, extreme values, and other data distortion issues.

In terms of methodology, as well as the more standard and traditional tools, the book comes up to date with extensive discussions of neural networks, support vector machines, bagging and boosting, and other tools.

The discussion of eight common misconceptions in Chapter 13 will be particularly useful to newcomers to the area, especially business users who are uncertain about the legitimacy of their analyses. And I was struck by the observation, also in this chapter, that for a successful business data mining exercise, the whole company has to buy into the exercise. It is not something to be undertaken by geeks in a back room. Neither is it a one-off exercise, which can be undertaken and then forgotten about. Rather it is an ongoing process, requiring commitment from a wide range of people in an organisation. More generally, data mining is not a magic wand, which can be waved over a miscellaneous and disorganised pile of data, to miraculously extract understanding and insight. It is an advanced technology of painstaking analysis and careful probing, using highly sophisticated software tools. As with any other advanced technology, it needs to be applied with care and skill if meaningful results are to be obtained. This book very nicely illustrates this in its mix of high level coverage of general issues, deep discussions of methodology, and detailed explorations of particular application areas.

An attractive feature of the book is its discussion of some of the most important data mining software tools and its illustrations of these tools in practice. Other data mining books tend to focus either on the technical methodological aspects, or on a more superficial presentation of the results, often in the form of screen shots, from a particular software package. This book nicely intertwines the two levels, in a way which I am sure will be attractive to readers and potential users of the technology.

The detailed case study of scoring methods in Chapter 12 is excellent, as are the other two application areas discussed in some depth – text mining and web mining. Both of these have become very important areas in their own right, and hold out great promise for knowledge discovery.

This book will be an eye-opener to anyone approaching data mining for the first time. It outlines the methods and tools, and also illustrates very nicely how they are applied, to very good effect, in a variety of areas. It shows how data mining is an essential tool for the data based businesses of today. More than that, however, it also shows how data mining is the equivalent of past centuries' voyages of discovery.

David J. Hand
Imperial College, London, and Winton Capital Management

Foreword from the French language edition

It is a pleasure for me to write the foreword to the third edition of this book, whose popularity shows no sign of diminishing. It is most unusual for a book of this kind to go through three editions in such a short time. It is a clear indication of the quality of the writing and the urgency of the subject matter.

Once again, Stéphane Tufféry has made some important additions: there are now almost two hundred pages more than in the second edition, which itself was practically twice as long as the first. More than ever, this book covers all the essentials (and more) needed for a clear understanding and proper application of data mining and statistics for decision making. Among the new features in this edition, I note that more space has been given to the free R software, developments in support vector machines and new methodological comparisons.

Data mining and statistics for decision making are developing rapidly in the research and business fields, and are being used in many different sectors. In the twenty-first century we are swimming in a flood of statistical information (economic performance indicators, polls, forecasts of climate, population, resources, etc.), seeing only the surface froth and unaware of the nature of the underlying currents.

Data mining is a response to the need to make use of the contents of huge business databases; its aim is to analyse and predict the individual behaviour of consumers. This aspect is of great concern to us as citizens. Fortunately, the risks of abuse are limited by the law. As in other fields, such as the pharmaceutical industry (in the development of new medicines, for example), regulation does not simply rein in the efforts of statisticians; it also stimulates their activity, as in banking engineering (the new Basel II solvency ratio). It should be noted that this activity is one of those which is still creating employment and that the recent financial crisis has shown the necessity for greater regulation and better risk evaluation.

So it is particularly useful that the specialist literature is now supplemented by a clear, concise and comprehensive treatise on this subject. This book is the fruit of reflection, teaching and professional experience acquired over many years.

Technical matters are tackled with the necessary rigour, but without excessive use of mathematics, enabling any reader to find both pleasure and instruction here. The chapters are also illustrated with numerous examples, usually processed with SAS software (the author provides the syntax for each example), or in some cases with SPSS and R.

Although there is an emphasis on established methods such as factor analysis, linear regression, Fisher's discriminant analysis, logistic regression, decision trees, hierarchical or partitioning clustering, the latest methods are also covered, including robust regression, neural networks, support vector machines, genetic algorithms, boosting, arcing, and the like. Association detection, a data mining method widely used in the retail and distribution industry for market basket analysis, is also described. The book also touches on some less familiar, but proven, methods such as the clustering of qualitative data by similarity aggregation. There is also a detailed explanation of the evaluation and comparison of scoring models, using the ROC curve and the lift curve. In every case, the book provides exactly the right amount of theoretical underpinning (the details are given in an appendix) to enable the reader to understand the methods, use them in the best way, and interpret the results correctly.

While all these methods are exciting, we should not forget that exploration, examination and preparation of data are the essential prerequisites for any satisfactory modelling. One advantage of this book is that it investigates these matters thoroughly, making use of all the statistical tests available to the user.

An essential contribution of this book, as compared with conventional courses in statistics, is that it provides detailed examples of how data mining forms part of a business strategy, and how it relates to information technology and the marketing of databases or other partners. Where customer relationship management is concerned, the author correctly points out that data mining is only one element, and the harmonious operation of the whole system is a vital requirement. Thus he touches on questions that are seldom raised, such as: What do we do if there are not enough data (there is an entertaining section on 'forename scoring')? What is a generic score? What are the conditions for correct deployment in a business? How do we evaluate the return on investment? To guide the reader, Chapter 2 also provides a summary of the development of a data mining project.

Another useful chapter deals with software; in addition to its practical usefulness, this contains an interesting comparison of the three major competitors, namely R, SAS and SPSS.

Finally, the reader may be interested in two new data mining applications: text mining and web mining.

In conclusion, I am sure that this very readable and instructive book will be valued by all practitioners in the field of statistics for decision making and data mining.

Gilbert Saporta
Chair of Applied Statistics
National Conservatory of Arts and Industries, Paris

List of trademarks

SAS®, SAS/STAT®, SAS/GRAPH®, SAS/Insight®, SAS/OR®, SAS/IML®, SAS/ETS®, SAS® High-Performance Forecasting, SAS® Enterprise Guide, SAS® Enterprise Miner™, SAS® Text Miner and SAS® Web Analytics are trademarks of SAS Institute Inc., Cary, NC, USA.

IBM® SPSS® Statistics, IBM® SPSS® Modeler, IBM® SPSS® Text Analytics, IBM® SPSS® Modeler Web Mining and IBM® SPSS® AnswerTree® are trademarks or registered trademarks of International Business Machines Corp., registered in many jurisdictions worldwide.

SPAD® is a trademark of Coheris-SPAD, Suresnes, France.

DATALAB® is a trademark of COMPLEX SYSTEMS, Paris, France.

1

Overview of data mining

This first chapter defines data mining and sets out its main applications and contributions to database marketing, customer relationship management and other financial, industrial, medical and scientific fields. It also considers the position of data mining in relation to statistics, which provides it with many of its methods and theoretical concepts, and in relation to information technology, which provides the raw material (data), the computing resources and the communication channels (the output of the results) to other computer applications and to the users. We will also look at the legal constraints on personal data processing; these constraints have been established to protect the individual liberties of people whose data are being processed. The chapter concludes with an outline of the main factors in the success of a project.

1.1 What is data mining?

Data mining and statistics, formerly confined to the fields of laboratory research, clinical trials, actuarial studies and risk analysis, are now spreading to numerous areas of investigation, ranging from the infinitely small (genomics) to the infinitely large (astrophysics), from the most general (customer relationship management) to the most specialized (assistance to pilots in aviation), from the most open (e-commerce) to the most secret (prevention of terrorism, fraud detection in mobile telephony and bank card applications), from the most practical (quality control, production management) to the most theoretical (human sciences, biology, medicine and pharmacology), and from the most basic (agricultural and food science) to the most entertaining (audience prediction for television). From this list alone, it is clear that the applications of data mining and statistics cover a very wide spectrum. The most relevant fields are those where large volumes of data have to be analysed, sometimes with the aim of rapid decision making, as in the case of some of the examples given above. Decision assistance is becoming an objective of data mining and statistics; we now expect these techniques to do more than simply provide a model of reality to help us to understand it. This approach is not completely new, and is already established in medicine, where some treatments have been developed on the basis of statistical analysis, even though the biological mechanism of the disease is little understood because of its

---

Data Mining and Statistics for Decision Making, First Edition. Stéphane Tufféry.

complexity, as in the case of some cancers. Data mining enables us to limit human subjectivity in decision-making processes, and to handle large numbers of files with increasing speed, thanks to the growing power of computers.

A survey on the www.kdnuggets.com portal in July 2005 revealed the main fields where data mining is used: banking (12%), customer relationship management (12%), direct marketing (8%), fraud detection (7%), insurance (6%), retail (6%), telecommunications (5%), scientific research (4%), and health (4%).

In view of the number of economic and commercial applications of data mining, let us look more closely at its contribution to 'customer relationship management'.

In today's world, the wealth of a business is to be found in its customers (and its employees, of course). Customer share has replaced market share. Leading businesses have been valued in terms of their customer file, on the basis that each customer is worth a certain (large) amount of euros or dollars. In this context, understanding the expectations of customers and anticipating their needs becomes a major objective of many businesses that wish to increase profitability and customer loyalty while controlling risk and using the right channels to sell the right product at the right time. To achieve this, control of the information provided by customers, or information about them held by the company, is fundamental. This is the aim of what is known as customer relationship management (CRM). CRM is composed of two main elements: operational CRM and analytical CRM.

The aim of analytical CRM is to extract, store, analyse and output the relevant information to provide a comprehensive, integrated view of the customer in the business, in order to understand his profile and needs more fully. The raw material of analytical CRM is the data, and its components are the data warehouse, the data mart, multidimensional analysis (online analytical processing[1]), data mining and reporting tools.

For its part, operational CRM is concerned with managing the various channels (sales force, call centres, voice servers, interactive terminals, mobile telephones, Internet, etc.) and marketing campaigns for the best implementation of the strategies identified by the analytical CRM. Operational CRM tools are increasingly being interfaced with back office applications, integrated management software, and tools for managing workflow, agendas and business alerts. Operational CRM is based on the results of analytical CRM, but it also supplies analytical CRM with data for analysis. Thus there is a data 'loop' between operational and analytical CRM (see Figure 1.1), reinforced by the fact that the multiplication of communication channels means that customer information of increasing richness and complexity has to be captured and analysed.

The increase in surveys and technical advances make it necessary to store ever-greater amounts of data to meet the operational requirements of everyday management, and the global view of the customer can be lost as a result. There is an explosive growth of reports and charts, but 'too much information means no information', and we find that we have less and less knowledge of our customers. The aim of data mining is to help us to make the most of this complexity.

It makes use of databases, or, increasingly, data warehouses,[2] which store the profile of each customer, in other words the totality of his characteristics, and the totality of his past and

---

[1] Data storage in a cube with n dimensions (a 'hypercube') in which all the intersections are calculated in advance, so as to provide a very rapid response to questions relating to several axes, such as the turnover by type of customer and by product line.

[2] A *data warehouse* is a set of databases with suitable properties for decision making: the data are thematic, consolidated from different production information systems, user-oriented, non-volatile, documented and possibly aggregated.

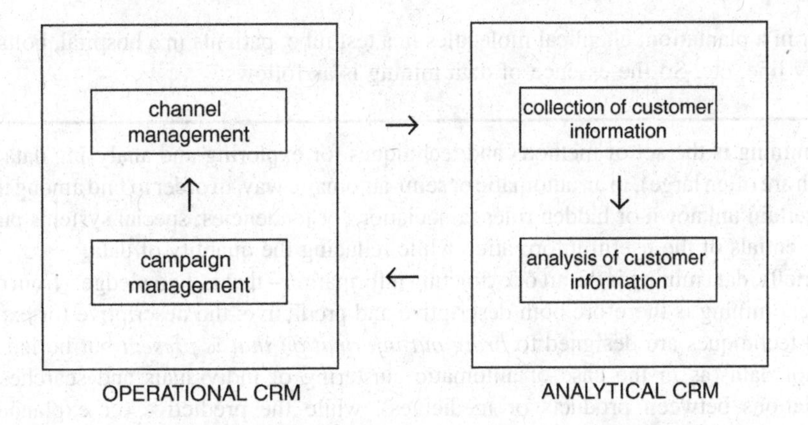

Figure 1.1 The customer relationship circuit.

present agreements and exchanges with the business. This global and historical knowledge of each customer enables the business to consider an individual approach, or 'one-to-one marketing',[3] as in the case of a corner shop owner 'who knows his customers and always offers them what suits them best'. The aim of this approach is to improve the customer's satisfaction, and consequently his loyalty, which is important because it is more expensive (by a factor of 3–10) to acquire a new customer than to retain an old one, and the development of consumer comparison skills has led to a faster customer turnover. The importance of customer loyalty can be appreciated if we consider that an average supermarket customer spends about €200 000 in his lifetime, and is therefore 'potentially' worth €200 000 to a major retailer.

Knowledge of the customer is even more useful in the service industries, where products are similar from one establishment to the next (banking and insurance products cannot be patented), where the price is not always the decisive factor for a customer, and customer relations and service make all the difference.

However, if each customer were considered to be a unique case whose behaviour was irreducible to any model, he would be entirely unpredictable, and it would be impossible to establish any proactive relationship with him, in other words to offer him whatever may interest him at the time when he is likely to be interested, rather than anything else. We may therefore legitimately wish to compare the behaviour of a customer whom we know less well (for a first credit application, for example) with the behaviour of customers whom we know better (those who have already repaid a loan). To do this, we need two types of data. First of all, we need 'customer' data which tell us whether or not two customers resemble each other. Secondly, we need data relating to the phenomenon to be predicted, which may be, for example, the results of early commercial activities (for what are known as propensity scores) or records of incidents of payment and other events (for risk scores). A major part of data mining is concerned with modelling the past in order to predict the future: we wish to find rules concealed in the vast body of data held on former customers, in order to apply them to new customers and take the best possible decisions. Clearly, everything I have said about the customers of a business is equally applicable to bacterial strains in a laboratory, types of

[3] Or, more modestly and realistically, 'one-to-few'.

fertilizer in a plantation, chemical molecules in a test tube, patients in a hospital, bolts on an assembly line, etc. So the essence of data mining is as follows:

> Data mining is the set of methods and techniques for exploring and analysing data sets (which are often large), in an automatic or semi-automatic way, in order to find among these data certain unknown or hidden rules, associations or tendencies; special systems output the essentials of the useful information while reducing the quantity of data.
>
> Briefly, data mining is the art of extracting information – that is, knowledge – from data.
>
> Data mining is therefore both descriptive and predictive: the descriptive (or exploratory) techniques are designed to *bring out information that is present* but buried in a mass of data (as in the case of automatic clustering of individuals and searches for associations between products or medicines), while the predictive (or explanatory) techniques are designed to *extrapolate new information based on the present information*, this new information being qualitative (in the form of classification or scoring[4]) or quantitative (regression).

The rules to be found are of the following kind:

- Customers with a given profile are most likely to buy a given product type.
- Customers with a given profile are more likely to be involved in legal disputes.
- People buying disposable nappies in a supermarket after 6 p.m. also tend to buy beer (a example which is mythical as well as apocryphal).
- Customers who have bought product A and product B are most likely to buy product C at the same time or *n* months later.
- Customers who have behaved in a given way and bought given products in a given time interval may leave us for the competition.

This can be seen in the last two examples: we need a history of the data, a kind of moving picture, rather than a still photograph, of each customer. All these examples also show that data mining is a key element in CRM and one-to-one marketing (see Table 1.1).

1.2 What is data mining used for?

Many benefits are gained by using rules and models discovered with the aid of data mining, in numerous fields.

1.2.1 Data mining in different sectors

It was in the *banking sector* that risk scoring was first developed in the mid-twentieth century, at a time when computing resources were still in their infancy. Since then, many data mining techniques (scoring, clustering, association rules, etc.) have become established in both retail and commercial banking, but data mining is especially suitable for retail banking because of

[4] The statistical technique is called 'classification' or 'discrimination'; the application of this technique to certain business problems such as the selection of customers according to certain criteria is called 'scoring'.

Table 1.1 Comparison between traditional and one-to-one marketing.

Traditional marketing	One-to-one marketing
Anonymous customer	Individually identified customer
Standard product	Personalized product and service
Serial production	Bespoke production
Mass advertising	Individual message
Unilateral communication	Interactive communication
Achievement of a sale, high take-up	Development of customer loyalty, low attrition rate
Market share	Customer share
Broad targets	Profitable niches
Segmentation by job and RFM	Statistical, behavioural segmentation
Traditional distribution channels, disconnected from each other	New, interconnected channels (telephone platforms, Internet, mobile telephones)
Product-oriented marketing	Customer-oriented marketing

the moderate unitary amounts, the large number of files and their relatively standard form. The problems of scoring are generally not very complicated in theoretical terms, and the conventional techniques of discriminant analysis and logistic regression have been extremely successful here. This expansion of data mining in banking can be explained by the simultaneous operation of several factors, namely the development of new communication technology (Internet, mobile telephones, etc.) and data processing systems (data warehouses); customers' increased expectations of service quality; the competitive challenge faced by retail banks from credit companies and 'newcomers' such as foreign banks, major retailers and insurance companies, which may develop banking activities in partnership with traditional banks; the international economic pressure for higher profitability and productivity; and of course the legal framework, including the current major banking legislation to reform the solvency ratio (see Section 12.2), which has been a strong impetus to the development of risk models. In banks, loyalty development and attrition scoring have not been developed to the same extent as in mobile telephones, for instance, but they are beginning to be important as awareness grows of the potential profits to be gained. For a time, they were also stimulated by the competition of on-line banks, but these businesses, which had lower structural costs but higher acquisition costs than branch-based banks, did not achieve the results expected, and have been bought up by insurance companies wishing to gain a foothold in banking, by foreign banks, or by branch-based banks aiming to supplement their multiple-channel banking system, with Internet facilities coexisting with, but not replacing, the traditional channels.

The *retail industry* is developing its own credit cards, enabling it to establish very large databases (of several million cardholders in some cases), enriched by behavioural information obtained from till receipts, and enabling it to compete with the banks in terms of customer knowledge. The services associated with these cards (dedicated check-outs, exclusive promotions, etc.) are also factors in developing loyalty. By detecting product associations on till receipts it is possible to identify customer profiles, make a better choice of products and arrange them more appropriately on the shelves, taking the 'regional' factor into account in the analyses. The most interesting results are obtained when payments are made with a loyalty card, not only because this makes it possible to cross-check the associations detected on the till receipts with sociodemographic information (age, family circumstances, socio-occupational category) provided by the customer when he joins the card scheme, but also because the use of the card makes it possible to monitor a customer's payments over time and to implement customer-targeted promotions, approaching the customer according to the time intervals and themes suggested by the model. Market baskets can also be segmented into groups such as 'clothing receipt', 'large trolley receipt', and the like.

In *property and personal insurance*, studies of 'cross-selling', 'up-selling' and attrition, with the adaptation of pricing to the risks incurred, are the main themes in a sector where propensity is not stated in the same terms as elsewhere, since certain products (motor insurance) are compulsory, and, except in the case of young people, the aim is either to attract customers from competitors, or to persuade existing customers to upgrade, by selling them additional optional cover, for example. The need for data mining in this sector has increased with the development of competition from new entrants in the form of banks offering what is known as 'bancassurance' (bank insurance), with the advantage of extended networks, frequent customer contact and rich databases. The advantages of this offer are especially great in comparison with 'traditional' non-mutual insurance companies which may encounter difficulties in developing marketing databases from information which is widely diffused and jealously guarded by their agents. Furthermore, the customer bases of these insurers, even if not divided by agent, are often structured according to contracts rather than customers. And yet these networks, with their lower loyalty rates than mutual organizations, have a real need to improve their CRM, and consequently their global knowledge of their customers. Although the propensity studies for insurance are similar to those for banking, the loss studies show some distinctive features, with the appearance of the Poisson distribution in the generalized linear model for modelling the number of claims (loss events). The insurers have one major asset in their holdings of fairly comprehensive data about their customers, especially in the form of home and civil liability insurance contracts which provide fairly accurate information on the family and its lifestyle.

The opening of the *landline telephone* market to European competition, and the development of the *mobile telephone* market through maturity to saturation, have revived the problems of 'churning' (switching to competing services) among private, professional and business customers. The importance of loyalty in this sector becomes evident when we consider that the average customer acquisition cost in the mobile telephone market is more than €200, and that more than a million users change their operator every year in some countries. Naturally, therefore, it is churn scoring that is the main application of data mining in the telephone business. For the same reasons, operators use *text mining* tools (see Chapter 14) for automatic analysis of the content of customers' letters of complaint. Other areas of investigation in the telephone industry are non-payment scoring, direct marketing optimization, behavioural analysis of Internet users and the design of call centres. The probability of a customer changing his mobile telephone is also under investigation.

Data mining is also quite widespread in the *motor industry*. A standard theme is scoring for repeat purchases of a manufacturer's vehicles. Thus, Renault has constructed a model which predicts customers who are likely to buy a new Renault car in the next six months. These customers are identified on the basis of data from concessionaires, who receive in return a list of high-scoring customers whom they can then contact. In the production area, data mining is used to trace the origin of faults in construction, so that these can be minimized. Satisfaction studies are also carried out, based on surveys of customers, with the aim of improving the design of vehicles (in terms of quality, comfort, etc.). Accidents are investigated in the laboratories of motor manufacturers, so that they can be classified in standard profiles and their causes can be identified. A large quantity of data is analysed, relating to the vehicle, the driver and the external circumstances (road condition, traffic, time, weather, etc.).

The *mail-order* sector has been conducting analyses of data on its customers for many years, with the aim of optimizing targeting and reducing costs, which may be very considerable when a thousand-page colour catalogue is sent to several tens of millions of customers. Whereas banking was responsible for developing risk scoring, the mail-order industry was one of the first sectors to use propensity scoring.

The *medical sector* has traditionally been a heavy user of statistics. Quite naturally, data mining has blossomed in this field, in both diagnostic and predictive applications. The first category includes the identification of patient groups suitable for specific treatment protocols, where each group includes all the patients who react in the same way. There are also studies of the associations between medicines, with the aim of detecting prescription anomalies, for example. Predictive applications include tracing the factors responsible for death or survival in certain diseases (heart attacks, cancer, etc.) on the basis of data collected in clinical trials, with the aim of finding the most appropriate treatment to match the pathology and the individual. Of course, use is made of the predictive method known as *survival analysis*, where the variable to be predicted is a period of time. Survival data are said to be 'censored', since the period is precisely known for individuals who have died, while it is only the minimum survival time that is known for those who remain. We can, for example, try to predict the recovery time after an operation, according to data on the patient (age, weight, height, smoker or non-smoker, occupation, medical history, etc.) and the practitioner (number of operations carried out, years of experience, etc.). *Image mining* is used in medical imaging for the automatic detection of abnormal scans or tumour recognition. Finally, the deciphering of the genome is based on major statistical research for detecting, for example, the effect of certain genes on the appearance of certain pathologies. These statistical analyses are difficult, as the number of explanatory variables is very high with respect to the number of observations: there may be several tens of millions of genes (genome) or pixels (image mining) relating to only a few hundred individuals. Methods such as partial least squares (PLS) regression or regularized regression (ridge, lasso) are highly valued in this field. The tracing of similar sequences ('sequence analysis') is widely used in genomics, where the DNA sequence of a gene is investigated with the aim of finding similarities between the sequences of a single ancestor which have undergone mutations and natural selection. The similarity of biological functions is deduced from the similarity of the sequences.

In cosmetics, Unilever has used data mining to predict the effect of new products on human skin, thus limiting the number of tests on animals, and L'Oréal, for example, has used it to predict the effects of a lotion on the scalp.

The *food industry* is also a major user of statistics. Applications include 'sensory analysis' in which sensory data (taste, flavour, consistency, etc.) perceived by consumers are correlated with physical and chemical instrumental measurements and with preferences for various products. Discriminant analysis and logistic regression predictive models are also used in the drinks industry to distinguish spirits from counterfeit products, based on the analysis of about ten molecules present in the beverage. Chemometrics is the extraction of information from physical measurements and from data collected in analytical chemistry. As in genomics, the number of explanatory variables soon becomes very great and may justify the use of PLS regression. Health risk analysis is specific to the food industry: it is concerned with understanding and controlling the development of microorganisms, preventing hazards associated with their development in the food industry, and managing use-by dates. Finally, as in all industries, it is essential to manage processes as well as possible in order to improve the quality of products.

Statistics are widely used in *biology*. They have been applied for many years for the classification of living species; we may, for example, quote the standard example of Fisher's use of his linear discriminant analysis to classify three species of iris. Agronomy requires statistics for an accurate evaluation of the effects of fertilizers or pesticides. Another currently fashionable use of data mining is for the detection of factors responsible for air pollution.

1.2.2 Data mining in different applications

In the field of customer relationship management, we can expect to gain the following benefits from statistics and data mining:

- identification of prospects most likely to become customers, or former customers most likely to return ('winback');
- calculation of profitability and *lifetime value* (see Section 4.2.2) of customers;
- identification of the most profitable customers, and concentration of marketing activities on them;
- identification of customers likely to leave for the competition, and marketing operations if these customers are profitable;
- better rate of response in marketing campaigns, leading to lower costs and less customer fatigue in respect of mailings;
- better cross-selling;
- personalization of the pages of the company website according to the profile of each user;
- commercial optimization of the company website, based on detection of the impact of each page;
- management of calls to the company's switchboard and direction to the correct support staff, according to the profile of the calling customer;
- choice of the best distribution channel;
- determination of the best locations for bank or major store branches, based on the determination of store profiles as a function of their location and the turnover generated by the different departments;
- in the retail industry, determination of consumer profiles, the 'market basket', the effect of sales or advertising; planning of more effective promotions, better prediction of demand to avoid stock shortages or unsold stock;

- telephone traffic forecasting;
- design of call centres;
- stimulating the reuse of a telephone card in a closely identified group of customers, by offering a reduction on three numbers of their choice;
- winning on-line customers for a telephone operator;
- analysis of customers' letters of complaint (using text data obtained by text mining – see Chapter 14);
- technology watching (use of text mining to analyse studies, specialist papers, patent filings, etc.);
- competitor monitoring.

In operational terms, the discovery of these rules enables the user to answer the questions 'who', 'what', 'when' and 'how' – who to sell to, what product to sell, when to sell it, how to reach the customer.

Perhaps the most typical application of data mining in CRM is propensity scoring, which measures the probability that a customer will be interested in a product or service, and which enables targeting to be refined in marketing campaigns. Why is propensity scoring so successful? While poorly targeted mailshots are relatively costly for a business, with the cost depending on the print quality and volume of mail, unproductive telephone calls are even more expensive (at least €5 per call). Moreover, when a customer has received several mailings that are irrelevant to him, he will not bother to open the next one, and may even have a poor image of the business, thinking that it pays no attention to its customers.

In *strategic marketing*, data mining can offer:

- help with the creation of packages and promotions;
- help with the design of new products;
- optimal pricing;
- a customer loyalty development policy;
- matching of marketing communications to each segment of the customer base;
- discovery of segments of the customer base;
- discovery of unexpected product associations;
- establishment of representative panels.

As a general rule, data mining is used to gain a better understanding of the customers, with a view to adapting the communications and sales strategy of the business.

In *risk management*, data mining is useful when dealing with the following matters:

- identifying the risk factors for claims in personal and property insurance, mainly motor and home insurance, in order to adapt the price structure;

- preventing non-payment of bills in the mobile telephone industry;
- assisting payment decisions in banks, for current accounts where overdrafts exceed the authorized limits;
- using the risk score to offer the most suitable credit limit for each customer in banks and specialist credit companies, or to refuse credit, depending on the probability of repayment according to the due dates and conditions specified in the contract;
- predicting customer behaviour when interest rates change (early credit repayment requests, for example);
- optimizing recovery and dispute procedures;
- automatic real-time fraud detection (for bank cards or telephone systems);
- detection of terrorist profiles at airports.

Automatic fraud detection can be used with a mobile phone which makes an unusually long call from or to a location outside the usual area. Real-time detection of doubtful bank transactions has enabled the Amazon on-line bookstore to reduce its fraud rate by 50% in 6 months. Chapter 12 will deal more fully with the use of risk scoring in banking.

A recent and unusual application of data mining is concerned with *judicial* risk. In the United Kingdom, the OASys (Offenders Assessment System) project aims to estimate the risk of repeat offending in cases of early release, using information on the family background, place of residence, educational level, associates, criminal record, social workers' reports and behaviour of the person concerned in custody and in prison. The British Home Secretary and social workers hope that OASys will standardize decisions on early release, which currently vary widely from one region to another, especially under the pressure of public opinion.

The *miscellaneous applications* of data mining and statistics include the following:

- road traffic forecasting, day by day or by hourly time slots;
- forecasting water or electricity consumption;
- determining whether a person owns or rents his home, when planning to offer insulation or installation of a heating system (Électricité de France);
- improving the quality of a telephone network (discovering why some calls are unsuccessful);
- quality control and tracing the causes of manufacturing defects, for example in the motor industry, or in companies such as the one which succeeded in explaining the sporadic appearance of defects in coils of steel, by analysing 12 parameters in 8000 coils during 30 days of production;
- use of survival analysis in industry, with the aim of predicting the life of a manufactured component;
- profiling of job seekers, in order to detect unemployed persons most at risk of long-term unemployment and provide prompt assistance tailored to their personal circumstances;

- pattern recognition in large volumes of data, for example in astrophysics, in order to classify a celestial object which has been newly discovered by telescope (the SKICAT system, applied to 40 measured characteristics);
- signal recognition in the military field, to distinguish real targets from false ones.

A rather more entertaining application of data mining relates to the prediction of the audience share of a television channel (BBC) for a new programme, according to the characteristics of the programme (genre, transmission time, duration, presenter, etc.), the programmes preceding and following it on the same channel, the programmes broadcast simultaneously on competing channels, the weather conditions, the time of year (season, holidays, etc.) and any major events or shows taking place at the same time. Based on a data log covering one year, a model was constructed with the aid of a neural network. It is able to predict audience share with an accuracy of ±4%, making it as accurate as the best experts, but much faster.

Data mining can also be used for its own internal purposes, by helping to determine the reliability of the databases that it uses. If an anomaly is detected in a data element X, a variable 'abnormal data element X (yes/no)' is created, and the explanation for this new variable is then found by using a decision tree to test all the data except X.

1.3 Data mining and statistics

In the commercial field, the questions to be asked are not only 'how many customers have bought this product in this period?' but also 'what is their profile?', 'what other products are they interested in?' and 'when will they be interested?'. The profiles to be discovered are generally complex: we are not dealing with just the 'older/younger', 'men/women', 'urban/rural' categories, which we could guess at by glancing through descriptive statistics, but with more complicated combinations, in which the discriminant variables are not necessarily what we might have imagined at first, and could not be found by chance, especially in the case of rare behaviours or phenomena. This is true in all fields, not only the commercial sector. With data mining, we move on from 'confirmatory' to 'exploratory' analysis.[5]

Data mining methods are certainly more complex than those of elementary descriptive statistics. They are based on artificial intelligence tools (neural networks), information theory (decision trees), machine learning theory (see Section 11.3.3), and, above all, inferential statistics and 'conventional' data analysis including factor analysis, clustering and discriminant analysis, etc.

There is nothing particularly new about exploratory data analysis, even in its advanced forms such as multiple correspondence analysis, which originated in the work of theoreticians such as Jean-Paul Benzécri in the 1960s and 1970s and Harold Hotelling in the 1930s and 1940s (see Section A.1 in Appendix A). Linear discriminant analysis, still used as a scoring method, first emerged in 1936 in the work of Fisher. As for the evergreen logistic regression,

[5] In an article of 1962 and a book published in 1977, J.W. Tukey, the leading American statistician, contrasts *exploratory* data analysis, in which the data take priority, with *confirmatory* data analysis, in which the model takes priority. See Tukey, J.W. (1977) *Exploratory Data Analysis*. Reading, MA: Addison-Wesley.

Pierre-François Verhulst anticipated this in 1838 and Joseph Berkson developed it from 1944 for biological applications.

The reasons why data mining has moved out of universities and research laboratories and into the world of business include, as we have seen, the pressures of competition and the new expectations of consumers, as well as regulatory requirements in some cases, such as pharmaceuticals (where medicines must be trialled before they are marketed), or banking (where the equity must be adjusted according to the amount of exposure and the level of risk incurred). This development has been made possible by three major technical advances.

The first of these concerns the storage and calculation capacity offered by modern computing equipment and methods: data warehouses with capacities of several tens of terabytes, massively parallel architectures, increasingly powerful computers.

The second advance is the increasing availability of 'packages' of different kinds of statistical and data mining algorithms in integrated software. These algorithms can be automatically linked to each other, with a user-friendliness, a quality of output and options for interactivity which were previously unimaginable.

The third advance is a step change in the field of decision making: this includes the use of data mining methods in production processes (where data analysis was traditionally used only for single-point studies), which may extend to the periodic output of information to end users (marketing staff, for example) and automatic event triggering.

These three advances have been joined by a fourth. This is the possibility of processing data of all kinds, including incomplete data (by using imputation methods), some aberrant data (by using 'robust' methods), and even text data (by using 'text mining'). Incomplete data – in other words, those with missing values – are found less commonly in science, where all the necessary data are usually measured, than in business, where not all the information about a customer is always known, either because the customer has not provided it, or because the salesman has not recorded it.

A fifth element has played a part in the development of data mining: this is the establishment of vast databases to meet the management requirements of businesses, followed by an awareness of the unexploited riches that these contain.

1.4 Data mining and information technology

An IT specialist will see a data mining model as an *IT application*, in other words a set of instructions written in a programming language to carry out certain processes, as follows:

- providing an output data element which summarizes the input data (e.g. a segment number);
- or providing an output data element of a new type, deduced from the input data and used for decision making (e.g. a score value).

As we have seen, the first of these processes corresponds to descriptive data mining, where the archetype is *clustering*: an individual's membership of a cluster is a summary of all of its *present* characteristics. The second example corresponds to predictive *data mining*, where the archetype is *scoring*: the new variable is a probability that the individual will behave in a certain way *in the future* (in respect of risk, consumption, loyalty, etc.).

Like all IT applications, a data mining application goes through a number of phases:

- development (construction of the model) in the decision-making environment;
- testing (verifying the performance of the model) in the decision-making environment;
- use in the production environment (application of the model to the production data to obtain the specified output data).

However, data mining has some distinctive features, as follows:

- The development phase cannot be completed in the absence of data, in contrast to an IT development which takes place according to a specification; the development of a model is primarily dependent on data (even if there is a specification as well).
- Development and testing are carried out in the same environment, with only the data sets differing from each other (as they must do!).
- To obtain an optimal model, it is both normal and necessary to move frequently between testing and development; some programs control these movements in a largely automatic way to avoid any loss of time.
- The data analysis for development and testing is carried out using a special-purpose program, usually designed by SAS, SPSS (IBM group), KXEN, Statistica or SPAD, or open source software (see Chapter 5).
- All these programs benefit from graphic interfaces for displaying results which justify the relevance of the developments and make them evident to users who are neither statisticians nor IT specialists.
- Some programs also offer the use of the model, which can be a realistic option if the program is implemented on a server (which can be done with the programs mentioned above).
- The conciseness of the data mining models: unlike the instructions of a computer program, which are often relatively numerous, the number of instructions in a data mining model is nearly always small (if we disregard the instructions for collecting the data to which the model is applied, since these are related to conventional data processing, even though there are special purpose tools), and indeed conciseness (or 'parsimony') is one of the sought-after qualities of a model (since it is considered to imply readability and robustness).

To some extent, the last two points are the inverse of each other. On the one hand, data mining models can be used in the same decision-making environment and with the same software as in the development phase, provided that the production data are transferred into this environment. On the other hand, the conciseness of the models means that they can be exported to a production environment that is different from the development environment, for example an IBM and DB2 mainframe environment, or Unix and Oracle. This solution may provide better performance than the first for the periodic processing of large bodies of data without the need for bulky transfers, or for calculating scores in real time (with inputting face to face with the customer), but it requires an export facility. The obvious advantage of the first

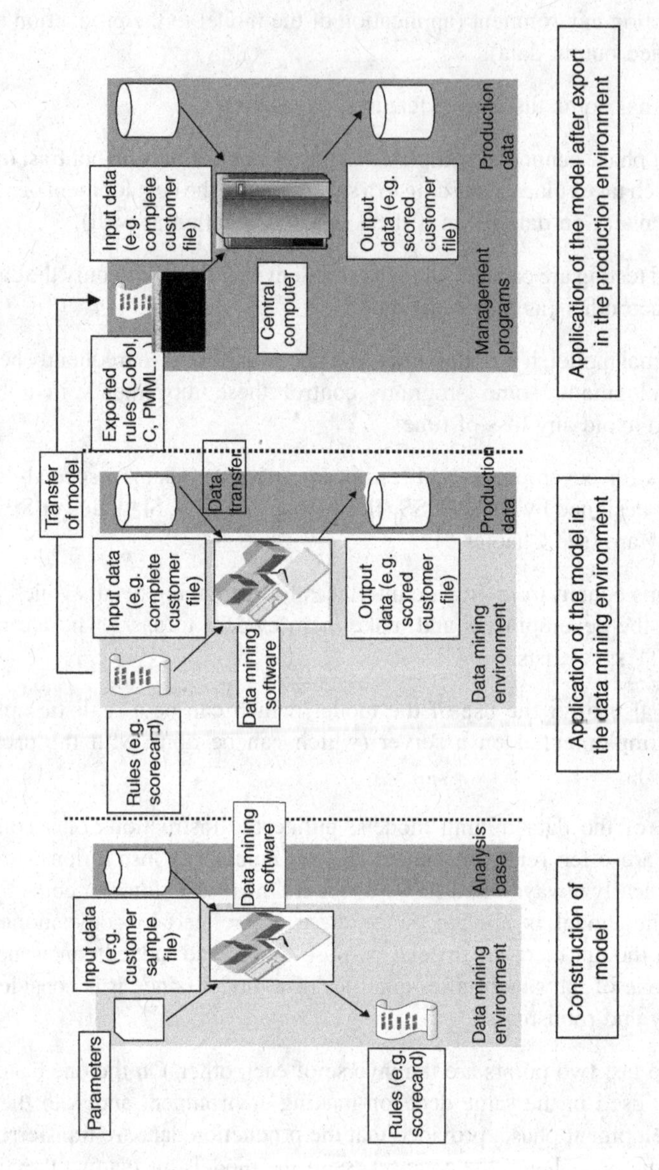

Figure 1.2 IT architecture for data mining.

solution is a gain in time in the implementation of the data mining processes. In the first solution, the data lead to the model; in the second, the model leads to the data (see Figure 1.2).

Some models are easily exported and reprogrammed in any environment. These are purely statistical models, such as *discriminant analysis* and *logistic regression*, although the latter requires the presence of an exponential function or the power function at least (which, it should be noted, is provided even in Cobol). These standard models are concise and high-performing, provided that they are used with care. In particular, it is advisable to work with a few carefully chosen variables, and to apply these models to relatively homogeneous populations, provided that a preliminary segmentation is carried out.

Here is an example of a logistic regression model, which supplies the 'score' probability of being interested in purchasing a certain product. The ease of export of this type of model will be obvious.

```
logit = 0.985 - (0.005*variable_W) + (0.019* variable_X) +
(0.122* variable_Y) - (0.002* variable_Z);
score = exponential(logit) / [1 + exponential(logit)];
```

Such a model can also be converted to a scoring grid, as shown in Section 12.8.

Another very widespread type of model is the *decision tree*. These models are very popular because of their readability, although they are not the most robust, as we shall see.

A very simple example (Figure 1.3) again illustrates the propensity to buy a product. The aim is to extend the branches of the tree until we obtain terminal nodes or *leaves* (at the end of the branches, although the leaves are at the bottom here and the root, i.e. the total sample, is at the top) which contain the highest possible percentage of 'yes' (propensity to buy) or 'no' (no propensity to buy).

The algorithmic representation of the tree is a set of rules (Figure 1.4), where each rule corresponds to the path from the root to one of the leaves. As we can see in this very simple example, the model soon becomes less concise than a statistical model, especially as real trees often have at least four or five depth levels. Exporting would therefore be rather more difficult if it were a matter of copying the rules 'manually', but most programs offer options for automatic translation of the rules into C, Java, SQL, PMML, etc.

Some clustering models, such as those obtained by the *moving centres* method or variants of it, are also relatively easy to reprogram in different IT environments. Figure 1.5 shows an example of this, produced by SAS, for clustering a population described by six variables into three clusters. Clearly, this is a matter of calculating the Euclidean distance separating each individual from each of the three clusters, and assigning the individual to the cluster to which he is closest (where CLScads[_clus] reaches a minimum).

However, not all clustering models can be exported so easily. Similarly, models produced by neural networks do not have a simple synthetic expression. To enable any type of model to be exported to any type of hardware platform, a universal language based on XML was created in 1998 by the Data Mining Group (www.dmg.org): it goes by the name of Predictive Model Markup Language (PMML). This language can describe the data dictionary used (variables, with their types and values) and the data transformations carried out (recoding, normalization, discretization, aggregation), and can use tags to specify the parameters of various types of model (regressions, trees, clustering, neural networks, etc.). By installing a PMML interpreter or relational databases, it is possible to deploy data mining models in an operating environment which may be different from the development environment. Moreover, these models can

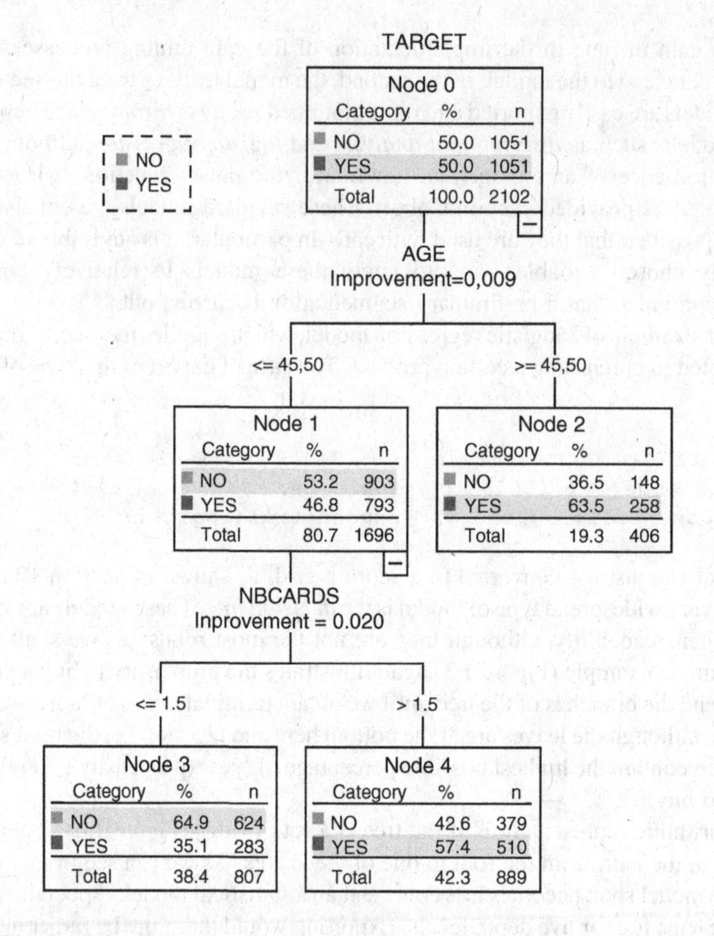

Figure 1.3 Example of a decision tree generated by Answer Tree.

be generated by different data mining programs (SAS, IBM SPSS, R, for example), since the PMML language tends to spread slowly, even though it remains less widespread and possibly less efficient than C, Java and SQL.

In R, for example, a decision tree is exported by using the *pmml* package (which also requires the *XML* package). The first step is to create the model, using the *rpart* tree function (Figure 1.6). The *pmml* package currently allows the export of models produced by linear regression, logistic regression, support vector machines, neural networks (the *nnet* package), decision tree (the *rpart* package), random forests and *k*-means.

I have mentioned three software packages in this section: SAS, IBM SPSS and R. These will be described in detail, with other data mining software, in Chapter 5.

1.5 Data mining and protection of personal data

Statisticians and data miners must comply with their national law with regard to the processing of personal data. Such data are defined as those which can be directly or indirectly related to an individual physical person by using his civil status, another identifier such as the

```
/* Node 3 */
DO IF (SYSMIS(age) OR (VALUE(age) LE 45.5)) AND
((VALUE(nbcards) LE 1.5) OR SYSMIS(nbcards)).
COMPUTE nod_001 = 3.
COMPUTE pre_001 = 'N'.
COMPUTE prb_001 = 0.649318.
END IF.
EXECUTE.

/* Node 4 */
DO IF (SYSMIS(age) OR (VALUE(age) LE 45.5)) AND
((VALUE(nbcards) GT 1.5) OR SYSMIS(nbcards)).
COMPUTE nod_001 = 4.
COMPUTE pre_001 = 'O'.
COMPUTE prb_001 = 0.573678.
END IF.
EXECUTE.

/* Node 2 */
DO IF (VALUE(age) GT 45.5).
COMPUTE nod_001 = 2.
COMPUTE pre_001 = 'O'.
COMPUTE prb_001 = 0.635468.
END IF.
EXECUTE.
```

Figure 1.4 Example of SPSS code for a decision tree.

telephone number of a customer or an assured party, or any other element belonging to him (voice, image, genetic or biometric fingerprints, address, etc.).[6]

Most countries have passed laws to restrict the collection, storage, processing and use of personal data, especially sensitive data, relating to health, sexual orientation, criminal convictions, racial origin, political opinions and religious faith. This is also the case in the European Union member states that have adopted European Directive 95/46/EC of 24 October 1995 into their national law. According to Article 6 of this directive, personal data must be:

(a) processed fairly and lawfully;

(b) collected for specified, explicit and legitimate purposes and not further processed in a way incompatible with those purposes. Further processing of data for historical, statistical or scientific purposes shall not be considered as incompatible provided that Member States provide appropriate safeguards;

[6] This does not apply to files on physical persons that are anonymized by the removal of all identifiers that could be used to trace them. Such files may be useful for statistical research.

```
*************************************;
*** begin scoring code for clustering;
*************************************;
label _SEGMNT_ = 'Cluster ID'
   Distance = 'Distance to Cluster Seed';
drop _nonmiss; _nonmiss = n(
  AGE , SAVINGS, NBPROD , EXPENDIT , INCOME , NBCARDS);
if _nonmiss = 0 then do;
  _SEGMNT_ _ = .; distance = .;
end;
else do;
  array _CLScads[3] _temporary_;
  drop _clus;
  do _clus = 1 to 3; _CLScads[_clus] = 0; end;
  if n(AGE) then do;
    _CLScads[1] + (AGE - 69.111111111 )**2;
    _CLScads[2] + (AGE - 70.095238095 )**2;
    _CLScads[3] + (AGE - 43.473900586 )**2;
  end;
  if n(SAVINGS) then do;
    _CLScads[1] + (SAVINGS - 383125.82523 )**2;
    _CLScads[2] + (SAVINGS - 256109.6931 )**2;
    _CLScads[3] + (SAVINGS - 14778.055064 )**2;
  end;
  if n(NBPROD) then do;
    _CLScads[1] + (NBPROD - 14.055555556 )**2;
    _CLScads[2] + (NBPROD - 15.476190476 )**2;
    _CLScads[3] + (NBPROD - 8.8776628878 )**2;
  end;
  if n(EXPENDIT) then do;
    _CLScads[1] + (DEPENSES - 5091.3631019 )**2;
    _CLScads[2] + (EXPENDIT- 3699.0411688 )**2;
    _CLScads[3] + (EXPENDIT - 2296.6205468 )**2;
  end;
  if n(INCOME) then do;
    _CLScads[1] + (INCOME - 3393.2086589 )**2;
    _CLScads[2] + (INCOME - 3247.8545619 )**2;
    _CLScads[3] + (INCOME - 1863.3223265 )**2;
  end;
  if n(NBCARDS) then do;
    _CLScads[1] + (NBCARDS - 1 )**2;
    _CLScads[2] + (NBCARDS - 1.2380952381 )**2;
    CLScads[3] + (NBCARDS - 1.4502304722 )**2;
  end;
_SEGMNT_ _ = 1; distance = _CLScads[1];
  do _clus = 2 to 3;
    if _CLScads[_clus]< distance then do;
     _SEGMNT_ = _clus; distance = _CLScads[_clus];
    end;
  end;
  distance = sqrt(distance*6/_nonmiss);
end;
*************************************;
*** end scoring code for clustering;
*************************************;
```

Figure 1.5 Example of SAS code generated by SAS Enterprise Miner.

```
> library(pmml)
Loading required package: XML
> pmml(titanic.rpart)
<PMML version=''3.2'' xmlns=''http://www.dmg.org/PMML-3_2''
xmlns:xsi=''http://www.w3.org/2001/XMLSchema-instance''
xsi:schemaLocation=''http://www.dmg.org/PMML-3_2
http://www.dmg.org/v3-2/pmml-3-
2.xsd''>
<Header copyright=''Copyright (c) 2010 Stéphane''
description=''RPart Decision Tree Model''>
<Extension name=''timestamp'' value=''2010-06-27 16:30:10''
extender=''Rattle''/>
<Extension name=''description'' value=''Stéphane''
extender=''Rattle''/>
<Application name=''Rattle/PMML'' version=''1.2.15''/>
</Header>
<DataDictionary numberOfFields=''4''>
<DataField name=''survived'' optype=''continuous''
dataType=''double''/>
<DataField name=''class'' optype=''continuous''
dataType=''double''/>
<DataField name=''age'' optype=''continuous'' dataType=''double''/>
<DataField name=''sex'' optype=''continuous'' dataType=''double''/>
</DataDictionary>
<TreeModel modelName=''RPart_Model'' functionName=''regression''
algorithmName=''rpart'' splitCharacteristic=''binarySplit''
missingValueStrategy=''defaultChild''>
<MiningSchema>
<MiningField name=''survived'' usageType=''predicted''/>
<MiningField name=''class'' usageType=''active''/>
<MiningField name=''age'' usageType=''active''/>
<MiningField name=''sex'' usageType=''active''/>
</MiningSchema>
<Node id=''1'' score=''0.323034984098137'' recordCount=''2201''
defaultChild=''2''>
<True/>
<Node id=''2'' score=''0.212016175621028'' recordCount=''1731''
defaultChild=''4''>
<SimplePredicate field=''sex'' operator=''greaterOrEqual''
value=''0.5''/>
<Node id=''4'' score=''0.202759448110378'' recordCount=''1667''>
<SimplePredicate field=''age'' operator=''greaterOrEqual''
value=''0.5''/>
</Node>
<Node id=''5'' score=''0.453125'' recordCount=''64''
defaultChild=''10''>
<SimplePredicate field=''age'' operator=''lessThan''
value=''0.5''/>
<Node id=''10'' score=''0.270833333333333'' recordCount=''48''>
<SimplePredicate field=''class'' operator=''greaterOrEqual''
value=''2.5''/>
```

Figure 1.6 Exporting a model into PMML in R software.

```
      </Node>
      <Node id=''11'' score=''1'' recordCount=''16''>
       <SimplePredicate field=''class'' operator=''lessThan''
value=''2.5''/>
      </Node>
     </Node>
    </Node>
    <Node id=''3'' score=''0.731914893617021'' recordCount=''470''
defaultChild=''6''>
    <SimplePredicate field=''sex'' operator=''lessThan''
value=''0.5''/>
     <Node id=''6'' score=''0.459183673469388'' recordCount=''196''>
      <CompoundPredicate booleanOperator=''surrogate''>
       <SimplePredicate field=''class'' operator=''greaterOrEqual''
value=''2.5''/>
       <SimplePredicate field=''age'' operator=''lessThan''
value=''0.5''/>
      </CompoundPredicate>
     </Node>
     <Node id=''7'' score=''0.927007299270073'' recordCount=''274''>
      <CompoundPredicate booleanOperator=''surrogate''>
       <SimplePredicate field=''class'' operator=''lessThan''
value=''2.5''/>
       <SimplePredicate field=''age'' operator=''greaterOrEqual''
 value=''0.5''/>
     </CompoundPredicate>
    </Node>
   </Node>
  </Node>
 </TreeModel>
</PMML>
>
```

Figure 1.6 (*Continued*).

(c) adequate, relevant and not excessive in relation to the purposes for which they are collected and/or further processed;

(d) accurate and, where necessary, kept up to date; every reasonable step must be taken to ensure that data which are inaccurate or incomplete, having regard to the purposes for which they were collected or for which they are further processed, are erased or rectified;

(e) kept in a form which permits identification of data subjects for no longer than is necessary for the purposes for which the data were collected or for which they are further processed.

Article 25 adds that personal data may be transferred to a third country only if the country in question ensures an adequate level of the protection of the data.

Unless an exception is made, for example in the case of a person working in the medical field, no statistician or other person is allowed to work on the aforementioned sensitive data unless they have been anonymized[7]. In some cases, the secrecy surrounding these data is reinforced by specific regulations, such as those relating to medical or banking confidentiality. Clearly, the disclosure of medical information about a patient may be harmful to him in terms of his image, but also, and more seriously, if it creates difficulties when he is seeking work or insurance cover. In another field, the abuse of Internet surfing data may be an intrusion into private life, because such data may reveal preferences and habits, and may lead to unwelcome selling operations. In banking, the disclosure of confidential data may expose their owner to obvious risks of fraud. Banking data, especially those relating to the use of bank cards, could also lead to drift, if they are used to analyse the lifestyle, movements and consumption habits of cardholders. This could be used to create a customer profile, which is exploited for gain, but which bears no relationship to the purpose for which these data were collected. As a general rule, therefore, all such statistical and computerized processing is highly restricted, in order to avoid cases in which customers innocently providing information simply for the purposes of managing their contracts find that, unbeknown to them, the information has been used in other ways for commercial purposes. Close attention is paid to the interconnection of computer files, which would make it possible to link different kinds of information, collected for different purposes, in such a way that disclosure would lead to the abusive or dangerous retention of files on individuals. Sadly, history has shown that the fear of 'Big Brother' is not unjustified. In this context, the sensitive nature of data on origin or political or religious opinions is evident, but legislators have generally considered that the storage of personal data on an individual could affect the freedom of the individual, even if none of the data are sensitive when taken in isolation.

This is particularly true of data relating to different aspects of an individual which can be used to create a 'profile' of the individual, using automatic processing, which may cause him to lose a right, a service or a commercial offer. In this area, risk scoring and behavioural segmentation are processes that are monitored and regulated by the authorities. The data miner must be aware of this, and must be careful to use only the data and processes that are legally available.

This fear of the power of files has even led some countries, including France, to restrict the dissemination of geomarketing data, such as those obtained from censuses, to a sufficiently coarse level of resolution to prevent these data from being applied too accurately to specific individuals (see Section 4.2.1). Geodemographic databases also contain hundreds of pieces of data on the income, occupational and family circumstances, consumption habits and lifestyles of the inhabitants of each geographical area. If each geographical area contains only a few tens of families, the data on a given family would clearly be very similar to the mean data in the geographical area, and the profile of each family could therefore be fairly accurately estimated.

The principles stated above are those that apply in the European Union, in the European Economic Area (EU, Iceland, Liechtenstein, Norway), in Switzerland, in Canada and in

[7] Current cryptographic methods, such as *hashing*, can be used for *secure matching* of files, and for reconciling anonymity with the need to cross-check files containing data on a single person or to update them subsequently. Each identifier is associated with a personal code, such that the identifier can be used to retrieve the personal code, but not vice versa. This personal code is then used to match the files. These mechanisms have been used in the medical field for several years.

Argentina. In the United States, personal data protection is much less closely regulated, and there is no equivalent to Directive 95/46/EC. Very few states have followed California in passing appropriate legislation, and data protection is based on the reactions of citizens rather than the law. However, some processes are subject to more restrictions than elsewhere in some areas; for example, the granting of credit has been covered by the Equal Credit Opportunity Act since 1974. This law prohibits the use of certain variables in the lending criteria, and therefore in scoring systems as well. Ethnic origin, nationality and religion must not be used. Neither must the sex of the applicant. Age can be used only if it does not penalize older persons. Income can be taken into account, but its origin (wages, pensions, social security, etc.) must not. No distinction must be made between married and unmarried applicants. However, occupation, seniority in employment, and owner-occupier status may be taken into account. Unless credit is refused, there is no obligation to tell an applicant what his score is (it is available only once per year on the Annualcreditreport.com site), but some organizations sell this information to applicants and also offer advice on improving their scores.

In another area, the *Safe Harbor Privacy Principles*, which US organizations are asked to comply with, were developed by the US Department of Commerce in response to Article 25 of Directive 95/46/EC, to protect the security of exchanges with those organizations that agree to observe these principles. However, it appears that these principles are not always followed in practice, and the exchange of data between Europe and the USA is a regular source of conflict. The history of Passenger Name Record (PNR) data is one example of this.

Following the attacks on the USA on 11 September 2001, the Information Awareness Office was established in January 2002, under the aegis of the US Department of Defense, to provide permanent automatic surveillance of all possible kinds of information which might be evidence of preparation for terrorist activities. This project, called *Total Information Awareness* (TIA), and renamed *Terrorism Information Awareness* in 2003, was intended to enable significant links to be created between police and judicial data and behaviour such as applications for passports, visas, work permits and driving licences, the use of credit cards, the purchase of airline tickets, the hiring of cars, and the purchase of chemical products, weapons, etc. TIA included other aspects such as automatic recognition of a human face in a crowd, automatic transcription of verbal and written communications in foreign languages, surveillance of medical databanks to detect a possible biological attack, and the surveillance of certain movements on the stock exchanges. This project led to protests by defenders of individual liberties, and its funding was stopped in September 2003. It reappeared shortly afterwards in other forms such as the CAPPS (Computer Assisted Passenger Prescreening System) and CAPPS2 of the Transportation Security Administration. This programme was more closely targeted than the preceding ones, since it was intended to be applied to users of air transport only: on embarking, each passenger supplied his name, address and telephone number, permitting automatic consultation of several hundred databases of various kinds (government and private). The result of this interrogation could trigger an alert and result in a passenger being searched. The initial aim was to extend this from the USA to Europe, but the US companies that had agreed to participate in the tests were boycotted by consumers, and the programme was finally wound up by Congress because of the dangers it posed to private life. Furthermore, opposition developed in Europe against the transmission of PNR data to the US customs and security services. These were data exchanged in standardized form between the stakeholders in air transport. They could be used by CAPPS2, and their extent and period of retention were critical. In 2004, CAPPS2 was replaced by *Secure Flight* and *VRPS* (Voluntary Registered Passenger System), and agreement was reached, with some difficulty, in August

2007 between the European Union and the United States on the transmission of PNR data to the authorities. The number and nature of these data (bank card number, telephone number, address, etc.) were always considered excessive by the Europeans, especially since their period of retention was 15 years, the right of access and correction was at the discretion of the Americans, access to sensitive data (ethnic origin, political opinions, state of health, etc.) was possible, the purposes of use could be extended, and the data could be transferred to third countries by the Americans: these were all infringements of Directive 95/46/EC. Further disputes between the two continents arose from the 'no-fly lists', or lists of passengers named as undesirable by the USA, supplied by the US authorities to the airlines who were asked to ensure that any passengers flying to the USA did not appear on these lists. The legal basis and quality of these lists were questioned by the Europeans, who pointed out, among other things, the risks of confusion of similar names in these lists, which are updated daily but still contain 65 000 names.

Any statistician or data miner dealing with personal data, even if these are rarely as sensitive as PNR data, must be careful to respect the private life of the persons concerned and must avoid using the data in a way which might cause undue offence. This is particularly true in a world in which the resources of information technology are so vast that enormous quantities of information can be collected and stored on each of us, almost without any technical limitation. This is even more important in the field of data mining, a method which is often used to help with decision making, and which could easily be transformed, if care is not taken, into a tool which takes decisions automatically on the basis of collected information. New problems have arisen concerning our individual data, which no longer simply form the basis for global, anonymous analysis, but can also be used to make decisions which can change the lives of individuals.

For further details on the protection of personal data and the problems it entails in modern society, see Chapter 4 of David Hand's book, *Information Generation: How Data Rule Our World*,[8] as well as the websites of the national data protection authorities.

1.6 Implementation of data mining

The main factors in the success of a project are:

(i) precise, substantial and realistic targets;

(ii) the richness, and above all the quality, of the information collected;

(iii) the cooperation of the departmental and statistical specialists in the organization;

(iv) the relevance of the data mining techniques used;

(v) satisfactory output of the information generated and correct integration into the information system where appropriate;

(vi) analysis of the results and feedback of experience from each application of data mining, to be used for the next application.

I will examine these matters in detail later on, especially in the next chapter and Chapter 13.

[8] Hand, D.J. (2007) *Information Generation: How Data Rule Our World.* Oxford: Oneworld Publications.

Data mining can be applied in different ways in a business. The business may entirely outsource the data mining operation, as well as its computer facilities management, supplying raw commercial files as required to specialist service providers. The service providers then return the commercial files supplemented with information such as customers' scores, their behavioural segments, etc. Alternatively, the business may subcontract the essentials of the data mining operation, with its service providers developing the data mining models it requires, but then take over these models and apply them to its own files, possibly modifying them slightly. Finally, the business may develop its own data mining models, using commercial software, possibly with the assistance of specialist consultants – and this book, of course! These different approaches are described more fully in Section 12.6.

2

The development of a data mining study

Before examining the key points of a data mining study or project in the following chapters, I shall briefly run through the different phases in this chapter. Each phase is mentioned here, even though some of them are optional (those covered in Sections 2.5 and 2.7), and most of them can be handed over to a specialist firm (except for those covered in Section 2.1 and parts of Sections 2.2 and 2.10). I will provide further details and illustrations of some of these steps in the chapter on scoring. We shall assume that statistical or data mining software has already been acquired; the chapter on software deals with the criteria for choice and comparison.

As a general rule, the phases of a data mining project are as follows:

- defining the aims;
- listing the existing data;
- collecting the data;
- exploring and preparing the data;
- population segmentation;
- drawing up and validating the predictive models;
- deploying the models;
- training the model users;
- monitoring the models;
- enriching the models.

Data Mining and Statistics for Decision Making, First Edition. Stéphane Tufféry.

2.1 Defining the aims

We must start by choosing the subject, defining the target population (e.g. prospects and customers, customers only, loyal customers only, all patients, only those patients who can be cured by the treatment under test), defining the statistical entity to be studied (e.g. a person, a household consisting of spouses only, a household including dependent children, a business with or without its subsidiaries), defining some essential criteria and especially the phenomenon to be predicted, planning the project, deciding on the expected operational use of the information extracted and the models produced, and specifying the expected results. To do this, we must arrange for meetings to be held between the clients (risk management and marketing departments, specialists, future users, etc., as appropriate) and the service providers (statisticians and IT specialists). As some data mining projects are mainly horizontal, operating across several departments, it will be useful for the general management to be represented at this stage, so that the inevitable arbitration can take place. The top management will also be able to promote the new data mining tools, if their introduction, and the changes in procedure that they entail, encounter opposition in the business. The contacts present at this stage will subsequently attend periodic steering committee meetings to monitor the progress of the project.

This stage will partly determine the choice of the data mining tools to be used. For example, if the aim is to set out explicit rules for a marketing service or to discover the factors in the remission of a disease, neural networks will be excluded.

The aims must be very precise and must lead to specific actions, such as the refining of targeting for a direct marketing campaign. In the commercial field, the aims must also be realistic (see Section 13.1) and allow for the economic realities, marketing operations that have already been conducted, the penetration rate, market saturation, etc.

2.2 Listing the existing data

The second step is the collection of data that are useful, accessible (inside or outside the business or organization), legally and technically exploitable, reliable and sufficiently up to date as regards the characteristics and behaviour of the individuals (customers, patients, users, etc.) being studied. These data are obtained from the IT system of the business, or are stored in the business outside the central IT system (in Excel or Access files, for example), or are bought or retrieved outside the business, or are calculated from earlier data (indicators, ratios, changes over time). If the aim is to construct a predictive model, it will also be necessary to find a second type of data, namely the historical data on the phenomenon to be predicted. Thus we need to have the results of medical trials, marketing campaigns, etc., in order to discover how patients, customers, etc. with known characteristics have reacted to a medicine or a mail shot, for example. In the second case, the secondary effects are rarer, and we simply need to know about the customer who has been approached has or has not bought the products being offered, or some other product, or nothing at all. The model to be constructed must therefore correlate the fact of the purchase, or the cure, with the other data held on the individual.

A problem may arise if the business does not have the necessary data, either because it has not archived them, or because it is creating a new activity, or simply because it has little direct contact with its customers. In this case, it must base its enquiries on samples of customers, if necessary by offering gifts as an incentive to reply to questionnaires. It could also use geomarketing, mega-databases (Acxiom, Wegener Direct Marketing) or tools such as 'first name scoring' (see Chapter 4). Finally, the business can use standard models designed by specialist companies. For example, generic risk scores are available in the banking sector (see Section 12.6.2).

Sometimes the business may have data that are, unfortunately, unsuitable for data mining, for example when they are:

- detailed data available on microfilm, useful for one-off searches but exorbitantly expensive to convert to a usable computerized format;
- data on individual customers that are aggregated at the end of the year to provide an annual summary (e.g. the number and value of purchases, but with no details of dates, products, etc.);
- monthly data of a type which would be useful, but aggregated at the store level, not the customer level;

In such cases, the necessary data must be collected as a matter of urgency (see also Section 4.1.7).

A note on terminology. *Variable* is taken to mean any characteristic of an entity (person, organization, object, event, case, etc.) which can be expressed by a numerical value (a measurement) or a coded value (an attribute). The different possible values that a variable can take in the whole set of entities concerned are called the *categories* of the variable. The statistical entity that is studied is often called the *individual*, even if it is not a person. According to statistical usage, we may also use the term *observation*.

2.3 Collecting the data

This step leads to the construction of the database that will be used for the construction of models. This *analysis base* is usually in the form of a table (DB2, Oracle, SAS, etc.) or a file (flat file, CSV file, etc.) having one record (one row) for each statistical individual studied and one field (one column) for each variable relating to this individual. There are exceptions to this principle: among the most important of these are time series and predictive models with repeated measurement. There are also spatial data, in which the full meaning of the record of each entity is only brought out as a function of the other entities (propagation of a polluting product, installation of parking meters or ticket machines, etc.). In some cases, the variables are not the same for all the individuals (for example, medical examinations differ with the patients). In general, however, the data can be represented by a table in which the rows correspond to the individuals studied and the columns correspond to the variables describing them, which are the same for all the individuals. This representation may be rather too simplistic in some cases, but has the merit of convenience.

The analysis base is often built up from snapshots of the data taken at regular intervals (monthly, for example) over a certain period (a year, for example). On the other hand, these data are often not determined at the level of the individual, but at the level of the product, account, invoice, medical examination, etc. In order to draw up the analysis base, we must therefore provide syntheses or aggregations along a number of axes. These include the individual axis, where data determined at a finer level (the level of the products owned by the individual, for example) are synthesized at the individual level, and the time axis, where n indicators determined at different instants are replaced by a single indicator determined for the whole period. For example, in the case of wages and other income paid into in a bank, we start by calculating the monthly sum of income elements paid into the different accounts of the household, before calculating the mean of these monthly income elements over 12 months. By observing the independent variables over a year, we can make allowance for a full economic cycle and smooth out the effect of seasonal variations.

As regards the aggregation of data at the individual level, this can be done (to take an example from the commercial sector) by aggregating the 'product' level data (one row per product in the files) into 'customer' level data, thus producing, in the file to be examined, one row per customer, in which the 'product' data are replaced by sums, means, etc., of all the customer's products. By way of example, we can move from this file:

customer no.	**date of purchase**	**product purchased**	**value of purchase**
customer 1	21/02/2004	jacket	25
customer 1	17/03/2004	shirt	15
customer 2	08/06/2004	T-shirt	10
customer 2	15/10/2004	socks	8
customer 2	15/10/2004	pullover	12
...	...	...	...
customer 2000	18/05/2004	shoes	50
...	...	...	...

to this file:

customer no.	**date of 1st purchase**	**date of last purchase**	**value of purchases**
customer 1	21/02/2004	17/03/2004	40
customer 2	08/06/2004	15/10/2004	30
...	...	...	...
customer 2000	18/05/2004	18/05/2004	50
...	...	...	...

Clearly, the choice of the statistical operation used to aggregate the 'product' data for each customer is important. Where risk is concerned, for example, very different results may be obtained, depending on whether the number of debit days of a customer is defined as the mean or the maximum number of debit days for his current accounts. The choice of each operation must be carefully thought out and must have a functional purpose. For risk indicators, it is generally the most serious situation of the customer that is considered, not his average situation.

When a predictive model is to be constructed, the analysis base will be of the kind shown in Figure 2.1. This has four types of variable: the identifier (or key) of the individual, the target variable (i.e. the dependent variable), the independent variables, and a 'sample' variable which shows whether each individual is participating in the development (learning) or the testing of the model (see Section 11.16.4).

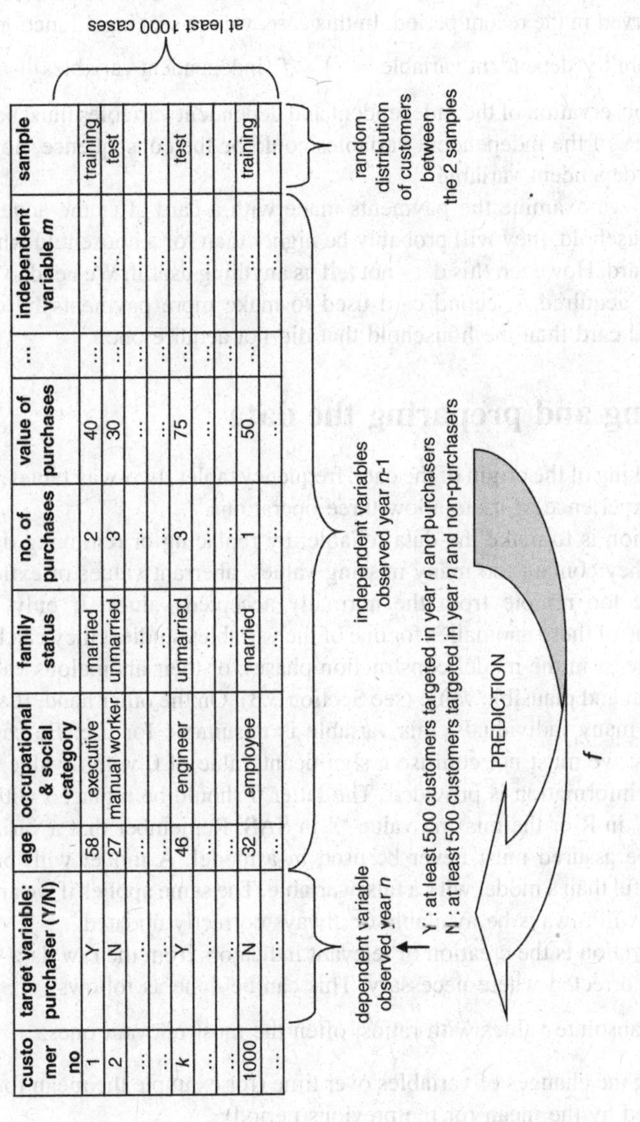

custo mer no	target variable: purchaser (Y/N)	age	occupational & social category	family status	no. of purchases	value of purchases	...	independent variable *m*	sample
1	Y	58	executive	married	2	40	...	...	training
2	N	27	manual worker	unmarried	3	30	...	...	test
...	...	...	...	...	...	...	...	...	...
k	Y	46	engineer	unmarried	3	75	...	...	test
...	...	...	...	...	...	...	...	...	...
1000	N	32	employee	married	1	50	...	...	training
...	...	...	...	...	...	...	...	...	...

Figure 2.1 Developing a predictive analysis base.

The analysis base is constructed in accordance with the aim of the study. The aim may be to find a function f of the independent variables of the base, such that f (independent variables of a customer) is the probability that the customer (or patient) will buy the product offered, or will fall ill, etc. The data are often observed over 18 months (better still, 24 months): the most recent 6 (or 12) months are the period of observation for the purchase of a product or the onset of a disease, and the previous 12 months are the period in which the customer's data are analysed to explain what has been observed in the recent period. In this case, we are seeking a function f such that

$$\text{Probability}(\text{dependent variable} = \text{x}) = f\,(\text{independent variables}).$$

The periods of observation of the independent and dependent variables must be separated; otherwise, the values of the independent variables could be the consequence, not the cause, of the value of the dependent variable.

For example, if we examine the payments made with a card after the acquisition of a second card in a household, they will probably be higher than for a household which has not acquired a second card. However, this does not tell us anything useful. We need to know if the household that has acquired a second card used to make more payments by card before acquiring its second card than the household that did not acquire one.

2.4 Exploring and preparing the data

In this step the checking of the origin of the data, frequency tables, two-way tables, descriptive statistics and the experience of users allow three operations.

The first operation is to make the data reliable, by replacing or removing data that are incorrect because they contain too many missing values, aberrant values or extreme values ('outliers') that are too remote from the normally accepted values. If only a few rare individuals show one of these anomalies for one of the variables studied, they can be removed from the study (at least in the model construction phase), or their anomalous values can be replaced by a correct and plausible value (see Section 3.3). On the other hand, if a variable is anomalous for too many individuals, this variable is unsuitable for use. Be careful about numerical variables: we must not confuse a significant value of 0 with a value set to 0 by default because no information is provided. The latter 0 should be replaced with the 'null' value in DB2, 'NA' in R or the missing value '.' in SAS. Remember that a variable whose reliability cannot be assured must never be used in a model. A model with one variable missing is more useful than a model with a false variable. The same applies if we are uncertain whether a variable will always be available or always correctly updated.

The second operation is the creation of relevant indicators from the raw data which have been checked and corrected where necessary. This can be done as follows:

- by replacing absolute values with ratios, often the most relevant ones;
- by calculating the changes of variables over time (for example the mean for the recent period, divided by the mean for the previous period);
- by making linear combinations of variables;
- by composing variables with other functions (such as logarithms or square roots of continuous variables, to smooth their distribution and compress a highly right-skewed distribution, as commonly found in financial or reliability studies);

- by recoding certain variables, for example by converting 'low, medium, high' into '1, 2, 3';
- by modifying units of measurement;
- by replacing the dates by durations, customer lifetimes or ages;
- by replacing geographical locations with coordinates (latitude and longitude).

Thus we can create relevant indicators, following the advice of specialists if necessary. In the commercial field, these indicators may be:

- the age of the customer when his relationship with the business begins, based on the date of birth and the date of the first purchase;
- the number of products purchased, based on the set of variables 'product P_i purchased (Yes/No)';
- the mean value of a purchase, based on the number and value of the purchases;
- the recentness and frequency of the purchases, based on the purchase dates;
- the rate of use of revolving credit, deduced from the ceiling of the credit line and the proportion actually used.

The third operation is the reduction of the number of dimensions of the problem: specifically, the reduction of the number of individuals, the number of variables, and the number of categories of the variables.

As mentioned above, the *reduction of the number of individuals* is a matter of eliminating certain extreme individuals from the population, in a proportion which should not normally be more than 1–2%. It is also common practice to use a sample of the population (see Section 3.15), although care must be taken to ensure that the sample is representative: this may require a more complex sampling procedure than ordinary random sampling.

The *reduction of the number of variables* consists of:

- disregarding some variables which are too closely correlated with each other and which, if taken into account simultaneously, would infringe the commonly required assumption of non-collinearity (linear independence) of the independent variables (in linear regression, linear discriminant analysis or logistic regression);
- disregarding certain variables that are not at all relevant or not discriminant with respect to the specified objective, or to the phenomenon to be detected;
- combining several variables into a single variable; for example, the number of products purchased with a two-year guarantee and the number of products purchased with a five-year guarantee can be added together to give the number of products purchased without taking the guarantee period into account, if this period is not important for the problem under investigation; it is also possible to replace the 'number of purchases' variables established for each product code with the 'number of purchases' variables for each product line, which would also reduce the number of variables;

- using factor analysis (if appropriate) to convert some of the initial variables into a smaller number of variables (by eliminating the correlated variables), these variables being chosen to maximize the variance (the new variables are also more stable in relation to sampling and random fluctuations).

Reducing the number of variables by factor analysis, for example by principal component analysis (PCA: see Section 7.1), is useful when these variables are used as the input of a neural network. This is because a decrease in the number of input variables allows us to decrease the number of nodes in the network, thus reducing its complexity and the risk of convergence of the network towards a non-optimal solution. PCA may also be used before automatic clustering.

Quite commonly, 100 to 200 or more variables may be investigated during the development of a classification or prediction model, although the number of ultimately discriminant variables selected for the calculation of the model is much smaller, the order of magnitude being as follows:

- 3 or 4 for a simple model (although the model can still classify more than 80% of the individuals);
- 5–10 for a model of normal quality;
- 11–20 for a very fine (or refined) model.

Beyond 20 variables, the model is very likely to lose its capacity for generalization. These figures may have to be adjusted according to the size of the population.

The *reduction of the number of categories* is achieved by:

- combining categories which are too numerous or contain too few observations, in the case of discrete and qualitative variables;
- combining the categories of discrete and qualitative variables which have the same functional significance, and are only distinguished for practical reasons which are unrelated to the analysis to be carried out;
- *discretizing* (binning) some continuous variables (i.e. converting them into ranges of values such as quartiles, or into ranges of values that are significant from the user's point of view, or into ranges of values conforming to particular rules, as in medicine).

Examples of categories which are combined because they are too numerous are the individual articles in a product catalogue (which can be replaced by product lines), social and occupational categories (which can be aggregated into about 10 values), the segments of a housing type (which can be combined into families as indicated in Section 4.2.1), counties (which can be regrouped into regions), the activity codes of merchants (the Merchant Category Code used in bank card transactions) and of businesses (NACE, the French name of the Statistical Classification of Economic Activities in the European Community, equivalent to the Standard Industrial Classification in the United Kingdom and North America), (which can be grouped into industrial sectors), etc. By way of example, the sectors of activity may be:

- Agriculture, Forestry, Fishing and Hunting
- Mining

- Utilities
- Construction
- Manufacturing
- Wholesale Trade
- Retail Trade
- Transportation and Warehousing
- Information
- Finance and Insurance
- Real Estate and Rental and Leasing
- Professional, Scientific, and Technical Services
- Management of Companies and Enterprises
- Administrative and Support and Waste Management and Remediation Services
- Education Services
- Health Care and Social Assistance
- Arts, Entertainment, and Recreation
- Accommodation and Food Services
- Other Services (except Public Administration)
- Public Administration

Extreme examples of variables having too many categories are variables which are identifiers (customer number, account number, invoice number, etc.). These are essential keys for accessing data files, but cannot be used as predictive variables.

2.5 Population segmentation

It may be necessary to segment the population into groups that are homogeneous in relation to the aims of the study, in order to construct a specific model for each segment, before making a synthesis of the models. This method is called the *stratification of models*. It can only be used where the volume of data is large enough for each segment to contain enough individuals of each category for prediction. In this case, the pre-segmentation of the population is a method that can often improve the results significantly. It can be based on rules drawn up by experts or by the statistician. It can also be carried out more or less automatically, using a statistical algorithm: this kind of 'automatic clustering', or simply 'clustering', algorithm is described in Chapter 9.

There are many ways of segmenting a population in order to establish predictive models based on homogeneous sub-populations. In the first method, the population is segmented according to general characteristics which have no direct relationship with the dependent variable. This pre-segmentation is *unsupervised*. In the case of a business, it may relate to its size, its legal status or its sector of activity. At the Banque de France, the financial health of businesses is modelled by sector: the categories are industry, commerce, transport, hotels, cafés and restaurants, construction, and business services. For a physical person, we may use sociodemographic characteristics such as age or occupation. In this case, this approach can be justified by the existence of specific marketing offers aimed at certain customer segments, such as 'youth', 'senior', and other groups. From the statistical viewpoint, this method is not always the best, because behaviour is not always related to general characteristics, but it may be very useful in some cases. It is usually applied according to rules provided by experts, rather than by statistical clustering.

Another type of pre-segmentation is carried out on behavioural data linked to the dependent variable, for example the product to which the scoring relates. This pre-segmentation is *supervised*. It produces segments that are more homogeneous in terms of the dependent variable. Supervised pre-segmentation is generally more effective than unsupervised pre-segmentation, because it has some of the discriminant power required in the model. It can be implemented according to expert rules, or simply on a common-sense basis, or by a statistical method such as a decision tree with one or two levels of depth. Even a decision tree with only one level of depth can be used to separate the population to be modelled into two clearly differentiated classes (or more with chi-square automatic interaction detection), without having the instability that is a feature of deeper trees. One common-sense rule could be to establish a consumer credit propensity score for two segments: namely, those customers who already have this kind of credit, and the rest. In this example, we can see that the rules can be expressed as a decision tree. This method often enables the results to be improved.

A third type of pre-segmentation can be required because of the nature of the available data: for example, they will be much less rich for a prospect than for a customer, and these two populations must therefore be separated into two segments for which specific models will be constructed.

In all cases, pre-segmentation must follow explicit rules for classifying every individual. The following features are essential:

- simplicity of pre-segmentation (there must not be too many rules);
- a limited number of segments and stability of the segments (these two aspects are related);
- segment sizes generally of the same order of magnitude, because, unless we have reasons for examining a highly specific sub-population, a 99%/1% distribution is rarely considered satisfactory;
- uniformity of the segments in terms of the independent variables;
- uniformity of the segments in terms of the dependent variable (for example, care must be taken to avoid combining high-risk and low-risk individuals in the same segment).

We should note that some operations described in the preceding section (reduction of the number of individuals, variables and categories) can be carried out either before any segmentation, or on each segment.

2.6 Drawing up and validating predictive models

This step is the heart of the data mining activity, even if it is not the most time-consuming. It may take the form of the calculation of a score, or more generally a predictive model, for each segment produced in the preceding step, followed by verification of the results in a test sample that is different from the learning sample. It may also be concerned with detecting the profiles of customers, for example according to their consumption of products or their use of the services of a business. Sometimes it will be a matter of analysing the contents of customers' trolleys in department stores and supermarkets, in order to find associations between products that are often purchased together. A detailed review of the different statistical and data mining methods for developing a model, with their principles, fields of application, strong points and limitations, is provided in Chapters 6–11.

In the case of scoring, the principal methods are logistic regression, linear discriminant analysis and decision trees. Trees provide entirely explicit rules and can process heterogeneous and possibly incomplete data, without any assumptions concerning the distribution of the variables, and can detect non-linear phenomena. However, they lack reliability and the capacity for generalization, and this can prove troublesome. Logistic regression directly calculates the probability of the appearance of the phenomenon that is being sought, provided that the independent variables have no missing values and are not interrelated. Discriminant analysis also requires the independent variables to be continuous and to follow a normal law (plus another assumption that will be discussed below), but in these circumstances it is very reliable. Logistic regression has an accuracy very similar to that of discriminant analysis, but is more general because it deals with qualitative independent variables, not just quantitative ones.

In this step, several models are usually developed, in the same family or in different families. The development of a number of models in the same family is most common, especially when the aim is to optimize the parameters of a model by adjusting the independent variables chosen, the number of their categories, the depth or number of leaves on a decision tree, the number of neurons in a hidden layer, etc. We can also develop models in different families (e.g. logistic regression and linear discriminant analysis) which are run jointly to maximize the performance of the model finally chosen.

The aim is therefore to choose the model with the best performance out of a number of models. I will discuss in Section 11.16.5 the importance of evaluating the performance of a model on a test sample which is separate from the learning sample used to develop the model, while being equally representative in statistical terms.

For comparing models of the same kind, there are statistical indicators which we will examine in Chapter 11. However, since the statistical indicators of two models of different kinds (e.g. R^2 for a discriminant analysis and R^2 for a logistic regression) are rarely comparable, the models are compared either by comparing their error rates in a confusion matrix (see Section 11.16.4) or by superimposing their lift curves or receiver operating characteristic (ROC) curves (see Section 11.16.5). These indicators can be used to select the best model in each family of models, and then the best model from all the families considered together.

The ROC curve can be used to visualize the separating power of a model, representing the percentage of correctly detected events (the Y axis) as a function of the percentage of incorrectly detected events (the X axis) when the score separation threshold is varied. If this curve coincides with the diagonal, the performance of the model is no better than that of a random prediction; as the curve approaches the upper left-hand corner of the square, the

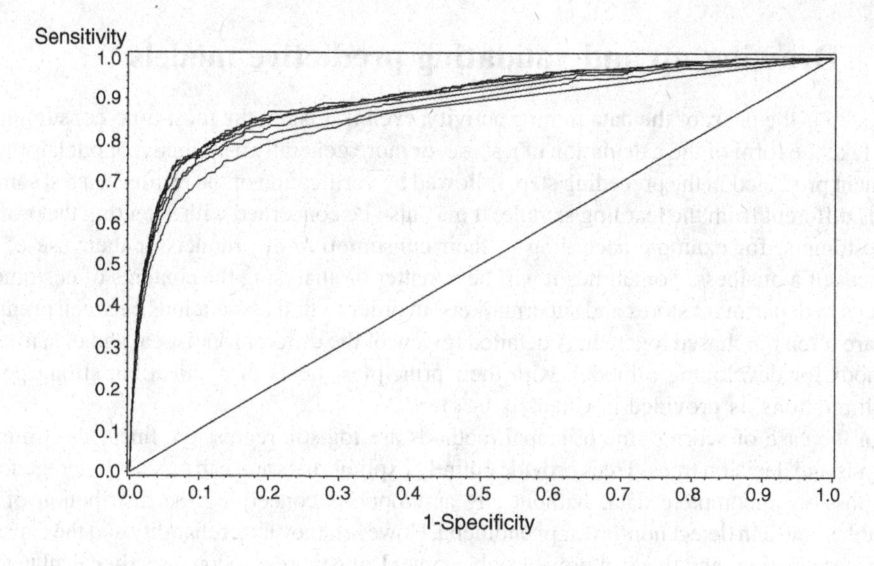

Figure 2.2 ROC curve.

model improves. Several ROC curves can be superimposed to show the progressive contribution of each independent variable in the model, as in Figure 2.2. This shows that a single variable, if well chosen, already explains much of the phenomenon to be predicted, and that an increase in the number of variables added to the model decreases the marginal gain they provide.

In addition to the qualities of mathematical accuracy of the model, other selection criteria may sometimes have to be taken into account, such as the readability of the model from the user's viewpoint, and the ease of implementing it in the production IT system.

2.7 Synthesizing predictive models of different segments

This small step only takes place if customer segmentation has been carried out. We need to compare the scores of the different segments, and calculate a unique normalized score that is comparable from one segment to another. This new score allows for the proportion of customers in each segment, and each score band, who have already achieved the aim, i.e. bought the product, repaid the loan, etc.

The synthesis of the scores for the different segments is easy in a logistic regression model, since the score is a probability which is inherently normalized and is suitable for the same interpretation in all the segments (provided that the sampling has been comparable in the different segments).

In all cases, it is possible to draw up a table showing the score points (for propensity in this case) of the different segments, indicating the subscription rate corresponding to this score and this segment on each occasion (Table 2.1). The rows of this table are then sorted by subscription rate, from lowest to highest. The rows sorted in this way are then combined into ten groups, each covering about 10% of the customers, with the rows of each group following each other and therefore having similar subscription rates. One point is given to the customers of the first group, and so on up to the customers of the tenth group, who have 10 points.

Table 2.1 Synthesizing the score points of the different segments.

Segment	Score	Number of customers	Subscription rate
1	1	$n_{1,1}$	$t_{1,1}$
1	2	$n_{1,2}$	$t_{1,2}$
...	...	...	...
1	10	$n_{1,10}$	$t_{1,10}$
2	1	$n_{2,1}$	$t_{2,1}$
2	2	$n_{2,2}$	$t_{2,2}$
...	...	...	...
5	1	$n_{5,1}$	$t_{5,1}$
...	...	...	...
5	10	$n_{5,10}$	$t_{5,10}$

2.8 Iteration of the preceding steps

The procedure described in Sections 2.4 to 2.7 is usually iterated until completely satisfactory results are obtained. The input data are adjusted, new variables are selected or the parameters are changed in the data mining tools, which, if operating in an exploratory mode, as mentioned above, may be highly sensitive to the choice of initial data and parameters, especially in neural networks.

Before accepting a model, it is useful to have it validated by experts in the field, for example credit analysts in the case of scoring, or market research managers. These experts can be shown some real cases of individuals to which different competing models have been applied, and they can then be asked to say which model appears to provide the best results, giving reasons for their response. Thus they can detect the characteristics of a model which would not have appeared during modelling. This may be done in the field of marketing and propensity scores. The method of modelling on historic data tends to reproduce the commercial habits of the business, even if precautions are taken to avoid this (by neutralizing the effect of sales campaigns, for example). Thus, recent customers may be rather too well scored, since they are usually most likely to be approached by sales staff. In the test phase, we may discover that older customers are not described well enough by the propensity score, and that the use of this score would result in their not being targeted. This phase could thus lead to a reduction in the effect of the variables relating to customer lifetime; these variables may even be removed from the model. In the field of risk, such variables may be those which forecast the risk too late and mask variables which would allow earlier prediction; or they may simply be variables which are not obvious to experts in the field, and which are discarded from the model.

2.9 Deploying the models

Deployment involves the implementation of the data mining models in a computer system, before using the results for action (adapting procedures, targeting, etc.) and making them available to users (information on the workplace, etc.). It is harder to update the 'workplace' applications to incorporate the score, especially if we wish to allow users to input new information in order to upload them to the central production files or to execute the model in real time. For an initial test, therefore, we may limit ourselves to creating a computer file containing the data yielded by data mining, for example the customers' score points; this file may or may not be uploaded to the computerized production system, and may be used for targeting in direct marketing. A spreadsheet file can be used to create a mail merge, using word processing software.

In this step we must decide on the output of the data at different levels of fineness. There may be fine grades in the production files (from 1 to 1000, for example), which are aggregated at the workstation (from 1 to 10) or even regrouped into bands (low/medium/high).

The confidentiality of the score output will be considered in this step, if not before: for example, is a salesman allowed to see the scores of all customers, or only the ones for which he is responsible? The question of the frequency of updating of the data must also be considered: this is usually monthly, but a quarterly frequency may be suitable for data that do not change very often, while a daily frequency may be required for sensitive risk data.

Before deploying a data mining tool in a commercial network, it is obviously preferable to test it for several months with volunteers. There will also be a phase of take-over of the tool, possibly followed by a phase in which the initial volunteers train other users.

2.10 Training the model users

To ensure a smooth take-over of the new decision tools by the future users, it is essential to spend some time familiarizing them with these tools. They must know the objective (and stick to it), the principles of the tools, how they work (without going into technical details, which would subsequently have to be justified individually), their limits (noting that these are only statistical tools which may not be useful in special cases: the difficulty in educational terms is to maintain a minimum of vigilance and critical approach without casting doubt on the reliability of the tools), the methods for using them (while pointing out that these are *decision support* tools, not tools for automatic decision making), what the tools will contribute (the most important point), and how the users' work patterns will change, in both operational and organizational terms (adaptation of procedures, increased delegation, rules for responsibility if the user does or does not follow the recommendations of the tool, in risk scoring for example).

2.11 Monitoring the models

Two forms of monitoring will be distinguished in this section. The first is one-off. Whenever a new data mining application is brought into use, the results must be analysed. Let us take the example of a major marketing campaign for which a propensity score has been developed to improve response rates. After the campaign, it is important to ensure that the response rates match the score values, and that the customers with the highest scores have in fact responded best. If a category of customers with a high score has not responded well to the marketing offer, this must be examined in detail to see if it is possible to segment the category or apply a decision tree to it, in order to discover the profiles of the non-purchasers and take this into account in the next campaign. It will be necessary to check whether the definition of this profile uses a variable which is not allowed for in the score calculation. It may be useful to complement this quantitative analysis with a qualitative analysis based on the opinions of the sales staff who have used the score: they may have some idea of the phenomena which were not taken into account or were underestimated by the score; they may also have intuitively identified the poorly scored populations. The analysis of the low score results is also useful, especially when the score has been developed in the absence of data on customers' refusal to purchase, and when 'negative' profiles for the score modelling therefore had to be defined in advance, instead of being determined on the basis of evidence, resulting in a degree of arbitrariness. An examination of the feedback from the campaign will make it possible to corroborate or to reject certain hypotheses on the 'negative' profiles.

The second form of monitoring is continuous. When a tool such as a risk score is used in a business, it is used constantly and both its correct operation and its use must be checked. Its operation can be monitored by using tables like the following:

month score	M-1	M-2	M-3	M-4	M-5	M-6	M-7	M-8	M-9	M-10	M-11	M-12
1												
2												
3												
...												
TOTAL												

The second column of this table shows, for the month $M-1$ before the analysis and for each score value, the number of customers, the percentage of customers in the total population, and the default rate. The sum of the losses of the business can be added if required. Clearly, the rate of non-payments will be replaced by the subscription rate in the case of a propensity score, and by the departure rate in the case of an attrition score. The third column contains the same figures for the month $M-2$, and so on. If required, a column containing the figures at the time when the score was created may be added. These figures can be used to check whether the score is still relevant in terms of prediction, and especially whether the rate of non-payments (or subscription, or attrition) is still at the level predicted by the score. It is even better to calculate the figures in column M-n in the following way. If the goal of the score is (for example) to give a 12 months prediction, we measure for each value of the score, the number of customers at M-n-12 and the number of defaults between M-n-12 and M-n.

The use of the score is monitored by using a table similar to the preceding one, in which the rate of non-payments is replaced by the number and value of transactions concluded in a given month and for a given score value. Thus it is possible to see if sales staff are taking the risk score into account to limit their involvement with the most high-risk customers. Indicators such as 'infringements' of the score by its users can be added to this table.

These tables are useful for departmental, risk and marketing experts, but may also be distributed in the business network to enable the results of actions to be measured. They also enable the business to quantify the prime target customers with which it conducts its transactions; these are preferably customers whose scores have a certain given value (low for risk, or high for propensity).

score M–1 → **↓ score M–2**	**1**	**2**	**3**	**...**		
1						
2						
3						
...						

A 'transition matrix' like the one above enables the stability of the score to be monitored.

		Original	**Now**	**χ^2 probability**
variable 1	category 1	number	number	
	category 2	number	number	
	...			
	...			
	TOTAL	total number	total number	
variable 2	category 1	number	number	
	...			
	...			
...				
variable k	...	...	...	
	...			

It is also worth using charts to monitor the distribution of the variables explaining the score, comparing the actual distribution to the distribution at the time of creation of the score (using a χ^2 or Cramér's *V* test, for example: see Appendix A), to discover if the distribution of a variable has changed significantly, or if the rate of missing values is increasing, which may mean that the score has to be revised.

2.12 Enriching the models

When the results of the use of a first data mining model are measured, the model can be improved, as we have seen, either by adding independent variables that were not considered initially, or by using feedback from experience to determine the profiles to be predicted (e.g. 'purchaser/non-purchaser') if these results were not available before.

At this point, we can also supplement the results of step 2.2) if they did not provide enough historic details of the variable to be predicted.

When we have done this, steps 2.3, 2.4, ... can be repeated to construct a new model which will perform more satisfactorily than the first one. This iterative process must become permanent if we are to achieve a continuous improvement of the results. There is one proviso, however: in the learning phase of a new model, we must never rule out individuals whose previous levels were considered to be low. Otherwise this model can never be used to question the validity of previous models, which may sometimes be necessary. For example, if we use a calculated scoring model to target only the individuals with the best score points, without ever approaching the others, there is a risk that we will fail to detect a change in behaviour of some of the initially low-scoring individuals. So it is always desirable to include a small proportion of individuals with medium or low scores in any targeting which is based on a model. We will return to this matter in Section 12.4.

2.13 Remarks

The procedure described in this chapter can use a number of data mining techniques with different origins: for example, multiple correspondence analysis for the transformation of variables, agglomerative hierarchical clustering for segmentation, and logistic regression for scoring. There are four important points to be considered:

- For step 2.3, it is essential to collect the data to be studied for several months or even several years, at least where predictive methods such as scoring are concerned, in order to provide valid analyses.
- Steps 2.3 and 2.4 are much more time-consuming than steps 2.5–2.7, but are very important.
- The data mining process is highly iterative, with many possible loopbacks to one or more of steps 2.4–2.7.
- If the population is large, it will be useful to segment it before calculating a score or carrying out classification, so that we can work on homogeneous groups of individuals, making processing more reliable. I will discuss this matter further in the chapter on automatic clustering.

2.14 Life cycle of a model

Data mining models (especially scores) go through a small-scale trial phase, for adjustment and validation, and to enable their application to be tested. When models are in production, they must be applied regularly to refreshed data. However, these models must be reviewed regularly, as described in the chapter on scoring.

2.15 Costs of a pilot project

The factors determining the number of man-days for a data mining project are so numerous that I can only give very approximate figures here. They will have to be modified according to the context. These numbers are mainly useful because they demonstrate the proportions of the different stages. Furthermore, these figures do not relate to major projects, only to small- and medium-scale ones such as pilot projects intended to initiate and promote the use of data mining in businesses. When the processes are fully established, the tools are in full production and the data are thoroughly known, a reduction in the costs shown in the table in Figure 2.3 may be expected. These costs relate to the data mining team only, excluding costs for the other personnel involved.

No.	Step	Costs (in days)		Remarks
		Small-scale project	Medium-scale project	
1	Definition of the target and objectives	4	8	Preliminary analyses and provision of numerical bases for decision making.
2	Listing the existing usable data	7	10	The cost depends on the organization's knowledge of the data
3, 4	Collection, formatting and exploration of the data	15	28	This cost is largely determined by the number of variables, their quality, the number of tables accessed, etc.
5–8	Developing and validating the models (clustering/scoring)	15	25	Must be multiplied if more than one population segment is to be processed
9	Complementary analyses and delivery of results.	9	12	For example, setting the scoring thresholds
10	Documentation and presentations	5	7	
11	Analysis of the first tests conducted	5	10	Statistical study leading to subsequent action
	TOTAL	60	100	

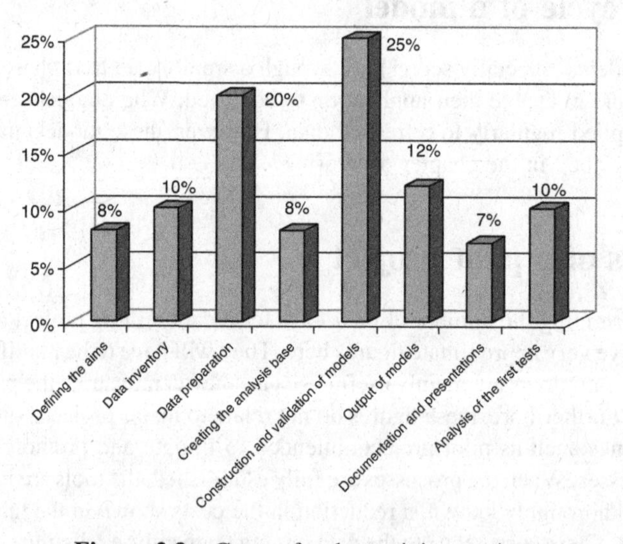

Figure 2.3 Costs of a data mining project.

3

Data exploration and preparation

This chapter starts with a brief survey of the different types of data. It then goes on to detail the tasks of exploring, analysing and preparing the data collected, using univariate, bivariate and multivariate methods and statistical tests. These tasks are fundamental, as the quality of the results generally depends more on the care taken here than on the modelling method which uses these input data. The chapter concludes with a discussion of sampling.

3.1 The different types of data

The data may be quantitative, qualitative, or, for some special problems, textual.

Quantitative (or *numerical*) data may be continuous or discrete. *Continuous* (or *scale* or *interval*) data are those whose values belong to an infinite subset of the set **R** of real numbers (e.g. wages, amount of purchases). *Discrete* data are those whose values belong to a finite or infinite subset of the set **N** of natural integers (e.g. number of children, number of products bought). When the set of values is infinite, or just large (several tens), discrete data are considered to be continuous data (e.g. age).

Qualitative (or *categorical*) data are data whose set of values is finite. These values are numerical or alphanumeric, but when they are numerical they are merely codes, not quantities (e.g. socio-occupational category, department number).

The values of *text* data are uncoded texts, written in natural language: examples are letters of complaint, reports and press despatches.

What distinguishes continuous and discrete data from other types is that they are concerned with *quantities*, *so* we can perform arithmetical operations on them; moreover, they are ordered (we can compare them by an order relationship '$\leq$'). Qualitative data are not quantities, but they may be ordered; in this case we speak of *ordinal* qualitative data (e.g. 'low, medium, high'). Non-ordered qualitative data are called *nominal*. Ordinal data can be classed in the family of discrete data and treated in the same way.

The analysis of text data requires a mixture of linguistics and statistics: this is *text mining*, which is discussed in Chapter 14. Note that text data in natural language may contain elliptical

Data Mining and Statistics for Decision Making, First Edition. Stéphane Tufféry.

Table 3.1 Data conversion.

Initial type	Final type	Operation	Principle
Continuous	Discrete	Discretization (syn : binning)	Splitting the whole set of values into ranges
Discrete or qualitative	Continuous	MCA (see Section 7.4)	Multiple correspondence analysis yields continuous factors based on the initial data

matter, abbreviations which may be more or less personal, faults in spelling or syntax, and ambiguities (where the meaning of terms depends on a context which is not easy to detect by automatic methods), making their analysis problematic.

Not all statistical and data mining methods can accept all types of input data. However, there are operations that can be used to switch from one type to another, as shown in Table 3.1.

3.2 Examining the distribution of variables

The first step in any investigation of data is to examine the *univariate statistics* of the variables, so that we can:

- detect any anomalies in their distribution (especially outliers or missing values);
- get an idea of some orders of magnitude (such as the average ages and income of the population) which will be useful in the subsequent analysis;
- see how to discretize the continuous variables if this is necessary.

For qualitative or discrete variables, we must look at the frequency of appearance of each category (frequency tables). For continuous variables, we can obtain a quick view of the distribution by using *box plots* (see Appendix A), which are output directly by some statistical software, or can be produced by calculating the most significant *quantiles*, i.e. 1%, 10%, 25%, 50% (median), 75%, 90% and 99%. In particular, the last percentile (99%) may include individuals selected erroneously because of poor coding: for example, a very high figure for income may be recorded for a professional who has been incorrectly coded as 'private individual'. We must therefore analyse the nature of the values in the extreme percentiles, and not eliminate them automatically.

We also have to ensure that the distribution of the values of a variable is indeed homogeneous and does not have a singularity, which might be an erroneous value, assigned by default where no information was acquired.

Finally, some variables such as age, age at commencement of working life, and time elapsed under a contract must lie between certain limits: we must check that they do not stray beyond these boundaries. Errors in the input of dates are quite common.

The second step is to use *bivariate statistics*, to detect:

- incompatibilities between variables;

Table 3.2 Contingency table.

(frequency in thousands)	Single	Married	Widowed	Divorced	TOTAL
<20 years	15 144 *100%*	6 *0%*	0 *0%*	0 *0%*	15 150
20–64 years	10 935 *32%*	20 048 *59%*	844 *2%*	2 423 *7%*	34 250
≥65 years	696 *8%*	4 755 *54%*	3 079 *35%*	328 *4%*	8 858
TOTAL	26 775 *46%*	24 809 *43%*	3 923 *7%*	2 751 *5%*	58 258

Matrimonial status at 1 January 1996 (source: INSEE [French National Institute of Statistics and Economic Studies], La situation démographique en 1995, p. 34).

- links between the dependent ('target') variable and the independent variables and their interactions, in order to eliminate the variables having no effect on the dependent variable;
- links between the independent variables, which must be avoided in some methods, such as linear and logistic regression.

These statistics can be based on detailed *association tests* detailed in Appendix A and used in Section 3.8 below, or, very simply (but less precisely), on the examination of two-way frequency tables such as Table 3.2.

These tables are called *contingency tables*. Note that numerical variables can be set out in a contingency table, provided that they are discretized. However, the intersection of more than two variables (multivariate statistics) can only be observed by a difficult and time-consuming process. Although some graphics tools can provide a three-dimensional display, sometimes with a facility to rotate axes, the most powerful multivariate analysis tool is still factor analysis, which I will discuss later on.

3.3 Detection of rare or missing values

This step must be carried out for all types of models.

Rare values can create bias in factor analysis and other analyses, by appearing to be more important than they really are. It is better to remove the observations in question, or to replace the rare value with a more frequent value when such a replacement makes sense in the context.

As for missing values, these are obviously far more troublesome, since most statistical methods are unable to handle them, and the corresponding observations must be eliminated from the study. Missing values can occur, for example, in reported variables where the input was optional and was not done; in responses to surveys; or in chemical analyses where the concentrations of some elements are below the detection thresholds. This poses two problems: on the one hand, there may be only 1% of missing values for each variable, but more than 10% of observations in which one of the variables has a missing value; on the other hand, if the values are not missing by chance and there are systematic differences between the complete and incomplete observations, the removal of incomplete observations introduces a bias into

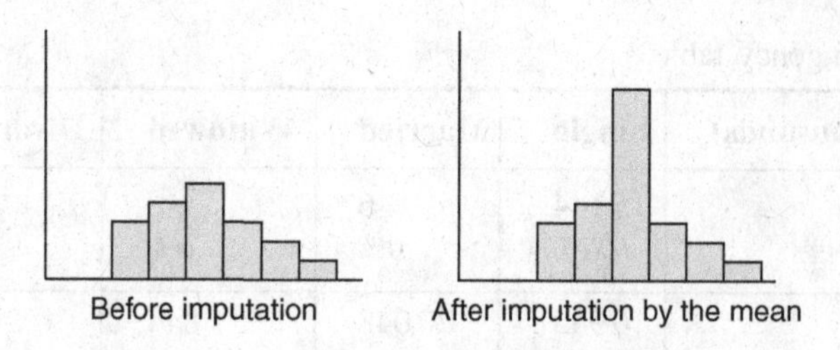

Figure 3.1 Effect of imputation by the mean.

the analysis. Before reaching this point, therefore, we must be quite sure that we cannot use other solutions, such as:

- not using the variable concerned if its contribution to the analysis of the problem does not appear to be essential, or replacing it with a similar variable that has no missing value;
- replacing the missing value with a value determined by a statistical method, by the investigator's knowledge of the data, or by an external source;
- treating the missing value as a whole separate value, containing a certain data element (this is not possible for numerical variables, since the '0' has another meaning, but we can divide them into classes and add a 'missing value' class).

The second and third solutions above should be avoided for any variable having a proportion of missing values in excess of 15–20%.

Statistical replacement of the missing values uses a process called 'imputation'; here we must take a cautious approach. The simplest method is to replace the missing value with the most frequent value (for qualitative variables) or the mean or median (for numerical variables). This is a hazardous procedure, because individuals having missing values tend to be atypical individuals (for example, because their risk level is above average, and they may conceal some information), rather than 'average' individuals. In any case, this procedure distorts the distribution of the imputed variable (see Figure 3.1, taken from the article by Jean-Pierre Nakache and Alice Guéguen cited below). There are more refined methods of imputation, based for example on a regression (Figure 3.2, *loc. cit.*), a clustering method, or a decision tree, for examining the individuals having the same profile as those with missing values. These other individuals may include some with a non-missing value that can be used as a reference. A fairly popular method is to classify individuals by the moving centres or *k*-means method (see Section 9.9.1), and then to impute each missing value with the mean of the variable in the class of the individual.

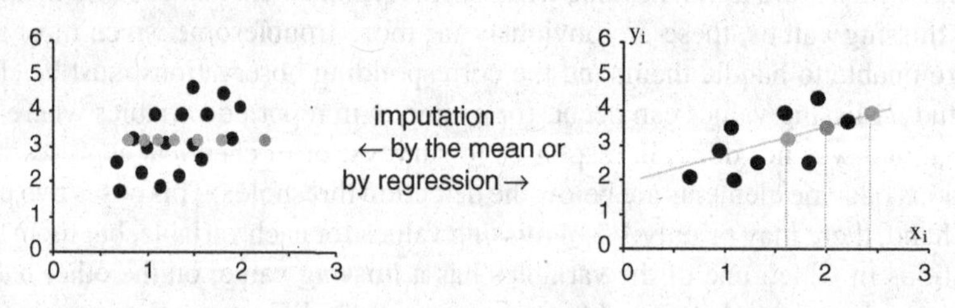

Figure 3.2 Imputation by the mean and by a regression.

In the most widespread imputation model, each missing value is replaced with an assumed value: this is *simple imputation*, which has the drawback of underestimating the variability of the imputed data and the confidence intervals of the estimated parameters. To overcome this, Rubin (1978)[1] and others have proposed a *multiple imputation* method in which each missing value is replaced with a number of plausible values (five is often enough), several complete data tables without missing values are thus obtained, and the desired statistical analyses (such as linear or logistic regression) are performed on each table, after which the results are combined into a set of estimated parameters with their standard deviations. This algorithm is implemented in the SAS/STAT MI procedure (for multiple implementation) and MIANALYSE (for combining the results). The interested reader should refer to *Analysis of Incomplete Multivariate Data*, by Joseph L. Schafer,[2] or the 2005 article by Jean-Pierre Nakache and Alice Guéguen in *Revue de Statistique Appliquée*.[3]

Of course, the MI procedure also allows single imputation, and its imputation using Markov chains (Markov chain Monte Carlo (MCMC) method) or other methods provides a closer approach to reality than an ordinary imputation by the mean or median. It can even impute qualitative values. However, we must remember that MCMC imputation and imputation by regression assume the normality of the imputed variables; if this is not the case, the TRANSFORM instruction can be used to apply a number of transformations (Box–Cox, power, exponential, logarithm, logit) to the variables before imputing them.

The command for single imputation of the variables Var1, Var2 and Var3 is given in the following syntax by the option 'NIMPUTE = 1'. As for the MIN and MAX options, these are used to limit the imputed values, by specifying limits which may be different for the different variables (be careful to specify the limits in the order in which the variables are cited in the 'VAR' command). If we do not wish to carry out the imputation, but simply need to obtain descriptive statistics and the pattern of missing values (Figure 3.3), we choose the 'NIMPUTE = 0' option. Where graphic representations are required, the *VIM* ('Visualization and Imputation of Missing values') package in R can be used to explore the missing data, and, in the later versions, carry out imputation by robust methods.

Note that, if more than one variable is named simultaneously in the VAR command, they are imputed with respect to each other, which presupposes the presence of a real correlation between these variables and enables the imputation to be enriched. Conversely, if we have reason to think that the variables must be independent, it is better to execute the MI procedure for one variable at a time.

```
PROC MI DATA = mytable.before_imputation
OUT = mytable.after_imputation
NIMPUTE = 1 MIN = 0 -1 . MAX = 100 1 .;
VAR var1 var2 var3 ;
RUN;
```

The pattern of missing values is described in a table in which the variables are grouped as a function of their missing values, the variables of a single group being simultaneously filled in or

[1] Rubin, D.B. (1978) Multiple imputations in sample surveys – a phenomenological Bayesian approach to nonresponse. *Proceedings of the Survey Research Methods Section*, American Statistical Association, pp. 20–34.

[2] Schafer, J.L. (2000) *Analysis of Incomplete Multivariate Data*. Boca Raton, FL: Chapman & Hall.

[3] Nakache, J.-P. and Guéguen, A. (2005) Analyse multidimensionnelle de données incomplètes. *Revue de Statistique Appliquée*, 53(3), 35–62.

Characteristics of the missing data								
						Group means		
Group	Var1	Var2	Var3	Freq	Percent	Var1	Var2	Var3
1	X	X	X	6557	80.79	12.217310	0.245615	3.102462
2	X	.	X	3	0.04	0	.	0.166667
3	.	X	X	1108	13.65	.	-0.075471	0.595276
4	.	X	.	353	4.35	.	0.160265	.
5	.	.	X	91	1.12	.	.	0.000916
6	O	O	O	4	0.05	.	.	.

Figure 3.3 Pattern of missing values.

missing for the same variables. A cross 'X' (or a point '.') for a variable signifies that it is filled in (or missing) in the corresponding group. The absence of all the variables is denoted by a circle 'O' on the whole line. Thus, in the fourth group of observations, only the variable Var2 is filled in, and its mean value is 0.16. For four individuals, the three variables Var1, Var2, Var3 are missing.

The *VIM* package in R carries out the same type of calculation and also provides a graphic representation of it, as shown in the following example.[4] In this example, the AGE variable has six missing values, the VISION variable has four missing values, and the COURSE variable has five missing values. Thirty observations have no missing value (Figure 3.4, right-hand graphic, bottom line), while in four observations it is only the VISION variable that has a missing value (top line), and so on.

```
Summary method:

    Missings per variable:
    Variable Count
      AGE     6
      VISION 4
      COURSE 5
    Missings in combinations of variables:
    Combinations   Count    Percent
      0:0:0          30     66.666667
      0:0:1           5     11.111111
      0:1:0           4      8.888889
      1:0:0           6     13.333333
```

Another VIM graphic displays the histogram of each variable (the AGE variable in Figure 3.5), showing the rectangle corresponding to the missing values of this variable on the far right, and showing the rectangles for the missing values of another variable in different colours. In our example, a darker grey indicates the four missing values of the VISION variable, corresponding to ages of 42, 44, 69 and 72 years.

Plenty of other graphics are offered in this useful package.

---

[4] Taken from Cody, R.P. and Smith, J.K. (2005) *Applied Statistics and SAS Programming Language*, 5th edn. Upper Saddle River, NJ: Prentice Hall.

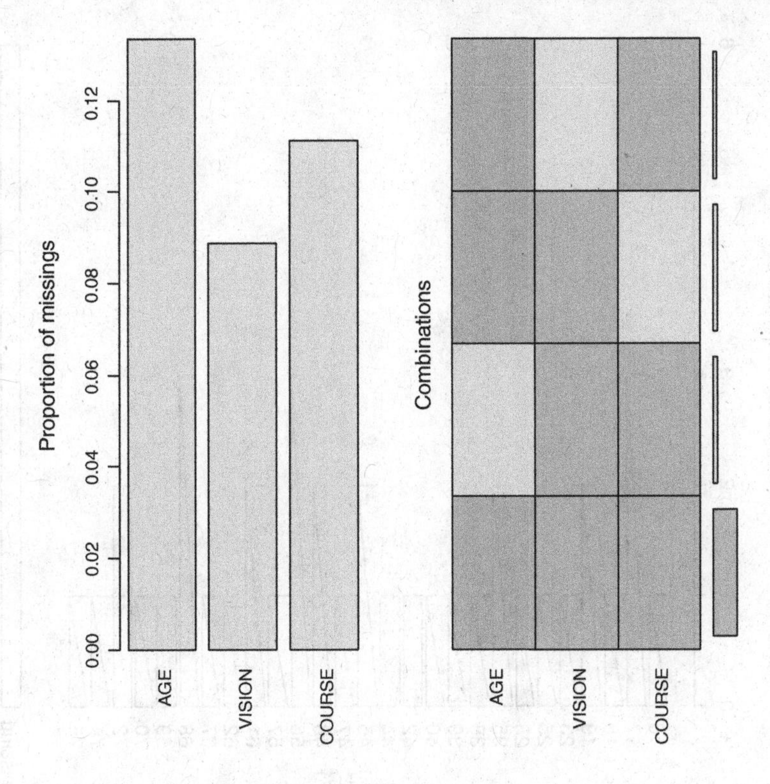

Figure 3.4 Graphic of missing values.

3.4 Detection of aberrant values

This step must be carried out for all types of models.

An aberrant value is an erroneous value corresponding to an incorrect measurement, a calculation error, an input error or a false declaration. While extreme values are not always aberrant, aberrant values are not always extreme, and this makes them harder to detect, possibly requiring a thorough knowledge of the data. However, it is essential to recognize them.

Commercial files can contain considerable numbers of aberrant values:

- incoherent dates, such as unknown birth dates that have been replaced by 'round numbers', subscription dates before the customer's date of birth, dates of last updates in the year 2050, 29 February in a non-leap year, etc.;
- customers declared as 'private' when they are 'business', with obvious effects on the values of their financial data;
- amounts input in cents when they should have been in euros;
- 'sex' codes taking more than two different values to encode specific provident or sickness benefit arrangements;

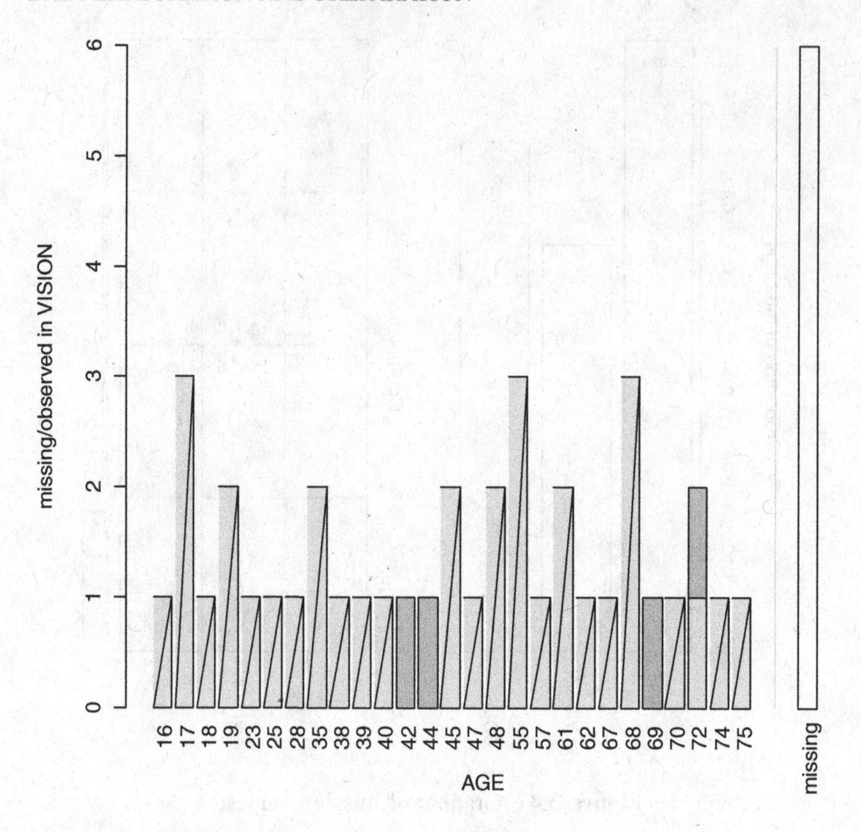

Figure 3.5 Histogram of missing values.

- profession not updated since the start of the customer's relationship with the business, and still appearing as 'student';
- telephone numbers that are not numbers at all, but codes indicating that the customer does not wish to be contacted, that he is not to be sent reminders if there is a delay in payment, that he is ex-directory, etc.

Luckily, aberrant values can often be detected using simple frequency tables or univariate statistics. When this has been done, there are several further steps we can take:

- We can delete the observations in question, if they are not too numerous and if their distribution is suitably random, or if it is clear that they should never have been included in the sample.
- We can keep the observations but remove the variable from the rest of the analysis if it is not considered essential, or replace it with a variable that is similar but has no aberrant values.
- We can keep the observations and the variable, but replace the aberrant value with another value assumed to be as close as possible to its true value (this brings us back to the problem of imputation of missing values).

- We can keep the observations and the variable, and use the variable as it is, tolerating a small margin of error in the results of the models.

The following parameters should be considered when we decide how to act:

- If the anomaly arises because the selected observation is outside the range of the study, as in the case above where business clients are confused with private customers, then clearly we must delete the observation.
- If the variable appears to be poorly discriminant for the phenomenon that is to be predicted, it will be better to keep the observation but eliminate the variable, or treat its aberrant values as missing values.
- On the other hand, if the variable appears to be discriminant and contains very few aberrant values, it would be a shame to lose it, and it would be better to eliminate the defective observations.
- If the variable appears to be discriminant, but contains a large proportion of non-correctable aberrant values, we will be unable to eliminate all the defective observations, and it would be better to avoid using the variable if we want the results to be reliable.
- If the variable appears to be discriminant and contains a reasonable proportion of aberrant values that looks as though it could be reduced, we can try to correct the aberrant values before using the variable.
- If the variable appears to be discriminant and contains a limited proportion of aberrant values, and if it does not look as though the errors can be corrected in a reliable and permanent way, it may be better to use the variable as it stands, without trying to improve it; or we could divide it into classes and assign all the missing values to a specific class.

The correction of an aberrant value, like the correction of a missing value, can be carried out in two ways:

- by using other data sources inside or outside the business (a simple example would be using the telephone directory to find reliable addresses and telephone numbers), or by cross-checking between several variables to establish the reliability of the initial variable (for example, by checking the declared number of children against the number that can be deduced from certain purchases);
- by replacing the aberrant value with a statistically imputed value.

As ever, the imputation must be done with care: if an individual has aberrant characteristics, this may be because his circumstances are very unusual, and it will not be acceptable to replace specific but significant values with a mean or median that smoothes over any differences. Corrections must be made with a light touch, or we may find that the remedy is worse than the disease. We must avoid solutions that would be reliable in the short term, but not in the medium or long term, or would require repeated or costly updates: this is the case with some pointlessly complex systems based on numerous parameters which are too dependent on the context in which they were established.

3.5 Detection of extreme values

This step must be carried out for all types of models. It is optional for decision trees, which can easily tolerate extreme values. However, it is recommended, since extreme values may be aberrant values.

An extreme value is not necessarily an aberrant value, though. It may relate to a specific profile or a specific category of individuals, which may or may not have to be retained in the study; we will have to decide this on a case-by-case basis. An extreme value may relate to a rare profile which is worth detecting. If we exclude it from the population studied we might reduce the usefulness of the learning sample and the resulting model. This is particularly true for the prediction of risks of non-payment or fraud, since the risk profiles often have extreme values for some of their financial ratios.

Leaving aside these cases of searches for atypical patterns, the extreme values of a continuous variable, even if they are not anomalies, will affect some methods, especially logistic regression, discriminant analysis and methods based on variance calculation. It may, therefore, be useful to take the following action:

- exclude the individuals ('outliers') with these extreme values from the learning sample of the model (the model can be tested on these temporarily excluded individuals afterwards), while ensuring that not more than 1–2% of individuals are excluded;
- or divide the continuous variable into classes, the extreme values being placed in the first or last class, but being 'neutralized' to some degree;
- or Winsorizing the variable – Winsorization means replacing the values of the variable beyond the 99th percentile with this percentile, while the values falling before the first percentile are replaced with this first percentile.

We can use standard deviations instead of percentiles. For a normal distribution (Figure 3.6), the values of a variable separated from the mean by more than three standard deviations are very rare, and can generally be considered extreme.

A tool which is very suited to the detection of these extreme values is the 'box plot' described in Appendix A.

Some software, such as SAS Enterprise Miner and IBM SPSS Modeler, can filter the extreme values according to standard deviations or percentiles (Figure 3.7).

3.6 Tests of normality

Tests of normality must be performed for a Fisher discriminant analysis or linear regression, because of the assumptions of these models. It is preferable to deal with the extreme values beforehand.

The normal law, or Laplace–Gaussian law, is frequently encountered in natural, biological and medical phenomena; it is the limit law of the binomial, Poisson and χ^2 distributions, the basis of many statistical indices (such as the kurtosis and skewness), the context of many statistical tests (Student's t, ANOVA, Pearson's correlation) and a basic assumption of Fisher discriminant analysis (X_i/Y distribution), linear regression (residual distribution), etc.

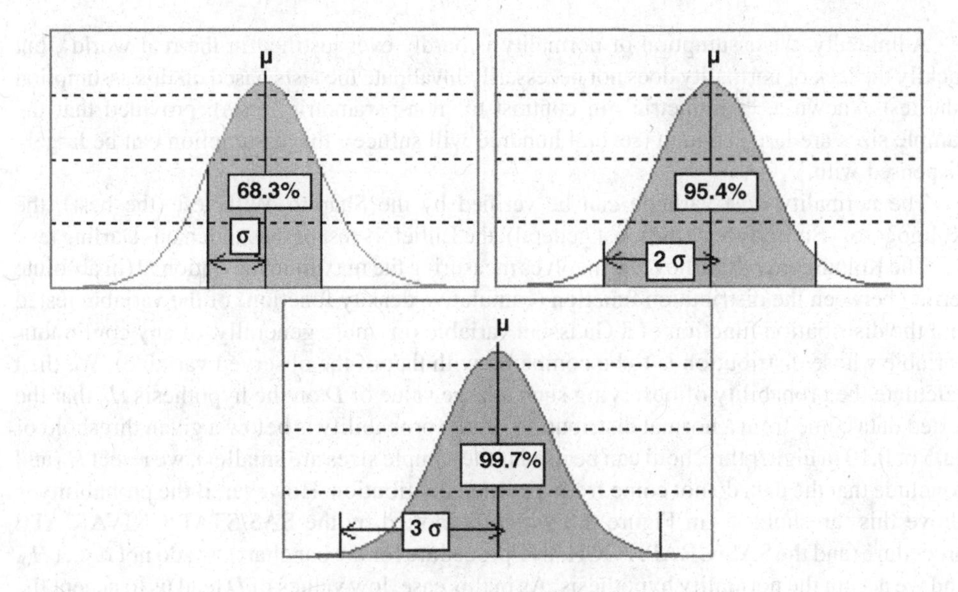

Figure 3.6 Normal distribution.

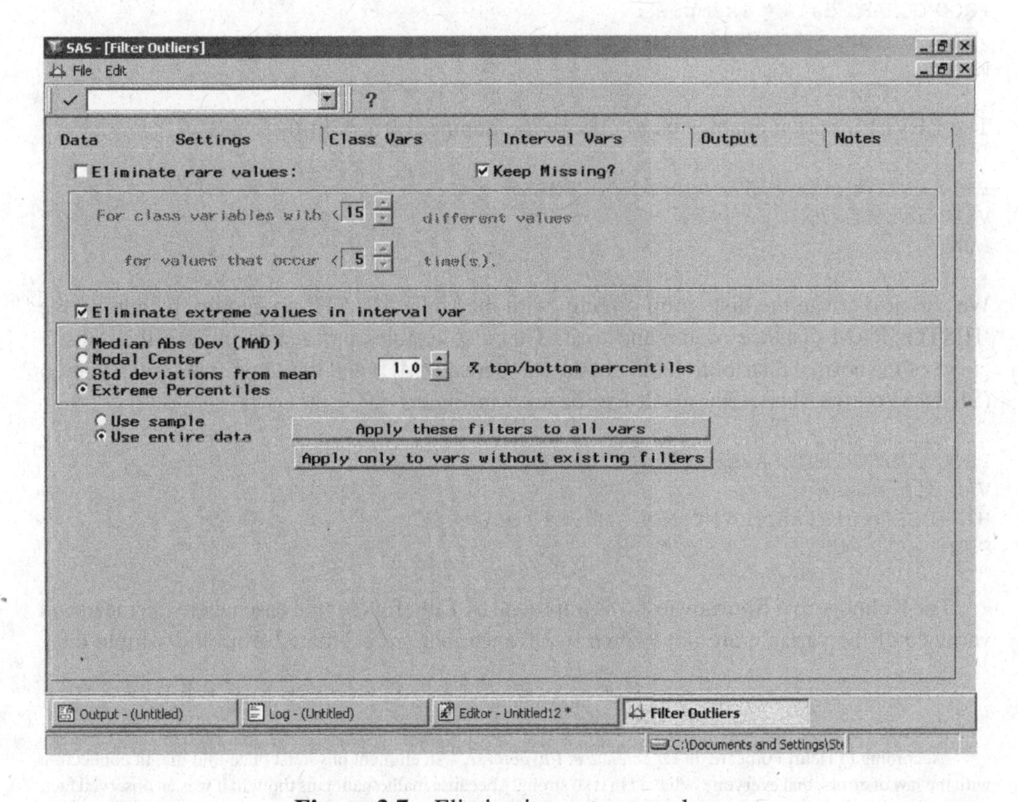

Figure 3.7 Eliminating extreme values.

Admittedly, the assumption of normality is hardly ever justified in the real world[5], but luckily the lack of normality does not necessarily invalidate the tests based on this assumption (the tests known as 'parametric', in contrast to 'non-parametric' tests); provided that the sample sizes are large enough (several hundred will suffice), this assumption can be largely dispensed with.

The normality of a variable can be verified by the Shapiro–Wilk test (the best), the Kolmogorov–Smirnov test (the most general), the Lilliefors test or the Anderson–Darling test.

The Kolmogorov–Smirnov test involves measuring the maximum deviation D (in absolute terms) between the distribution function (cumulative density function) of the variable tested and the distribution function of a Gaussian variable (or, more generally, of any continuous variable whose distribution is to be compared with that of the observed variable). We then calculate the probability of observing such a large value of D on the hypothesis H_0 that the tested data come from a normal distribution. If this probability is below a given threshold of 0.05 or 0.10 (a higher threshold can be used if the sample sizes are smaller), we reject H_0 and conclude that the data do not come from a normal distribution. However, if the probability is above this threshold, as in Figure 3.8 which is plotted by the SAS/STAT UNIVARIATE procedure (and the SAS/GRAPH GCHART procedure for the bar chart), we do not reject H_0, and we accept the normality hypothesis. As in this case, low values of D lead us to accept the hypothesis of normality.

The SAS syntax is as follows:

```
PROC GCHART data = example ;
VBAR distance / LEVELS = 8;
RUN ; QUIT ;
```

The LEVELS option enables us to choose the number of bars in the chart.

```
PROC UNIVARIATE DATA=example NORMAL;
VAR distance;
RUN;
```

We can also create the histogram directly with the UNIVARIATE procedure, by adding the 'HISTOGRAM distance' command to it. This also enables us to superimpose the density curve of the normal distribution having the same mean (MU) and the same standard deviation (SIGMA) on the chart. The result can be seen in Figure 3.9.

```
PROC UNIVARIATE DATA=example NORMAL;
VAR distance;
HISTOGRAM distance / NORMAL (MU=EST SIGMA=EST);
RUN;
```

The Kolmogorov–Smirnov test was improved by Lilliefors for the case where the mean and variance of the variable are not known in advance, but are estimated from the sample data.

[5] According to Henri Poincaré, in *La Science et l'hypothèse*, 'An eminent physicist once told me, in connection with the law of errors, that everyone believed in it so strongly because mathematicians thought it was an observed fact, while observers thought it was a mathematical theorem'.

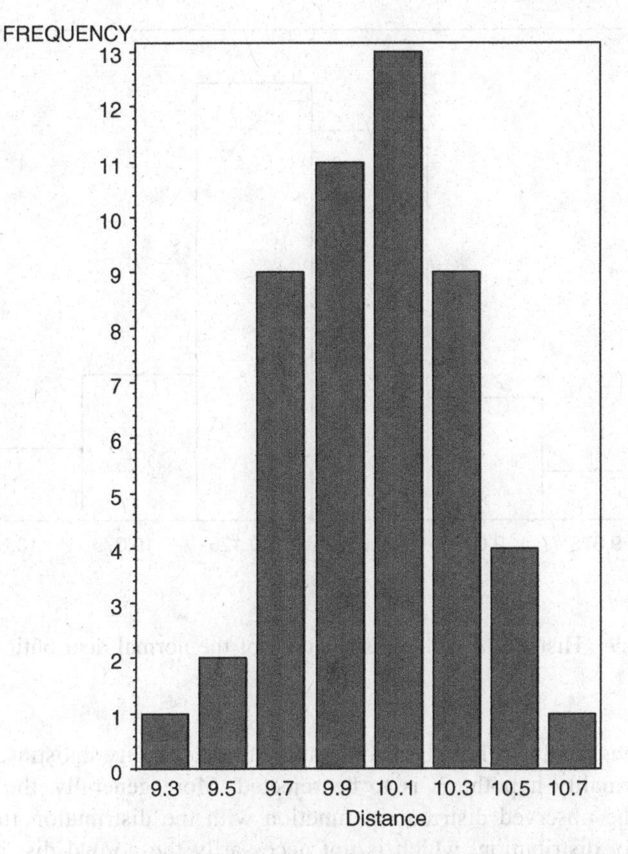

Tests for Normality				
Test	**Statistic**		**p Value**	
Shapiro-Wilk	**W**	0.992264	**Pr < W**	0.9845
Kolmogorov-Smirnov	**D**	0.051999	**Pr > D**	>0.1500
Cramer-von Mises	**W-Sq**	0.02621	**Pr > W-Sq**	>0.2500
Anderson-Darling	**A-Sq**	0.162071	**Pr > A-Sq**	>0.2500

Figure 3.8 The Kolmogorov–Smirnov test.

The original Kolmogorov–Smirnov test was not powerful enough in this case (it rejected the null hypothesis less often when it was false). Another weak point of the Kolmogorov–Smirnov test is that it is less sensitive in the tails of the distribution than in the centre. This aspect has been corrected by the Anderson–Darling test, and by the Cramér–von Mises test which is a special case of the Anderson–Darling test.

In the Shapiro–Wilk test, the cumulative distribution of the data is shown on a normal probability scale, called a P-P (probability–probability) plot, where a normal distribution is shown by a straight line with a slope of 1 (Figure 3.10). Thus the Shapiro–Wilk statistic is a way of measuring how far the graphic representation of the data deviates from the

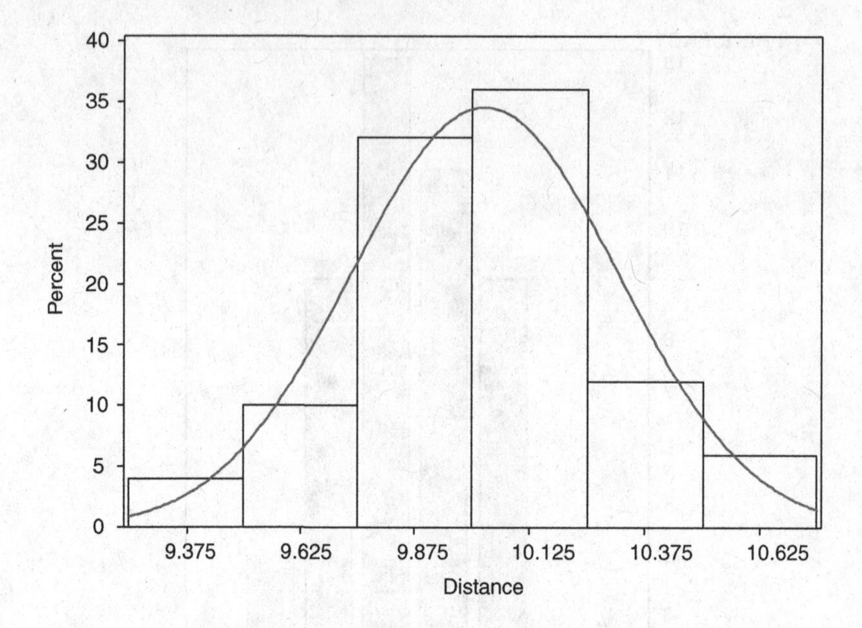

Figure 3.9 Histogram with density curve of the normal distribution.

straight line. It ranges from 0 to 1, and, in contrast to other normality statistics, a low value means that the normality hypothesis is to be rejected. More generally, the P-P plot is used to compare the observed distribution function with the distribution function of a specified probability distribution, which is not necessarily the normal distribution. The probability distribution must be fully specified, and, in the example of a normal distribution, the mean and the standard deviation are specified by the user or calculated by the procedure on the basis of the sample data. The match between the observed distribution function and the theoretical distribution function appears in the form of a diagonal on the P-P plot.

The SAS syntax for generating the P-P plot in Figure 3.10 is:

```
SYMBOL V=plus;
PROC CAPABILITY DATA=example NOPRINT;
     PPPLOT distance / NORMAL (COLOR = black)
                      CFRAME = white
                      SQUARE;
```

This test is more powerful than the Kolmogorov–Smirnov test for small samples (less than 2000 individuals); in these circumstances, it is the best test of normality.

Another useful test, the Jarque–Bera test,[6] specifically uses the characteristics of the normal distribution. It operates by jointly testing the coefficient of skewness S and the

[6] Jarque, C.M. and Bera, AK. (1980) Efficient tests for normality, homoscedasticity and serial independence of regression residuals. *Economics Letters*, 6(3), 255–259.

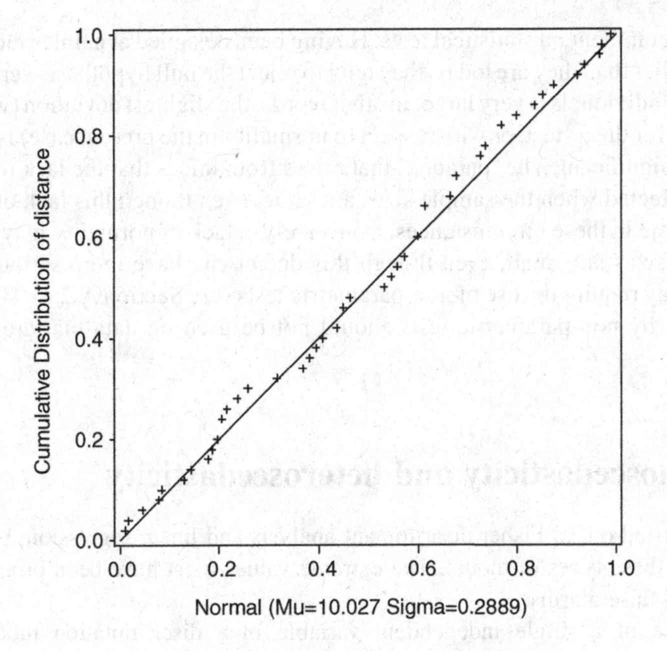

Figure 3.10 Shapiro–Wilk test.

coefficient of kurtosis K of the data set (see Section A.2.1). More specifically, the Jarque–Bera test does not directly test the normality of the data, but tests whether or not the skewness and kurtosis calculated from these data are those of a normal distribution (having the same expectation and the same variance):

H_0: $S=0$ and $K=3$,

H_1: $S\neq 0$ and $K\neq 3$.

According to H_0, the following statistic follows a χ^2 distribution with two degrees of freedom, enabling the hypothesis to be tested:

$$JB = \frac{n}{6}\left(S^2 + \frac{(K-3)^2}{4}\right).$$

This test is asymptotic and is preferably reserved for cases where the sample sizes are fairly large. It is run in R by the jarque.bera.test function in the *tseries* package.

Normality tests are widely available in statistical software. For example, SAS offers them in its procedures NPAR1WAY (Kolmogorov–Smirnov, Cramér–von Mises – see the syntax for this procedure in Section 3.8.3) and UNIVARIATE (Kolmogorov–Smirnov, Cramér–von Mises, Shapiro–Wilk for fewer than 2000 observations, and Anderson–Darling). The Kolmogorov–Smirnov, Lilliefors and Shapiro–Wilk tests are also provided in SPSS (the EXAMINE procedure) and R (the ks.test, lillie.test and shapiro.test functions); R also includes the Anderson–Darling and Cramér–von Mises tests (ad.test and cvm.test functions).

A general comment on statistical tests. Having been designed at a time when volumes of data were smaller than they are today, they tend to reject the null hypothesis very easily when the number of individuals is very large; in other words, the slightest deviation (with respect to independence for the χ^2 test, or with respect to normality in the present case) is detected and appears to be significant. The 'paradox' that arises from this is that the lack of normality is very easily detected when the sample sizes are large, even though this lack of normality is less troublesome in these circumstances. Conversely, a lack of normality may be unnoticed when sample sizes are small, even though this defect can have more serious effects; for purists, this may require the use of non-parametric tests (see Section A.2.4). However, there is no reason why non-parametric tests should not be used on data that are presumed to be normal.

3.7 Homoscedasticity and heteroscedasticity

This test is carried out in Fisher discriminant analysis and linear regression, because of the hypotheses of these types of model. The extreme values must have been processed beforehand, to avoid false alarms.

In the case of a single independent variable in a discrimination model, the term 'homoscedasticity' means the equality of the variances of the variable in a number of samples, for example in the different groups of a population. In the case of more than one independent variable, homoscedasticity means the equality of the covariance matrices of the variables in a number of samples. In the case of linear regression, we speak of homoscedasticity when the variance of the residuals does not depend on the value of the predictors; when this occurs, solutions for correcting it are shown in Section 11.7.1. When a score is calculated to separate k types of individuals, we are particularly concerned with the equality of the covariance matrices of the independent variables in the k sub-populations to be discriminated. This is a basic hypothesis in Fisher's linear discriminant analysis. We will see that, in the absence of homoscedasticity (i.e. when heteroscedasticity is present), we should theoretically abandon linear discriminant analysis in favour of quadratic discriminant analysis. In practice, the use of quadratic discriminant analysis is not often necessary, and only provides a slight improvement in the results, while increasing the number of parameters to be estimated, so adding to the complexity of the model. Heteroscedasticity is considered to be less troublesome when all the samples are of the same size.

Homoscedasticity can be verified by the Levene test (which is the best, because it has low sensitivity to non-normality), the Bartlett test (best if the distribution is normal) or the Fisher test (the least robust if normality is not present), which test the null hypothesis that the variances are equal,

$$H_0 : \sigma_1^2 = \sigma_2^2 = \ldots = \sigma_k^2,$$

against the alternative hypothesis

$$H_1 : \sigma_i^2 \neq \sigma_j^2 \text{ for at least one pair } (i,j).$$

Test of homogeneity of variance

Power

Levene statistic	df1=	df2	Significance
50.448	2	396	.000

Figure 3.11 Levene test.

The Levene test used by SPSS (Figure 3.11) shows that the variance differs significantly between the two groups, the calculated probability being less than 0.05.

The Levene test, together with the Bartlett test that is not included in SPSS, is also available in R (levene.test and bartlett.test functions) and SAS (ANOVA and GLM procedures).

3.8 Detection of the most discriminating variables

This investigation must be carried out when a predictive model is being constructed, except in the case of a decision tree, which detects the most discriminating variables automatically and is not affected by the presence of non-discriminating variables. Most of the other methods give much better results when they are used on a list of variables limited to the most relevant ones. This selection phase is crucial, therefore, and it is also essential to have a thorough knowledge of the data and their functional significance. The problem here is that we must exclude the less discriminating variables while being aware that some variables are less discriminating when taken in isolation and more discriminating when considered with others, because they are less correlated with them. We must therefore mark and retain these variables for the subsequent modelling procedure. There are some helpful techniques, such as variable clustering (see Section 9.14) for ensuring that we have selected at least one representative of each class of variables. Although some software, such as SAS Enterprise Miner, offers automatic detection of discriminating variables, it is also useful if the operator makes a brief inspection to avoid excluding a variable unnecessarily, given the importance of this selection.

For cases where the dependent variable and the independent variables are all quantitative, see Appendix A for information on the Pearson and Spearman correlation coefficients. In the case of a quantitative dependent variable and a qualitative independent variable, the aim is to determine whether the values of a quantitative variable differ significantly between a number of groups: this requires a test of the equality of the mean, known as 'single factor analysis of variance' (see Appendix A). When there are *n* qualitative independent variables, an '*n*-factor ANOVA' is carried out. Single-factor analysis is much more common, and I will discuss below the case of a qualitative dependent variable and a quantitative independent variable, which is symmetrical and which we shall return to: we will check to see if the dosage of a medicine affects recovery (the qualitative dependent variable), in order to find out whether the fact of belonging to the 'recovered' group has a link with the dosage (the quantitative dependent variable).

So for now we simply need to examine the case of a qualitative dependent variable, as found in scoring for example. We will do this by making a distinction based on the nature of the independent variables.

3.8.1 Qualitative, discrete or binned independent variables

For variables that are qualitative, discrete, or continuous but divided into classes, we test their link with the dependent variable by calculating the values of χ^2, the associated probability, and Cramér's *V* coefficient. The advantage of Cramér's V is that it integrates the sample size and the number of degrees of freedom of the contingency table, using χ^2_{max} (see Section A.2.12). It is more readable than the probability associated with χ^2 and, above all, it provides an absolute measure of the strength of the link between two qualitative or discrete variables, regardless of the number of their categories and the size of the population.

Note that, in contrast to the probability associated with the χ^2 of a table, χ^2_{max} depends only on the minimum of the number of rows and the number of columns, not on both of these numbers at once. In particular, if one of the two cross-tabulated variables (for example, the dependent variable) is binary, χ^2_{max} is equal to the size of the population and does not depend on the number of categories of the second variable.

Let us take the case of a two-column table, to see why χ^2_{max} does not depend on the number of rows. In this case, χ^2_{max} is found when one or other of the columns *X* and *Y* is 0 in each row. Thus we obtain a table in the following form:

		V1		
		X	Y	*TOTAL*
V2	A	50	0	*50*
	B	0	25	*25*
	C	0	25	*25*
	TOTAL	*50*	*50*	*100*

In this table, χ^2 does not change when we aggregate all the rows for which $Y = 0$ as well as all the rows for which $X = 0$, in other words when we return to a 2×2 table (for which $\chi^2 = \chi^2_{max} =$ sample size):

		V1		
		X	Y	*TOTAL*
V2	A	50	0	*50*
	B + C	0	50	*50*
	TOTAL	*50*	*50*	*100*

Thus χ^2_{max} for an $l \times 2$ table depends not on *l*, but only on the total sample size. This is why Cramér's *V* for two tables, $l \times 2$ and 2×2, can be compared, and Cramér's *V* for two variables V2 can be compared with a binary dependent variable V1, even if the two variables V2 do not have the same number of categories (particularly if V2 is formed from a single continuous variable discretized in several ways). In our example, Cramér's *V* is 1 and χ^2 is 100 for both tables, but the probability associated with χ^2 is 1.93×10^{-22} for the first table and

1.52×10^{-23} for the second. Clearly, it is meaningless to claim that V2 is more discriminating in the second case, as its power to predict V1 is exactly the same.

Two other examples demonstrate the superiority of Cramér's *V* over the χ^2 probability for the process of selecting variables. The two variables V3 and V4 shown below have exactly the same power to predict V1, since V4 is obtained by aggregating the rows of V3, and rows A and B on the one hand, and C and D on the other hand, are linearly related. V3 and V4 have the same Cramér's $V = 0.19$, and the same $\chi^2 = 5.357$, but the probability of the χ^2 of V3 is 0.15 (a non-significant link), while for V4 it is 0.02 (a significant link). At the 5% threshold we will accept V4 but not V3, although in fact the number of degrees of freedom changes (1 instead of 3), but not the discriminating power of the variables.

		V1		
		X	Y	*TOTAL*
V3	A	30	20	*50*
	B	30	20	*50*
	C	10	15	*25*
	D	10	15	*25*
	TOTAL	*80*	*70*	*150*

		V1		
		X	Y	*TOTAL*
V4	A + B	60	40	*100*
	C + D	20	30	*50*
	TOTAL	*80*	*70*	*150*

Another, less trivial, example is provided by the variable V5 below:

		V1		
		X	Y	*TOTAL*
V5	A	30	20	*50*
	B	25	15	*40*
	C	5	5	*10*
	D	20	30	*50*
	TOTAL	*80*	*70*	*150*

V5 is a refinement of the division of V4, the first category of which is divided into three subcategories. Since the rows of V5 are linearly independent, we can say that V5 is slightly more discriminating than V4: its power to predict V1 is slightly greater. In fact, its Cramér's *V* is slightly greater than that of V4: 0.20 instead of 0.19. On the other hand, χ^2 probability is much higher for V5 (at 0.12) than for V4 (0.02), suggesting that V5 ought to be excluded from the selection variables in favour of V4.

Even more annoying is the fact that, if we rely on the χ^2 probability for the selection of variables, we may reach opposite conclusions depending on whether we are working on a whole population or a sample. Suppose that V4 and V5 are observed in 1% of the population, and the contingency tables for the whole population are as follows:

		V1		
		X	Y	*TOTAL*
V4	A + B	6000	4000	*10 000*
	C + D	2000	3000	*5000*
	TOTAL	*8000*	*7000*	*15 000*

		V1		
		X	Y	*TOTAL*
V5	A	3000	2000	*5000*
	B	2500	1500	*4000*
	C	500	500	*1000*
	D	2000	3000	*5000*
	TOTAL	*8000*	*7000*	*15 000*

The Cramér's V values are unchanged, because they are insensitive to the sample size. However, the χ^2 probabilities are now 1.6×10^{-118} for V4 and 1.1×10^{-126} for V5. This time, therefore, the variable selection process will favour V5 rather than V4, by contrast with the result found for the 1% sample.

In conclusion, the selection of variables that are qualitative, discrete, or continuous but divided into classes must be based on Cramér's *V* rather than the χ^2 probability. If we rely on this probability, we run the risk of failing to select variables that would in fact be found to be discriminating after a regrouping of some of their categories.

For ordinal independent variables, we can use Kendall's tau as described in Appendix A.

3.8.2 Continuous independent variables

For continuous independent variables, we can carry out a parametric test of variance (ANOVA) with one factor, or a non-parametric test, unless we conduct both tests to give us a better chance of capturing all the independent variables that are potentially useful. The ANOVA test is appropriate if the independent variables are normal and have the same variance regardless of the category of the dependent variable. If one of the two hypotheses of normality and homoscedasticity is not satisfied, it will be preferable to use a more robust non-parametric test, which could be either the Wilcoxon–Mann–Whitney test (binary dependent variable) or the Kruskal–Wallis test (two or more categories). We can also use the Welch ANOVA test where there is normality without homoscedasticity. There is another test, known as the median test, but it is less powerful than the Wilcoxon–Mann–Whitney and Kruskal–Wallis tests. Overall, I prefer the Kruskal–Wallis test, because of its generality and the readability of the result, which takes the form of a χ^2 with one degree of freedom, increasing as the continuous independent variable is linked more closely with the dependent variable.

The ANOVA test yields the R^2 indicator, which is the inter-class sum of squares *ESS* divided by the total sum of squares *TSS*, and which is closer to 1 as the strength of the link between the dependent qualitative variable and the continuous variable tested increases (see Section A.2.13). We are mainly interested in the classification of the variables according to the value of R^2, rather than the significance level (the probability associated with this value), since in some cases most of the variables would be selected at the 5% threshold. This indicator R^2 is supplied by the SAS/STAT ANOVA procedure using the following syntax, in which it should be noted that the qualitative dependent variable becomes an independent variable of the (means of the) continuous variables indep_var_x. Note that the single-factor ANOVA test is also executed by the GLM procedure, which is slower but can also be used for multiple-factor ANOVA tests in both balanced and unbalanced designs, whereas the ANOVA procedure can only be used for multiple-factor ANOVA in the case of balanced designs (with the same number of observations for each combination of factors) and in some other very specific cases.

```
PROC ANOVA DATA=table;
 CLASS dependent;
 MODEL indep_var_1 indep_var_2 ... = dependent;
RUN;
```

The outputs of the ANOVA procedure shown below demonstrate that the first independent variable is clearly better than the second in terms of explaining the dependent variable.

```
R-Square   Coeff Var   Root MSE   indep_var_1 Mean
0.018519   532.7779    4.218779    0.791846

R-Square   Coeff Var   Root MSE   indep_var_2 Mean
0.001116   31.41090    0.295972    0.94225
```

Warning: it is sometimes better to run these tests after excluding abnormal individuals from the sample, because extreme values bias the results to a significant degree, and may make a variable appear to be more predictive than it really is (or vice versa).

The values of R^2 and χ^2 in the Kruskal–Wallis test increase with the strength of the link; the independent variables are therefore listed by decreasing values of the preceding indicators, so that the most predictive variables are at the head of the list. This list can be obtained simply, even when there are many independent variables, by using the programming languages of software such as SAS and SPSS, as mentioned in Section 3.8.4. To this list we can add the variables which would not be detected by the above methods, because of a non-monotonic effect on the dependent variable for example, but which are detected by a CHAID or CART decision tree. If we use a tree, we must restrict ourselves to the first split and measure all the best splitting variables detected by the software, in addition to the first variable selected, and not take into account the first splitting variables of all the subsequent splits, which are most discriminating variables for a sub-population only. This list can be written in summary form as in Table 3.3.

We may need to supplement the list with variables that were not previously selected, but are identified as useful by experts in the field. In fact, some variables are found to be

Table 3.3 Table for selecting variables.

variable	KW χ^2	KW rank	R^2	R^2 rank	CHAID χ^2	no. of CHAID nodes	df	CHAID prob.	CHAID rank	CART rank
X	2613.076444	1	0.0957	1	1023.9000	5	4	2.3606E-220	1	1
Y	2045.342644	2	0.0716	2	949.0000	5	4	4.0218E-204	2	11
Z	1606.254491	3	0.0476	5	902.3000	4	3	2.8063E-195	4	12
...	1566.991315	4	0.0491	4	920.4000	4	3	3.3272E-199	3	4

powerfully discriminating in risk or attrition scores, because they alert us to the event late and without much risk of error: in other words, they sound the fire alarm when the smoke is already visible. On the other hand, some variables are earlier indicators and apparently less discriminating with respect to the risk, but allow more effective preventive action to be taken by flagging up problems sooner.

There is a happy medium, as suggested above, by which we can avoid losing variables which would only be moderately discriminating individually, but are globally useful because they are less closely linked with other independent variables. In this method, we quickly identify groups of variables and then look for at least one variable in each group. This factor analysis is carried out on numerical data, discretized previously (see Section 3.10) if some of them have missing values, using either the SAS VARCLUS algorithm (see Section 9.14), or the ACP Varimax method (see Section 7.2.1). The latter method operates as follows.

For each of the factor axes used (the ones with eigenvalues greater than 1, using the Kaiser criterion), we can obtain the coefficients of correlation with the set of variables. Each axis is strongly correlated with some variables and weakly correlated with others: for each axis, we will examine the variables that are most closely correlated with it. Variables with a correlation coefficient (coordinate of the variable on the axis) above 0.7 will be assigned to each axis: these are the variables that contribute more to this axis than to all the others added together (see Section 7.1.2: for a given variable, the sum of squares of its coefficients of correlation with the set of axes is equal to 1). In this way we obtain several groups of variables: the number will be equal to the number of axes used, less a few axes without any variable assigned, plus any group of variables that is not assigned to any axis. If the last group is too large, the threshold of 0.7 on the correlation coefficient is reduced. We can also choose to assign each variable to the axis with which it is most closely correlated, without specifying a threshold for the correlation coefficient.

If the volume of data is not too great (a few tens of thousands of observations), we can use another technique to refine the selection of variables by marking the most robust ones. With this method, we construct a certain number (50–100) of samples found by random sampling with replacement; these are called 'bootstrap' samples. We launch a modelling process (logistic regression or discriminant analysis) on each bootstrap sample, with stepwise selection of the variables, and then count how many times each variable is selected in a model. Depending on the number of variables tested, we accept only those variables which have appeared at least two or three times, or all those which have appeared at least once. This selection method decreases the sensitivity to sampling, which makes a variable appear sufficiently discriminating on one sample but not on another, even if both samples were constructed by random sampling from the same number of individuals using the same rule.

By means of this exhaustive analysis, we can expect to find all the discriminating variables, regardless of their type, which may be that of the variable X in Figure 3.12 or Y in Figure 3.13. X is discriminating and can be used as it stands in any classification method. Y is discriminating, but must be used in a decision tree or must be divided into classes in advance if it is to be used in a classification method such as logistic regression. Z in Figure 3.14 is not discriminating and should be rejected.

3.8.3 Details of single-factor non-parametric tests

The aforementioned generalizations of Student's t test are shown, with others, in Table 3.4.

The non-parametric tests (Wilcoxon–Mann–Whitney, Kruskal–Wallis, median, and Jonckheere–Terpstra) operate on the ranks of the values, rather than on the values themselves,

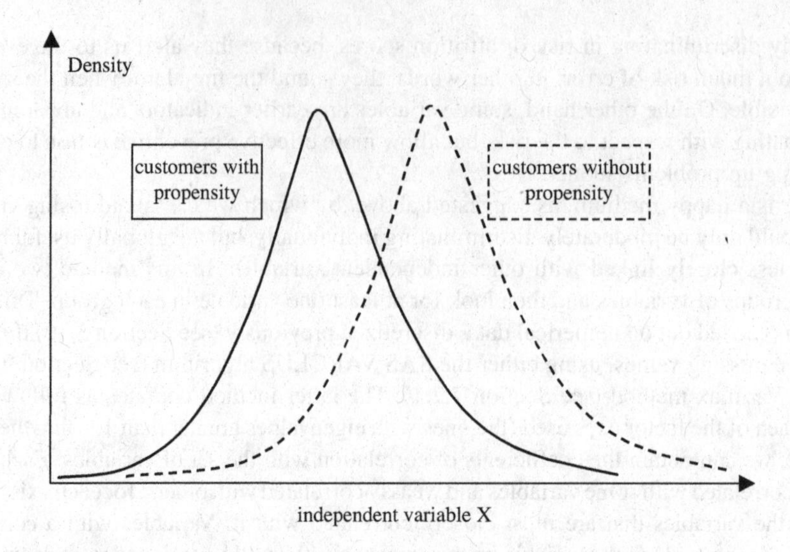

Figure 3.12 Linear discrimination.

and are therefore more robust and can also be applied to purely ordinal variables. Note that the *t* and ANOVA tests are less affected by a lack of normality than by heteroscedasticity.

I will now describe the principle of the Wilcoxon–Mann–Whitney test for two groups. The observations of the two groups are combined and ordered, and a mean rank is assigned to them for tied values. The observations for the two groups must be independent and the number of tied values must be small with respect to the total number of observations. The number of times that a value of group 1 is smaller (i.e. its rank is lower) than a value in group 2 is calculated and denoted by U_1. The number of times that a value in group 2 is smaller (i.e. its

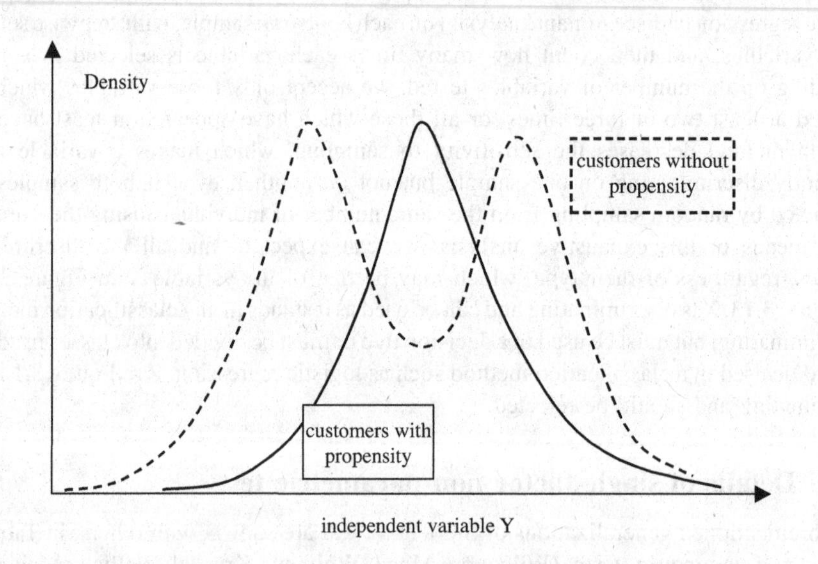

Figure 3.13 Non-linear discrimination.

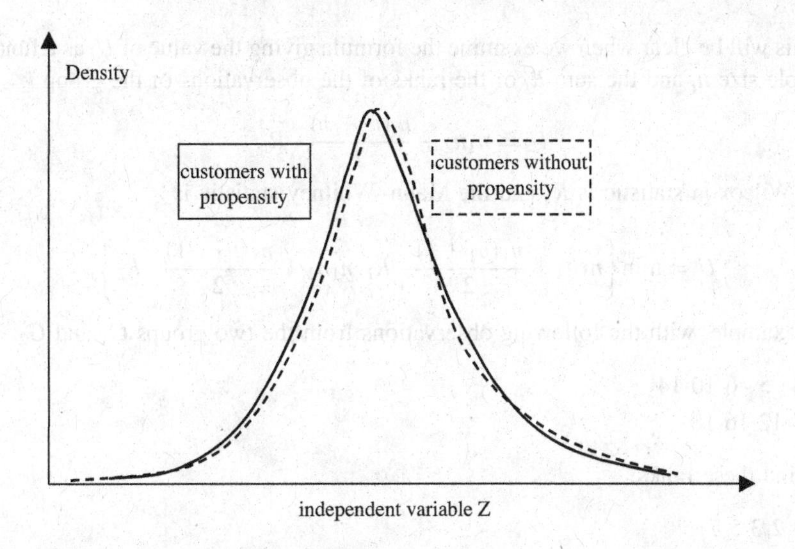

Figure 3.14 Non-discrimination.

Table 3.4 Mean comparison tests.

Form of distribution	**Two samples**	**Three or more samples***
normality and homoscedasticity	Student's t test	ANOVA
normality and heteroscedasticity	Welch's t test	Welch – ANOVA
non-normality and heteroscedasticity	Wilcoxon–Mann–Whitney	Kruskal–Wallis
non-normality and heteroscedasticity	median test	median test
non-normality and heteroscedasticity		Jonckheere–Terpstra test (ordered samples)

*Do not compare all the pairs by t tests, because significant differences will be incorrectly detected (at the 95% threshold, for four equal means, at least one significant difference will be detected in 27% of cases). Such a test, which rejects the null hypothesis too readily, is described as 'liberal' or 'powerful'.

rank is lower) than a value in group 1 is calculated and denoted by U_2. The Mann–Whitney U statistic is the smaller of these two numbers.

Wilcoxon's rank-sum statistic[7] S is the sum of the ranks of the smaller (or first) sample. Evidently, as U decreases, the groups become more significantly different, and the same applies if S is very large or very small. Each of these statistics is associated with a test in which the null hypothesis is that the ranks of group 1 do not differ from the ranks of group 2 (both groups are taken from the same population). The alternative hypothesis is that either the ranks of the two groups are different (two-tailed test) or that the ranks of one group are greater than those of the other group (one-tailed test). The commonly used term, 'Wilcoxon–Mann–Whitney test', is justified by the fact that both statistics lead to equivalent

[7] Not to be confused with the Wilcoxon signed-rank statistic for two matched samples.

tests. This will be clear when we examine the formula giving the value of U_i as a function of the sample size n_i and the sum R_i of the ranks of the observations of the group i:

$$U_i = n_1 n_2 + \frac{n_i(n_i+1)}{2} - R_i.$$

The Wilcoxon statistic is R_1 and the Mann–Whitney statistic is

$$U = \min\left\{ n_1 n_2 + \frac{n_1(n_1+1)}{2} - R_1, n_1 n_2 + \frac{n_2(n_2+1)}{2} - R_2 \right\}.$$

For example, with the following observations from the two groups G_1 and G_2:

G_1: 3 5 6 10 14
G_2: 8 12 16 18

we find these ranks:

G_1: 1 2 3 5 7
G_2: 4 6 8 9

and therefore $R_1 = 18$, $R_2 = 27$, $U_1 = 20 + 15 - 18 = 17$, $U_2 = 20 + 10 - 27 = 3$ and $U = U_2 = 3$ (note: it is a general property that $U_1 + U_2 = n_1 n_2$). The three cases ($U_2 = 3$) in which a value of group G_2 is smaller than a value of group G_1 are those of the pairs (10, 8), (14, 8) and (14, 12).

The fact, as mentioned above, that the null hypothesis is more easily rejected as U becomes smaller is explained as follows. U is compared to a critical value found in a table for a fixed threshold, and if U is below this critical value, the null hypothesis is rejected. This exact test has a corresponding asymptotic test: when the sample sizes n_1 and n_2 are greater than 8, the U statistic tends, according to the null hypothesis, towards a normal distribution with a mean $\mu = n_1 n_2/2$ and variance $\sigma^2 = n_1 n_2(n_1 + n_2 + 1)/12$. The asymptotic test is conducted by calculating the quotient $Z = (U - \mu)/\sigma$ and comparing the calculated value $|Z|$ with the theoretical value found in the table for the standard normal distribution. If $|Z| \geq |Z_{table}|$, we reject the null hypothesis; if $|Z| < |Z_{table}|$, we accept it and conclude that there is no significant difference between the groups. At the usual threshold of 5%, the null hypothesis will be rejected if $|Z| \geq 1.65$ for a one-tailed test, and if $|Z| \geq 1.96$ for a two-tailed test.

Similarly, the exact distribution of the Wilcoxon statistic can be found in a table for small sample sizes, or calculated by software that can carry out an exact test, but there is also an asymptotic test available in the same conditions as for the Mann–Whitney U statistic: when n_1 and $n_2 > 8$, the S statistic tends under the null hypothesis towards a normal distribution with a mean $\mu = n_1(n_1 + n_2 + 1)/2$ and variance $\sigma^2 = n_1 n_2(n_1 + n_2 + 1)/12$ (note that this is equal to the variance of the Mann–Whitney statistic). In the example below, this variance is 4.082, while the mean μ is 10 for Mann–Whitney and 20 for Wilcoxon (based on the smaller group).

The SAS/STAT NPAR1WAY procedure (single-factor non-parametric tests) calculates the Wilcoxon statistic instead of the Mann–Whitney statistic, based on the smaller group; in this case the result is $S = 27$. With the EXACT instruction, the NPAR1WAY procedure (see below for the full syntax) carries out an exact test; we may note here that it is considered non-significant at the 5% threshold and does not prove the existence of a significant difference between the two grounds. For its part, the one-tailed asymptotic test ('normal approximation') is considered significant, but the two-tailed test is not (since $1.96 > Z \geq 1.65$). The value of Z used for the test is the quotient $(S - \mu)/\sigma = (27 - 20)/4.082 = 1.7146$. This difference between the conclusions of the asymptotic and exact tests is commonly found.

Wilcoxon Scores (Rank Sums) for Variable x Classified by Variable a

a	N	Sum of Scores	Expected Under H0	Std Dev Under H0	Mean Score
1	5	18.0	25.0	4.082483	3.600
2	4	27.0	20.0	4.082483	6.750

Wilcoxon Two-Sample Test

Statistic (S)	27.0000
Normal Approximation	
Z	1.7146
One-Sided Pr > Z	0.0432
Two-Sided Pr > |Z|	0.0864
t Approximation	
One-Sided Pr > Z	0.0624
Two-Sided Pr > |Z|	0.1248
Exact Test	
One-Sided Pr >= S	0.0556
Two-Sided Pr >= |S - Mean|	0.1111

The non-parametric Kruskal–Wallis test is used for $k \geq 2$ groups, with the same null hypothesis as before, in other words that the groups all come from the same population and that there is no difference between their ranks (since the alternative hypothesis is that the ranks of at least two groups differ, the Kruskal–Wallis test is essentially two-tailed). Let N be the number of observation, n_i the size of the group i, and R_i the sum of the ranks of the observations of the group i. The test statistic is

$$H = \frac{12}{N(N+1)} \sum_{i=1}^{k} \frac{R_i^2}{n_i} - 3(N+1).$$

A correction must be made where there are equalities of ranks. If the sizes are large or if $k > 6$, H tends under the null hypothesis towards a χ^2 distribution with $k - 1$ degrees of freedom; otherwise, we must look up the critical values in a table. Even in a case of normality and homoscedasticity, this test is almost as powerful as ANOVA. Here, the test is not considered significant at the 5% threshold, and we find that the associated probability (0.0864) is equal to that of the two-tailed Wilcoxon test. This is by chance, but in any case these probabilities are always similar. No exact Kruskal–Wallis test is implemented in the SAS/STAT NPAR1WAY procedure.

Kruskal-Wallis Test	
Chi-Square	2.9400
DF	1
Pr > Chi-Square	0.0864

The median test is designed to test the null hypothesis that the samples come from populations having the same median. When there are two samples, the median test is a χ^2 test (at least if the sample sizes are large) applied to a 2×2 contingency table whose columns correspond to the samples and whose rows contain the number of observations over (or under) the median.

The Jonckheere–Terpstra test is used when there is a natural a priori classification (ascending or descending) of the k (≥ 3) populations from which the samples are taken. Suppose we wish to determine the effect of temperature on the growth rate of trout, where this is measured in three samples at water temperatures of 16, 18 and 24 °C. If we are attempting to reject the null hypothesis of the equality of growth rates in the three samples, the Jonckheere–Terpstra test is preferable to the ANOVA test because it takes the increasing order of the temperatures into account.

These tests can be carried out by most statistical programs such as SAS and SPSS. The SAS/STAT procedure, as mentioned above, is:

```
PROC NPAR1WAY WILCOXON DATA=table CORRECT=no;
/* suppress continuity correction */
CLASS a; /* classification variable */
VAR x; /* numeric variable*/
EXACT; /* optional exact test */
RUN;
```

The WILCOXON option on the first line launches the Kruskal–Wallis test, and the Wilcoxon–Mann–Whitney test if the number of groups (= number of categories of the variable 'a') is 2. By replacing WILCOXON with ANOVA, MEDIAN or EDF we will respectively obtain an ANOVA test, a median test, or tests of the equality of the distribution of the variable 'x' in the groups defined by the variable 'a' (Kolmogorov–Smirnov and Cramér–von Mises tests, and, if 'a' has only two categories, the Kuiper statistic). The instruction EXACT is optional, and launches the determination of an exact test to be used for small samples. The Jonckheere–Terpstra test is not run by the SAS NPAR1WAY procedure, but by the FREQ procedure, using the option JT on the TABLES line.

3.8.4 ODS and automated selection of discriminating variables

I have already mentioned the programming language of certain software, which enables the list of independent variables, classed according to their discriminating power, to be drawn up automatically. In this section, we will see how the SAS Output Delivery System (ODS) can be used for this purpose, noting that the SPSS Output Management System can also yield the results shown here, although the programming would be rather more complex.

For continuous independent variables, let us take the example of an automatic method of this kind, using the Kruskal–Wallis test and its implementation in the SAS NPAR1WAY procedure which we looked at before. In the example of syntax shown below, the dependent variable is the 'target' qualitative variable specified on the CLASS line, and the continuous independent variables are specified on the VAR line. They may be very numerous, and it would be tedious to scan tens of pages of results to find all the Kruskal–Wallis χ^2 values, given that only the highest ones are useful. Rather than read the results directly in the SAS Output window, we will start by sending them to an SAS file, using the ODS. By placing the instruction 'ODS OUTPUT KruskalWallisTest = kruskal' in front of the NPAR1WAY procedure, we specify that the Kruskal–Wallis test results are to be retrieved, and that we wish to retrieve them into the file 'kruskal'. The first qualification is required because we have seen that the NPAR1WAY procedure runs not only the Kruskal–Wallis test, but also the Wilcoxon–Mann–Whitney test when the number of categories of the dependent variable is 2. Of course, if we wished to retrieve the result of the Wilcoxon test, we would have to replace 'KruskalWallisTest' with 'WilcoxonTest'.

The data set 'kruskal' contains three observations for each variable mentioned on the VAR line: one of these (characterized by the variable _NAME1_ = '_KW_') contains the χ^2 value, another (with the variable _NAME1_ = 'DF_KW') contains the number of degrees of freedom of the χ^2 test (still 1 in this case), and a third (with the variable _NAME1_ = 'P_KW') contains the probability associated with the value of χ^2 (the content here being equal to that of the χ^2 value, since the number of degrees of freedom is always the same). In our case, we simply need to retrieve the first type of observation '_KW_', which we can do with the condition WHERE. We then sort the file by decreasing values 'nValue1' of χ^2 ('cValue1' is the χ^2 value in alphanumeric format) and the first 30 lines of it are printed, assuming that we are only interested in the 30 variables most closely linked to the dependent variable. We could also export this file in Excel format.

```
ODS OUTPUT KruskalWallisTest = kruskal ;

PROC NPAR1WAY WILCOXON DATA = table ;
CLASS target;
VAR var1 var2 var3 var4 var5 ...;
DATA kruskal (keep = Variable cValue1 nValue1);
  SET kruskal;
WHERE name1 = '_KW';

PROC SORT DATA = kruskal;
BY DESCENDING nValue1;

PROC PRINT DATA = kruskal (obs = 30);
RUN;
```

The result is as follows:

Obs	Variable	cValue1	nValue1
1	var3	5821.0000	5821.000000
2	var2	1039.1879	1039.187933
3	var4	1032.8305	1032.830498
4	var5	803.8562	803.856189
5	var1	693.4167	693.416713
6	...	633.9656	633.965609

Otherwise, for independent variables that are qualitative, discrete, or continuous and divided into classes, where the link with the dependent variable is tested by calculating the value of Cramér's *V* coefficient, the ODS 'KruskalWallisTest' is replaced by the 'ChiSq' table and the above syntax becomes:

```
ODS OUTPUT ChiSq = Chi2;

PROC FREQ DATA=table;
TABLES target * (var1 var2 var3 var4 var5 ...) / CHISQ;

DATA Chi2 (keep = Table Value);
SET Chi2;
WHERE Statistic = 'Cramers V';

PROC SORT DATA = Chi2;
BY DESCENDING Value;

PROC PRINT DATA = Chi2 (obs = 30);
RUN;
```

Note that it is the FREQ procedure that, used with the option CHISQ, yields Cramér's *V*, together with the χ^2, the Mantel–Haenszel χ^2 and the coefficient Φ.

One last comment. The normal output to the Output window is specified in ODS by the keyword LISTING (the default output) and it is possible to request the cancellation or reactivation of this output. By writing the following syntax, we prevent the results of PROC NPAR1WAY from being displayed in the Output window (using the ODS LISTING CLOSE instruction), but we allow the results of subsequent procedures to be displayed there (using the ODS LISTING instruction). This option is useful if the number of variables is very large, because the outputs of PROC NPAR1WAY could fill several hundred pages in the output window unnecessarily.

```
ODS LISTING CLOSE ;
ODS OUTPUT KruskalWallisTest = kruskal ;

PROC NPAR1WAY WILCOXON DATA = table ;
CLASS target;
VAR var1 var2 var3 var4 var5 ...;
RUN;

ODS LISTING ;
```

However, this solution is not suitable if we wish to send the outputs simultaneously to the Output window and an HTML or RTF document, which we can do simply by bracketing the processes between a first line ODS RTF FILE (for an RTF document) and a last line ODS RTF CLOSE (to close the file), enabling us to present SAS outputs in a Word document in a much more elegant way than by copying and pasting from the Output window. By adding these two extra lines, and leaving the ODS LISTING CLOSE instruction, we will block the display of the outputs of PROC NPAR1WAY in the Output window, but not in the RTF file, which will be created with all the outputs as if no ODS LISTING CLOSE instruction had been given. To block all the ODS outputs of all kinds, we must use the ODS EXCLUDE ALL instruction; afterwards, we can return to the normal display with the ODS SELECT ALL instruction. The syntax for placing the list of 30 variables most closely linked to the dependent variable in a file 'sas_ods.doc' is as follows:

```
ODS RTF FILE = "c:\sas_ods.doc" ;
ODS EXCLUDE ALL ;
ODS OUTPUT KruskalWallisTest = kruskal ;

PROC NPAR1WAY WILCOXON DATA = table ;
CLASS target;
VAR var1 var2 var3 var4 var5 ...;
RUN;

ODS SELECT ALL ;

DATA kruskal (keep = Variable cValue1 nValue1);
 SET kruskal;
WHERE name1 = 'KW';

PROC SORT DATA = kruskal;
BY DESCENDING nValue1;
PROC PRINT DATA = kruskal (obs = 30);
RUN;
ODS RTF CLOSE;
```

3.9 Transformation of variables

We often need to carry out this step, regardless of the type of model. The particular type of transformation known as *normalization* of the variables is widely used in Fisher discriminant analysis, because of the assumption of multivariate normality in this type of model, and also in cluster analysis. Another transformation is even more widely used, but we will save this for the next section: this is *discretization*, i.e. the division of continuous variables into classes. Another fairly common transformation is one in which the original variables are replaced with their factors, continuous variables which are produced by *factor analysis* (see Chapter 7) and which are remarkable in that a few of the factors (which are sorted) contain the essentials of the information. This method will be discussed later. Finally, there is the transformation known as *creation of indicators*, such as the ratios X/Y or $X(\text{period } t)/X(\text{period } t{-}1)$ from the raw data, which is preferably done in cooperation with specialists in the field under investigation.

If the number of variables is large, this step may sometimes be carried out only after the step of selecting the most discriminating variables, to avoid having to transform all the variables instead of concentrating on the discriminating variables only. However, some transformations can be executed automatically, such as the one in which a value of 'missing' is substituted for one which is known to be equivalent (e.g. '999999999'), or a qualitative variable coded in n characters is replaced with one coded in fewer characters, with which a label is then associated. Thus 'accommodated free of charge' is replaced with 'F', and the appropriate label is associated with the code 'F': this reduces computation time and the amount of data stored.

Now let us examine the normalization of a continuous variable. This is done by transforming the variable with a mathematical function, which compresses its distribution, brings it closer towards a normal distribution, and if necessary decreases its heteroscedasticity and increases its discriminating power in a linear model. Financial data, whose distributions are generally more 'peaked' than the normal curve and more elongated towards the right, are processed in this way. The same applies to reliability data, product concentrations, cell counts in biology or percentages ranging from 0 to 100. The smaller degree of flattening compared with the normal curve is revealed by the *kurtosis*, while the elongation of the distribution to the right is shown by a coefficient of skewness greater than 0 (see Section A.2.1).

The transformation function is frequently the Napierian logarithm (if $V \geq 0$, V is replaced with $\log(1 + V)$) or the square root, if the skewness is positive. If this coefficient is negative, the transformation function is often V^2 or V^3. For a percentage from 0 to 100, the arcsine function ($\sqrt{V/100}$) is most suitable. Some software can carry out an appropriate transformation automatically, using the Box–Cox transformation function. Examples are the box.cox function in the *car* package and the boxcox function in the *MASS* package in R. The Box–Cox transformation is based on a parameter λ which is estimated in such a way as to optimize the result. Demonstrations can be found on the Internet at:

```
http://wiki.stat.ucla.edu/socr/index.php/SOCR_
EduMaterials_Activities_ PowerTransformFamily_Graphs
```

Following the 1964 paper by Box and Cox,[8] variants have been proposed by other authors, including Manly (1971),[9] John and Draper (1980),[10] Bickel and Doksum (1981)[11] and Yeo and Johnson (2000).[12]

The example in Figure 3.15 of transformation of the variable 'household income' shows that the logarithm is not always the most suitable transformation function. This variable is visibly 'peaked' and skewed to the right, as confirmed by the values of the skewness (2.38) and kurtosis (11.72) and the P-P plot of the cumulative distributions (Figure 3.16). Looking at the logarithm of the income (actually the logarithm of '1 + income') in Figure 3.17, we see that the situation has not improved, because the distribution is even more 'peaked' (kurtosis = 12.03) and it is skewed to the left this time (skewness = −2.03). However, the square root of the incomes shows a distribution much closer to normality (Figure 3.18), with kurtosis 1.76 and skewness 0.64. The diagram of cumulative distributions (Figure 3.19) is also very close to a straight line.

3.10 Choosing ranges of values of binned variables

Discretization is unnecessary if we use a decision tree, which can itself divide continuous variables into classes: the CHAID, CART, C4.5 and C5.0 trees can also be used very successfully for this purpose (see below). However, when we start with continuous variables, this step must always be carried out before multiple correspondence analysis (MCA), DISQUAL discriminant analysis (a linear discriminant analysis on the factors of an MCA)

[8] Box, G.E.P. and Cox, D.R. (1964) An analysis of transformations. *Journal of the Royal Statistical Society, Series B*, 26(2): 211–252.

[9] Manly, B.F.J. (1976) Exponential data transformations. *The Statistician*, 25, 37–42.

[10] John, J.A. and Draper, N.R. (1980) An alternative family of transformations. *Applied Statistics*, 29, 190–197.

[11] Bickel, P.J. and Doksum, K.A. (1981). An analysis of transformations revisited. *Journal of the American Statistical Association*, 76, 296–311.

[12] Yeo, I.K. and Johnson, R.A. (2000). A new family of power transformations to improve normality or symmetry. *Biometrika*, 87, 954–959.

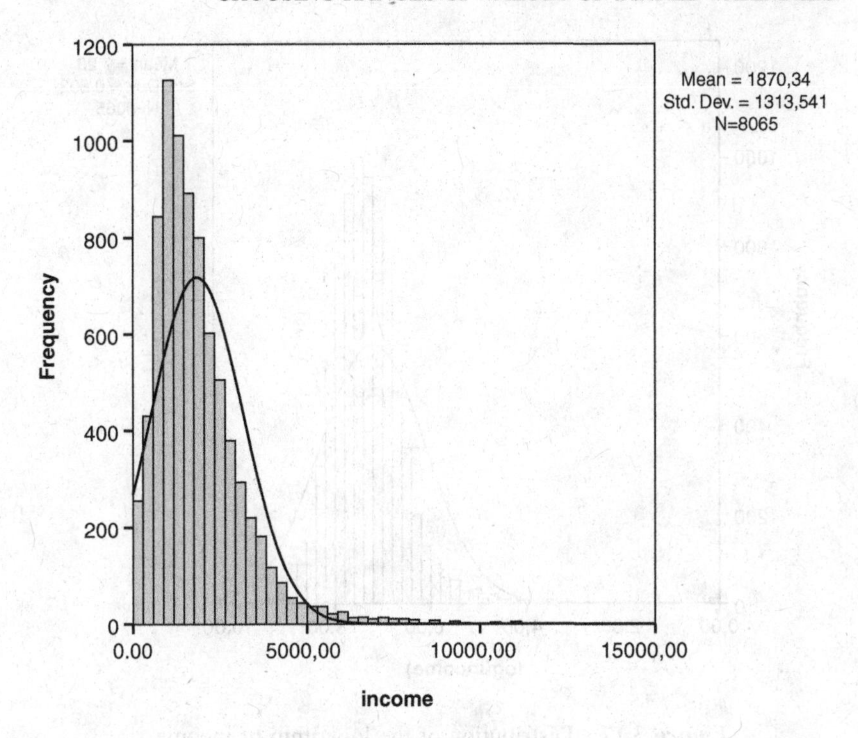

Figure 3.15 Income distribution.

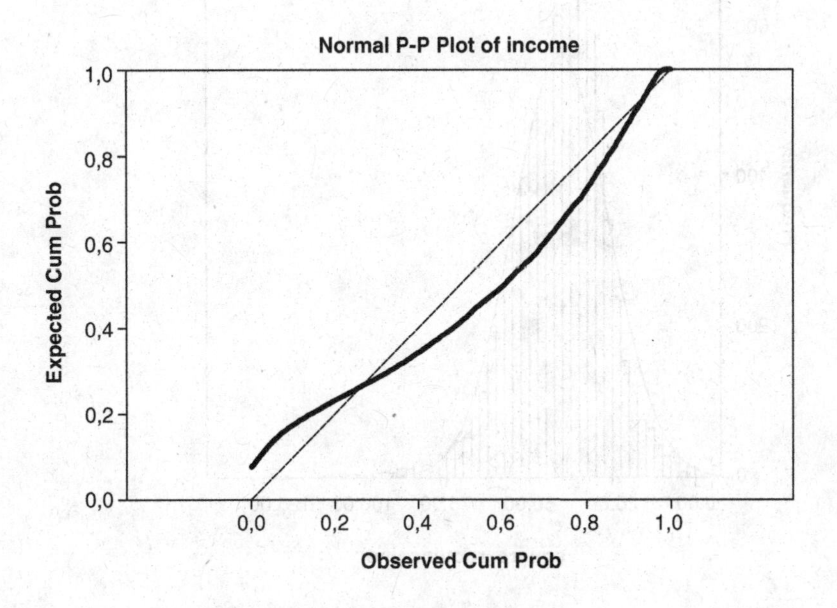

Figure 3.16 P-P plot of income.

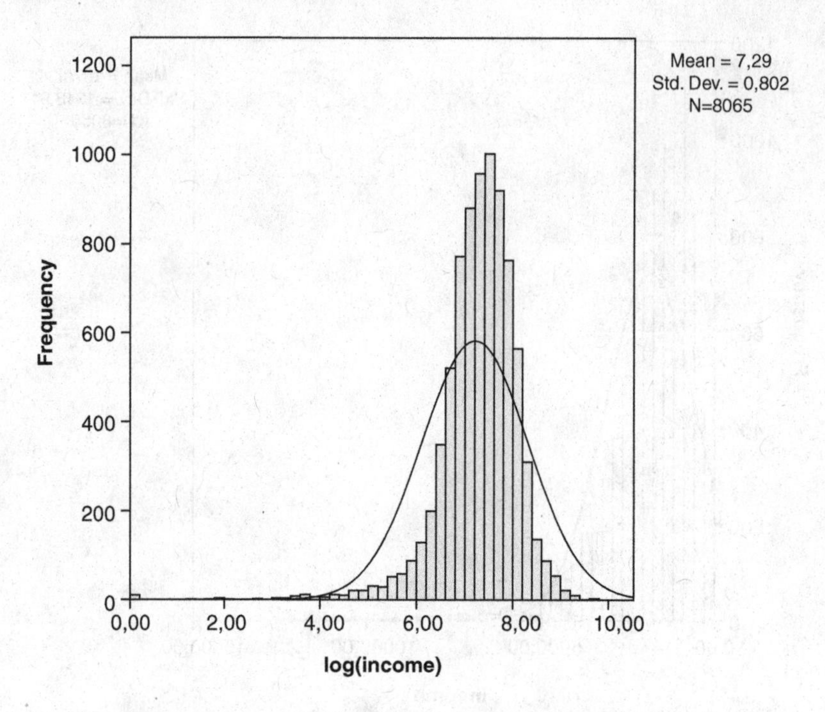

Figure 3.17 Distribution of the logarithm of income.

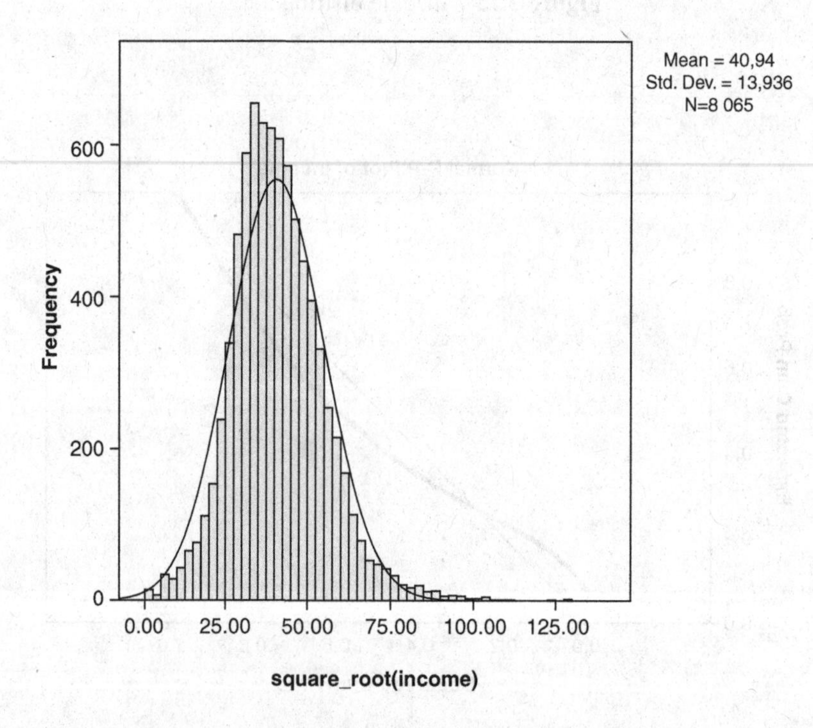

Figure 3.18 Distribution of the square roots of income.

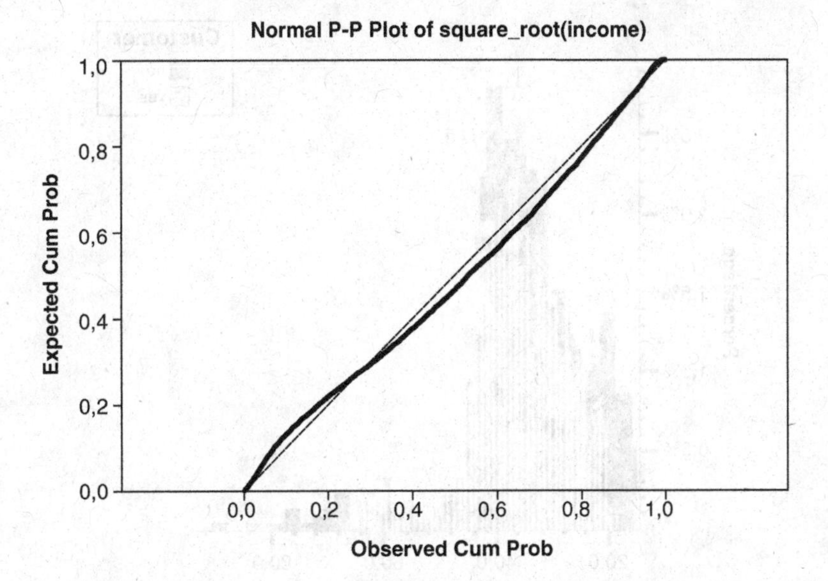

Figure 3.19 P-P plot of the square roots of income.

or clustering by similarity aggregation (the Condorcet method). It is unnecessary before factor analysis of mixed data, since this handles both quantitative and qualitative data. The question of discretization arises when logistic regression is to be undertaken, since logistic regression can incorporate quantitative and qualitative variables into the same model. It is true that logistic regression on discretized continuous variables is often better than logistic regression on continuous initial variables (see Section 11.8.4), but this is not always the case.

As a general rule, and especially before carrying out a logistic regression, we need to decide whether to divide each independent variable into classes, by checking to see if the phenomenon to be predicted is a linear function of the independent variable, or at least monotonic (in other words, always increasing or decreasing). An example of a non-linear response might relate to an illness where the frequency of occurrence increases more rapidly at some ages than at others. An example of a non-monotonic response is attendance at health spas, which will be lower for young and retired persons and higher for active persons. Non-linear responses are quite common, unlike non-monotonic responses.

Where we have a non-linear monotonic response as a function of X, we can sometimes preserve a continuous form, by modelling $f(X)$ instead of X, where f is a monotonic function such that $f(x) = x^2$ for example, if the response is quadratic.

However, a non-monotonic response requires a deeper transformation of the independent variable, and in this case we usually discretize it, in other words divide it into classes or 'ranges'. We can then model it by logistic regression or DISQUAL discriminant analysis.

The division into classes is carried out naturally by taking the points of inflection or intersection of the distribution curves as the class limits ('purchasers exceed non-purchasers'). However, ranges of values having the same behaviour with respect to the dependent variable are grouped together. For example, in Figure 3.20, we can distinguish the range '45–65 years'.

There are three other situations where continuous variables may usefully be discretized. The first of these relates to the presence of *missing values,* where these are infrequent enough

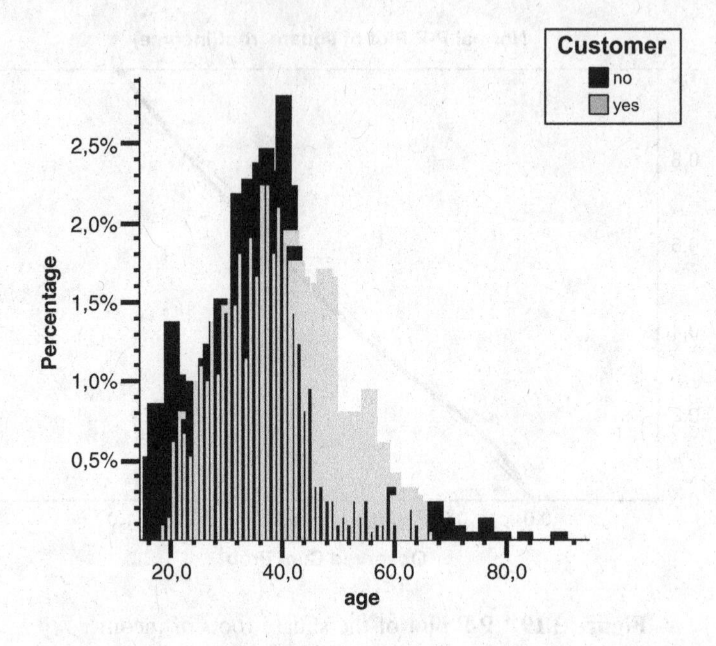

Figure 3.20 Division of a continuous variable into classes.

to make it unnecessary to exclude the variable (see Section 3.3). Since the imputation of missing values is always a tricky matter, the preferred solution is usually to divide the variable into ranges, then to add another range corresponding to all the missing values; this range can be examined and joined to another range if appropriate. Note in passing that division into classes may help to decrease the rate of missing values at source: in a questionnaire, respondents are more willing to provide details of wage ranges than their exact pay.

Then there are *extreme values* which are not easy to correct: clearly, discretization eliminates this problem, and we no longer have 150-year-old customers, but simply customers in the '80 years and above' range. The extreme values will be assigned to the first or last class. Clearly, discretization is not the only solution; for instance, another one is Winsorization (Section 3.5).

Finally, when using logistic regression, the determination of the coefficients may be uncertain if the number of individuals is small, because there will not be enough points to estimate the maximum likelihood. In this case, discretization increases the robustness of the model by grouping individuals together.

This all goes to show that dividing continuous variables into classes does not necessarily lead to a loss of information, but is often beneficial to the modelling procedure. It also enables us to go on to process quantitative variables in the same way as qualitative variables.

Having said this, we still need to know the best way of discretizing the variables. There is no universal recipe, and certainly no fully automatic method, even though some data mining software has started to incorporate this kind of functionality.

Among other decision trees, CHAID (see Section 11.4.7) can be very helpful, as the following example will show. Suppose that we wish to predict a target variable from a number of variables including age, and that we wish to discretize the age variable. We start by dividing the ages into 10 ranges (or more, if the number of individuals is large), then inspect the percentage of individuals in the target for each age class.

			target		
			no	yes	Total
age	18–25 years	Count	127	81	208
		% within age	61.1%	38.9%	100.0%
	25–29 years	Count	104	126	230
		% within age	45.2%	54.8%	100.0%
	29–32 years	Count	93	101	194
		% within age	47.9%	52.1%	100.0%
	32–35 years	Count	113	99	212
		% within age	53.3%	46.7%	100.0%
	35–38 years	Count	93	94	187
		% within age	49.7%	50.3%	100.0%
	38–40 years	Count	149	123	272
		% within age	54.8%	45.2%	100.0%
	40–42 years	Count	108	72	180
		% within age	60.0%	40.0%	100.0%
	42–45 years	Count	116	97	213
		% within age	54.5%	45.5%	100.0%
	45–51 years	Count	77	113	190
		% within age	40.5%	59.5%	100.0%
	> 51 years	Count	71	145	216
		% within age	32.9%	67.1%	100.0%
Total		Count	1051	1051	2102
		% within age	50.0%	50.0%	100.0%

We will subsequently group together the classes which are close in terms of percentage in the dependent variable, namely range 2 and 3, ranges 4 to 8, and ranges 9 and 10. If we now launch the CHAID tree on the variable 'age', we can see that it automatically does what we have done manually, by dividing the continuous variable into deciles, then using the χ^2 criterion to group together the classes which are closest in terms of the dependent variable (Figure 3.21). The number of classes produced by CHAID is not usually specified by the user directly, but is based on the minimum size of a node and the threshold of χ^2, these values being included in the input parameters of the tree algorithm (see Section 11.4.7).

We do not always use such a purely statistical method for the discretization; sometimes we may combine statistical criteria with 'professional' ones which may result in the choice of thresholds which are significant for the problem under investigation (for example, 18 years for an age criterion).

Certain basic principles are generally followed for clustering or MCA, regardless of the discretization method used:

1. Avoid having too many differences in the numbers of classes between one variable and another.

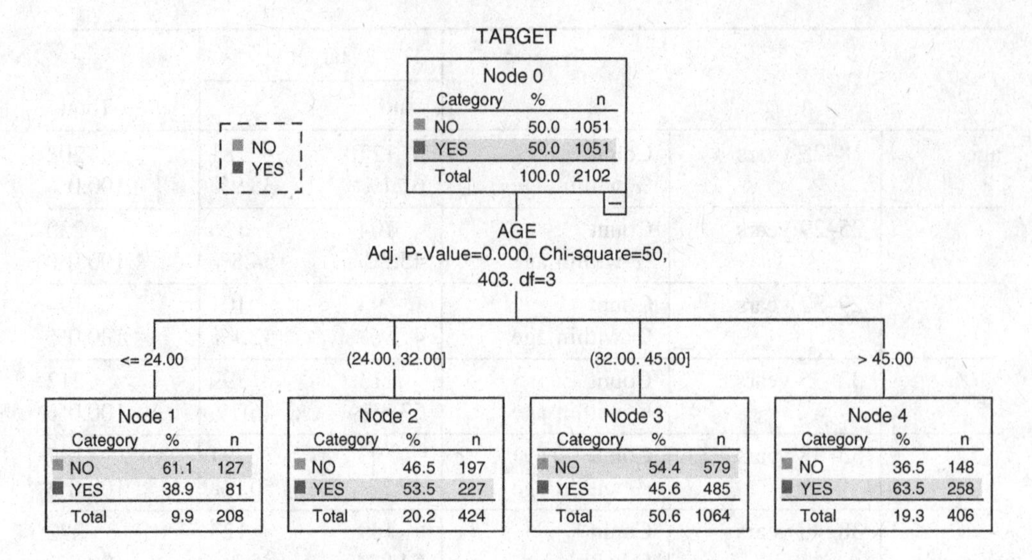

Figure 3.21 Automatic discretization using CHAID.

2. Avoid having too many different class sizes for each variable.

3. Especially avoid having classes that are too small.

4. About 4 or 5 classes is often found to be a good number.

To speed up the process, we can divide continuous variables into quartiles or quintiles, and then adjust the thresholds so that they 'fall just right' or correspond to the reference thresholds of the problem.

For prediction, the number of classes is determined partly by the sample size (there can be more classes with a larger size, since there has to be a minimum of individuals in each class), and partly by the distribution of the dependent variable as a function of the independent variable. In the example in Figure 3.20, there will be either three or two natural classes (if we group the two extreme classes together), and in the example in Figure 3.21 there will be either four or three natural classes (if the two central classes, which are quite similar, are grouped together). Quite often, a continuous variable can be usefully divided into two classes; this operation is called *binarization*. This gives us two classes of the type '0/>0' (absent/present) or '$\leq$0/>0', or possibly with other thresholds. As a general rule, we try to limit the number of classes, to ensure that the percentages in the target are clearly different from one class to the next, and to reduce the number of parameters in the model, thereby increasing its robustness.

We have seen how the CHAID tree can be used to automate the division into k classes. Other algorithms such as ChiMerge[13] are based on the χ^2 criterion, or on the Pearson ϕ (see Section A.2.12), for example StatDisc.[14] Other approaches are based on entropy and

[13] Kerber, R. (1992) ChiMerge: Discretization of numeric attributes. In *Proc. Tenth National Conference on Artificial Intelligence*, pp. 123–128. Cambridge, MA: MIT Press.

[14] Richeldi, M. and Rossotto, M. (1995) Class-drivenstatistical discretization of continuous attributes. In *Machine Learning: ECML-95 Proceedings European Conference on Machine Learning*, Lecture Notes in Artificial Intelligence 914, pp. 335–338. Berlin: Springer.

minimize the sum of entropies of the categories, as in the C4.5 tree. For continuous variables, these have the advantage of finding the optimal threshold (according to the chosen criterion), since all possible split thresholds are tested. An example of this approach is implemented in the SPSS Data Preparation module (see Chapter 5 on software). The CART tree (see Section 11.4.7) can also be used to carry out binarization in an optimal way, since it optimizes the Gini index of purity by testing all the possible split thresholds for the variable. These algorithms can also be used generally for the optimal supervised grouping of the categories of a categorical variable, and are increasingly common in data mining software.

Of course, each continuous variable can be divided in several different ways, for example in binary and non-binary ways, and all the forms of division can be input into the predictive model for testing. We can then keep the most appropriate division.

When we have a number of variables to be discretized, rather than just one, we should not attempt to complete the operation in one step with just one tree. What we need to do when using this method is to request, on each occasion, the construction of a tree with a single level and a single variable to be divided, identified as an independent variable. Multiple-level trees will not be suitable here, because the division of a variable at a level of the order of $n \geq 2$ depends on the division at the level $n-1$ and therefore is not globally, but only locally, optimal (see Section 11.4.9). Only a tree with one level of depth provides an overall optimum method, in this case for the division of the variable. Thus, in order to discretize a set of p variables, we start the tree algorithm p times, once for each of the variables. A macro-language such as that provided by some software is very useful for automating the procedure. My recent book *Étude de cas en statistique décisionnelle* proposes a macro-program based on the CHAID algorithm and the SAS TREEDISC macro for automatic discretization of a set of variables.[15]

Since these processes may be rather complex, if the number of variables is high and if there is no automatic procedure that is fast enough, this step is sometimes left until after the selection of the most discriminating variables, in order to concentrate on the truly discriminating variables only. Warning: the classification will affect the discriminating power, so we may miss out on an interesting variable!

3.11 Creating new variables

This step must be considered for all types of models. Like the transformation of variables carried out before it, this step is not universal; some models only use the initial variables. However, it is often useful in non-scientific areas, where the variables have not been collected with statistical analysis in mind, and may not always be best suited to the investigation of the problem.

In this case, new variables can be created from the initial variables and may be more discriminating. Here are a few common examples, in no particular order:

- the date of birth and date of first purchase are combined to give the customer's age at the commencement of his relationship with the business;
- the set of variables 'product P_i purchased (Yes/No)' can be used to find the number of products purchased;
- the number of purchases and the total value of these purchases give the mean value of a purchase;

[15] Tufféry, S. (2009) *Étude de Cas en Statistique Décisionnelle*. Paris: Technip.

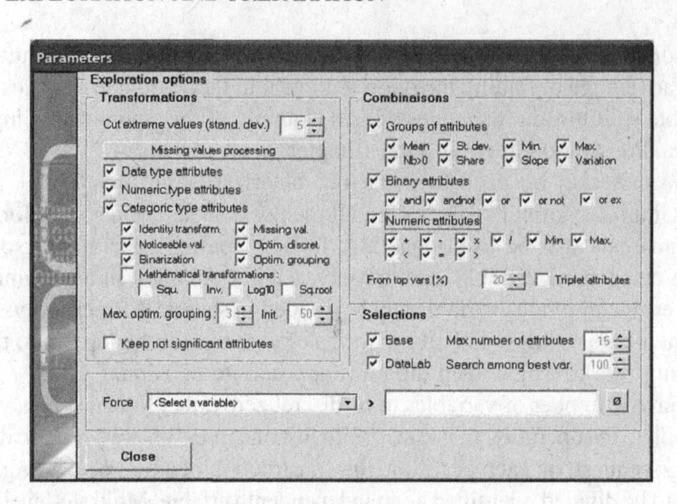

Figure 3.22 Generation of variables by DataLab.

- the dates of purchase tell us how recent and frequent the purchases were;
- the credit limit and the amount actually used can be combined to find the rate of use of the credit.

Thus new variables are generally created by the modeller, but there are automatic variable creation methods, such as the genetic algorithms described in Section 11.13.

Another useful program is DataLab, developed by Complex Systems. One of its functions, Data Scanning (Figure 3.22), is designed for data preparation as a preliminary to predictive modelling, and for generating combinations of variables. The user specifies the dependent variable and the independent variables. DataLab then automatically transforms the variables, Winsorizing and discretizing the continuous variables, binarizing the categorical variables or grouping their categories, and combining variables with each other. It automatically creates a large number of variables of the form V^2, $1/V$, $\log(V)$, $\text{root}(V)$, V_1/V_2, V_1V_2, $V_1 + V_2$, $(V_1 > V_2)$, etc. Finally, it uses a linear regression (for a numeric dependent variable) or logistic regression (for a categorical dependent variable) to select the most discriminating variables. The resulting model can be exported in SAS, SPSS, SQL or other code. This tool should not be relied on unthinkingly for the automatic selection of all variables and the construction of the final model, but DataLab is very useful for suggesting combinations of variables which might not have been thought of.

3.12 Detecting interactions

One phenomenon commonly encountered in the real world is that the simultaneous action of two variables is not the sum of the independent actions of the variables. We speak of 'interactions' between the variables.

This step of detecting the interactions has to be performed for some models such as linear models (linear discriminant analysis or logistic regression), which are additive, meaning that when they are expressed in the form

$$Y = \alpha_1 X_1 + \ldots + \alpha_k X_k,$$

the effect of each variable X_i is independent of the effect of the other variables. In other words, each coefficient α_i is uniquely determined. If the reality to be modelled is more complex, and if, for example, a person's weight has different effects in a model depending on his size or age, we can sometimes take the quotient 'weight/size' if the relationship between the two variables is linear. But if the relationship between them is more complex, or if the variables are not numeric, the interactions must be introduced, by taking the product of indicators, such as:

$$1_{age\in A1} \cdot 1_{weight\in P1}, 1_{age\in A2} \cdot 1_{weight\in P2}, 1_{age\in A2} \cdot 1_{weight\in P1}, 1_{age\in A1} \cdot 1_{weight\in P2}, \ldots$$

or

$$1_{\text{married}} \cdot 1_{\text{no children}}, 1_{\text{unmarried}} \cdot 1_{\text{no children}}, 1_{\text{married}} \cdot 1_{\text{with children}}, 1_{\text{unmarried}} \cdot 1_{\text{with children}}.$$

The interactions of variables are used when the sample sizes are large enough. They are less useful for continuous or discrete variables, and more useful for qualitative variables (e.g. 'one-parent family with three children').

The interactions can be detected by two-way tables or by decision trees (incidentally, part of the name of the first tree, AID, stands for 'interaction detection'). It is increasingly common for logistic regression algorithms or those of other generalized linear models (see Section 11.9.6) to enable the contribution of interactions to be tested in a model. Thus, if we take the example of predicting deaths in the sinking of the *Titanic*, we see that the class*sex interaction has more effect than age (in binary form: 0 = child/1 = adult).

Analysis of Maximum Likelihood Estimates

Parameter	DF	Estimate	Standard Error	Wald Chi-Square	Pr > ChiSq
Intercept	1	−5.9403	0.5645	110.7220	<.0001
CLASS	1	1.7191	0.1898	82.0339	<.0001
AGE	1	1.0205	0.2275	20.1137	<.0001
SEX	1	6.0881	0.5148	139.8860	<.0001
CLASS*SEX	1	−1.5505	0.1947	63.4238	<.0001

Now, when we relate sex (0 = F, 1 = M), class (1st, 2nd, 3rd, 0 = crew) and survival (yes/no) to each other, we see (Figure 3.23) that the rate of survival does indeed decrease with the class number for women, but that the survival rate for men is slightly higher in third class than in second. The interaction between class and sex is not very pronounced, but it is there. There is also an interaction between age and sex, because the survival rate is higher for women than for men, and higher for children than for adults, but lower for girls than for women. To investigate this interesting example further, you can download the data file from:

http://www.amstat.org/publications/jse/datasets/titanic.dat

We can make the SAS/STAT LOGISTIC procedure test all the possible interactions between the three independent variables beyond these three variables, by writing

```
PROC LOGISTIC DATA=titanic;
MODEL survived = class | age | sex / SELECTION=forward;
RUN;
```

Sex				Survived no	Survived yes	Total
female	Class	crew	Count	3	20	23
			% within Class	13,0%	87,0%	100,0%
		first	Count	4	141	145
			% within Class	2,8%	97,2%	100,0%
		second	Count	13	93	106
			% within Class	12,3%	87,7%	100,0%
		third	Count	106	90	196
			% within Class	54,1%	45,9%	100,0%
	Total		Count	126	344	470
			% within Class	26,8%	73,2%	100,0%
male	Class	crew	Count	670	192	862
			% within Class	77,7%	22,3%	100,0%
		first	Count	118	62	180
			% within Class	65,6%	34,4%	100,0%
		second	Count	154	25	179
			% within Class	86,0%	14,0%	100,0%
		third	Count	422	88	510
			% within Class	82,7%	17,3%	100,0%
	Total		Count	1364	367	1731
			% within Class	78,8%	21,2%	100,0%

Figure 3.23 Survival rates on the *Titanic.*

To test all the possible second-order interactions, we write

```
PROC LOGISTIC DATA=titanic;
MODEL survived = class | age | sex @2 / SELECTION=forward;
RUN;
```

Finally, to test the class*sex interaction only (which gives the above coefficients), we write

```
PROC LOGISTIC DATA=titanic;
MODEL survived = class age sex class*sex /
SELECTION=forward; RUN;
```

Note that the HIERARCHY = SINGLE option (the default option) of LOGISTIC offers an interesting possibility: we can decide to input an interaction into a model only if its component variables have already been selected as principal effects, and we can retain a variable as long as it has an effect in an interaction. We can avoid this hierarchical structure by using the HIERARCHY = NONE option.

3.13 Automatic variable selection

This step is not essential if the discriminating variables have already been carefully chosen, as mentioned in Section 3.8. However, it is strongly advisable to include this step for certain models such as linear models (linear discriminant analysis or logistic regression). All serious programs offer stepwise automatic selection, at the very least. To describe this selection mode, I will use the example of the prediction of a binary variable by logistic regression, although the technique can be applied more generally.

The main stepwise selection methods are as follows:

- Forward stepwise selection, in which there is no variable in the model at the outset, and those which are sufficiently closely linked to the dependent variable, and which contribute most to the model (in a sense which can vary: e.g. likelihood ratio test, score test in the SAS and IBM SPSS LOGISTIC procedures, etc.), are added one by one, with allowance for the previously selected variables.
- Backward stepwise selection, in which we start by entering all the variables into the model before removing, one by one, those which contribute least to the model (in a sense which can vary: the SAS and IBM SPSS LOGISTIC procedures use the Wald test). This method is not recommended if the initial number of variables is very large (especially if the number of observations is also large), but it can enable more useful variables to be detected; in logistic regression, however, it has the drawback of being sensitive to possible problems of complete separation (see Section 11.8.7).
- In combined stepwise selection, each forward selection step is followed by one or more backward selection steps, and the process is interrupted when no further variables can be added to the model or when the addition or removal of a variable results in a model which has already been evaluated; this method is most demanding in terms of computation time, but it is the most reliable and the most widely used method.

Two more remarks may be added:

- It is possible to combine forward and backward selection with the aim of retaining only those variables which appear both times.
- Automatic stepwise selection procedures are often insufficiently selective, and the statistician must often exclude other variables manually, for example according to the correlation coefficients between independent variables or evidence of overfitting (see Section 11.3.4) measured in the test sample.

The detailed algorithm for stepwise selection, as implemented for example in SAS or IBM SPSS, is as follows:[16]

Step 1. Estimate the constant by searching for the maximum likelihood.

Step 2. Calculate the score statistic for each variable not included in the model.

[16] I will not go into the details of the options for specifying a minimum or maximum number of variables in the model.

Step 3. Choose the variable V with the lowest p-value for this statistic. If this p-value is higher than the input threshold (the SLENTRY parameter in SAS; by default this is 0.05), the algorithm ends; otherwise, go to the next step.

Step 4. Add the variable V to the model. If this model is then identical to a previously calculated model, this means that V would be excluded subsequently, and the algorithm ends without the addition of V (otherwise, the algorithm would 'loop'). Otherwise, go to the next step.

Step 5. Calculate the Wald statistic and the corresponding p-value for each variable (V and the others) in the resulting model.

Step 6. Choose the variable W having the highest p-value. If this p-value is below the output threshold (the SLSTAY parameter in SAS; by default this is 0.05), go back to step 2. Otherwise, if the model produced by excluding the variable W is a previously calculated model, this means that V would be added subsequently: this model is therefore retained, without W, and the algorithm ends, to prevent it from 'looping'. Otherwise, i.e. if the p-value is greater than or equal to the output threshold and if the model without W is not a previous model, W is excluded, and we go back to step 5.

Thus we obtain a model in which the variables V_{i1}, V_{i2}, ..., V_{ip} are a subset of the set of tested variables V_1, ..., V_k, chosen because they optimized a certain statistical criterion (the score statistic in this case) at each iteration, with allowance for the choice of the variables V_{i1}, V_{i2}, ..., $V_{ip'}$ ($p' < p$) previously entered into the model. Clearly, this model is not necessarily the best one overall, since the best model with $p' + 1$ variables is not necessarily found by adding a variable to the best model with p' variables (just as the best clustering into k classes is not necessarily derived from the best clustering into $k + 1$ classes). Suppose, for example, that we have three variables V_1, V_2 and V_3, such that the variable V_1 optimizes the statistical criterion and V_2 is strongly correlated with V_1. The one-variable model will be $\{V_1\}$ and the two-variable model will be $\{V_1,V_3\}$, since V_1 eliminates V_2 which is highly correlated with it. However, it is possible that the model $\{V_2,V_3\}$ is better than the model $\{V_1,V_3\}$. Although this is not the most common example, it is not exceptional. This is what makes global selection methods useful.

In the case of continuous independent variables, SAS, SPAD and R (with the *leaps* package) have also implemented a global method which is better than the stepwise methods, namely the *leaps and bounds* algorithm of Furnival and Wilson,[17] which attempts to calculate the best regressions for a subset of 1, 2, ..., k independent variables, by comparing some of all the possible models and eliminating the least useful ones immediately. By optimizing the exploration of all the possible cases, we can keep the computation time within acceptable limits, at least if $k \leq 40$.

3.14 Detection of collinearity

This step of detecting linear links between independent variables is essential for Fisher discriminant analysis, logistic regression and linear regression. However, collinearity does

[17] Furnival, G.M. and Wilson, R.W. (1974) Regression by leaps and bounds. *Technometrics*, 16, 499–511.

not affect decision trees, neural networks, or PLS regression. This section is concerned with the links that may exist between continuous independent variables, but logistic regression can also have qualitative or discrete independent variables, and the absence of excessively strong associations between these is tested by the χ^2 criterion and Cramér's V, as mentioned in Sections 3.8.1 and A.2.10–A.2.12.

The simplest and most usual way of detecting collinearity is to calculate the correlation coefficients of the variables in pairs (Pearson's coefficient for continuous variables that are fairly close to a normal distribution, Spearman's in other cases). This can be done automatically for all the variables using the programming language of a package such as SAS or SPSS.

In the SAS example shown below, we are focusing on the correlation coefficients of the variables VAR1, VAR2... contained in the TEST file, and we wish to detect the strongest correlations. The results of the analysis are shown in the form of a list of the pairs (VAR*i*,VAR*j*) with their correlation coefficients (Pearson's, see below), starting with the strongest correlations in absolute terms and ending with the weakest. These calculations are performed by the CORR procedure, which outputs them to the file specified by OUTP. In this file the correlations are shown in a matrix, with one row (of the 'CORR' type) and one column per variable, so that the correlation coefficient of (VAR*i*,VAR*j*) is read on the *j*th column of the *i*th row. The TRANSPOSE procedure is used to transform this matrix display into the required list. The file called PEARSON below then contains the name of the first variable of each pair in VARIABLE1, the name of the second variable in VARIABLE2, the correlation coefficient and its absolute value in ABSCORRELATION. Before printing the list (using PRINT), we eliminate the unwanted terms and keep only those in which VARIABLE1 < VARIABLE2.

```
PROC CORR DATA = test PEARSON SPEARMAN OUTP=pearson
OUTS=spearman NOPRINT;
VAR var1 var2 var3 var4 var5 ... ;

DATA pearson (DROP = _TYPE_ RENAME = (_NAME_ =
variable1)); SET pearson;
WHERE _TYPE_ = "CORR";

PROC TRANSPOSE DATA = pearson NAME=variable2
PREFIX=correlation OUT = pearson ;
VAR var1 var2 var3 var4 var5 ... ;
BY variable1 NOTSORTED ;

DATA pearson;
 SET pearson;
WHERE variable1 < variable2;
abscorrelation = ABS(correlation1);

PROC SORT DATA=pearson;
BY DESCENDING abscorrelation ;
PROC PRINT DATA=pearson;
RUN;

... (same for Spearman)
```

As an empirical rule, we can consider that the correlation is unacceptable when the correlation coefficient exceeds 0.9, very risky when the coefficient exceeds 0.8, and needs to be treated with caution when it exceeds 0.7. Obviously, we will be more tolerant if the available variables are less numerous and if it will be difficult to achieve a sufficiently accurate prediction.

If we use a method that is sensitive to the collinearity of the independent variables, it is not always sufficient to verify the absence of collinearity of the variables examined in pairs. In other words, it is not enough to calculate the correlation coefficients of all the pairs of variables. This is because there may be a linear relationship between three variables even when there is no linear relationship between any two of the three. We must therefore check that there is no multicollinearity between the variables. This can be done in two ways.

One way is to calculate an index called the 'tolerance' or its inverse, the 'variance inflation factor' (VIF). The VIF is so called because it acts as a multiplier of the variance of the estimator of the coefficient of the variable in a linear regression (see Section 11.7.1). The tolerance of a variable is the proportion of the variance of this variable that is not explained by the other variables, i.e. $1 - R^2$, where R is the multiple correlation coefficient of the tested variable with the other independent variables. It is often considered that the tolerance should be greater than 0.2 or at least 0.1 (VIF $\leq$ 10).

Alternatively, we examine the correlation matrix and calculate its condition indices (according to Belsley, Kuh and Welsch),[18] defined as the square root[19] of the ratio of the largest eigenvalue to each of the eigenvalues:

$$\eta_k^2 = \mu_{\max} / \mu_k.$$

According to Belsey,the multicollinearity is moderate if some indices η_k are greater than 10, and high if some indices η_k are greater than 30. If this is the case, we check to see if we can link the corresponding eigenvalue to a strong contribution (in excess of 50%) of the principal component (the eigenvector associated with the eigenvalue) to the variance of two or more variables. In other words, we find out whether two or more columns in Table 3.5 contain values greater than 0.5. In our case, the table shows that there is no problem of collinearity, as the collinearity present between variables 4 and 6 is only moderate.

In SAS and IBM SPSS, only the linear regression procedure REG offers the possibility of measuring the tolerance and VIF, although these measurements are equally useful in logistic regression or discriminant analysis. However, all we need to do is to launch the REG procedure, specifying as the dependent variable that of the problem (event if it is binary). In any case, the choice of dependent variable has no effect on the result, as this will only be based on the correlations between the independent variables, not the correlations with the dependent variable. We must be careful about using these criteria when the independent variables are not continuous, for example if we recode qualitative variables: recoding a binary variable in 0/1 form will not lead to the same result as recoding in 1/2. The VIF (and the condition indices) will be substantially higher with 1/2 coding.

The SAS syntax for producing these indices is shown in Section 11.7.8.

Note that we can get an idea of the risks of multicollinearity and the natural groupings of variables by carrying out an agglomerative hierarchical clustering of the variables, rather than the individuals, as mentioned in Section 9.14.

---

[18] Belsley, D.A., Kuh, E. and Welsch, R.E. (1980) *Regression Diagnostics: Identifying Influential Data and Sources of Collinearity.* New York: John Wiley & Sons, Inc.

[19] Some authors do not use the square root.

Table 3.5 Multicollinearity.

	Eigenvalue	Condition index	Variance Proportions						
			(cst)	var 1	var 2	var 3	var 4	var 5	var 6
1	3.268	1.000	.01	.00	.03	.02	.01	.01	.02
2	1.022	1.788	.00	.56	.01	.02	.00	.33	.00
3	.976	1.830	.00	.42	.00	.10	.00	.42	.01
4	.811	2.008	.00	.02	.07	.81	.00	.14	.00
5	.636	2.266	.01	.00	.78	.04	.02	.09	.00
6	.221	3.842	.01	.00	.11	.01	.20	.00	.73
7	.065	7.099	.97	.00	.00	.00	.76	.00	.24

3.15 Sampling

3.15.1 Using sampling

Sampling is an indispensable procedure in statistics and data mining, especially in prediction and classification, where most algorithms use a *training sample* for developing the model and a *test sample* for validating the model, or multiple samples for cross-tabulated validation. Also, if the class to be predicted is rare, it may be necessary to sample the population and adjust the sample in order to increase the frequency of that class. Some classification methods require sampling. It is often the case that the overall numbers to be processed are so great that we have to sample in order to reduce the number of observations to be handled. Finally, Monte Carlo methods and resampling (see Section 11.15) are increasingly used to enhance or at least estimate the robustness of models, either by aggregation of models or by calculating confidence intervals for the estimated parameters.

Sampling is only applicable if, on the one hand, we can control the representativeness of the sample,[20] and, on the other hand, if we are not looking for excessively unusual phenomena. This can easily be understood if we consider the investigation of types of fraud or narrow segments with high added value. We also need to be sure that the accuracy of the results only increases as the square root of the sample size (i.e. it does not depend on the size of the total population):[21] a multiplier of 10 only divides the confidence interval by 3.2 ($=\sqrt{10}$). As a general rule, optimal sampling requires a thorough knowledge of the population under investigation, which is not always available, especially if we are studying a constantly changing population such as a customer base.

[20] For example, we can conduct a Student test or a non-parametric test on the means of the variables used (see Appendix A).

[21] See Appendix A.

Consider an example from the world of banking: the creation of a sample of current account holders. Do we start by taking accounts at random, according to a probability distribution, and then find their holders, or should we start by picking holders at random and then find their accounts? In the first case, there will be a bias towards holders who have several accounts (and are therefore more likely to be chosen); in the second case, there will be a bias towards accounts with more than one holder, i.e. joint accounts. In one case, there is a bias towards professionals and well-off customers, while in the other case there is a bias towards couples; in all cases, the sampling is biased and must be adjusted. We must therefore know the population so that we can find the distribution of single/joint accounts and customers as a function of their numbers of current accounts.

Clearly, sampling is required for the development of a predictive model, if only for the creation of the training and test samples for optimizing the selection of the independent variables. This selection will be carried out in such a way that the performance of the model on the test sample is maximized. Once this has been done, we must still return to the whole population to recalculate the parameters of the model (such as the logistic regression coefficients), not only on the learning sample, but on the totality of individuals, in order to provide the best estimate of the parameters of the model (unless we use a *bagging* method as described in Section 11.15.2).

3.15.2 Random sampling methods

The main methods of random sampling are simple random sampling, systematic sampling, stratified sampling and cluster sampling.

Simple random sampling involves drawing n individuals at random without replacement from a population of N, each individual having a probability of $1/N$ of being drawn. For this purpose, many programs have a 'random' function which randomly outputs a number from 0 to 1 according to a statistical distribution that can be specified (uniform, normal, etc.). In some variants, the drawing can be with replacement and/or with unequal probability (not equal to 1/N for each individual). Such an unequal probability is used in boosting. As for the sampling with replacement, it is the foundation of bootstrap (see Section 11.15).

In *systematic sampling,* the individuals are drawn not at random, but in a regular way. If we carry out a 'one in a hundred' sampling, we take the first individual, then the 101st, then the 201st, and so on. We must pay attention to cyclical data with this form of sampling: if we use customer numbers, the hundreds number may be a family number, and if we take one customer in every hundred, we will never choose two individuals in the same family. However, this sampling mode can also provide a degree of comprehensiveness.

In *stratified sampling*, we divide the population, for example by dividing the customers into age ranges, and then draw customers at random from each stratum to obtain a sub-sample for each stratum; we can then bring all these sub-samples together. In *proportional* stratified sampling, the relative size of each sub-sample is equal to the relative size of the corresponding division: for example, if 30% of the customers in the population are aged over 60, then 30% of any stratified sample by age must be customers aged over 60. It may be useful to carry out a *non-proportional* stratified sampling procedure to take into account the variability of the phenomena studied in each stratum: thus we can underrepresent the strata in which the variability is low (where the interesting information is concentrated in a few individuals) and overrepresent the strata in which the variability is high (which require a larger number of individuals to establish the information).

Stratified sampling is required when we need to construct two samples for the training and testing of a classification model, since it enables us to control the distribution in the two samples of the dependent variable. In the aforementioned case of a rare class to be predicted and the necessity of increasing its frequency in the sample, this can be done by non-proportional stratified sampling.

Cluster sampling is a matter of drawing families of individuals (the 'clusters') at random and choosing all the individuals in each cluster, this being known as a *census*. We may, for example, choose certain urban districts at random and then ask questions of all the customers from these districts. Or we can choose a family name at random and then carry out a census of all the customers whose family name starts with the letter drawn at random. We must be careful about the representativeness of the sample, if some initials are chosen at random, for example 'S' for 'Smith'! It may be better to work with the second and third letters of the family name, after conducting a test of representativeness.

By contrast with stratified sampling, cluster sampling improves as the clusters resemble each other more closely, and as the individuals in each cluster differ from each other. This is because the basic principle of this form of sampling is that the cluster is a miniature version of the population under investigation. If we wish to measure a certain indicator in the sample, we will choose a cluster division criterion having no relationship with the indicator to be measured. We often choose a geographical criterion (although its relevance must still be checked), especially as the use of this criterion limits travelling time and expense in the case of a field study.

4

Using commercial data

This chapter describes the main kinds of data that are generally studied in commercial data mining applications, grouped into types. Particular attention will be paid to geodemographic and profitability data, which are very useful in certain contexts. Finally, there is a detailed examination of the data used in banking, personal and property insurance, the telephone industry and mail order.

4.1 Data used in commercial applications

Before looking at the details of the data used in four of the main commercial sectors where large-scale data mining projects are typically found, I will list the main kinds of data common to all these sectors, divided into a few major families. These data are aggregated by household or used at the individual level, depending on the studies concerned and the possible options.

4.1.1 Data on transactions and RFM Data

In many investigations, and in all of those that relate to a propensity to consume, the most important data are those on commercial transactions. We ask the following questions: 'where?' (geographical locations, businesses where the transactions took place, Internet, etc.), 'when?' (frequency and recency of the transactions), 'how?' (method of payment), 'how much?' (number and value of transactions), 'what?' (what has been purchased).

A typical recency, frequency, monetary value (RFM) analysis is conducted by cross-tabulating the recency of the last purchase in the period studied (e.g. quarter $T-1$, $T-2$, $T-3$, $T-4$), with the frequency of purchases in that period (in our example, the number of quarters between 1 and 4 when a purchase was made), and then examining the distribution of purchases in each intersection. We can denote these by 0 and 1 to indicate quarters without a purchase and with purchases, respectively; for example, 1001 signifies a customer who has made a purchase in the preceding quarter $T-1$ and in the quarter $T-4$, and no purchases in the other two quarters. The results will be as shown in Table 4.1.

Data Mining and Statistics for Decision Making, First Edition. Stéphane Tufféry.

Table 4.1 RFM segmentation.

frequency recency	4	3	2	1
$T-1$	1111	1110 1101 1011	1100 1010 1001	1000
$T-2$		0111	0110 0101	0100
$T-3$			0011	0010
$T-4$				0001

When the subscription rates are measured for each cell of the table, we often find a segmentation having the following general form:

- Very good customers: 1111, 1110, 1101, 1011;
- Good customers: 0111, 1100, 1010, 1001;
- Average customers: 0110, 0101, 0011, 0100, 0010, 0001;
- New active customers: 1000.

4.1.2 Data on products and contracts

Other important data relate to the ownership of products: these include the numbers, types, options, prices, date of purchase or subscription, date and reason for cancellation or return of products, mean product life or expiry date, payment date and method, discount granted to the customer, and profit margin on this product for the business.

4.1.3 Lifetimes

The first of the variables relating to customer lifetime is, of course, *age*. For obvious reasons of market segmentation, this is a very important variable in marketing. If we do not know the ages of our customers, we can sometimes estimate them with a reasonable degree of accuracy by an alternative method. This is 'first name scoring', and it will be described later.

The second major lifetime relating to a customer is his *lifetime as a customer of the business*. This variable is used, together with others including *length of time at present address* and *length of time in present job* (rather than length of time at work, i.e. number of years since first job), as a risk indicator.

For propensity studies, we must also consider the period of time that has elapsed since the subscription to a contract or the purchase of a product, which is important, especially when there is a life cycle for the product concerned, as in the case of vehicles, credit, and some managed savings products.

For risk studies, we may also consider the time since the last claim (related to no-claims bonuses in car insurance, for example), since the last dispute, since the last non-payment, etc.

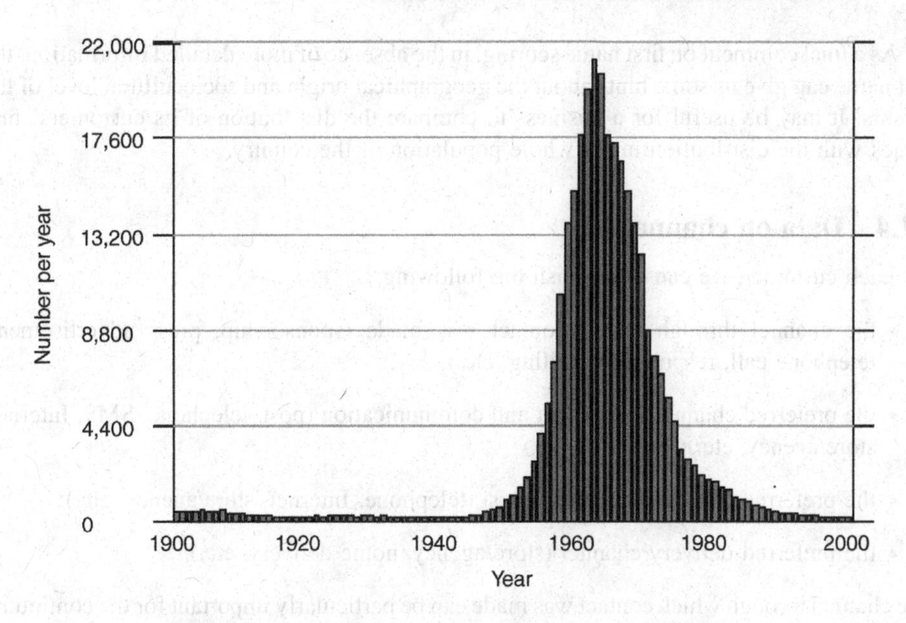

Figure 4.1 Distribution of the first name 'Pascal'.

As I have mentioned, we can attempt to estimate customers' ages when they are not directly known, by using the first name. This first name scoring method is based on the fact that certain first names go in and out of fashion through the years. The usefulness of this method is evident for first names such as the one in the graph in Figure 4.1, based on the INSEE data for France. Data of this type are provided by the Office for National Statistics in the United Kingdom and by the Social Security Administration in the United States.[1] These are names that are sufficiently popular to have a statistical value, but not as widespread as first names which are present at all periods (John, David, Lucy...), meaning that they cannot be used as predictors of age (although if we know two first names at the same address we can try to cross-tabulate them). Such names are John and David in the United Kingdom, James and Mary in the USA, and Pierre, Paul and Marie in France. In the USA, Mary was always one of the top two girls' names from 1910 to 1965, James was always one of the top five boys' names from 1910 to 1980, John and Robert stayed in the top five until 1971, and Michael has become increasingly popular since the 1940s. However, some first names are highly predictive: in France, these are Émile, Jeanne and Germaine for the 1920s, Joseph, Lucien, Roger and Thérèse for the 1930s, Claude for 1935–1945, Monique for 1940–1950, Daniel, Bernard, Michel and Françoise for 1945–1955, Philippe, Patrick and Catherine for 1955–1965, Éric, Thierry, Pascal (with a peak in 1962, as seen in Figure 4.1), Christine, Isabelle and Nathalie for 1960–1970, Julien and Nicolas for 1980–1990, and so on. In the USA and UK, we find George and Dorothy at the start of the twentieth century, Margaret in 1910–1930, Richard in 1930–40 (USA) and 1950–1970 (UK), Karen in the 1960s, Christopher in the 1970s and 80s, Jessica in the 1990s, and so on. In the UK, Margaret was one of the two commonest names between 1914 and 1934, then declined from 4th to 39th place from 1944 to 1954.

[1] See, for example, http://www.nameplayground.com/ and http://www.ssa.gov/OACT/babynames/ for the USA, http://www.babynames.co.uk/popular-baby-names for the United Kingdom, http://meilleursprenoms.com for France, and www.beliebte-vornamen.de for Germany.

As a final comment on first name scoring, in the absence of more detailed information, the first name can give us some hints about the geographical origin and sociocultural level of the parents. It may be useful for a business to compare the distribution of its customers' first names with the distribution in the whole population of the country.

4.1.4 Data on channels

For each customer, we can distinguish the following:

- the channel through which contact was made (sponsorship, press advertisement, telephone call, response to mailing, etc.);
- the preferred channel for contact and communication (post, telephone, SMS, Internet, store/agency, etc.);
- the preferred channel for orders (post, telephone, Internet, store/agency, etc.);
- the preferred delivery channel (store/agency, home delivery, etc.).

The channel through which contact was made can be particularly important for the continuing relationship between the business and the customer: customers recruited by sponsorship are considered to be most loyal.

4.1.5 Relational, attitudinal and psychographic data

Relational data are: responses to marketing campaigns and offers, rejection of direct marketing, preference for a contact channel, preference for a delivery channel, response to questionnaires (on order or guarantee forms, for example), responses to courtesy or customer satisfaction calls, calls to the customer service or after-sales service, and complaints (but remember that fewer than 10% of dissatisfied customers actually make complaints).

Relational data are less commonly available than other data, but are very important. For example, a customer's propensity may depend on the *distribution channel*; a customer with a high propensity score may refuse an offer because it is made by telephone, where he would have accepted it by post. If these data are available, it may be helpful to use them to construct 'distribution channel' propensity scores which are then cross-tabulated with the 'product' propensity scores; we can then avoid contacting customers who do not wish to be contacted.

Attitudinal data may have a significant effect on the customer's loyalty, which does not only depend on his satisfaction. This is because, as we know, a satisfied customer may change brands, but a customer who is well treated after a complaint is generally more loyal than before the incident. Customers who complain most are sometimes those who are most attached to the brand. The attitudinal factors to be incorporated into loyalty modelling are: the image and prominence of the brand for the customer (low, medium or high), the predisposition to purchase (is unaware of the product/has heard of it/knows about it/is interested/would like it/expects to buy it), the attitude to the product (enthusiastic/positive/indifferent/negative/hostile), the reasons for buying (quality, price, service, etc.), the attractiveness of the competition, the customer's propensity or disinclination to change, any special barriers to change existing in the market (making the customer more or less 'tied'), etc. We need to distinguish voluntary loyalty from captivity.

Psychographic data are lifestyle, personality (shy, authoritarian, prudent, ambitious, etc.), values (pro-modern, conservative, politically involved, hedonistic, materialistic, critical, etc.), risk aversion (trustful, mistrustful, anxious, demanding), knowledge, focus of interest, opinions and behaviour. These data are widely used for customer segmentation. They include the well-known social styles of the Centre de Communication Avancée, the criteria for which are more concerned with psycho-sociology.

4.1.6 Sociodemographic data

This list of data used in data mining will end with sociodemographic data, even though these are the best-known type. This is because they are rarely the most discriminating data in practice. Perhaps their main advantage is that they are universal and easily understood. However, they can also suffer from input errors, intentional false statements, and, above all, frequent lack of updating.

The main sociodemographic data are:

- personal (sex, level of education);
- family (family situation, number and ages of children, number of dependants);
- occupational (income, occupation and social category, number of working and retired people in the household);
- wealth (fixed and movable property, owner-occupier or tenant, value of residence, possession of a second home, etc.);
- geographical (length of time at the address, region of residence, place of residence (commune, district, municipality), number of inhabitants of the place of residence, ZIP code, ZIP+4 code and Block Group (USA), Super Output Area and Census Output Area (UK), IRIS and INSEE block (France), type of housing (geodemographic segment) deduced from the preceding geographical area;
- environmental and geodemographic (competition, population, working population, customer population, unemployment rates, economic potential, product ownership rates, etc., in the area of residence of the customer or prospect).

4.1.7 When data are unavailable

Some geodemographic data (see Section 4.2.1) and relational data (survey responses) are useful in the absence of other, more precise data, especially in a population of prospects.

We can also use *behavioural mega-databases*. These mega-databases contain hundreds of indicators of the consumption habits and lifestyles of several tens of millions of households with their personal details (names, telephone numbers, postal address, and e-mail address if any). There are two main types of mega-databases.

In the first type, the databases are created by the sharing of files from partners in retailing, mail order, the Internet sector, the press, the community sector, etc. In France, the leading mega-database is Apollinis, belonging to WDM (formerly Wegener Direct Marketing), with 10 million postal addresses and 3 million e-mail addresses. A good example in the UK is the Data Locator Group and its Data Rental based containing 2000 data items on 24 million individuals. In the second type of megadatabase, the data are collected via large-scale surveys, where the households agree to respond and allow the resale of these data in exchange for discount vouchers or free samples for a number of products. The American Acxiom company is a specialist in these databases, and has become even more dominant since 2004 when it purchased the major European company Claritas and the leading French provider, Consodata.

The owners of these mega-databases may conduct studies for businesses (drawing up profiles of their customers) or may sell them address files which are 'qualified', in other words enriched with a certain amount of information or created by using certain selection criteria. Naturally, the amount of accompanying information will increase the price of each address: this ranges from around €0.20 to €0.30 for a plain address (€0.50 with the telephone number added), plus about €0.05 for each lifestyle criterion, plus about €0.10 to €0.15 for each precise consumption habit. So the price will often be more than €0.50 per qualified address, which is more expensive than the addresses sold by La Redoute or the Bottin international directory, but is justified by the accuracy of the information. A business can even add its own questions to a supplier's surveys, although this will obviously cost more, and be the only business to know what the answers are.

4.1.8 Technical data

These are data which are not generally used in data mining analysis, but are required for the selection of individuals admitted to the analysis base, or for the implementation of data mining in targeted marketing. These data are:

- date of death (the fact that this is filled in);
- the type of customer (private, business, company, etc.);
- non-acceptance of direct marketing (shown in the Robinson lists used in France, Belgium, Switzerland and Germany);
- bad payer status;
- status as employee of the business or of a subsidiary;
- title, surname, forename, telephone number and full address;
- the 'not at this address' indicator (in direct marketing, a good file should contain fewer than 3% 'not at this address' responses).

4.2 Special data

4.2.1 Geodemographic data

Geodemographic data have the distinctive feature of not relating directly to individuals, but to their geographical environment. For each individual, customer or consumer, it is possible to discover the details of his place of residence in terms of economics (number of businesses, working population, population on benefits, unemployment, local businesses and services, consumption habits, etc.), sociodemographics (population, wealth, average number and ages of children, family structures, social and occupational level, etc.), housing (age, type and convenience of housing, proportion of tenants and owner-occupiers, etc.) and competition (presence of the business, presence of its competitors, market share, etc.). This information is supplied by the general census (sources: the U.S. Census Bureau in the USA, the Office for National Statistics in the UK, INSEE in France), statistics on civil status, electoral registers, tax authorities, social services organizations, postal services, housing files, etc. In France, the sources include the SIRENE file of businesses, the Revenue Department, the family allowance office, the annual statement of social data, the Banque de France (the file of bank branches), etc. Sometimes this information can be supplemented with street-level information such as that available in the files of postal services.

The existence of these data has led to the emergence of geomarketing, which Franck Bleuzen defines as 'modelling and analysis of all the correlation factors between the consumer's place of residence and his mode of consumption'.[2] The underlying assumption of geomarketing is that the choice of a place of residence is dictated to a household by social and financial considerations, resulting in a degree of uniformity of the information mentioned above, at least if it is examined with a sufficiently fine degree of resolution, and leading to similar behaviour in terms of consumption and purchasing.

Of course, geodemographic data are not directly related to separate individuals, and are therefore less precise than some other data that the business may hold on its customers. However, they have the major advantage of being available for individuals who are not (yet) customers of the business. This enables the business to supplement the information it holds on its prospects, to investigate a geographical area where it has no market presence with a view to establishing itself there, to identify the catchment areas and population movements, to find areas with high commercial potential, to prepare for the targeting of prospects or the marketing of a new product, to allocate the territories of its sales staff, to analyse the customer profiles by local market, and so on.

By identifying the catchment areas of its points of sale, a business can limit their overlap, or, conversely, find areas of poor coverage, decide on the areas for distribution of its advertising flyers and posters, adapt its offer to the customer base, measure the effects of its competitors, etc. When it is established in an area, the business can measure its commercial performance by comparing it, on the one hand, to its performance in other areas with the same socioeconomic and demographic characteristics (in the same block groups, see below), and, on the other hand, to the expected potential performance allowing for the household resources as shown in the geodemographic databases for this area. If the business considers that its trading performance is inadequate, it can use geomarketing either to focus its efforts on areas with major potential for the proposed product(s), or it can preferentially target areas geographically close to areas where its penetration rate is high.

Geodemographic data can be used for propensity studies. The customer's potential for the business will be roughly equal to the difference between the 'shopping basket' of consumers in his area and what he has purchased already.

Geodemographic data are available at different geographical levels, ranging from the whole country to the district, via regions, departments, cantons and communes (in France), or from the state to the county and the municipality (in the USA), or again from the region to the country and district (Ward) in the UK. The nomenclature of territorial units for statistics (NUTS), established by Eurostat in 1981, defines territorial divisions which are comparable between the countries of the European Union, for the purpose of drawing up regional statistics

[2] Cited by Laure Gontard (http://mvmemoire.free.fr/m%E9moires/Les%20m%E9moires/GONTARD.pdf).

and implementing the EU's regional policies. The NUTS nomenclature has three levels: NUTS 1, NUTS 2 and NUTS 3. Each member state is divided into one or more NUTS 1 regions, which are divided into one or more NUTS 2 regions, which are again divided into one or more NUTS 3 regions. NUTS 1 in France is a group of regions (there are nine of these, including one for overseas territories); in Germany it is a *Land* (there are sixteen of these), and in the UK it is a region (there are nine of these). A NUTS 2 territory is a region in France and a county in the UK. NUTS 3 is a département in France, and a district in the UK and Germany. The NUTS nomenclature of 2006 divides the European territory into 97 NUTS 1 regions, 271 NUTS 2 regions and 1303 NUTS 3 regions.

In France, the district level has been defined by INSEE, for all municipalities with more than 5000 inhabitants, as sets of blocks grouped according to statistical information (IRIS) with an average of 2000 inhabitants each. There are approximately 16 000 of these IRIS districts. The usefulness of the IRIS district is that it relates to an area of uniform population, instead of a purely administrative division. This makes it suitable for geomarketing studies. The 35 000 municipalities with less than 5000 inhabitants each are not divided into IRIS districts.

If the base unit for general census data on the population is taken to be the minimum geographical area for the free distribution of all[3] the 17 000 sociodemographic indicators, below which level the Data Protection Act (CNIL) considers (since the 1990 census) that there is a risk of supplying information which could be too closely applicable to individual cases (this risk is judged to be unacceptable because it is compulsory to respond to the census questions), the 51 000 base units are the 35 000 municipalities with less than 5000 inhabitants, plus the 16 000 IRIS districts of the 1800 municipalities with more than 5000 inhabitants.

Some information, generally relating to the files of the Post Office or the Revenue Agency, is compiled at the address and street level. Between the street and the municipality or IRIS level, there is an intermediate level, with about 120 inhabitants: this is the block level, and is also the area covered by a census taker in the general census. Municipalities with more than 10 000 inhabitants (numbering about 900) are divided into blocks, as are all municipalities in urban areas with more than 50 000 inhabitants, and there are about 222 000 of these blocks, containing more than half of French households. Oddly enough, municipalities with 5000 to 10 000 inhabitants are divided into IRIS districts, not into blocks, although the IRIS districts are defined as groups of blocks.

INSEE has distributed a maximum of 15 items of information at block level since 1990. These items are the distribution of the population by sex and five age ranges (0–19, 20–39, 40–59, 60–74, 75 and above), the number of people in primary homes, and the division of housing into four categories (primary homes, second homes, occasional accommodation, vacant accommodation). However, some specialist companies, which already held databases at block level before 1990, have continued to update these after each general census, by interpolation of the information supplied at IRIS level. In the USA, the smallest unit used by the Census Bureau for data collection is the Census Block. It is similar to the French block, but at an even finer level of detail, representing an area bounded by a street, a road or a watercourse, even if there are no inhabitants. Of the 8.2 million Census Blocks, about 2.7 million are uninhabited. The Census Blocks are grouped into Census Block Groups (more than 200,000 in each), which are themselves grouped into Census Tracts which are roughly equivalent to municipalities, and then into counties and states. With an average of 500

[3] Except for indicators of nationality and immigration.

households each, the Census Block Groups are the equivalent of the French IRIS, and the Census Bureau distributes information at this level. In the United Kingdom, the smallest unit use by the Office for National Statistics is the Census Output Area, which is the equivalent of the French block, the number of OAs being almost the same as the number of blocks. This unit is then aggregatèd into Super Output Areas, and subsequently into Census Tracts, districts and counties.

There are two further resources which can increase the relevance of geomarketing studies.

The first resource is provided by data mining itself, using the clustering techniques which, in France, enabled the research company COREF to develop a typology of French municipalities (in about 40 clusters called *geotypes*™), and then a typology of blocks (in 31 clusters called *îlotypes*™, which have then been refined into about 50 *sous-îlotypes*), after creating databases of several hundred variables on these two types of entity.

This clustering is based on factor analysis and dynamic cloud clustering (see Section 9.9.2). This approach has the advantage of summarizing numerous data and substituting a single indicator, the block cluster, or a few factors, for several hundreds of sociodemographic, behavioural, economic and other indicators (age, family circumstances, number of children, income, socio-occupational class, housing, distance from shops, etc.). This is always useful, especially when there is no preconceived idea of the phenomenon to be examined, and when we are not interested in a single specific data element.

COREF was purchased in 1996 by CCN, which then merged with other companies to create Experian. Experian eventually abandoned the COREF typologies, and developed its own geodemographic typologies, now in use worldwide, starting with MOSAIC.

This typology is available in an international version, MOSAIC GLOBAL, developed for 380 million households and 880 million consumers in 25 countries, including the USA, Canada, Western and Northern Europe, Japan, Australia and New Zealand. It is based on sociodemographic data, lifestyles, behaviour and preferences, and contains 10 types which are universal and found to varying degrees in the 25 countries covered by this typology. For example, "sophisticated singles" include 29% of Finns but only 1.3% of Irish citizens. The types are distributed along two major discriminant axes: the income level (from low to high) and the housing type (urban to rural).

<table>
<tr><td>High</td><td>A
Sophisticated
Singles</td><td>B
Bourgeois
Prosperity</td><td>C
Career and
Family</td><td>D
Comfortable
Retirement</td></tr>
<tr><td rowspan="2">Affluence</td><td colspan="2">E
Routine Service
Workers</td><td>F
Hard Working Blue
Collar</td><td rowspan="2">J
Rural
Inheritance</td></tr>
<tr><td>G
Metropolitan
Strugglers</td><td>H
Low Income
Elders</td><td>I
Post Industrial
Survivors</td></tr>
<tr><td>Low</td><td colspan="3">Urban →</td><td>Rural</td></tr>
</table>

This global typology can be used for marketing analyses at the international level, for comparing customers in different countries, or for finding identical behaviours in different countries. This information is useful for developing coherent marketing approaches, applying the knowledge of a customer in one country to another country, etc.

The MOSAIC GLOBAL typology is also subdivided in each country, where there is a specific typology related to the global typology by means of common variables (proportion of the population aged over 65, or of school age, etc.). Of course, these MOSAIC types are always calculated by statistical methods (k-means on standardized variables; see Section 9.9.2), but they are also validated by experts in the sociology of consumption, economics and human geography. Their variables are carefully chosen: they must have balanced weights, without any one characteristic taking precedence; they must not be redundant, must be related to consumer behaviour, and must have categories which are not too rare and which are well distributed over the whole territory. In the USA, we find variables relating to age, ethnic origin, educational level, family circumstances, size of household, occupation, income, the status of the housing and age of the building, the means of transport to work, and the number of vehicles owned.

In the USA, the MOSAIC typology is based on 300 variables, including the data from the 10-yearly general Census of the population with annual updates by AGS Demographics, and consists of 60 types aggregated into 12 groups, available at the ZIP+4 code, the ZIP code, or the Block Group level. The ZIP+4 level is very detailed, since it corresponds to a street or about ten households, but not all addresses have a ZIP+4 code; conversely, not all ZIP+4 codes relate to addresses (they may be box numbers), and moreover the ZIP+4 code created to optimize the routing of the mail is not as stable over time as the Block Group. This last unit is therefore preferred as the elementary unit for analysis and storage of geodemographic data. In the United Kingdom, MOSAIC is constructed from 400 variables, 54% taken from the Census and 46% from other sources. It consists of 61 types aggregated into 11 groups: Ties of Community, Suburban Comfort, Blue Collar Enterprise, Happy Families, etc (Table 4.2). It was in the UK that MOSAIC was first developed, as a result of the work done by Richard Webber, Professor of Geography at Kings College, University of London with Experian.

In France, MOSAIC consists of 52 "portraits" aggregated into 14 "landscapes", including "working class tradition", "future executives", "popular middle class", "culture and leisure", "urban seniors", etc.

Other companies are now offering these typologies at the national and international level, including Acxiom, with Personicx, a typology with about 70 segments, defined at the level of the Block Group or ZIP Code (USA), the postcode (UK) or IRIS (France). It can be used on national statistical data, and also on the Acxiom megadatabase (see Section 4.1.7).

In the UK, Richard Webber, creator of MOSAIC, has also developed the ACORN (A Classification Of Residential Neighbourhoods) typology.

In the USA, the Experian MOSAIC typology is used alongside the very popular PRIZM housing typology of Nielsen Claritas, with its 14 groups and 66 segments, including Money & Brains, Home Sweet Home, Old Glories, and Young & Rustic. In 1974 this was the first major commercial geodemographic segmentation. Like those of MOSAIC, the PRIZM segments are distributed along two major discriminant axes, namely the income level (from low to high) and the housing type (from rural to urban).

The second resource of geomarketing is the existence, alongside geodemographic databases, of GIS (Geographical Information Software), enabling all the manipulated data to be linked to their geographical coordinates to provide true spatial databases and analyses. For example, we may wish to study the propagation of a phenomenon, or calculate the distance between each customer and the nearest point of sale. To do this, the software incorporates mapping data and socioeconomic and demographic indicator bases, to which data specific to the business can be added. All these data can be recalculated for areas defined by the user. This software also allows for changes of scale: we can examine the figures at county level, then 'zoom in' on a particular county, municipality, district, and so on. Finally, we can export certain data calculated for an area defined on a map by the user.

Table 4.2 The Experian MOSAIC groups.

Group	Percentage of UK Households	Social Groups	Description
Symbols of Success	9.62%	Upper Middle and Middle Middle class	This group represents the wealthiest 10% of people in Britain, Set in their careers and with substantial equity and net worth. These people tend to be white British but with some Jewish, Indian and Chinese Minorities. Tends to contain older people advanced in their careers.
Happy Families	10.76%	Lower middle class and Middle middle class	Families from Middle England, focussed on children, home and career. Tends to be in new suburbs in more prosperous areas of the UK. Mostly white with few minorities
Suburban Comfort	15.10%	Lower Middle Class	People in comfortable homes in mature suburbs built between 1918 & 1970, moderate incomes. Includes Middle class Asian Enterprise
Ties of Community	16.04%	Lower middle class and Skilled working class	People focussed on local communities, families concentrated near Industrial areas, Includes lower income Asians
Urban Intelligence	7.19%	Mixture of Middle classes	Young educated people in urban areas starting out in life, Includes significant minority presence and students

(*continued*)

Table 4.2 *(Continued)*

Group	Percentage of UK Households	Social Groups	Description
Welfare Borderline	6.43%	Working class and Poor	Poorest people in the UK, Urban with significant ethnic minority presence
Municipal Dependency	6.71%	Working class and Poor	Poor people in council houses and dependent on benefits, Mostly white British with few immigrants
Blue Collar Enterprise	11.01%	Skilled Working Class	Enterprising rather than well educated, includes White Van Man, Few Ethnic minorities
Twilight Subsistence	3.88%	Working class pensioners	Poorer pensioners in council houses, few ethnic minorities
Grey Perspectives	7.88%	Middle Class pensioners	Pensioners in comfortable retirement and traditional values
Rural Isolation (K)	5.39%	Mixed Rural	People with relatively low incomes but high non liquid assets, traditional values, very few ethnic minorities

To study a customer base, the first step is to 'geocode' it, in other words link each standardized[4] address with its IRIS code and its block code, and possibly its coordinates in the form of longitude and latitude. Geocoding has to be done regularly (at least once or twice yearly) to allow for new customers and changes of customers' addresses. In France, the Institut Géographique National supplies maps for geocoding (from the *Géoroute* road database), and has worked in partnership with INSEE to create the *Base-îlots* database containing a map of streets and blocks 90 and 99, the addresses at the ends of street sections, and a few details of cladding or marking. INSEE has brought this database closer to Directory of Location Below the Municipal Level (known as REPLIC), which supplies, for each block, the reference, the reference of the IRIS district to which it belongs, the type of road, the starting address, the end address and the side of the road, thus enabling address files to be divided by block.

But even with all these resources, geomarketing is limited, especially in France. This is because it is illegal for anyone except a few public authorities to distribute data on areas of habitation below the IRIS level, apart from the 15 types of data available at block level (see above). This means that geomarketing data are less precise in France than in the United Kingdom, for example, where many types of data are available at the level of the Census Output Area, equivalent to the block. Because of this, the predictive power of these data is rather limited (this can be tested with scoring tools), and their use is reserved for cases where more precise data are not on hand.

4.2.2 Profitability

Profitability (or *economic value*) is a factor in many analyses, which may be concerned with the profitability of markets, customer segments or individual customers, not to mention profitability per product, per territory or per distribution channel. It is the difference between the profits to the business from a customer, segment, or market, etc., and the costs incurred, namely the acquisition and structural costs, commercial costs, operation processing costs and the cost of finance. While the profits are not always easily estimated, it is even harder to ascertain the costs in a business using multi-channel distribution, since we have to trace all contacts and interactions between a business and its customer, according to the types of channel and the frequency and duration of interactions. And although, fortunately, information technology always allows us to see what product was sold to what customer for what price at what date, it is not always capable of showing us who sold it (a physical agency, a travelling salesman, a call centre, a website...), or how (spontaneous request by a customer, response to a direct marketing effort, etc.), and it is even less likely to reveal how long it took (unless we plug sales representatives' diaries into the databases), which also has a considerable effect on costs. Even if the business can answer these questions, this is not the end of the process of determining cost and profit, because cost accounting is not always sufficiently detailed to show the cost of each operation.

[4] Standardization of addresses is highly recommended in all customer files. It enables us to reduce the number of 'not at this address' replies, avoid duplication in the files, and ensure that two people in the same family do in fact live at the same address, while benefiting from the reduced postal rates offered in France for mailings using standardized addresses.

Finally, even if the business can achieve a precise calculation of the profitability of each customer, it must be wary of drawing conclusions too quickly: the method of calculating profitability is not always a direct reflection of a customer's behaviour, a risk, or a propensity to buy, but may reflect the behaviours of financial markets and stock exchanges which affect the products owned by the customer. Thus, the profitability of a product such as a home-ownership savings scheme may be positive or negative according to the year of joining. Since the profitability of a customer is affected by the profitability of the products, we must be cautious when interpreting this. In any case, before focusing too strongly on our customers who are most profitable at any given time, we should note that a customer who remains loyal throughout his life to the same company, the same brand, the same trademark, generally moves through several segments of differing profitability. This naturally brings us to the most interesting aspect of profitability, namely *lifetime value* (LTV), or the updated net value of the expected future financial transactions with a customer (income relating to the customer minus acquisition and service costs), in other words the updated net value of profitability. In the airline industry, for example, the LTV of a student travelling in economy class may be greater than that of a company director travelling in business class. This is because the student will become an executive in a few years' time, just when the company director retires and starts travelling in economy class.

This information is much richer than the simple profitability at a given time, but is also much harder to calculate, especially if the business does not have regular financial transactions with the customer. This would be the case for a motor manufacturer, by contrast with a telephone service provider where the billing and revenues are regular. The LTV must include elements of propensity – including the propensity to buy a new product, to upgrade or to cross-purchase – as well as the margins on each product, the costs (structural, acquisition, operation processing, etc.), and, last but not least, customer attrition and product lifetime, allowing for possible contract breaks (for example, early repayment of bank loans or disconnections due to unpaid telephone bills). The calculation of LTV is a combination of the main predictive indicators (propensity, attrition, risk) that can be established for a customer. It enables us to target promotional and advertising investment on customers who are loyal to a brand, rather than using a scatter-gun approach with a large number of customers who in some cases are only 'promotional buyers'. This is also true of banking, where it may be tempting to cut margins to offer credit at very favourable rates to customers who will switch to another provider at the first opportunity. If we are unable to carry out this complex LTV calculation, we can examine the cross-tabulation of profitability and loyalty (the inverse of attrition), and the distribution of customers on these two axes.

Profitability	+	customers to be made loyal	customers to be retained
	–	customers to be let go	customers to be made profitable
		–	+
			Loyalty

4.3 Data used by business sector

4.3.1 Data used in banking

We will start this survey of the specific types of data used in data mining with the banking sector, which is distinctive because of the variety of problems and the richness of the data that can be used.

A retail bank will keep the following data:

- personal and family data (age, sex, family situation, number and ages of children, number of dependants);
- occupational data (occupation and social category, years in employment, number of working people in the household);
- geographical (length of time at the address, code of the municipality of residence, area of residence (district), type of residence deduced from the area of residence, other geodemographic data;
- assets (income and savings kept at the bank, home owner or tenant status, possession of second home, etc.);
- data on bank products held (number, type, date of commencement, expiry date, liabilities, net banking income, profitability);
- data on the use of bank products, mainly credit and payment methods;
- data on the operation of current accounts (number and value of credit and debit entries, distinguishing between transfers between a person's accounts from external movements; highest credit entries for the month; average credit and debit balances; authorized and unauthorized overdrafts in terms of value and number of days; statements; debits; credit transfers, etc.)
- data on the characteristics of credit (fixed term or revolving credit, reason for credit, period of credit, nominal amount, available amount, monthly instalments, outstanding capital, type of interest rate, value of rate, indexing of rate, early repayment, number and value of outstanding payments, number and nature of guarantees, normal or questionable accounting situation, etc.);
- data on risk (disputes, outstanding credit repayments or dishonoured cheques, value of unpaid amounts, blocking by courts or banks, over-indebtedness);
- relational data (reactions to marketing initiatives and commercial offers, refusal of direct marketing, preference for a contact or distribution channel, responses to courtesy calls or satisfaction surveys, complaints, multibanking);
- event data (birthdays, start of working life, marriage, birth of children, retirement, expiry of a savings or credit product).

A business bank will keep the following data:

- data on the banking behaviour of businesses (operation of accounts, use of credit, etc.);
- accounting data obtained from balance sheets and used to construct economic and financial ratios relating to the equilibrium of the balance sheet, profitability, solvency, increase in activity, productive structure, indebtedness, inter-business credit, etc.;
- risk data (unpaid items, receivership, liquidation, etc.);
- data supplied by the Banque de France (from the FIBEN database) and rating agencies such as Standard & Poor's, Moody's, Fitch, and Dun & Bradstreet.

4.3.2 Data used in insurance

The list below covers the main requirements of cross-selling, up-selling and attrition studies which may be carried out in a general insurance company.

First of all, we have data on the customers: age, sex, family situation, number of children, municipality code, pricing zone, lifetime as a customer of the insurance company, relationship between the insured object and the customer (home-owner/tenant, main driver, etc.), socio-professional category, and geodemographic and environmental data.

The contract data are the number of insurance policies, cover/options, situation of the contract, reason for cancellation, original company, age of product, level of reduction, payment frequency, amount of premium, and discount offered to customers.

The claims data are the limited cost and actual cost, number of claims in the current year, number of claims in other years, cover used, rate of liability, and history of claims.

Finally, data on the insured property are particularly useful for up-selling.

For home insurance, we have the following data: nature of residence (main or secondary), type of residence (apartment or house), number of rooms in the residence, data of construction of the residence, insured capital, and insured capital for valuables.

For motor insurance, we have the following data: vehicle make/type/model, vehicle segment/class, type of gearbox, taxable capacity, date of first use, date of purchase, date of birth of the main driver, date of licence of main driver, whether or not a young driver is to be covered, no-claims bonus, number of years with 50% bonus, use of vehicle, and type and amount of excess.

4.3.3 Data used in telephony

The data used in the mobile or fixed telephone industry come from various sources. Mostly they are obtained from the management and use of telephone lines, but some are collected by polls or surveys from customer panels.

There are 'customer' data, namely: subscriber's address, sex of subscriber, type of residence, ownership of a computer, availability of Internet access, geodemographic and environmental data, first subscription date, number, types and references of lines, previous telecommunications company, phone number portability.

There are line data: type of line, status of line, subscription start date, commitment end date, reason for termination (house move, competition, etc.), type of subscription, ISDN subscription, options taken up (call diversion, call waiting, call transfer, etc.), pricing options (local package, mobile package, etc.), start date and average reduction for each pricing option, inclusion on telephone directory, switching type and model.

There are 'billing' data, as follows: date of bill, amount of bill, due date, payment method (cheque, direct debit, etc.), total call duration, number of calls (per types), number of different called parties, average distance of called parties, and average duration and frequency of call per called party. We also have the numbers, dates, durations and prices of calls of the following types: local, national, international, mobile, SMS, Internet, and customer service.

Finally, there are the 'call' data: type of 'call' (voice, SMS, Internet), calling number, number called, date and time of start and end of call, duration of the call (voice), number of characters (SMS), origin and destination of the call, distance of called party, call successful (yes/no), call billed (yes/no), call pricing (local, trunk, etc.), pricing option for the call (package, reduction period, special offer, etc.) and use of voice messaging.

4.3.4 Data used in mail order

The data that are used in mail order are:

- personal and family data (age, birthday, sex, forename, number of children and their approximate ages);
- sociodemographic data (address, area of residence, type of residence deduced from the area of residence, catchment area, change of address);
- commercial activity (recency, frequency, value of orders, season of each order) – this is essential information, because a customer who has already ordered products is more likely to order again;
- purchasing habits (product types, product style);
- the channel used for orders (post, telephone, Internet, store);
- the method of payment (cheque, bank card, store card, on-line payment, cash, interest-free instalments, COD);
- the delivery channel (store, home, 24-hour home delivery);
- incidents (returns, refusals).

It is worth noting that RFM analysis is well established in data mining for the prediction of mail-order buying behaviour. The recency and frequency are analysed by season, in other words by half-years, and cover the last two years. That is to say, the recency is the half-year in which the last order was given, and the frequency is the number of half-years of activity in the last two years.

The forename is used not only to estimate the customer's age when this is not known (using the forename scoring method described above), but also in 'loyalty/profitability' typologies.

5

Statistical and data mining software

The statistical and data mining software market may be dominated by a few products, but there are many other packages, which are often much less expensive and, in some cases, may offer a comprehensive functionality. Some highly useful free software products are also being developed in this field. The most popular of these is R. However, the choice of software is not a simple matter, since statistical functionality is not the only criterion. We must also consider how the software performs with high volumes of data, the ease of access to different databases, the simplicity of deployment of the product models (model production and export functions) and the possibility of automating common tasks. The computing power may be surprising: some microcomputer packages can process hundreds of thousands of lines or more. Given that the price of commercial software ranges from €1500 to €150 000 (even if there are big discounts for teaching and research), while the software houses' brochures only highlight the benefits, we can easily become confused. The aim of this chapter is to help you make your choice by summarizing the points of comparison between products, showing you the range of current software on offer, and providing details on the three leading packages, namely SAS, IBM SPSS, and R. I will conclude with some advice about optimization to reduce machine processing time.

5.1 Types of data mining and statistical software

With the advent of microcomputing, numerous statistical and data mining programs have been developed for computers. These are relatively inexpensive, easily installed, and generally user-friendly; they contain good algorithms and can process tens or hundreds of thousands of individuals. They include Insight's S-PLUS, Neuralware's Predict, R (free software based on the S engine, like the commercial S-PLUS product) and TANAGRA (freeware). Most of them, however, cannot fully process very large databases, and often use only one or two

Data Mining and Statistics for Decision Making, First Edition. Stéphane Tufféry.

techniques – although newer, more powerful versions may be on their way. Some very highly developed products, such as S-PLUS, R, TANAGRA, Weka and JMP (pronounced 'Jump'), are an exception to this rule and use multiple techniques. JMP was developed by John Sall, one of the founders of SAS. It can read and write SAS tables, although there is no need to install SAS; it is also much less expensive. It has an interactive, simple, intuitive interface, and produces excellent graphics. It has its own algorithms for clustering, regression, generalized linear models, decision trees, neural networks, survival analysis, time series and many others, but it can also call SAS routines by entering their options in the parameters.

With these software well established, others have been designed to use large data volumes and cover a wide range of techniques. They can operate in microcomputer (or 'local') mode, and also in client–server mode, if the databases are very large or if they require operation on an industrial scale, for example with secure data sharing and protection. In this case, the server can process millions or tens of millions of lines, while the client is used for a quality display. The price of this software is at least five to ten times as great as that of the equivalent microcomputer software, and depends mainly on the configuration (the number of processors) of the server. Some packages, such as SPAD, are cheaper. The range of software is summarized in Table 5.1. You can now choose which section of the table your new system is to come from.

Single-technique software will only meet a limited, one-off requirement, and a professional statistician will not be interested in it, except for a very specific application that his usual software cannot handle: for example, a statistician may use SAS/STAT, but create his decision trees with IBM SPSS Answer Tree, and use DataLab for the transformation and selection of variables.

Table 5.1 Chart of statistical and data mining software.

Multi-technique software	TIBCO Software – S-PLUS R Weka University of Lyon – TANAGRA SAS – JMP	SAS – SAS/STAT SAS – Enterprise Miner IBM – IBM SPSS Statistics IBM – IBM SPSS Modeler Coheris SPAD – SPAD Statsoft – Statistica Data Miner TIBCO Software – Insightful Miner KXEN Oracle – Oracle Data Mining Microsoft – Analysis Services
Single-technique software	Salford Systems – CART Neuralware – Predict Complex Systems – DataLab	Isoft – Alice IBM – SPSS Answer Tree
↑ Statistical resources Computing power →	**Microcomputer software**	**Client–server software**

Multi-technique microcomputer software will satisfy the more demanding statistician, at a price that is still reasonable. It can also handle the data volumes required by small, medium and some large businesses. All statistical algorithms are available in this sector.

A business or an organization is likely to be persuaded to opt for client–server software, not because of a need for statistical techniques, but rather because of a need for industrialization. This software enables several users to work cooperatively on the same machine and even on the same data if necessary, with more flexible licence management, and it is possible to schedule back-ups, automatic data transfers and mass processing on a Unix or mainframe server, or even to export the models provided by the software into a management information application. If we need to carry out frequent, secure, automatic statistical processing with output of the results to numerous users, we will choose this category of product.

But even when we have chosen a category, that is not the end of the matter. For multi-technique software, especially for client–server systems, we often have to choose between two types: these are known as 'statistical' and 'data mining' software (Table 5.2).

Table 5.2 Statistical software vs. data mining software.

Trade designation	Statistical software	Data mining software
Platform	Microcomputer or client–server	Microcomputer or client–server
Graphic interface	Programming windows or scrolling menus	Icons which can be moved and linked with arrows
Algorithms	Those which are currently used, except for decision trees (these may be present in special-purpose software)	As for statistical software - without a number of statistical algorithms (e.g. non-parametric tests) and data analysis algorithms (e.g. linear discriminant analysis) which have to be called by lines of script - with the addition of decision trees, neural networks, detection of association rules - may sometimes provide higher performance in managing large databases
Price	A significant price	A high price
Examples	SAS/STAT IBM SPSS Statistics S-PLUS (TIBCO Software) Statistica Base (Statsoft)	SAS Enterprise Miner IBM SPSS Modeler Insightful Miner (TIBCO Software) Statistica Data Miner (Statsoft)

This double list is only a few years old. Several well-known statistical software developers have extended their product range to include data mining software, which is quite separate from their statistical products. What are the differences between statistical and data mining software, which sometimes cohabit and interact with each other? Table 5.2 provides a summary. In view of the price difference, the choice should be straightforward, but marketing can still work miracles in every field. We should also consider the attractions of neural networks (currently available in a module of IBM SPSS Statistics) and association rules detection algorithms.

5.2 Essential characteristics of the software

5.2.1 Points of comparison

We must bear a number of factors in mind when comparing and choosing statistical or data mining software. First of all, we must look for a wide range of data mining and data preparation techniques. The second point may be more or less important, depending on whether or not the user has other data request, analysis and preparation software. However, even if he has this software, it is always more convenient to have all the tools in a single package. This will avoid data transfers which may be complicated by different native data formats.

For constructing statistical models, in most cases we need to have software that can provide logistic regression, Fisher discriminant analysis, decision trees and cluster analysis. For other common applications, the software must also be capable of executing linear regression and general linear models (GLMs), to enable us to process quantitative and qualitative predictors simultaneously while controlling random effects.

A software package should also have advanced statistical functionality, covering the following tasks:

- carrying out tests on the distribution of variables, which are essential for choosing the correct algorithms and selecting the right variables;
- transforming the variables in the best possible way (binning, normalization, etc.);
- detecting correlations of variables with each other;
- carrying out factor analysis (PCA and MCA);
- sampling data for validating and establishing the reliability of models (cross-tabulated validation, bootstrapping, stratified sampling, etc.).

We must then check the quality of the implemented data mining algorithms. This may not be an easy matter, since the marketing literature of software houses is not always explicit, especially as regards their weak points. We usually have to rely on other users to tell us if the learning technique of a supplier's neural network is based on the rather outdated gradient back-propagation algorithm, or if his decision tree is unreliable because it lacks an automatic validation procedure, or if the sampling method is a little rough-and-ready.

A third criterion, which may be decisive, is the computing power and the capacity to handle large data volumes. The importance of this factor is directly related to the size of the business and the number of customers. Most microcomputer software can handle several tens of thousands or even hundreds of thousands of individuals, especially for a discriminant analysis not requiring excessive amounts of computing power. So small and medium enterprises without very complex requirements can use these products quite happily, especially if the models generated by the software can be exported, in C or another language for example, in such a way that they can be imported into the central computers of the business. Large enterprises should be aware that computing power, even in client–server systems, depends on the power of both the hardware and the software, and the ratios between processing speeds of different programs can be as much as 1 to 5. Processing speed can be crucial in certain areas of work where large amounts of calculation, repeated tests and multiple-sample models (e.g. bootstrapping) are required in order to comply with regulations. Even if initially large populations can be broken down by clustering and segmentation operations, a large amount of computing power is still needed for the cluster analysis itself, as well as for modelling in each of the segments, which, even if smaller, may require highly complex analysis.

A fourth criterion relates to the types of data handled: for example, if the business already has an SAS Infocentre, it will clearly be beneficial to choose SAS software. In any case, the software must be able to import data in different formats.

Another criterion is the user-friendliness of the software and the ease of producing reports summarizing the operations and the results. Although this factor should be borne in mind because it may have an effect on the user's productivity and avoid the need for repetitive tasks, it should not be overestimated, and we should always remember that greater user-friendliness will never be a substitute for the basic experience of statistics and data which a user is expected to have.

The final criterion is . . . the price! It is not always easy to compare the prices of different suppliers, who may either sell or lease their software, may offer maintenance for technical support and the delivery of later versions, may charge according to a price list or by the volume of data processed, and so on.

5.2.2 Methods implemented

The more numerous and varied the methods implemented by the software, the more likely the statistician is to be able to deal with all the problems he may encounter. The main methods are:

(i) prediction (linear regression, general linear model, robust regression, non-linear regression, PLS regression, decision trees, neural networks, *k* nearest neighbours, etc.);

(ii) classification (linear discriminant analysis, binary logistic regression, polytomous logistic regression (ordinal or nominal), generalized linear model, decision trees, neural networks, *k* nearest neighbours);

(iii) cluster analysis (moving centres, *k*-means, agglomerative or divisive hierarchical clustering, hybrid methods, density estimation methods, Kohonen maps);

(iv) association rules detection;

(v) survival analysis;

(vi) time series analysis.

Some programs offer automatic chaining of a number of methods, with stopping points between the methods if required (if a method does not provide suitable results, the next one may or may not be executed).

5.2.3 Data preparation functions

The data preparation phase requires relatively extensive functionality, not provided in all software, if we wish to avoid a high workload and excessive prolongation of this stage, which is in any case the longest part of a study. We should therefore ensure that the following functions are present:

(i) file handling (merging, aggregation, transposition, etc.);

(ii) data display, colouring of individuals according to a criterion;

(iii) detection, filtering and Winsorization of outliers;

(iv) analysis and imputation of missing values;

(v) transformation of variables (recoding, standardization, automatic normalization, discretization, etc.);

(vi) creation of new variables (predetermined logical, chain, statistical, mathematical, and other functions);

(vii) selection of the best independent variables, discretizations and interactions.

5.2.4 Other functions

The following functions are fundamental for any proper study:

(i) statistical functions (determination of central tendency, dispersion, and shape characteristics; statistical tests of the mean, variance, distribution, independence, heteroscedasticity, etc.);

(ii) data sampling and partitioning functions, for creating training, test and validation samples, bootstrap functions, and jackknife functions (for cross-tabulated validation);

(iii) functions for exploratory data analysis, especially factor analysis (principal component analysis, PCA with rotation of axes, correspondence factor analysis, multiple correspondence analysis);

(iv) display of results, table manipulation, 2D, 3D and interactive graphics library, navigation through decision trees, display of statistical parameters and performance curves (ROC, lift, Gini index), facility for incorporating these elements into a report, etc.;

(v) advanced programming language.

5.2.5 Technical characteristics

The technical characteristics are the main factors in determining the price of the software, as well as the productivity of the statistician and the possibility of group working on the same data. It is important, therefore, to make a careful estimate of the requirements for:

(i) the hardware platform (Unix, Windows, Sun, IBM MVS, etc.);

(ii) the databases accessed (Oracle, Sybase, DB2, SAS, SQL Server, Access, etc.);

(iii) native access (more reliable and much faster) or ODBC access (easier to program) to these databases;

(iv) client–server or standalone architecture;

(v) the algorithms, parallel or not;

(vi) the maximum data volume which can be processed (in a reasonable time);

(vii) execution in deferred (batch) mode or interactive (transactional) mode;

(viii) the possibility of exporting models (C, PMML, Java, SQL, etc.).

An algorithm is parallelized when it is divided into a number of tasks that can be executed simultaneously on several processors of a computer, or on more than one computer (grid computing). This enables the execution speed of an algorithm to be increased significantly, and is particularly useful for neural networks or even simple sorting.

5.3 The main software packages

5.3.1 Overview

Table 5.3 lists many of the existing software packages, divided into three main families according to the approximate volumes of data that they can handle.

The two leading products competing in the market for data mining in large systems are SAS and SPSS. These are also the most widely used packages in all systems.[1] To this we should add Statsoft's Statistica Data Miner, which is very comprehensive and well developed, making good use of the richness of its underlying S-PLUS statistical software. KXEN, which operates as a modelling engine for implementation in IT applications or other statistical software, should be considered separately. SPAD has always been popular in the French academic world because of the quality of its algorithms, especially for factor analysis. From Version 6 onwards, it has become more useful for businesses, with the development of a client–server architecture and improvements to reduce processing time. New algorithms have

[1] Source: KDnuggets survey, October 2000 (covering 698 users), June 2002 (551 users) and May 2004 (650 users).

Table 5.3 The main statistical and data mining software packages.

Data volume	Product	Speciality*	Producer
Low (tens of thousands of records)	NeuralWorks Predict	Neural networks	Neuralware
	NeuroOne	Neural networks	Netral
	Wizwhy	Associations	Wizsoft
	Weka		Open Source (University of Waikato, New Zealand)
	R		Open Source (initially at the University of Auckland)
	DataLab	Data preprocessing	Complex Systems
Medium (hundreds of thousands of records)	Alice	Decision trees	Isoft
	KnowledgeSEEKER	Decision trees	Angoss
	KnowledgeSTUDIO		Angoss
	C5.0 (Unix) See5 (Windows)	Decision trees	RuleQuest Research
	Data Mining Suite		Salford Systems
	CART	Decision trees	Salford Systems
	Polyanalyst		Megaputer
	TANAGRA		University of Lyon, France
	JMP		SAS
	S-PLUS		TIBCO Software
High (millions of records)	KXEN		KXEN
	Oracle Data Mining		Oracle
	SPAD		Coheris SPAD
	IBM SPSS Statistics		IBM
	IBM SPSS Modeler		IBM
	Statistica Data Miner		Statsoft
	Insightful Miner		TIBCO Software
	SAS/STAT		SAS
	Enterprise Miner		SAS

* If no 'speciality' is shown in this column, the software is a suite containing a number of algorithms.

also appeared, for logistic regression (not included in earlier versions), the PLS approach and (starting with Version 7) interactive hierarchical descending clustering ('interactive clustering tree').

All these products are available in client–server versions and can therefore handle large volumes of data.

Alongside these data mining package with user-friendly graphic interfaces, we find less expensive statistical software. Some of these, such as SAS/STAT and SPSS, are also offered in client–server versions. Traditionally, SAS has been more widely used in pharmacy, finance and insurance, SPSS has been used in human sciences and distribution, and S-PLUS in industry.

Leaving aside these 'juggernauts' (I will return to SAS and IBM SPSS shortly), there are many specialist packages on the market for decision trees and neural networks.

Some free software has also spread beyond university laboratories: these products include the excellent TANAGRA, the successor of SIPINA, Weka[2] (Waikato Environment for Knowledge Analysis), and, above all, R, a very popular free software which I will go on to discuss more fully below.

Finally, some major database producers (Oracle and Microsoft) have recently appeared on the market, with data mining algorithms incorporated into their databases. The value of this approach is that complex calculations can be carried out as close to the data as possible, without costly exchanges between a voluminous database and the data mining software. The statistical model is implemented directly in the database, and can be applied very efficiently to all new data. The database can also execute a model sent to it in PMML format by data mining software which has a function for export in this format.

5.3.2 IBM SPSS

SPSS (Statistical Package for the Social Sciences) appeared in 1968, and has been used very widely in the social sciences, marketing and health. Its name was changed to IBM SPSS Statistics in 2009, after the acquisition of the SPSS business by the IBM Group. Its functionality is accessed via a proprietary L4G (fourth generation) programming language, called the 'syntax language', or via a graphic interface of scrolling menus which generate a syntax invisibly to the user, unless he requests the 'pasting' of the syntax into a command window. The advantage of the graphic interface is its user-friendliness and simplicity of use for a less experienced user; on the other hand, the syntax language can be used to chain lengthy and complex operations together, and to repeat common tasks by recording their syntax. Note that part of the functionality can only be accessed by using the syntax, not the graphic interface. The commands can be launched in interactive mode or in batch mode, using the Production Facility module. There is also a macro language for automating repeated commands which may depend on parameters. A script language is available for constructing dialogue boxes.

The main window of IBM SPSS Statistics, the Data Editor, looks like a spreadsheet, and direct input is possible, as in Excel (Figure 5.1). The SPSS commands are executed line by line and update the table or add results to the Output Editor window. This window also provides an option for storing the executed syntaxes with their execution times; it is therefore related to both the Output and Log windows of SAS. If it is necessary to input the syntax, this takes place in the Syntax Editor, similar to the SAS Editor window. By contrast with Excel, these software do not require the user to input syntax in the cells which will contain the results. IBM SPSS Statistics can read from and write to ASCII files, some

[2] 'Weka' is the name of an emblematic bird of New Zealand.

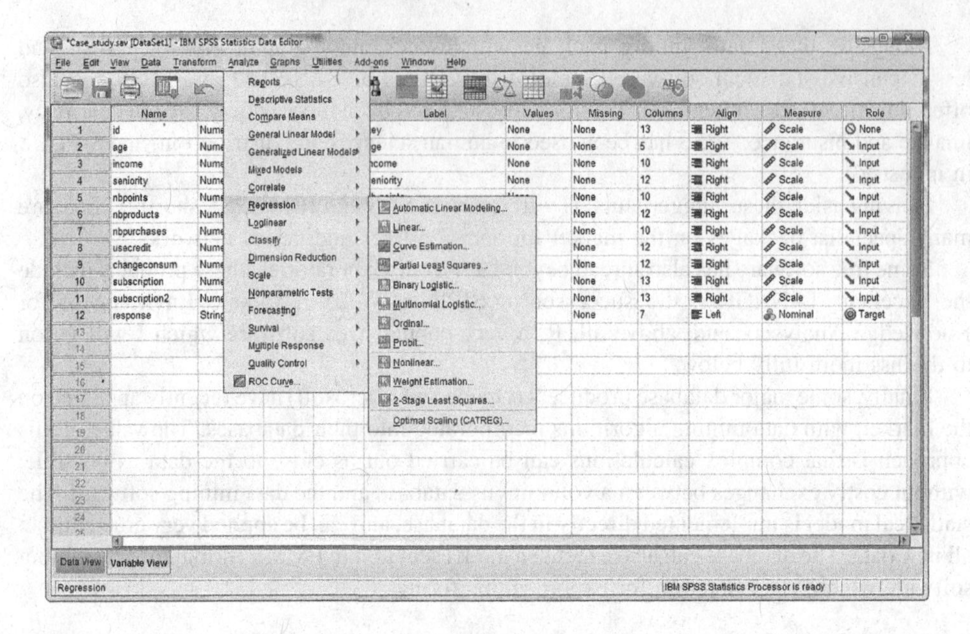

Figure 5.1 IBM SPSS Statistics.

databases and some tables of other statistical software. The basic data management functions (sorting, aggregation, transposition, table merge, etc.) are provided. Similar to the ODS in SAS, but less flexible and less concise, the Output Management System (OMS) in IBM SPSS Statistics can be used to send the output to a file, instead of the Output Editor. The file can be in SPSS (allowing it to be reinserted into another process), TXT, HTML or XML format. The OMS can also store the results of a large number of iterated calculations in a single file (by creating a loop with a macro), so that they can be compared subsequently and the best ones can be saved.

IBM SPSS Statistics is available in several environments, including Windows, Mac OS X and Unix, and a new version appears roughly once per year. Although IBM SPSS Statistics is primarily a microcomputer tool and the management of large data volumes is not its strong point, a client–server version has been created with some functionality not provided in the microcomputer version, such as the application of score functions. Finally, a range of modules which can be purchased separately allow the user to access some statistical tools that are more advanced than those of the basic module, starting with logistic regression which is available in the Regression module.

From Version 15 onwards, the Data Preparation add-on module has included an effective algorithm called Optimal Binning, for discretizing (binning) continuous variables. Each continuous variable is divided into categories in such a way that the measurement of the association between the discretized continuous variable and the class variable is optimized. In the Optimal Binning algorithm, this measurement is based on entropy (Fayyad and Irani, 1993),[3] and the algorithm minimizes the sum of the entropies of the categories, as in the C4.5 tree (see Section 11.4.7).

[3] Fayyad, U.M. and Irani, K.B. (1993) Multi-interval discretization of continuous-valued attributes for classification learning. In *Proc. 13th International Joint Conference on Artificial Intelligence*, pp. 1022–1027. Los Altos, CA: Morgan Kaufmann Publishers.

New modules appear regularly, including a neural network module in Version 16, a module for multiple imputation of missing values and an RFM analysis module in Version 17, a bootstrap module and a direct marketing module (for RFM, segmentation, profile, postcode and other analyses) in Version 18. Additionally, the Categories module has been enhanced with the addition of ridge, lasso and elastic net regularization (see Section 11.7.2.). The basic module has been extended with a nearest-neighbour analysis algorithm. Now it can also be used to modify or create new dialogue boxes. But perhaps the most spectacular improvement in recent years is the considerable extension of the programming possibilities in IBM SPSS Statistics. This is based on the use of the Python programming language, enabling new functions and procedures to be developed. Since Version 16, it has also been possible to call and use R software within IBM SPSS Statistics, thus providing access to all the R packages and a vast amount of additional functionality.

Since the weak point of R is the difficulty of handling large volumes of data (because these are loaded into RAM), whereas IBM SPSS Statistics does not have these problems, the integration of R is based on the following procedure. IBM SPSS Statistics reads the data, transforms them if necessary, selects them, and only sends the subset of useful observations and variables to R. A very simple example is shown below.

```
GET FILE = 'mytable.sav'.
SELECT IF (condition=1).
BEGIN PROGRAM R.
mytable <- spssdata.GetDataFromSPSS (variables=c("V1 to V5") row.
label=V1)
regression <- lm(V2 ~ V3+V4+V5, data=mytable)
print (summary(regression))
spsspivottable.Display (anova(regression))
END PROGRAM.
```

IBM SPSS Statistics can also display a graphic produced by R in its Output window. It can integrate R functions, which may be native or user-constructed, into its syntax, and it can also integrate them in the form of dialogue boxes, to supplement the IBM SPSS Statistics menus with new functionality.

In 1998, SPSS Inc. bought ISL, producer of the Clementine data mining software, and the integration of this package with SPSS is bound to gather pace in future years. Decision trees, which up to 2005 were offered in the standalone Answer Tree software of SPSS Inc., are now integrated into IBM SPSS Statistics in the form of a module called IBM SPSS Decision Trees, enabling them to be incorporated into IBM SPSS Statistics syntax, macro programs, etc. It is also possible to program decision tree bagging or boosting. The tree algorithms are CART, CHAID and QUEST, as in Answer Tree, but IBM SPSS Decision Trees has the drawback of not allowing interactive manipulation of the trees. In 2009, Clementine was renamed 'IBM SPSS Modeler'.

In conclusion, I should mention the free PSPP software, distributed under the terms of the GNU General Public Licence, which is claimed to be a 'clone' of SPSS. It does not have all the functionality of the latter product, but its syntax and database format are compatible with those of SPSS, and it can handle large volumes. On the other hand, it does not exist in a Windows version, and it can only be run on this platform by installing a Unix emulator such as Cygwin. The address of the official PSPP website is http://www.gnu.org/software/pspp/, and detailed documentation can be found at http://cict.fr/~stpierre/doc-pspp.pdf.

5.3.3 SAS

SAS (Statistical Analysis System) was founded in 1976[4] in the IBM mainframe world, and still retains its original capacity for handling large data volumes, a capacity which increased still further with the implementation of a parallel architecture from 1996. With successive versions, the language used by SAS changed from Fortran to PL/I and then C. The first version for PC/DOS appeared in 1985; it was followed in 1986 by the first SAS/STAT module. As in SPSS, specialist modules multiplied over the years, around the central module called SAS/BASE. Some modules are more concerned with statistics, namely IML (Interactive Matrix Language), STAT, ETS (Econometrics and Time Series), OR (Operational Research), and QC (Quality Control). Other modules are dedicated to reporting (AF, EIS), but these are giving way to the new SAS Business Intelligence platform. The SAS data mining module, Enterprise Miner, a competitor of IBM SPSS Modeler, appeared in 1998. In 2000, the SAS Enterprise Guide module enabled SAS users to benefit from the user-friendly interface which was already provided in SPSS but not in SAS, where the mainframe origins were evident in a traditionally austere interface. However, I should mention the useful SAS/INSIGHT module which appeared in 1991, and could be used for displaying data in 2D or 3D, plotting various graphs, carrying out calculations such as PCA, and even carrying out interactive operations on data, for example in order to exclude some outliers from the analysis.

Version 7 brought the ODS (see the description of the OMS above), which enabled the output of a procedure to be sent, not to the Output window, but to a file which could be in SAS, RTF, PDF, HTML or XML format. I have given a few examples of the use of ODS (see Section 3.8.4, for instance), a system which enables us to retrieve the results of statistical tests conducted on an indefinite number of variables in SAS files, for subsequent formatting and presentation in a summary file. The same approach can be used to calculate a large number of models from a set of data, or from bootstrap samples, before comparing them and selecting the best model or models. ODS enables us to conduct and evaluate tests automatically in large numbers, which would be impossible otherwise. An extremely valuable tool, then. Another aid to productivity offered by the ODS is the facility for automatic creation of a Word report containing the results.

Starting with Version 8, SAS has been running on Microsoft Windows, Unix, and z/OS for IBM mainframe, using code which is portable from one version to another. The OS/2 and Apple Macintosh versions which appeared with Version 6 of SAS have since disappeared. Version 8 brought a number of improvements to SAS/STAT, for logistic regression for example, and new procedures, such as GAM for generalized additive models and MI for imputing missing values. Version 9.1, which came out in 2003, brought new procedures such as ROBUSTREG, together with ODS GRAPHICS which is an extension of ODS for creating high-quality graphics. They are directly integrated with the outputs of the SAS procedures, not produced by supplementary programming of SAS/GRAPH procedures. However, they can be modified, either by subsequent modification in the editor of ODS GRAPHICS (Statistical Graphics Editor), or by preliminary modification of their graphic model using the new Graph Template Language (GTL).

[4] To be precise, the SAS company was set up in 1976, but the software was developed progressively from 1966 onwards based on the work of Anthony J. Barr, followed up by his PhD students, Jim Goodnight and John Sall. For more details, see the SAS website (http://www.sas.com/presscenter/bgndr_history.html) and Wikipedia.

Version 9.2 was first distributed in 2008. It provided improvements to older procedures, two new procedures, SEQDESIGN and SEGTEST, for analysing clinical trials, and the integration of three statistical procedures which had previously been available by downloading for SAS 9.1.3. These are GLMSELECT (which enriches the GLM procedure with advanced selection methods, such as Tibshirani's lasso and Efron et al.'s least angle reression (LAR)), GLIMMIX (for generalized linear models with random effects) and QUANTREG (for modelling conditional quantiles, rather than a conditional expectation as in conventional regression). Another new feature of SAS 9.2 is a worthy successor of SAS/INSIGHT, namely SAS/IML Studio (briefly called SAS Stat Studio), which elegantly combines the functionality of SAS/STAT with the interactivity of SAS/INSIGHT. Thus graphics such as those in Figure 5.2 are dynamically related to the data in the table under study, and these data can be located on either a microcomputer or a server. Following the example of IBM SPSS Statistics, SAS decided to integrate the R software, but only in SAS/IML Studio, which can manage the objects manipulated by R.

Like IBM SPSS Statistics, SAS has an L4G programming language, although it is more concise and efficient, with a more flexible and powerful macro language and an SQL procedure which substitutes SQL syntax for SAS syntax in certain cases, with a possible gain in performance (see below). SAS also has a matrix language, IML (Interactive Matrix Language). An SAS program is built up from DATA steps, procedure steps, and macros if required. Several tens of procedures provide the very comprehensive range of functions

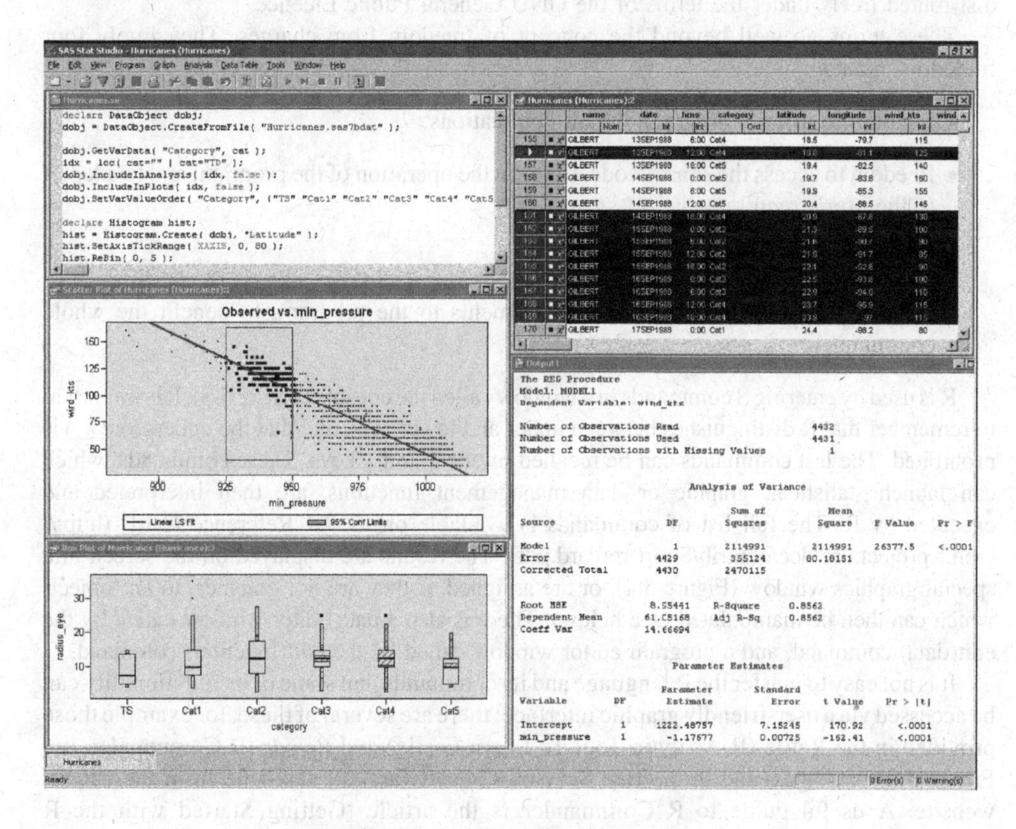

Figure 5.2 SAS/IML STUDIO.

(statistics, graphics, utilities, etc.), while the DATA step enables the user to open files (or import databases), read each record in turn, write to another file (or export to a database), merge a number of files if necessary, and close the files, while concentrating more on the content of the data than on their physical storage. It should be noted that the import/export function with Excel 2003 (and later versions) is very well managed, taking into account the tabs of an Excel sheet and specific Excel formats. Plenty of clear and accurate information can be found on the SAS technical support website (http://support.sas.com), in the articles by Olivier Decourt on his website (see Section B.8 in Appendix B) and in books on SAS (see Section B.6).

5.3.4 R

The R software is based on the same language as the commercial S-PLUS product. This is the S language, developed by John Chambers and others at Bell Laboratories in the 1970s. This language, similar to Scheme, is interpreted (its commands are directly executed, unlike those of a compiled language) and object-oriented (the data, functions and outputs are stored in the computer's RAM in the form of objects, each of which has a name). It differs from the SAS and SPSS languages in its greater concision: R code is more compact and closer to mathematical language. R was created by Robert Gentleman and Ross Ihaka at the University of Auckland, for teaching statistics, and since 1997 its source code has been available and distributed freely under the terms of the GNU General Public Licence.

These terms go well beyond the concept of freedom from charges. They imply four freedoms, namely:

- freedom to run the program, for all applications;
- freedom to access the source code, to study the operation of the program, and to adapt it to the user's requirements;
- freedom to redistribute copies;
- freedom to publish one's own improvements to the program, to benefit the whole community.

R is used by entering S commands in a window called the console (Figure 5.3). It is important to remember that R distinguishes between upper and lower case, and that the underscore (_) is prohibited. The last commands can be recalled by the ↑ and ↓ keys. These commands, which can launch statistical, graphic or data management functions, are then interpreted and executed by R. The full list of commands is available on the 'R Reference Card' (http://cran.r-project.org/doc/contrib/Short-refcard.pdf). The results are displayed on the screen in a special graphics window (Figure 5.4), or are assigned, if they are not graphics, to an 'object' which can then be manipulated (see below). There is also a data editor window called by the edit(data) command, and a program editor window called by the edit(function) command.

It is not easy to master the R language and its commands, but some of its functionality can be accessed via a user-friendly graphic interface: there are several of these, for example those provided in the Rattle (R Analytic Tool To Learn Easily) and *Rcmdr* (R Commander, see Figure 5.5) packages, and in the free SciViews R software, all available from the CRAN website. A useful guide to R Commander is the article 'Getting Started with the R Commander' by John Fox, available on the Internet. The R Commander interface is

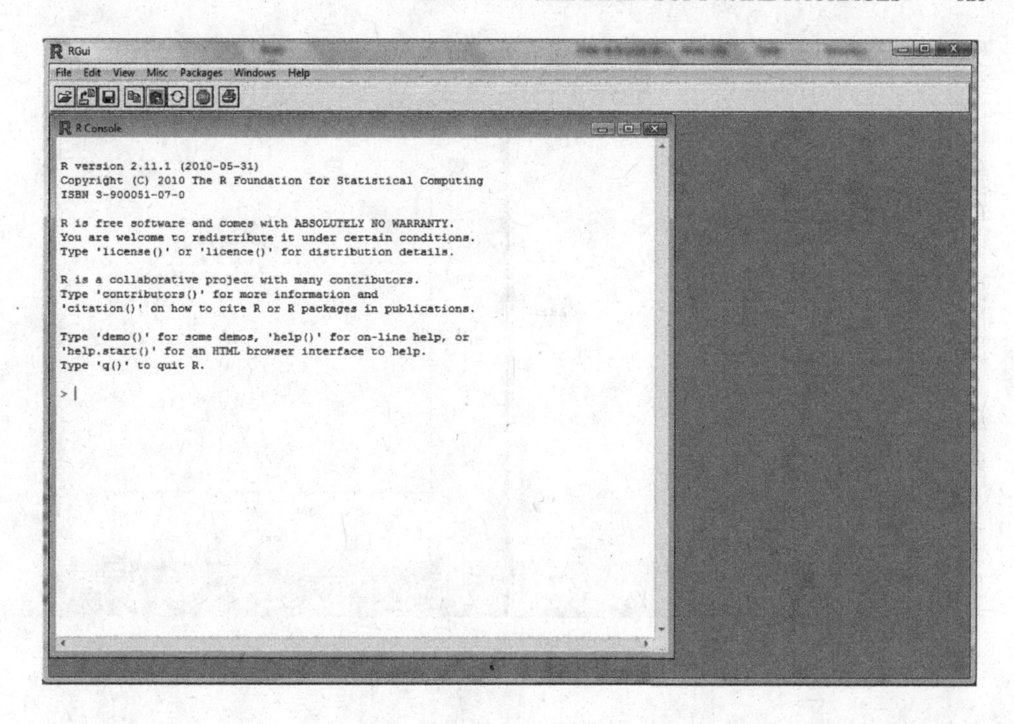

Figure 5.3 The R console.

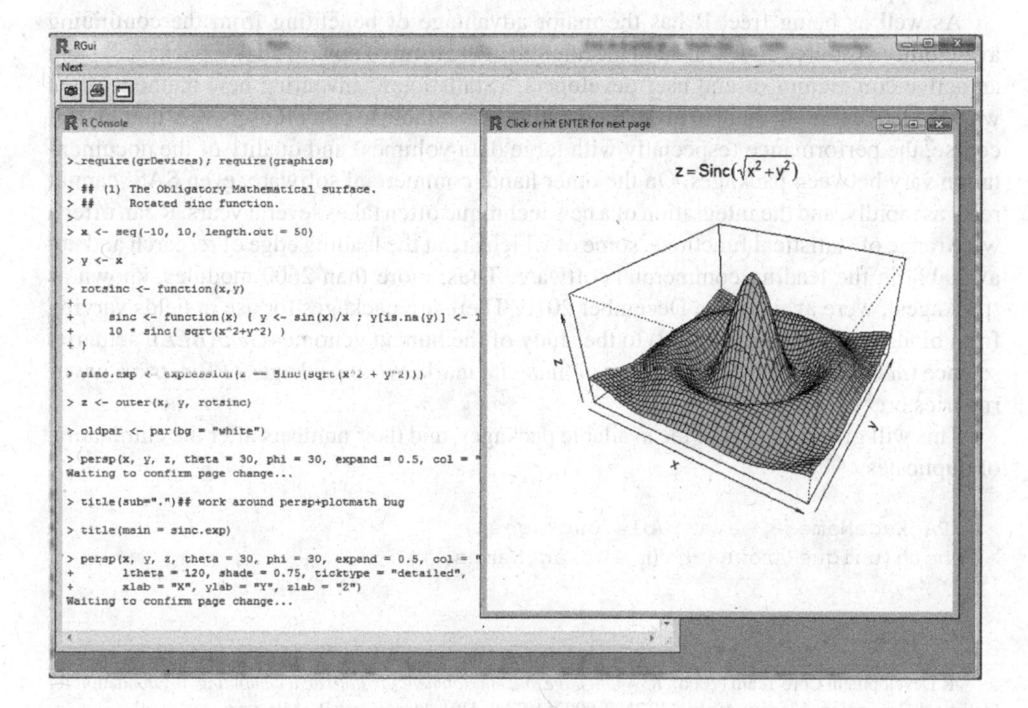

Figure 5.4 Graphic window in R.

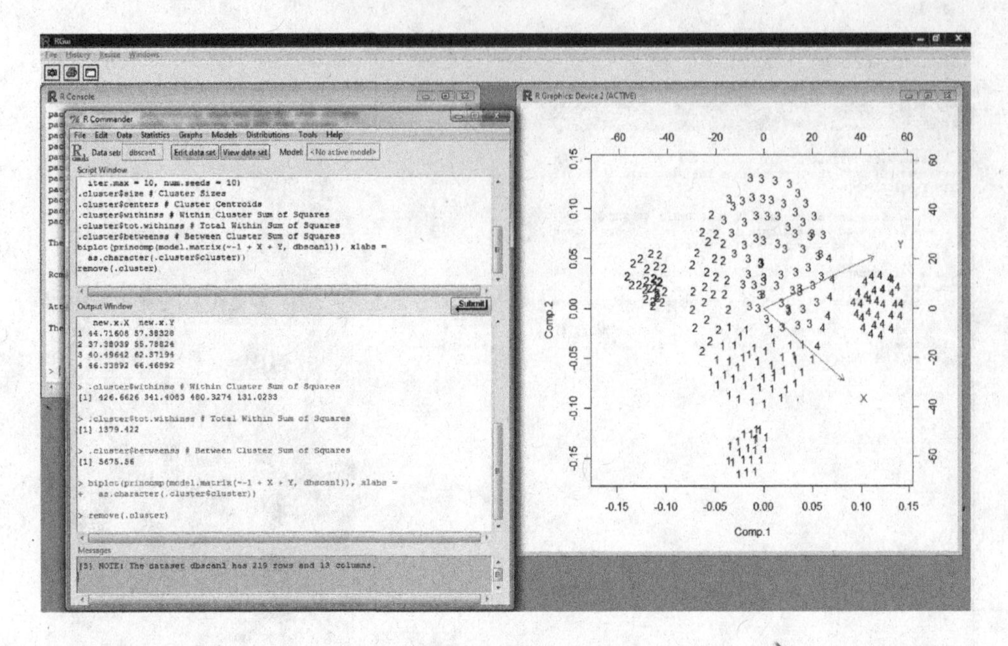

Figure 5.5 R Commander.

comprehensive and is the most widely used, but Rattle is useful for more typical data mining applications.

As well as being free, R has the major advantage of benefiting from the continuing availability of source codes and enhancements (in the form of downloadable packages) from an active community of end user developers.[5] Statisticians inventing new methods often work with R, enabling them to make their methods available to other R users very quickly. Of course, the performance (especially with large data volumes) and quality of the documentation vary between packages. On the other hand, commercial software, even SAS, cannot react as rapidly, and the integration of a new technique often takes several years. R can offer a wide range of statistical functions, some of which are at the leading edge of research and not available in the leading commercial software. Thus, more than 2600 modules, known as 'packages', were available in December 2010. There are packages for use in fields varying from biodiversity (*BiodiversityR*) to the study of the human genome (*GenABEL*), actuarial science (*actuar*) and the econometrics of financial markets (all packages of *Rmetrics*: www.rmetrics.org[6]).

This will give you a list of the available packages, and their numbers after the elimination of duplicates:

```
> myPackageNames <- available.packages()
> length(unique( rownames(myPackageNames) ))
```

[5] R Development Core Team (2006). *R: A language and environment for statistical computing.* R Foundation for Statistical Computing, Vienna, Austria. ISBN 3-900051-07-0, URL http://www.R-project.org.

[6] See also the document by Daniel Herlemont at http://www.yats.com/doc/r-trading-projet-index.pdf

The writing of R functions and packages is facilitated by the fact that they are written in the same language as the programming language which uses functions already integrated into R. This contrasts with SAS, which requires SAS/IML, or IBM SPSS Statistics, which uses Python, a language external to SPSS. To consider just one example, an operation such as variable selection does not follow the same rules in SAS/BASE and SAS/IML. Such differences are not found in R. Consequently, the learning curve for this software is different from that of SAS, not to mention that of IBM SPSS Statistics: although it is more complicated at first, the use of R becomes easier in later stages, for programming repeated tasks or new functions or algorithms, because all of these actions are based on the same language in R.

R is also attractive because of its graphic resources (see http://addictedtor.free.fr/graphiques/) and the many platforms on which it is provided (Unix, Linux, Windows, Mac/OS X). With the *foreign* package, R can read SAS, SPSS or Stata tables. But R can manipulate more varied types of data (vectors, matrices, etc.) than the rectangular tables used in these software products.

R also has a matrix language which is as highly developed as that of MATLAB, which it can also emulate via a special package (*matlab*). The Internet resources, FAQ, manuals and discussion lists are numerous, and most of them are brought together on the website of the *Comprehensive R Archive Network* (CRAN, http://cran.r-project.org/ or http://www.r-project.org/, the site at the top of the page when you type 'r' into Google). Another advantage of R is the provision of an 'R help list' (http://www.r-project.org/mail.html) to which users can subscribe for discussions of problems and possible solutions, for news about R, updates to documentation, etc. Compared with the technical support offered by commercial software developers, the information is less focused, and several solutions may be offered by different contributors when a problem is posted, but the response is very quick, often within an hour.

For all these reasons, R has experienced spectacular growth in a few years, to the point that it is competing with SAS not only in academic settings, but also in industry.

Compiled ('binary') files for installing R and its packages are distributed from the CRAN website, which also contains source codes and instructions for the installation of each platform. Version 2.12.1 of R is available in December 2010. Once on the CRAN website, you choose a mirror site (by country), choose a platform, decide whether to install R or one of its packages, and then download the compiled file or read the installation instructions. Installation is quick and easy. When R is installed, an additional package can be installed by going to the Packages menu (on the Windows platform) and selecting installation from a zip file on a local disk or installation directly from a CRAN website. The *library()* command provides a list of installed packages. The *library(toto)* function loads an installed package called *toto* into RAM. A package must be loaded into memory to be used, but the basic packages are loaded by default. These packages which are loaded when R starts are listed by the *search* command. Loading a package such as *Rcmdr* results in the display of the graphical user interface of R Commander.

When you close R, by typing *q()*, it asks if you wish to save an image of the work session. If you reply 'yes', R saves two files: Rdata, which contains the working environment and the set of objects created; and Rhistory, which contains the set of commands entered during the session. These files are stored in the R working directory, which is displayed by the command *getwd()*. This directory can be modified by the command *setwd(<directory>)*, but can also be modified continuously in Windows by right-clicking on the R icon, then clicking on Properties and modifying the directory displayed in 'Start in'.

A suitable function for a problem can be found quickly in R by using the *help.search* command. For example, *help.search("anova")* displays all the R functions containing the string "anova" in the function description. The search can be extended to the Internet by the RSiteSearch("anova") command or by clicking on 'Search' at www.r-project.org. This is useful, because adding the letter 'R' in a search engine rarely limits the search to R software. Note that the Firefox browser has an 'Rsitesearch' plug-in. The command *help(tutu)*, or its synonym *?tutu*, displays the help for the function *tutu()* and the name of the package containing it. For help on an operator, put the operator in quotation marks, thus: *help("<-")*. The option *try.all.packages = TRUE* extends the search for the function to all the installed packages, even those which have not been loaded into memory. The function *help(package = toto)* provides a list of the functions of the package *toto*. R help in HTML format can be accessed by the command *help.start()*. This leads to documentation, manuals and FAQs. Demonstrations can be launched with commands such as *demo(lm.glm, package="stats")*, and examples are provided by the example command: *example(lm)*.

R is not suitable for large data volumes, as it loads all the data it requires into RAM, and does not use a temporary file on a local disk for its calculations. To give you an idea of the sizes involved, with 1 GB of RAM it is possible to carry out a logistic regression on a file of several tens of thousands of observations for a few tens of variables, but not much more than that. It also tends to be slower than SAS, and the repetition of many modelling operations, in bagging for example, may take much longer in R than in SAS. The difference may vary from several hours to a few minutes (see O'Kelly, 2009).[7]

The main statistical functions of R (followed by the name of the package containing them, in brackets) are as follows:

- table (base): frequency table (one variable) or contingency table (two variables)
- mean (base): mean
- median (stats): median
- range (base): range
- var, sd (stats): variance, standard deviation
- quantile (stats): quantiles
- summary (base): basic statistics describing a quantitative variable (minimum, maximum, mean, quartiles), a qualitative variable (frequencies), a model, etc.
- cov (stats): covariance of two variables; variance–covariance matrix
- cor (stats): Pearson, Spearman, and Kendall correlation coefficient for two variables; correlation matrix
- chisq.test (stats): χ^2 test
- fisher.test (stats): Fisher's exact test

---

[7] O'Kelly, M. (2009) R vs. SAS in model based drug development. Paper presented to UserR! The R User Conference, Rennes, France. http://www.agrocampus-ouest.fr/math/useR-2009/slides/OKelly.pdf

- t.test (stats): Student's test
- aov, anova (stats): analysis of variance
- manova (stats): multivariate analysis of variance
- var.test (stats): Fisher's variance test
- bartlett.test (stats): Bartlett's variance test
- levene.test (car): Levene's variance test
- wilcox.test (stats): Wilcoxon test
- kruskal.test (stats): Kruskal–Wallis test
- binom.test, prop.test (stats): tests of proportion
- shapiro.test (stats): Shapiro–Wilk test for normality
- ks.test (stats): Kolmogorov–Smirnov test for normality
- lillie.test (nortest): Lilliefors test for normality
- cvm.test (nortest): Cramér–von Mises test for normality
- ad.test (nortest): Anderson–Darling test for normality
- friedman.test (stats): Friedman rank test (ANOVA on paired samples)
- mcnemar.test (stats): McNemar test (χ^2 on paired samples)
- density (stats): density estimation
- boot (boot): bootstrap
- princomp (stats): PCA
- varimax (stats): PCA varimax.

The main clustering functions in R (more numerous than in SAS and IBM SPSS Statistics) are as follows:

- hclust (stats): agglomerative hierarchical clustering
- cutree (stats): cuts a tree diagram produced by agglomerative hierarchical clustering (similar to PROC TREE in SAS)
- kmeans (stats): *k*-means algorithm
- agnes (cluster): agglomerative nesting
- clara (cluster): clustering large applications
- daisy (cluster): dissimilarity matrix calculation

- diana (cluster): divisive analysis clustering
- fanny (cluster): fuzzy analysis clustering
- mona (cluster): monothetic analysis clustering of binary variables
- pam (cluster): partitioning around medoid
- Mclust (mclust): MCLUST probabilistic clustering based on a search for Gaussian models
- pop (amap): clustering by aggregation of similarities
- som (class): Kohonen maps.

The main modelling functions in R are as follows:

- lm (stats): linear regression, analysis of variance and covariance
- lm.ridge (MASS),[8] ridge (survival): ridge regression
- lars (lars): lasso regression
- glmnet (glmnet): lasso and elastic net regressions
- nls (stats): non-linear regression (by least squares)
- loess (stats): LOESS regression
- spline (stats): spline interpolation
- glm (stats): generalized linear model for the following Y/X=x distributions: normal (regression), binomial (logistic regression), Poisson, gamma
- lme (nlme): mixed-effects linear model
- nlme (nlme): mixed-effects non-linear model
- clogit (survival): conditional logistic regression
- gam (mgcv): generalized additive model
- gamm (mgcv): generalized additive mixed-effects model
- lda (MASS): linear discriminant analysis
- qda (MASS): quadratic discriminant analysis
- rpart (rpart), tree (tree): CART decision trees
- bagging (ipred): bagging a CART tree constructed by *rpart*
- randomForest (randomForest): random forests

[8] The MASS package is grouped with the class, nnet and spatial packages in the VR package.

- ada (ada): Discrete AdaBoost, Real AdaBoost, LogitBoost and Gentle AdaBoost
- adaboost (boost): AdaBoost
- logitboost (boost): LogitBoost
- gbm (gbm): boosting (generalized boosted regression modelling)
- knn (class): *k*-nearest-neighbour classification
- nnet (nnet): neural networks
- ksvm (kernlab), lssvm (kernlab), svmpath (svmpath): support vector machines
- ts (stats): time series
- arima (stats): ARIMA model
- survreg (survival): parametric survival model
- coxph (survival): Cox proportional hazards regression model
- survfit (survival): survival curve for censored data.

There is a package called *tseries* for time series. Here are some of its useful functions:

- arima.sim: for simulating ARIMA trajectories
- ARMAacf: for calculating the theoretical autocovariance function
- ar, arima, arima0, arma: for adjusting an AR, ARMA, etc. model with a choice of different methods; returns the residuals
- acf, pacf: for plotting the autocorrelation and partial autocorrelation functions
- diff: for differentiating the series with different orders
- predict: for forecasting with different horizons
- acf2AR: can be used for one-step forecasting
- pp.test: Phillips–Perron test (unit root)
- Box.test: decorrelation test (portmanteau test combining a number of tests, such as the Ljung–Box and Box–Pierce tests; the term 'portmanteau' alludes to an object for containing several different kinds of garments)
- garch: for adjusting ARCH/GARCH models.

There is also a *tm* package for text mining.

Several user-friendly interfaces have been created for factor analysis, including those of the *ade4* package (http://pbil.univ-lyon1.fr/ADE-4) developed at the University of Lyon 1 (France) for analysing ecological and environmental data, and the interface of the *FactoMineR* package. The *FactoMineR* interface is incorporated in the interface of the *Rcmdr* package: it is obtained simply by connecting to the Internet (first ensuring that port 80 is not protected by a firewall) and typing the following line into the R console:

```
> source("http://factominer.free.fr/install-facto.r")
```

This only has to be done once. Then the *FactoMineR* scrolling menu will be included in *Rcmdr* whenever the package is loaded.

This menu (Figure 5.6) displays the methods handled by *FactoMineR*. Some of these are conventional, including principal component analysis (PCA), correspondence factor analysis (CA) and multiple correspondence analysis (MCA). Some of them are more advanced and can deal with structures on variables or individuals. These are multiple factor analysis (MFA), hierarchical multiple factor analysis (HMFA), dual multiple factor analysis (DMFA), factor analysis for mixed data (FAMD) and generalized Procrustes analysis (GPA). Unlike PCA and MCA, factor analysis for mixed data (FAMD) deals with both quantitative and qualitative variables. When using these data, it is possible to execute both a PCA and an MCA and to allow for their common structure as it is manifested in a correlation of certain axes of the PCA and MCA. FAMD constructs new axes and breaks down the inertia by axis and group (PCA or MCA).We can thus identify the axes corresponding to directions of inertia that are important for both groups, and others relating to only one of the groups. In MFA, devised by Escofier and Pagès, we consider data structured in groups of variables, these groups being hierarchically arranged in HMFA (as in the example of an enquiry structured in themes and sub-themes). The advantage of multiple factor analysis is that it does not start by combining all the groups of variables, which would make the groups with higher variance eclipse the others. Instead, it carries out separate analyses on each group of variables; these analyses are PCA if the group is made up of quantitative variables, MCA if the variables are qualitative, and FAMD if the

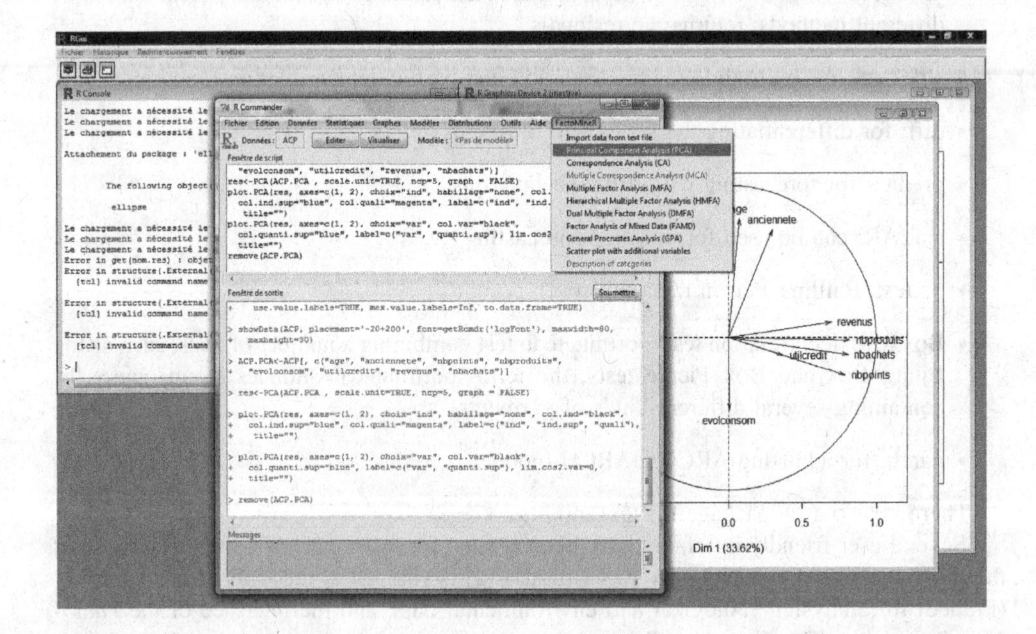

Figure 5.6 FactoMineR.

variables are mixed. It then carries out a global analysis, but only after each group of variables has been weighted by dividing its elements by the square root of the first eigenvalue of the group. Thus we can analyse each group in isolation (in the usual way) and then analyse all the variables globally. In this global analysis, we are particularly interested in factors which are common to the groups of variables and are also directions with high inertia. We can also represent each group of variables in the form of a point in a graphic.

Here I should also mention another method called STATIS, developed by Escoufier and others. This is similar to MFA, but weights the groups of variables differently, by under-weighting the groups whose structure is farthest from the structure of the set. We speak of seeking a 'compromise', and we disregard the groups that are a long way from the compromise. STATIS is not implemented in *FactoMineR*, but is in the *ade4* package.

Like MFA, Procrustes analysis deals with several groups of variables, provided that they are all quantitative. Another generalization of MFA is dual multiple factor analysis, in which the data are structured in groups of variables and groups of individuals.

The *FactoMineR* interface can be used to save the outputs (such as eigenvalues, contributions, and factor coordinates) to an R file (named 'res' by default) or to a CSV file which can be read in Excel. Like *Rcmdr*, it can be used to save the R code that is generated in a script window. For example, the eigenvalues can be read in res$eig. This interface also produces carefully designed graphics. Supplementary individuals and variables can be handled.

5.3.5 Some elements of the R language

Here is a small sample of the R language.

To read a CSV file, containing the name of the variables on the first line, with the separator ';', with the decimal separator ',', and adding 'blanks' if the lines do not all have the same number of variables:

```
read.table(file, header = TRUE, sep = ";", dec=",", fill = TRUE)
```

To assign the Napierian logarithm of the weight to the object lweight:

```
lweight <- log(weight)
```

To calculate the specified quantiles for the variable 'weight' of the file *File*, disregarding observations containing missing values (represented by NA: 'not available'):

```
quantile(File$weight, c( 0, .25, .5, .75, 1), na.rm=TRUE)
```

To display the descriptive statistics of the file *File*:

```
summary(File)
```

To carry out the test of variance on the qualitative variable 'sex' and the variable 'size':

```
var.test(size,sex)
```

To carry out the χ^2 test on the contingency table of the variables 'situation' and 'purchase':

```
chisq.test(table(situation,purchase))
```

To carry out a linear regression of the weight on size and age, and display the results of this regression (coefficients, R^2, F-ratio, RSE, etc.) and the main graphics (residual plot, Q-Q diagram, Cook's distance):

```
Linear.Model <- lm(formula = weight ~ (size + age), data = File)
Summary(Linear.Model)
Plot(Linear.Model)
```

This final example shows how models are specified in the R syntax. The response is the term on the left of the symbol '~', and the expression $\mathrm{y} \sim \mathrm{x}_1 + \mathrm{x}_2$ specifies a model of the form $\mathrm{y} = \alpha_0 + \alpha_1\mathrm{x}_1 + \alpha_2\mathrm{x}_2$. To specify a model

of the form	we write:
$y = \alpha_0 + \alpha(x_1 + x_2)$	$\mathrm{y} \sim \mathrm{I}\ (\mathrm{x}_1 + \mathrm{x}_2)$
$y = \alpha_1 x_1 + \alpha_2 x_2$	$\mathrm{y} \sim \mathrm{x}_1 + \mathrm{x}_2 - 1$
$y = \alpha_0 + \alpha_1 x + \alpha_2 x^2$	$\mathrm{y} \sim$ poly(x, 2)
with interaction between x_1 and x_2:	$\mathrm{x}_1 : \mathrm{x}_2$
with additive and interactive effects:	$\mathrm{x}_1 * \mathrm{x}_2$ (equivalent to $\mathrm{x}_1 + \mathrm{x}_2 + \mathrm{x}_1 : \mathrm{x}_2$).

This example also provides an illustration of object-oriented language. The result of the linear regression is copied into an object assigned to the variable 'Linear.Model', to which the 'summary' function can be applied to provide the main outputs of the linear regression, in the same way that this function can be applied to a file to provide the main descriptive statistics of the variables of this file. It is a general principle of R that the functions act specifically in accordance with the nature of the objects entered into the argument. Instead of immediately displaying all the results of the linear regression, as most software do, R stores them in an object from which the desired results are extracted subsequently, using the 'summary' function or the 'print' function to provide the minimum information, or the 'plot' function to generate graphics, etc. This procedure is unusual, but may be useful for comparing the results of a large number of tests, where it is undesirable to display all the results. Admittedly, the outputs in software such as SAS or IBM SPSS Statistics can be limited to what is strictly necessary, but if the user forgets to request the display of even one result, all the calculations have to be repeated. In this respect, the operation of R is somewhat similar to that of the ODS (see Section 3.8.4), which enables the results of the calculations to be sent to SAS files instead of being sent directly to the output window.

However, it should be noted that the outputs of R are much more rough-and-ready, using a notepad-style character set without formatting. On the other hand, the *prettyR* package can be used to export outputs in HTML.

There are other differences between the R language and the SAS and SPSS languages. In these languages, there are procedures which process a file 'vertically' by analysing all the observations ('cases' in the SPSS language) to produce results, while the functions process a file 'horizontally' by creating new variables from the existing ones, this being done for each observation independently of the others (with exceptions such as the 'delay' or 'lag' function).

In R, the functions do both, according to the arguments sent to them. For example, the apply function can apply the same function to the margins of the table, either horizontally or vertically. Thus *apply(x,1, mean)* calculates the means of each row of a table x, and *apply(x,2, sum)* calculates the sums of each column of x.

Like the macro language of SAS and SPSS, R offers solutions to those who need to execute identical tasks several times but with parameters which may be different. Instead of typing the commands into the R console one by one, we can write them to a file saved in ASCII format with the extension '.R'. A first method of writing syntax to execute repeated commands is to use vector writing, placing the values of the parameters in vectors of mode character, and then using indexation to execute the command with different values of the parameters. Another method, similar to the SAS or SPSS macro language, is that of defining a function. Emmanuel Paradis, in his excellent brief guide *R for Beginners*, available on the Internet, gives the example of a scatter diagram to be produced for three data sets contained in three different files. Instead of writing the same commands three times:

```
layout(matrix(1:3, 3, 1)) # partitions the graphic
data <- read.table("Swal.dat") # reads the data
plot(data$V1, data$V2, type="l")
title("swallow") # adds the title
data <- read.table("Wren.dat")
plot(data$V1, data$V2, type="l")
title("wren")
data <- read.table("Dunn.dat")
plot(data$V1, data$V2, type="l")
title("dunnock")
```

he defines the function:

```
myfun <- function(S, F)
{
data <- read.table(F)
plot(data$V1, data$V2, type="l")
title(S)
}
```

which is called in this way:

```
layout(matrix(1:3, 3, 1))
myfun("swallow", "Swal.dat")
myfun("wren", "Wrenn.dat")
myfun("dunnock", "Dunn.dat")
```

or as follows:

```
layout(matrix(1:3, 3, 1))
species <- c("swallow", "wren", "dunnock")
file <- c("Swal.dat" , "Wren.dat", "Dunn.dat")
sapply(species, myfun, file)
```

These instructions can be written to a file MyBirds.R, which will be called by the command

```
source("Mybirds.R")
```

An R function can include loops (for, repeat, while), conditions (if, ifelse), etc. The general syntax of a function is:

```
function_name <- function (arguments)
{body of the function}.
```

Other illustrations of the R language will be provided later in this book, especially in the section on regression.

5.4 Comparison of R, SAS and IBM SPSS

Several comparisons of software packages are available in the literature, but as far as I know none of these deals with the details of their functionality. Table 5.4 is an attempt to fill this gap, for the three leading packages, namely:

- the IBM SPSS suite (IBM SPSS® Statistics 19, IBM SPSS® Modeler 14);
- the SAS suite (SAS® 9.2, SAS® Enterprise Miner™ 6.1);
- R and its packages.

The functions, grouped by type, are shown in the rows, and the packages are shown in the columns. Of course, the information given is for guidance only, based on what was available to the author at the time of writing, but the reader can contact the developers to obtain the latest official documentation, and visit the R project site for the R software.

Note that only the native functions are shown, not those which can be obtained by special-purpose developments or macros which are more or less accessible and more or less maintained by authors who are not necessarily linked to the software developers. Some SAS macros are available on the INSEE website at www.insee.fr/fr/nom_def_met/outils_stat/macro.htm, and quite a large number of SPSS macros can be found at www.spsstools.net/Macros.htm. This selective approach was necessary in order to fix the boundary of the functionality (and limit the number of lines in the table!), even though it has to be admitted that this is rather restrictive in the case of R, which allows functions to be written quickly to meet specific needs.

In the SPSS column, 'MOD' indicates proprietary functions of IBM SPSS Modeler. In the SAS column, 'SEM' means proprietary functions of Enterprise Miner. Since SEM is an 'upper layer' of SAS, a function is indicated in SAS itself even if it is also available in SEM. Similarly, a function is shown in IBM SPSS Statistics even if it is also available in IBM SPSS Modeler. For R, all the packages are at the same level. Their names are shown in the rightmost column, unless they are included in the basic packages. The name of the R function is usually shown in brackets when it is different from the package name, and if the package contains more than one function. Remember that the case is significant in the name of an R object: *lmrob* and *lmRob* are names of different functions. The last column of the table also shows the names of the SAS and IBM SPSS procedures.

Table 5.4 Comparison of IBM SPSS, SAS and R.

Functions			SPSS	SAS	R	Procedures[a]
Data reading	File formats	flat	x	x	x	
		CSV	x	x	x	
		Excel	x	x	x	R: xlsReadWrite, gdata (read.xls)
		Access	x	x	x	R: via ODBC
		SAS	x	x	x	R: foreign, Hmisc (sasxport.get)
		SPSS	x	x	x	R: foreign, Hmisc (spss.get)
	DBMS	Oracle	x	x	x	R: ROracle
		SQL Server	x	x	x	
		DB2	x	x		
		Teradata	x	x		
		MySQL	x	x	x	R: RMySQL
		others, if ODBC driver	x	x	x	R: RODBC
	Automatic variable type recognition		x	SEM		and SAS® Enterprise Guide
	Variable labels import		x	x	x	
Data writing	File formats	flat	x	x	x	
		CSV	x	x	x	
		Excel	x	x	x	R: xlsReadWrite
		Access	x	x	x	R: via ODBC
		SAS	x	x	x	R: foreign (write.foreign)
		SPSS	x	x	x	R: foreign (write.foreign)
	DBMS	Oracle	x	x	x	R: ROracle
		SQL Server	x	x	x	
		DB2	x	x		
		Teradata	x	x		

(*continued*)

Table 5.4 (*Continued*)

Functions			SPSS	SAS	R	Procedures[a]
		MySQL	x	x	x	R: RMySQL
		others, if ODBC driver	x	x	x	R: RODBC
Data preparation	File handling	selection	x	x	x	
		deduplication	x	x	x	R: base (unique)
		aggregation	x	x	x	
		merge	x	x	x	
		transposition	x	x	x	SAS: TRANSPOSE – SPSS: FLIP
		transposition by groups of observations		x		SAS: TRANSPOSE
		SQL language	x	x	x	R: RODBC, RMySQL
		copying the data dictionary from one file to another	x			
	Data display	navigation in spreadsheet mode in the file	x	x	x	R: Rcmdr
		2D and 3D display of individuals (scatter plot)	x	x	x	R: lattice, rgl
		display of data in n dimensions (Radviz method[b])			x	R: dprep (radviz2d)
		colouring of individuals according to a variable different from the axes	x	x	x	R: lattice
		interactive selection of observations	MOD	SEM	x	+ SAS/IML Studio – SPSS: the Modeler graphics allow this selection – R: rggobi

	Cleaning	outlier detection	MOD	SEM	x	R: dprep (baysout, lofactor, mahaout, maxlof, robout), mvoutlier
		outlier filtering	MOD	SEM	x	R: outlier (rm.outlier)
		Winsorization		SEM	x	R: psych (winsor)
	Automatic management of missing values	detection of missing values	x	x	x	R: VIM, dprep (imagmiss, clean)
		linking of missing values	x	x	x	SPSS: MVA – SAS: MI – R: VIM
		elementary imputation (mean, median, maximum, etc.)	x	x	x	SAS: STANDARD, STDIZE, PRINQUAL – R: dprep (ce.impute, ce. mimp), Hmisc (impute)
		imputation by EM (expectation maximization) algorithm	x	x	x	SAS: MI, MIANALYSE – R: mix, cat
		imputation by regression	x	x	x	SAS: MI, MIANALYSE – R: mi (mi. continuous, mi.method), mice, Hmisc (transcan)
		imputation by logistic regression	x	x	x	SAS: MI, MIANALYSE – R: mic, Hmisc (transcan)
		imputation by discriminant analysis		x	x	SAS: MI, MIANALYSE – R: mice
		imputation by *k* nearest neighbours			x	R: dprep (ce.impute)
		imputation by Markov chains	x	x	x	SAS: MI, MIANALYSE – SPSS: MULTIPLE IMPUTATION (Missing Values module) – R: mix
		imputation by decision tree	MOD	SEM	x	R: Hmisc (transcan)

(*continued*)

Table 5.4 (*Continued*)

Functions			SPSS	SAS	R	Procedures[a]
		multiple imputation	x	x	x	SAS: MI, MIANALYSE – SPSS: MULTIPLE IMPUTATION (Missing Values module) – R: mitools, mice, mix, cat, Hmisc (transcan), Amelia
	Transformation of variables	recoding	x	x	x	R: dprep (baysout, rangenorm)
		automatic normalization		x	x	SAS: TRANSREG – R: car, MASS (Box–Cox transformation)
		automatic standardization	x	x	x	SAS: STANDARD, STDIZE, DISTANCE
		automatic transformation of a numeric variable by a spline function		x	x	SAS: TRANSREG, PRINQUAL – R: Hmisc (transcan), acepack (ace)
		optimal automatic digitization of categorical variables	x	x	x	SAS: TRANSREG, PRINQUAL – SPSS: CATREG, CATPCA, OVERALS – R: Hmisc (transcan), acepack (ace)
		creation of new variables	x	x	x	
		character functions	x	x	x	
		logic functions	x	x	x	
		mathematical functions	x	x	x	
		statistical functions	x	x	x	
		financial functions	x	x	x	R: Rmetrics
		automatic discretization (in equal frequency or equal width ranges)	x	x	x	R: dprep (disc.ef, disc.ew)

		optimal automatic discretization (according to a dependent variable)	x	SEM	x	SPSS: OPTIMAL BINNING (module Data Preparation) – R: dprep (chiMerge, disc.lr, disc.mentr)
		random number generation	x	x	x	
		date management	x	x	x	
	Selection of variables	automatic detection of interactions		SEM		
		step-by-step algorithms	x	x	x	SAS: STEPDISC – R: stats (step), MASS (stepAIC), pps (pps1)
		global algorithms (Furnival and Wilson)		x	x	SAS: REG, PHREG, LOGISTIC – R: leaps
	Random sampling	simple (srs: simple random sampling)	x	x	x	R: stats (sample), sampling (srswor, srswr)
		simple (pps: probability proportional to size)	x	x	x	R: stats (sample), pps (pps1, ppss, ppswr), sampling (UPbrewer, UPmaxentropy, UP ...), sampfling
		stratified	x	x	x	SAS: SURVEYSELECT – SPSS: Complex Samples module – R: sampling (strata), pps (ppssstrat)
		cluster	x	x	x	SAS: SURVEYSELECT – SPSS: Complex Samples module – R: sampling (cluster)
		bootstrap	x	x	x	SAS: SURVEYSELECT – SPSS: Complex Samples module – R: boot
		according to the a posterior distribution of the parameters of an adjusted model			x	R: lme4 (mcmcsamp)

(*continued*)

Table 5.4 (*Continued*)

Functions			SPSS	SAS	R	Procedures[a]
		determination of sample sizes for significant results		x	x	SAS: POWER, GLMPOWER – R: SampleSizeProportions, samplesize, binomSamSize
		automatic division of data into training, test and validation samples	MOD	SEM		
Data analysis	Descriptive statistic	*n*-variable contingency table	x	x	x	
		single-variable frequency table	x	x	x	
		central tendency characteristics	x	x	x	
		dispersion characteristics	x	x	x	R: stats
		shape characteristics	x	x	x	R: e1071 (skewness, kurtosis)
		Winsorized means	x	x	x	R: psych (winsor)
		Durbin–Watson statistic	x	x	x	SAS: GLM, REG – R: lmtest
	Statistical tests	tests of means (parametric or nonparametric)	x	x	x	SAS: MEANS, UNIVARIATE, TTEST, GLM, NPAR1WAY MEANS, T-TEST, ONEWAY, NPAR TESTS – R: stats
		tests of variance (parametric or nonparametric)	x	x	x	SAS: ANOVA NPAR1WAY – SPSS ONEWAY – R: stats (var.test. fligner, test)
		tests of distribution (parametric or nonparametric)	x	x	x	SAS: NPAR1WAY – R: stats, nortest

		tests of normality	x	x	x	SAS: NPAR1WAY, UNIVARIATE – SPSS: EXAMINE, NPAR TESTS – R: stats (Shapiro.test), nortest, dprep (mardia), tseries (jarque.bera.test)
		tests on independent samples	x	x	x	R: stats
		tests on paired samples	x	x	x	R: stats
		tests of correlations (parametric or nonparametric)	x	x	x	SAS: CORR – SPSS: CORRELATIONS, NONPAR COR – R: stats
		tests of independence	x	x	x	SAS: FREQ – SPSS: CROSSTABS – R: stats
		tests of multicollinearity (VIF…)	x	x	x	SAS: REG - SPSS: REGRESSION – R: car
		Levene's test of homoscedasticity	x	x	x	SAS: ANOVA, GLM –SPSS: EXAMINE, UNIANOVA, GLM – R: car (levene.test)
		Bartlett's test of homoscedasticity		x	x	SAS: ANOVA, GLM – R: stats (bartlett.test)
		White's test (equality of the variance of residuals)		x		SAS: REG and MODEL
		exact tests	x	x	x	SAS: FREQ, NPAR1WAY – R: stats
		tests by the Monte Carlo method	x	x	x	SAS: FREQ, NPAR1WAY – R: stats . . .
	Graphics	bar, curve, area, and pie charts, etc.	x	x	x	
		histograms	x	x	x	
		scatter plots	x	x	x	
		box plot	x	x	x	SAS: BOXPLOT

(*continued*)

Table 5.4 (*Continued*)

Functions			SPSS	SAS	R	Procedures[a]
		stem and leaf	x	x	x	R: aplpack (stem.leaf)
		P-P plot	x	x	x	R: e1071 (probplot)
		Q-Q plot	x	x	x	R: stats (qqplot)
		3D graphics	x	x	x	
		rotating graphics	x	x	x	
		time series graphics	x	x	x	R: stats (plot.ts, monthplot, lag.plot, tsdiag, ts.plot), tseries (seqplot.ts)
		colouring of selected individuals in the file	x	x		
	Factor analysis	PCA	x	x	x	SAS: PRINCOMP, FACTOR – SPSS: FACTOR – R: stats (princomp, prcomp), FactoMineR (PCA), ade4 (dudi.pca)
		PCA with orthogonal rotation	x	x	x	SAS and SPSS: FACTOR – R: stats (varimax, factanal), psych (fa), GPArotation (GPA)
		PCA with oblique rotation	x	x	x	SAS and SPSS: FACTOR – R: stats (varimax, factanal), psych (fa), GPArotation (GPA)
		PCA on transformed numeric variables[c]		x		SAS: PRINQUAL
		PCA on categorical variables[d]	x	x		SAS: PRINQUAL – SPSS: CATPCA
		CA	x	x	x	SAS: CORRESP - SPSS: CORRESPONDENCE – R: MASS (corresp), FactoMineR (CA), ade4 (dudi.coa), ca (ca)

		MCA	x	x	x	SAS: CORRESP - SPSS: HOMALS[e] - R: MASS (mca), FactoMineR (MCA), homals, ade4 (dudi.acm), ca (mjca)
		linear canonical correlation analysis (continuous or binary variables) with two groups of variables		x	x	SAS: CANCORR – R: CCA (cc)
		regularized (ridge) linear canonical correlation analysis with two groups of variables			x	R: CCA (rcc)
		non-linear canonical correlation analysis[f] with two groups of variables	x	x		SPSS: OVERALS – SAS: TRANSREG
		non-linear canonical correlation analysis with more than two groups of variables	x			SPSS: OVERALS
	Factor analysis of multiple tables	multiple factor analysis			x	FactoMineR (MFA), ade4 (mfa)
		hierarchical multiple factor analysis			x	FactoMineR (HMFA)
		dual multiple factor analysis			x	FactoMineR (DMFA)
		factor analysis for mixed data			x	FactoMineR (FAMD)

(*continued*)

Table 5.4 (*Continued*)

Functions			SPSS	SAS	R	Procedures[a]
		generalized Procrustes analysis			x	FactoMineR (GPA)
		STATIS method			x	ade4 (statis)
	Positioning analysis	calculation of distances, similarities, dissimilarities	x	x	x	SAS: DISTANCE – R: diss
Modelling/ prediction	Calculation of confidence intervals of estimators		x	x	x	
	Simple and multiple linear regression	ordinary least squares	x	x	x	SAS: REG, GLM – SPSS: REGRESSION – R: stats (lm), biglm[g]
		median least squares		x	x	SAS: QUANTREG – R: quantreg (rq)[h]
		quantile regression		x	x	SAS: QUANTREG – R: quantreg (rq)
		weighted least squares	x	x	x	SAS: REG, GLM – SPSS: WLS – R: MASS
		least trimmed squares		x	x	SAS: ROBUSTREG - R: robustbase (ltsReg)
		regression on principal components		x	x	SAS: REG – R: pls
		ridge regression		x	x	SAS: REG – R: MASS (lm.ridge), penalized, survival
		lasso regression		x	x	SAS: GLMSELECT – R: lars, lasso2 (l1ce), penalized, biglars[i], relaxo
		elastic net regression			x	R: elasticnet, glmnet
		multivariate regression[j]		x		SAS: TRANSREG

	Regression on transformed variables	quadratic or cubic transformation	x	x		SAS: TRANSREG, REG, RSREG, GLM – SPSS: CURVEFIT
		general polynomial transformation		x	x	SAS: TRANSREG – R: mda (polyreg)
		logarithmic, logistic, exponential, power, inverse hyperbolic, arcsine transformations	x	x		SAS: TRANSREG – SPSS: CURVEFIT
		spline transformation		x		SAS: TRANSREG
		regression on categorical variables by optimal coding[k]	x	x		SAS: TRANSREG – SPSS: CATREG
	Non-parametric regression	locally weighted least squares on neighbours (LOESS)		x	x	SAS: LOESS – R: stats (loess, lowess[l])
		spline regression		x	x	SAS: TPSPLINE – R: spline, stats (smooth.spline)
		kernel method		x	x	SAS: KDE – R: ks
		Theil–Sen robust regression[m]			x	R: mblm
		other robust regression		x	x	SAS: ROBUSTREG - R: MASS (rlm, lqs), robustbase (lmrob), robust (lmRob)
	Non-linear regression	without constraint	x	x	x	SAS: NLIN – SPSS: NLR – R: stats (nls)
		with constraints	x	x	x	SAS: NLIN – SPSS: CNLR – R: stats (nls)
		robust			x	R: robustbase (nlrob)
		quantile			x	R: quantreg (nlrq)
		with mixed effects			x	R: nlme (nlme), lme4 (nlmer)

(continued)

Table 5.4 (*Continued*)

Functions			SPSS	SAS	R	Procedures[a]
		Friedman's 'projection pursuit regression'			x	R: stats (ppr)
	PLS regression	classical	x	x	x	SAS: PLS – SPSS: PLS[n] – R: pls, ade4 (nipals)
		penalized			x	R: ppls
	General linear model	ANOVA	x	x	x	SAS: ANOVA, GLM, MIXED – SPSS: GLM, MANOVA – R: stats (lm, anova)
		MANOVA	x	x	x	SAS: ANOVA, GLM, MIXED – SPSS: GLM, MANOVA – R: stats (lm, anova)
		ANCOVA	x	x	x	SAS: GLM, MIXED – SPSS: GLM, MANOVA – R: stats (aov)
		MANCOVA	x	x	x	SAS: GLM, MIXED – SPSS: GLM, MANOVA, UNIANOVA – R: stats (aov)
		ANOVA, MANOVA, ANCOVA, MANCOVA for noisy data		x		SAS: ORTHOREG
		linear models with lasso penalization		(x)	x	SAS: GLMSELECT[o] - R: lasso2 (l1ce), penalized
		fixed effect models	x	x	x	SAS: GLM, MIXED – SPSS: GLM – R: nlme (lme)
		random effect models	x	x	x	SAS: GLM, MIXED, VARCOMP – SPSS: GLM – R: nlme (lme)
		mixed models	x	x	x	SAS: GLM, MIXED, VARCOMP – SPSS: MIXED, GLM – R: nlme (lme), lme4 (lmer)

		models with repeated measurements	x	x	x	SAS: ANOVA, GLM, CATMOD, MIXED – SPSS: MIXED, GLM – R: stats (aov), nlme (lme)
		mixed models with repeated measurements	x	x	x	SAS: MIXED – SPSS: MIXED – R: nlme (lme)
		hierarchical linear models	x	x	x	SAS: NESTED, GLM, MIXED – SPSS: MIXED – R: nlme (lme)
	Decision trees	CHAID	x	SEM		SPSS: TREE
		CART or equivalent	x	SEM	x	SPSS: TREE – R: rpart (recommended package), tree
		C4.5/C5.0 or equivalent	MOD	SEM	x	R: RWeka (J48)
		CTree (conditional inference algorithm)			x	R: party (ctree)
		LMT (logistic model trees)			x	R: RWeka (LMT)
		multivariate tree (several responses to predict)			x	R: mvpart, party (ctree)
		automatic pruning by validation on a test sample	x	SEM		SPSS: TREE
		automatic pruning by cross-validation	x	SEM	x	SPSS: TREE – R: rpart
		interactive construction with choice of variable for development of a node	MOD	SEM		
		interactive construction with choice of categories for development of a node	MOD	SEM		

(*continued*)

Table 5.4 (*Continued*)

Functions			SPSS	SAS	R	Procedures[a]
	Neural networks	MLP	MOD	SEM	x	SPSS: MLP (Neural Networks module) – R: nnet
		RBF	MOD	SEM		SPSS: RBF (Neural Networks module)
		gradient back-propagation	x	SEM		
		quick propagation	MOD	SEM		
		conjugate gradient descent	x	SEM		
		Quasi-Newton		SEM	x	R: nnet
		Levenberg–Marquardt		SEM		
		choice of transfer function	x	SEM	x	R: nnet
		interactive learning control	MOD	SEM		
	k nearest neighbours		x	SEM	x	SPSS: KNN- R: knnflex, kknn
Modelling/ classification	Fisher's discriminant analysis	linear	x	x	x	SAS: DISCRIM, CANDISC – SPSS: DISCRIMINANT – R: MASS (lda), candisc
		quadratic	x	x	x	SAS: DISCRIM – SPSS: DISCRIMINANT – R: MASS (qda)
		regularized[p]			x	R: klaR (rda)
		flexible[q]			x	R: mda (fda)
		Gaussian mixture[r]			x	R: mda (mda)
		DISQUAL				can be programmed

	Non-parametric Bayesian discriminant analysis				x	SAS: DISCRIM
	Logistic regression (logit)	binary	x	x	x	SAS: LOGISTIC, GENMOD, CATMOD, PROBIT – SPSS: LOGISTIC – R: stats (glm), elrm, biglm
		nominal polytomous	x	x		SAS: CATMOD, LOGISTIC – SPSS: NOMREG
		ordinal polytomous	x	x	x	SAS: CATMOD, GENMOD, LOGISTIC, PROBIT – SPSS: PLUM – R: MASS (polr)
		PLS logistic regression			x	R: gpls
	Other logistic regressions (binomial distribution)	probit model	x	x	x	SAS: GENMOD, PROBIT, LOGISTIC – SPSS: PROBIT, PLUM – R: stats (glm)
		log-log model	x	x	x	SAS: GENMOD, PROBIT, LOGISTIC – SPSS: PLUM – R: stats (glm)
		cauchit model	x		x	SPSS: PLUM – R: stats (glm)
		ordinal probit model	x	x	x	SAS: GENMOD – SPSS: PLUM – R: stats (polr)
		ordinal log-log model	x	x		SAS: GENMOD – SPSS: PLUM
		ordinal cauchit model	x			SPSS: PLUM
	Other counting models (discrete distributions)	Poisson model	x	x	x	SAS: GENMOD – SPSS: GENLOG – R: stats (glm)
		multinomial model	x	x		SAS: GENMOD, CATMOD - SPSS: GENLOG, HILOGLINEAR
		log-linear model (Poisson distribution)	x	x		SAS: GENMOD – SPSS: GENLOG

(*continued*)

Table 5.4 (*Continued*)

Functions			SPSS	SAS	R	Procedures[a]
		log-linear model (multinomial distribution)	x	x		SAS: GENMOD, CATMOD - SPSS: GENLOG, LOGLINEAR, HILOGLINEAR
	Other generalized linear models (continuous distributions)	gamma model	x	x	x	SAS: GENMOD – SPSS: GENLIN – R: stats (glm)
		lognormal model		x		SAS: GENMOD
		random effects models and mixed models		x		SAS: NLMIXED (non-linear random effects), GLIMMIX
		mixed models with repeated measurements		x		SAS: GLIMMIX
		free choice of specified distribution and link function	x	x		SAS: GENMOD – SPSS: GENLIN
		free specification of distribution or link function		x	x	SAS: GENMOD, NLMIXED – R: stats (power)
	Penalized ridge models (logistic, Poisson, Cox)				x	R: penalized
	Penalized lasso models (logistic, Poisson, Cox, gamma, inverse Gaussian)				x	R: penalized, lasso2 (gl1ce)
	Robust models (logistic, Poisson)				x	R: robustbase (glmrob), robust (glmRob)
	Models with repeated measurements			x	x	SAS: GENMOD, CATMOD, GLIMMIX – R: repolr, mprobit

	Decision trees	see above (Prediction) + QUEST	x		x	SPSS: TREE – R: LohTools
	Neural networks	see above (Prediction)				
	naive Bayesian classifier		x		x	SPSS: NAIVEBAYES – R: e1071 (naiveBayes), klaR (NaiveBayes), predbayescor
	k nearest neighbours		x	SEM	x	SPSS: KNN – R: class (knn), knncat, klaR (sknn), knnflex, knnTree, kknn
Advanced modelling	generalized additive model			x	x	SAS: GAM – R: mgcv (gam, gamm)
	vector generalized additive model/vector generalized linear model				x	R: VGAM
	MARS (multivariate adaptive regression splines)				x	R: mda (mars), polspline (polymars), earth
	structural equation models		AMOS	x	x	SAS: CALIS – R: sem
	genetic algorithms			x	x	SAS/OR (proc GA) - SAS/IML – R: gafit, rgenoud
	SVM (support vector machines)		MOD		x	R: kemlab, svmpath, e1071 (svm)
	automatic optimization of the parameters of a predictive model (linear, tree, SVM, *k* nearest neighbours, etc.)				x	R: e1071 (tune)
Ensemble methods	bagging		MOD	SEM	x	R: adabag, ipred (bagging)
	arcing x4			SEM		
	discrete adaboost		MOD		x	R: ada, adabag, boost (adaboost), gbm
	real adaboost, gentleadaboost				x	R: ada
	logitboost				x	R: ada, caTools (LogitBoost), boost (logitboost)
	gradient boosting			SEM	x	R: gbm, mboost

(*continued*)

Table 5.4 (*Continued*)

Functions			SPSS	SAS	R	Procedures[a]
	boosting for generalized linear models				x	R: GLMBoost, mboost (glmboost)
	boosting for generalized additive models				x	R: GAMBoost, mboost (gamboost)
	random forests				x	R: randomForest, party (cforest)
Cluster analysis	Agglomerative hierarchical methods	hierarchical clustering (distance based)	x	x	x	SAS and SPSS: CLUSTER – R: stats (hclust), flashClust[s], *cluster (agnes), FactoMineR (HCPC)*
		hierarchical clustering (density based)		x		SAS: CLUSTER
		Wong's hybrid method		x		SAS: CLUSTER
		hybrid BIRCH method	x			SPSS: TWOSTEP
		hierarchical clustering of variables	x	x	x	SAS: VARCLUS – R: Hmisc (varclus)
	Divisive hierarchical methods				x	R: cluster (diana, mona)
	Partitioning methods	*k*-means, dynamic clouds	x	x	x	SAS: FASTCLUS - SPSS: QUICK CLUSTER – R: stats (kmeans)
		k-means for truncated data			x	R: trimcluster
		k-means, dynamic clouds: searching for strong forms				
		non-parametric methods (density estimation)		x	x	SAS: MODECLUS – R: fpc (dbscan), RWeka (DBScan)
		k-medoids (PAM, CLARA)			x	R: cluster (pam, clara), fpc (pamk)
		Gaussian mixture models			x	R: mclust

		k-modes			x	R: klaR (kmodes)
		aggregation of similarities			x	R: amap
		Kohonen maps	MOD	SEM	x	R: class (SOM), kohonen
Survival analysis	Cox proportional hazards regression model		x	x	x	SAS: PHREG and TPHREG - SPSS: COXREG – R: survival (coxph), coxrobust
	Kaplan–Meier model for estimating the survival function		x	x	x	SAS: LIFETEST - SPSS: KM – R: survival (survfit), rms
	tobit model			x	x	SAS: LIFEREG – R: survival (tobin)
	other			x	x	R: survival, timereg
Time series	decomposition, filtering, deseasonalization		x	x	x	SAS/ETS: X11, X12, TIMESERIES - SPSS: SEASON – R: forecast (seasadj), stats (decompose, stl, filter)
	X12 deseasonalization			x	x	SAS/ETS: X12 – R: x12
	automatic generation of time series models		CLEM	x	x	with SAS High-Performance Forecasting or SAS/ETS Time Series Forecasting System – R: forecast (auto. arima, ets)
	spectral analysis		x	x	x	SAS/ETS and SPSS: SPECTRA – R: stats (spectrum, spec.ar, spec.pgram)
	non-parametric models (simple and Holt-Winters exponential smoothing)		x	x	x	SAS/ETS: FORECAST - SPSS: EXSMOOTH – R: forecast (ets[t]), stats (HoltWinters)
	parametric models (ARMA)		x	x	x	SAS/ETS and SPSS: ARIMA – R: tseries (arma)

(continued)

Table 5.4 (*Continued*)

Functions			SPSS	SAS	R	Procedures[a]
	semi-parametric models (ARIMA, SARIMA)		x	x	x	SAS/ETS and SPSS: ARIMA – R: stats (arima), forecast (Arima)
	vector autoregressive models (VAR, VECM, etc.)			x	x	SAS/ETS: VARMAX – R: vars
	autoregressive models		x	x	x	SAS/ETS: AUTOREG - SPSS: AREG – R: stats (ar, ar.ols)
	non-linear processes – generalized autoregressive models (ARCH, GARCH)			x	x	SAS/ETS: AUTOREG – R: tseries (garch)
	long memory models			x	x	SAS/ETS: AUTOREG – R: longmemo
Association Rules detection			MOD	SEM	x	R: arules, RWeka
Optimization				x	x	SAS/OR (OPTMODEL procedure) – R: stats (optim[a], optimize, nlm), gasl, maxLik
Matrix calculation			x	x	x	SAS/IML – SPSS 'Matrix' instruction – R: matlab, Matrix, sparseM
Text Mining			x	x	x	SAS Text Miner IBM SPSS Text Analytics R: tm
Visualization of results	Reporting	editable and reformattable tables	x	x		
		library of model tables	x	x		
		graphics	x	x	x	R: lattice
		interactive graphics	x	x	x	R: iplots, rgl, Rcmdr (Scatter3DDialog)

		3D graphics	x	x	x	R: rgl
		graphics gallery	x	x	x	R: R Graph Gallery
		library of model graphics	x	x	x	
		Control of graphics creation in R and inclusion in Output Viewer	x	x	ns	SPSS: using a plug-in Python – SAS/ IML Studio
		incorporation into a report	x	x		
		export to Word or Excel	x	x		
		saving in PostScript, JPEG, TIFF, PNG, Windows metafile, bitmap	x	x	x	
		saving in HTML, RTF (or DOC) and PDF	x	x		SAS: Output Delivery System – SPSS: Output Management System
	Performance evaluation	confusion matrix	x	x	x	
		lift curve	MOD	SEM	x	IBM SPSS Modeler: with graphic selection of the score zone chosen for filtering the observations – R: ROCR
		ROC curve	x	x	x	SAS: ROCPLOT macro, LOGISTIC procedure – R: ROCR, Epi (ROC)
		Gini index/area under the ROC curve	x	x	x	SAS: ROC macro, LOGISTIC procedure – R: ROCR, Epi (ROC)
		ROC curves with superimposition of models	x	x	x	SAS: ROCCOMP macro – R: ROCR

(continued)

Table 5.4 (*Continued*)

Functions			**SPSS**	**SAS**	**R**	**Procedures**[a]
	Automatic report creation			SEM		
	Statistical help	contextual interpretation of the results (on-line help by right clicking on a result)	x			
		case studies – examples with comments	x	x		
		detailed manual on statistics supplied by the developer with the software, with examples, theory and details of algorithms	x	x		SPSS manual is less detailed and is prèsented in two documents: 'Command Syntax Reference' and 'Algorithms'
Industrialization	Metadata management	declaration of the role of a variable (identifier, dependent, independent, illustrative, useless, etc.)	x	x		
		creation of subgroups of the set of variables	x			SPSS: use of subgroups in the dialogue boxes
	Programming language	script language to create a graphic interface	x	x	x	
		library of predefined functions	x	x	x	
		user definition of functions	x	x	x	
		programming of complex operations	x	x	x	

		calling C and Fortran	x	x	x	SPSS: by DLL declaration
		integration into MS Excel for data exchange and use of functions			x	R: RExcelInstaller (installs the RExcel add-in)
		colouring of syntax in the programming window to help with syntax checking	x	x	x	R: in various editors, e.g.: Tinn-R,[v] Emacs, WinEdt
		colouring of messages in the 'log' window to highlight errors and warnings		x		
		debugger	x	x	x	
		production (batch) functions	x	x	x	
	Platforms	Windows	x	x	x	
		AIX	x	x		
		HP UX	x	x		
		Linux	x	x	x	
		Solaris	x	x		
		Mac OS	x		x	
		z/OS		x		
	Export of models	C-C + +	MOD	x		
		Java	x	x		SPSS: Java class supplied for PMML interpretation
		XML-PMML	x	x	x	R: pmml, XML

(*continued*)

Table 5.4 (*Continued*)

Functions		SPSS	SAS	R	Procedures[a]
Computing power	client–server architecture	x	x		
	working in client–server mode and on local workstation in the same session		x		
	parallelization of the data mining algorithms (more specific than the parallelization of the SQL database processes)	x	x	x	SAS: use of SPDE (scalable parallel data engine) in some procedures, and possibility of grid computing – SPSS: sort and multinomial logistic regression – R: a list of packages can be found under 'High-Performance and Parallel Computing'

[a] See also Kleinman, K. and Horton, N.J. (2009) *SAS and R: Data Management, Statistical Analysis, and Graphics*. Boca Raton, FL: Chapman & Hall.

[b] In this method of 'radial coordinate visualization', the observations described by n variables are represented in a plane, inside a circle in which a polygon is inscribed, with each vertex corresponding to one variable. The variables are normalized between 0 and 1. An observation is closer to one of the vertices of the polygon when the value of the variable associated with this vertex is greater than the others for this observation. An observation is close to the centre if all the variables have approximately the same values for this observation.

[c] The PRINQUAL procedure uses the alternating least squares method to find the transformations of the initial variables which maximize the variance of the first axes. These transformations may or may not be linear, and use spline functions and the power, exponential, logarithm, rank, logistic and arcsine functions.

[d] The PRINQUAL (SAS) and CATPCA (SPSS) procedures use the alternating least squares method to find the numeric codings of the initial categorical variables which maximize the variance of the first axes ('optimal scoring').

[e] The principle of HOMALS is different from that of an MCA procedure like SAS CORRESP, since HOMALS uses the alternating least squares method (with longer calculation time) instead of simple diagonalization. In practice, however, we often have to go to the fourth decimal place to find any difference between the two procedures.

[f] Optimal transformation of the variables by alternating least squares.

[g] The *biglm* package should be used for data volumes which are too large to be stored in memory.

[h] Median regression is a special case of quantile regression, processed by the *quantreg* package.

[i] The *biglars* package should be used for data volumes which are too large to be stored in memory.

[j] There are several dependent variables Y_i for the same independent variables X and the aim is to find regressions Y_i/X which have the same constants, or the same slopes, or which satisfy certain conditions. Multivariate regression can be applied to transformed variables (splines, etc.).

[k] The TRANSREG (SAS) and CATREG (SPSS) procedures use the alternating least squares method to find the numeric codings of the initial categorical variables which optimize the adjustment of the regression. This method is also used by the TRANSREG procedure to find the best spline transformation of numeric variables. This method involves alternating a search for the best adjustment (in the sense of the least squares) of the parameters with allowance for the coding of the data with a search for the best coding with allowance for the preceding adjustment of the parameters.

[l] Note the syntax lowess(x,y) which is different from the syntax loess(y~x) which is used more widely in R.

[m] This regression involves calculating the slopes of the straight lines passing through all possible pairs of points, and taking the median of these as the estimator.

[n] The PLS procedure in SPSS requires the preliminary installation of a Python editor programming module, downloadable from the SPSS website.

[o] The GLMSELECT procedure is designed purely for the selection of variables, not for modelling: it does not provide regression diagnostics, for example. It implements the LAR method as well as the lasso.

[p] Regularized discriminant analysis (Friedman, J. H. (1989) Regularized discriminant analysis. *Journal of the American Statistical Association*, 84, 165–175) is a compromise between linear discriminant analysis and quadratic discriminant analysis. It operates on a similar principle to ridge regression, and is useful when the predictors are collinear. It is controlled by a parameter in the range from 0 to 1, which progressively limits the covariance matrices by classes towards a common covariance matrix as in linear discriminant analysis.

[q] Flexible discriminant analysis (Hastie, T., Tibshirani, R. and Buja, A. (1994) Flexible discriminant analysis by optimal scoring. *Journal of the American Statistical Association*, 89, 1255–1270) makes use of the link between linear discriminant analysis and multiple linear regression, and applies generalizations of linear regression, such as spline regression or the MARS method, to discriminant analysis. As in the case of support vector machines, this represents a move into a larger variable space.

[r] In discriminant analysis by Gaussian mixtures (Hastie, T. and Tibshirani, R. (1976) Discriminant analysis by Gaussian mixtures. *Journal of the Royal Statistical Society, Series B*, 58, 155–176), the conditional law $P(x/G_i)$ (see Section 11.6.4) is assumed to be not a Gaussian (of dimension p), but a mixture $\sum_k \pi_{ik}\varphi(x;\mu_{ik},\sum)$ of Gaussians. A simplification is provided by the assumption that all the Gaussians have the same covariance matrix Σ. The Gaussian mixture can be used to model non-homogenous classes. The parameters of the Gaussians are found by maximizing the likelihood, which does not appear to be easy, but can be done with the EM algorithm of Dempster *et al.* (Dempster, A.P., Laird, N.M. and Rubin, D.B. (1977) Maximum likelihood from incomplete data via the EM algorithm (with discussion). *Journal of the Royal Statistical Society, Series B*, 39, 1–38).

[s] The *flashClust* package is faster than *hclust*.

[t] The ets function is preferable to the forecast HoltWinters function in the same package.

[u] The optim function implements the conjugate gradient, Nelder–Mead, quasi-Newton and simulated annealing methods.

[v] http://www.sciviews.org/Tinn-R/.

To sum up, the advantages of IBM SPSS Statistics are its ease of installation and use, as well as its practical and user-friendly data import assistant. And before SAS Enterprise Guide came on the scene, SPSS was significantly ahead in terms of user-friendliness, with an interface closer to the Microsoft standards. This also explains its success in sectors such as marketing and the social sciences. As mentioned above, IBM SPSS Statistics has a graphic interface with scrolling menus which make it unnecessary to know a programming language, even where such a language exists and is useful for repeated operations. However, R does not have a universal graphic interface, and by no means all of its functions have been integrated into the various graphic interfaces distributed in its packages.

IBM SPSS Statistics also has an advantage over SAS in its lower price and the possibility of obtaining Answer Tree for decision trees without having to buy the data mining suite, whereas anyone wanting to construct decision trees with SAS has to buy Enterprise Miner. For decision trees, IBM SPSS is also more competitive than R, which does not offer many tree algorithms. Most of the packages only implement CART, and their interface is very unfriendly. However, it is worth noting the useful CTree algorithm (Hothorn, Hornik and Zeileis, 2006),[9] implemented in the *party* package.

For its part, SAS is more comprehensive and flexible in terms of file management; for example, it has an improved TRANSPOSE procedure. The SQL procedure can be used to manipulate not only relational tables, but also SAS tables, with better performance in some cases than when using native SAS instructions. For example, if there is no index on the join keys, it will be quicker to merge two tables by using an SQL 'join' rather than an SAS 'merge' (equivalent to 'match files' in IBM SPSS Statistics), which requires a preliminary sort of the tables to be merged. SAS is also unequalled in its processing speed for large volumes. As regards performance, SAS is probably the most stable of the three systems: it is very difficult to make SAS crash, but it is by no means unusual for IBM SPSS Statistics to freeze when processing large volumes.

In terms of these 'data management' aspects, IBM SPSS performs less well than SAS but better than R. A major drawback of R is that most of its functions have to load all the data into memory before execution, which sets a serious limit on the volumes that can be handled. However, some packages are beginning to break free of this constraint: one example is the *biglm* package for linear models.

The technical documentation of SAS is very comprehensive (almost 8000 pages covering the procedures of the SAS/STAT module alone) and is not dispersed in the same way as the documentation for R.

Lastly, SAS is much more widely used than IBM SPSS Statistics, and therefore has more sources and resources devoted to it, such as forums, user clubs, trainers, websites, macro libraries, books, etc. From this point of view, R is also well served, with a very comprehensive website, many contributions, conferences and articles, an increasing number of books, and even an on-line journal (http://journal.r-project.org/index.html).

SAS offers many more predefined functions, such as mathematical and financial functions, than IBM SPSS Statistics. These include depreciation, compound interest, cash flow, hyperbolic functions, factorials, combinations and arrangements, and others. There are also more predictive and descriptive algorithms in R and SAS than in IBM SPSS Statistics; the

---

[9] Hothorn, T., Hornik, K. and Zeileis, A. (2006) Unbiased recursive partitioning: a conditional inference framework. *Journal of Computational and Graphical Statistics*, 15(3), 651–674. Preprint available from http://statmath.wu-wien.ac.at/~zeileis/papers/Hothorn+Hornik+Zeileis-2006.pdf

statistical indicators are more detailed and the parameter setting possibilities are much greater. Finally, the SAS macro language is more flexible and complete than that of SPSS.

SAS offers better aids to productivity, for example with the possibility of saving the logistic regression model obtained for a sample in an SAS 'model' file, so that it can subsequently be applied to any other sample. Another very useful SAS procedure that has no equivalent in SPSS is FORMAT, which is used to associate labels with data values. FORMAT enables formats to be defined globally, without reference to any given table, unlike SPSS where labels are defined only for the working table and are attached to this table. When a SAS format has been defined, it can be used with any table without modifying it, to format the display of its data in procedures such as PRINT, TABULATE, GCHART and GPLOT. Formats can even be stored in a permanent catalogue and re-used without any need to redefine them for each session. And this is not all: unlike the SPSS labels which are only applied to precise values, the FORMAT procedure in SAS is applied to ranges of values, as in the example below. Moreover, the FORMAT procedure operates not only in display but also in the calculation of data. Thus, FORMAT can be used to discretize a continuous variable 'logically', simply by applying a suitable format to it, without any need to modify it physically. This eliminates the step of transforming and recording the data, and, by way of example, a logistic regression can be calculated on continuous variables as if they had been discretized. Clearly, this makes it much quicker to test more than one discretization. The syntax of the FORMAT procedure is:

```
PROC FORMAT;
VALUE age
0-<18 = '< 18 years'
18-<25 = '18-24 years'
25-<35 = '25-34 years'
35-<45 = '35-44 years'
45-<55 = '45-54 years'
55-<65 = '55-64 years'
65-<75 = '65-74 years'
75-high = '>= 75 years' ;
```

In automatic clustering, the hybrid Wong method implemented by SAS (chaining the FASTCLUS and CLUSTER procedures) is superior to the IBM SPSS Statistics 'two-step' method because of its capacity to detect clusters of different shapes (non-spherical) and optimize the choice of the number of clusters. Moreover, SAS has implemented algorithms for clustering by density estimation. R has its own useful clustering algorithms, but none of them appears to compare with SAS in terms of handling large volumes.

But the superiority of SAS over IBM SPSS is possibly seen most clearly in the area of prediction. The links between SAS and a large number of universities have enabled it to implement methods and algorithms not found in IBM SPSS Statistics: These include generalized additive models, numerous refinements of generalized linear models, genetic algorithms, and boosting. The new versions of SAS regularly implement some of the very latest statistical discoveries, on an experimental basis in some cases. However, we should note the initiative taken by SPSS in implementing in an IBM SPSS Statistics module the neural network algorithms which are traditionally reserved for data mining software such as IBM SPSS Modeler or SAS Enterprise Miner, and which are rather poorly provided for in R (the *nnet* package is practically the only one, and is not very comprehensive). Furthermore, its IBM

SPSS Data Preparation module includes much advanced data preparation functionality, particularly an efficient algorithm for automatic discretization of continuous variables.

The advantages of R have been mentioned already, but let me summarize them here. First of all, there is the price: R is free! Then there are the number and richness of the packages, unequalled in some fields (econometrics, actuarial science, biostatistics, etc.). These packages are also updated frequently and the user only has to connect to the Internet and run the Update packages command to benefit automatically from all the latest updates.

This richness is particularly evident as regards regression methods where the explained variable is continuous, robust and non-linear methods, and ensemble methods, especially boosting. R also has the benefit of some very well-designed graphics and advanced display functions.

The user will also appreciate the flexible programming and matrix calculation offered by R, even if it requires a rather longer learning period than SAS or IBM SPSS. It is certainly easier for an SAS user to switch to IBM SPSS, or vice versa, than to R. But anyone who is willing to spend some time learning R will have the satisfaction of programming his own functions or even creating his own packages.

For a user who is just looking for the simplest possible way of using R, the RExcel add-in integrates R with Excel, offering the benefit of the familiar Excel interface for reading or writing R data, and the possibility of calling thousands of R functions, either as macros, or directly in the cells of Excel.

5.5 How to reduce processing time

Processing times can be shortened – drastically in some cases – by following a few rules:

- Work on structured files (SAS, SPSS, DB2, etc.) rather than flat files.
- Limit the analysed file to the lines and variables relevant to the current process (by careful selection and using the KEEP and DROP commands).
- Recode the variables and make them smaller by using formats. Formats enable numerous categories of variables to be replaced with codes which are much more compact and economical in terms of disc space.
- Create Booleans such as alphanumeric variables of length 1, rather than numerical variables.
- Clearly define the length of the variables used, limiting it to the minimum possible (e.g. use the LENGTH command in SAS).
- Remove intermediate files which are no longer required, and especially (in SAS) clear out the temporary WORK directory as often as possible (PROC DATASETS LIB = WORK KILL NOLIST), since it is not automatically purged until the end of the SAS session.
- Keep enough free space on the hard disk: at least four times the size of the file to be analysed.
- Defragment the hard disk if necessary.

- Do not place the analysed file or the temporary workspace on a remote network.
- Increase the amount of RAM.
- (IBM SPSS Statistics) Avoid unnecessary EXECUTE commands.
- (IBM SPSS Statistics) Use the PRESORTED option in aggregations.
- (IBM SPSS Statistics) Clean out the log file regularly.
- (SAS) Use BY rather than CLASS in the MEANS procedure.
- (SAS) If a request uses a variable at least three times in a WHERE filter or in a BY, create an index on this variable, which will optimize the WHERE (making it unnecessary to read the whole table) and ensure that the BY is not preceded by a SORT which is more time-consuming; a single index can avoid several sorts; a simple index or a compound index can be created on several variable; an index also reduces the execution times for table joining in PROC SQL.
- (SAS) Use compression to reduce the disk space occupied by a file and save time for its processing if necessary (but this is not always possible: check the message in the log); this option is written in the form COMPRESS = YES if the variables are mostly alphanumeric, or COMPRESS = BINARY if the variables are mostly numeric; it is possible to write either

  ```
  DATA Table 2 (COMPRESS = YES) ;
  SET Table 1;
  RUN ;
  ```

  or

  ```
  OPTIONS COMPRESS = YES ;
  DATA Table 2;
  SET Table 1 ;
  RUN ;
  ```

  according to whether you need to compress a specific table (DATA SET option) or all tables (SYSTEM option).
- (SAS) For copying tables, use PROC COPY or PROC DATASETS (COPY instruction) rather than a DATA SET step.
- (SAS) For sorting a large table with a small sort key, the TAGSORT option can save a lot of time by avoiding loading into memory the data which do not appear in the BY. The saving can be more than 40%, as shown for example in the book by Olivier Decourt and Hélène Kontchou (see Section B.6). This option is useful if there is insufficient memory, but it requires many read/write operations and a lot of processor capacity; it should therefore be kept for cases in which saving takes a relatively long time with respect to the sort key.

- (SAS) On the other hand, if there is insufficient memory for the sort, the memory allocated to sorting can be maximized by the SORTSIZE = MAX option.
- (SAS) Since SAS 9.2, the PRESORTED option has been available for making SAS analyse the table at the outset and only sort it if this has not been done already; this option is recommended if there is a suspicion that the table has been already been sorted.
- (SAS) Other PROC SORT options can provide a significant time saving. The NOEQUALS option tells SAS that there is no need to keep the same order in the table for observations which have the same values for all the variables of the BY. The THREADS option, which appeared with SAS 9, enables the sort calculations to be parallelized on a multi-processor machine. If this option is not specified in the PROC SORT, then the one defined in the SYSTEM options is used (another option for specifying the number of processors). For sorting, the THREADS option is incompatible with TAGSORT. Note that parallelization is available for procedures other than sorting, i.e. SUMMARY, MEANS, REPORT, TABULATE, and SQL, as well as some SAS/STAT procedures. By default, in the SYSTEM options, THREADS is activated and CPUCOUNT is equal to the number of processors of the computer.
- (SAS) With the MULTIPASS option, the LOGISTIC procedure (since SAS 9.2) extends its capacities by rereading the input data if necessary, instead of trying to store them in memory or on disk.

6

An outline of data mining methods

This chapter introduces the five chapters which form the core technical content of this book. They are rather more accessible than some specialist books on statistics, data analysis and neural networks, and I hope that they will be enjoyable to read. However, a reader who is only interested in the applications of data mining and the procedures for implementing it in a business may omit these chapters. On the other hand, they are essential for anyone wishing not only to understand the working of the tools, in order to use them more successfully, but also to know when and where to use any particular algorithm. In this first technical chapter, I shall outline the descriptive and predictive methods of data mining and statistics as a whole, and compare their main features, which will be discussed in detail in the following chapters.

It is important to note that the logarithms used in this book are Napierian (natural) logarithms in all cases.

6.1 Classification of the methods

As mentioned in Chapter 1, the main data mining and data analysis methods can be divided into two large families: *descriptive methods* and *predictive methods*. In descriptive methods, for reducing, summarizing and grouping data, there is no dependent variable, i.e. no privileged variable. In predictive methods, which explain data, there is a dependent variable, in other words a variable to be explained, or a privileged variable.

A more detailed version of this classification is shown in Table 6.1, where methods forming part of conventional statistics and data analysis have been given grey backgrounds.

Considering predictive methods only (Table 6.2), we can be more precise by distinguishing the differences relating to the type of variable, namely independent (in the rows) and dependent (in the columns). Clearly, the rows 'n quantitative (representing different quantities)' and 'n qualitative' are only relevant if the dependent variables are correlated with each other. Otherwise, it is sufficient to carry out n analyses of the '1 quantitative' or '1 qualitative' type.

Data Mining and Statistics for Decision Making, First Edition. Stéphane Tufféry.

Table 6.1 Classification of methods.

Type	Family	Sub-family	Algorithm
descriptive methods	geometrical models	factor analysis (projection and visualization in a space of lower dimension)	principal component analysis (PCA) (continuous variables)
			correspondence analysis (CA) (qualitative and binary variables)
			multiple correspondence analysis (MCA) (qualitative and binary variables)
		cluster analysis (grouping in homogeneous clusters in the whole space)	partitioning methods (moving centres, *k*-means, dynamic clouds, *k*-medoids, etc.)
			hierarchical methods (agglomerative, divisive)
		cluster analysis + dimension reduction	neural clustering (Kohonen maps)
	combinatory models		clustering by aggregation of similarities (qualitative variables)
	logical rule-based models	link detection	search for association rules
			search for similar sequences

predictive methods	logical rule-based models	decision trees	decision trees (dependent variable is numeric or qualitative)
	models based on mathematical functions	neural networks	supervised learning networks (perceptron, radial basis function network, etc.)
		parametric or semi-parametric models	linear regression, ANOVA, MANOVA, ANCOVA, MANCOVA, general linear model (GLM), PLS regression (continuous dependent variable)
			Fisher's discriminant analysis, logistic regression, PLS logistic regression (qualitative dependent variable)
			log-linear model (dependent variable = counting = number of individuals have a given combination of categories of qualitative variables)
			generalized linear model (GLM), generalized additive model (GAM) (dependent variable continuous, discrete, counting or qualitative)
	prediction without model	probabilistic analysis	k nearest neighbours

Table 6.2 Predictive methods.

Independent → **↓ Dependent**	**1 quantitative (covariable)**	***n* quantitative (covariables)**	**1 qualitative (factor)**	***n* qualitative (factors)**	**Combination**
1 quantitative	simple linear regression, spline regression, robust regression, decision trees, MARS, SVR (support vector regression), *k* nearest neighbours	multiple linear, regression, spline regression, robust regression,* PLS regression, decision trees, MARS, neural networks, SVR, *k* nearest neighbours	ANOVA, decision trees, MARS, SVR, *k* nearest neighbours	ANOVA, decision trees, MARS, neural networks, SVR, *k*-nearest neighbours	ANCOVA, univariate GLM, decision trees, MARS, neural networks, SVR, *k* nearest neighbours
***n* quantitative (representing different quantities)**	multivariate regression, PLS2 regression	multivariate regression, PLS2 regression, neural networks	MANOVA	MANOVA, neural networks	MANCOVA, multivariate GLM, neural networks
1 qualitative nominal or binary	Fisher's discriminant analysis, logistic regression, regularized generalized linear models, decision trees, MARS, SVM, naive Bayesian classifier, *k* nearest neighbours	Fisher's discriminant analysis, logistic regression, PLS logistic regression, regularized generalized linear models, decision trees, MARS, neural networks, SVM, naive Bayesian classifier, *k* nearest neighbours	logistic regression, DISQUAL discriminant analysis, regularized generalized linear models, decision trees, MARS, SVM, naive Bayesian classifier, *k* nearest neighbours	logistic regression, DISQUAL discriminant analysis, regularized generalized linear models, decision trees, MARS, neural networks, SVM, naive Bayesian classifier, *k* nearest neighbours	logistic regression, regularized generalized linear models, decision trees, MARS, neural networks, SVM, naive Bayesian classifier, *k* nearest neighbours

***n* qualitative nominal or binary (representing different characteristics)**	decision trees, vector generalized linear model, vector generalized additive model	decision trees, vector generalized linear model, vector generalized additive model, neural networks	decision trees, vector generalized linear model, vector generalized additive model	decision trees, vector generalized linear model, vector generalized additive model, neural networks	decision trees, vector generalized linear model, vector generalized additive model, neural networks
1 quantitative asymmetrical	gamma and log-normal regressions	gamma and log-normal regressions	gamma and log-normal regressions	gamma and log-normal regressions	gamma and log-normal regressions
1 discrete (counting)	Poisson regression, log-linear model	Poisson regression, log-linear model	Poisson regression, log-linear model	Poisson regression, log-linear model	Poisson regression, log-linear model
1 qualitative ordinal (at least 3 groups)	ordinal logistic regression	ordinal logistic regression	ordinal logistic regression	ordinal logistic regression	ordinal logistic regression
***n* quantitative or qualitative (representing repeated measurements of the same characteristic)**	generalized linear models with repeated measures	generalized linear models with repeated measures	generalized linear models with repeated measures	generalized linear models with repeated measures	generalized linear models with repeated measures

*LOESS, ridge, lasso, LARS, and other robust regressions.

Table 6.3 Comparison of methods.

Method	**Absence of assumptions concerning the problem to be solved**	**Exhaustive processing of databases**	**Heterogeneous or incomplete data processed**
		Clustering	
moving centres method and its variants	no (fixed number of initial clusters and centres)	yes	numerical variables and variables without missing values
hierarchical clustering	yes, but the clusters at level n are determined by those at level $n-1$	no (non-linear algorithm), impossible to process more than several thousand observations	yes (possible to process non-numeric variables with an *ad hoc* distance)
neural clustering (Kohonen)	no (fixed number of clusters)	yes	the variables $\in [0,1]$ must be transformed
clustering by aggregation of similarities	yes	in principle yes, but depends on the implementation	qualitative variables
		Classification and prediction	
decision trees	as for hierarchical clustering (a kind of 'reverse tree')	no (but does not reach the limit as soon as hierarchical clustering)	some trees, such as CHAID, must discretize continuous variables
neural networks perceptrons	yes (but the number of hidden neurons must be specified)	no (no learning on several hundred variables)	the variables $\in [0,1]$ must be transformed
radial basis function networks	as for perceptrons	yes	the variables $\in [0,1]$ must be transformed
discriminant analysis	no (assumptions on the conditional distributions X_i/Y)	yes	numerical variables and variables without missing values

discriminant analysis on factorial coordinates of MCA (DISQUAL method)	yes (assumptions on conditional distributions X_i/Y can generally be dispensed with)	yes	yes (missing values are treated as entirely separate values)
linear regression	no (linearity in x of E(Y∣X=x) + assumptions on the residuals)	yes	numerical variables and variables without missing values
logistic regression, generalized linear model	no (linearity in x of g(E(Y∣X=x)) + non-complete separation (see Section 11.8.7))	yes (provided that a sufficiently powerful machine is used, if the number of observations is very large)	yes (continuous variables with missing values are divided into classes)
		Associations	
search for association rules	yes	depends on the parameter settings	yes
similar sequences	yes	yes (same remarks apply)	yes

As for the descriptive methods of clustering, these are detailed in a summary table at the end of Chapter 9.

6.2 Comparison of the methods

Table 6.3 summarizes the advantages and disadvantages of the various data mining methods in relation to these three essential qualities that are expected:

- the absence of restrictive assumptions concerning the problem to be solved;
- the capacity of treating the data exhaustively within a reasonable period in all cases;
- the possibility of handling incomplete and heterogeneous data which may or may not be numerical (in the case of independent variables for the classification and prediction methods).

7

Factor analysis

In multivariate analysis, the factor methods described in this chapter are much appreciated by statisticians, who use them as a way of representing the individuals of a population in two or three dimensions as faithfully as possible, while also detecting the links between the variables as well as the variables which separate the individuals most clearly. These methods are based on linear algebra, and also on a tool which is very useful for clustering and pattern recognition: the human eye. A simple glance is enough to locate large clusters of individuals, detect exceptional individuals and find any isolated groups of individuals. Thus factor analysis is also a powerful resource for reducing the dimensions of a problem, decreasing the number of variables to be studied while losing as little information as possible. In some cases, it is very useful as a preliminary process before using certain algorithms, such as neural networks, which are sensitive to the number of input variables; it may also be useful before clustering. Transformation of qualitative variables into continuous variables by multiple correspondence analysis is quite widely used, especially in discriminant analysis on qualitative variables (DISQUAL). Finally, principal component analysis with rotation can be used to create groups of variables based on their correlations, and is the foundation for an effective variable clustering algorithm (VARCLUS).

7.1 Principal component analysis

7.1.1 Introduction

When the p variables describing the n individuals of a population are all numerical, each individual can be represented by a point in a p-dimensional space $\mathcal{R}^p$. The set of individuals is a 'cloud of points'. When $p \leq 2$, the distances between individuals can be seen clearly by simple observation of the cloud; this observation becomes more difficult when $p = 3$, and is impossible when $p > 3$. Evidently, it would be desirable to reduce the space $\mathcal{R}^p$ to $\mathcal{R}^2$ or $\mathcal{R}^3$ – we speak of the *projection* of the variables of $\mathcal{R}^p$ on to $\mathcal{R}^2$ or $\mathcal{R}^3$. The problem is that the choice of two or three variables, such as age, wages, or length of service, is intrinsically

Data Mining and Statistics for Decision Making, First Edition. Stéphane Tufféry.

arbitrary, and may result in a considerable loss of information from the data, since there is no way of knowing in advance whether these are the most discriminating variables. At one extreme, if all the individuals in a study have the same age, the same wages and the same length of service, the projection in these three axes will shrink to a point, even if some individuals may be very different in other respects. The projection of the cloud of individuals from the initial p-dimensional space into a space with fewer dimensions automatically decreases the distances between individuals: clearly, we must try to decrease these distances as little as possible, if we want to distinguish between the individuals and understand what they have in common and what separates them.

Principal component analysis (PCA), which will be examined in this section, is a method for projecting the cloud of individuals on to subspaces with fewer dimensions while maintaining the distances between individuals as much as possible. We begin by systematically centring all the variables, by subtracting their means, so that we are working on variables with a zero mean. This simplifies the calculation and geometrical representations, because the centre of gravity of the cloud of individuals then coincides with the origin 0 of the axes and subspaces.

The determination of these subspaces is carried out for each axis in turn. Each individual x_i has a weight p_i. This weight is generally $p_i = 1/n$ for every i, but different weights can be given to individuals belonging to different sub-populations. The sum of squares of the distances of the individuals x_i from their centre of gravity, multiplied by their weight p_i, is called the *total inertia*:

$$I = \sum_{i=1}^{n} p_i d(0, x_i)^2.$$

We can say that the aim of PCA is to find the axis for which the inertia projected on this axis is maximized. The inertia projected on an axis is, by definition, the sum of squares of the coordinates v_i of the individuals on the axis, these squares being weighted by p_i. In other words, we look for the axis for which the sum

$$\sum_{i} p_i v_i$$

reaches a maximum and is thus as close as possible to I. This is equivalent to minimizing the difference between each individual and its projection, in other words providing maximum elongation of the projection of the cloud of individuals on the axis. Having done this, we look for a second axis which, out of all the axes orthogonal (i.e. perpendicular and therefore not correlated) to the first axis, will be the one which maximizes the inertia projected on it. This inertia projected on the second axis is, by construction, less than that projected on the first axis. A number of factor axes can thus be determined in succession with decreasing projected inertias. Because of their orthogonality, the total inertia of the cloud of individuals is broken down into the sum of inertias projected on each axis.

As we have seen, the concept of *distance* is used here. In the space of the individuals, the simplest distance is the Euclidean distance, according to which the distance of two individuals $x = (x_1, x_2, \ldots, x_p)$ and $y = (y_1, y_2, \ldots, y_p)$ is:

$$d(x, y) = (x_1 - y_1)^2 + (x_2 - y_2)^2 + \ldots + (x_p - y_p)^2.$$

This distance is very useful in the physical world, but less so in the world of economics and social sciences, where the data $x_1, x_2, \ldots, x_p$ to be manipulated may be as unlike and non-comparable as age, income, turnover, number of children, etc.

In practice, the 'inverse of variances' is practically always used as the distance. It is defined thus:

$$d(x,y) = ((x_1-y_1)/\sigma_1)^2 + ((x_2-y_2)/\sigma_2)^2 + \ldots + ((x_p-y_p)/\sigma_p)^2,$$

where each σ_i is the standard deviation of the ith variable. With this new distance, which is a way of *reducing* the variables (dividing them by their standard deviation), the distance between two individuals no longer depends on the unit of measurement, and the more dispersed variables are not favoured. Even if the units of measurement are not the same for all variables, the 'inverse of variances' distance brings them all to the same level. We speak of *normalized* PCA as opposed to *non-normalized* PCA, in which the variables are centred but not reduced. In my discussion of this topic I will assume that we are dealing with this 'inverse of variances' distance, but will occasionally point out certain special features which arise from the use of simple Euclidean distance.

Before examining the cloud of variables, let us recall that the *covariance* cov(X,Y) of two numeric variables X and Y is an indicator of their simultaneous variation, which is positive if Y increases whenever X increases, and is zero if X and Y are independent, although the opposite is false (it is possible to have dependence and zero covariance) as for the linear correlation coefficient. If the standard deviations of the variables X and Y are denoted σ_X and σ_Y, their means are denoted μ_X and μ_Y, their values are denoted $(x_i)_i$ and $(y_i)_i$, and their linear correlation coefficient is r_{XY}, then the covariance is

$$\mathrm{cov}(X,Y) = \frac{1}{n}\sum_{i=1}^{n}(x_i-\mu_X)(y_i-\mu_Y),$$

and we find that

$$\mathrm{cov}(X,Y) = \sigma_X \cdot \sigma_Y \cdot r_{XY}. \quad (7.1)$$

If we use σ_{ij} as a simpler notation for the covariance of X_i and X_j, the *covariance matrix* (also called the *variance–covariance matrix*) is given by

$$M_{\mathrm{cov}} = \begin{pmatrix} \sigma_1^2 & \sigma_{12} & \cdots & \sigma_{1n} \\ \cdots & \sigma_2^2 & \cdots & \sigma_{2n} \\ \cdots & \cdots & \cdots & \cdots \\ \sigma_{n1} & \sigma_{n2} & \cdots & \sigma_n^2 \end{pmatrix}.$$

This is a matrix in which the diagonal terms are the variances of the variables, and in which the *trace*, i.e. the sum of the diagonal terms, is the sum of the variances of the variables. This matrix is also positive, semi-definite and symmetric, meaning that it is diagonalizable with orthogonal eigenvectors and eigenvalues, all non-negative. By changing the variables, therefore, it is possible to find a base in which the non-diagonal terms are all zero.

When the variables are reduced, formula (7.1) shows that the covariance matrix becomes

$$M_{\text{corr}} = \begin{pmatrix} 1 & r_{12} & \dots & r_{1n} \\ \dots & 1 & \dots & r_{2n} \\ \dots & \dots & \dots & \dots \\ r_{n1} & r_{n2} & \dots & 1 \end{pmatrix},$$

which is the matrix of the linear correlation coefficients (r_{XY}) and is therefore called the *correlation matrix*. Its trace is equal to the number p of variables.

In the variable space, the norm $\|V\|$ of a variable V (its 'length') is equal to its standard deviation, and is therefore 1 when V is reduced. The scalar product of two centred variables is equal to their covariance $\text{cov}(X,Y)$, and the cosine of their angle is equal to their scalar product divided by the product of their norms; in other words, it is equal to their linear correlation coefficient $r_{XY} = \text{cov}(X,Y)/\sigma_X \cdot \sigma_Y$.

More particularly, since it is generally assumed that $\|V\| = 1$, the coordinate of the projection of a variable on an axis is equal to its linear correlation coefficient with this axis (norm of the variable multiplied by its cosine with the axis).

Given this definition of the norm of a variable, we can see that the total inertia of the cloud of variables, which is the sum of squares of the norms of the variables, i.e. the sum of their variances, is equal to the total inertia of the cloud of individuals defined previously, assuming that $p_i = 1/n$. This inertia is equal to the number of variables in normalized PCA.

In the cloud of variables, we aim to maximize the projected inertia, as for the cloud of individuals. However, in this case we are not looking for the direction of maximum elongation of the cloud, since all the variables have the same norm of 1 and are therefore on a hypersphere with radius 1. The aim is to maximize the sum of squares of the coordinates of the projections of the variables on an axis, in other words to maximize the sums of the squared cosines of the angles formed by the axis with the variables. This is equivalent to maximizing the sum of squares of the correlation coefficients of the variables and the axis which we are seeking, where this axis gives the direction of maximum inertia. The axis having this property is called the *factor axis*. In PCA, we determine a first factor axis in this way, then a second factor axis which is the one out of all the axes orthogonal to the first that has the maximum inertia projected on it. This inertia projected on the second axis is, by construction, less than that projected on the first axis. A number of factor axes can thus be determined in succession with decreasing projected inertias. Because of their orthogonality, the total inertia of the cloud of variables is broken down into the sum of inertias projected on each axis. This total inertia is equal to the sum of the variances of the variables, and therefore to the trace of the covariance matrix (equal to the correlation matrix in normalized PCA).

The sum of the projected inertias is expressed as the trace of a matrix because of the way in which the inertias are found, as the eigenvalues of the covariance matrix (for non-normalized PCA) or the correlation matrix (for normalized PCA). As for the factor axes, these are the eigenvectors of the matrix in question. Thus the diagonalization of the matrix enables us to find the axes and their inertia, bearing in mind that, in practice, it is usually the correlation matrix that is diagonalized.

A subspace generated by two factor axes is a *factorial plane* (or factorial plot).

The objective of PCA can be expressed in the form of dual conditions on the variables and the individuals ('dual' because they originate from the same data table showing 'individuals × variables'). The aim is to find the axis which:

- in the cloud of variables, maximizes the squares of the correlation coefficients of the variables with this axis (which defines a variable C because we are in the variable space):

$$\sum_i r(X_i, C)^2 = \lambda;$$

- in the cloud of individuals, maximizes the weighted squares of the coordinates of the individuals on this axis:

$$\sum_i p_i v_i^2 = \lambda, \tag{7.2}$$

bearing in mind that, generally, all the individuals x_i have the same weight $p_i = 1/n$, and (7.2) can be written more simply as

$$\frac{1}{n}\sum_i v_i^2 = \lambda. \tag{7.3}$$

The left-hand term of equation (7.3) is the variance of c, where c is the variable defined by the axis, in other words the variable which, for each individual, associates its projection on the axis.

The aim is therefore to find the variable, the linear combination of the variables analysed, which is both most closely correlated with the set of variables analysed and has the maximum variance, these two conditions being equivalent and based on a projected inertia $\lambda \geq 0$, the same in the cloud of variables and the cloud of individuals. This variable c, called the *principal component,* is a linear combination of the centred and reduced analysed variables, and it can be shown (see Escofier and Pagès, *Analyses factorielles simples et multiples,* Section 5.4.1) that the coefficients of this linear combination are equal to $1/\sqrt{\lambda}$ times the coordinates of the variables on the factor axis, in other words their linear correlation with the axis.

This last point can be explained as follows. In the variable space, the projection of the p variables on the jth factor axis defines a numerical value x_{jm} for each of the p variables: these values form what is called the jth factor F_j and correspond to a point $(x_{j1}, \ldots, x_{jp})$ in the individual space. We can show that this point is located in $\mathcal{R}^p$ on the jth factor axis u_j of the cloud of individuals, and we have

$$u_j = \frac{1}{\sqrt{\lambda_j}} F_j. \tag{7.4}$$

Finally, the above set of inertias λ corresponding to each factor axes is added up, giving the total inertia, which as we have seen is the same in the cloud of variables and the cloud of individuals. In both clouds, we have the same total inertia and the same inertia projected on

Eigenvalues of the Correlation Matrix				
	Eigenvalue	Difference	Proportion	Cumulative
1	2.69628418	1.21286011	0.3370	0.3370
2	1.48342407	0.54475789	0.1854	0.5225
3	0.93866619	0.00785658	0.1173	0.6398
4	0.93080961	0.26102963	0.1164	0.7561
5	0.66977998	0.10079635	0.0837	0.8399
6	0.56898363	0.17882698	0.0711	0.9110
7	0.39015665	0.06826097	0.0488	0.9598
8	0.32189568		0.0402	1.0000

Figure 7.1 Eigenvalues of the correlation matrix.

each factor axis, this inertia decreasing with the rank of the axis, as can be seen in the example in Figure 7.1, which also shows that the sum of the eigenvalues is 8, which is the number of variables (we shall return to this example in the following pages).

Figure 7.1 is one of the tables provided by the PRINCOMP procedure, the PCA procedure of the SAS package, which has the following basic syntax:

```
PROC PRINCOMP DATA=sasuser.case_study OUT=individuals OUTSTAT=stat;
VAR age seniority income nbproducts nbpurchases nbpoints changeconsum
usecredit;
WEIGHT weight;
RUN;
```

Note that the correlation matrix (normalized PCA) is used by default, and if we wish to use the covariance matrix (for non-normalized PCA) we must include the COV option on the first line. The variable following the keyword WEIGHT allows us to specify a weight other than 1/ n for an individual. This weight must be non-negative and can be zero for an individual who is to be represented in the cloud but not taken into account in the calculations: such an individual is called *illustrative* or *supplementary*.

A parameter $N = p$ can be added on the first line to calculate only the first p factor axes.

The OUTSTAT data set contains a column for each variable analysed, and one or more rows for each type of statistic (the type is indicated by the variable _TYPE_): i.e. mean of variables, standard deviation, correlation coefficients and eigenvalues, and a row (with _TYPE_ = SCORE) for each factor axis requested, containing not the coordinate of each variable on the factor axis, but the coefficient of each variable in the expression of the principal component as a linear combination of the initial centred and reduced variables. According to the above formula (7.4), this coefficient has to be multiplied by the square root of the eigenvalue to give the coordinate of each variable on the factor axis, in other words its correlation coefficient with the axis.

Thus we cannot directly trace the cloud of variables by a graphic procedure (PLOT or GPLOT) applied to the observations PRIN1, PRIN2... in the OUTSTAT data set (where each observation PRINx contains the coefficients a_i of the different variables V_i in the expression

PRINx $= \sum_i a_i V_i$). We must first multiply each a_i by the square root of the eigenvalue of PRINx, or use a macro such as the %ACP macro of INSEE (see the following section) which carries out this operation automatically.

As for the OUT data set, this is used to trace the cloud of individuals directly (the GPLOT procedure is preferable because of its higher resolution), as it contains one observation for each individual, as well as the DATA input data set, with the same number of variables plus a variable for each factor axis, which contains the coordinate of the individual on this axis. An example of a cloud is shown in Figure 7.8.

7.1.2 Representation of variables

The cloud of variables is analysed more often than the cloud of individuals because, especially when the individuals are numerous (more than a few hundred), the projections of the individuals on the factor axes and their contributions are not usually of interest. In the individual space the distances between points are important, whereas in the variable space the angles between the variables are most significant. This is due to the property mentioned above, namely that the cosine of the angle between two centred and reduced variables is equal to their linear correlation coefficient, and that a search for two positively correlated variables is equivalent to a search for two variables at an acute angle.

The *quality of representation* of a variable X on a factor axis is the square of its cosine with this axis, this definition being based on its relationship with the correlation coefficient. The sum of the qualities of representation on the set of factor axes is 1:

$$\sum_i r(X_i, C)^2 = 1. \tag{7.5}$$

This equality is due to the fact that $\|X\| = 1$ and $\{C_i\}_i$ is an orthonormal basis.

The quality of representation of a variable on a factor plane (C_i, C_j) is the sum of the squares of the cosines of the variable with the axes C_i and C_j. This quality varies between 0 (if the variable is entirely uncorrelated with C_i and C_j) and 1 (if the variable belongs to the factor plane and is independent of the other C_k). In the variable space, the intersection of the unit sphere, which contains all the variables (which are assumed to be reduced and therefore with a norm of 1), with the plane (C_i, C_j) is called the *circle of correlation.* It therefore follows that:

- the variable is perfectly represented on the plane (quality = 1) if and only if its projection is on the circle of correlation;
- the quality of representation of the variable is strictly between 0 and 1 if and only if its projection is within the circle of correlation;
- the variable is not represented on the plane at all (quality = 0) if and only if its projection is in the centre of the plane and of the circle of correlation.

In PCA, the choice of the first two factor axes is that in which the projection of the variables is as close as possible to the circle of correlation.

Two variables which are close together on the factor plane may actually be poorly correlated if they are distant from the circle of correlations and close to the centre of the plane.

Variables actives			_AXE1_					_AXE2_				
Ident.	CONTR	POIDS	COORD	CTR	RCTR	CO2	QLT	COORD	CTR	RCTR	CO2	QLT
age	12.50	12.50	0.06	0.1	7	0.3	0.3	0.78	40.6	1	60.3	60.6
seniority	12.50	12.50	0.30	3.3	6	9.0	9.0	0.74	36.7	2	54.5	63.5
income	12.50	12.50	0.73	19.6	4	52.9	52.9	0.10	0.6	7	1.0	53.9
nbproducts	12.50	12.50	0.81	24.5	1	66.1	66.1	0.00	0.0	8	0.0	66.1
nbpurchases	12.50	12.50	0.81	24.1	2	65.1	65.1	-0.11	0.9	5	1.3	66.4
nbpoints	12.50	12.50	0.79	22.9	3	61.7	61.7	-0.23	3.7	4	5.5	67.1
changeconsum	12.50	12.50	0.05	0.1	8	0.2	0.2	-0.50	16.7	3	24.8	25.0
usecredit	12.50	12.50	0.38	5.3	5	14.2	14.2	-0.10	0.7	6	1.1	15.3

Figure 7.2 Interpreting the INSEE PCA macro.

Consider, for example, the projections of the two poles of a sphere on the plane passing through the equator: the projections meet in the centre, although the poles are diametrically opposed. On the other hand, if two variables are both close to each other and close to the circle of correlation, their correlation coefficient is close to 1. If they are diametrically opposed on the circle of correlation, their correlation coefficient is close to -1.

Note that, unfortunately, the SAS PRINCOMP procedure does not display the quality of representation of the variables, although this is done in SPAD or IBM SPSS Statistics. The FACTOR procedure in IBM SPSS Statistics displays the quality of representation corresponding to the number k of axes selected (i.e. the sum of the first k squared cosines). To perform the same calculation with SAS, we must write a macro to supplement the PRINCOMP procedure; some specialists, particularly at INSEE, have already done this. Figure 7.2 shows the output from the PCA macro developed at INSEE (available for download from www.insee.fr). This shows, for each of the first k axes selected:

- the coordinate COORD of each active variable on the axis;
- the contribution CTR of the variable to the axis, i.e. (see below) COORD^2 divided by the eigenvalue of the axis, expressed as a percentage;
- the rank RCTR of the variable, the variables being classified by decreasing contribution;
- the quality CO2 of representation of the variable on the axis, i.e. COORD^2, expressed as a percentage;
- the sum QLT of the qualities of the variable on this axis and on all the preceding axes.

Thus, overall, we have:

- the contribution CONTR of the variable to the total inertia of the cloud of variables;
- the weight WEIGHT of the variable in the cloud.

Eigenvectors									
		Prin1	Prin2	Prin3	Prin4	Prin5	Prin6	Prin7	Prin8
age	age	0.035722	0.637571	0.258608	0.314857	0.332034	-.485766	0.260887	0.109258
seniority	client seniority	0.182484	0.605982	0.187981	0.057860	-.389761	0.622209	-.049299	-.138971
income	client income	0.443030	0.080052	-.040509	-.273237	0.651810	0.117833	-.318435	-.425293
nbproducts	nb of products	0.495244	0.001202	0.117938	0.022297	-.306377	-.337970	-.616976	0.389326
nbpurchases	nb of purchases	0.491426	-.093575	0.003426	-.304492	0.130701	0.220890	0.542087	0.545237
nbpoints	nb of points	0.478177	-.191761	0.030312	0.110554	-.381114	-.308847	0.389827	-.573440
changeconsum	change in consumption	0.029868	-.408820	0.770733	0.384360	0.184274	0.234905	-.032310	0.005507
usecredit	use of credit	0.229862	-.084950	-.535980	0.754779	0.151093	0.222308	-.024057	0.100762

Figure 7.3 Coefficients of the principal components.

The best way to interpret the factor axes is to observe their correlations with the variables analysed. So now we change our point of view. Instead of fixing a variable, we fix a factor axis C_k and calculate the sum of squares of the linear correlation coefficients,

$$\sum_{j=i}^{p} r(X_j, C_k)^2,$$

where $\{X_j\}_j$ is the set of analysed variables. By the definition of the projected inertia, this sum is equal to the eigenvalue λ_k of C_k. We say that $r(X_j, C_k)^2/\lambda_k$ is the *contribution* of X_j to the axis C_k. It is also, according to equation (7.4) above, the square of the centred and reduced coefficient of X_j in the expression of C_k as the linear combination of the X_j. Since the sum of contributions of the p variables X_j to the axis is equal to 1, the mean contribution of a variable is $1/p$ and a contribution of more than $1/p$ is considered to be important.

Figure 7.3, calculated by the SAS PRINCOMP procedure, provides an example of the coefficients of the principal components as linear combinations of the centred and reduced analysed variables. It therefore contains the quantities $r(X_j, C_k)/\sqrt{\lambda_k}$. If, for example, we require the correlation coefficients of the first axis with the variables analysed, in other words the coordinates of these variables on the axis, then we must multiply all the terms in the Prin1 column by $\sqrt{\lambda_1}$. The first principal component is written:

$$(0.035722*\text{age}_{\text{centred-reduced}}) + (0.182484*\text{seniority}_{\text{centred-reduced}}) + \ldots$$

The sum of squares of each column is 1, because it is the sum of contributions of the variables to the corresponding axis. Equation (7.5) shows that the sum of squares of a row is 1. We can also see that the scalar product of two columns is 0, corresponding to the value 0 of the correlation coefficient of two different axes.

The FACTOR procedure in IBM SPSS Statistics does not display this table, but displays two others. The first of these, called the 'component matrix', contains the coordinates $r(X_j, C_k)$ of the variables on the factor axes (Figure 7.4). The second table, called the 'matrix of the coefficients of the coordinates of the components', contains, oddly enough, the

	Component matrix[a]							
	Component							
	1	2	3	4	5	6	7	8
age	.059	.777	.251	.304	.272	−.366	.163	.062
seniority	.300	.738	.182	.056	−.319	.469	−.031	−.079
income	.727	.097	−.039	−.264	.533	.089	−.199	−.241
nbproducts	.813	.001	.114	.022	−.251	−.255	−.385	.221
nbpurchases	.807	−.114	.003	−.294	.107	.167	.339	.309
nbpoints	.785	−.234	.029	.107	−.312	−.233	.243	−.325
changeconsum	.049	−.498	.747	.371	.151	.177	−.020	.003
usecredit	.377	−.103	−.519	.728	.124	.168	−.015	.057

Extraction method: Principal component analysis.
[a]8 components extracted.

Figure 7.4 Coordinates of the variables on the axes.

quantities $r(X_j, C_k)/\lambda_k$ (without the square root of the eigenvalue!), which have no obvious use.

Figure 7.5 shows the factor plane for the above data, plotted with the GPLOT procedure of SAS/GRAPH, the circle being constructed according to the instructions in an ANNOTATE table used by the GPLOT procedure, as in the example described in Section 1.3.4 of my book *Étude de cas en statistique décisionnelle*.[1] The result is satisfactory, but not very easy to obtain. It is fairly similar to the output of the *FactoMineR* package in R.

A more elegant result can be obtained using a rather more concise syntax, which uses GTL, a new language that appeared with SAS 9.1.3. An application of GTL to a multiple correspondence analysis graph is described in Section 12.7.

We can also construct a graph (not quite as pleasing to look at, but still better than a simple output from the PLOT procedure) by using the %PLOTACP macro developed by INSEE, if we have used the %ACP macro beforehand. For an example of application, see Section 1.7 of *Étude de cas en statistique décisionnelle*.

We can also use the SAS %PLOTIT macro (see Section 7.3.2) after the PRINCOMP procedure, specifying the name DATA of the data set containing the coordinates PRIN1 and PRIN2 of the variables on the two factor axes, the label LABELVAR of the points (identical to the name of the variables in this case), the plotting of the horizontal axis 0 and vertical axis 0, the frame colour, COLOR, and the internal graph colour, COLORS. The result is shown in Figure 7.6.

```
%PLOTIT(DATA=sortie, plotvars=prin2 prin1, labelvar=_name_, href=0,
vref=0, color=black, colors=black)
```

We find that all the variables are on the same side of the first factor axis: this is because they are all positively correlated with each other. This is what is called the *size effect* (or size

[1] Tufféry, S. (2009) *Étude de Cas en Statistique Décisionnelle*. Paris: Technip.

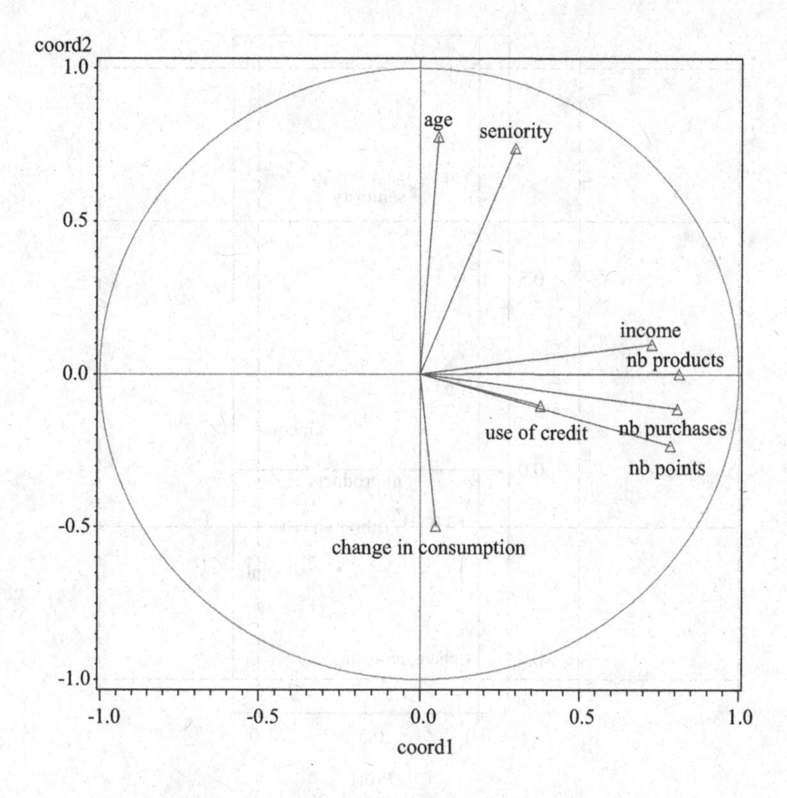

Figure 7.5 Representation of the variables.

factor), a reference to the size that has an effect on a whole set of measurements, such that some individuals have high values for the set of variables while others have low values for the set of variables. In this case, the first axis is seen as a factor summarizing the strength of an underlying factor based on a number of different measurements. It is then useful to examine the plane intersecting the second and third factor axes.

7.1.3 Representation of individuals

Usually, when the cloud of individuals is represented, we are interested less in the individuals in isolation than as a set. The only relatively common exception relates to the individuals which, because of their abnormally high contribution to the factor axes, may be 'outliers', or extremes, which we may prefer to omit from the analysis to avoid falsifying the results.

For a more precise concept of the contribution of an individual to an axis, let us recall that, by definition of the projected inertia (see the start of this section), if p_i denotes the weight of the individual i and v_{ik} is the individual's coordinate on the kth axis (with an inertia λ_k), then

$$\sum_{i=1}^{n} p_i v_{ik}^2 = \lambda_k.$$

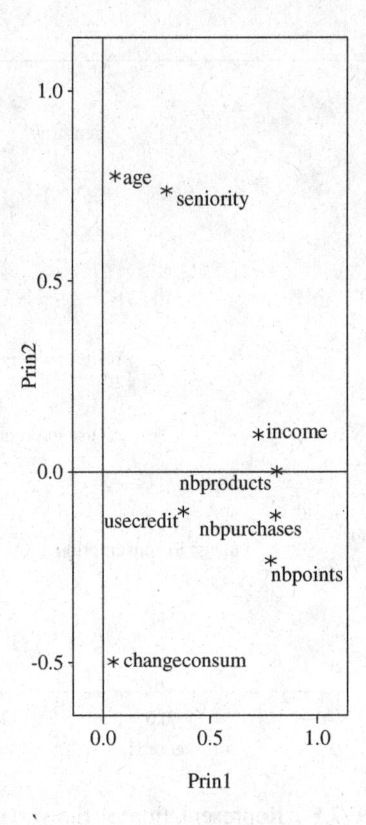

Figure 7.6 Representation of variables with the PLOTIT macro.

By analogy with the previously introduced concept of the contribution of a variable to the kth axis, we can then define the *contribution* of the individual i to the kth axis by

$$\frac{p_i v_{ik}^2}{\lambda_k},$$

in such a way that the sum of the n contributions is 1. A contribution that exceeds the weight p_i of the individual is considered to be important, while a contribution exceeding 0.25 is dangerously high for the stability of the PCA. If this case arises, the WEIGHT instruction of the SAS PRINCOMP procedure (see the example of syntax above) can be used to assign a weight of 0 to this individual (and a weight of 1 to the others) to transform it into an *illustrative* (or supplementary) individual, which is not used in the analyses but whose factor coordinates are calculated so that it can be represented.

If we look at the two clouds of points (individuals and variables) simultaneously, we will see that an individual is on the side of the variables for which it has high values and is on the opposite side from the variables for which it has low values. The values increase with the distance of the individual from the origin (the centre of the cloud); if the individual has mean values for all variables, it is at the origin. However, do not be led astray by the superimposition of the two clouds of points, which are located in different spaces which have different distances: they are not directly comparable. Furthermore, the cloud of variables is contained in a hypersphere with radius 1, but the cloud of individuals is not.

7.1.4 Use of PCA

Although there are software packages offering user-friendly and relatively simple PCA procedures, there are still some pitfalls to avoid when interpreting PCA. Some of the ones I have encountered are as follows:

- The individual space and the variable space must not be superimposed.
- In a factor plane, the proximity of two variables does not mean anything unless they are near the circle of correlation.
- The first factor plane (intersection of the first two factor axes) is not the only one that offers useful information; it is also profitable to intersect the first and third axes, or the second and third axes, and so on.
- Avoid letting an individual (or a small group of individuals) have too great a contribution to the first axes; in the worst case, an axis may be almost entirely accounted for by a single individual.

Having said this, the representation of the variables in a factor plane is the most intuitive and most practical method of identifying which variables are interrelated or opposed to each other; it is far more satisfactory than the correlation matrix. PCA, in the form using rotation (Section 7.2.1), is also the essential ingredient of one of the best methods of variable clustering, used in the SAS VARCLUS procedure (see Section 9.14).

PCA can also be used to represent on the factor plane variables which were not used in the construction of the axes. Just as some individuals are defined as *illustrative* (see above) when they are to be excluded from the construction of the axes because they are suspected of being the result of measurement errors or because they contribute too strongly to an axis, some continuous variables may be defined as *supplementary* or *illustrative*, in contrast to the *active* variables.

Supplementary variables may be, for example, variables that are to be related to the active variables but not to each other, or variables that are to be accounted for by the active variables, or possibly variables that are to be used to reinforce the interpretation of the axes without using the variables that were used to determine the axes.

Unfortunately, the SAS PRINCOMP procedure is rather outdated, and cannot generate supplementary variables, unlike its most recent competitor, the CORRESP procedure for multiple correspondence analysis (see below). However, we can position supplementary variables on the circle of correlation, making use of the fact that the coordinates of these variables on the axes are their correlation coefficients with the axes. SAS macros have been developed in various places to compensate for these deficiencies (and the small number of printed outputs) of the PRINCOMP procedure; once again, the INSEE ACP macro is useful here (see above).

Users of R have access to the *FactoMineR* package (Figure 7.7), described in Chapter 5 on software. This includes all the requisite functionality, even if the graphs are lacking in readability when the number of variables is large (the SAS GTL language is better for these cases).

Another advantage of PCA is that its graphic representation can be used to check the outcome of a clustering procedure (see Chapter 9) carried out either independently or based on the principal components. This representation enables us to:

- check the relevance of the clustering visually (the human eye is a very efficient instrument for detecting clusters);

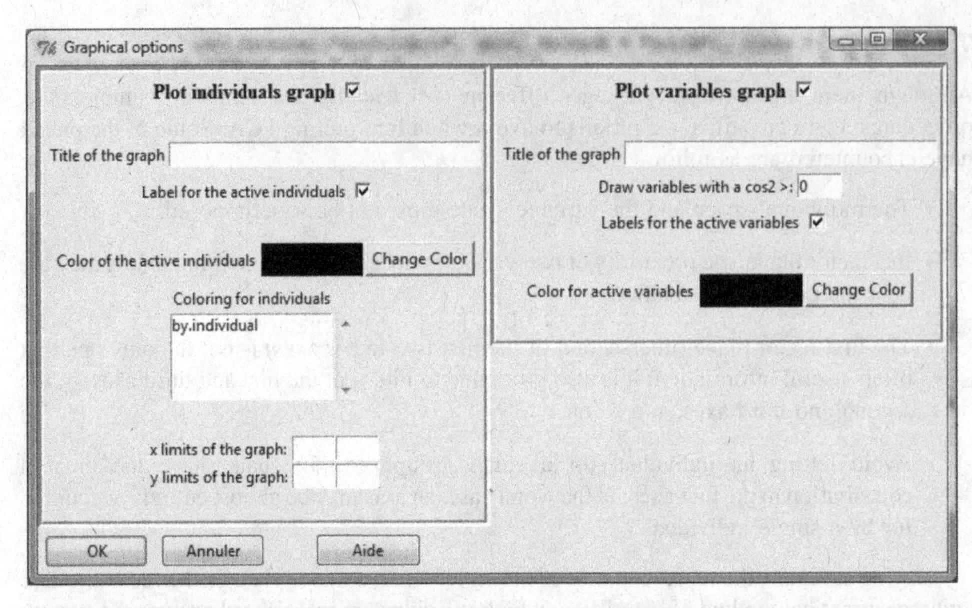

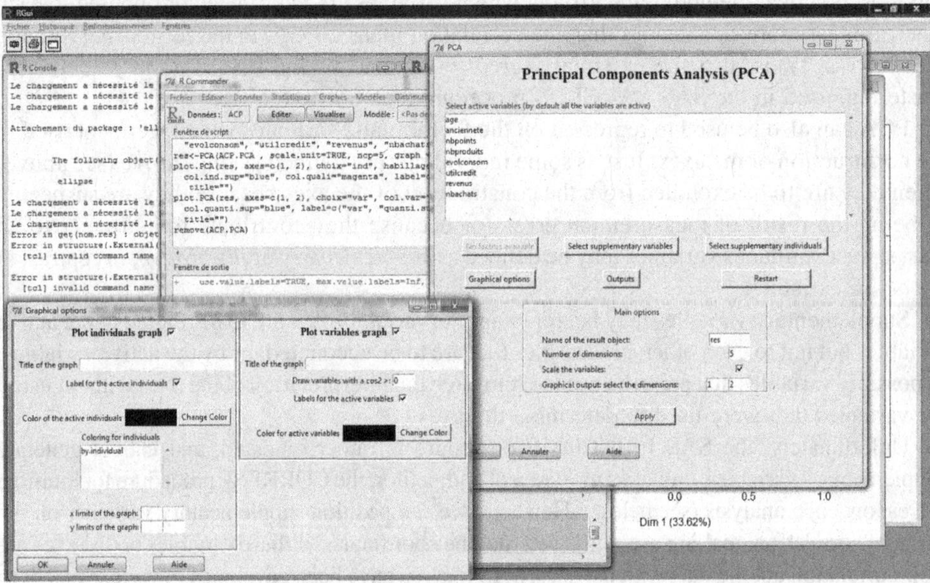

Figure 7.7 PCA with *FactoMineR*.

- choose the most appropriate number of clusters where necessary;
- easily isolate certain individuals if their data appear to be extreme, possibly due to a measurement or input error;
- select the most typical individuals in a cluster, or conversely those which are similar to an adjacent cluster towards which they may develop, this being of interest to a commercial business if the second cluster is a more profitable customer segment.

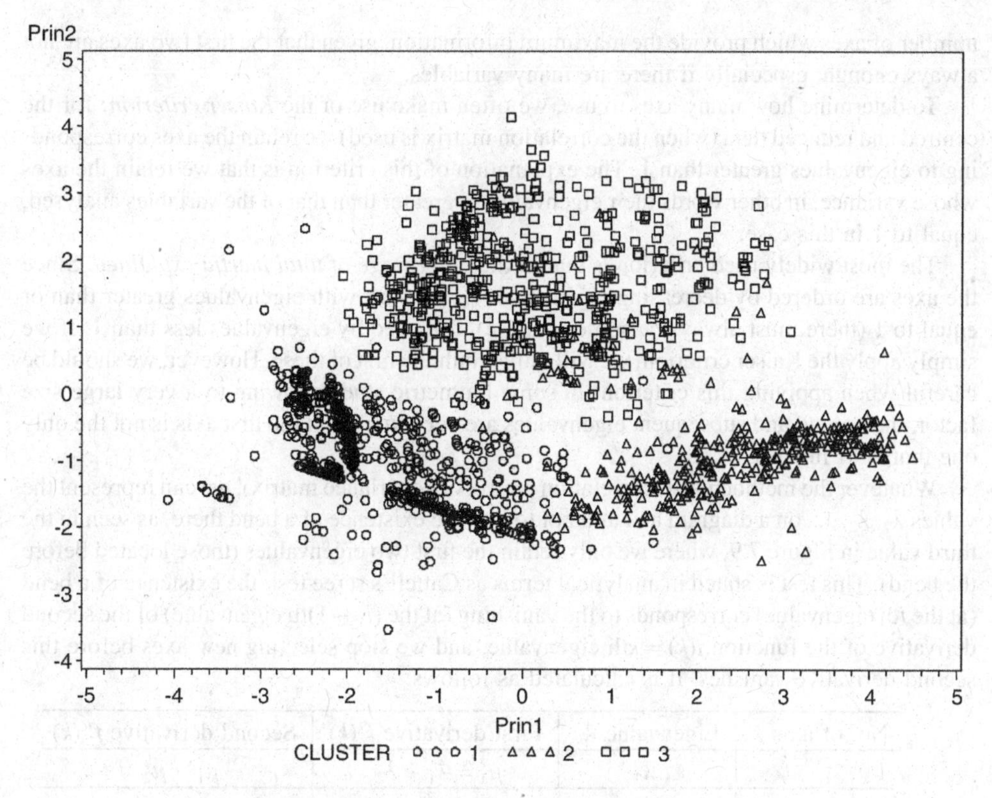

Figure 7.8 Representation of PCA-based clustering.

This is done by 'colouring' (or representing by different symbols) the points in a cloud of individuals according to the clusters they belong to (Figure 7.8). Suppose that this cluster is contained in the variable CLUSTER, and that PRIN1 and PRIN2 are the first two principal components. In SAS, we can write:

```
SYMBOL1 V=CIRCLE C=BLACK;
SYMBOL2 V=TRIANGLE C=BLACK;
SYMBOL3 V=SQUARE C=BLACK;
PROC GPLOT;
PLOT PRIN2*PRIN1=CLUSTER;
RUN;
QUIT;
```

7.1.5 Choosing the number of factor axes

The inertia projected on a factor axis, in other words its eigenvalue, corresponds to the share of information carried by this axis. According to the statistical principle of parsimony, we wish to 'summarize' the information as neatly as possible, or in other words to find a minimum number of axes which provide the maximum information, given that the first two axes are not always enough, especially if there are many variables.

To determine how many axes to use, we often make use of the *Kaiser criterion:* for the centred and reduced data (when the correlation matrix is used) we retain the axes corresponding to eigenvalues greater than 1. The explanation of this criterion is that we retain the axes whose variance, in other words their eigenvalue, is greater than that of the variables analysed, equal to 1 in this case.

The most widely used criterion is that of the *percentage of total inertia explained.* Since the axes are ordered by decreasing eigenvalues λ_i, we start with eigenvalues greater than or equal to 1 (there must always be some of these), followed by eigenvalues less than 1. If we simply apply the Kaiser criterion, we will only retain the first of these. However, we should be careful when applying this criterion: in some biometric studies, owing to a very large size factor, the second and subsequent eigenvalues are very small, but the first axis is not the only one that is of interest.

Whatever the method used (correlation matrix or covariance matrix), we can represent the values $\lambda_1, \lambda_2, \ldots$ on a diagram and attempt to find the existence of a bend there, as seen in the third value in Figure 7.9, where we only retain the first two eigenvalues (those located before the bend). This test is stated in analytical terms as Cattell's scree test: the existence of a bend (at the ith eigenvalue) corresponds to the vanishing (at the $(i + 1)$th eigenvalue) of the second derivative of the function $f(k) = k$th eigenvalue, and we stop selecting new axes before this second derivative vanishes. It is calculated as follows:

No. of axis k	Eigenvalue λ_k	First derivative $f'(k)$	Second derivative $f''(k)$
1	λ_1	$\mu_1 = \lambda_1 - \lambda_2$	$\mu_1 - \mu_2$
2	λ_2	$\mu_2 = \lambda_2 - \lambda_3$	$\mu_2 - \mu_3$
3	λ_3	$\mu_3 = \lambda_3 - \lambda_4$	$\mu_3 - \mu_4$
4	λ_4	$\mu_4 = \lambda_4 - \lambda_5$	$\mu_4 - \mu_5$
...	...	...	...

Taking the example of Figure 7.1, we have:

No. of axis k	Eigenvalue λ_k	First derivative $f'(k)$	Second derivative $f''(k)$
1	2.69628418	1.21286011	0.66810222
2	1.48342407	0.54475789	0.53690131
3	0.93866619	0.00785658	–0.25317305
4	0.93080961	0.26102963	...
...	...	...	...

We find that the bend in the third eigenvalue appears in the form of an inflection point in the fourth eigenvalue.

A third criterion requires the calculation of successive cumulative values $\lambda_1/\sum_i\lambda_i$, $(\lambda_1 + \lambda_2)/\sum_i\lambda_i$, $(\lambda_1 + \lambda_2 + \lambda_3)/\sum_i\lambda_i, \ldots$, 1, to see what proportion of the sum of variances $\sum_i\lambda_i$ (equal to the total inertia) is provided by the first axis, the first two axes, the first three,

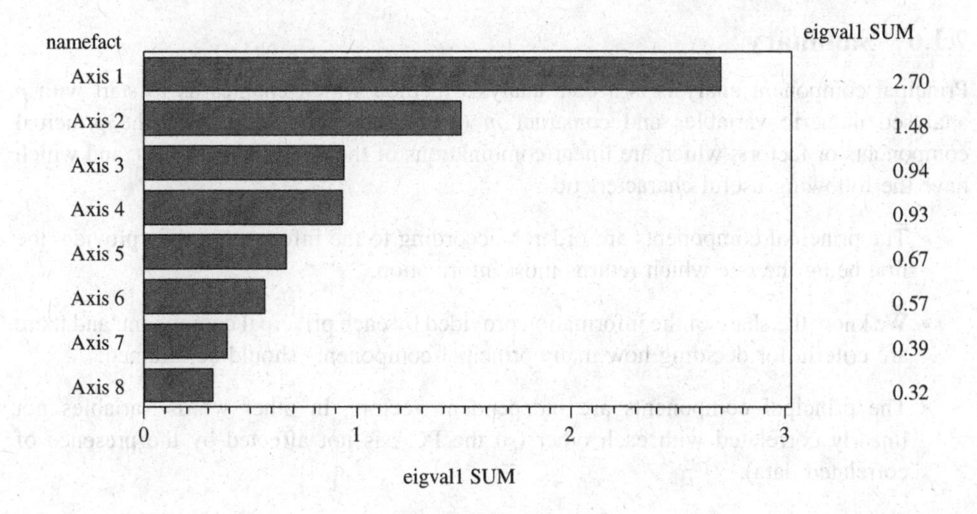

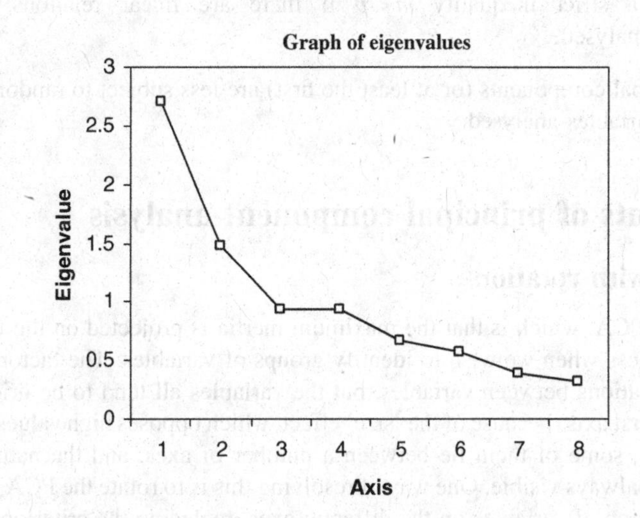

Figure 7.9 Choosing the number of factor axes.

etc. If the first p axes provide a proportion of the total inertia that is considered satisfactory, we stop at that point and do not select any further axes. For example, if the first two axes provide 80% of the total inertia, it is clear that the cloud of individuals is virtually flat, in a two-dimensional subspace, and that a projection on the factor plane will be highly satisfactory. This is more likely to happen as the correlations among the analysed variables increase, or in other words if the correlation matrix has terms greater than 0.

When this criterion is used to determine the number of axes to be retained, we must be careful, because the fact that 40% of the inertia is provided by the first axis does not have the same meaning regardless of whether we are dealing with 10 or 50 variables (it is much more significant in the second case).

7.1.6 Summary

Principal component analysis is a data analysis method which enables us to start with p analysed numeric variables and construct m $(\leq p)$ other variables, called the principal components or factors, which are linear combinations of the analysed variables, and which have the following useful characteristics:

- The principal components are ordered according to the information they provide, the first being the one which returns most information.
- We know the share of the information provided by each principal component, and there are criteria for deciding how many principal components should be retained.
- The principal components are independent vectors, in other words variables not linearly correlated with each other (so the PCA is not affected by the presence of correlated data).
- There is a strict inequality $m < p$ if there are linear relations between the variables analysed.
- The principal components (or at least the first) are less subject to random fluctuations than the variables analysed.

7.2 Variants of principal component analysis

7.2.1 PCA with rotation

The strength of PCA, which is that the maximum inertia is projected on the first axis, may become a weakness when we wish to identify groups of variables. The factor plane clearly shows the correlations between variables, but the variables all tend to be orientated in the direction of the first axis (because of the 'size' effect, which opposes high values to low values of the variables), some of them lie between a number of axes, and the natural groups of variables are not always visible. One way of resolving this is to rotate the PCA axes to obtain the best distribution of variables on the different axes, replacing the criterion of maximum inertia provided on the first axis with another criterion which makes for easier interpretation. This criterion depends on the method, but in any case the total inertia does not change after rotation – only its breakdown changes.

Rotation may be *orthogonal* or *oblique*. In the first case, the factors are not correlated, allowing easier interpretation. In the second case, the factors are no longer orthogonal, making interpretation more difficult, but it has the advantage that the eigenvalues are stronger and the correlation of the factors with the variables is stronger.

The main oblique forms of PCA are *oblimin* and *promax* PCA; the second of these is faster and is used for large volumes of data.

The main orthogonal forms of PCA are *varimax*, *quartimax* and *equamax* PCA, the last of which is a compromise between the first two (while *orthomax* PCA is a generalization of the first three). In quartimax PCA, all the variables have a high contribution to the same factor, and each variable has a non-zero contribution to another factor and practically zero contributions to all the other factors.

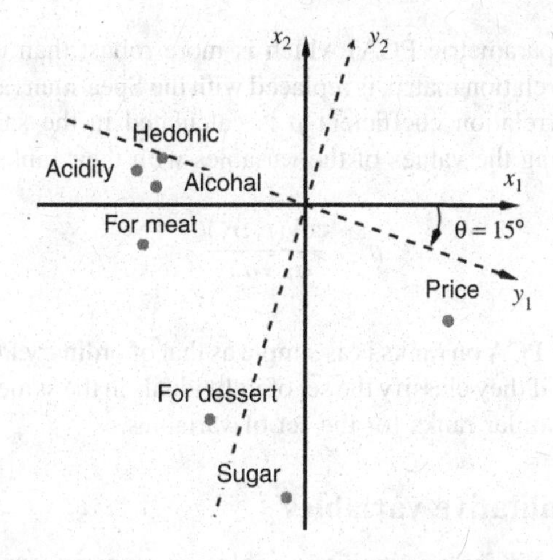

Figure 7.10 Example of VARIMAX rotation.

The most widely used form of PCA with rotation is varimax PCA. This is based on the principle of maximizing, for each factor, not the sum of squares of the correlation coefficients of this factor with the set of variables, but the variance of these correlation coefficients, with the result that each factor is strongly correlated with some variables and weakly correlated with the others. Thus some variables have a high contribution to each axis, while the others have a very low contribution, and the axes are easy to interpret.

Hervé Abdi (www.utdallas.edu/~herve/Abdi-rotations-pretty.pdf) gives an example of five wines, described in terms of acidity, sugar and alcohol content, matching with meat and desserts, the hedonic dimension, and price. He shows (Figure 7.10) that varimax PCA provides the best interpretation of the price and sugar axes.

Promax PCA is a hybrid method, consisting of a varimax rotation followed by an oblique rotation such that the high and low factor coordinates of the variable space correspond to the same variables, but with low values of coordinates which are even weaker.

These variants of PCA are provided in IBM SPSS Statistics and in the SAS/STAT FACTOR procedure, which is more generally applicable but more complex and slower than PRINCOMP which is used for ordinary PCA. Quartimax PCA also forms the basis of the VARCLUS procedure in SAS/STAT, used for clustering numeric variables (see Section 9.14). However, the forms of PCA with rotation preferred in English-speaking countries are not included in the SPAD software which relates more to French-style data processing.

7.2.2 PCA of ranks

If extreme values (outliers) or totally asymmetric distributions are present, the reduction of variables provided by PCA may be insufficient to yield good results, and it may be helpful to work on the ranks of the variables rather than on the variables themselves. This gives us a form of PCA on ranks (non-parametric PCA) which is more robust than ordinary PCA. In this method, the Pearson correlation matrix is replaced with the Spearman rank correlation matrix. In fact, Spearman's correlation coefficient ρ is calculated in the same way as Pearson's coefficient, after replacing the values of the variables with their ranks (see Appendix A):

$$\rho = \frac{\text{cov}(r_x, r_y)}{\sigma_{r_x} \cdot \sigma_{r_y}}.$$

The interpretation of PCA on ranks is as simple as that of ordinary PCA. Two variables are close on the factor plane if they classify the set of individuals in the same way. Two individuals are close if they have similar ranks for the set of variables.

7.2.3 PCA on qualitative variables

This form of PCA is used for qualitative variables or numeric variables with non-linear relationships, and consists of a PCA applied to variables which have undergone an optimal transformation into numeric variables. This method is described by Gifi (1990);[2] it is implemented in the CATPCA procedure (an update of PRINCALS) in IBM SPSS Statistics, and in the SAS PRINQUAL procedure.

7.3 Correspondence analysis

7.3.1 Introduction

You will recall that the *contingency table* of two qualitative (or discretized) variables A and B, with categories $(a_k)_k$ and $(b_l)_l$, is the table (x_{ij}) in which the value x_{ij} is the number of individuals x such that $A(x) = a_i$ and $B(x) = b_j$.

The χ^2 test enables us to detect a dependence between the two variables. The frequencies and the contribution to the χ^2 of each cell of the contingency table show the association between the categories of the two variables: a high frequency denotes a strong positive relationship, a low frequency denotes a strong negative relationship, and an intermediate frequency denotes a weak relationship. By examining this table we can obtain a good description of the relationships between the two variables, but if there are numerous categories it is tiresome to have to inspect all the cells. It would be even harder to read the table if there were more than two variables to be cross-referenced.

Correspondence analysis (CA) or *binary correspondence analysis* overcomes this problem by providing a two-dimensional view of contingency tables, thus:

- two positively related categories of A and B (high frequency) are close;
- two negatively related categories of A and B (low frequency) are opposed;
- the strongest oppositions are on the first axis (horizontal);
- the categories not related to others are in the centre.

[2] Gifi, A. (1990) *Nonlinear Multivariate Analysis*. Chichester: John Wiley & Sons, Ltd.

		colour of hair				mean profile
		dark	brown	red	fair	
colour of eyes	brown	11	20	4	1	37
	hazel	3	9	2	2	16
	green	1	5	2	3	11
	blue	3	14	3	16	36
mean profile		18	48	12	21	100

Figure 7.11 Example of CA.

The idea behind CA is clearly attractive, which explains its popularity among data analysts. I will not describe the theoretical principles of CA, since it can be considered as a form of PCA with a specific distance, namely the χ^2 *distance*. This distance is used to compare two categories of a single variable A by comparing their relative frequencies on the set of categories of B, while complying with two properties:

- On the one hand, the distance of the two categories of A must be independent of their total frequencies, which are independent of B and do not correspond to the desired aim.
- On the other hand, the distance of the categories of A, which is calculated on the basis of their frequencies on each of the categories of B, must not be affected by the high-frequency categories of B; this is achieved by giving each category of B a weight equal to the inverse of the frequency of this category.

Thus two categories of a single variable are close if the two groups of individuals which possess them also have the same characteristics (with respect to the other variables).

To see how the χ^2 distance is calculated, let us take an example (Figure 7.11) from Section 1.3.2 of Lebart *et al.* (2006).[3] The distance (hazel eyes; blue eyes) is equal to the following sum:

$$\frac{1}{18}\left(\frac{3}{16}-\frac{3}{36}\right)^2+\frac{1}{48}\left(\frac{9}{16}-\frac{14}{36}\right)^2+\frac{1}{12}\left(\frac{2}{16}-\frac{3}{36}\right)^2+\frac{1}{21}\left(\frac{2}{16}-\frac{16}{36}\right)^2.$$

This formula, which enables us to evaluate the proximity of two categories of a single variable, is clearly inapplicable when there are two different variables. However, we may, as suggested above, show the categories of A and B simultaneously in a graph, thus giving a meaning to the distance of a category A_i of A and a category B_j of B. Without going into the rather complicated details and the concept of axial expansion (see Lebart *et al.*), we can say that the idea is to represent the categories of A, then position each category B_j of B as the centre of gravity of the categories A_i, each A_i being weighted by the relative frequency of A_i, given B_j. In the example above, the relative frequency of brown eyes, given that the hair colour is brown, is 20/48.

In the same way as in PCA, we diagonalize a matrix to find eigenvalues for which we examine the proportion of the total inertia that they provide, and we obtain the appropriate number of factor axes by observing the decrease in the eigenvalues. The total inertia,

[3] Lebart, L., Morineau, A. and Piron, M. (2006) *Statistique Exploratoire Multidimensionnelle: Visualisations et Inférences en Fouille de Données*, 4th edn. Paris: Dunod.

measured by the χ^2 metric as the weighted mean of the squares of the distances of the individuals from their centre of gravity, is here equal to χ^2 divided by the number of individuals, which explains the name given to this measurement system. As in PCA, the total inertia is the sum of the eigenvalues. It measures the scatter of the cloud of points and the relation between the two variables. As in PCA, we must avoid interpreting the proximities of categories on the factor plane before we are sure that this plane represents the categories correctly, in other words that the squared cosines of the angles of the categories with the plane are sufficiently close to 1. In this case, though, we do not speak of the 'circle of correlation', since the variables are qualitative, not continuous. However, it is worth noting that we can superimpose categories and individuals in a single plane here, which is not the case in PCA. This is due to the fact that the point representing a category is actually the centre of gravity of the individuals having this category. More precisely, its coordinate on an axis is the mean of the coordinates on this axis of the individuals having this category, this mean being divided by the square root of the eigenvalue of the axis. Thus two categories of different variables are close if they tend to relate to the same individuals. We have the dual property: the coordinate of an individual on an axis is the mean of the coordinates of the categories of the individual on this axis, this mean being divided by the square root of the eigenvalue of the axis. I will show later, when discussing MCA, that this property can be used profitably to calculate the coordinates of an individual from the coordinates of its categories, found by analysing the Burt table. This property can also be expressed as follows: an axis is the sum of the indicators of the categories, divided by the number of variables and the root of the eigenvalue of the axis. In particular, a factor axis is a linear combination of the indicators of the categories of the variables analysed. This is also true of MCA. Two individuals are close if they tend to have the same categories for the set of variables.

In this factor plane, the centre is the mean profile for the two variables. A category A_i is distant from the centre if the distribution of the categories of the other variable differs widely between the set of individuals for which A_i is true and the set of all the individuals. Two categories A_i and A_j of the same variable A are close if the other variable B has the same restricted distribution at A_i and A_j. The proximity of two categories A_i and B_j of two variables A and B generally corresponds to an excess frequency of the intersection $A_i \times B_j$.

The eigenvalues are all less than or equal to 1. If we have k eigenvalues equal to 1, the rows and columns of the contingency table can be divided into $k + 1$ groups and each group of rows is associated with one group of columns only (and vice versa), so that the table (x_{ij}) is split into $k + 1$ blocks, and the cloud of points is split into $k + 1$ groups. Each factor axis associated with an eigenvalue of 1 is a perfect reflection of this association between groups I_1 and I_2 of rows and groups J_1 and J_2 of columns (I_k being associated with J_k):

$$\begin{array}{cccl} I_1 & 0 & I_2 & \text{axis associated with } \lambda = 1 \\ \hline j_1 & & j_2 & \end{array}$$

Within each group (e.g. I_1), the variables have exactly the same coordinate on the associated factor axis.

If all the eigenvalues are equal (or close) to 1, each category of A corresponds to a single (or almost a single) category of B. If A and B are ordered and if these order structures are associated, so that the table can be reordered to form a table with '0' throughout except in a strip surrounding the diagonal, the first factor axis opposes the extremes to each other (scale factor), while the second axis opposes the extremes and the means, and the cloud of points takes a parabolic form called a 'horseshoe' (the Guttman effect). More generally, the rank factor r is a polynomial function of degree r of the first factor. If the Guttman effect is very pronounced (a very clear parabola on the first factor plane), factors with a rank of 3 or greater will be disregarded. This effect is particularly likely to occur in CA or MCA when the qualitative variables being examined are the result of a clustering of initially continuous variables, and when these continuous variables are related by a size effect (see Section 7.1 on PCA).

A classic example of this situation is one in which A and B are the educational qualifications and the posts of a population of employees: the parabola opposes top executives and postgraduate degrees, on the one hand, to manual workers and non-graduates or those with vocational qualifications on the other, while the base of the parabola corresponds to intermediate technical and business diplomas and posts (middle management, engineers). Brigitte Escofier and Jérôme Pagès give a comprehensive discussion of this subject.[4]

7.3.2 Implementing CA with IBM SPSS Statistics

Since the SAS procedure used for CA, CORRESP, is the same as that used for MCA which will be examined below, I will simply mention it here and refer the reader to Section 7.4.3 for a detailed example of implementation.

For the time being, I will concentrate on the CORRESPONDENCE procedure of IBM SPSS Statistics, which is reserved for CA, and which provides similar results to those of the SAS CORRESP procedure in this context.

For this purpose, we will use a data set which is also used in the section on MCA. It relates to visits to a department store by 582 customers, whose ages and the departments they visited are known (it is assumed that there is only one department per customer).

As we shall see in Section 7.4.3, the SAS syntax for CA is as follows:

```
PROC CORRESP DATA=sasuser.survey OUT=output ALL ;
TABLES age , department ;
RUN ;
%PLOTIT(DATA=output, datatype=corresp, plotvars=Dim2 Dim1, symvar=_
type_, href=0, vref=0, tsize=0.5, color=black, colors=black)
```

This provides a number of outputs which will be studied later, similar to the outputs of IBM SPSS Statistics reviewed below. Figure 7.12 has been produced by the %PLOTIT macro developed and supplied by SAS for creating high resolution graphics, or more specifically clouds of points labelled by the value of a variable. This macro has filled an important gap, because:

- the GPLOT procedure produces high resolution graphics, but is not good at controlling the automatic label positioning (this requires the additional use of the ANNOTATE instruction, which complicates the syntax);
- and the PLOT procedure can only produce low resolution graphics.

The %PLOTIT macro can be used to provide displays after the CORRESP, PRINCOMP, PRINQUAL, MDS and TRANSREG procedures. It makes use of the GANNO procedure and requires the SAS/GRAPH module. It provides optimal label positioning by an iterative process, hence the name: PLOT ITeratively.

[4] Escofier, B. and Jérôme Pagès, J. (2008) *Analyses Factorielles Simples et Multiples*, 4th edn. Paris: Dunod.

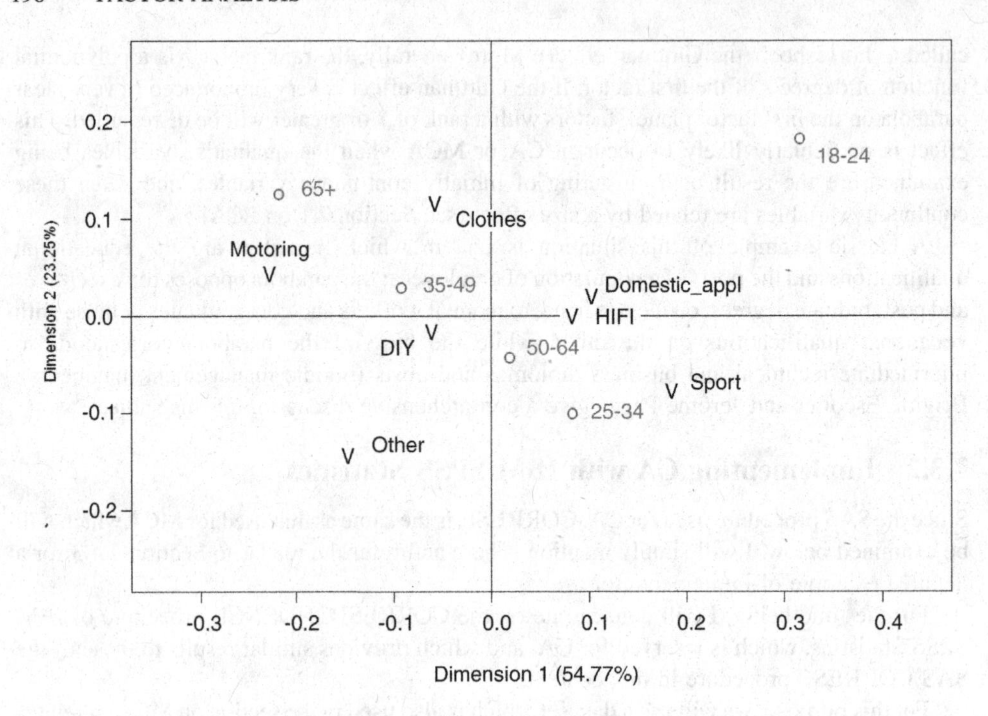

Figure 7.12 Factor plane of a CA.

The %PLOTIT macro is therefore called at the end of the CORRESP procedure in this case, by specifying the name DATA of the data set containing the coordinates DIM1 and DIM2 of the categories to be represented on the two factor axes, the type DATATYPE of the data to be plotted (originating from PROC CORRESP in this case, resulting in the display of the proportion of inertia in the description of each axis), the variable SYMVAR whose first character (more characters can be requested with SYMLEN=n) is the symbol representing each point on the graphic, the size TSIZE of the texts and symbols in the graphic, the plotting of the horizontal axis 0 and vertical axis 0, and the frame colours COLOR and the internal graphic colours COLORS.

In IBM SPSS Statistics, the syntax is:

```
CORRESPONDENCE
TABLE = age(1 5) BY department (1 7)
/PRINT = TABLE RPOINTS CPOINTS
/PLOT = NDIM(1,MAX) BIPLOT(20)
```

The CORRESPONDENCE procedure is applicable to numeric variables only, for which the ranges of values are to be specified. However, they are treated as qualitative variables. Alphanumeric variables must therefore be recoded as numeric variables, but may have value labels associated with them:

```
VALUE LABELS AGE 1 '18-24' 2 '25-34' 3 '35-49' 4 '50-64' 5 '65+'.
VALUE LABELS DEPARTMENT 1 'Motoring' 2 'Sport' 3 'Clothes' 4 'Domestic
appliances' 5 'HIFI' 6 'DIY' 7 'Other'.
```

The procedure begins by producing the contingency table of the two variables being analysed. For example, we may find 46 individuals aged 18–24 out of 582, i.e. 7.9%, which is shown in the row point table below as a 'mass' of 0.079.

Correspondence table

Age category	Primary Department							
	Motoring	Sport	Clothes	Domestic appliances	Hi-Fi	DIY	Other	Active margin
18–24	2	6	10	15	8	3	2	46
25–34	8	16	21	31	21	10	20	127
35–49	22	17	47	54	34	21	35	230
50–64	12	17	27	40	17	11	23	147
65 +	4	3	9	6	3	2	5	32
Active margin	48	59	114	146	83	47	85	582

The program then displays the table showing the inertia of each axis, with its proportion in the total inertia and its cumulative proportion. We can see that the first two axes account for 78% of the total inertia. The latter value is equal to the χ^2 divided by the total population, i.e. $0.028 = 16.341/582$. The table also contains the singular value of each axis, which is the square root of its inertia.

Summary

Dimension	Singular value	Inertia	Chi square	Sig.	Proportion of inertia		Confidence Singular Value	
					Accounted for	Cumulative	Standard deviation	Correlation 2
1	.124	.015			.548	.548	.039	−.139
2	.081	.007			.232	.780	.039	
3	.068	.005			.163	.943		
4	.040	.002			.057	1.000		
Total		.028	16.341	.875[a]	1.000	1.000		

a. 24 degrees of freedom

The next two tables show, in the right-hand columns for 'contribution of dimension (1, 2) to inertia of point', the squares of the cosines of the categories with each axis. The 'total' column is the sum of the squares of the cosines on each axis, which is called the 'quality' of the factor representation. We can see that the categories are represented well on the factor plane, except for the 50–64 category of the 'Age' variable, and the Hi-Fi and DIY categories of the 'Department' variable. This must be taken into account in the interpretation of the proximity of the categories on the factor plane.

The columns showing the 'contribution of point to inertia of dimension (1, 2)' contain the proportion of the inertia of each category in the inertia of each axis: this is the contribution of the category to the axis.

The 'Score in dimension (1, 2)' columns contain the coordinates of each category on each axis (which are not calculated in the same way as in the SAS CORRESP procedure, although the other results are identical).

The content of the 'mass' column has already been mentioned.

Finally, the 'Inertia' column contains the contribution of each category to the total inertia, a quantity that is analysed less than the contribution to each axis. This inertia is the sum of contributions to the total χ^2, calculated for all the cells of the contingency table corresponding to the category analysed and divided by the total number of individuals. In the example of the 18–24 category, the addition takes place on the first row of the contingency table. Neither the CORRESPONDENCE procedure nor the IBM SPSS Statistics CROSSTABS procedure yield these contributions to the total χ^2, but all that needs to be done is to calculate the following ratio for each cell (see Appendix A):

$$\frac{(\textit{observed frequency} - \textit{theoretical frequency})^2}{\textit{theoretical frequency}}$$

Overview row points[a]

		Score in Dimension			Contribution				
					Of Point to Inertia of Dimension		Of Dimension to Inertia of Point		
Age	Mass	1	2	Inertia	1	2	1	2	Total
18–24	.079	.898	−.644	.011	.514	.406	.749	.251	1.000
25–34	.218	.237	.355	.004	.099	.340	.347	.506	.852
35–49	.395	−.259	−.102	.005	.214	.051	.651	.066	.717
50–64	.253	.053	.151	.003	.006	.071	.032	.170	.202
65 +	.055	−.615	−.441	.005	.168	.132	.482	.162	.644
Active Total	1.000			.028	1.000	1.000			

a. Symmetrical normalization

Overview column points[a]

		Score in dimension			Contribution				
					Of Point to Inertia of Dimension		Of Dimension to Inertia of Point		
Primary department	Mass	1	2	Inertia	1	2	1	2	Total
Motoring	.082	−.643	−.168	.005	.275	.029	.939	.042	.980
Sport	.101	.525	.276	.005	.225	.096	.636	.115	.751
Clothes	.196	−.163	−.400	.004	.042	.388	.180	.703	.883
Domestic appliances	.251	.291	−.067	.004	.172	.014	.741	.026	.766
Hi-Fi	.143	.235	.010	.004	.063	.000	.270	.000	.270
DIY	.081	−.175	.065	.001	.020	.004	.271	.025	.295
Other	.146	−.415	.509	.006	.203	.469	.502	.492	.995
Active Total	1.000			.028	1.000	1.000			

a. Symmetrical normalization

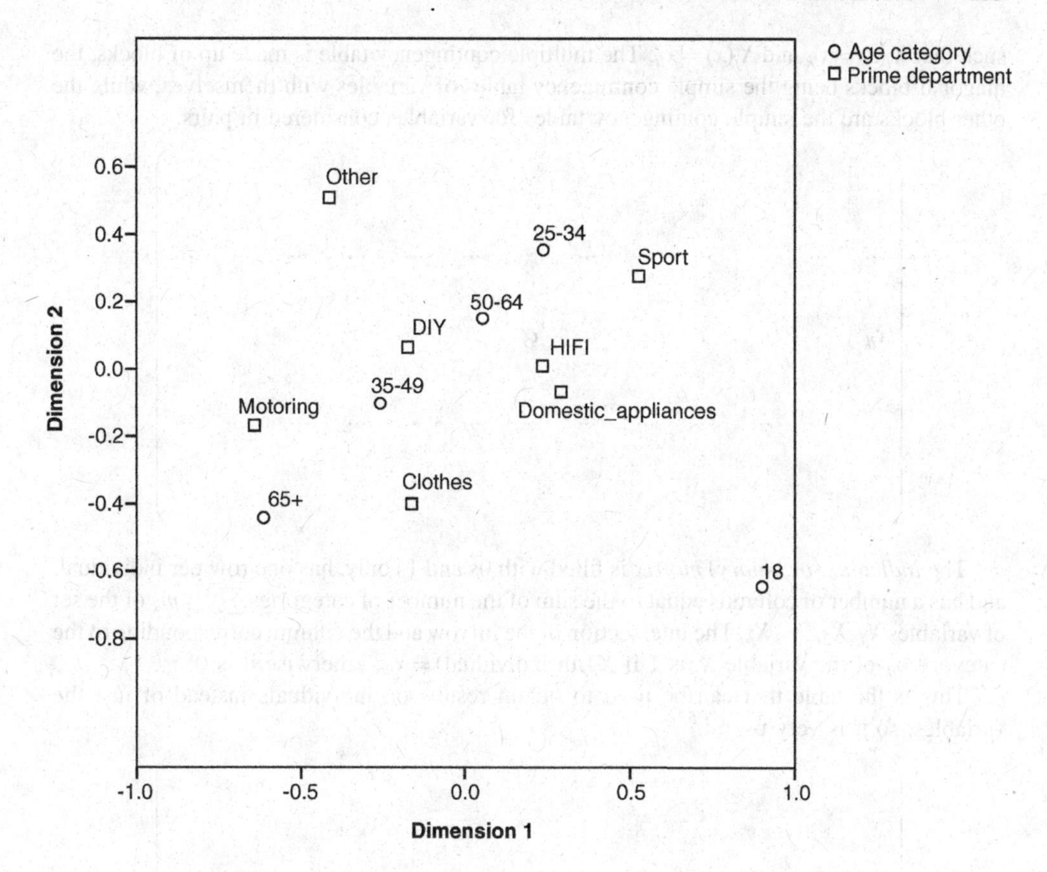

Figure 7.13 Factor plane of a CA (in IBM SPSS Statistics).

The CORRESPONDENCE procedure also plots the factor plane shown in Figure 7.13. The logical proximity of some categories, such as 25–34 and Sport, can be seen here.

7.4 Multiple correspondence analysis

7.4.1 Introduction

Multiple correspondence analysis (MCA) is an extension of correspondence analysis, applied to more than two qualitative variables. In this case, we cannot use the simple contingency table, which must either be generalized or replaced with the *indicator matrix*.

The generalization of the simple contingency table is the *multiple contingency table*, also called the *Burt table*. In this symmetrical square table, instead of having rows showing the categories of a single variable and columns showing the categories of another variable, both the rows and the columns correspond to the categories of the set of variables $X_1, X_2, \ldots, X_p$. At the intersection of two identical categories, we find the number of individuals having this category. At the intersection of two different categories, belonging to different variables, we find the number of individuals having both the first and the second category. At the intersection of the kth category x_{ik} of X_i and the lth category x_{jl} of X_j, we find the number α of individuals x

such that $X_i(x) = x_{ik}$ and $X_j(x) = x_{jl}$. The multiple contingency table is made up of blocks, the diagonal blocks being the simple contingency tables of variables with themselves, while the other blocks are the simple contingency tables for variables considered in pairs.

$$
\begin{pmatrix}
 & & & \vdots & & x_{jl} & & \vdots & & & \\
 & & & \vdots & & . & & \vdots & & & \\
\cdots & \cdots & \cdots & \cdots & \cdots & . & \cdots & \cdots & \cdots & \cdots & \cdots \\
 & & & \vdots & & . & & \vdots & & & \\
x_{ik} & . & . & . & . & \alpha & & \vdots & & & \\
 & & & \vdots & & & & \vdots & & & \\
\cdots & \cdots & \cdots & \cdots & \cdots & \cdots & \cdots & \cdots & \cdots & \cdots & \cdots \\
 & & & \vdots & & & & \vdots & & & \\
 & & & \vdots & & & & \vdots & & &
\end{pmatrix}
$$

The *indicator (or binary) matrix* is filled with 0s and 1s only, has one row per individual, and has a number of columns equal to the sum of the number of categories $\sum_{k=1}^{p} m_k$ of the set of variables $X_1, X_2, \ldots, X_p$. The intersection of the ith row and the column corresponding to the category x_{jk} of the variable X_j is 1 if X_j(*i*th individual) $= x_{jk}$; otherwise it is 0.

This is the table that can be used to obtain results on individuals instead of just the variables, so it is very useful.[5]

$$
\begin{pmatrix}
 & & & \vdots & & & X_j & & & \vdots \\
 & & & \vdots & & & x_{jk} & & & \vdots \\
 & & & \vdots & & & . & & & \vdots \\
 & & & \vdots & & & . & & & \vdots \\
 & & & \vdots & & & . & & & \vdots \\
i & . & . & \vdots & 0 & 0 & 1 & 0 & 0 & \vdots \\
 & & & \vdots & & & & & & \vdots \\
 & & & \vdots & & & & & & \vdots \\
 & & & \vdots & & & & & & \vdots
\end{pmatrix}
$$

The possibility of carrying out an MCA based on a complete indicator matrix arises because a CA provides the same factor axes regardless of whether it is calculated on the contingency table or on the complete indicator matrix. The principles of CA, especially the use of the χ^2 metric, can be applied to the complete indicator matrix. However, the eigenvalues differ considerably depending on the table from which they are calculated (if λ is the eigenvalue of the complete indicator matrix, then λ^2 is the eigenvalue of the Burt table), as

[5] The use of the complete indicator matrix was first reported by Jean-Pierre Nakache in 1970 and is described in Nakache, J.-P. (1973) Influence du codage des données en analyse factorielle des correspondances. Étude d'un exemple pratique medical. *Revue de Statistique Appliquée*, 21(2). Subsequently, Ludovic Lebart described MCA as CA carried out on a complete indicator matrix. Anyone interested in the history of data analysis will enjoy the flamboyant style of Benzécri, J.-P. (1982) *Histoire et Préhistoire de l'Analyse des Données*, new edn. Paris: Dunod.

does the sum of these eigenvalues, which is equal to the total inertia, as in the case of principal component analysis. In CA, this total inertia is equal to the χ^2 divided by the total number of individuals in the population when the contingency table is used; with the complete indicator matrix, it is

$$\frac{m_1+m_2}{2}-1,$$

where m_1 and m_2 are the numbers of categories of the two variables X_1 and X_2. In a MCA on the complete indicator matrix, we find a total inertia of

$$\frac{1}{p}\left(\sum_{i=1}^{p} m_i\right)-1,$$

where p is the number of variables and m_i is the number of categories of the ith variable. As in PCA, and unlike CA, this sum of eigenvalues does not depend on the structure of the data (in CA, it depends on χ^2, i.e. on the association between variables). When it is not related to χ^2, the total inertia does not depend on the association between variables, so it has no particular statistical significance.

The inertia of a category with frequency n_j, in other words its contribution to the total inertia, is

$$\frac{1}{p}\left(1-\frac{n_j}{n}\right),$$

which, incidentally, shows that we must avoid having categories with frequencies that are too low, in order not to unbalance the results.

The inertia of a variable with m_i categories is therefore

$$\sum_{j=1}^{m_i}\frac{1}{p}\left(1-\frac{n_j}{n}\right)=\frac{m_i-1}{p},$$

and, since it depends on its number of categories, we can see that it is preferable to avoid disparities between the numbers of categories of the different variables.

These points should be taken into account in the data preparation phase, as already mentioned in Chapter 3 on data.

Of course, we still have the problem of knowing how many factor axes to use, and how to interpret them, just as in PCA. First of all, we need to know that the number of eigenvalues which are not simply equal to 0 or 1, in other words the number of factor axes, is

$$\sum_{i=1}^{p} m_i-p,$$

and therefore, in view of the value of the sum of eigenvalues shown above for the complete indicator matrix, the mean value of the eigenvalues is $1/p$. In PCA, we would use the Kaiser criterion, but in this case we must retain only the axes whose eigenvalues are greater than $1/p$. A second criterion, as in PCA, is the presence of a bend in the bar chart of eigenvalues. By contrast with PCA, however, the percentage of total inertia explained by the first axes is not necessarily significant; it is often rather low, because of the large number of categories found. The case of an eigenvalue equal to 1, sometimes encountered in CA (see above), is the exception here. The percentage of inertia explained is even smaller when the MCA is carried out on the complete indicator matrix, because of the number of columns created by the indicator coding. A solution to this problem was proposed by J.-P. Nakache and others in 1977.[6] This involves considering the squares or other special functions of the eigenvalues, rather than the eigenvalues themselves (see Sections 1.4.8 and 4.15 of the book by Lebart *et al.* cited earlier in this chapter). As a result of these transformations, the first eigenvalues represent a higher percentage of inertia than the first eigenvalues of the complete indicator matrix. Because of the low percentage of inertia explained and its dependence on the method used, the eigenvalues and the percentages of inertia are rarely important in the interpretation of a MCA. They are pessimistic measurements of the quality of a MCA, and it is incorrect to speak of a proportion of information delivered when dealing with percentages of inertia. In practice, we rarely go beyond the first five axes.

When we have decided which factor axes to keep, how do we interpret them? The best way is to find the categories that make the strongest contribution to each factor axis. This contribution is

$$\frac{1}{\lambda}\frac{n_j}{n \cdot p}(v_j)^2$$

where λ is the eigenvalue of the axis, v_j is the coordinate of the category on this axis, and the other quantities are as described above. We will generally prefer the categories whose contribution is greater than the weight

$$\frac{n_j}{n \cdot p},$$

in other words those whose coordinate v_j is greater than $\sqrt{\lambda}$. An axis is accounted for by categories with strong contributions.

When we look at the representation of a category on an axis, as in CA, we must check the quality of this representation. It is measured by the square of the cosine of the angle of the category with the axis (Table 7.1). This squared cosine is the percentage accounted for by the axis in the scatter of the category, and the representation of the category improves as this value approaches 1. The proximity of two categories on an axis should not be evaluated unless they both have a reasonably large squared cosine on this axis. If two categories have high coordinates on the same axis (meaning that they are distant from the centre), but one has a higher squared cosine than the other, then both are different from the mean profile (represented by the centre) but the difference from the mean is accounted for by this axis, more for one than for the other, and not by other characteristics. Note that the sum of the

[6] Nakache, J.-P., Lorente, P., Benzécri, J.-P. and Chastang C. (1977) Aspects pronostiques et thérapeutiques de l'infarctus myocardique compliqué. *Cahiers de l'Analyse des Données*, 2(4), 415–434.

Table 7.1 Squared cosines at the end of a MCA.

Squared cosines for the column points

	Dim1	Dim2	Dim3	Dim4
Female	0.4328	0.0983	0.0452	0.0328
Male	0.4328	0.0983	0.0452	0.0328
18–24	0.0001	0.2284	0.0003	0.0935
25–34	0.0019	0.1601	0.1011	0.0030
35–49	0.0580	0.0205	0.1706	0.0008
50–64	0.0111	0.1569	0.0252	0.1008
65 +	0.1472	0.0523	0.0002	0.1617
1 per month	0.0564	0.0111	0.3032	0.0200
1 per week	0.0226	0.1777	0.0008	0.2069
1st time	0.0002	0.2353	0.0153	0.0034
$<$ 1 per month	0.0064	0.0162	0.2210	0.0002
$>$ 1 per week	0.1398	0.0732	0.0184	0.2040
Motoring	0.0532	0.0328	0.0021	0.0332
Other	0.0092	0.0374	0.0457	0.0634
DIY	0.2397	0.0038	0.0238	0.0031
Domestic_appl	0.1378	0.0769	0.0003	0.1725
Hi-Fi	0.0115	0.0659	0.1047	0.0001
Clothes	0.0570	0.0336	0.0429	0.1990
Sport	0.0334	0.0001	0.3639	0.0315

squared cosine values over all the axes is 1. As for the supplementary variables, their squared cosine can also be analysed, even if they do not contribute to the axes.

In the example of Table 7.1, the '18–24' and '1st time' categories are very poorly represented on the first axis, and therefore we should not place any reliance on their proximity on the first axis (moreover, they are near the origin). However, their proximity on the second axis is much more significant. Figure 7.14 superimposes the categories of the supplementary variable 'satisfaction' (represented by 'o') on those of the active variables (represented by 'x'). All these variables relate to visits to a department store by 582 customers, and include the age, sex, departments visited, frequency of visits and satisfaction of each customer. They are mentioned in Section 7.3.2 and will be analysed in greater detail in Section 7.4.3 to illustrate the implementation of MCA with the SAS software, which provided the initial results above.

7.4.2 Review of CA and MCA

CA and MCA provide many benefits:

- The factors are the numeric variables that provide the best separation of the categories of the qualitative variables under examination.

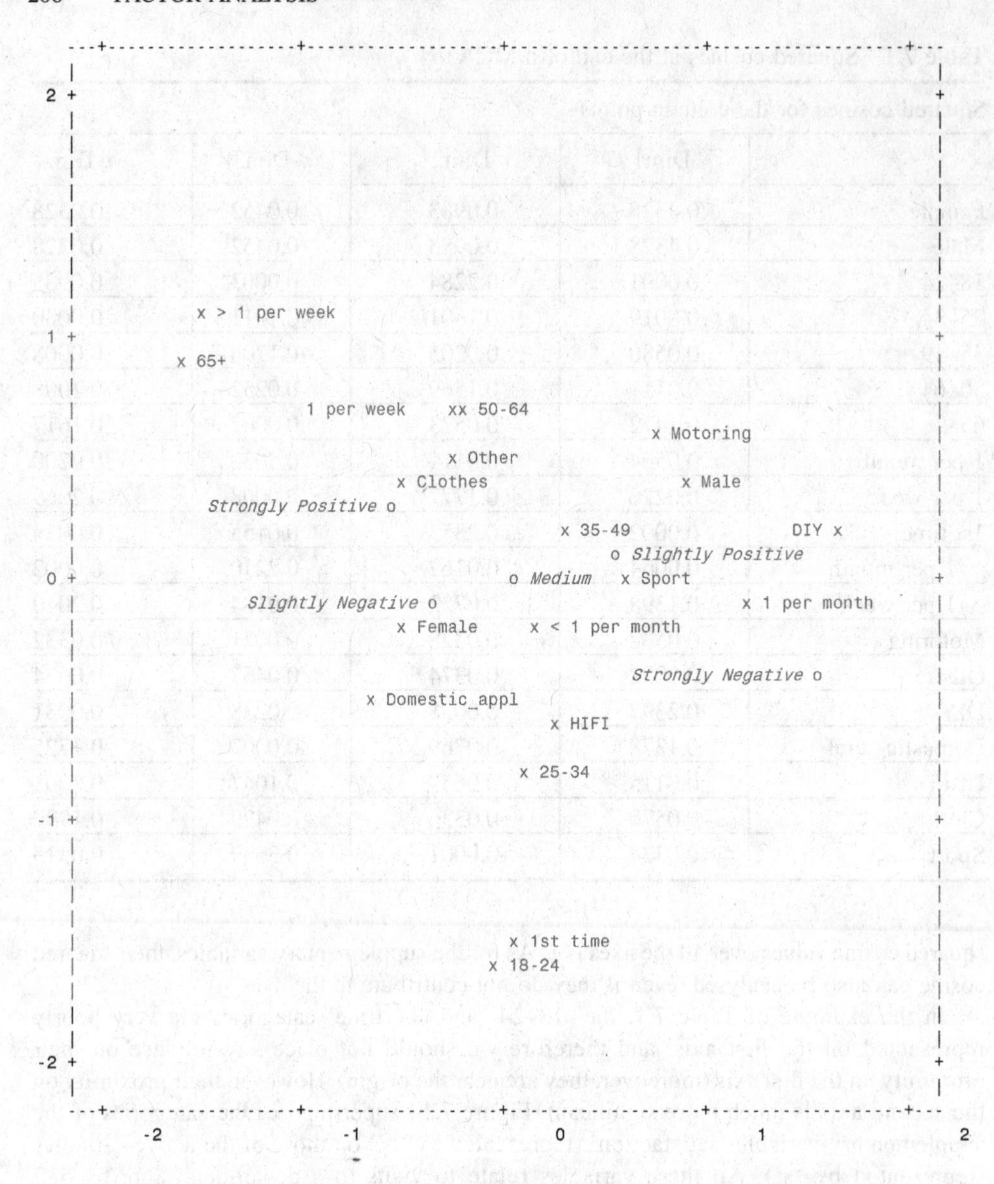

Figure 7.14 Multiple correspondence analysis on commercial data.

- CA and MCA can be used to transform qualitative (initial) variables into quantitative variables (the projections of the categories of the initial variables on the factor axes).
- They can be used to find non-linear relationships (of degree greater than 1) between the previously discretized continuous variables, and, on the other hand, to detect dependences between variables whose linear correlation coefficient is close to 0 (relationships of degree greater than 1; see Section A.2.7).

- They can be used to represent individuals and categories simultaneously on the same plane (using squared cosines to ensure the quality of the projection).
- They can be used to filter random fluctuations of data by replacing the original variables with the first factor axes, which is useful before clustering or neural network processing.
- Some software enable supplementary variables to be displayed (see above), as for PCA, without taking them into account in the calculation of correspondences.

In the graphic representation of CA and MCA:

- two individuals are close together if they have almost the same categories;
- two categories of two different variables are close together if they are possessed by almost the same individuals (a high frequency in the contingency table); more particularly, they merge together if they are possessed by exactly the same individuals;
- two categories of the same variable are close together if the two groups of individuals which possess them resemble each other in respect of the other variables.

Also, the distance of a category from the centre increases as its frequency decreases, since the square of distance to the centre, d^2, is inversely proportional to the frequency. We have $d^2 = (n/n_j) - 1$. Such categories can be enough to determine the first factor axes almost exclusively, completely hiding any interesting general phenomena behind specific phenomena which only relate to a few individuals. This is why we should avoid having categories with excessively low frequencies.

To summarize the benefits of this powerful method, we can say that examining a factorial plane is both more efficient than examining all the planes (x, y) of the original variables, and much faster than scanning all the contingency tables.

7.4.3 Implementing MCA and CA with SAS

IBM SPSS Statistics, with its HOMALS procedure, is not really suitable for MCA as practised by its founders, so I will concentrate on the SAS CORRESP procedure which is dedicated to factor analysis. The CORRESP procedure is quite comprehensive and can be used to process both types of data encountered in this form of analysis.

In the first type, which is more common, the data set contains 'raw' data: each row of the data set corresponds to one individual, each column corresponds to a qualitative variable characterizing the individuals, and each row/column intersection contains the category of the variable for the individual.

In the second type, the data in the data set are already in the form of a contingency table, Burt table, or indicator matrix.

I will illustrate this with a small example. When the data are arranged in the data set as in the following table, they are of the first type:

Name	Sex	Status
Mr Smith	Male	Married
Mrs Smith	Female	Married
Mr Brown	Male	Unmarried
Mr Wood	Male	Married
Mrs Wood	Female	Married
Mrs Black	Female	Widowed

The contingency table for these data is:

	Married	Unmarried	Widowed
Female	2	0	1
Male	2	1	0

The Burt table is:

	Female	Male	Married	Unmarried	Widowed
Female	3	0	2	0	1
Male	0	3	2	1	0
Married	2	2	4	0	0
Unmarried	0	1	0	1	0
Widowed	1	0	0	0	1

The indicator matrix is:

	Female	Male	Married	Unmarried	Widowed
1	0	1	1	0	0
2	1	0	1	0	0
3	0	1	0	1	0
4	0	1	1	0	0
5	1	0	1	0	0
6	1	0	0	0	1

These three tables illustrate the second type of data processed by the CORRESP procedure.

With the first type of data, CORRESP must be used with the instruction TABLES, which creates a contingency table, a Burt table (multiple contingency table) or an indicator matrix, depending on the options, and then carries out the CA or MCA. The SAS syntax is in the following form:

```
PROC CORRESP DATA=survey MCA DIMENS=4 OUT=output ALL;
TABLES sex age frequency department satisfaction;
SUPPLEMENTARY satisfaction;
RUN;
```

The variables following the TABLES instruction can be numeric or alphanumeric, but in any case they are considered to be qualitative.

The MCA option creates a Burt table (use the BINARY option to obtain an indicator matrix) of the variables written after the TABLES instruction. In CA, when there are only two variables to examine, neither the MCA nor the BINARY option is used, and the contingency table is created by writing the two variables, separated by a comma, the variable on the left of the comma being the row variable and the variable on the right being the column variable:

```
TABLES sex , status;
```

We can also analyse the simple contingency table of the variables $X_1 \ldots X_M$ (in the rows) by $Y_1 \ldots Y_N$ (in the columns) by writing the following syntax:

```
TABLES X1 ... XM , Y1 ... YN;
```

In particular, the instruction

```
TABLES X1 ... XM;
```

with the MCA option is equivalent to the instruction

```
TABLES X1 ... XM , X1 ... XM;
```

without the MCA option.

The SUPPLEMENTARY row contains the supplementary variables, which must also be entered in the TABLES row. The desired number of factor axes is specified by the DIMENS command (default value 2). The OUT data set contains the inertia, coordinates on the axes, contributions to the axes and squared cosines for the rows (_type_ = 'OBS') and columns (_type_ = 'VAR'). The coordinates of these can be used to draw the factor plane as in Figure 7.14 of Section 7.4.1. The SAS macro %PLOTIT (see Section 7.3.2 on CA) is commonly used for this purpose. The DATA data set of the %PLOTIT macro must be the OUT data set at the output of the CORRESP procedure, and the parameter PLOTVARS specifies the axes to be represented:

```
%PLOTIT(DATA=output, DATATYPE=corresp, PLOTVARS=Dim2 Dim1)
```

The 'VAR' type entries are always present and correspond to the categories following the instruction TABLES (and any which may follow SUPPLEMENTARY), except for the categories to the left of the comma, if any. On the other hand, the 'OBS' type entries are not always present. With the MCA option, these entries do no exist, because the Burt table is symmetrical. With the BINARY option, the 'OBS' entries correspond to the individuals analysed. In the case of CA, these entries correspond to the categories of the row variable (to the left of the comma). If we wish to find results on the individuals themselves, not just on their variables, we must use the BINARY option. This enables us to represent the individuals and the categories on the same plane.

One drawback of the CORRESP procedure, not found in PRINCOMP which is used for PCA, is that the output data set OUT contains the inertia, the coordinates on the axes, the contributions and the squared cosines for each observation, but not the variables of the input data set DATA, nor even the identifier of this data set which would enable the data set DATA to be matched with the OUT data set to enrich it. If we wish to enrich the input data set with the coordinates of the individuals on the factor axes, in order to transform qualitative to quantitative data before clustering or discriminant analysis for example (see Section 12.9), we can start by executing an MCA on the Burt table, as indicated in the SAS syntax example above, with the option MCA. The advantage of MCA on the Burt table is that the calculations are faster than for the indicator matrix. It is then simply necessary to recover the coordinates of the categories of the variables on the axes, then calculate the sum of the coordinates of the categories of each individual, divided by the number of variables and by the square root of the eigenvalue of the axis. The result of this calculation is the coordinate of the individual on the axis, found without using the indicator matrix! This is based on the fact that, as seen in Section 7.3.1, the coordinate of an individual on an axis is the mean of the coordinates of the individual's categories on this axis, divided by the square root of the eigenvalue of the axis (meaning that an axis is a linear combination of the indicators of the categories of the analysed variables, a property which is used in DISQUAL discriminant analysis as described in Section 11.6.7).

Figure 7.15 illustrates this result with the example of customers' visits to a department store which we have looked at before. Columns B to E on the left of the spreadsheet contain the categories of the variables of each customer (the customers are numbered in column A). The coordinates of these categories on the first two factor axes are shown in columns L and M, obtained from the OUT data set which is the output of the CORRESP procedure. These are the coordinates reproduced in Figure 7.17. Column F contains the mean of the coordinates on the first axis of the categories of sex, age, department and frequency of visits, this mean being divided by the square root of the eigenvalue of the axis (in cell N2). This eigenvalue is shown in Figure 7.16. The same applies to column G and the second factor axis. Thus we find that the first customer is male (coordinate 0.87891 on the first axis), aged between 35 and 49 years (coordinate 0.29799), visits the DIY department (coordinate 1.65167), and comes to the store less than once per month (coordinate 0.13411). The mean of these four coordinates is 2.96268/4 = 0.74067. After dividing by the square root of 0.31833, we have 1.312761918, as shown in cell F2. This is the coordinate of the first customer on the first axis of the factor plane of individuals.

If we wish to use the BINARY option but without representing the individuals, we specify this when calling the %PLOTIT macro:

```
%PLOTIT(DATA=output (WHERE = (_type_ ne 'OBS')), DATATYPE=corresp,
PLOTVARS=Dim2 Dim1)
```

If we are interested in the results for the categories of the variables (_type_ = 'VAR' in the OUT data set), we can find them with either the MCA or the BINARY option.

For the second type of data processed by CORRESP, we have seen that the data in the data set are already in the form of a contingency table,[7] a Burt table or an indicator matrix: each line of the data set corresponds to one row of the table and each column corresponds to a category specified in the VAR instruction, and this category has to be numeric. Each intersection of a row and column contains the frequency of a cell of the table, this intersection being a numeric data element greater than or equal to 0 (if it is less than 0 or missing, the line is not included in the analysis). The Burt table is symmetrical and its rows do not have to be specified, because

[7] The TRANSREG procedure can be used to construct an indicator matrix and store in a macro variable the list of indicators created.

F2 =MOYENNE(RECHERCHEV(B2;$J$3:$M$21;3;FAUX);RECHERCHEV(C2;$J$3:$M$21;3;FAUX);RECHERCHEV(D2;$J$3:$M$21;3;FAUX);RECHERCHEV(E2;$J$3:$M$21;3;FAUX))/RACINE($N$2)

obs	sex	age	department	frequency	Dim1	Dim2
1	Male	35-49	DIY	< 1 per month	1,312761918	0,263258425
2	Male	18-24	Other	1 per month	0,418028896	-0,397102461
3	Female	35-49	Motoring	< 1 per month	0,314228415	0,148569378
4	Male	35-49	Clothes	< 1 per month	0,366571755	0,336385518
5	Female	50-64	Clothes	1 per month	-0,367998535	0,300056167
6	Male	35-49	Other	1 per month	0,563897239	0,409521532
7	Female	50-64	Clothes	1 per week	-0,630109592	0,695891741
8	Male	35-49	Motoring	1 per week	0,745275573	0,866265866
9	Female	50-64	Other	1 per week	-0,51828458	0,738765192
10	Male	35-49	DIY	< 1 per month	1,312761918	0,263258425
11	Male	35-49	Motoring	1st time	0,882273065	-0,155581871
12	Male	50-64	DIY	1 per month	1,18584937	0,518527427
13	Female	50-64	Other	< 1 per month	-0,341673996	0,312667055
14	Male	35-49	Other	1 per month	0,563897239	0,409521532
15	Female	35-49	Domestic_appl	< 1 per month	-0,311007082	-0,334565481
16	Male	35-49	Clothes	1 per week	0,189961171	0,762483654
17	Female	35-49	Other	> 1 per week	-0,853724167	0,66737784
18	Female	25-34	Domestic_appl	1 per month	-0,321034423	-0,721042143
19	Female	50-64	Domestic_appl	< 1 per month	-0,523420103	-0,109559043
20	Female	25-34	Other	1 per week	-0,401399373	0,097019529
21	Male	35-49	Domestic_appl	1 per month	0,382151132	-0,012704567
22	Female	25-34	HIFI	< 1 per month	-0,005817896	-0,818400698
23	Female	65+	HIFI	1 per month	-0,661662705	-0,027269317
24	Female	65+	Clothes	1 per month	-0,992458609	0,419179322
25	Female	50-64	Clothes	1 per month	-0,367998535	0,300056167
26	Female	50-64	Domestic_appl	1 per month	-0,437919631	-0,07929648
27	Female	50-64	Domestic_appl	1 per month	-0,437919631	-0,07929648
28	Female	35-49	Motoring	1 per month	0,399728888	0,178831941
29	Male	50-64	Domestic_appl	1 per week	-0,092372945	0,608137445

TYPE	_NAME_	Inertia	Dim1	Dim2	Contr1	Contr2
INERTIA		3,75	.	.	0,31833	0,31408
VAR	18-24	0,0614	-0,0312	-1,6313	0,00006	0,16741
VAR	25-34	0,05212	0,0824	-0,7573	0,00116	0,0996
VAR	35-49	0,04032	0,29799	0,17696	0,02756	0,00985
VAR	50-64	0,04983	-0,1814	0,68136	0,00653	0,09333
VAR	65+	0,063	-1,5907	0,9484	0,10926	0,03937
VAR	Female	0,02394	-0,4925	-0,2347	0,12207	0,02811
VAR	Male	0,04273	0,87891	0,41894	0,21786	0,05017
VAR	Motoring	0,06117	0,76953	0,60391	0,03836	0,02394
VAR	Other	0,05693	-0,2314	0,46737	0,00614	0,02539
VAR	DIY	0,06128	1,65167	0,20733	0,17302	0,00276
VAR	Domestic_appl	0,04994	-0,6415	-0,4791	0,08108	0,04584
VAR	HIFI	0,05716	0,26283	-0,6296	0,00774	0,04499
VAR	Clothes	0,05361	-0,4837	0,37126	0,03599	0,02149
VAR	Sport	0,05991	0,54389	0,02415	0,02355	0,00005
VAR	1 per month	0,04364	0,32707	-0,1452	0,02901	0,0058
VAR	1 per week	0,0504	-0,2645	0,74211	0,0134	0,10696
VAR	1st time	0,06071	0,04471	-1,5486	0,00014	0,17055
VAR	< 1 per month	0,04914	0,13411	-0,2131	0,00371	0,0095
VAR	> 1 per week	0,06277	-1,5009	1,08648	0,10335	0,05489

Figure 7.15 Coordinates of categories and individuals on the factor axes.

Inertia and Chi-Square Decomposition					
Singular Value	Principal Inertia	Chi-Square	Percent	Cumulative Percent	2 4 6 8 10 ----+----+----+----+----+---
0.56420	0.31833	759.10	8.49	8.49	*********************
0.56043	0.31408	748.97	8.38	16.86	*********************
0.53721	0.28860	688.21	7.70	24.56	*******************
0.52911	0.27996	667.61	7.47	32.03	*******************
0.52401	0.27459	654.80	7.32	39.35	******************
0.51099	0.26111	622.66	6.96	46.31	*****************
0.50151	0.25151	599.77	6.71	53.02	*****************
0.49404	0.24408	582.04	6.51	59.53	****************
0.49052	0.24061	573.77	6.42	65.94	****************
0.48516	0.23538	561.31	6.28	72.22	****************
0.47999	0.23039	549.40	6.14	78.36	***************
0.47222	0.22299	531.75	5.95	84.31	***************
0.46499	0.21622	515.61	5.77	90.08	**************
0.44178	0.19517	465.41	5.20	95.28	*************
0.42071	0.17700	422.09	4.72	100.00	************
Total	3.75000	8942.50	100.00		
Degrees of freedom = 324					

Figure 7.16 Eigenvalues of the MCA.

they are equal to its columns. On the other hand, the rows of a contingency table or indicator matrix must be specified: this is done with the ID instruction.

CORRESP has to be used with the VAR instruction which directly executes the CA or MCA. To tell SAS that the input is a Burt table, we use the MCA option below, indicating the number of variables with the NVARS command. This is because SAS does not know this number of variables, because it is the categories of these variables, rather than the variables themselves, that appear after the VAR instructions (care must be taken to list them correctly in the order of the data set).

```
PROC CORRESP DATA=table MCA NVARS=2 OUT=output ALL;
VAR male female married unmarried widowed;
RUN;
```

If the input is an indicator matrix or a simple contingency table, the syntax is simplified by removing the MCA and NVARS instructions, as the distinction between the two types of table (indicator matrix or contingency table) is shown by the ID instruction, followed by the name of the row variable for a contingency table, or by a variable identifying the individuals for an indicator matrix.

Column Coordinates				
	Dim1	Dim2	Dim3	Dim4
Female	-0.4925	-0.2347	-0.1592	-0.1355
Male	0.8789	0.4189	0.2842	0.2418
18-24	-0.0312	-1.6313	0.0627	1.0440
25-34	0.0824	-0.7573	0.6018	-0.1037
35-49	0.2980	0.1770	-0.5109	-0.0344
50-64	-0.1814	0.6814	0.2733	-0.5462
65+	-1.5907	0.9484	-0.0621	1.6670
1 per month	0.3271	-0.1452	-0.7581	0.1949
1 par week	-0.2645	0.7421	-0.0498	-0.8008
1st time	0.0447	-1.5486	0.3944	0.1858
< 1 per month	0.1341	-0.2131	0.7872	0.0211
> 1 per week	-1.5009	1.0865	0.5446	1.8132
Motoring	0.7695	0.6039	-0.1511	0.6080
Other	-0.2313	0.4674	0.5169	-0.6090
DIY	1.6517	0.2073	-0.5205	-0.1879
Domestic_appl	-0.6415	-0.4791	-0.0306	-0.7177
HIFI	0.2628	-0.6295	-0.7936	0.0237
Clothes	-0.4837	0.3713	-0.4197	0.9038
Sport	0.5439	0.0242	1.7960	0.5286

Figure 7.17 Coordinates of the categories on the factor axes.

In the case of the Burt table, the OUT data set only contains results (coordinates, contributions, etc.) for the categories following the VAR instruction, with _type_ = 'VAR' entries in OUT. In other cases, the OUT data set also contains results on the lines corresponding to the ID instruction, with _type_ = 'OBS' entries in OUT: thus there are results for the categories of the row variable (contingency table) and results for individuals (indicator matrix). These _type_ = 'OBS' entries in the OUT data set contain the variable ID. As before, it is the indicator matrix that is used to obtain results on individuals and represent them on the same plane as the categories. As before, the results on the categories of the variables are the same with the indicator matrix and the Burt table. We can also add a SUPPLEMENTARY line containing supplementary variables, which must also appear on the VAR line.

A contingency table is indicated by a command of the following type:

```
PROC CORRESP DATA=table OUT=output ALL;
VAR married unmarried widowed;
ID sex;
RUN;
```

Summary Statistics for the Column Points			
	Quality	Mass	Inertia
Female	0.6092	0.1602	0.0239
Male	0.6092	0.0898	0.0427
18-24	0.3223	0.0198	0.0614
25-34	0.2660	0.0546	0.0521
35-49	0.2498	0.0988	0.0403
50-64	0.2941	0.0631	0.0498
65+	0.3615	0.0137	0.0630
1 per month	0.3908	0.0863	0.0436
1 par week	0.4080	0.0610	0.0504
1st time	0.2541	0.0223	0.0607
< 1 per month	0.2438	0.0657	0.0491
> 1 per week	0.4354	0.0146	0.0628
Motoring	0.1213	0.0206	0.0612
Other	0.1556	0.0365	0.0569
DIY	0.2703	0.0202	0.0613
Domestic_appl	0.3875	0.0627	0.0499
HIFI	0.1823	0.0357	0.0572
Clothes	0.3325	0.0490	0.0536
Sport	0.4289	0.0253	0.0599

Figure 7.18 Quality of representation on the first four factor axes.

An indicator matrix will be indicated by a command of the following type:

```
PROC CORRESP DATA=table OUT=output NOROW=print;
VAR male female married unmarried widowed;
ID label;
RUN;
```

Here, the NOROW = print command is used to prevent the display of the factor coordinates, contributions and squared cosines of the observations (of which there may be several thousand or more) while retaining the corresponding display for the variables.

For the complete indicator matrix, the variable *label* specified in the ID *label* instruction is the identifier of the individual. There is one line per individual in the processed data set, each one marked by the variable *label* in the OUT data set, and the variables of the VAR line are dichotomous variables (0/1).

With the MISSING option, the missing values of the qualitative variables are treated as if they formed a specific category and the observations are not excluded.

Partial Contributions to Inertia for the Column Points				
	Dim1	Dim2	Dim3	Dim4
Female	0.1221	0.0281	0.0141	0.0105
Male	0.2179	0.0502	0.0251	0.0188
18-24	0.0001	0.1674	0.0003	0.0769
25-34	0.0012	0.0996	0.0685	0.0021
35-49	0.0276	0.0099	0.0894	0.0004
50-64	0.0065	0.0933	0.0163	0.0673
65+	0.1093	0.0394	0.0002	0.1364
1 per month	0.0290	0.0058	0.1720	0.0117
1 par week	0.0134	0.1070	0.0005	0.1397
1st time	0.0001	0.1705	0.0120	0.0028
< 1 per month	0.0037	0.0095	0.1411	0.0001
> 1 per week	0.1034	0.0549	0.0150	0.1715
Motoring	0.0384	0.0239	0.0016	0.0272
Other	0.0061	0.0254	0.0338	0.0484
DIY	0.1730	0.0028	0.0190	0.0025
Domestic_appl	0.0811	0.0458	0.0002	0.1154
HIFI	0.0077	0.0450	0.0778	0.0001
Clothes	0.0360	0.0215	0.0299	0.1429
Sport	0.0236	0.0000	0.2833	0.0253

Figure 7.19 Contributions of the categories to the factor axes.

The WEIGHT *variable* command is used to assign weights to the observations, telling SAS that the weight of each observation is different from 1 and is contained in the variable *variable.* An observation with a non-positive weight is a supplementary observation that is delivered but not taken into account in the calculations. This option cannot be used when the input to the procedure is a Burt table (VAR and MCA options), because this table is symmetrical and the supplementary variables are indicated by SUPPLEMENTARY.

Returning to the example in the previous section (visits to departments of a store), we obtain the following tables at the output of PROC CORRESP. They are the same regardless of whether the table constructed was a Burt table (MCA option) or the indicator matrix (BINARY option).

The table of eigenvalues is shown in Figure 7.16, with the values of the indicator matrix in the second column and their square roots in the first column. The eigenvalues of the Burt table, which are the squares of the second column, are not shown. According to the formulae in the previous section, the sum of the eigenvalues is 3.75, i.e.

$$\frac{\text{total number of categories}}{\text{number of variables}} - 1 = \frac{19}{4} - 1.$$

The usual rule for selecting the axes would be to keep all those whose eigenvalues exceeded 0.25 (the inverse of the number of variables). In this case, this would result in the selection of the first seven axes, but I have chosen a smaller number, four, for this description. These first four eigenvalues represent 32% of the inertia, while we can see that the first four eigenvalues of the Burt table represent 38% of the inertia, illustrating the usefulness of the proposal of J.P. Nakache mentioned above.

Figure 7.17 shows the coordinates of the categories on the factor axes, which are very useful, not only for graphic representation, but also whenever we need to transform qualitative variables into continuous variables.

Then we have the squares of the cosines which measure the quality of the representation of a category on a factor axis. The table produced by CORRESP was shown as Table 7.1 in Section 7.4.1. It will be recalled that the representation of the category on the axis improves as the square of the cosine approaches 1. Additionally, the sum of squares of the cosines for all the axes is 1 for each category. As in this case, I have not kept all the axes, but only four of them, and the sum of the first four squares of the cosines of a variable appears in Figure 7.18 in the 'quality' column. The categories having low qualities are those which are poorly represented on the four chosen factor axes, and which are better represented on axes which have not been retained.

Figure 7.19 shows the contribution of the categories to the factor axes. It is a kind of inverse of the squared cosines which measures the explanation of the category per axis. This contribution is the proportion of the inertia of the category in the inertia of the axis, and the sum of contributions of all the categories for a given axis is, of course, 1. An axis is accounted for by its categories having high contributions.

8

Neural networks

Data mining would be not be the same without neural networks, which lie at the root of certain descriptive and predictive methods of data mining. These networks have become widely used, owing to their modelling power (they can approximate any sufficiently regular function), with excellent results across a broad range of problems, even when faced with complex phenomena, irregular forms, and data that are difficult to grasp and follow no particular probability law. In some cases, however, their use is impeded by certain difficulties in implementation, such as the 'black box' nature of the networks, the delicacy of the necessary adjustments, the amount of computing power required, and especially the risks of overfitting and convergence to a globally non-optimal solution.

This chapter has been placed before the chapters on clustering, classification and prediction methods, because neural networks are used both for clustering (Kohonen networks) and classification and prediction (perceptrons, radial basis function networks). Any reader not interested in the details of these methods may skip this chapter.

8.1 General information on neural networks

Following the initial description of a *formal neuron* by McCulloch and Pitts in 1943, the first *neural networks* appeared in 1958 with the 'perceptron' of Rosenblatt. They were developed rapidly in the 1980s and have been used widely in industry since the 1990s. A neural network has an architecture based on that of the brain, organized in neurons and synapses, and takes the form of a set of interconnected *units* (or *formal neurons*), with each continuous input variable corresponding to a unit at a first level, called the *input layer*, and each category of a qualitative variable also corresponding to a unit of the input layer. In some cases, when the network is used in a predictive technique, there may be one or more dependent variables: in this case each of them corresponds to one unit (or several units in the case of qualitative variables – see below) at a final level, called the *output layer*. Predictive networks are called 'supervised learning' networks, and descriptive networks are called 'unsupervised learning' networks.

Data Mining and Statistics for Decision Making, First Edition. Stéphane Tufféry.

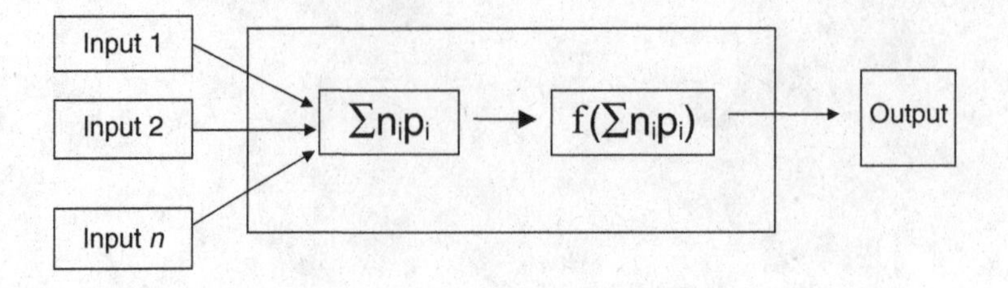

Figure 8.1 Unit of a neural network.

Units belonging to an intermediate level, the *hidden layer*, are sometimes connected between the input layer and the output layer. There may be several hidden layers.

A unit receives values at its input and returns 0 to n values at the output. All these values are normalized so that they lie between 0 and 1 (or sometimes between −1 and 1, depending on the limits of the transfer function described below). A *combination function* calculates a first value from the units connected at the input and the weight of the connections. Thus, in the most widely used networks, this is the weighted sum $\sum_i n_i p_i$ of the input values n_i of the units. To determine an output value, a second function, called the *transfer function* (or *activation function*), is applied to this value. The units in the input layer are simple, in the sense that they do not create any combinations but only transmit the values of the variables corresponding to them.

Thus a perceptron unit takes the form shown in Figure 8.1. The notation used in this diagram is as follows:

- n_i is the value of unit i at the preceding level (the summation over i corresponds to all the units at the preceding level connected to the unit being observed);
- p_i is the weight associated with the connection between unit i and the observed unit;
- f is the transfer function associated with the observed unit.

The learning of the neural network takes place on the basis of a sample of the population under study; it uses the individuals in the sample to adjust the weights of the connections between the units. In the course of learning, the value delivered by the output unit is compared with the actual value, and the weights p_i of all the units are adjusted so as to improve the prediction, by a mechanism which depends on the type of neural network. One mechanism which is still widely used is 'gradient back-propagation', but there are more recent and more effective ones such as the Levenberg–Marquardt, quasi-Newton, conjugate gradient, quick propagation, and genetic algorithms (see Section 11.13). The network runs through the learning sample many (often several thousand) times. Learning is completed when an optimal solution[1] has been found and the weights p_i are no longer modified significantly, or when a previously specified number of iterations have been run. At the end of the learning phase, the network forms a function which associates the variables with each other. In the perceptron, the transfer function may be the linear function $f(x) = x$, but it is best to choose a function that behaves linearly in the neighbourhood of 0 (when the weights of the units are small) and non-linearly at the limits, so that both linear and non-linear phenomena can be modelled. In almost

[1] This optimum may be global or only local: see Section 8.7.1 on the multilayer perceptron.

all cases, a *sigmoid* function is chosen, more specifically the logistic sigmoid $s(x) = 1/(1+e^{-x})$ (which we will meet again in the section on logistic regression), or the tangent sigmoid, which is simply the hyperbolic tangent function:

$$\tanh(x) = \frac{e^x - e^{-x}}{e^x + e^{-x}}$$

There are other sigmoid functions, but the performance is very similar regardless of which is chosen. Although these functions are constantly increasing, they can approach any continuous function when they are combined with each other (see Section 8.7.1), in other words when the activation of a several units is combined.

The capacity to handle non-linear relations between the variables is a major benefit of neural networks.

The necessity of normalizing the values of the input data can be seen with the logistic function (Figure 8.2). If this were not done, the data with large values would 'crush' the others, and the adjustments of the weights would have no effect on the value $1/(1+\exp(-\sum n_i p_i))$, as this value does not vary greatly around 0 or 1 when the absolute value of $\sum n_i p_i$ is large. Also, the fact that all the values lie between 0 and 1 (or −1 and 1) means that a unit can receive the output of a preceding unit at its input without encountering problems due to excessively large values.

As a general rule, the stages in the implementation of a neural network for prediction or classification are:

(i) identification of the input and output data;

(ii) normalization of these data;

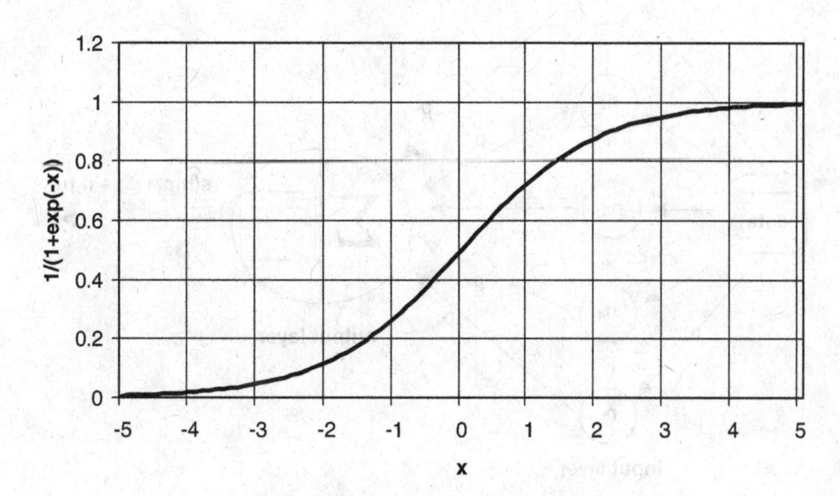

Figure 8.2 The logistic function.

(iii) establishment of a network with a suitable structure;

(iv) learning;

(v) testing;

(vi) application of the model generated by learning;

(vii) denormalization of the output data.

8.2 Structure of a neural network

The structure of a neural network, also referred to as its 'architecture' or 'topology', consists of the number of layers and units, the way in which the different units are interconnected (the choice of combination and transfer functions) and the weight adjustment mechanism. The choice of this structure will largely determine the results that will be obtained, and is the most critical part of the implementation of a neural network.

The simplest structure is one in which the units are distributed in two layers: an input layer and an output layer. Each unit in the input layer has a single input and a single output which is equal to the input (see Figure 8.3). The output unit has all the units of the input layer connected to its input, with a combination function and a transfer function. There may be more than one output unit. In this case, the resulting model is a linear or logistic regression, depending on whether the transfer function is linear or logistic, and the weights of the network are the regression coefficients.

The predictive power can be increased by adding one or more hidden layers between the input and output layers (Figure 8.4). Although the predictive power increases with the number

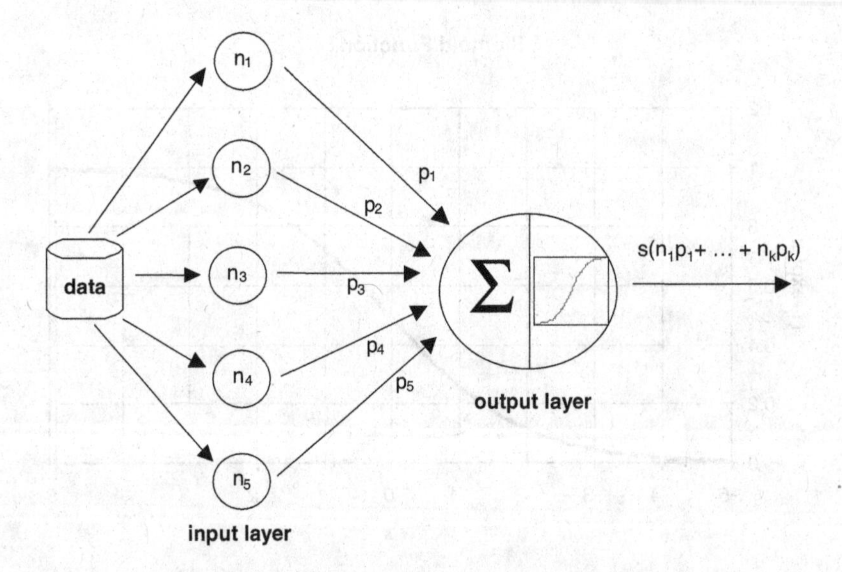

Figure 8.3 Neural network with no hidden layer.

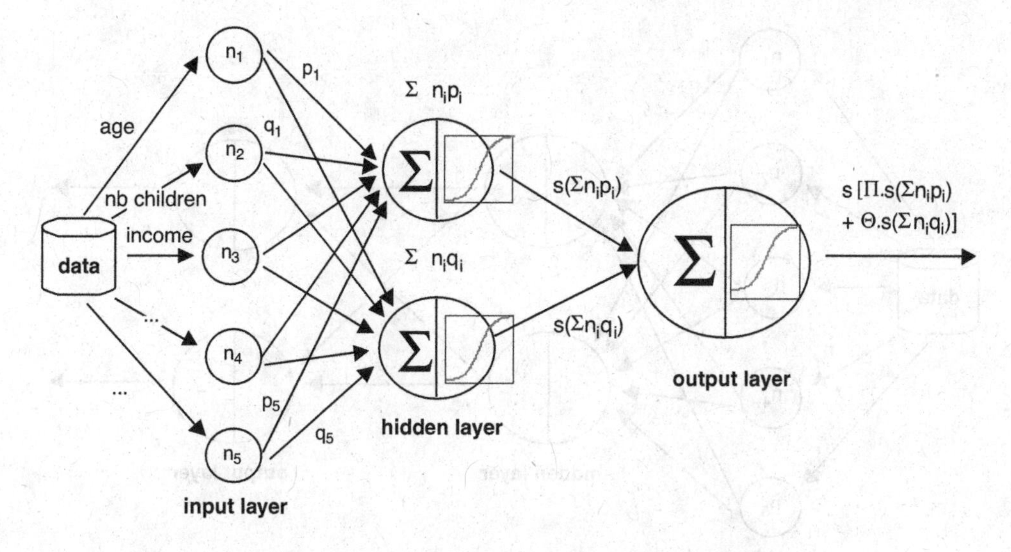

Figure 8.4 Neural network with a hidden layer.

of hidden layers and units in these layers, this number must nevertheless be as small as possible, to ensure that the neural network does not simply store all the information from the learning set but can generalize it, thus avoiding what is known as 'overfitting' (see Section 11.3.4), which occurs when the weights simply make the system learn the details of the learning set, instead of discovering general structures. This happens when the size of the learning set is too small in relation to the complexity of the model, which in this case means the complexity of the network topology. This is discussed further in Section 8.4 below.

Whether or not a hidden layer is present, the output layer of the network can sometimes have a number of units, when there are a number of classes to predict (Figure 8.5).

8.3 Choosing the learning sample

The learning of the neural network will be improved if it takes place on a sample that is sufficiently rich to represent all the possible values of all the layers of the network, in other words all the possible categories of each variable, at the input or at the output. A network can only learn from the configurations that it has encountered during its learning: customers with overdrafts of more than €1000 may be very much at risk, but if the network's learning sample did not include any of these, then the network will not be able to predict anything about them. However, we should remember that the learning time increases greatly with the size of the sample, as the neural network runs through its learning sample many times.

For the output variables, the learning sample must include all the categories in equal proportions, even if some categories are more frequent in the real population (for example, we must have as many 'negative' events as 'positive' ones, even if the 'negative' events are much rarer in reality).

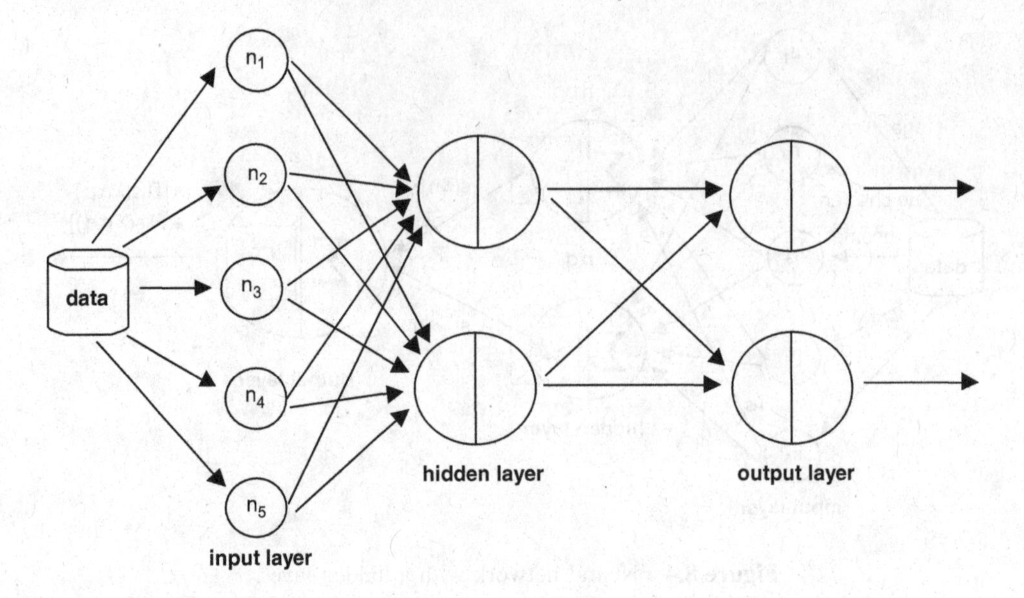

Figure 8.5 Neural network with more than one output unit.

8.4 Some empirical rules for network design

In a back-propagation network, at least 5–10 individuals will be needed to adjust each weight. To increase the robustness of the network, it is advisable to have a single hidden layer for a radial basis function network, and one, or in exceptional cases two, for the multilayer perceptron: instead of adding a third hidden layer, it is better to modify other parameters, retest with other initial weights, or reprocess the input data.

A network with n input units, a single hidden layer, m units in the hidden layer and k output units has $m(n + k)$ weights. We therefore need a sample of at least $5m(n + k)$ individuals for the learning process. If the number of input units has to be reduced, because the learning sample is too small, then the number of predictive variables must also be reduced. Suppose that we wish to reduce 20 predictive variables to 10. We can test all the combinations of 10 variables, changing only two or three of them each time. This is time-consuming, but takes into account the fact that some variables only reveal their predictive nature when combined with certain other variables. Another procedure, which is fast and elegant, involves carrying out a principal component analysis (see Section 7.1) and substituting the first principal components for the variables at the input of the network. A minor drawback of this technique is that it is inherently linear, and may conceal important non-linear structures.

The value of m generally lies between $n/2$ and $2n$. Some authors suggest extending the range to $3n$; others recommend $3n/4$, and yet others prefer a value between $\sqrt{nk}/2$ and $2\sqrt{nk}$. The interested reader should consult the report by Iebeling Kaastra and Milton Boyd.[2] For

---

[2] Kaastra, I. and Boyd, M. (1996) Designing a neural network for forecasting financial and economic time series. *Neurocomputing*, 10, 215–236.

classification, m is generally at least equal to the number of classes to be predicted. It is best to proceed by conducting a number of tests, measuring the error rate on the test sample each time, and stopping the increases in m as soon as this rate reaches a minimum, to avoid overfitting.

8.5 Data normalization

You will recall that the data used in a neural network must be numeric and their categories must lie within the range [0,1]; if this is not already the case, the data must be normalized. To ensure that the normalization process described below is correct, the learning data set must of course cover all the values found in the whole population, particularly the extreme values of continuous variables.

8.5.1 Continuous variables

Even when *continuous variables* are normalized, the extreme values may still tend to 'bury' the normal values. Thus most monthly income levels are in the range from €0 to €10 000, but if an income exceeds €100 000, the standard normalization of the 'income' variable, i.e. its replacement with the variable

$$\frac{\text{income} - \text{minimum income}}{\text{maximum income} - \text{minimum income}}$$

will make the difference between €5000 and €10 000 almost imperceptible, placing it on the same level as the much less significant difference between €95 000 and €100 000.

There are several ways of normalizing this type of variable correctly. The variable can be discretized, and replaced with its quartiles, for example. We could normalize the logarithm of the variable, instead of the variable itself; this would 'stretch' the lower part of the scale. We could normalize the variable in a linear way, as mentioned above, in respect of its values in the range from -3 to $+3$ times the standard deviation[3] σ about the mean μ, then change values lower than $\mu - 3\sigma$ to 0 and change values greater than $\mu + 3\sigma$ to 1. In this variant, we can divide the range $[\mu - 3\sigma, \mu + 3\sigma]$ in two if necessary, by setting the mean μ to the centre of the range, 0.5, and applying the two half-ranges in a linear way.

8.5.2 Discrete variables

To normalize *discrete variables* for which the difference between 0 and 1 is greater than between 1 and 2, 2 and 3, etc., we can carry out the following translation:

- $0 \rightarrow 0$
- $1 \rightarrow {}^1\!/_2$
- $2 \rightarrow {}^1\!/_2 + {}^1\!/_4$

[3] Note that, when a variable follows a normal distribution with a mean μ and standard deviation σ, we find 68% of the observations in the range $[\mu - \sigma, \mu + \sigma]$, 95% of the observations in the range $[\mu - 2\sigma, \mu + 2\sigma]$ and 99.7% of the observations in the range $[\mu - 3\sigma, \mu + 3\sigma]$.

- ...
- $n \rightarrow \sum_{k=1}^{n} 2^{-k}$

8.5.3 Qualitative variables

The normalization of *qualitative variables* poses a problem: it makes an order relationship appear among its categories, which is often artificial and leads the neural network astray.

A common way of overcoming this difficulty is to make the number of units equal to the number of categories of the qualitative variables, by creating binary variables (called 'indicator variables') whose value of 1 or 0 signifies that the qualitative variable does or does not have this category. The drawback of this solution is that it requires a larger number of units, resulting in a more complex network with a longer learning time, as well as an increase in the size of the sample required for learning.

Before using a neural network on qualitative data, therefore, we should reduce the number of categories as much as possible.

8.6 Learning algorithms

At the present time, the Levenberg–Marquardt algorithm is often favoured by experts, because it converges more quickly, and towards a better solution, than the gradient back-propagation algorithm. However, it needs a large amount of computer memory, proportional to the square of the number of units. It is therefore limited to small networks with only a few variables. It is also restricted to a single output unit.

The gradient back-propagation algorithm is the oldest and most widely used method, especially for large data volumes. But it lacks reliability because of its sensitivity to local minima.

The conjugate gradient descent algorithm is a good compromise, because its performance approaches that of the Levenberg–Marquardt algorithm in terms of convergence, but it can be used on more complex networks with more than one output if necessary.

Finally, I should also mention the quasi-Newton algorithm and the genetic algorithms which will be examined in Section 11.13.

8.7 The main neural networks

There are various neural network models. The main ones are the multilayer perceptron (MLP), the radial basis function (RBF), and the Kohonen network, which are described below. More recently, the density estimation networks of Specht (1990)[4] have been used both for classification (probabilistic neural networks) and for prediction (general regression neural networks). There are also networks similar to RBF networks but based on the mathematical theory of wavelets.

The Kohonen network is an unsupervised learning network used for clustering, while the other networks mentioned above (MLP, RBF, etc.) are supervised learning networks, used with one or more dependent variables at the output.

[4] Specht, D.F. (1990) Probabilistic neural networks. *Neural Networks*, 3, 109–118.

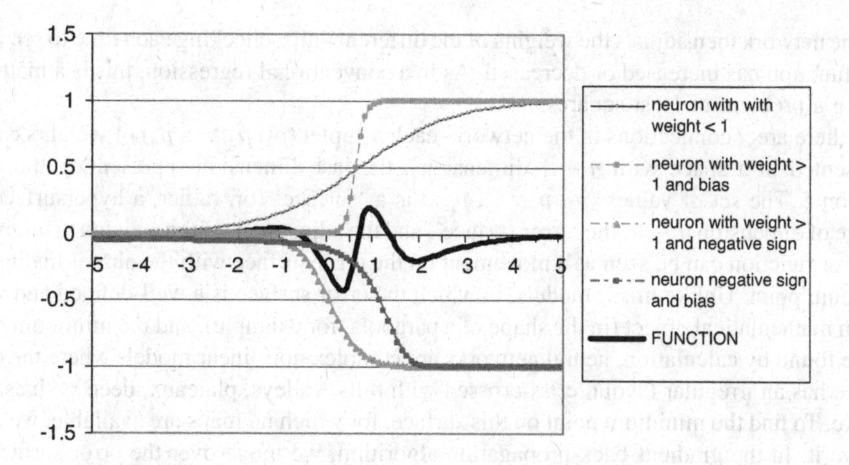

Figure 8.6 Approximation of a function by a sum of sigmoids.

8.7.1 The multilayer perceptron

The archetypal neural network is the multilayer perceptron. It is particularly suitable for the discovery of complex non-linear models. Its power is based on the possibility of approximating any sufficiently regularly function with a sum of sigmoids (Figure 8.6). As its name indicates, this network is made up of several layers: the input variables, the output variable or variables, and one or more hidden levels. Each unit at a level is connected to the set of units at the preceding level.

The number of input units is always equal to the number of variables in the model; if necessary, these variables may be the 'indicator' variables substituted for the original qualitative variables (see Section 8.5.3). There is usually just one output unit. For the choice of the number of units in the hidden layer, see Section 8.4.

To explain the operation of the MLP, let us consider the special, but quite common, case of an MLP using gradient back-propagation.

Each connection has an associated weight, which changes in the course of learning. The network starts its learning by assigning a random value to each of the weights and calculating the output value on the basis of a set of records for which the expected output value is known: this is the learning sample. The network then compares the calculated output value with the expected value, and calculates an *error function* ε, which can be the sum of squares of the errors occurring for each individual in the learning sample:

$$\sum_i \sum_j (E_{ij} - O_{ij})^2,$$

where the first summation is performed on the individuals of the learning set, the second summation is performed on the output units, and E_{ij} (O_{ij}) is the expected (obtained) value of the jth unit for the ith individual.

The network then adjusts the weights of the different units, checking each time to see if the error function has increased or decreased. As in a conventional regression, this is a matter of solving a problem of least squares.

If there are n connections in the network, each n-tuplet $(p_1, p_2, \ldots, p_n)$ of weights can be represented in a space with $n + 1$ dimensions, the last dimension representing the error function ε. The set of values $(p_1, p_2, \ldots, p_n, \varepsilon)$ is a 'surface' (or, rather, a hypersurface) in a space of dimension $n + 1$, the 'error surface', and the adjustment of the weights to minimize the error function can be seen as a movement on the error surface with the aim of finding the minimum point. Unlike linear models, in which the error surface is a well-defined and well-known mathematical object (in the shape of a parabola, for example), and the minimum point can be found by calculation, neural networks are complex non-linear models where the error surface has an irregular layout, criss-crossed with hills, valleys, plateaux, deep ravines, and the like. To find the minimum point on this surface, for which no maps are available, we must explore it. In the gradient back-propagation algorithm, we move over the error surface by following the line with the greatest slope, which offers the possibility of reaching the lowest possible point. We then have to work out how quickly we should travel down the slope. If we go too quickly, we may pass over the minimum point or set off in the wrong direction; if we go too slowly, we will need too many iterations in the network to find a solution. When we speak of an 'iteration', this means inputting the whole learning set into the network, comparing the expected and obtained outputs, and calculating the error function. The range of possible iterations is very wide, but the order of magnitude is 10 000.

The correct speed is proportional to the slope of the surface and to another important parameter, namely the *learning rate*. This rate, between 0 and 1, determines the extent of the modification of the weights during learning. It is useful to vary this rate, which will be high at the outset (between 0.7 and 0.9) to allow a speedy exploration of the error surface and a fast approximation to the best solutions (the minima of the surface), and then decrease at the end of the learning to bring us as close as possible to an optimal solution. In a situation such as that shown in Figure 8.7, this decrease in the learning rate will ensure that we do not go from the local optimum A straight to the local optimum C, possibly with oscillations between A and C, without reaching the global optimum B.

A second important parameter affects the performance of a multilayer perceptron: this is the *moment (of a neural network)*, which makes the weights tend to keep the same direction of change, increasing or decreasing, because a factor incorporates the preceding weight adjustments. The moment limits oscillations which could be caused by irregularities in the

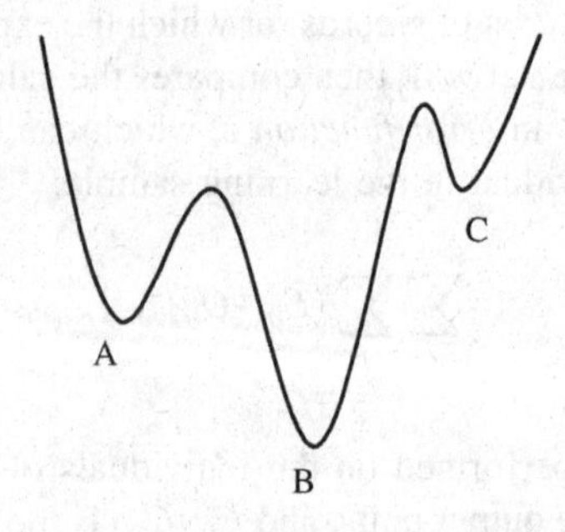

Figure 8.7 Local optimum and global optimum.

learning examples. The effect of the moment is that, if we move several times successively in the same direction over the error surface, we tend to continue the movement without being 'trapped' by the local minima (such as point A in Figure 8.7) and pass over them to reach the global minima (such as point B in the same figure). Just as the learning rate decreases as learning continues, the moment often increases during learning, to enable the network to make a smooth approach to a globally optimal solution.

To sum up, the learning rate controls the extent of modification of the weights during the learning process; a higher rate means faster learning, but there is a greater risk that the network will converge towards a solution other than the globally optimal one. The moment acts as a damping parameter, reducing oscillations and helping to achieve convergence; with a smaller moment, the network is better at 'adapting to its environment', but extreme data have more effect on the weights. To some extent, the learning rate controls the speed of movement and the moment controls the speed of the changes of direction on the error surface; at the start of the process, we move quickly in all directions, but at the end we slow down and change direction less often.

The main danger of neural network modelling is obvious: the network may converge towards a solution that is locally, but not globally, optimal. This risk has led to the development of graphic tools for real-time display of the error rate in learning and validation, enabling the learning to be interrupted as soon as there is any sign of overfitting and an increased error rate in validation (Figure 8.8).

8.7.2 The radial basis function network

An RBF network is an supervised learning network, like the multilayer perceptron, which it resembles in some ways. However, it works with only one hidden layer, and, when calculating

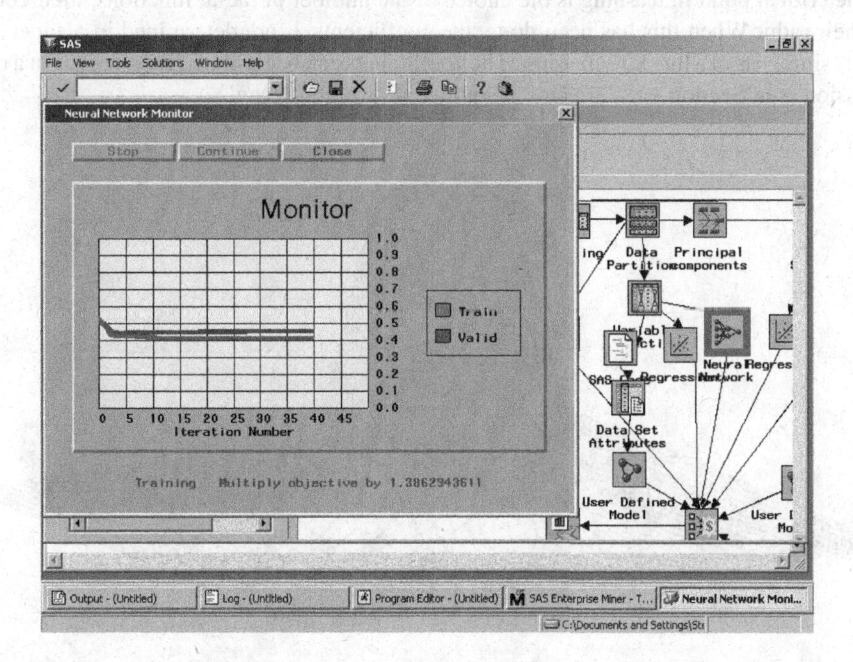

Figure 8.8 Graphic monitoring of a neural network with SAS Enterprise Miner.

the value of each unit in the hidden layer for an observation, it uses the distance in space between this observation and the centre of the unit, instead of the sum of the weighted values of the units of the preceding level. Unlike the weights of a multilayer perceptron, the centres of the hidden layer of an RBF network are not adjusted at each iteration during learning (but some may be added if the space is not sufficiently covered). In a perceptron, the modification of a synaptic weight makes it necessary to re-evaluate all the others, but in an RBF network the hidden neurons share the space and are virtually independent of each other. This makes for faster convergence of RBF networks in the learning phase, which is one of their strong points.

Now, the response surface (the set of values) of a unit of a hidden layer of a multilayer perceptron, before the application of the (generally non-linear) transfer function, is a hyperplane $\sum_i p_i X_i = K$, and similarly the response surface of a unit of the hidden layer of an RBF network is a hypersphere $\sum_i (X_i - \omega_i)^2 = R^2$, and the response of the unit to an individual (x_i) is a decreasing function Γ of the distance between the individual and this hypersphere. As this function Γ is generally a Gaussian function, the response surface of the unit, after the application of the transfer function, is a Gaussian surface, in other words a 'bell-shaped' surface (Figure 8.9). We speak of a *radial function* for Γ, i.e. a function symmetrical about a centre.

Comparing the MLP and RBF networks, we find the differences listed in Table 8.1.

Finally, the global response of the network to each individual (x_i) presented to it is:

$$\sum_{k=1}^{\text{no. of hidden units}} \lambda_k \exp\left[-\frac{1}{2\sigma_k^2} \sum_{i=1}^{\text{no. of input units}} (x_i - \omega_i^k)^2 \right]$$

The learning of an RBF is a matter of determining the number of units in the hidden layer, i.e. the number of radial functions, their centres $\Omega_k = (\omega_i^k)$, their radii σ_k, and the coefficients λ_k. The critical point in learning is the choice of the number of radial functions, their centres and their radii. When this has been done, the coefficients λ_k are determined in a supervised way, as simply as in a linear regression. The coefficients can be limited if required, as in a ridge regression (see Section 11.7.2). This is known as 'weight decay'.

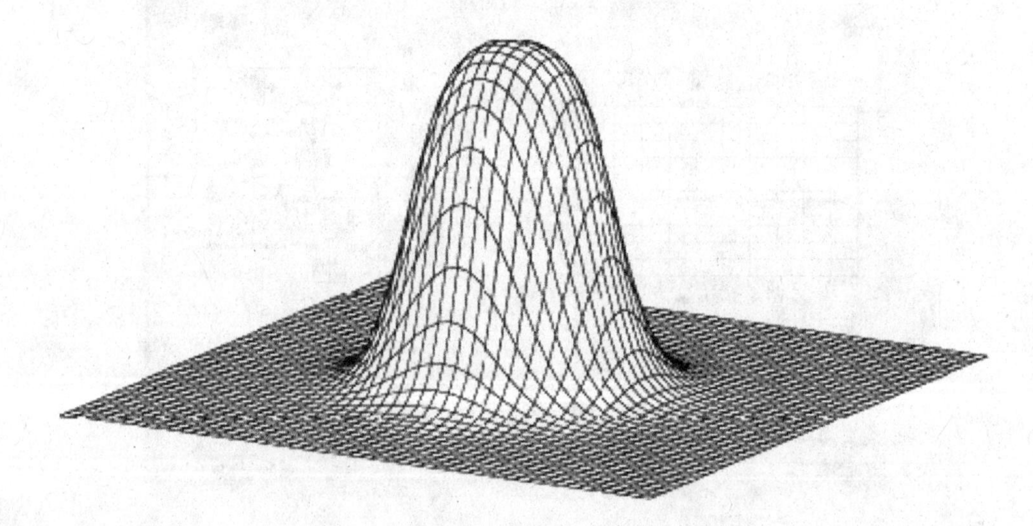

Figure 8.9 Response surface of a radial unit.

Table 8.1 Comparison of MLP and RBF networks.

	Network →	**MLP**	**RBF**
	'Weight'	Weight p_i	Centre ω_i
Hidden layer(s)	Combination function	Scalar product $\sum_i p_i x_i$	Euclidean distance $\sum_i (x_i - \omega_i)^2$
	Transfer function	Logistic $s(X) = 1/(1 + \exp(-X))$	Gaussian $\Gamma(X) = \exp(-X^2/2\sigma^2)$
	Number of hidden layers	≥ 1	$= 1$
Output layer	Combination function	Scalar product $\sum_k p_k x_k$	Linear combination of Gaussians $\sum_k \lambda_k \Gamma_k$ (see below)
	Transfer function	Logistic $s(X) = 1/(1 + \exp(-X))$	Linear function $f(X) = X$
	Speed	Faster in 'model application' mode	Faster in 'model learning' mode
	Advantage	Better generalization	Less risk of non-optimal convergence

The number of units is generally specified by the user, even if the network can create others to improve the accuracy of the results. A sufficiently high number must be provided, generally more than in a multilayer perceptron, to enable the data structure to be modelled correctly. The units may commonly number several hundred. This is because the fast decrease of the Gaussian means that the RBF network has a lower extrapolation capacity when farther from the centres of the units of the hidden layer. These units must therefore be sufficiently numerous to ensure that at least one unit is activated for each observation; in other words, at least one radial function must have a non-negligible value in any region where data are present. This is an evident drawback of the RBF network as compared with the multilayer perceptron, even if it is also better protected against certain risky extrapolations found with the multilayer perceptron. In fact, the complexity of the RBF network increases exponentially with the number of input variables, because the radial function space has to be filled. As the number of variables increases, therefore, the calculation time of the RBF network increases, together with the number of observations needed for learning. This is one of its major weaknesses. It is therefore essential to select the input variables of a RBF network with great care.

When the number of units has been chosen, we must consider their centres. Some networks position the centres in a random way. However, the results can be improved by using the moving centres method (see Section 9.9.1) or Kohonen networks (see below) to divide the space into clusters (partitions) according to the distribution of the data. Thus, if the data are distributed in packets, the centres of these packets will be chosen as the centres of the RBF network. Also, more centres will be positioned in areas with a high observation density (input adaptation), or in areas where the result to be predicted varies more rapidly (output adaptation), in order to reduce the output error. Output adaptation is not very often used in learning, but if it is used it gives rise to the problem of not always being compatible with the input adaptation. This is because the response distribution (the dependent variable) may not coincide with the data density distribution, and may lead to the determination of other centres for the radial function.

With the exception of the possible application of output adaptation, the search for the centres is unsupervised and can also be carried out on observations for which the response is not always known, since it is then simply a matter of estimating the data probability density. It may be useful to be able to train the RBF network on observations for which the dependent variable is not always known: this enables us to use a larger number of observations. In this case, we speak of *semi-supervised learning*. This is used whenever the collection of labelled observations is more difficult or costly than the collection of unlabelled observations.

However, this semi-supervised learning has the drawback of being more sensitive to noise. When areas of high density are sought in order to determine their centres without considering the dependent variable, the input variables which are not related to the dependent variable are not distinguished from the related variables, and may introduce noise.

The final aspect of parameter setting relates to the radii of the units of the hidden layer, which are the standard deviations of the Gaussian distributions. A simple solution is to choose radii equal to twice the mean distance between centres. If they are too large, the network will lack structural detail and its precision will be reduced. If they are too small, the space will be poorly covered by the Gaussian surfaces, and the network will have to interpolate between these surfaces, which will decrease the capacity for generalizing the results of the learning phase. As for the centres, the radii will be chosen so that they are smaller in areas with a high density of observations, or in areas in which the result to be predicted varies more rapidly. There are several ways of determining the radii as precisely as possible: a useful method finds the k nearest neighbours (see Section 11.2), and examines each unit centre to see where its k nearest neighbours are located (k is chosen appropriately by the user), and the mean distance to these k nearest neighbours is taken to be the radius. This method has the merit of adapting to the structure of the data. The radii are not necessarily equal to each other, but this is sometimes assumed to decrease the number of network parameters.

Compared with the MLP network, the RBF network has the major advantage of needing only a single hidden layer, and using *linear* combination and transfer functions in the output layer in most cases – except in certain sophisticated variants (see Table 8.1). This makes for faster learning and far fewer problems of complicated parameter adjustment for the user. It also avoids the risk, inherent in the back-propagation mechanism of the multilayer perceptron, of convergence towards a locally, but not globally, optimal solution. From this point of view, the sequential search for the centres of the radial functions, their radii, and then the coefficients of their linear combination is an advantage: it provides greater simplicity and faster learning, and decreases the risk of overfitting by comparison with a search for global optimization by gradient descent.

The weakness of the radial basis function, compared with the multilayer perceptron, is that it may need a large number of units in its hidden layer, which increases the execution time of the network without always yielding perfect modelling of complex structures and irregular data. This happens when the number of input variables is too large, and it is desirable to reduce this number as far as possible. This problem is due to the fact that the RBF network, even more than the multilayer perceptron, requires a learning set which covers all the configurations and all the categories of variables which may be found when it is applied to the whole population to be studied. The advantages and disadvantages of the RBF network tend to be those that are generally found in networks for probability density estimation. The MLP network offers the best generalization capacity, especially for noisy data.

8.7.3 The Kohonen network

The Kohonen network is the most widely used unsupervised learning network. It can also be called a self-adaptive or self-organizing network, because it 'self-organizes' around the data. Other synonyms are 'Kohonen map' and 'self-organizing map'.

Like any neural network, it is made up of layers of units and connections between these units. The major difference from the networks described above is that there is no variable to be predicted. The purpose of the network is to 'learn' the structure of the data so that it can distinguish clusters in them.

The Kohonen network is composed of two levels (Figure 8.10):

- the input layer, with a unit for each of the n variables used in the clustering;
- an output layer, whose units are arranged as a generally square or rectangular (sometimes hexagonal) grid of $l \times m$ units (in some cases l and $m \neq n$), each of these $l \times m$ units being connected to each of the n units of the input layer, the connection having a certain weight p_{ijk} ($i \in [1,l]$, $j \in [1,m]$, $k \in [1,n]$).

The units of the output layer are not interconnected, but a distance is defined between them, such that we can speak of the 'neighbourhood' of a unit.

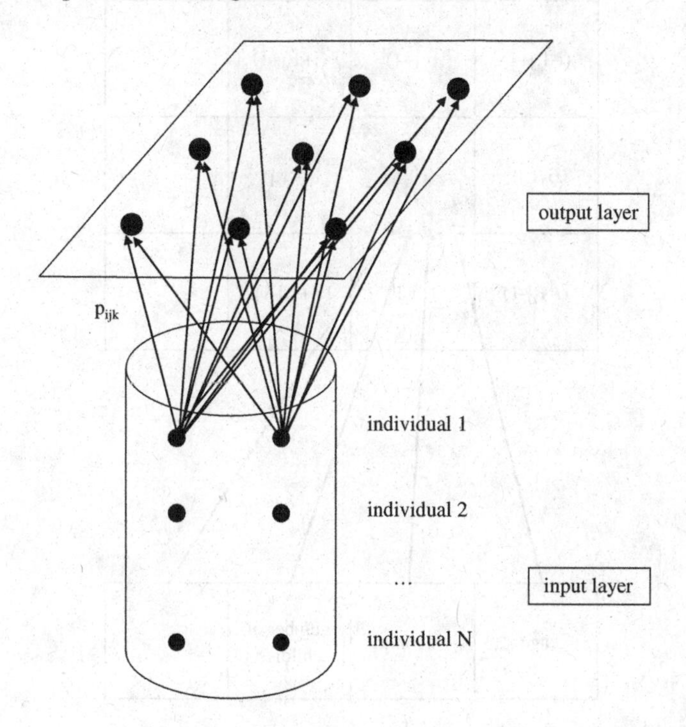

Figure 8.10 Kohonen network.

The units of the input layer correspond to the variables of the individuals to be clustered, and this layer is used to present the individuals; the states of its units are the values of the variables characterizing the individuals to be clustered. This is why this layer contains n units, where n is the number of variables used in the clustering.

The grid on which the output units are placed is called the 'topological map'. The shape and size of this grid are generally chosen by the user, but they may also change in the course of learning. Each output unit (i,j) is associated with a weight vector $(p_{ijk})_{k\in[1,n]}$, and therefore the *response* of this unit to an individual $(x_k)_{k\in[1,n]}$ is, by definition, the Euclidean distance

$$d_{ij}(x) = \sum_{k=1}^{n} (x_k - p_{ijk})^2.$$

So how does a Kohonen network learn? First of all, the weights p_{ijk} are initialized randomly. Then the responses of the $l \times m$ units of the output layer are calculated for each individual (x_k) in the learning sample. The unit chosen to represent (x_k) is the unit (i,j) for which $d_{ij}(x)$ has the minimum value. We say that this unit is 'activated' (Figure 8.11). This unit and all the neighbouring units have their weights adjusted to bring them closer to the

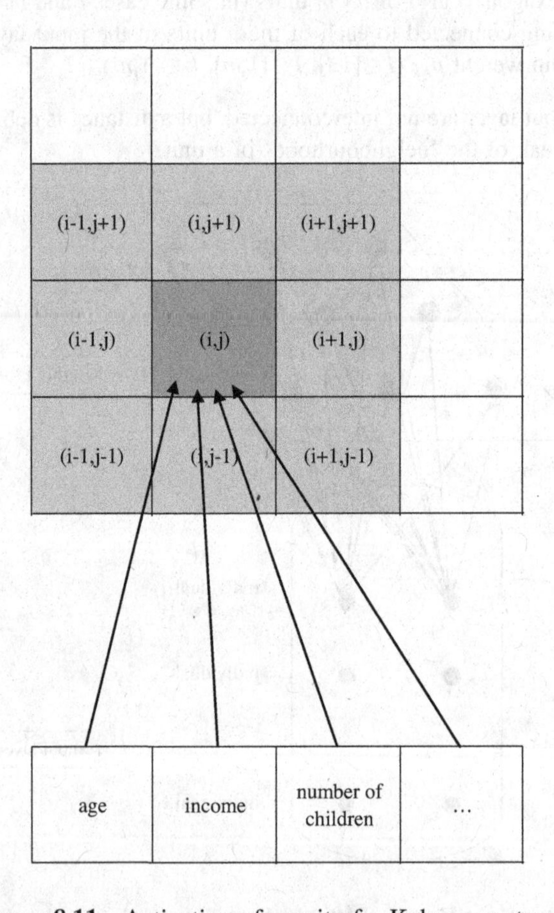

Figure 8.11 Activation of a unit of a Kohonen network.

individual at the input. For example, the neighbouring units of (i,j) are the eight units $(i-1,j)$, $(i+1,j)$, $(i,j-1)$, $(i,j+1)$, $(i+1,j+1)$, $(i+1,j-1)$, $(i-1,j+1)$, $(i-1,j-1)$. The size of the neighbourhood generally decreases during learning: at the beginning, the neighbourhood can be the whole grid; by the end, it may be reduced to the unit itself. These adjustments form part of the network parameters.

The new weights of a neighbour (I,J) of the 'winner' (i,j) are

$$p_{IJk} + \Theta \cdot f(i,j;I,J) \cdot (x_k - p_{IJk}) \text{ for every } k \in [1,n],$$

where $f(i,j;I,J)$ is a decreasing function of the distance between the units (i,j) and (I,J), such that $f(i,j;i,j) = 1$. It may also be a Gaussian function: $\exp(-\text{distance}(i,j;I,J)^2/2\sigma^2)$.

The parameter $\Theta \in [0,1]$ is a learning rate which, as in the case of a multilayer perceptron, changes during learning by decreasing linearly or exponentially.

It is the extension of the weight adjustment to the whole neighbourhood of the 'winning' unit that brings the neighbouring units of (i,j) close to the individual (x_k) at the input, and enables the individuals that are close together in variable space to be represented by identical or neighbouring units in the layer, just as neighbouring neurons respond to nearby stimuli in the cerebral cortex. The whole process takes place as though the Kohonen network was made of rubber and was deformed to make the cloud of individuals pass over it while approaching as closely as possible to the individuals. By contrast with the factor plane (see Section 7.1), the projection concerned is non-linear.

When all the individuals in the learning sample have been presented to the network and all the weights have been adjusted, the learning is complete.

To summarize, during the network's learning:

- For each individual, only one output unit (the 'winner') is activated.
- The weights of the winner and its neighbours are adjusted.
- The adjustment is such that two closely placed output units correspond to two closely placed individuals.
- Groups (clusters) of units are formed at the output.

In the application phase, the Kohonen network operates by representing each input individual by the unit of the network which is closest to it in terms of the distance defined above. This unit will be the cluster of the individual.

This algorithm has some similarities with the moving centres and k-means methods (see Section 9.9.1). However, there is an important difference. In the k-means method, the introduction of a new individual into a cluster only results in the recalculation of the centre of gravity of the cluster, without any effect on the other centres of gravity. But the introduction of a new individual into a Kohonen network results in the adjustment of not just the unit nearest to the individual, but also the neighbouring units. The neighbourhood of the 'winner' unit is significant, while the neighbourhood of the 'winner' centre of gravity is not.

Another major difference between Kohonen networks and the moving centres of k-means methods is that, unlike these methods, the Kohonen clustering takes place by reducing the number of dimensions of the variable space, as in factor analysis, the new working space generally being of dimension 2, as in my description, or, exceptionally, of dimension 3 or 1.

9

Cluster analysis

Clustering, also known as 'segmentation', which I will describe more fully below, is the most widespread descriptive method of data analysis and data mining. It is used when there is a large volume of data and the aim is to find homogeneous subsets, which can be processed and analysed in different ways. Such a requirement is present in a wide range of contexts, especially in the social sciences, medicine and marketing, where human factors mean that the data are numerous and difficult to understand. In clustering, unlike classification (which will be dealt with in a later chapter), there is no particular dependent variable, and it is harder to compare two forms of clustering objectively. There is no privileged criterion for comparing two algorithms, such as we find in the rate of correct classification, and many competing algorithms vie for our attention. Only the main ones will be described here, but it is worth noting that the theory is constantly and rapidly evolving in this field, especially now that text mining and web mining are beginning to give rise to problems of document clustering.

9.1 Definition of clustering

Clustering is the statistical operation of grouping objects (individuals or variables) into a limited number of groups known as clusters (or segments), which have two properties. On the one hand, they are not defined in advance by the analyst, but are discovered during the operation, unlike the classes used in *classification*. On the other hand, the clusters are combinations of objects having similar characteristics, which are separated from objects having different characteristics (resulting in internal homogeneity and external heterogeneity). This can be measured by criteria such as the between-cluster sum of squares (see below). As with classification, the essence of clustering is the distribution of objects into groups. However, this distribution is not carried out on the basis of a predefined criterion, and is not intended to combine the objects having the same value for such a criterion. In other words, the cluster to which each object belongs is not known in advance, in contrast to the classification process. Even the number of clusters is not always fixed in advance. This is because there is no

Data Mining and Statistics for Decision Making, First Edition. Stéphane Tufféry.

dependent variable: clustering is descriptive, not predictive. It is widely used in marketing, medicine, the social sciences, and similar fields. In marketing, it is often referred to as 'segmentation' or 'typological analysis'. In medicine, the term used is 'nosology'. In biology and zoology, we speak of numerical taxonomy. Finally, neural network experts use the term *unsupervised pattern recognition.*

9.2 Applications of clustering

In marketing, clustering is particularly useful for finding the different customer profiles which make up a customer base. After it has detected the clusters which 'sum up' its customer base, the business can develop a specific offer and communications for each of them. It can also follow the development of its customers over the months, and see which customers and how many of them move from one cluster to another every month. If the business also wishes to follow certain customers in detail, it can set up a *customer panel*, based on the clustering, to ensure that all the clusters are well represented.

In the retail sector, clustering is used to divide up all the stores of a particular company into groups of establishments which are homogeneous in terms of the type of customer, turnover, turnover per department (according to the type of product), size of store, etc.

In the medical field, clustering can be used to discover groups of patients suitable for particular treatment protocols, each group comprising all the patients who react in the same way.

In sociology, clustering is used to divide the population into groups of individuals who are homogeneous in terms of social demographics, lifestyle, opinions, expectations, etc.

More generally, clustering is useful as a preliminary to other data mining operations.

In the first place, most predictive algorithms are not good at handling an excessively large number of variables, because of the correlations between the variables which can affect their predictive power. However, it is difficult to describe a heterogeneous population correctly with a small number of variables. The groups formed by clustering are useful because they are homogeneous and can be described by a small number of variables which are specific to each group.

Secondly, it is sometimes helpful if an algorithm can process missing values in a clustering process without replacing them with *a priori* values such as the means (or minima or maxima) for the whole population. We can then wait for an individual to be placed in its cluster in order to replace its missing values, taking the mean values (or the minima or maxima) not for the population as a whole, but for this cluster.

9.3 Complexity of clustering

To gain some idea of the complexity of the problem, let us recall that the number of (non-overlapping) partitions of n objects is the Bell number,

$$B_n = \frac{1}{e}\sum_{k=1}^{\infty}\frac{k^n}{k!}$$

For example, for $n = 4$ objects, $B_n = 15$, with:

- 1 partition with 1 cluster (*abcd*);

- 7 partitions with 2 clusters (*ab,cd*), (*ac,bd*), (*ad,bc*), (*a,bcd*), (*b,acd*), (*c,bad*), (d,abc);
- 6 partitions with 3 clusters (*a,b,cd*), (*a,c,bd*), (*ad,bc*), (*a,d,bc*), (*b,c,ad*), (*b,d,ac*), (*c,d,ab*);
- 1 partition with 4 clusters (*a,b,c,d*).

For $n = 30$ objects, $B_{30} = 8.47 \times 10^{23}$, an enormous number, greater than Avogadro's number (6.022×10^{23}), which is the number of molecules in one mole of any gas. As a general rule, $B_n > \exp(n)$, which shows how necessary it is to define the correct criteria for clustering and use efficient algorithms, because it would be out of the question to test all the possible combinations.

9.4 Clustering structures

9.4.1 Structure of the data to be clustered

The data are set out in the form of a rectangular matrix where the rows are the individuals and the columns are the variables, or else in the form of a square matrix of similarities, showing the distances between individuals or between variables (for example, a matrix in which all the coefficients are 1, minus the correlation matrix). These structures enable individuals or variables to be clustered. The distance used is most commonly the Euclidean distance,

$$L_2 = \sqrt{\sum_{i=1}^{p} (x_i - y_i)^2},$$

but the Manhattan distance,

$$L_1 = \sum_{i=1}^{p} |x_i - y_i|,$$

is sometimes used, especially to reduce the effect of extreme individuals whose coordinates are not squared.

9.4.2 Structure of the resulting clusters

There are three possible cases. In the first, two clusters are always separated: we are concerned here with partitioning methods. The number of clusters is generally defined *a priori*, but some methods can dispense with this constraint (for example, clustering by similarity aggregation, non-parametric methods using density estimation such as the SAS/STAT MODECLUS procedure). The main partitioning methods are:

- moving centres, *k*-means and dynamic clouds;
- *k*-medoids, *k*-modes, *k*-prototypes;
- methods based on a concept of density;
- Kohonen networks;
- clustering by similarity aggregation.

In the second case, two clusters are separate or one contains the other: the methods used here are hierarchical ascendant methods (known as 'agglomerative') or descendant methods (known as 'divisive'). Agglomerative methods are based on a concept of distance or density. This type of clustering can be combined with the type used in the first case, in what are known as 'mixed' or 'hybrid' methods.

In the third case, two clusters can have a number of objects in common ('overlapping' clusters), and we speak of 'fuzzy' clustering, in which each object has a certain probability of belonging to a given cluster. This form of clustering is very rarely used and is not described here.

9.5 Some methodological considerations

9.5.1 The optimum number of clusters

The definition of natural clusters is a tricky matter, because they are not always as obvious as in some of the textbook cases presented in this section. The results do not always seem as natural as these, and may also differ according to the algorithm used to compute them. In practice, the determination of the 'real' number of clusters is empirical as much as theoretical. Clustering is often required to be readable and interpretable in concrete terms, rather than theoretically optimal. However, if the data are naturally structured into three clusters but four clusters are requested, there is bound to be an arbitrary element in the division of the clusters.

The question of the number of clusters to be discovered is therefore particularly important. In some methods, such as the moving centres method and its variants (see below), this number has to be determined *a priori*, which obviously has a highly detrimental effect on the quality of clustering if this number does not correspond to the actual distribution of individuals.

Other methods, such as clustering by similarity aggregation, allow the algorithm itself to determine the optimum number of clusters automatically. We can then advantageously begin by carrying out an initial clustering without fixing the number of clusters and then examining the results to see that we have n clusters of significant size while the others are very small (perhaps less than 1% of the population in each case, for example). Finally, we recompute the clustering with the number of clusters to be discovered fixed at n.

Agglomerative hierarchical clustering can also be used to choose an optimum number of clusters in a simple way (see below).

In addition to these technical aspects, practical matters must often be borne in mind when choosing the number of clusters. If the clustering is to be used for research or direct marketing, it can be fairly complex. If it is to be used by sales personnel in the field, when dealing with customers, it must be relatively simple, containing not more than seven or eight segments. There have been cases in which typologies were never used because they were too complex, with about 15 segments.

9.5.2 The use of certain types of variables

The variables must be standardized if they were not all measured in the same units and if they have different means or variances. It is preferable to isolate the outliers, as in the example in Figure 9.1, in which a single individual in the upper right-hand corner of the diagram, measured at more than 30 on the vertical axis, is enough to invalidate the entire clustering (a 'single linkage' agglomerative hierarchical clustering method), because the extreme individual represents a cluster on its own, whereas the cloud of points is composed of two

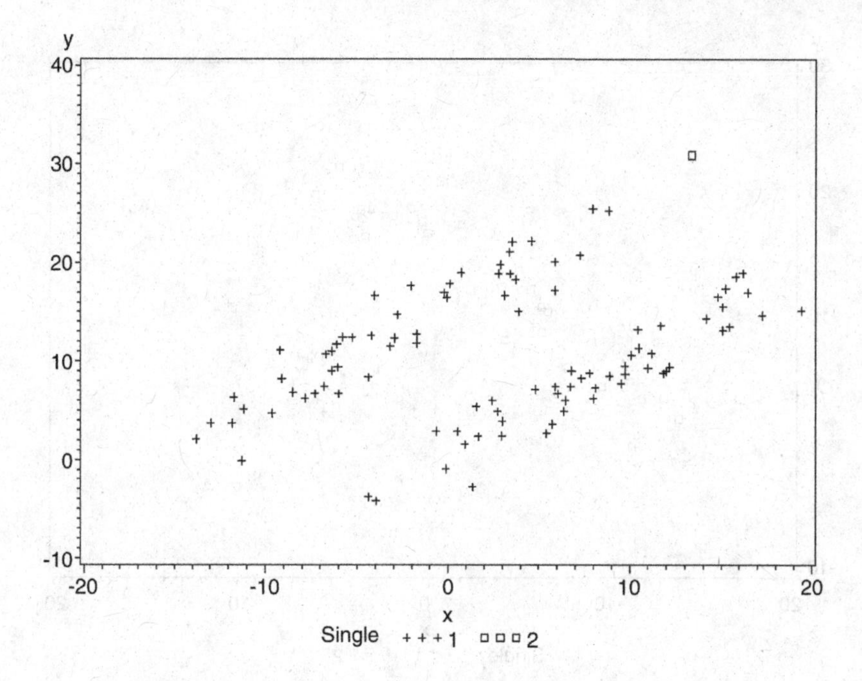

Figure 9.1 Effect of an extreme individual on clustering.

clearly visible clusters. And yet this individual is not so far away from the rest. The effects of an individual placed completely outside the normal range can be imagined. Admittedly, in this case the 'single linkage' method amplifies the influence of this outlier individual, because of the chain effect used in this method (see Section 9.10.2).

When the extreme individual is removed, the agglomerative hierarchical clustering detects the two natural clusters of the cloud in a satisfactory way (Figure 9.2).

When using qualitative variables, it is possible to use clustering on continuous variables by multiple correspondence analysis (MCA), by acting on the factors associated with each variable category.

9.5.3 The use of illustrative variables

In order to distinguish true clusters in the data, we often have to interpret the first results before transforming, adding or excluding variables, and then restart the clustering. Excluding a variable does not necessarily mean deleting it from the analysis base. Instead, we cease to take it into account in the clustering operation, while retaining it as an inactive variable to observe the distribution of its categories in the various clusters. It is no longer an 'active' variable, but becomes an 'illustrative' variable (also called a 'supplementary' variable), as seen in factor analysis.

9.5.4 Evaluating the quality of clustering

There are statistical quality measurements for each clustering methods, and these are detailed below. They are very useful for choosing the correct number of clusters for agglomerative hierarchical clustering.

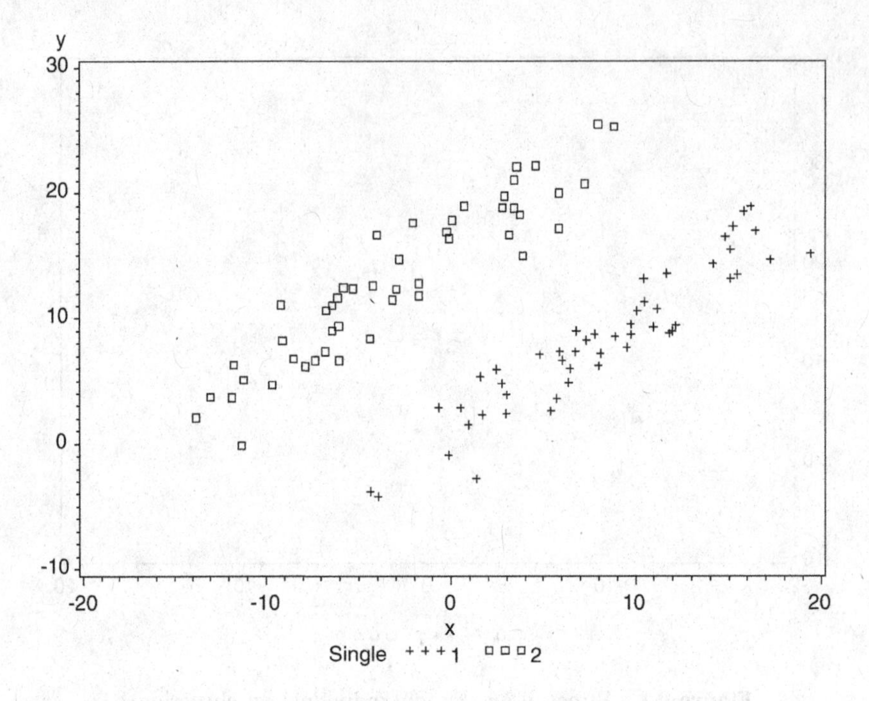

Figure 9.2 Removal of the extreme individual in clustering.

A more intuitive validation of a clustering can be achieved by using another statistical method to visualize the clusters. For example, the individuals can be represented on the factor plane of a principal component analysis (carried out on the clustering variables) and can be 'coloured' according to their clusters, as in the example in Figure 7.8 which shows three clear clusters.

9.5.5 Interpreting the resulting clusters

A major difference between classification and clustering methods is that, in the latter type, there is no objective universal scale for comparison. Any predictive classification models can be compared by measuring the rate of correctly classed individuals or the area under the ROC curve. This is not true of clustering, since even an indicator like within-cluster sum of squares depends on the number of clusters and cannot be used universally to prove the superiority of one form of clustering over another which does not have the same number of clusters.

In any case, experience shows that most users prefer a clustering method to be intuitive and easy to understand, rather than perfect in statistical terms. It will be necessary to explain to them that, in statistical clustering, since the clusters are not predefined, a 'rather young' cluster may include a number of elderly people. The extreme cases might need to be monitored and reallocated in a few cases.

An interesting way of interpreting a clustering is to draw up a decision tree after determining the clustering, taking as the dependent variable of the tree the number of the cluster (and considering the variables used in the course of clustering as the independent variables). If we obtain a tree with a sufficiently low error rate (less than 10–15%, for

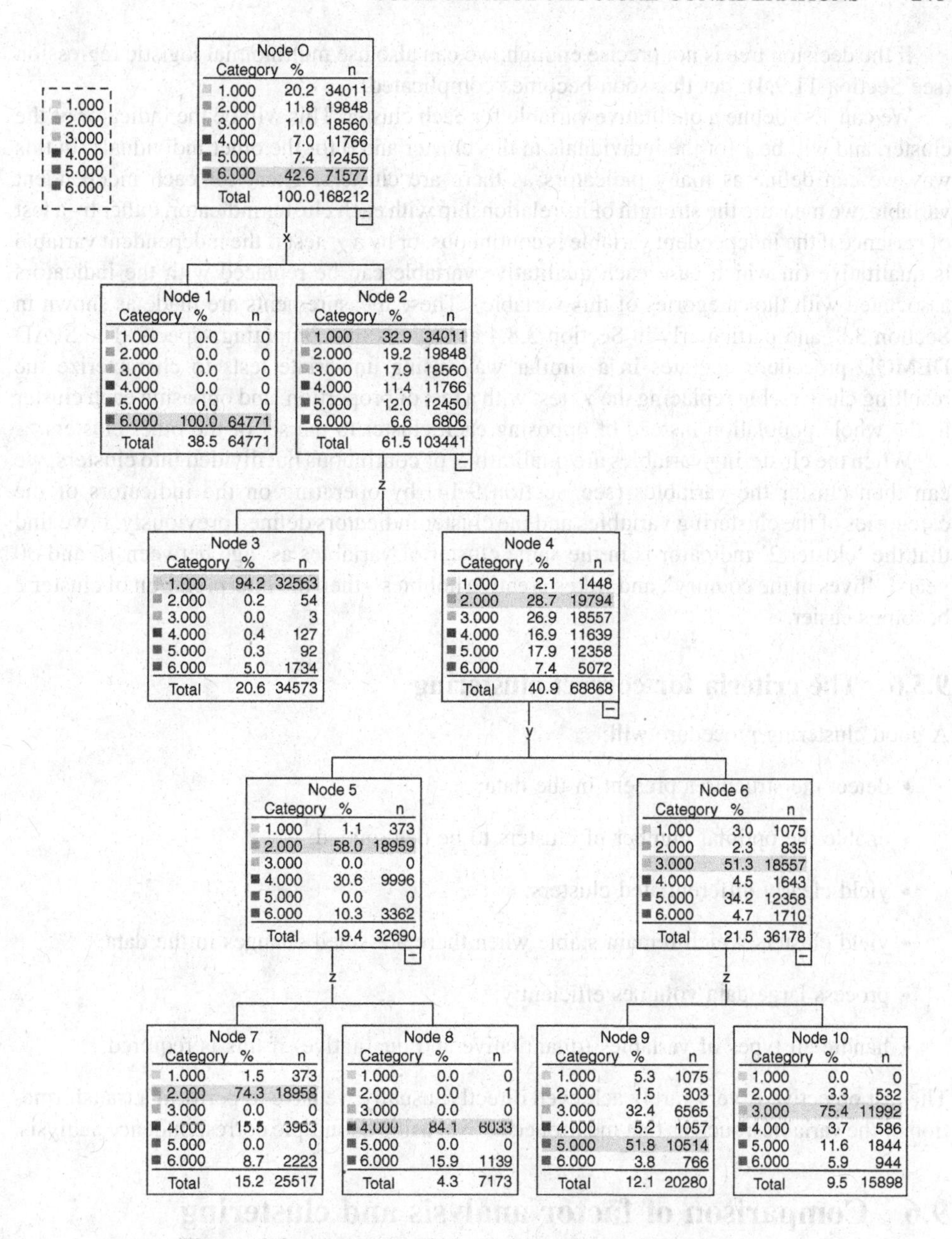

Figure 9.3 Analysis of clustering using a decision tree.

instance), we can even replace the initial clustering (which may not have been very explicit) with a clustering defined by a limited set of rules, namely those deduced from the tree.

The result is not always as convincing as the example in Figure 9.3, where six leaves of the tree enable us to find the six clusters in a reasonably exact way: the only clusters that are poorly described by the tree are the third and fourth.

If the decision tree is not precise enough, we can also use multinomial logistic regression (see Section 11.9.4), but this soon becomes complicated.

We can also define a qualitative variable for each cluster. This will be the indicator of the cluster, and will be 1 for the individuals in this cluster and 0 for the other individuals. In this way we can define as many indicators as there are clusters. Then, for each independent variable, we measure the strength of its relationship with each cluster indicator, either by a test of variance if the independent variable is continuous, or by a χ^2 test if the independent variable is qualitative (in which case each qualitative variable can be replaced with the indicators associated with the categories of this variable). These measurements are made as shown in Section 3.8, and particularly in Section 3.8.4 as regards the computing aspects. The SPAD DEMOD procedure operates in a similar way, using univariate tests to characterize the resulting clusters, but replacing the χ^2 test with a test of proportion, and opposing each cluster to the whole population instead of opposing each cluster to the set of the other clusters.

When the clustering variables are qualitative, or continuous but divided into clusters, we can then cluster the variables (see Section 9.14) by operating on the indicators of the categories of the clustering variables and the cluster indicators defined previously. If we find that the 'cluster 2' indicator is in the same cluster of variables as 'age between 18 and 30 years', 'lives in the country', and 'likes creative hobbies', then the interpretation of cluster 2 becomes easier.

9.5.6 The criteria for correct clustering

A good clustering procedure will:

- detect the structures present in the data;
- enable the optimal number of clusters to be determined;
- yield clearly differentiated clusters;
- yield clusters which remain stable when there are small changes in the data;
- process large data volumes efficiently;
- handle all types of variables (quantitative and qualitative) if this is required.

The last objective is very rarely achieved directly; usually we have to perform a transformation of the variables, such as that mentioned above, using a multiple correspondence analysis.

9.6 Comparison of factor analysis and clustering

Factor analysis, in its various forms (PCA, CA, MCA) is the ideal method for providing an overview of the data and continuous spatial visualization of the individuals, and even, in some cases, detecting the natural number of clusters. By interpreting the factor axes, we can use it to show up very clearly marked tendencies. However, it has certain limitations which make it useful to apply clustering methods on a complementary basis. The first limitation is the lack of readability of the principal planes when the volume of data is large, with several hundreds, or even several thousands, of individuals or more appearing as points on the plane. Another difficulty is that the projection on a subspace of lower dimension can make individuals appear closer than they would if all the dimensions were taken into account. A third problem arises from the fact that the first components are the ones which are predominantly taken into account in a factor analysis, the others being difficult to interpret. This means that we lose some useful information: an important dimension may be missing from the first main axes. For example, the 'risk of debt default' will not be fully shown on the 'income' and 'capital' axes, because it is linked with other characteristics of the customer.

Clustering offers solutions to these problems. First of all, we should note that it works in a different way, by partitioning the individual space instead of representing it continuously. It generally requires the use of an algorithm, i.e. a sequence of elementary operations performed in a repetitive and recursive way, whereas factor methods find the solution of a calculation. Clustering methods do not have one of the benefits of factor methods, namely the parallel processing of individuals and variables. However, they have the major advantage of taking into account all the dimensions (all the variables) of a problem, without projection on to a subspace of lower dimension. Thus there is no loss of information. Moreover, we can obtain a direct description of the clusters, which makes it simpler to interpret and use the results, in comparison to factor methods. Finally, the substitution of the centres of gravity of the clusters for the original individuals makes the graphic representation much more readable.

9.7 Within-cluster and between-cluster sum of squares

Before looking at clustering by moving centres and hierarchical clustering, we must define the between-cluster and within-cluster sum of squares of a population (Figure 9.4).

As mentioned with reference to factor analysis, the *total sum of squares* (or *inertia*) *I* of the population is the weighted mean (usually weighted by the inverse of the total frequency) of the squares of the distances of the individuals from the centre of gravity of the population. This can be written $\sum_{i \in I} p_i (x_i - \bar{x})^2$, where $\bar{x}$ is the mean of the x_i. The sum of squares of a cluster is calculated in the same way with respect to its centre of gravity and can be written as

$$\sum_{i \in I_j} p_i (x_i - \bar{x}_j)^2.$$

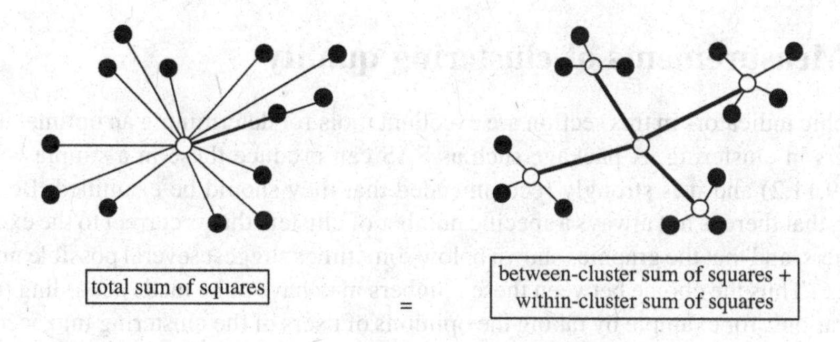

Figure 9.4 Within-cluster, between-cluster and total sum of squares.

If the population is segmented into k clusters, with sums of squares $I_1, \ldots, I_k$, the *within-cluster sum of squares* is, by definition,

$$I_A = \sum_{j=1}^{k} I_j.$$

A cluster becomes more homogeneous as its sum of squares decreases, and the clustering of the population becomes better as I_A diminishes.

Finally, the *between-cluster sum of squares* I_R of the clustering is defined as the mean (weighted by the sum of the weights of each cluster, $p_j = \sum_{i \in I_j} p_i$) of the squares of the distances of the centres of gravity of each cluster from the global centre of gravity. This can be written as

$$\sum_{j \in \text{clusters}} \left(\sum_{i \in I_j} p_i \right) (\bar{x}_j - \bar{x})^2.$$

As I_R increases, the separation between the clusters also increases, indicating satisfactory clustering.

Thus there are two criteria for correct clustering: I_R should be large and I_A should be small. Now, if these two criteria depend on the clustering, *Huygens' formula*,

$$I = I_A + I_R,$$

shows that their sum depends on the global population only, and that the two preceding criteria (minimization of the within-cluster sum of squares and maximization of the between-cluster sum of squares) are therefore equivalent. Using the above notation, this formula can be written as

$$\sum_{i \in I} p_i (x_i - \bar{x})^2 = \sum_{j \in \text{clusters}} \left(\sum_{i \in I_j} p_i (x_i - \bar{x}_j)^2 \right) + \sum_{j \in \text{clusters}} \left(\sum_{i \in I_j} p_i \right) (\bar{x}_j - \bar{x})^2$$

It is important to note that clustering into $k + 1$ clusters will have a higher between-cluster sum of squares than clustering into k clusters, and will therefore be 'better'; so two clusterings having different numbers of clusters cannot be compared according to the inertial criterion alone. This is the intrinsic drawback of clustering methods using a purely inertial criterion: if the number of clusters is not fixed in advance, then the optimization of this criterion results in the isolation of all the individuals in the same number of separate clusters, each containing one individual.

9.8 Measurements of clustering quality

The graphic indicators in this section are excellent tools for determining an optimal number of clusters in clustering. A package such as SAS can produce these in a simple way (see Section 9.11.2) and it is strongly recommended that they should be examined. Be aware, however, that there is not always a specific number of clusters that is correct to the exclusion of all others, and that the graphics shown below sometimes suggest several possible numbers of clusters. Thus the choice between these numbers may have to be made according to other considerations, for example by taking the opinions of users of the clustering into account. It is also possible, in a specific situation, that one graphic is more readable and provides easier separation than another: so it is useful to examine all the available graphics, to take

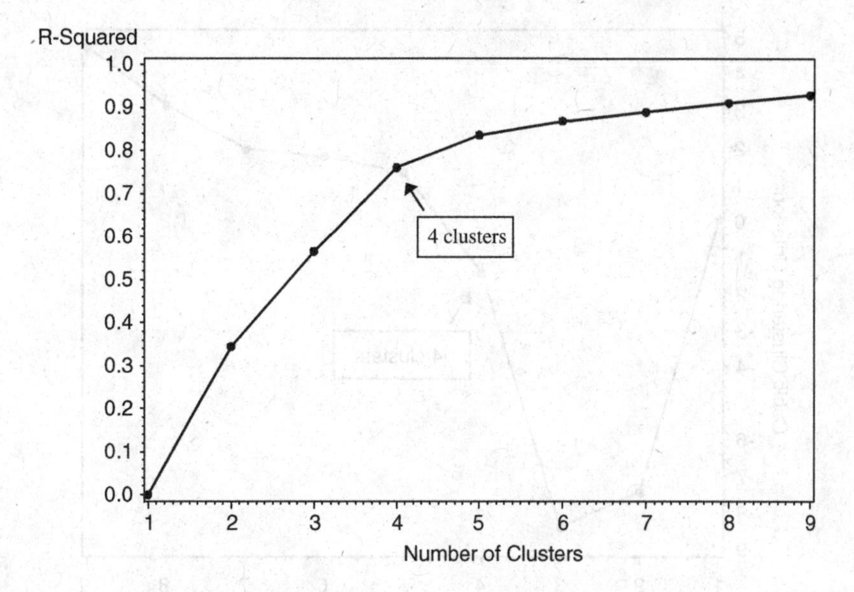

Figure 9.5 R^2 as a function of the number of clusters.

advantage of the information provided in them, which may not always be identical even if it is rarely conflicting.

9.8.1 All types of clustering

R^2 (RSQ) is the proportion of the sum of squares explained by the clusters (between-cluster sum of squares / total sum of squares). The nearer it is to 1, the better the clustering will be, but we should not aim to maximize it at all costs, because this would result in the maximum number of clusters: there would be one cluster per individual. So we need an R^2 that is close to 1 but without too many clusters. A good rule is that, if the last significant rise in R^2 occurs when we move from k to $k + 1$ clusters, the partition into $k + 1$ clusters is correct. In the example in Figure 9.5, the best number of clusters is obviously four.

The *cubic clustering criterion* (CCC), identified by Sarle (1983),[1] indicates whether the clustering is good (CCC > 2), requires examination (CCC between 0 and 2), or may be affected by outliers (CCC < 0). If the CCC is slightly negative, the risk of outliers is low, and this slightly negative value may indicate the presence of small clusters (see Figure 9.22). The risk of outliers becomes high only if CCC is markedly negative, at less than −30 or thereabouts. A good partition into $k + 1$ clusters will show a dip for k clusters and a peak for $k + 1$ clusters, followed either by a gradual decrease in the CCC, or a smaller rise (more generally, an inflection point) where isolated points or groups are present (as in the case of Figure 9.6, which still has an optimum at four clusters). The CCC should not be used with the 'single linkage' hierarchical method, but should preferably be used with the Ward or moving centres method (see below), and each cluster must have at least 10 observations.

The *pseudo F* measures the separation between all the clusters. It must be high. If n is the number of observations and c is the number of clusters, then

[1] Sarle, W.S. (1983) *Cubic Clustering Criterion*, SAS Technical Report A-108. Cary, NC: SAS Institute Inc.

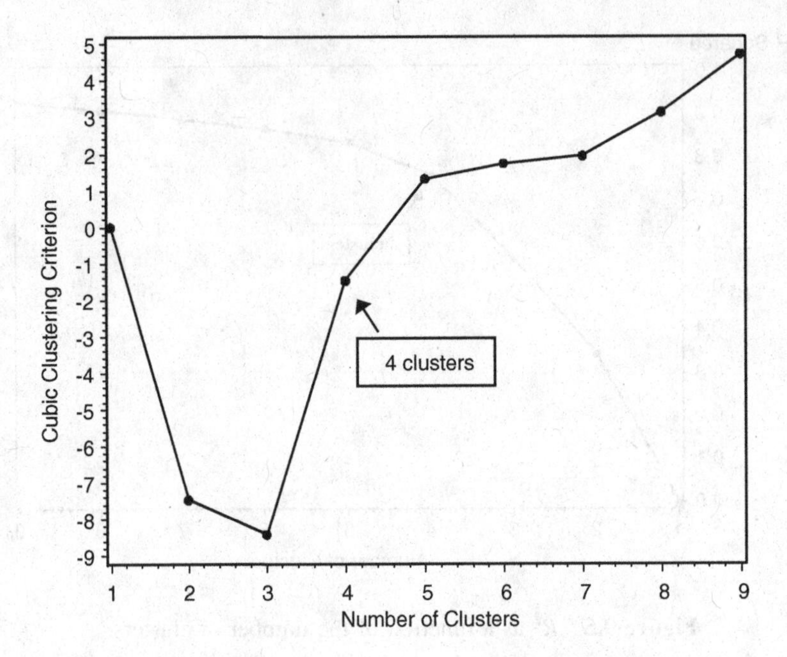

Figure 9.6 CCC as a function of the number of clusters.

$$pseudo\ F = \frac{R^2/(c-1)}{(1-R^2)/(n-c)}.$$

We speak of 'pseudo *F*' because, despite the analogy with the *F*-ratio of the analysis of variance, it does not follow a Fisher distribution. Like the CCC, the pseudo *F* should not be used with the 'single linkage' hierarchical method.

9.8.2 Agglomerative hierarchical clustering

The *semi-partial* R^2 (SPRSQ) measures the loss of between-cluster sum of squares caused by grouping two clusters together, in other words the decrease in R^2. The SPRSQ shown for k clusters is the loss of between-cluster sum of squares due to the change from $k + 1$ clusters to k clusters. Since the aim is to have maximum between-cluster sum of squares, we look for a low SPRSQ followed by a high SPRSQ on the following aggregation. In other words, a peak for k clusters and a dip for $k + 1$ clusters indicate a satisfactory clustering into $k + 1$ clusters. In the example shown in Figure 9.7, the optimum is four clusters (this is the same example as for the R^2 and the CCC, and I will describe this example more fully in Section 9.11.2; the values of the indicators are given in Figure 9.19). The loss of between-cluster sum of squares is only 0.07 when we move from 5 to 4 clusters, as against a loss of 0.20 when moving from 4 to 3 clusters. The curve is generally decreasing, but not always, because a merging of clusters may cause a smaller loss of between-cluster sum of squares than the previous merge.

The *pseudo* t^2 (PST2) measures the separation between the two clusters aggregated most recently: a peak for k clusters and a dip for $k + 1$ clusters indicate a satisfactory clustering into $k + 1$ clusters.

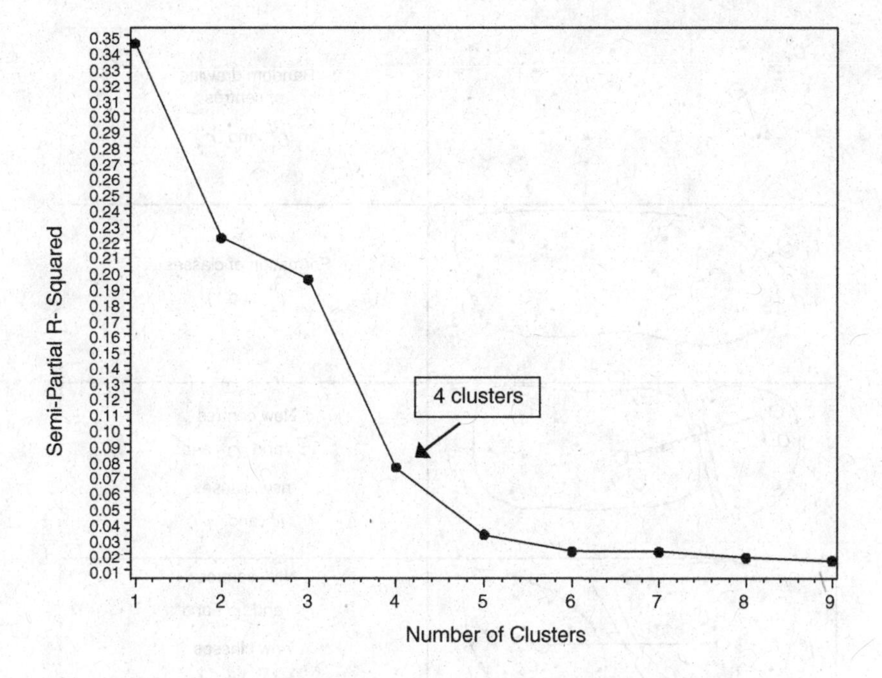

Figure 9.7 SPRSQ as a function of the number of clusters.

9.9 Partitioning methods

9.9.1 The moving centres method

This method, proposed by Forgy (1965),[2] with variants such as the *k-means* and *dynamic clouds* methods (see below), proceeds in the following way:

1. *k* individuals are chosen as the initial centres of the clusters (either by picking them at random, or taking the first *k*, or taking 1 out of *n/k*, although some packages such as the SAS one provide refinements as described in Section 9.11.2).

2. The distances between each individual and each centre c_i of the preceding step are calculated, and each individual is assigned to the nearest centre, thus defining *k* clusters.

3. The *k* centres c_i are replaced with the centres of gravity of the *k* clusters defined in step 2 (these centres of gravity are not necessarily individuals in the population).

4. A check is made to see if the centres have remained sufficiently stable (by comparing their movement to the distances between the initial centres) or if a fixed number of iterations has been completed:

   - if the answer is yes, the process stops (usually after at least 10 iterations);

   - if no, return to step 2.

[2] Forgy, E.W. (1965) Cluster analysis of multivariate data: efficiency versus interpretability of classification. *Biometrics*, 21, 768–769.

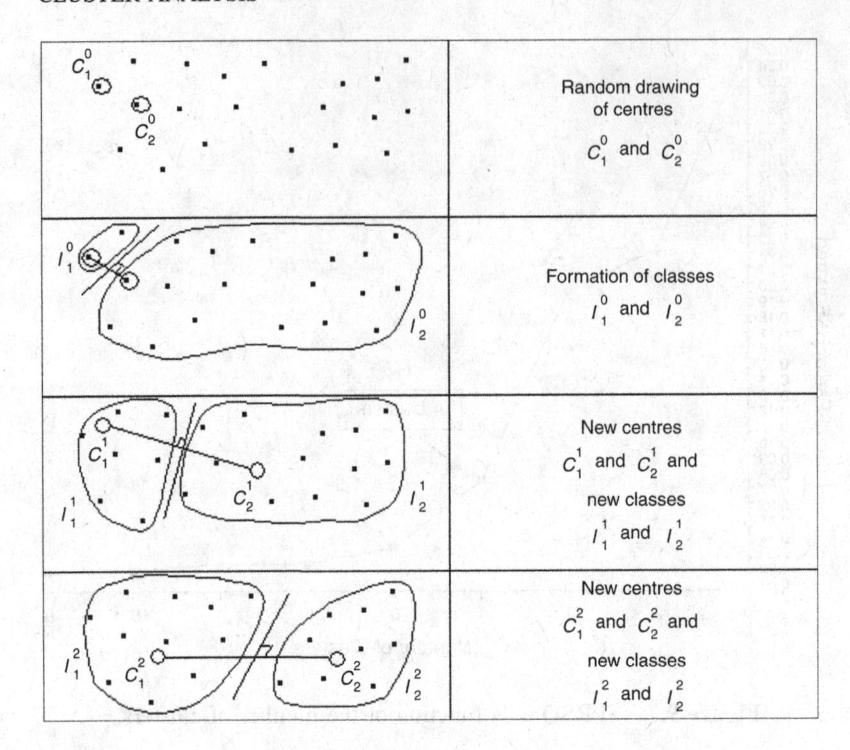

Figure 9.8 Moving centres.

Figure 9.8 shows how Ludovic Lebart illustrates this method.[3]

9.9.2 *k*-means and dynamic clouds

In MacQueen's *k*-means variant (1967),[4] the centre of gravity of each group is recalculated for each new individual introduced into the group, instead of waiting for the assignation of all the individuals before recalculating the centres of gravity. The convergence is faster and may even be completed in a single iteration, but the result depends on the order of the individuals in the data set.

Diday's *dynamic clouds* method (1971)[5] differs from the moving centres method mainly in the fact that each cluster is not represented by its centre of gravity (which may lie outside the population) but by a subset of the cluster, called the *kernel*, which, if it is well structured (for example, if it contains the most central individuals), will be more representative of its cluster than the centre of gravity.

[3] Lebart, L., Morineau, A. and Piron, M. (2006) *Statistique Exploratoire Multidimensionnelle: Visualisations et Inférences en Fouille de Données*. Paris: Dunod.

[4] MacQueen, J.B. (1967) Some methods for classification and analysis of multivariate observations. *Proceedings of the Fifth Berkeley Symposium on Mathematical Statistics and Probability*, Vol. 1, pp. 281–297. Berkeley, University of California Press.

[5] Diday, E. (1971) La méthode des nuées dynamiques. *Revue de Statistique Appliquée*, 19(2), 19–34.

In all these methods, there is one individual that is more representative than the others in the cluster, namely the one closest to the centre of gravity: this individual is called the *paragon*.

9.9.3 Processing qualitative data

The above methods are only applicable to continuous data, and therefore, if the data being investigated are non-continuous, then either they must be transformed into continuous data by multiple correspondence analysis, or other partitioning methods must be used.

The advantage of MCA is that it enables quantitative variables (binned into categories) and qualitative variables to be processed simultaneously, by applying the above methods to the factors resulting from the MCA.

Another approach is to use variants of the k-means method, such as *k-modes* (Huang, 1998)[6] which are applicable to qualitative data, and *k-prototypes* (Huang, 1997)[7] which combine k-means and k-modes and can be used on mixed data.

9.9.4 *k*-medoids and their variants

k-medoids, which are used in the PAM (Partitioning Around Medoids) algorithm of Kaufman and Rousseeuw (1990),[8] have the advantage of being more robust in relation to outliers than k-means. This robustness is due to the principle of the algorithm: a medoid is the representative of a cluster, chosen as its most central object, which is tested by systematic permutation of one representative and another object of the population chosen at random, to see if the quality of the clustering increases, in other words if the sum of the distances of all the objects from their representatives decreases. The algorithm stops when no further permutation improves the quality. The main disadvantage of this algorithm is its complexity, of the order of $O(ik(n-k)^2)$, where i is the number of iterations, k is the number of clusters and n is the number of objects.

To reduce the computing time, which can quickly become prohibitive, we can use certain devices such as that provided by the CLARA (Clustering LARge Applications) algorithm of Kaufman and Rousseeuw (1990),[9] which acts on a number of samples instead of on the total population, applies the PAM algorithm to them on each occasion, and finally accepts the best result. Both PAM and CLARA are implemented in R (in the *cluster* package) and S-PLUS. Since the quality of the clusters produced by CLARA depends on the sampling, a variant, known as CLARANS (Clustering LArge applications upon RANdomized Search), devised by Ng and Han (1994),[10] extends the area of the search for the objects and products clusters of better quality than PAM and CLARA, but does not have the capacity of CLARA to handle large databases.

---

[6] Huang, Z. (1998) Extensions to the k-means algorithm for clustering large data sets with categorical values. *Data Mining and Knowledge Discovery*, 2, 283–304.

[7] Huang, Z. (1997) Clustering large data sets with mixed numeric and categorical values. In *Proceedings of the First Pacific Asia Knowledge Discovery and Data Mining Conference*, pp. 21–34. Singapore: World Scientific.

[8] Kaufman, L. and Rousseeuw, P. (1990) *Finding Groups in Data: An Introduction to Cluster Analysis*. New York: John Wiley and Sons, Inc.

[9] Ibid.

[10] Ng, R.T. and Han, J. (1994) Efficient and effective clustering methods for spatial data mining. In *Proceedings of 20th International Conference on Very Large Data Bases*, Santiago de Chile, pp. 144–155.

9.9.5 Advantages of the partitioning methods

The main advantage of some of these methods (moving centres, k-means, k-modes, k-prototypes) is that their complexity is linear; in other words, their execution time is proportional to the number n of individuals (since the nk distances between the individuals are calculated at each step), so that they can be used with large volumes of data. Furthermore, the number of iterations needed to minimize the within-cluster sum of squares is generally small, making these methods even more suitable for such applications.

The second advantage, in the SAS implementation at least, is that it is possible to detect outliers, which appear in the form of clusters reduced to one element. These can be excluded from the set of initial centres by the DELETE option in the SAS FASTCLUS procedure (see Section 9.11.2). We can also use an option ('strict=s' in FASTCLUS) which assigns the cluster number $-k$ to each observation that is closer to the kth cluster than the others, but is separated from this cluster by a distance greater than the stated threshold s. This threshold is specified by inspecting the maximum distance (_radius_) between an individual and the centre of its cluster, and setting the value '_radius_' of the clusters with high frequencies slightly above this value (see Figure 9.9).

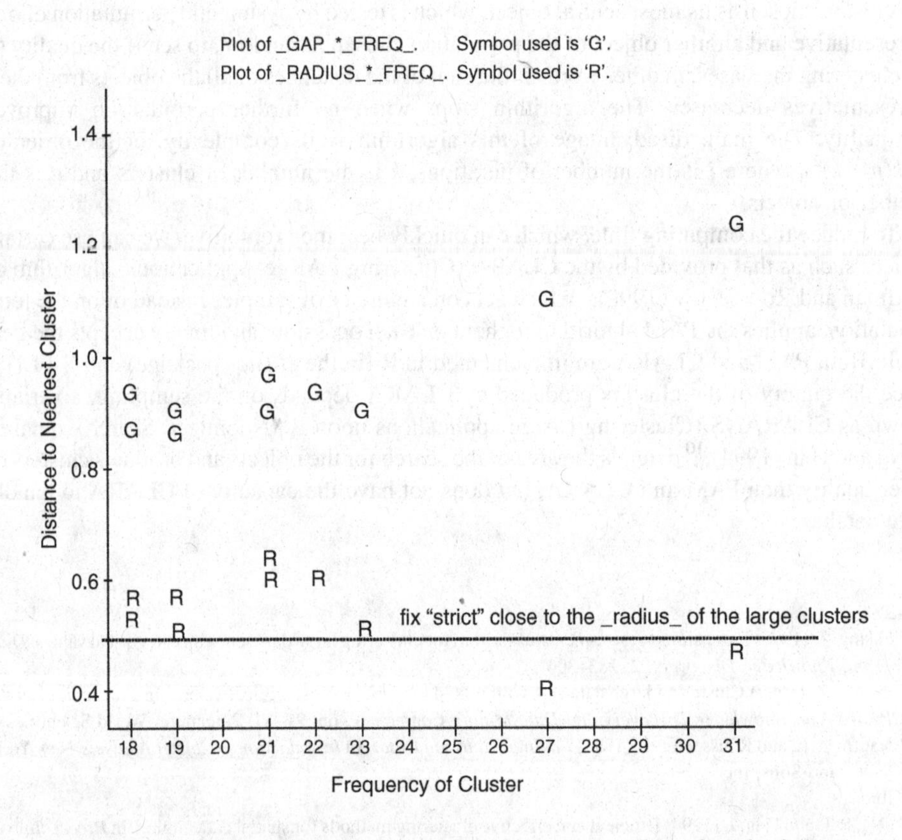

Figure 9.9 Cross-tabulation of the frequency and radius of each cluster.

There is also a third advantage. Unlike hierarchical methods, in which the clusters are not altered once they have been constructed, the reassignment algorithms constantly improve the quality of the clusters, which can thus reach a high level when the form of the data (spherical) is suitable.

9.9.6 Disadvantages of the partitioning methods

The first disadvantage of partitioning methods is that the final partition depends greatly on the more or less arbitrary initial choice of the centres c_i (see Section 9.9.7). Consequently we do not have a global optimum, but simply the best possible partition based on the starting partition.

To mitigate this problem, some algorithms make several random selections of the c_i, compare the resulting clusterings, and cross-tabulate the clusters to establish *strong forms*. Note that, if we perform p clusterings with k clusters each, we can have k^p strong forms, each strong form $\{f_1, f_2, \ldots, f_p\}$ containing the individuals classed firstly in cluster f_1 (there are k possibilities), and secondly in cluster f_2, and so on. In particular, the number of strong forms is k^2 when two clusterings are performed (this may be easier to understand): the individuals can be distributed in a two-way table where the intersection of the ith row and the jth column contains the individuals classed in cluster i on one occasion and in cluster j on another occasion. Since this number of strong forms is much too large to be usable, we only keep those having a significant frequency, by setting a threshold corresponding to a switch of the frequency, and reassigning the individuals who do not belong to one of the selected strong forms.

Another solution is to carry out a number of clusterings on different randomly chosen samples of the global population, and use the centres of the best clustering obtained as the initial centres of the algorithm applied to the whole population. A simpler solution is to perform two clusterings: the first of these is simply intended to provide the final centres which are the initial centres of the second clustering. All software packages allow the user to choose the initial centres instead of picking them at random. The following SAS syntax is an example of this (see also Section 9.11.2):

```
PROC FASTCLUS DATA=test SUMMARY MAXC=10 MAXITER=50
OUTSEED=centres DELETE=2; VAR &var;
DATA=centres; SET centres; WHERE _FREQ__ > 2;
PROC FASTCLUS DATA=test SEED=centres SUMMARY MAXC=10
MAXITER=50
STRICT=0.6 OUT=partition; VAR &var;
RUN;
```

The first run of the FASTCLUS procedure excludes the centres with two or fewer individuals attached (DELETE = 2) and places the final centres in the OUTSEED data set. This file is then read as the SEED data set of the initial centres during the second execution of FASTCLUS. Note the use of the STRICT = 0.6 option mentioned above for processing individuals at a distance of more than 0.6 from any cluster. We can see the effect of this option on individuals 102 and 103 who are given a negative cluster number of −1 (−3) because they are separated by a distance of 0.65 (0.64) from the centre of their cluster 1 (3).

102	−0.59774	−1.05012	4	−1	0.65400
103	−0.80046	−0.97533	4	−3	0.64009

Note also that, between the two runs of FASTCLUS, we exclude the clusters with a frequency (_FREQ_) of 2 or less from the CENTRES file. The DELETE option could have left some of them there, namely those created in the last iteration, after which there is no further exclusion of centres (to prevent the elimination of individuals in the corresponding cluster).

The second drawback of partitioning methods is that the number of clusters, k, is fixed in these methods, and is not less than k unless certain clusters are empty or are excluded by the DELETE option from the SAS FASTCLUS procedure (see Section 9.11.2). If this number does not correspond to the actual configuration of the cloud of individuals, the quality of the clustering may be adversely affected. We can try to mitigate this problem by testing different values of k, but this increases the duration of the processing. We can also use PCA to visualize the individuals and attempt to identify the clusters.

The third drawback of these methods is that they are only good at detecting spherical forms. Even convex forms such as ellipses cannot be detected well if they are not sufficiently separated.

9.9.7 Sensitivity to the choice of initial centres

An illustration of the effect of the choice of initial centres is provided in Figures 9.10 and 9.11.

If the initial centres are A, B and C, the first clusters are $\{A\}$, $\{B\}$ and $\{C,D,E,F,G\}$. The centres of gravity of these clusters are A, B, and a point I located near the centre of the graphic. In the next iteration, all the points remain in their cluster, except for C which is closer to B than to the centre of gravity I. The final clusters are therefore $\{A\}$, $\{B,C\}$ and $\{D,E,F,G\}$.

If the initial centres are A, D and F, the first clusters are $\{A,B,C\}$, $\{D,E\}$ and $\{F,G\}$. The centres of gravity of these clusters are such that no point changes its cluster in the next iteration.

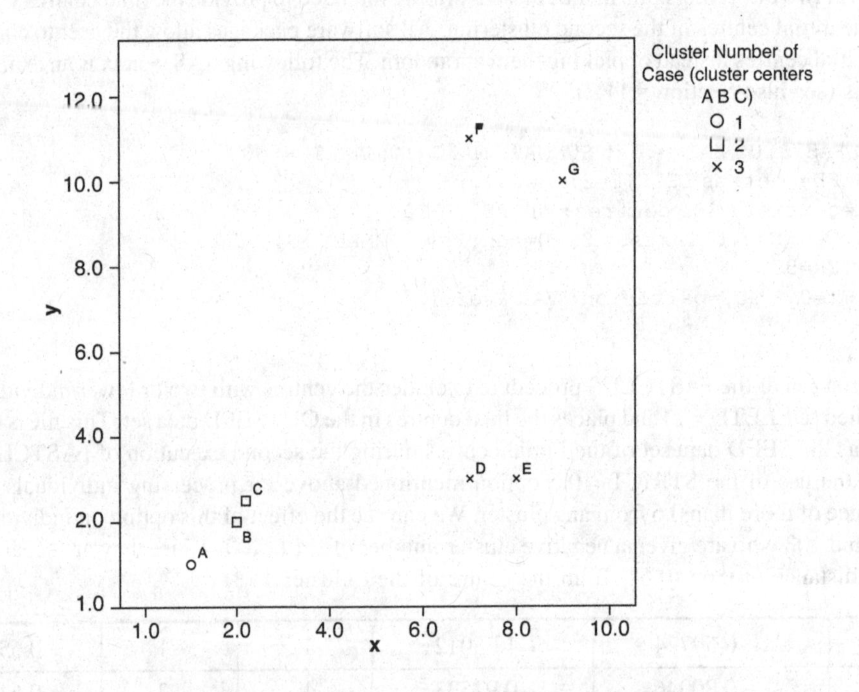

Figure 9.10 Poor choice of initial centres.

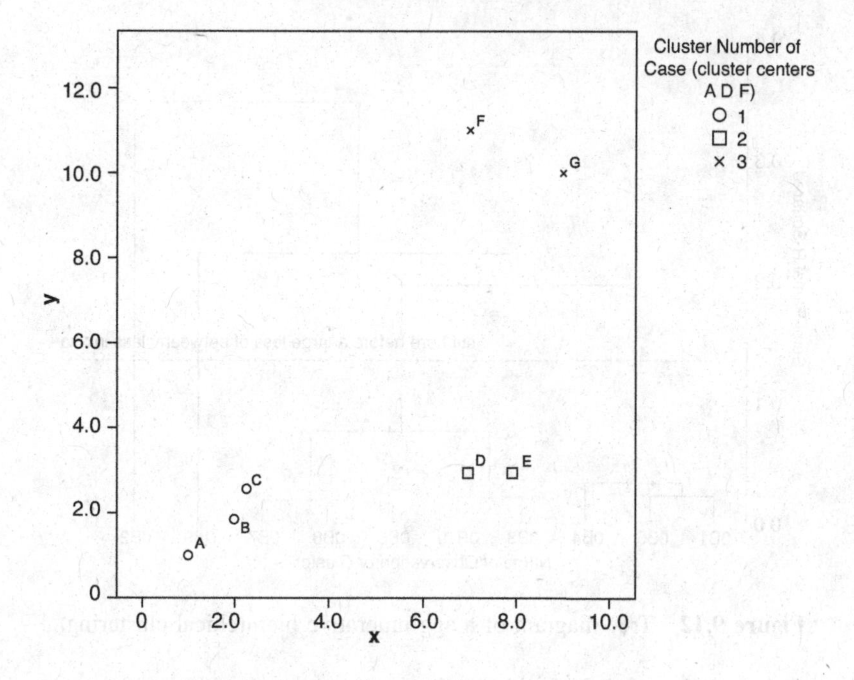

Figure 9.11 Good choice of initial centres.

9.10 Agglomerative hierarchical clustering

9.10.1 Introduction

Unlike the methods considered above, which are *non-hierarchical* and produce a partition into a certain number (which may or may not be fixed) of clusters, agglomerative hierarchical clustering (AHC) produces sequences of nested partitions of increasing heterogeneity, between partition into *n* clusters where each object is isolated and partition into one cluster which includes all the objects. AHC can be used if there is a concept of distance, which can be in either an individual space or a variable space. We must have defined the distance of two objects, generally natural, and the distance of two clusters, which gives us more possibilities, as we will see in the next section.

The general form of the algorithm is as follows:

Step 1. The initial clusters are the observations.

Step 2. The distances between clusters are calculated.

Step 3. The two clusters which are closest together are merged and replaced with a single cluster.

Step 4. We start again at step 2 until there is only one cluster, which contains all the observations.

The sequence of partitions is represented in what is known as a *tree diagram* (see Figure 9.12), also known as a *dendrogram*. This tree can be cut at a greater or lesser height to obtain a

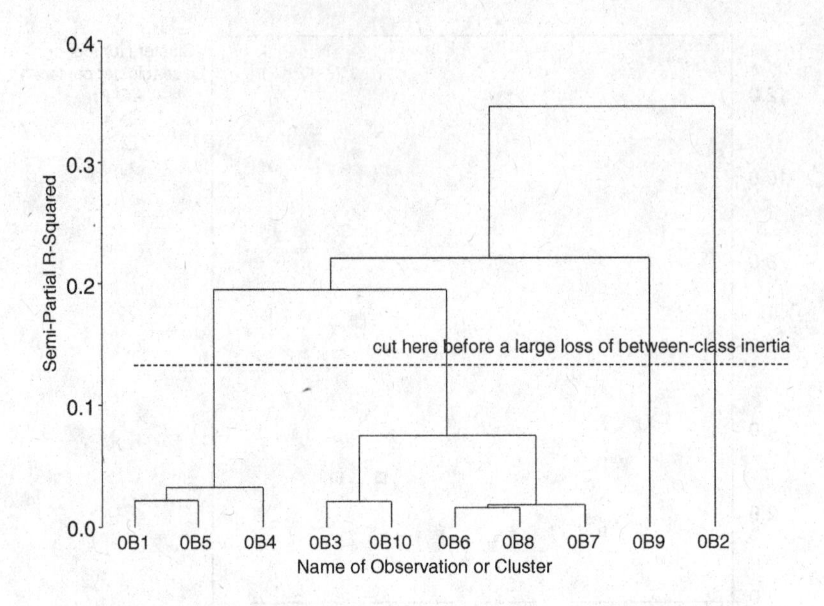

Figure 9.12 Tree diagram of a agglomerative hierarchical clustering.

smaller or larger number of clusters; this number can be chosen by the statistician to optimize certain statistical quality criteria (Section 9.8). The main criterion is the loss of between-cluster sum of squares (semi-partial R^2), represented in Figure 9.12 by the height of the two connected branches: since this loss must be as small as possible, the tree diagram is cut at a level where the height of the branches is large.

9.10.2 The main distances used

Thus the AHC algorithm works by searching for the closest clusters at each step and merging them, and the critical point of the algorithm is the definition of the distance between two clusters A and B. When each of the two clusters is reduced to one element, the definition of their distance is natural (it is usually the Euclidean distance between the two members), but as soon as a cluster has more than one element the concept of the distance between two clusters is less obvious. It can be defined in many ways, but the most usual definitions are the ones described below.

The *maximum distance* between two observations $a \in A$ and $b \in B$ tends to generate clusters of equal diameter. By definition, it is highly sensitive to outliers, and is therefore little used. The corresponding form of AHC is called 'farthest-neighbor technique', 'diameter criterion' or 'complete linkage' AHC.

The *minimum distance* between two observations $a \in A$ and $b \in B$ defines what is known as 'nearest-neighbor technique' or 'single linkage' AHC. Its weak point is that it is sensitive to the 'chain effect' (or chaining): if two widely separated clusters are linked by a chain of individuals who are close to each other, they may be grouped together.

Thus, in Figure 9.13, we can distinguish two natural sets, corresponding to clusters 1 and 2 on one hand, and 4 on the other hand, cluster 3 being the 'umbilical cord' between the two sets. These two sets are detected well by most of the methods (see below), including the Ward, average linkage and complete linkage methods. However, if the single linkage method is used to find two clusters, it isolates cluster 2 and groups all the rest together (see Figure 9.14). This

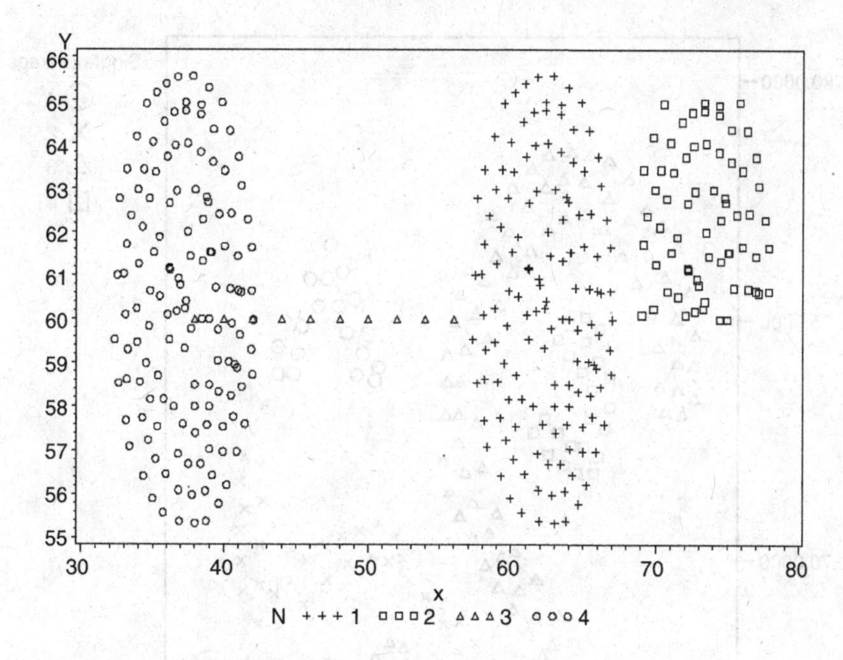

Figure 9.13 Illustration of the chain effect.

is due to the fact that the closest points of clusters 1 and 2 are separated by a distance that is greater than the shortest distance between two points of cluster 1, of clusters 1 and 3, of cluster 3, and of clusters 3 and 4. Since the distance between two clusters is the shortest distance

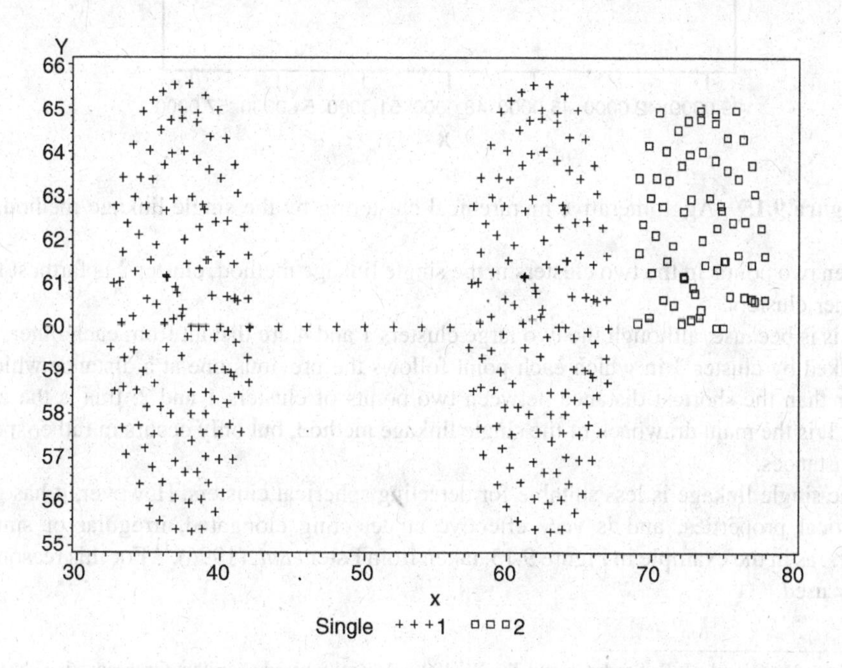

Figure 9.14 Sensitivity of single linkage to the chain effect.

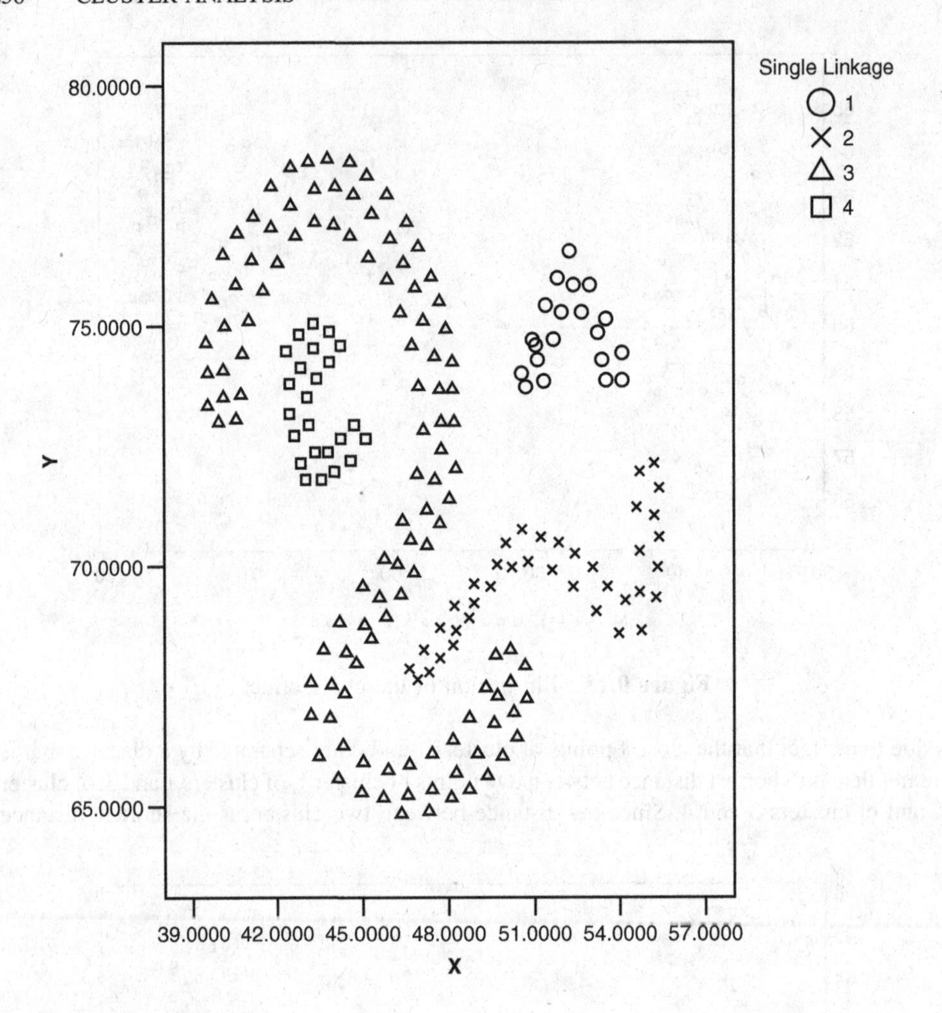

Figure 9.15 Agglomerative hierarchical clustering by the single linkage method.

between two points in the two clusters in the single linkage method, cluster 2 is farthest from the other clusters.

This is because, although the two large clusters 1 and 4 are distant from each other, they are linked by cluster 3 in which each point follows the previous one at a distance which is shorter than the shortest distance between two points of clusters 1 and 2: this is the chain effect. It is the main drawback of the single linkage method, but only occurs in rather special circumstances.

The single linkage is less suitable for detecting spherical clusters. However, it has good theoretical properties, and is very effective at detecting elongated, irregular or sinuous clusters, as in the example in Figure 9.15, taken from Ester *et al.* (1996).[11] For this reason it is widely used.

---

[11] Ester, M. Kriegel, H.-P., Sander, J. and Xu, X. (1996) A density-based algorithm for discovering clusters in large spatial databases with noise. In *Proceedings of the 2nd ACM SIGKDD*, Portland, OR, pp. 226–231.

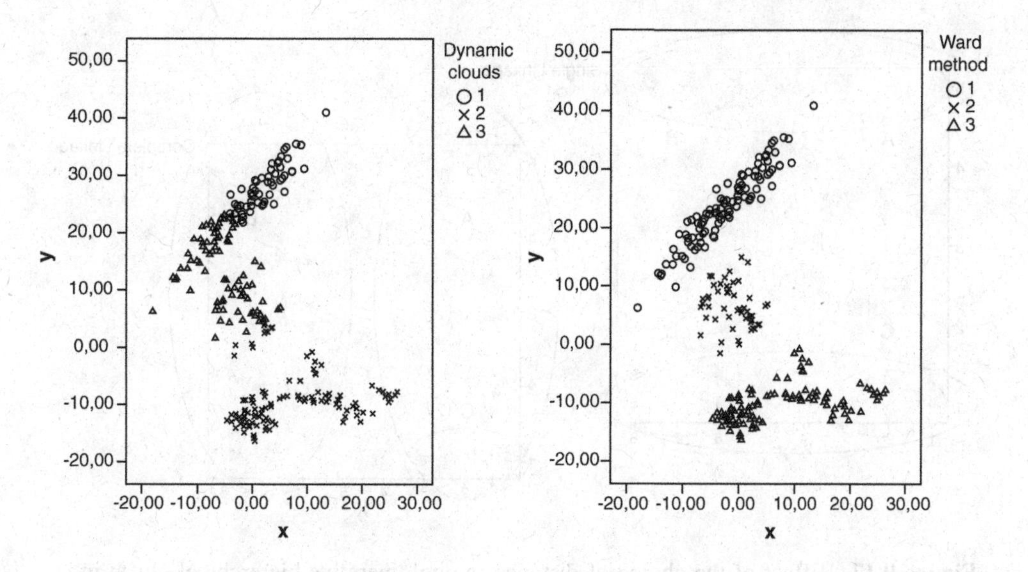

Figure 9.16 The Ward method and the dynamic clouds method.

The *mean distance* between two observations $a \in A$ and $b \in B$, which defines what is known as 'average linkage' AHC, is intermediate between the maximum distance and minimum distance methods, and is less sensitive to noise. It tends to produce clusters having the same variance.

The *distance between the centres of gravity* of A and B, which defines what is known as the 'centroid' method of AHC (the centre of gravity is sometimes referred to as the centroid), is more robust to outliers but less precise. This is the simplest in terms of calculation.

The *Ward method* (Figure 9.16) is one of those which match the purpose of the clustering most closely. Since effective clustering is clustering in which the between-cluster sum of squares is high, and since a change from clustering into $k+1$ clusters to clustering into k clusters (grouping of two clusters) can only reduce the between-cluster sum of squares, the aim is to merge the two clusters which cause the smallest decrease in the between-cluster sum of squares. The concept of distance corresponding to this objective is the Ward distance between two clusters, defined as the reduction in between-cluster sum of squares (or the increase in the within-cluster sum of squares) due to their merging.

The Ward distance between two clusters A and B having centres of gravity a and b, and frequencies n_A and n_B, is

$$d(A,B) = \frac{d(a,b)^2}{n_A^{-1} + n_B^{-1}}.$$

We can see that it is a function of the distance between centres of gravity. Although the Ward method tends to produce spherical clusters with the same frequencies (owing to the form of the above formula), is relatively ineffective for elongated clusters (for which single linkage is preferable) and very sensitive to outliers, it is by far the most popular method for agglomerative hierarchical clustering, because it is effective when applied to real problems.

The choice of distance is not of purely academic interest. Its effect on the shape of the detected clusters is shown in Figure 9.17. When four clusters are selected, the single linkage

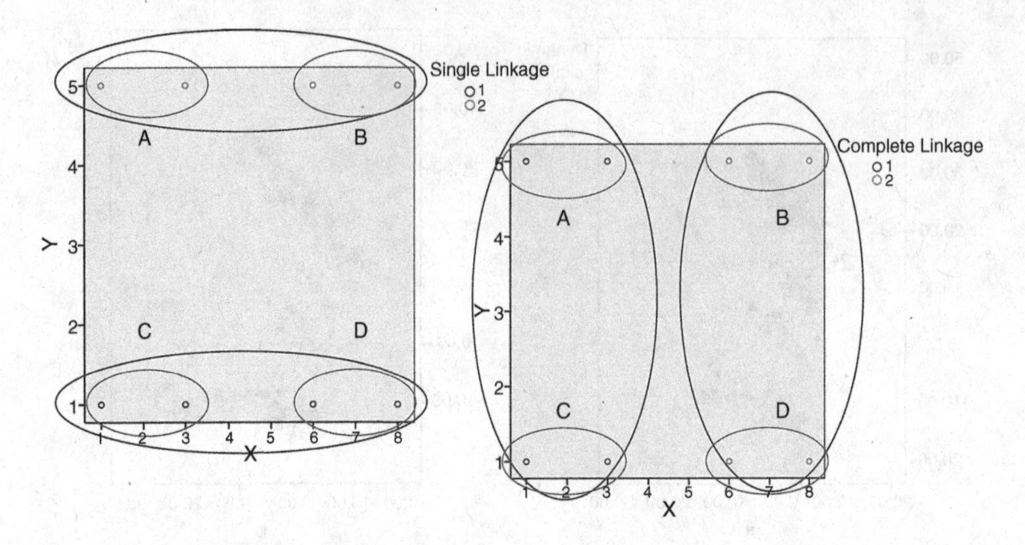

Figure 9.17 Effect of the choice of distance in agglomerative hierarchical clustering.

and the complete linkage methods both result in the same clusters, A, B, C and D. However, when two clusters are selected, the two methods do not lead to the same result. This is because $d_{\min}(A,B) = d_{\min}(C,D) = 3$, $d_{\min}(A,C) = d_{\min}(B,D) = 4$, and $d_{\min}(A,D) = d_{\min}(B,C) = 5$. Furthermore, $d_{\max}(A,B) = d_{\max}(C,D) = 7$, $d_{\max}(A,C) = d_{\max}(B,D) = \sqrt{2^2 + 4^2} = 4.47$, and $d_{\max}(A,D) = d_{\max}(B,C) = \sqrt{4^2 + 7^2} = 8.06$. Because of this, the single linkage method groups clusters A and B together on the one hand, and C and D on the other hand, whereas the complete linkage method groups clusters A and C on the one hand and B and D on the other.

9.10.3 Density estimation methods

In addition to the above methods, three others will be mentioned. These *density estimation* methods are often among the most suitable for detecting the structure of rather complex clusters. We can picture the data space as a landscape of peaks and valleys, where the mountains are the clusters and bottoms of the valleys are their boundaries. In this picture, the mountains are regions of high density. Density is generally defined as the number of objects in a certain neighbourhood. It is estimated by one of three methods:

- the k-nearest-neighbours method (the density at a point x is the number k of observations in a sphere centred on x, divided by the volume of the sphere);
- the uniform kernel method (in which the radius of the sphere is fixed, not the number of neighbours);
- the Wong hybrid method (which uses the k-means algorithm in a preliminary analysis).

A distance d_P between two clusters is then defined as inversely proportional to the density in the middle of these two clusters (it is assumed that $d_P = \infty$ if the two clusters are not adjacent). The last step is to apply a single linkage AHC method to the d_P.

These methods are effective for detecting all types of cluster, especially clusters having irregular shapes and unequal sizes and variances. This arises from the principle of these methods, which does not specify any shape for the cluster in advance: a cluster grows in any direction where the density is great enough. This principle is different from that of partitioning methods such as moving centres, k-medoids and their derivatives, which are based on the reassignment of objects to different clusters whose number is predetermined. Density estimation methods operate by specifying not the number of clusters, but a *smoothing parameter*, which, depending on circumstances, can be:

- the number k of neighbours of each point x;
- the radius r of the sphere surrounding x;
- the number of clusters of the preliminary k-means (Wong method).

The problem with these methods is the difficulty of finding a good value of the smoothing parameter (see Figure 9.18). Their constraints are that, on the one hand, it is better to standardize the continuous variables and exclude the outliers, and, on the other hand, that they require sufficiently high frequencies. It should be noted that these methods are not only associated with hierarchical algorithms, but are also found in partitioning algorithms such as the popular DBSCAN[12] and its extension OPTICS,[13] which are very suitable for finding different shapes of clusters.

Here is the SAS syntax for clustering by the 20 nearest neighbours:

```
PROC CLUSTER DATA=test OUTTREE=tree METHOD=density k=20 CCC PSEUDO
PRINT=10;
VAR &var; RUN;
```

If we replace the instruction 'k = ' by 'r = ' or 'hybrid', we obtain the uniform kernel method or the Wong hybrid method.

The following is the 'two-stage' variant which stops large clusters merging with each other if not all the small clusters have merged with other clusters:

```
PROC CLUSTER DATA=test OUTTREE=tree METHOD=twostage k=20 CCC PSEUDO
PRINT=10;
VAR &var; RUN;
```

This two-stage variant is rather more effective than the standard 'density' method.

9.10.4 Advantages of agglomerative hierarchical clustering

This type of clustering does not suffer from the two major drawbacks of the moving centres method, namely its dependence on the choice of initial centres and the fixed number of clusters chosen in advance. Instead, it allows the optimal number of clusters to be chosen, using indicators such as the pseudo t^2 and semi-partial R^2.

The second advantage of AHC is that it can detect clusters of different shapes, according to the distance chosen (Ward, single linkage and similar methods). The best possible distance

[12] Ibid.

[13] Ankerst, M., Breunig, M., Kriegel, H.-P. and Sander, J. (1999) OPTICS: Ordering points to identify the clustering structure. In *Proc. 1999 ACM-SIGMOD Int. Conf. Management of Data (SIGMOD'99)*, pp. 49–60.

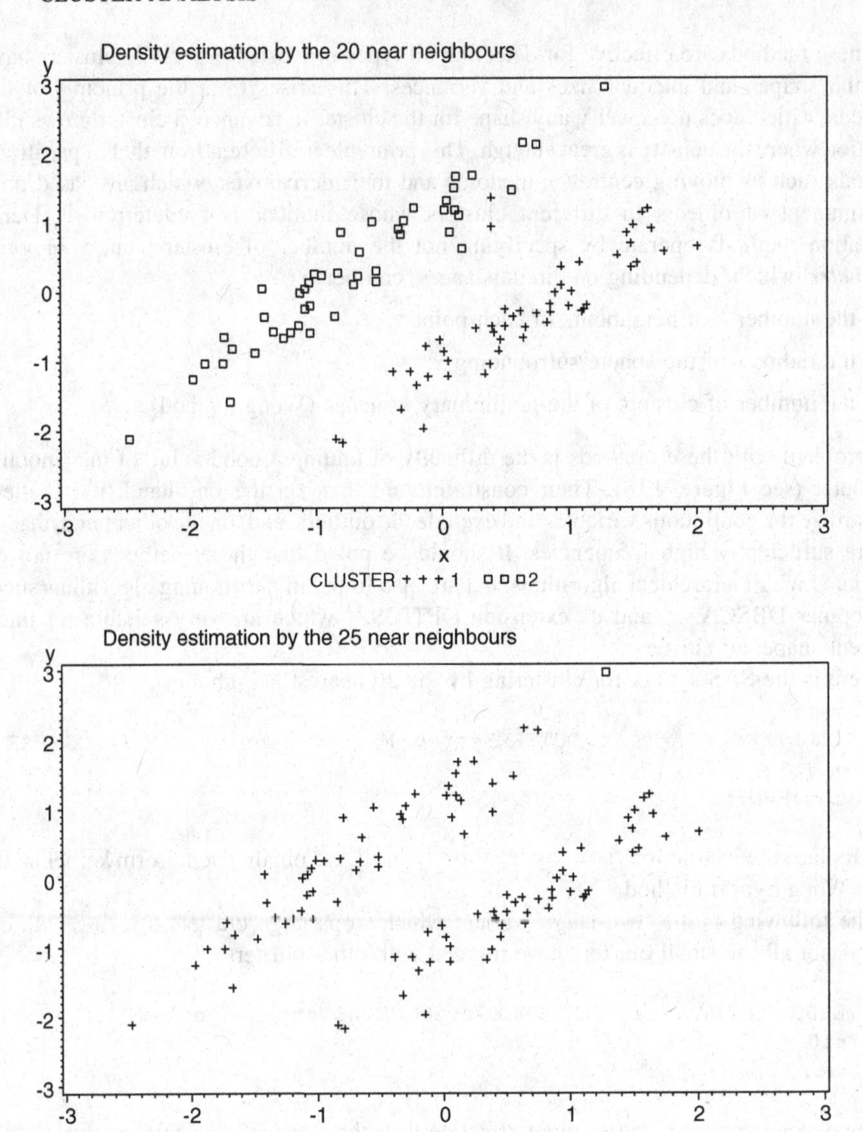

Figure 9.18 Density estimation and effect of the smoothing parameter.

can be chosen according to previous knowledge of the shape of the clusters to be detected (elongated, curved, with outliers, etc.), but the choice is mainly made by carrying out a number of trials and observing the results. We may decide to use a given distance because the resulting clusters are more easily described (see Section 9.5.5) or because the number of clusters appears more natural with respect to the graphic indicators (see Section 9.8).

The third advantage of AHC is that it enables us to cluster individuals, variables or centres of clusters obtained by using a moving centres algorithm (if centres are clustered, we improve the results if we know not only the centres of the clusters but also the within-cluster sum of squares and the frequencies of the clusters).

9.10.5 Disadvantages of agglomerative hierarchical clustering

The main disadvantage of AHC is its algorithmic complexity, which is non-linear: in order to move from $k + 1$ clusters to k clusters, we must calculate $(k + 1)k/2$ distances and combine the two closest clusters. If n is the number of individuals to be clustered, the complexity of the basic algorithm is of the order of n^3, and it will soon exceed the capacity of even a powerful computer. The difference from the moving centres method is that we are comparing individuals with each other instead of comparing them with the centres of the clusters only.

The situation is improved to a certain extent by the nearest-neighbour algorithm, which reduces a complexity of n^3 to n^2, by a judicious combination of more than two observations (not only the two nearest) on each iteration. It can be used with the Ward, average linkage, single linkage and complete linkage methods.

A second drawback is that, at each step, the partitioning criterion is not global, but depends on the clusters obtained so far: two individuals placed in different clusters will no longer be compared. In other words, this type of clustering into n clusters is not necessarily the best possible outcome, but only the best of those obtained by combining the clusters of a clustering procedure into $n + 1$ clusters. Thus some natural clusters may be hidden by an earlier branching.

9.11 Hybrid clustering methods

9.11.1 Introduction

Hybrid methods combine the strengths of both hierarchical methods and partitioning methods (moving centres, *k*-means, etc.), namely the precision and the lack of *a priori* specification of the former type, and the speed of the latter type.

In these methods, a first clustering is carried out on the n observations by a moving centres or *k*-means method, specifying a number of clusters between 10 and 100 (or the Wong limit, $n^{0.3}$). This value is set at a high enough level to limit the risk of merging natural clusters and greatly increase the between-cluster sum of squares. An agglomerative hierarchical clustering is then performed on the centres of these clusters (or of their strong forms), not on the initial observations (note: the centres do not have to form part of the initial observations), and the tree diagram is cut at a height that is considered to be optimal. In this way we obtain the final clusters, to which the initial observations are assigned by means of their pre-clusters which resulted from the first step. In some software (SPAD), the AHC is followed by optimization carried out by performing a clustering of the moving centres on the centres of the clusters resulting from the AHC.

Here are some examples of hybrid methods:

- The Wong density estimation method must be preceded by a *k*-means procedure (the other density estimation methods are not hybrid).
- Ward, average linkage and centroid agglomerative hierarchical clusterings may be preceded by a *k*-means procedure (the first method is widely used).

- The 'two-step cluster component' algorithm of IBM SPSS Statistics is based on the BIRCH algorithm (Zhang *et al.*, 1996),[14] which has the advantage of handling both quantitative and qualitative variables, and has linear complexity (effective with large volumes), but which, in the case of quantitative variables, has the same disadvantage as the k-means method, in that it can only detect spherical clusters satisfactorily.

- The BRIDGE algorithm (Dash *et al.*, 2001)[15] links a k-means procedure with a DBSCAN procedure, thus providing both high speed and excellent quality of detection of clusters of arbitrary shape.

9.11.2 Illustration using SAS Software

I have reproduced one of the data sets from the article by Ester *et al.* cited in Section 9.10.2. We start by standardizing the continuous variables x and y, in other words centring them (so that they have mean 0) and reducing them (so that they have standard deviation 1). This step is strongly recommended if there are substantial differences between the variances or means of the variables. We can use the conventional STANDARD procedure or the more recent STDIZE procedure. The first line of the syntax associates the variables x, y, which are analysed into the macro-variable VAR, which is then called &VAR in the rest of the syntax, and enables the analysed variables to be changed by replacing their name only on the line for assignment to the macro-variable.

```
%LET var = x y;
PROC STANDARD DATA=dbscan OUT=test MEAN=0 STD=1;
VAR &var;
RUN;
```

We then launch a moving centres procedure to obtain a maximum number of clusters specified by the parameter MAXC (10 in this case), in a maximum number of iterations MAXITER (50 in this case). MAXITER is a maximum which will not be reached if the algorithm converges previously. The default value of this parameter is 1, but it should preferably be at least 10, and if it is given the value 0 each individual is directly assigned to one of the initial centres and the final clusters are immediately produced. The choice of the parameter CONVERGE = 0.02 stops the iterations when no centre is displaced through a distance greater than 0.02 times the minimum distance between the initial centres. This option is not applicable if MAXITER $\leq$ 1. Another option, DELETE = 2, excludes the centres with two or fewer individuals attached (this is to prevent having an outlier as the centre). This exclusion takes place after each iteration except the last, and the excluded centres are not replaced. The parameter RADIUS = d specifies the minimum distance (default value 0) between two initial centres at the time of their selection. If the initial centres are to be chosen at random, for example in order to find the strong forms, we can set the parameter REPLACE = RANDOM. We can also choose the parameter REPLACE = NONE, to accelerate the selection of the initial centres by preventing the substitution of the nearest centres (set 'radius' to a large

[14] Zhang, T., Ramakrishnan, R. and Linvy, M. (1996). BIRCH: An efficient data clustering method for large databases. In *Proc. of 1996 ACM-SIGMOD International Conference on Management of Data*, Montreal, Quebec.

[15] Dash, M., Liu, H. and Xu, X. (2001). '1 + 1 > 2': Merging distance and density based clustering. *Proceedings of the 7th International Conference on Database Systems for Advanced Applications (DASFAA)*, pp. 32–39.

enough value), but it is best to leave the default option REPLACE = FULL. Note that the DRIFT option can be used to replace the moving centres method with the k-means method.

```
PROC FASTCLUS DATA=test SUMMARY MAXC=10 MAXITER=50 CONVERGE=0.02
MEAN=centres OUT=partitio CLUSTER=presegm DELETE=2;
VAR &var;
RUN;
```

At the output of the algorithm, each individual is assigned in the OUT data set to a cluster whose number is specified by the variable CLUSTER = presegm. These clusters are described in the MEAN data set, which contains one observation per cluster, making 10 in this case. The list of statistics displayed can be limited by the SHORT and SUMMARY options. If this is not done, the statistics are displayed for each variable, revealing the variables which make the greatest contribution to the clustering: these are the ones with the highest R^2 (between-cluster sum of squares/total sum of squares) and the highest $R^2/(1 - R^2)$ (the ratio of between-cluster sum of squares to the within-cluster sum of squares). In this case, the two variables have very high and similar values of R^2:

Statistics for Variables

Variable	Total STD	Within STD	R-Square	RSQ/(1-RSQ)
X	1.00000	0.24548	0.942225	16.308599
Y	1.00000	0.23181	0.948484	18.411491
OVER-ALL	1.00000	0.23874	0.945355	17.299831

The table also contains the global standard deviation, 'Total STD', and the intra-cluster standard deviation 'Within STD' of each variable, and $R^2 = 1-(\text{Within STD}/\text{Total STD})^2$. All these results are provided for each variable and also globally (the 'OVER-ALL' row).

The MEAN data set contains one row for each resulting cluster, with the following information:

Obs	presegm	_FREQ_	_RMSSTD_	_RADIUS_	_NEAR_	_GAP_	X	Y
1	1	21	0.26371	0.62286	7	0.90908	0.03354	0.78624
2	2	31	0.22412	0.45198	6	1.23883	1.25483	1.60977
...	...	...	...	...	...	...	...	...

- CLUSTER = presegm: number of the cluster
- _FREQ_: number of individuals in the cluster
- _RMSSTD_: within-cluster sum of squares ('root-mean-square standard deviation'); the lower this value is, the more homogeneous the cluster, which is preferable
- _RADIUS_: maximum distance between an individual in the cluster and its centre
- _NEAR_: number of the nearest cluster
- _GAP_: the distance between the centre of the cluster and the nearest centre of another
- X, Y: values of the initial variables &VAR for the centre of the cluster.

Remember that the centre of the cluster, i.e. its centre of gravity, is not necessarily an individual in the cluster. The variables _FREQ_ and _RMSSTD_, and the initial variables &VAR, will be used for the subsequent agglomerative hierarchical clustering.

The OUT data set contains the initial variables for each individual (N is the cluster number, which is known *a priori* in our example), together with the following variables:

Obs	X	Y	N	presegm	DISTANCE
1	−1.92073	−1.49419	1	9	0.38436
2	−1.88339	−1.36798	1	9	0.33055

- CLUSTER = presegm: number of the cluster
- DISTANCE: the distance between the individual and the centre of his cluster.

The statistics at the output of the procedure, namely the pseudo *F*, CCC, and the expected R^2 from the clustering, are then inspected, to ascertain that the expected R^2 approaches 1 and that CCC > 2.

CCC > 2 is satisfied in this case, but its value is really only valid if the variables are not correlated. Similarly, the expected R^2 is calculated under the null hypothesis of the linear independence of the variables. It is also less than the observed R^2, which is 0.945355.

As a general rule, we aim to maximize these indicators by testing different values of the number of clusters.

Pseudo F Statistic = 401.74

Approximate Expected Over-All R-Squared = 0.90858

Cubic Clustering Criterion = 8.533

When this step of partitioning by the moving centres method is complete, we can move on to the agglomerative hierarchical clustering. As mentioned above, the most widely used method is the Ward method. This is applied to the 10 centres at the output of the FASTCLUS procedure, contained in the ‘centres’ data set.

```
PROC CLUSTER DATA=centres OUTTREE=tree METHOD=ward CCC PSEUDO
PRINT=10; VAR &var; COPY presegm; RUN;
```

Because the input data set DATA contains variables named _FREQ_ and _RMSSTD_ created by the previous procedure, SAS knows that the input data are centres of clusters, not individuals, and it uses the frequency _FREQ_ and the within-cluster sum of squares _RMSSTD_ of each cluster to optimize the AHC. This is because the proximity of two centres must be assessed with allowance for the size and dispersion (within-cluster sum of squares) of the corresponding clusters. These two variables are used whenever the AHC is preceded by a moving centres procedure. They can also be used if the AHC is used directly on centres of clusters, even if they are not defined by a previous FASTCLUS procedure, in which case the two variables must be entered under their names _FREQ_ and _RMSSTD_ by the user.

The output data set, OUTTREE, is used to construct the tree diagram. It contains all the information needed to plot the tree diagram, as well as information for the choice of the best number of clusters to use, in other words the level at which the tree should be cut. This data set contains an observation for each node of the tree of the cluster hierarchy, i.e. normally $2n-1$ observations (n = number of objects to be clustered), corresponding either to the objects to be clustered (nodes at the first level of the tree $\Leftrightarrow$ clusters reduced to one object), or to the subsequent clusters resulting from the grouping of objects and clusters.

The clustering is based on the variables &VAR, and the instruction COPY presegm is added, to copy into the output data set the variable presegm which is the pre-cluster number calculated by the first step of the moving centres procedure.

The CLUSTER procedure can also be used to perform an AHC by the Wong hybrid method. Select the instruction HYBRID and choose one of the two methods DENSITY and TWOSTAGE:

```
PROC CLUSTER DATA=centres OUTTREE=tree METHOD=density HYBRID CCC
PSEUDO PRINT=10...

PROC CLUSTER DATA=centres OUTTREE=tree METHOD=twostage HYBRID CCC
PSEUDO PRINT=10...
```

The hierarchical structure is displayed at the output of the CLUSTER procedure (see Figure 9.19). Thus we see that cluster 9 is formed by joining the objects (clusters reduced to a single element, and therefore to a centre obtained from the moving centres procedure) OB6 and OB8, that it contains 19 + 18 individuals (attached to the 6_{th} and 9_{th} moving centre), that it is then joined to the object OB7, and so on.

This description is summarized in the OUTTREE data set (Figure 9.20). The information shown in this data set includes, for each object or cluster, the coordinates &VAR of its centre (which are the weighted means of the objects forming the cluster), its 'parent' _PARENT_ in the tree diagram, its height _HEIGHT_ in the tree diagram (the definition of this height

Cluster History										
NCL	**Clusters Joined**		**FREQ**	**SPRSQ**	**RSQ**	**ERSQ**	**CCC**	**PSF**	**PST2**	**Tie**
9	OB6	OB8	37	0.0160	.931	.906	5.04	350	51.2	
8	CL9	OB7	56	0.0181	.913	.893	3.38	312	31.2	
7	OB1	OB5	43	0.0198	.893	.877	2.41	292	69.4	
6	OB3	OB10	38	0.0199	.873	.855	2.37	290	62.0	
5	CL7	OB4	65	0.0315	.842	.824	1.95	281	56.3	
4	CL6	CL8	94	0.0739	.768	.778	−.88	235	84.1	
3	CL5	CL4	159	0.1980	.570	.700	−8.2	142	140	
2	CL3	OB9	186	0.2255	.344	.530	−7.4	113	98.0	
1	CL2	OB2	217	0.3442	.000	.000	0.00	.	113	

Figure 9.19 History of a agglomerative hierarchical clustering.

Obs	_NAME_	_PARENT_	_NCL_	_FREQ_	_HEIGHT_	_RMSSTD_	_SPRSQ_	_RSQ_	_PSF_	_PST2_
1	OB6	CL9	10	19	0.00000	0.26769	0.00000	0.94681	.	.
2	OB8	CL9	10	18	0.00000	0.25155	0.00000	0.94681	.	.
3	CL9	CL8	9	37	0.01599	0.40234	0.01599	0.93081	349.798	51.224
...										
15	CL3	CL2	3	159	0.19795	0.75796	0.19795	0.56970	141.663	140.372
...										

Figure 9.20 Extract from the CLUSTER output data set.

depends on the METHOD option; for the Ward method, the height is defined as the loss of between-cluster sum of squares due to the merger which created the cluster), its frequency _FREQ_ (for the clusters reduced to one of the centres produced by FASTCLUS, this frequency is provided by the DATA data set), and the largest number of clusters _NCL_ which can be specified so that it appears in the tree diagram. To some extent, _NCL_ is the level of the tree diagram where the cluster appears. Thus, _NCL_ = 10 for the objects OBx because they appear at the lowest level of the tree diagram, which is obtained when 10 clusters are requested. The clusters CLx appear at _NCL_ levels which become smaller (higher in the tree diagram) as the clusters are merged. For example, cluster CL3 is selected when we require clustering into three clusters. Thus we have _NCL_ = 3 on the CL3 row, because cluster 3 appears when we specify three clusters only, and ceases to appear if we allow four clusters.

The OUTTREE data set also contains the clustering quality indicators for the number of clusters _NCL_: these depend on the METHOD option, and in this case (with the Ward method) we have the within-cluster sum of squares _RMSSTD_ of the cluster (supplied in the DATA=CENTRES data set for clusters reduced to one centre, OBnn), the semi-partial R^2 _SPRSQ_, R^2 _RSQ_, pseudo F _PSF_, pseudo t^2 _PST2_, the cubic clustering criterion _CCC_, and the expected R^2 _ERSQ_.

In order to choose the level where the tree diagram will be cut, in other words the number of clusters, we use the statistical indicators contained in the OUTTREE data set and in the table describing the history of the mergers of the clusters. To represent these graphically, as in Section 9.8, we can run the following SAS syntax (these indicators are not included in IBM SPSS Statistics):

```
PROC SORT DATA=tree;BY ncl;RUN;
SYMBOL1 COLOR=black INTERPOL=join VALUE=dot HEIGHT=1; PROC
GPLOT DATA=tree;
PLOT (_sprsq_ _pst2_ _rsq_ _ccc_ _psf_) * _ncl_
RUN;
```

In this syntax, the TREE file is the output from the CLUSTER procedure (OUTTREE file), in which _ncl_ is the number of clusters on which the data set must be sorted before displaying the graphic. Note the INTERPOL = join instruction which connects the points with a continuous line.

The SPRSQ and PST2 must show a dip for k clusters and a peak for $k-1$ clusters for the clustering into k clusters to be satisfactory: this suggests an optimum $k=4$, since we go from SPRSQ = 0.074 to 0.198, showing that a change from 4 to 3 clusters greatly reduces the between-cluster sum of squares (Figure 9.21). According to the same criterion, the choice $k=5$ would also be acceptable, and it also meets the requirement for a large pseudo F and an R^2 close to 1. For $k=4$, we find that CCC is slightly negative, but not to any harmful extent. This CCC is due to the small isolated clusters which can be seen in Figure 9.22.

Having chosen the number of clusters, we can display the tree diagram and create a data set OUT containing the number (CLUSTER) and the name (CLUSNAME) of the cluster of each object (denoted OBx) for the clustering with the number of clusters specified by NCL (four in this case). This is done by the TREE procedure which takes as its input the file produced by the CLUSTER procedure and which produces at its output the data set printed in Figure 9.23.

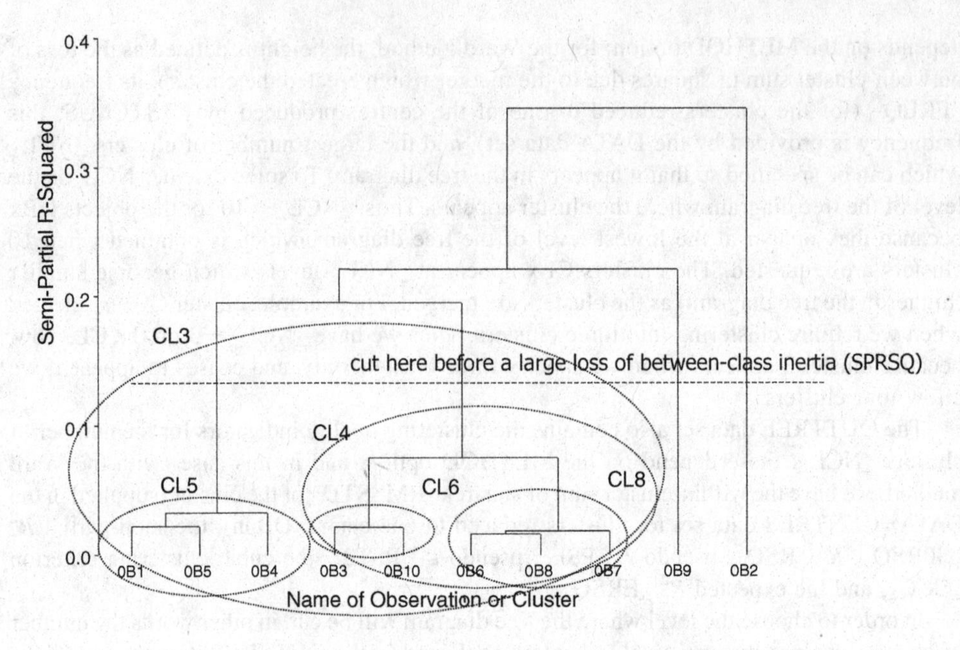

Figure 9.21 Example of a tree diagram.

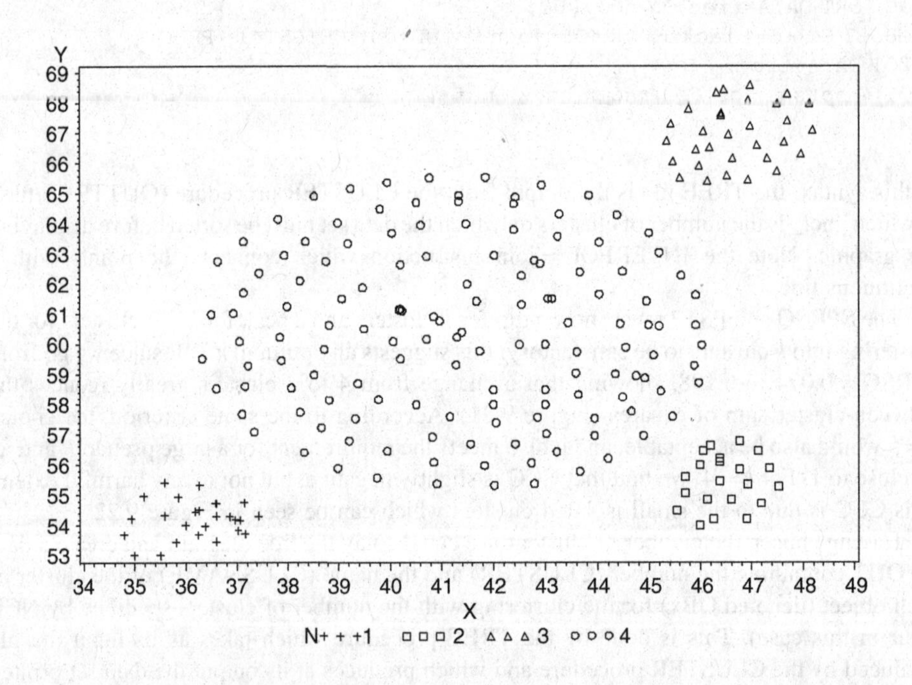

Figure 9.22 Representation of the clusters.

Obs	_NAME_	presegm	CLUSTER	CLUSNAME
1	OB6	6	1	CL4
2	OB8	8	1	CL4
3	OB7	7	1	CL4
4	OB3	3	1	CL4
5	OB10	10	1	CL4
6	OB1	1	2	CL5
7	OB5	5	2	CL5
8	OB4	4	2	CL5
9	OB9	9	3	OB9
10	OB2	2	4	OB2

Figure 9.23 Content of the output data set from the tree diagram cut at the chosen level.

```
PROC TREE DATA=tree NCL=4 OUT=segmhier; COPY presegm; RUN;
```

Finally, we simply have to match the following:

- the data set PARTITIO created in the moving centres step and containing the pre-cluster PRESEGM (between 1 and 10) of each individual;
- and the data set SEGMHIER created by AHC and containing the final cluster CLUSTER (between 1 and 4) of each pre-cluster PRESEGM, according to the correspondence shown in Figure 9.23.

The match is carried out by a 'merge' on the two data sets:

```
PROC SORT DATA = partitio; BY presegm;
PROC SORT DATA = segmhier; BY presegm;
DATA segm; MERGE partitio segmhier; BY presegm; RUN;
```

Thus we know the final cluster of each individual in the study population, which can be shown in Figure 9.24. We have not carried out a final consolidation of the clustering by reapplying the moving centres algorithm, although this is automatically suggested by SPAD, and we have not used strong forms. But, since the result is not perfect, we can try to improve it by a final consolidation programmed in SAS.

This consolidation is therefore carried out by applying the moving centres method to the centres of the clusters produced by the preceding AHC (see Figure 9.25). We can see that these centres are well positioned, even if they are outside the population in almost all cases. These are:

- either centres produced in the first step of the moving centres method, when the centres have not been grouped with others by the AHC (as is the case with OB2 and OB9 here) – these centres therefore have the same coordinates at the output of the AHC as at the input;

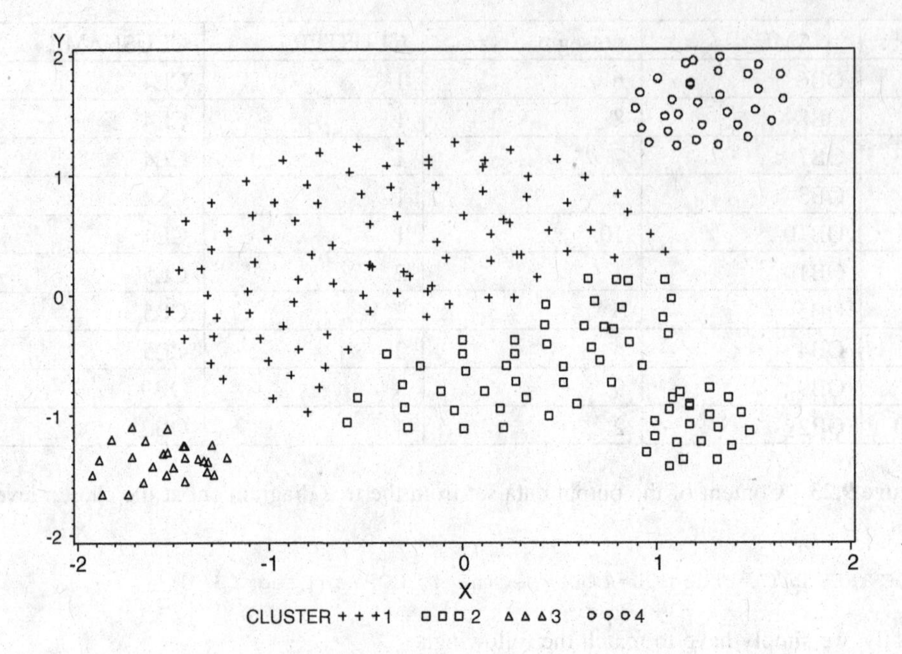

Figure 9.24 Clusters detected by the hybrid Ward method.

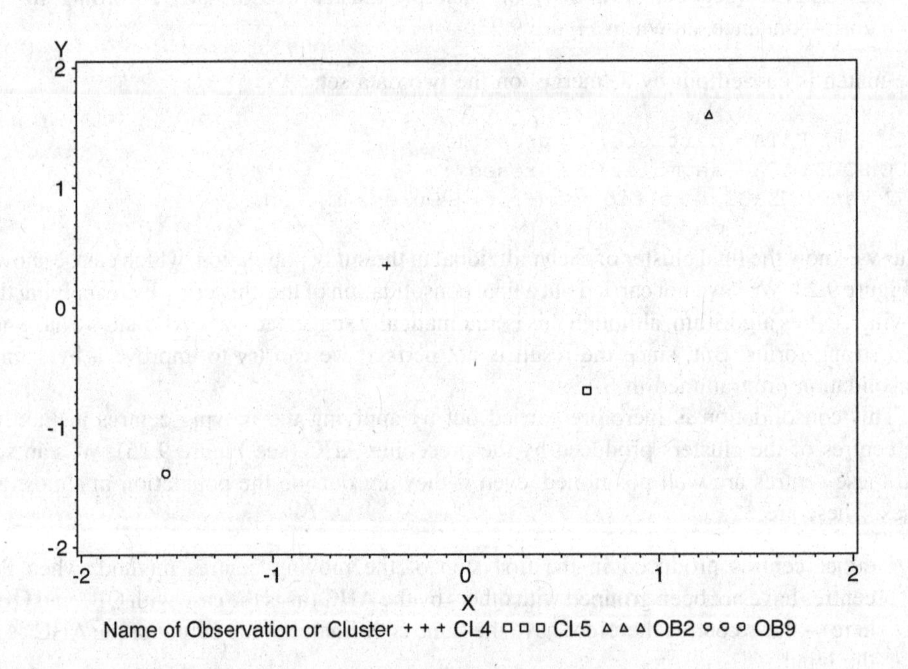

Figure 9.25 Centres of final clusters produced by AHC.

- or centres produced from groupings into clusters (CL4 and CL5 here) carried out by the AHC – the coordinates of these centres are the means of the coordinates of the objects in the cluster, for example the coordinates (0.63057; −0.70409) of CL5 are the weighted means of the coordinates 21/65 (−0.03354; −0.78624), 22/65 (1.16729; −1.07084) and 22/65 (0.72778; −0.25893) of the moving centres OB1, OB4 and OB5; these coordinates are contained in the OUTTREE data set output by the CLUSTER procedure.

We create a data set CENTRECAH containing the centres of the four clusters produced by the AHC, selecting them by their name found in Figure 9.23. Then we launch the FASTCLUS procedure, using the SEED command to indicate that it must not determine the initial centres of the algorithm, but must read them in the specified data set. Clearly, the desired number of clusters is set to four.

```
DATA centrescah (keep = x y _name_) ; SET tree;
WHERE _name_ in ("OB2" "OB9" "CL4" "CL5") ; RUN;
PROC FASTCLUS DATA=test SEED=centrescah MAXC=4 MAXITER=10
CONVERGE=0 MEAN=centres2 OUT=presegm2 CLUSTER=presegm;
VAR &var;
RUN;
```

The 'consolidated' clusters are shown in Figure 9.26. We can see that this consolidation is not really an improvement over the clustering shown in Figure 9.24.

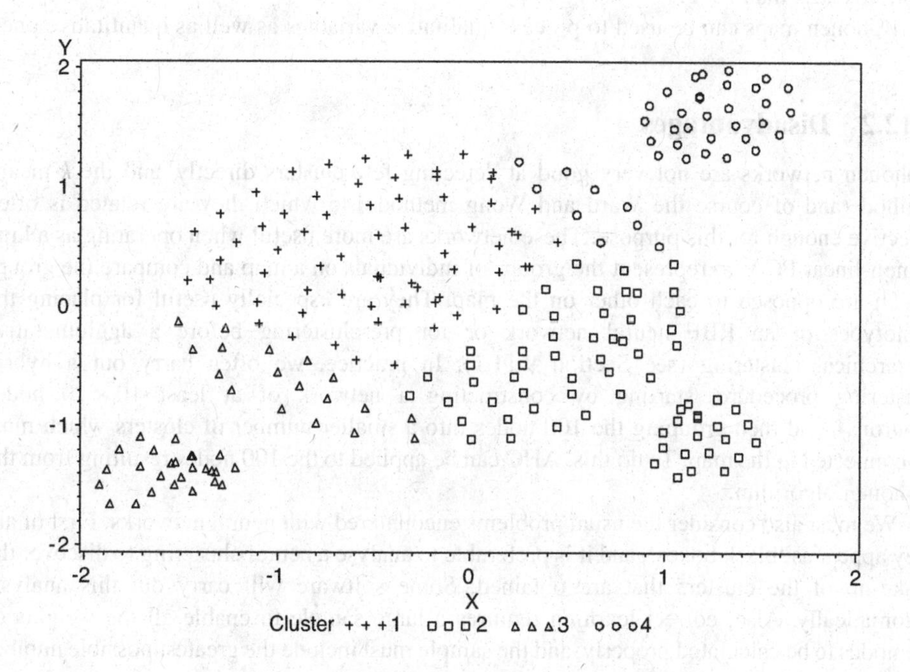

Figure 9.26 Final clusters after consolidation.

This shows that the moving centres method fails to discover clusters of dissimilar size, even though they are all convex. They are rather limited as to the type of clusters that they can detect, and their main advantage is their speed, enabling them to be used as preliminary operations, as in the hybrid clustering described here. The methods operating by density estimation are much more effective in pattern recognition, and in this case the four clusters could be detected perfectly by the 5-nearest-neighbours method (or by direct AHC with the single linkage method). Some other highly effective methods are described in Section 9.15 below.

9.12 Neural clustering

As we have seen, unsupervised learning neural networks, such as Kohonen maps (see Section 8.7.3), are intended to be used for clustering. These networks will not be described further at this point.

9.12.1 Advantages

The strength of Kohonen maps lies in their ability to model non-linear relationships between data. These networks are also helpful for determining the optimum number of clusters automatically, instead of fixing it in advance. This is because if the numbers r and c of rows and columns in the grid of the Kohonen map are fixed, the network allows for the configuration of the data in the determination of the clusters, which can finally be present in a number which is strictly less than $r \times c$.

Kohonen maps can be used to process qualitative variables as well as quantitative ones.

9.12.2 Disadvantages

Kohonen networks are not very good at detecting few clusters directly, and the k-means method (and of course the Ward and Wong methods) to which they are related is often effective enough for this purpose. These networks are more useful when operating as a kind of non-linear PCA, to represent the groups of individuals on a map and compare the groups which are opposed to each other on the map. They are especially useful for placing the prototypes of an RBF neural network or for pre-clustering before a agglomerative hierarchical clustering (see Section 9.11.1). In practice, we often carry out a hybrid clustering procedure, starting by constructing a network of at least 10×10 nodes (neurons), and then grouping the 100 nodes into a smaller number of clusters which must be connected in the map. To do this, AHC can be applied to the 100 nodes resulting from the Kohonen algorithm.

We must also consider the usual problems encountered with neural networks. First of all, they appear as 'black boxes', and it is preferable to analyse a neural clustering to discover the make-up of the clusters that are obtained. Some software will carry out this analysis automatically. Also, correct learning requires a large sample to enable all the weights of the nodes to be calculated properly, and the sample must include the greatest possible number of categories of the variables. Finally, Kohonen networks, like other clustering algorithms, are sensitive to outliers.

9.13 Clustering by similarity aggregation

The algorithm which I will examine here is based on the work of Pierre Michaud[16] and Jean-François Marcotorchino.[17] It is also called *relational clustering* because of the relational analysis used by the authors; sometimes it is known as the *voting method* or *Condorcet method*. The principles of this clustering method have also been implemented in the POP freeware by Michel Petitjean of the University of Paris 7 (see http://petitjeanmichel.free.fr), and this program has now been incorporated in the *amap* package of R by Antoine Lucas.

9.13.1 Principle of relational analysis

Relational analysis is based on the representation of data in the form of equivalence relations – hence the name. A clustering is actually an equivalence relation $\mathcal{R}$, where $i\mathcal{R}j$ if i and j are in the same cluster. As for any binary relation defined for a set of n objects, we can associate $\mathcal{R}$ with an $n \times n$ matrix, which is defined by $m_{ij} = 1$ if $i\mathcal{R}j$, and $m_{ij} = 0$ otherwise. The three properties of an equivalence relation, namely reflexivity, symmetry and transitivity, are shown by the following relations:

- $m_{ii} = 1$,
- $m_{ij} = m_{ji}$,
- $m_{ij} + m_{jk} - m_{ik} \leq 1$.

The clustering procedure is therefore a matter of finding a matrix $M = (m_{ij})$ which meets the above conditions.

In relational analysis, all the variables of the individuals of the population to be clustered must be qualitative; if they are not, then they must be discretized, into deciles for example. Each of p variables has its own natural clustering: each cluster consists of the individuals having the same category for the variable in question. The aim of relational analysis is to find a clustering which is a good compromise between the initial p natural clusterings. To do this, we assume that m_{ij} is the number of times that the individuals i and j have been placed in the same cluster (i.e. the number of variables for which i and j have the same category), and that

[16] Marcotorchino, J.-F. and Michaud, P. (1979) *Optimisation en Analyse Ordinale des Données*. Paris: Masson, Chapter X.

[17] Marcotorchino, J.-F. (1981) Agrégation des similarités en classification automatique. Doctoral Thesis in Mathematics, University of Paris VI. See also M. Petitjean, M. (2002) Agrégation des similarités: une solution oubliée. *RAIRO Oper. Res.*, 36(1), 101–108.

$M' = (m'_{ij}) = 2(m_{ij}) - p$. Then $m'_{ij} > 0$ if i and j are in the same cluster (they 'coincide') for a majority of variables, $m'_{ij} < 0$ if i and j are in different clusters for a majority of variables, and $m'_{ij} = 0$ if the number of variables for which i and j are grouped together is the same as the number of variables for which i and j are separated. It is natural to place i and j in the same final cluster if m'_{ij} is positive, and to separate them if m'_{ij} is negative. But this criterion is not sufficient, because of the non-transitivity of the majority rule (*Condorcet's paradox*): there may be a majority for joining i and j, j and k, but not for joining i and k. We must therefore add equivalence relation constraints of the kind stated above (reflexivity, symmetry and transitivity) to find a clustering that is closest to the majority of the p initial clusterings. This brings us to a linear programming problem which can be resolved correctly, as shown in the work of Marcotorchino and Michaud.

9.13.2 Implementing clustering by similarity aggregation

To give a better picture of the working of clustering based on relational analysis, I shall describe the stages of clustering by similarity aggregation, using an approach which is intuitive rather than absolutely rigorous.

For each pair of individuals (A,B), let $m(A,B)$ be the number of variables having the same value for A and B, and $d(A,B)$ be the number of variables having different values for A and B, given that, for continuous variables,

- either we consider that they have the 'same value' if they are in the same decile,
- or we define their contribution to $c(A,B)$ below as being equal to

$$1-2\left(\frac{|v(A)-v(B)|}{v_{max}-v_{min}}\right),$$

where v_{min} and v_{max} are the outlying values of the variable V.

The *Condorcet criterion* for two individuals A and B is defined as

$$c(A,B) = m(A,B) - d(A,B).$$

We then define the Condorcet criterion of an individual A and a cluster S as

$$c(A,S) = \sum_i c(A,B_i),$$

the summation being over all the $B_i \in S$.

Given the above, we start to construct the clusters by placing each individual A in the cluster S for which $c(A,S)$ is maximum and at least 0. Sometimes we can replace this value 0 with a larger value, to strengthen the homogeneity of the clusters. We can also have an effect on this homogeneity by introducing a factor $\sigma > 0$ into the definition of the Condorcet criterion, which becomes $c(A,B) = m(A,B) - \sigma.d(A,B)$. A large value of σ will be a high cluster homogeneity factor. If $c(A,S) < 0$ for every existing S, then A forms the first element of a new cluster.

We therefore take a first individual A, which is compared with all the other individuals, and group it with another individual B_A if necessary. We then take the second individual B, which is compared with the other individuals, as well as with the cluster $\{A, B_A\}$, if it exists. And so on. This step is the first iteration of the clustering.

We can perform a second iteration by taking each individual again and reassigning it, if necessary, to another cluster taken from those defined in the first iteration. In this way we perform a number of iterations, until:

- the specified maximum number of iterations is reached, or
- the global Condorcet criterion ceases to improve sufficiently (by more than 1% for example, a value which can be set in advance) from one iteration to the next, this global Condorcet criterion being

  $$\sum_A c(A, SA),$$

  where the summation is performed on all the individuals A and the clusters S_A to which they have been assigned.

In practice, two iterations (or three if absolutely necessary) will be enough to provide good results.

9.13.3 Example of use of the R amap package

We begin by loading the package:

```
> library(amap)
```

If the variables of the data frame (data table) to be processed are not factors (qualitative variables), they must be transformed in advance as follows:

```
> for (i in 1:17) credit[,i] <- factor(credit[,i])
```

In this example, 17 variables are assumed. The variables can be numeric at the outset, but the number of categories must be small.

We then calculate the dissimilarity matrix, using the diss function in the package. However, this function only processes whole numbers, and we must therefore transform the variables in advance, thus:

```
> creditn <-
matrix(c(lapply(credits,as.integer),recursive=T),ncol=17)
```

We obtain the following matrix:

```
> matrix <- diss(creditn)
> matrix
```

	[,1]	[,2]	[,3]	[,4]	[,5]	[,6]	[,7]	[,8]	[,9]	[,10]	[,11]	[,12]	[,13]	[,14]	[,15]	[,16]	[,17]	[,18]	[,19]	[,20]
[1,]	17	−3	−3	−5	−3	−3	1	−5	−1	−3	−9	−5	−1	5	−3	−3	11	−1	−5	1
[2,]	−3	17	−1	−3	−3	−5	−1	3	3	1	5	3	7	−3	5	7	−3	−5	3	−1
[3,]	−3	−1	17	1	−5	3	−1	−3	5	−5	−7	−7	−5	−1	−3	−5	−1	−3	−9	−1
[4,]	−5	−3	1	17	1	1	3	−3	−1	−11	−5	1	−5	−5	3	−7	−3	−5	−7	−3
[5,]	−3	−3	−5	1	17	−1	−3	−7	−9	−3	−1	1	−7	3	1	−5	−3	−7	−5	−5
[6,]	−3	−5	3	1	−1	17	−1	1	1	−13	−9	−7	−3	−7	−1	−5	−3	−7	−3	1
[7,]	1	−1	−1	3	−3	−1	17	−3	3	−7	−1	3	−1	−1	1	−3	7	−3	−5	7
[8,]	−5	3	−3	−3	−7	1	−3	17	−3	−1	1	−3	5	−5	5	−1	−7	−5	5	1
[9,]	−1	3	5	−1	−9	1	3	−3	17	−7	−5	−3	−1	−3	1	1	1	−5	−7	−1
[10,]	−3	1	−5	−11	−3	−13	−7	−1	−7	17	1	−5	−3	7	−3	1	−1	−7	1	−7
[11,]	−9	5	−7	−5	−1	−9	−1	1	−5	1	17	9	5	−1	7	1	−7	−5	1	−3
[12,]	−5	3	−7	1	1	−7	3	−3	−3	−5	9	17	−1	−1	7	1	−3	−3	−1	−5
[13,]	−1	7	−5	−5	−7	−3	−1	5	−1	−3	5	−1	17	−5	5	3	−3	−5	1	3
[14,]	5	−3	−1	−5	3	−7	−1	−5	−3	7	−1	−1	−5	17	1	3	5	−3	−3	−5
[15,]	−3	5	−3	3	1	−1	1	5	1	−3	7	7	5	1	17	3	−3	−1	−3	−3
[16,]	−3	7	−5	−7	−5	−5	−3	−1	1	1	1	1	3	3	3	17	−3	−5	1	−1
[17,]	11	−3	−1	−3	−3	−3	7	−7	1	−1	−7	−3	−3	5	−3	−3	17	−1	−7	1
[18,]	−1	−5	−3	−5	−7	−7	−3	−5	−5	−7	−5	−3	−5	−3	−1	−5	−1	17	−13	−5
[19,]	−5	3	−9	−7	−5	−3	−5	5	−7	1	1	−1	1	−3	−3	1	−7	−13	17	−3
[20,]	1	−1	−1	−3	−5	1	7	1	−1	−7	−3	−5	3	−5	−3	−1	1	−5	−3	17

Finally, we can use the function for clustering by similarity aggregation:

```
> pop(matrix)
Upper bound (half cost)            : 189
Final partition (half cost)        : 129
Number of classes                  : 6
Forward move count                 : 879424708
Backward move count                : 879424708
Constraints evaluations count      : 1758849416
Number of local optima             : 4

          Individual          class
1             1                 1
2             2                 2
3             3                 3
4             4                 3
5             5                 4
6             6                 3
7             7                 1
8             8                 2
9             9                 3
10            10                5
11            11                2
12            12                2
13            13                2
14            14                5
15            15                2
16            16                2
17            17                1
18            18                6
19            19                2
20            20                1
```

In this example, the function has detected six clusters among the 20 individuals.

Note that we have only processed a very small number of individuals, because the pop function unfortunately only has processing capacity for a few tens of individuals. However, this drawback is inherent in this implementation, rather than in the clustering method itself, which was implemented in a very high-performance version by IBM in its Intelligent Miner software in the 1990s, capable of processing millions of individuals.

9.13.4 Advantages of clustering by similarity aggregation

The method is useful because it enables us:

(i) to determine the optimum number of clusters automatically, instead of fixing them in advance;

(ii) to process missing values without transforming them;

(iii) to compare all the individuals in pairs at each step, thus building up a global clustering, instead of local clustering as in hierarchical clustering methods.

9.13.5 Disadvantages of clustering by similarity aggregation

Clustering by similarity aggregation has a number of drawbacks. Essentially, they show that this method has to be confined to nominal variables.

1. Clustering by similarity aggregation is inherently restricted to nominal variables, and therefore it is necessary to discretize continuous variables, a task that software packages do not always perform in an optimal way, especially as the resulting discrete variables are ordinal, and clustering by similarity aggregation cannot handle their ordinal properties.

2. These methods are sensitive to the number of categories of the variables. Variables with only a few categories have a higher weight than the others (this is the inverse of factorial methods, in which the number of categories increases the contribution to the sum of squares), because it is easier for two individuals to be similar on a variable having few categories. These variables therefore tend bring the individuals together and have a strong cluster shaping effect. The binary variables tend to 'overwhelm' the others.

3. In contrast to factorial methods, the method of similarity aggregation is affected by the presence of redundant variables, which bias the results of the clustering in favour of those variables which will become most discriminating in the description of the clusters, while a principal component analysis, for example, supplies linearly independent principal components.

4. The principle of the method means that individuals which are close on a majority of variables are brought together. In the case of two variables, for example, we can see that if

   a. two individuals x and y are in the same decile of the first variable, and are very different on the second variable, and

   b. x and another individual z are 'fairly close', because they are in the same quintile but not the same decile for the two variables, then x will be aggregated with y, and not with z! The result is a division into clusters which may be surprising.

9.14 Clustering of numeric variables

So far, we have only considered the clustering of individuals. However, methods have been developed for clustering variables as this is more important than the clustering of individuals in some fields: for example, sensory analysis (creation of groups of descriptors) and medicine (identifying syndromes on the basis of a set of symptoms). The methods described here are only applicable to numeric variables, but if we have qualitative variables we can always transform them into numeric variables by using multiple correspondence analysis.

One method involves applying agglomerative hierarchical clustering to the variables, as we would to individuals, the difference being that the Euclidean distance is replaced by Pearson's correlation coefficient r (or rather, $1 - r$). The 'complete linkage' method is preferably used, and two groups of variables V and W will be close if every variable $v \in V$ is sufficiently correlated with every variable $w \in W$.

The most successful method consists of a divisive hierarchical clustering (DHC) by iterative applications of an oblique PCA (see Section 7.2.1) and this is provided by the SAS/STAT VARCLUS procedure. We start with the set of variables. A PCA is then performed and the first two factor axes, with eigenvalues λ_1 and λ_2, are inspected. By default, VARCLUS operates with the correlation matrix to give the same weight to all the variables, and it will be assumed henceforth that this parameter setting is used (specify the COVARIANCE option to use the covariance matrix). If $\lambda_2 > 1$, this means that the second axis – not just the first – is important (the Kaiser criterion is chosen by default, but the value 1 can be replaced with another value λ, by entering MAXEIGEN $= \lambda$). We then perform an oblique quartimax rotation and redistribute the variables into the group V_1 of variables most highly correlated with the first axis and the group V_2 of variables most highly correlated with the second axis. If $\lambda_2 \leq 1$, the second axis contains too little information and the set of variables is not divided. This step is repeated for each of the clusters V_1 and V_2.

We stop when there are no more clusters with $\lambda_2 > 1$ (or $\lambda_2 > \lambda$) or when a specified number of clusters (set by the option MAXCLUSTERS $=$ k, which neutralizes MAXEIGEN $= \lambda$) has been reached. It is not essential to specify a number of clusters, because the first criterion for stopping is always effective in the absence of any other setting. Clearly, as the threshold λ increases, it will be harder to surmount, and there will be a smaller number of clusters.

Sometimes, after a cluster has been divided, we may find that the frequency of the cluster is not equal to the sum of the frequencies of the two sub-clusters. This is due to the fact that, following each division, VARCLUS is able to reassign each variable to another cluster formed previously, in an attempt to maximize the variance accounted for. This reassignment of variables can be prevented, if we wish to reduce the computation time or avoid destroying the structure of the clustering tree. To do this, we have to specify the option HIERARCHY, or specify a data set OUTTREE $=$ ARBRE which will cause the information required to display the divisive hierarchical clustering tree diagram to be stored in the data set ARBRE. The diagram will then be displayed by the TREE procedure, as in the case of a tree diagram produced by AHC (see Section 9.10.1). Figure 9.27 shows on the horizontal axis the proportion of variance explained as a function of the number of clusters, and we can see that three clusters of variables explain 61% of the variance, as described more fully below. The horizontal orientation of the tree diagram is provided by the option HORIZONTAL, and the option HEIGHT _PROPOR_ means that the horizontal axis of the tree diagram represents the proportion of the variance explained at each step of the DHC.

The result of the VARCLUS procedure is a partitioning of the variables such that two variables of the same cluster are intercorrelated as much as possible, and that two variables of different clusters are correlated as little as possible.

The VARCLUS procedure differs from a standard PCA. In a PCA, all the principal components are calculated from the same variables (the initial variables), whereas in VARCLUS the initial variables are separated iteratively into subgroups (by quartimax) and the principal components are calculated on these subgroups, not on the set of initial variables.

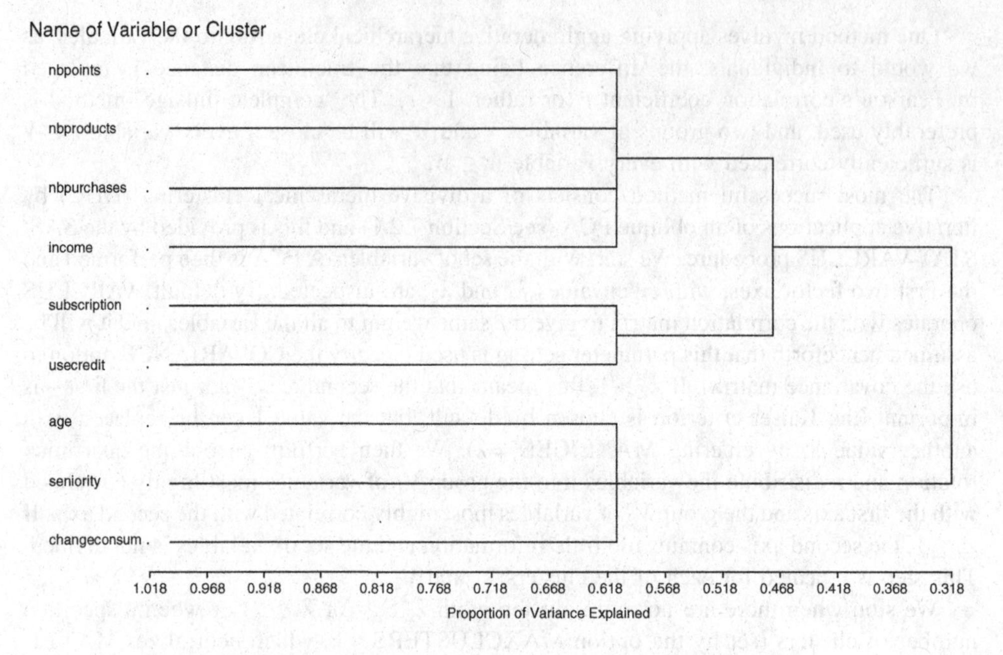

Figure 9.27 Tree diagram of variable clustering.

If the number of clusters of variables is limited to k, so that we keep only k components for each VARCLUS procedure, these k components may explain less variance than the k first principal components of the PCA, but they are easier to interpret.

VARCLUS is an effective method for clustering variables, but requires more computation time than PCA; care must be taken if there are more than 30 variables.

To illustrate this procedure, let us reconsider the example of Figure 7.5.

```
PROC VARCLUS DATA=file_customer OUTTREE=tree;
VAR age seniority nbpoints nbproducts nbpurchases income
subscription changeconsum usecredit;
RUN;
PROC TREE DATA=tree HORIZONTAL;
HEIGHT _PROPOR_;
RUN;
```

Looking at the SAS/STAT outputs, we find that the first component, with its eigenvalue $\lambda_1 = 2.86$, explains 31.82% of the total variance. This variance is 9, the number of variables, since these are reduced (with variance = 1) when the correlation matrix is used:

Cluster summary for 1 cluster

Cluster	Members	Cluster Variation	Variation Explained	Proportion Explained	Second Eigenvalue
1	9	9	2.863822	0.3182	1.4884

Total variation explained = 2.863822 Proportion = 0.3182

For the second eigenvalue, we find $\lambda_2 = 1.49 > 1$, meaning that we must divide the set of variables into two clusters, namely one cluster of six variables and another of three variables:

Cluster summary for 2 clusters

Cluster	Members	Cluster Variation	Variation Explained	Proportion Explained	Second Eigenvalue
1	6	6	2.808456	0.4681	1.2806
2	3	3	1.464485	0.4882	0.9239

Total variation explained = 4.272942 Proportion = 0.4748

The resulting two clusters explain 47.48% of the total variance, with a first eigenvalue of 2.81 for cluster 1 (i.e. 46.81% of the variance of this cluster and 31.2% of the total variance), and a first eigenvalue of 1.46 for cluster 2 (i.e. 48.82% of the variance of this cluster and 16.27% of the total variance). This second cluster will not be divided, because its second eigenvalue is less than 1, but the first cluster will be. However, when we look at the table showing the two clusters we can see that the 'change in consumption' variable is located in the second cluster, and this variable is the one least correlated with its cluster (coefficient $= 0.2151$), a sign that the clustering of the variable is not satisfactory.

The 'R^2 with next closest' column in the table below shows the correlation coefficient between the variable and the cluster (other than its own) which is closest to it. The next column is the ratio between $1 - R^2$ (own cluster) and $1 - R^2$ (next closest), and this ratio is smaller when the variable is well clustered (it is even possible to have a ratio of more than 1 if the variable is particularly poorly clustered).

Cluster	Variable	R-squared with		1-R**2 Ratio	Variable Label
		Own Cluster	Next Closest		
Cluster 1	**nbpoints**	0.6546	0.0011	0.3458	nb of points
	nbproducts	0.6189	0.0183	0.3882	nb of products
	nbpurchases	0.5950	0.0007	0.4053	nb of purchases
	income	0.4551	0.0234	0.5580	client income
	subscription	0.2537	0.0042	0.7495	subscription other service
	usecredit	0.2312	0.0002	0.7689	use credit
Cluster 2	**age**	0.6033	0.0000	0.3967	age
	seniority	0.6461	0.0336	0.3662	client seniority
	changeconsum	0.2151	0.0027	0.7870	change in consumption

After cluster 1 has been divided, we obtain the following result, in which we find that the three resulting clusters explain 61.05% of the total variance, and that the second eigenvalues λ_2 of the three clusters are all less than 1, meaning that the process of cluster division stops.

Cluster summary for 3 clusters

Cluster	Members	Cluster Variation	Variation Explained	Proportion Explained	Second Eigenvalue
1	4	4	2.537365	0.6343	0.6969
2	3	3	1.464485	0.4882	0.9239
3	2	2	1.492515	0.7463	0.5075

Total variation explained = 5.494365 Proportion = 0.6105

The distribution of the variables in the three clusters, with the two slightly separate variables 'subscription to other service' and 'payments using credit', shows a result which is consistent with that shown in Figure 7.5. We can see that there is now only one poorly clustered variable, 'change in consumption', which perhaps ought to be separated from the rest of its cluster. To force SAS/STAT to do this, there are two options:

- set a parameter MAXEIGEN = 0.9 (less than the value λ_2 of cluster 2);
- or directly specify a number of clusters, MAXCLUSTERS = 4.

In this way, we could explain a total of 71.27% of the variance.

Cluster	Variable	R-squared with		1-R**2 Ratio	Variable Label
		Own Cluster	Next Closest		
Cluster 1	nbpoints	0.6290	0.1399	0.4314	nb of points
	nbproducts	0.6643	0.0679	0.3601	nb of products
	nbpurchases	0.6952	0.0344	0.3156	nb of purchases
	income	0.5489	0.0234	0.4619	client income
Cluster 2	age	0.6033	0.0012	0.3972	age
	seniority	0.6461	0.0379	0.3679	client seniority
	changeconsum	0.2151	0.0030	0.7872	change in consumption
Cluster 3	subscription	0.7463	0.0743	0.2741	subscription other service
	usecredit	0.7463	0.0643	0.2712	use of credit

With the division into three clusters, SAS/STAT provides the correlation coefficients between each variable and each cluster, in other words between the variable and the component representing the cluster.

Cluster Structure

Cluster		1	2	3
age	age	−.005609	0.776713	0.034486
seniority	client seniority	0.194737	0.803797	0.053219
nbpoints	nb of points	0.793065	−.033371	0.373993
nbproducts	nb of products	0.815057	0.135163	0.260543
nbpurchases	nb of purchases	0.833807	0.027315	0.185432
income	client income	0.740852	0.152848	0.147148
subscription	subscription other service	0.272504	0.064823	0.863862
changeconsum	change in consumption	0.054681	−.463802	0.008141
usecredit	use of credit	0.253561	0.012789	0.863862

We can see the negative correlation of the variable 'change in consumption' with its cluster, according to Figure 7.5. Furthermore, looking at the correlations between the clusters as shown below, we see that the correlations between cluster 2 and the other two are very weak.

Inter-Cluster Correlations

Cluster	1	2	3
1	1.00000	0.08659	0.30448
2	0.08659	1.00000	0.04492
3	0.30448	0.04492	1.00000

Finally, SAS/STAT summarizes its clustering process in the form of a table showing the progressive refinement of the clustering, with a regularly increasing proportion of the variance explained.

Number of Clusters	Total Variation Explained by Clusters	Proportion of Variation Explained by Clusters	Minimum Proportion Explained by a Cluster	Maximum Second Eigenvalue in a Cluster	Minimum R-squared for a Variable	Maximum 1-R**2 Ratio for a Variable
1	2.863822	0.3182	0.3182	1.488449	0.0025	
2	4.272942	0.4748	0.4681	1.280620	0.2151	0.7870
3	5.494365	0.6105	0.4882	0.923863	0.2151	0.7872

The procedure also supplies the score functions of each cluster, enabling us to calculate the component of the cluster as a linear combination of the centred and reduced variables in the cluster.

Table 9.1 A survey of clustering methods.

Algorithm (date)	Parameters	Most suitable for:	Shape of clusters
		Partitioning methods (with reassignment of objects)	
Moving centres, *k*-means (1965, 1967)	Number of clusters	Separate clusters, large frequencies	Spherical
k-modes (1998)	Number of clusters	Qualitative variables, large frequencies	
k-prototypes (1998)	Number of clusters	Quantitative and qualitative variables, large frequencies	
PAM (1990)	Number of clusters	Separate clusters, more robust to outliers than those of *k*-means, small frequencies	Spherical
CLARA (1990)	Number of clusters	Fairly large frequencies	Spherical
CLARANS (1994)	Number of clusters, maximum number of neighbours	Better-quality clusters than in PAM and CLARA, small frequencies	Spherical
Kohonen maps (1982)	Number of clusters	Separate clusters, fairly large frequencies	Spherical
		Hierarchical methods	
AHC, single linkage (1951)	Cut-off level in the tree diagram	Small frequencies, clusters of irregular shape	Elongated
AHC, Ward and others (1963, and 1948–1967)	Cut-off level in the tree diagram	Small frequencies, better quality clusters than in *k*-means	Spherical
CURE (1998)	Number of clusters, number of representatives per cluster	Clusters of any shape	Arbitrary
ROCK (1999)	Number of clusters	Qualitative variables, small frequencies	
BIRCH (1996)	Maximum number of sub-clusters of an intermediate node, maximum diameter of sub-clusters of terminal nodes	Large frequencies, algorithm generalized to hybrid variables (quantitative and qualitative)	Spherical

CHAMELEON (1999)	k nearest neighbours, size of sub-clusters created, parameter α	Clusters of any shape	Arbitrary
Density estimation methods			
Wong's hybrid method (1982)	Initial number of clusters, cut-off level in the tree diagram	Large frequencies	Spherical
DBSCAN (1996)	Radius of clusters, minimum number of objects per cluster	Any shape of cluster (more effective than CLARANS), fairly large frequencies	Arbitrary
OPTICS (1999)	Radius of clusters, minimum number of objects per cluster	Any shape of cluster, fairly large frequencies	Arbitrary
BRIDGE (2001)	Radius of clusters, minimum number of objects per cluster	Any shape of cluster, large frequencies	Arbitrary
DENCLUE (1998)	Radius of clusters, minimum number of objects per cluster	Any shape of cluster, fairly large frequencies	Arbitrary
Grid-based methods			
STING (1997)	Number of cells at the lowest level, number of objects per cell	Large frequencies	Rectangular
WaveCluster (1998)	Number of cells for each dimension, wavelet, number of transformation applications	Large frequencies, high-quality clusters, effective allowance for outliers	Arbitrary
CLIQUE (1998)	Grid size, minimum number of objects per cell	Large frequencies, numerous variables	Arbitrary
MAFIA (1999)	Cluster dominance factor, minimum number of objects per cell	Large frequencies, numerous variables, better-quality clusters than in CLIQUE	Arbitrary
Other methods			
Similarity aggregation (1979)	Similarity threshold	Qualitative variables, large frequencies	
CACTUS (1999)	Support threshold, validation threshold	Qualitative variables, large frequencies, not too many variables	
VARCLUS (1976)	Number of clusters or threshold of second eigenvalue	Clustering of variables	

Standardized Scoring Coefficients				
Cluster		**1**	**2**	**3**
age	age	0.000000	0.530366	0.000000
seniority	client seniority	0.000000	0.548860	0.000000
nbpoints	nb of points	0.312555	0.000000	0.000000
nbproducts	nb of products	0.321222	0.000000	0.000000
nbpurchases	nb of purchases	0.328611	0.000000	0.000000
income	client income	0.291977	0.000000	0.000000
subscription	subscription other service	0.000000	0.000000	0.578796
changeconsum	change in consumption	0.000000	−.316700	0.000000
usecredit	use of credit	0.000000	0.000000	0.578796

9.15 Overview of clustering methods

Table 9.1 sums up the main characteristics of the best-known and most widely used clustering algorithms. Further details can be found in the standard work by Leonard Kaufman and Peter J. Rousseeuw describing their PAM, CLARA, AGNES, DIANA and other algorithms (see Section 9.9.4 above), the more recent work of A.D. Gordon,[18] Brian S. Everitt, Sabine Landau and Morven Leese,[19] Jean-Pierre Nakache and Josiane Confais,[20] and Periklis Andritsos.[21] There was a proliferation of algorithms in the late 1990s, but unfortunately these are not yet available in most commercial software, or even in R. WaveCluster, a highly effective algorithm, is just one that ought to be more widespread.

In addition to the previously mentioned partitioning, hierarchical and density estimation methods, I have alluded to grid-based algorithms, in which the data space is divided into small cubes. The density of each cube is estimated, and adjacent dense cubes are then grouped into clusters. In this case there is no reassignment of objects, as in the moving centres method, and these methods are more closely related to the hierarchical methods, except for the fact that the aggregation of the clusters depends, not on distance, but on another criterion. These methods were developed for the clustering of highly dimensional spatial data (describing objects in space). Some algorithms such as CLIQUE are also density estimation methods.

[18] Gordon, A.D. (1999) *Classification*, 2nd edn. Boca Raton, FL: Chapman & Hall/CRC.

[19] Everitt B.S., Landau S., Leese M. (2001) *Cluster Analysis*, 4th edition. London: Arnold.

[20] Nakache, J.-P. and Confais, J. (2004) *Approche pragmatique de la classification*. Paris: Technip.

[21] Andritsos, P. (2002) Data clustering techniques. Tech. Report CSRG-443, University of Toronto.

10

Association analysis

The detection of association rules is another descriptive method which is very popular in data mining, especially in such areas as web mining, where it is used to analyse the pages visited by a web user, and the retail industry, where it can analyse the products bought by a customer on a single visit. This explains the alternative name for this method: *market basket analysis*. Of course, this method can be usefully applied to other activities as well. It does not have the same theoretical difficulties as clustering and classification methods; instead, the difficulties arise from the need to process enormous volumes of data (up to several million till receipts, for example) and to pick out new and interesting associations from the overwhelming majority of irrelevant or previously known associations.

10.1 Principles

Finding *association rules* is a matter of finding rules of the following type: 'If, for any one individual, variable $A = x_A$, variable $B = x_B$, and so on, then, in 80% of cases, variable $Z = x_Z$, and this configuration is found for 20% of the individuals.' In other words, the aim is to find the most frequent combined values of a set of variables of a data set. In market basket analysis, the variables are the indicators of the products, and the rules are applied to indicators equal to 1, in other words the products bought. Note that some recent research has been carried out on 'negative' rules, where we are interested in the products that are *not* bought.

The value of 80% is called the *index of confidence* and the value of 20% is called the support index of the rule $\{A = x_A, B = x_B, \ldots\} \Rightarrow \{Z = x_Z\}$. The first part of the rule is called the 'antecedent' or 'condition'; the second part is called the 'consequent' or 'result'; and expressions of the form $\{A = x_A\}$ are called 'items'. In an association rule, an item can never be in both the condition and the result simultaneously.

Data Mining and Statistics for Decision Making, First Edition. Stéphane Tufféry.

A rule is therefore an expression of the form:

If *Condition*, then *Result*.

Here is an example taken from marketing (mythical, if not veracious):

If *Nappies* and *Saturday*, then *Beer*.

The support index is the probability

$$\text{Prob}(\textit{Condition} \text{ and } \textit{Result}).$$

The confidence index is the probability

$$\text{Prob}(\textit{Condition} \text{ and } \textit{Result})/\text{Prob}(\textit{Condition}).$$

Naturally, the aim is to find association rules for which the support and confidence are above specified minimum thresholds.

For example, in the transactions shown in Table 10.1, where each row corresponds to a market basket T_X, and each column corresponds to a product $A, B, \ldots$, the confidence index of the association $B \Rightarrow E$ is $\frac{3}{4}$ and its support index is $\frac{3}{5}$. Similarly, the confidence index of the association $C \Rightarrow B$ is $\frac{2}{3}$ and its support index is $\frac{2}{5}$. One thing is evident: B is present in almost all the transactions, or more precisely the *a priori* probability of having B there is 0.8. This probability is greater than the confidence index for $C \Rightarrow B$, and therefore the rule $C \Rightarrow B$ is not helpful for predicting B. If we say that a transaction taken at random contains B, there is only one chance in five that we will be wrong, as against one chance in three if we follow the rule $C \Rightarrow B$.

The improvement brought by a rule, by comparison with a random response, is called the lift (or simply the 'improvement'), and is as follows:

$$\text{lift(rule)} = \frac{\text{confidence\_index(rule)}}{\text{Prob(Result)}} = \frac{\text{Prob(Condition and Result)}}{\text{Prob(Condition)} \times \text{Prob(Result)}}.$$

When the 'result' is independent of the 'condition', the lift is clearly equal to 1. If the lift is less than 1, the rule does not help. Thus we find that lift$(C \Rightarrow B) = \frac{5}{6}$ (useless rule) and lift $(B \Rightarrow E) = \frac{5}{4}$ (useful rule). But note that, if the lift of the rule

$$\text{Condition} \Rightarrow \text{Result}$$

Table 10.1 Set of transactions.

T_{26}	A	B	C	D	E	
T_{163}	B	C	E	F		
T_{1728}	B	E				
T_{2718}	A	B	D			
T_{3141}	C	D				

is less than 1, then the lift of the inverse rule, i.e. the rule

$$\text{Condition} \Rightarrow \text{NOT Result.}$$

is greater than 1, since

$$\text{confidence index(inverse rule)} = 1 - \text{confidence index(rule)}$$

and

$$\text{Prob(NOT Result)} = 1 - \text{Prob(Result)}.$$

If a rule is not useful, we can try using the inverse rule, in the hope that it will be helpful for business or marketing purposes.

The main algorithm for detecting association rules is the *Apriori* algorithm proposed by Agrawal and other researchers.[1]

Apriori operates in two steps, which have become standard for this type of algorithm:

- It starts by searching for the subsets of items having a probability of appearance (support) above a certain threshold.
- Then it attempts to break down each subset in a form $\{\text{Condition} \cup \text{Result}\}$ such that the quotient Prob(Condition and Result)/Prob (Condition), i.e. the confidence index, is above a certain threshold.

In the first step, Apriori starts by making a first pass through the data, to eliminate all the items which are less frequent than the specified minimum support. It then performs a second pass, in order to construct all the sets of items with two elements, formed from the items retained previously. Of these sets, it only retains those whose frequency exceeds the specified minimum support. On each pass, Apriori retains only the sets of items which are more frequent than the support threshold, out of all those constructed on the basis of the sets from the previous pass and the items selected in the first pass. The frequent items with a size of n which are useful for our purposes are those constructed from sets with a size of $n-1$ which are themselves frequent. The first optimization of Apriori is that only a single pass is required for each value of n.

The difficulty of implementing the search for rules is due to the exponential growth of the number of rules with the number of items. For each subset of items E with n elements, there are 2^{n-1}–1 rules of the form $A \Rightarrow \{E - A\}$, and therefore the same number of possible breakdowns in the second step. Another improvement provided by the designers of Apriori is a way of quickly identifying the rules which may exceed the fixed threshold of the confidence index.

Because of these advantages, the Apriori algorithm is the most widespread and most commonly implemented algorithm for detecting association rules.

In practice, however, there are still a very large number of rules remaining, and most packages offer an option for storing these rules in a file, in which the Condition ⇒ Result rules can be filtered up to a certain value of the support index, and can be sorted according to their

[1] Agrawal, R., Imielienski, T. and Swami, A.N. (1993). Mining association rules between sets of items in large databases. In *Proceedings of the 1993 ACM SIGMOD International Conference on Management of Data*, pp. 207–216. New York: ACM Press.

Agrawal, R., Mannila, H., Srikant, R., Toivonen, H., and Verkamo, A. I. (1995). Fast discovery of association rules. In *Advances in Knowledge Discovery and Data Mining*, pp. 307–328. Cambridge, MA: AAAI Press/MIT Press.

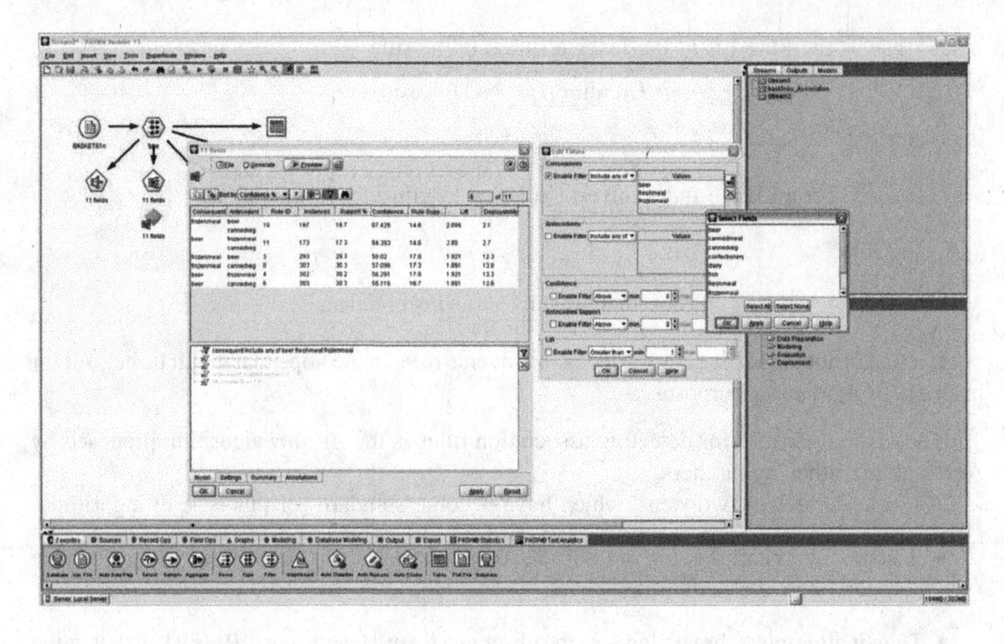

Figure 10.1 Parameter setting in IBM SPSS Modeler.

support, confidence or lift. This file is often a text file, but SAS Enterprise Miner can store the rules in an SAS table.

The requirement in respect of the confidence threshold is generally stricter than for the support threshold; a common example of a filter is 75% for confidence and 5% for support (and 1 for lift, of course).

However, even with these filters, the number of rules soon becomes dizzyingly high, up to several million for just a few hundred items and a few thousand observations. Indeed, this number increases exponentially with the decrease in the minimum support and an increase in the number of items in each rule. In fact, not only are almost all of these rules uninteresting or well known already (cheese goes with bread and wine, white wine goes with oysters, nails go with a hammer, and so on), but, purely in terms of computing power, it may be impossible to process and store so many rules. So some packages offer a useful option for adding a filter on the content of the rules, making it possible to retain only the rules which contain a certain item in their consequent or antecedent. This functionality is even more useful because we often seek rules that 'predict' a certain behaviour, where the consequent contains certain items specified in advance. Among the commercial software programs, IBM SPSS Modeler has this functionality (Figure 10.1).

The packages also enable us to set a limit to the size of the rules, in other words to the number of items they contain. We would rarely need to go beyond 10 items. Note that some packages, but not all, permit consequents with more than one item. This is the case with SAS Enterprise Miner and IBM SPSS Modeler, but not the freeware developed by Christian Borgelt.[2] However, this package is often mentioned and used, or implemented in other

[2] Downloadable from http://fuzzy.cs.uni-magdeburg.de/%7Eborgelt/software.html, or more directly from http://www.borgelt.net//apriori.html.

software (such as R and its *arules* package,[3] and also Tanagra[4]), because of its high speed, making it suitable for detecting a large number of rules.

Interesting rules are those which are non-trivial, usable in practice, and preferably explicable.

10.2 Using taxonomy

Products can be defined at a more or less fine level of detail. For example, we may consider:

- savings products in banking, finance, etc.;
- among the bank savings products, there are current accounts, passbooks, etc.;
- among passbooks, there are instant savings, building society savings, post office savings accounts, and so on.

The *taxonomy* of products is the set of these levels, with its hierarchy. The finest level enables us to undertake more accurate marketing operations. However, working at the finest level multiplies the rules, many of which will only have low support and must therefore be eliminated. Working at the most general level enables us to have stronger rules. Both viewpoints have their advantages and disadvantages. A good compromise is to adapt the level of generality to each product, based on its scarcity, for example.

Products which are scarcest and most expensive (e.g. microcomputers or hi-fi in a department store) will be coded at a finer level, whereas more common products (e.g. food products) will be coded at a more general level. By way of example, we can group all yogurts, cheeses, creams, etc., into 'dairy products', while making a distinction between DVD players and camcorders. Even in this example, we can see that the finest level that is of any use is most often the level of the product type (e.g. television), in other words the level of the department or sub-department, rather than the identification number of the product (such as the Efficient Article Numbering, or Stock Keeping Unit (SKU), which is the reference number of the product in the stores or in the catalogue). A level as fine as the SKU, which identifies everything down to the format and colour of the product, is rarely useful.

The value of this procedure is that it can provide more relevant rules, in which the commonest products do not hide the less common ones purely because of their frequency.

The best market basket analyses are therefore generally carried out on the basis of different levels of the product taxonomy. In all cases, even if just one level is used, the products in the transactions analysed must be carefully coded, to clearly distinguish a separate product from an option which is not to be taken into account. For each product, we must also ask what the most important property in the associations is to be: is it the type of product, its brand, or maybe its size (for clothes)?

[3] See: http://cran.univ-lyon1.fr/web/packages/arules/index.html and http://rss.acs.unt.edu/Rdoc/library/arules/html/apriori.html.

[4] See: http://eric.univ-lyon2.fr/~ricco/tanagra/fichiers/fr_Tanagra_Assoc_Rules_Comparison.pdf.

10.3 Using supplementary variables

In addition to the products in a market basket, events relating to customers, etc., the transaction lines analysed may include supplementary variables such as the date and time of the transaction, or the method of payment. These enable us to detect rules such as:

If *Nappies* and *Saturday*, then *Beer*.

By adding temporal variables, we can look for the sequence of events which ends with the purchase of a new product, the departure of the customer, or the like. In this case, we speak of *temporal* associations.

Other information may be found here, such as the name of a manufacturer which is included with some product types. Thus a market basket analysis can detect brand loyalty phenomena.

For this purpose, the data to be analysed are presented as follows:

	Product 1			Product 2			...
Customer A	Type	Brand	Purchase date	Type	Brand	Purchase date	...
Customer B	Type	Brand	Purchase date	Type	Brand	Purchase date	...
...	...	...	...	...	...	...	...
			...	...	...	...	
				...			

In mail order, insurance and banking, we can also add some information about the distribution channel: shop/agency, telephone, Internet, etc.

10.4 Applications

The method of finding association rules has been used widely since the 1960s in the retail industry for analysing market baskets, stocking departments, organizing promotions, managing stocks to prevent shortages and overstocks, etc. It is also useful for detecting associations of options chosen in packaged products (in banking, telephony, insurance, etc.) or associations of terms in a corpus of documents. It can be applied to any kind of items; for example, it can be used to detect rules in sports, for example: if player X is on the field and the match takes place in given circumstances, then the player Y scores more goals in 70% of cases.

As mentioned above, the main problem in implementing this method is the large number of irrelevant association rules which may submerge the relevant ones. This problem can be mitigated by using filters and taxonomies. However, some rules with high lifts and confidence indices may pass unnoticed because their support indices are below the threshold which had to be specified in order to prevent the numbers of rules becoming impossible to process. Hastie *et al.* (2009)[5] offer the light-hearted example of 'vodka ⇒ caviar' which is penalized by the scarcity of the consequent.

---

[5] Hastie, T., Tibshirani, R. and Friedman, J.H. (2009) *The Elements of Statistical Learning: Data Mining, Inference and Prediction*, 2nd edn. New York: Springer.

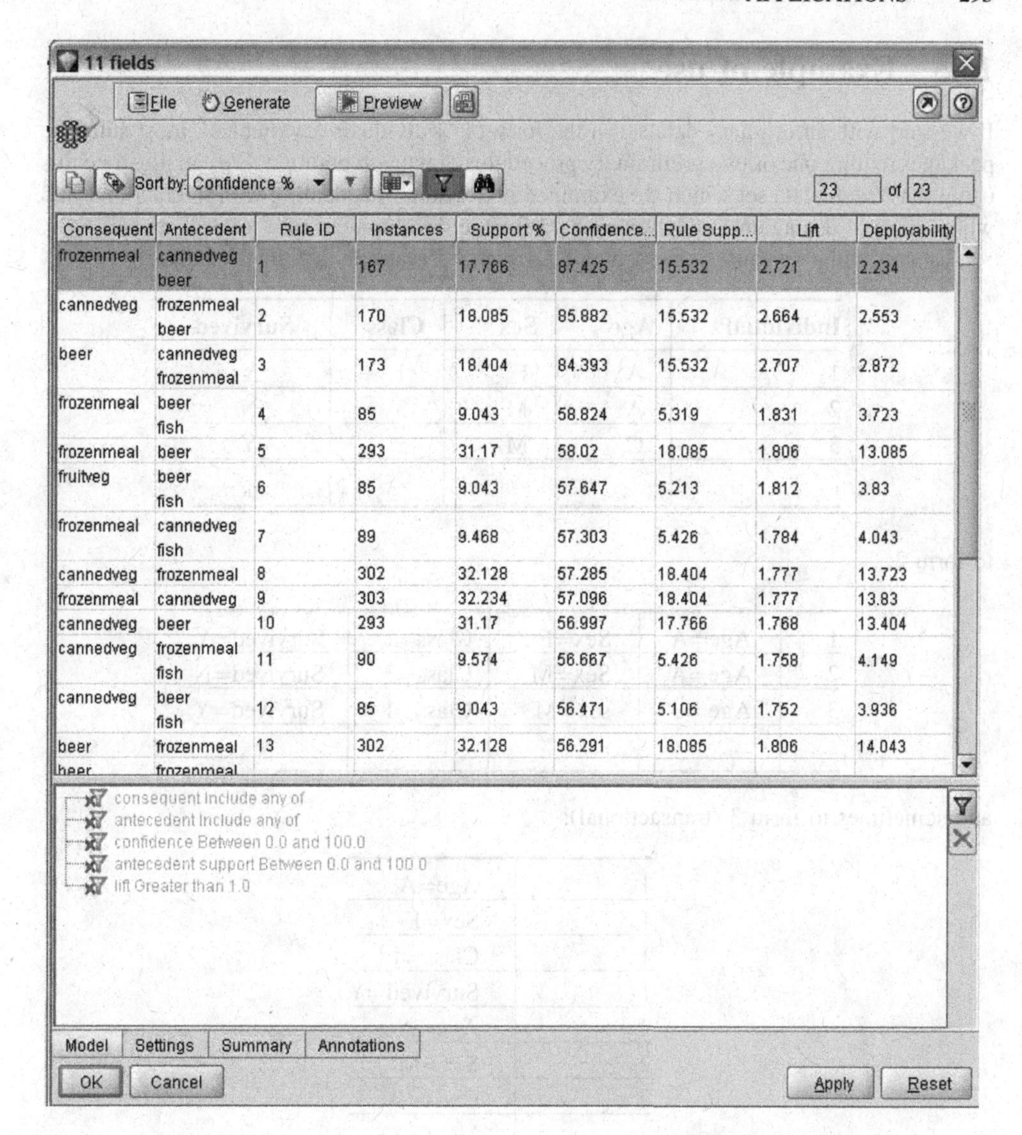

Consequent	Antecedent	Rule ID	Instances	Support %	Confidence...	Rule Supp...	Lift	Deployability
frozenmeal	cannedveg beer	1	167	17.766	87.425	15.532	2.721	2.234
cannedveg	frozenmeal beer	2	170	18.085	85.882	15.532	2.664	2.553
beer	cannedveg frozenmeal	3	173	18.404	84.393	15.532	2.707	2.872
frozenmeal	beer fish	4	85	9.043	58.824	5.319	1.831	3.723
frozenmeal	beer	5	293	31.17	58.02	18.085	1.806	13.085
fruitveg	beer fish	6	85	9.043	57.647	5.213	1.812	3.83
frozenmeal	cannedveg fish	7	89	9.468	57.303	5.426	1.784	4.043
cannedveg	frozenmeal	8	302	32.128	57.285	18.404	1.777	13.723
frozenmeal	cannedveg	9	303	32.234	57.096	18.404	1.777	13.83
cannedveg	beer	10	293	31.17	56.997	17.766	1.768	13.404
cannedveg	frozenmeal fish	11	90	9.574	56.667	5.426	1.758	4.149
cannedveg	beer fish	12	85	9.043	56.471	5.106	1.752	3.936
beer	frozenmeal	13	302	32.128	56.291	18.085	1.806	14.043
beer	frozenmeal							

Figure 10.2 Association rules detection in IBM SPSS® Modeler.

Naturally, a huge amount of computation power is needed to analyse the market baskets of a hypermarket with several tens of millions of products on its lists and several million transactions per year.[6] Association detection algorithms are provided in data mining programs available for a client–server system, such as SAS Enterprise Miner™ and IBM SPSS Modeler (Figure 10.2), as well as in freeware such as R, Tanagra, RapidMiner and Weka.

[6] The total number of transactions in all the Wal-Mart stores is more than 20 million per day!

10.5 Example of use

If we start with an ordinary data set in the form of 'individuals × variables', most software packages require one or two preliminary procedures of data preparation. We can illustrate this using the *Titanic* data set which we examined in Section 3.12, dealing with interactions, and which we will use again in Section 11.8.13 for the development of a logistic model.

To start with, we must convert the data from observations in form 1 (tabular):

Individual	Age	Sex	Class	Survived
1	A	F	1	Y
2	A	M	3	N
3	C	M	1	Y
...	...	...	...	...

to form 2:

1	Age=A	Sex=F	Class=1	Survived=Y
2	Age=A	Sex=M	Class=3	Survived=N
3	Age=C	sex=M	Class=1	Survived=Y
...	...	...	...	...

and sometimes to form 3 (transactional):

1	Age=A
1	Sex=F
1	Class=1
1	Survived=Y
2	Age=A
2	Sex=M
2	Class=3
...	...

To consider only three examples, the freeware by C. Borgelt processes form 2, but SAS Enterprise Miner requires form 3, while IBM SPSS Modeler can handle forms 1 and 3.

The following SAS code can be used to create form 3, used by SAS, directly. First, we must add a key, in other words a unique identifier of each individual, if the file does not already have one. The input file will then be transposed with respect to this key by an association rules detection program. Note that the variables are numeric in this example.

```
DATA titanic ;
SET sasuser.titanic ;
id = _n_ _ ;
RUN;
```

	CLASS	AGE	SEX	SURVIVED	ID
1	1	1	1	1	1
2	1	1	1	1	2
3	1	1	1	1	3
4	1	1	1	1	4
5	1	1	1	1	5
6	1	1	1	1	6
7	1	1	1	1	7
8	1	1	1	1	8
9	1	1	1	1	9
10	1	1	1	1	10

The following transposition transforms the 'individuals × variables' data set into a data set with one line per (individual, variable) pair with the name of the variable in _name_ ('name of the former variable') and its content in var1, where 'var' is the prefix specified in the TRANSPOSE procedure. Since the variable 'ID' has also been transposed (all the variables have been transposed, by the VAR _all_ instruction), the corresponding lines are deleted from the TRANSPO file.

```
PROC TRANSPOSE DATA=test OUT=transpo (WHERE = (_name_ NE "id"))
PREFIX= var ;
BY id ;
VAR _all_ ;
RUN;
```

	ID	NAME OF THE FORMER VARIABLE	var1
1	1	CLASS	1
2	1	AGE	1
3	1	SEX	1
4	1	SURVIVED	1
5	2	CLASS	1
6	2	AGE	1
7	2	SEX	1
8	2	SURVIVED	1
9	3	CLASS	1
10	3	AGE	1
11	3	SEX	1
12	3	SURVIVED	1

A step called DATA transforms the preceding data set into form 3 as mentioned above, by concatenating the name of each variable with its content:

```
DATA titanic_assoc (KEEP = id item)
SET transpo ;
LENGTH item $20. ;
item = CATX ('=',_name_, var1) ;
RUN;
```

	key	item
1	1	CLASS=1
2	1	AGE=1
3	1	SEX=1
4	1	SURVIVED=1
5	2	CLASS=1
6	2	AGE=1
7	2	SEX=1
8	2	SURVIVED=1
9	3	CLASS=1
10	3	AGE=1
11	3	SEX=1
12	3	SURVIVED=1

A data set in the above form (form 3) can be analysed in Enterprise Miner. Other packages require form 2, and the file can be transposed again from form 3 to form 2 by using the ID as a pivot.

```
PROC TRANSPOSE DATA=titanic_assoc OUT=titanic_assoc2 (DROP =_name_)
PREFIX=var ;
BY id ;
VAR item ;
RUN ;
```

	ID	var1	var2	var3	var4
1	1	CLASS=1	AGE=1	SEX=1	SURVIVED=1
2	2	CLASS=1	AGE=1	SEX=1	SURVIVED=1
3	3	CLASS=1	AGE=1	SEX=1	SURVIVED=1
4	4	CLASS=1	AGE=1	SEX=1	SURVIVED=1
5	5	CLASS=1	AGE=1	SEX=1	SURVIVED=1
6	6	CLASS=1	AGE=1	SEX=1	SURVIVED=1
7	7	CLASS=1	AGE=1	SEX=1	SURVIVED=1
8	8	CLASS=1	AGE=1	SEX=1	SURVIVED=1
9	9	CLASS=1	AGE=1	SEX=1	SURVIVED=1
10	10	CLASS=1	AGE=1	SEX=1	SURVIVED=1

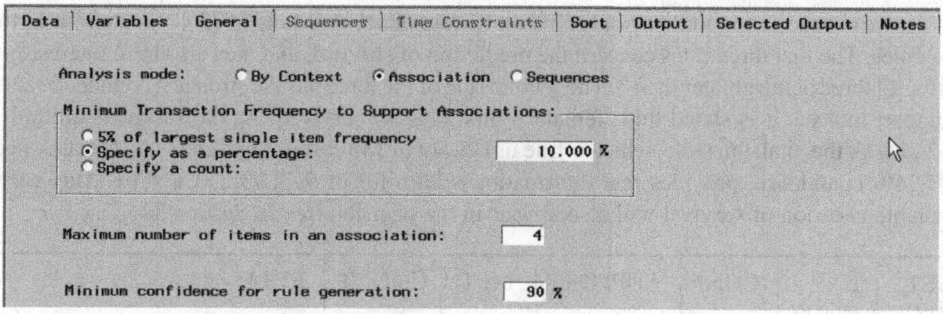

Data | Variables | General | Sequences | Time Constraints | Sort | Output | Selected Output | Notes

Analysis mode: ○ By Context ◉ Association ○ Sequences

Minimum Transaction Frequency to Support Associations:

○ 5% of largest single item frequency
◉ Specify as a percentage: 10.000 %
○ Specify a count:

Maximum number of items in an association: 4

Minimum confidence for rule generation: 90 %

Figure 10.3 Parameter setting for association detection in SAS Enterprise Miner.

Figure 10.3 shows the parameter setting screen of the Association node of SAS Enterprise Miner, which is applied to the data set in form 3 (transactional). This results in the 17 rules shown in Figure 10.4. As mentioned above, the package provides an option for storing these rules in an SAS data set or exporting them in another format.

The first rule is: male ⇒ adult (SEX=1 ⇒ AGE=1). It relates to 1667 individuals out of 2201 passengers on the *Titanic*, i.e. the support index is 75.74%. As there are 1731 males, of whom 1667 are adults, the confidence index is 96.30%. The lift of this rule is its confidence index divided by the probability of being an adult, which is 95.05% (2092 out of 2201 passengers). This is only 1.01, and the 96.30% is only very slightly greater than the 95.05% of confidence achieved by trivial prediction. This rule is therefore of low interest.

The rule with the strongest lift is 'SURVIVED=0 & CLASS=0 ⇒ SEX=1 & AGE=1': drowned + member of crew ⇒ male + adult. The lift is 99.55% (confidence) divided by 75.74% (this percentage of passengers are male adults), i.e. 1.31. But is this prediction really useful? What we need is rules in which survival or drowning appears in the consequents (results) and not in the antecedents (conditions). None of the above 17 rules meets this condition.

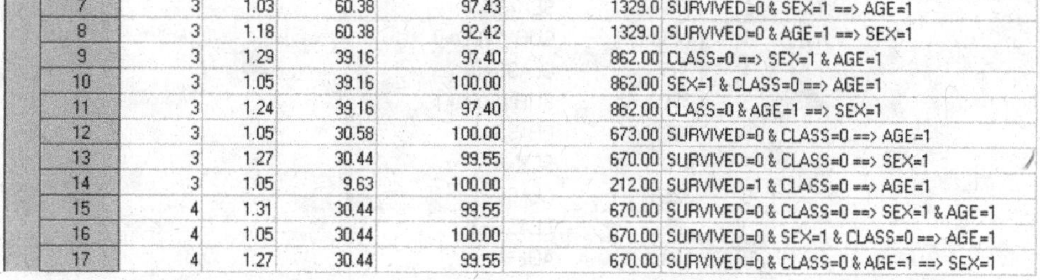

Rules | Frequencies | Code | Log | Notes

	Relations	Lift	Support(%)	Confidence(%)	Transaction Count	Rule
1	2	1.01	75.74	96.30	1667.0	SEX=1 ==> AGE=1
2	2	1.02	65.33	96.51	1438.0	SURVIVED=0 ==> AGE=1
3	2	1.16	61.97	91.54	1364.0	SURVIVED=0 ==> SEX=1
4	2	1.05	40.21	100.00	885.00	CLASS=0 ==> AGE=1
5	2	1.24	39.16	97.40	862.00	CLASS=0 ==> SEX=1
6	2	1.03	14.49	98.15	319.00	CLASS=1 ==> AGE=1
7	3	1.03	60.38	97.43	1329.0	SURVIVED=0 & SEX=1 ==> AGE=1
8	3	1.18	60.38	92.42	1329.0	SURVIVED=0 & AGE=1 ==> SEX=1
9	3	1.29	39.16	97.40	862.00	CLASS=0 ==> SEX=1 & AGE=1
10	3	1.05	39.16	100.00	862.00	SEX=1 & CLASS=0 ==> AGE=1
11	3	1.24	39.16	97.40	862.00	CLASS=0 & AGE=1 ==> SEX=1
12	3	1.05	30.58	100.00	673.00	SURVIVED=0 & CLASS=0 ==> AGE=1
13	3	1.27	30.44	99.55	670.00	SURVIVED=0 & CLASS=0 ==> SEX=1
14	3	1.05	9.63	100.00	212.00	SURVIVED=1 & CLASS=0 ==> AGE=1
15	4	1.31	30.44	99.55	670.00	SURVIVED=0 & CLASS=0 ==> SEX=1 & AGE=1
16	4	1.05	30.44	100.00	670.00	SURVIVED=0 & SEX=1 & CLASS=0 ==> AGE=1
17	4	1.27	30.44	99.55	670.00	SURVIVED=0 & CLASS=0 & AGE=1 ==> SEX=1

Figure 10.4 Result of association detection in SAS Enterprise Miner.

If we choose a support threshold of 5% and a confidence threshold of 75%, we go from 17 to 62 rules. The first three rules concern the prediction of survival, and they also have interesting lifts, all three being greater than 3. The second rule of the three has the strongest confidence and support indices. It is stated thus: female + first class ⇒ survived. As the survivors are only 32.30% of the total, thus rule, which is true in 141 out of 145 cases of first class and females, i.e. 97.24% confidence, provides real information with a lift of 97.24/32.30 = 3.01. This very reliable criterion of survival will also appear in the decision tree in Section 11.4.2.

SET_ SIZE	EXP_ CONF	CONF	SUPPORT	LIFT	COUNT	RULE
4	29.71	96.55	6.36	3.25	140.00	SEX=0 & CLASS=1 ⇒ SURVIVED=1 & AGE=1
3	32.30	97.24	6.41	3.01	141.00	SEX=0 & CLASS=1 ⇒ SURVIVED=1
4	32.30	97.22	6.36	3.01	140.00	SEX=0 & CLASS=1 & AGE=1 ⇒ SURVIVED=1
4	60.38	75.71	30.44	1.25	670.00	CLASS=0 ⇒ SURVIVED=0 & SEX=1 & AGE=1
3	61.97	75.71	30.44	1.22	670.00	CLASS=0 ⇒ SURVIVED=0 & SEX=1
4	61.97	75.71	30.44	1.22	670.00	CLASS=0 & AGE=1 ⇒ SURVIVED=0 & SEX=1
...	...	...	...	...	...	...
2	95.05	96.51	65.33	1.02	1438.0	SURVIVED=0 ⇒ AGE=1
2	95.05	96.30	75.74	1.01	1667.0	SEX=1 ⇒ AGE=1

As shown in Figure 10.5, SAS Enterprise Miner also displays the most frequent items, namely those whose frequency exceeds 5% (the support threshold that was set previously) of the number of individuals, which is 2201 in this case.

Rules | Frequencies | Code | Log | Notes

	Count	Item
1	2092	AGE=1
2	1731	SEX=1
3	1490	SURVIVED=0
4	885	CLASS=0
5	711	SURVIVED=1
6	706	CLASS=3
7	470	SEX=0
8	325	CLASS=1
9	285	CLASS=2
10	109	AGE=0

Figure 10.5 The most frequent items in SAS Enterprise Miner.

A note on *redundant rules*. In the example above, rule 13,

$$\text{SURVIVED} = 0 \,\&\, \text{CLASS} = 0 \Rightarrow \text{SEX} = 1,$$

and rule 15,

$$\text{SURVIVED} = 0 \,\&\, \text{CLASS} = 0 \Rightarrow \text{SEX} = 1 \,\&\, \text{AGE} = 1,$$

have exactly the same support (670 observations), because rule 14,

$$\text{SURVIVED} = 0 \,\&\, \text{CLASS} = 0 \Rightarrow \text{AGE} = 1,$$

is always true. Rule 13 is therefore redundant with respect to rule 15, because $15 \Rightarrow 13$.

11

Classification and prediction methods

This is the longest chapter in the book, covering the predictive methods used in statistics and data mining. These are the longest established, the most widely applied, and the most profitable techniques, and are used in various fields, such as hospitals, for calculating the probability of occurrence of a disease, the recovery of a patient or the effectiveness of a treatment; in research and industry, for calculating the probability of occurrence of a phenomenon; in meteorology, for forecasting the weather or peak pollution incidences; in agriculture, for predicting the yield of a crop; in banks and insurance, for calculating the probability of customer defaults or claims; in humanities and social sciences, for predicting types of behaviour; in archaeology, for dating excavated objects, etc. There are many predictive techniques. They are undergoing constant development and being extended to a wide range of problems. However, all these techniques operate within a precise theoretical framework, which we must be familiar with if we are to avoid using them inappropriately. Between the overview section outlining the qualities to be expected in a predictive technique (Section 11.3) and the guide to the application of these techniques at the end of the chapter (Section 11.16), we shall examine them one by one, pointing out their strong points and the restrictions on their use, and starting with the type of data to which they are applied.

11.1 Introduction

The predictive techniques used in data mining can be divided into two major operations: *classification* (or *discrimination*) and *prediction* (or *regression*). The aim of these two operations is to estimate the value of a variable (called the 'dependent', 'target', 'response', 'explained' or 'endogenous' variable) relating to an individual or object as a function of the value of a certain number of other variables relating to the same individual, identified as the 'independent' variables (also called the 'explanatory', 'control' or 'exogenous' variables).

Data Mining and Statistics for Decision Making, First Edition. Stéphane Tufféry.

What distinguishes them is the nature of the dependent variable: this is *qualitative* in the case of classification, but *continuous* in the case of prediction.[1]

Classification is therefore an operation that places each individual from the population under study in one of a number of specified classes, according to the characteristics of the individual which are identified as independent variables. An individual is generally assigned to a class on the basis of the explanatory characteristics by using a formula, an algorithm, or a set of rules, which forms a *model*, and which must be discovered.

In practice, there are often two classes to be predicted, such as

- purchasing or non-purchasing customer (a score called the *propensity to consume*),
- a patient with a good or a bad prognosis (*risk* score),
- a customer who is loyal or non-loyal to the business (*attrition* score),

and the model will enable us to determine

- the probability that a customer will buy a given product,
- the probability of a patient's recovery,
- the probability of losing a customer.

Prediction is a well-known operation in data analysis at least as regards its elementary aspects: for example, it may be used to estimate the price of an apartment as a function of its area, the storey it is situated on, and the district it is located in, or the consumption of electricity as a function of the outside temperature and the thickness of insulation, or the size of the wings of a species of bird as a function of age, and so on.

11.2 Inductive and transductive methods

Classification and prediction techniques fall into two types (Figure 11.1). In *inductive* techniques, a training phase (the inductive phase) is used to develop a model which in some way summarizes the relations between the variables, and which can then be applied to new data to deduce a classification or prediction from them (the deductive phase).

Transductive techniques have only one step (repeated if necessary) in which each individual is classified directly (or is predicted) with reference to the other individuals that have already been classified – no model is created. Consequently, there is no determination of parameters, because transductive methods are non-parametric.

The best-known of the *transductive techniques*, apart from the kernel method, is probably the k-nearest-neighbour method,[2] also called 'memory-based reasoning', in which each individual is classified by searching among previously classified individuals for the class of the k individuals which are its nearest neighbours, in terms of Euclidean distance or some

[1] 'Prediction' should not be confused with 'forecasting', which is a matter of evaluating the value of a variable at time t, given its values at previous points in time.

[2] Previously encountered in automatic clustering, in the context of estimating the probability density (see Section 9.10.3).

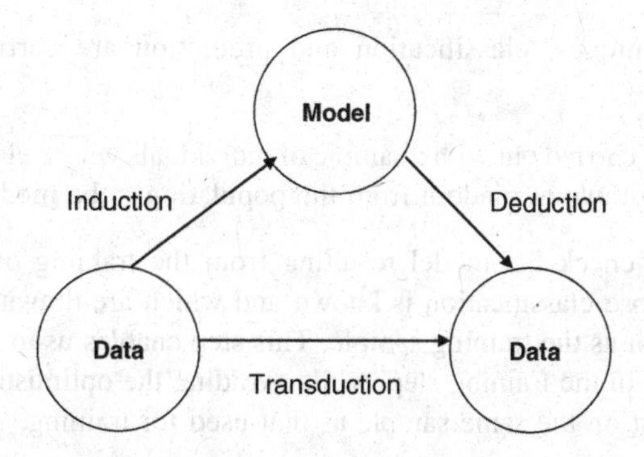

Figure 11.1 Inductive and transductive methods.

other distance metric. Thus, in the example of Figure 11.2, the individual '?' is assigned to '0', because it is mainly surrounded by 0s.

This principle can also be used for prediction, with a continuous predicted variable whose value is known for a certain number of individuals, and in this case we calculate the mean of the predicted variable in the neighbourhood of any new individual.

The value of k will be chosen so as to obtain the best possible classification (or prediction). The choice of the distance function and the optimal value of k is the main difficulty in using this technique. If k is too small, the small number of neighbours makes the prediction unreliable; if k is too large, the prediction becomes less accurate. Some software packages offer an automatic choice of k which optimizes the accuracy.

This algorithm is useful when the number of variables is not too large and when the distribution of the classes in Euclidean space is irregular. It works well with heterogeneous data or even text data.

However, it has the drawback of requiring the manipulation of a number of previously classified individuals whenever a new classification (or prediction) is carried out, even for only one individual. It therefore needs a lot of storage and computing capacity.

Since it is generally preferably to *summarize in a model* the information contained in the data, in order to be able to *control* this model and *rapidly apply it* to new data, transductive techniques are much less widely used than inductive techniques, which will now be discussed.

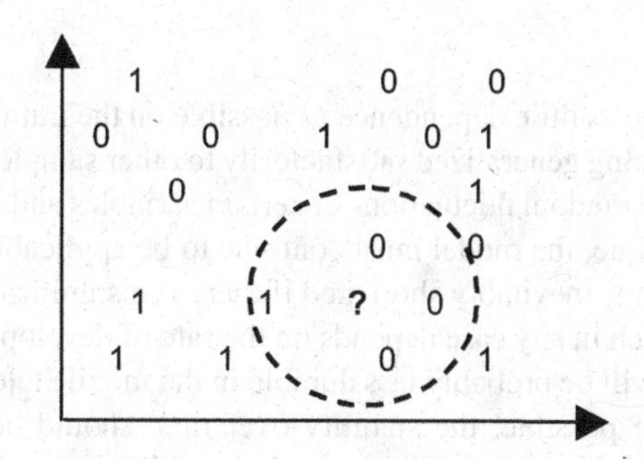

Figure 11.2 Classification by the k-nearest-neighbour method.

In inductive techniques, classification and prediction are carried out in three or four steps:

1. A *training* step, carried out with a sample of individuals whose classification is known, and which are drawn at random from the population to be modelled.

2. A *test* step, to check the model resulting from the training on another sample of individuals whose classification is known and which are drawn at random from the same population as the training sample. This step enables us to select the best of the models created in the training step, while avoiding the optimistic bias that would be caused by a test on the same sample as that used for training.

3. An optional step of *validation* on a third sample whose classification is known, to measure the performance of the best model selected in the previous two steps. The purpose of this step is to predict the quality of the results obtained when the model is applied.

4. An *application* step, in which the resulting model is applied to the whole of the population to be modelled.

In practice, the three samples – training, test and validation – may all be in the same physical data set, or may be in separate data sets.

11.3 Overview of classification and prediction methods

11.3.1 The qualities expected from a classification or prediction method

Precision

In a classification technique, the error rate, i.e. the proportion of incorrectly classified individuals, must be as low as possible. Quality indicators, such as the area under the ROC curve and the Gini index (which will be described below), must be as close as possible to 1.

In a prediction technique, there are other quality indicators, such as the R^2 of a linear regression, which must be as close as possible to 1.

Robustness

The model must also have as little dependence as possible on the training sample that is used, and must be capable of being generalized satisfactorily to other samples. It must have the least possible sensitivity to the random fluctuations of certain variables and to missing values. Even if the data change over time, the model must continue to be applicable to new samples for a reasonable period, which is inevitably shortened if there is a significant change in the law or legal conditions, but which in any case depends on the rate of development of product ranges and consumers: a score will be probably less durable in the mobile telephony industry than in banking, for example. If possible, the stability over time should be tested at the time of construction of the model, if an out-of-time sample is available for the tests. Such a sample covers a different period from that of the training sample.

The model should not be applied to variables which are doubtful, difficult to obtain, or unstable from one sample to another or from one period to another.

Concision (parsimony)

The rules of the model must be as simple as possible and the number of rules must be as small as possible. This will ensure that they are easier to understand and control, and more capable of generalization to populations other than that of the training sample. Concision is a factor in robustness.

Explicit results

The rules of the model should preferably be accessible and understandable. When expressed in the form of explicit conditions on the original variables, they have two advantages: they are immediately understandable to any user, and can easily be programmed by an IT worker, in the form of SQL requests for example, for integration into the information system of a business. Laws or regulations may require the results to be readable. This is also a requirement in the medical field.

The diversity of data types processed

Not all algorithms can handle data that are qualitative, discrete, continuous, or simply missing.

Model development speed

Even if the application of a given model is always relatively fast (as it has to be, for some real-time applications), its training, in other words its construction, may take too long when large volumes are data are involved (more than several hundred thousand observations). If many tests and adaptations are required to refine the model, its training needs to be reasonably speedy.

The possibilities of parameter setting

In a classification, it may be useful to weight the classification errors, for example in order to show that it is more serious to classify an patient who is ill as 'not ill' than vice versa. It may also be useful to specify the *a priori* distribution of the individuals into the classes to be predicted. Other parameter settings are possible, depending on the techniques used.

11.3.2 Generalizability

A small training sample can easily yield a low error rate in the training phase, while resulting in a relatively high error rate in the test phase, because the model is poorly generalizable owing to incomplete training. Conversely, a large training sample may make the model appear to be less effective during training, because it will not be able to learn all the specific cases of this more complex sample, but this model will perform better in the test phase because it is more generalizable. However, the error rate in the test phase does not decrease

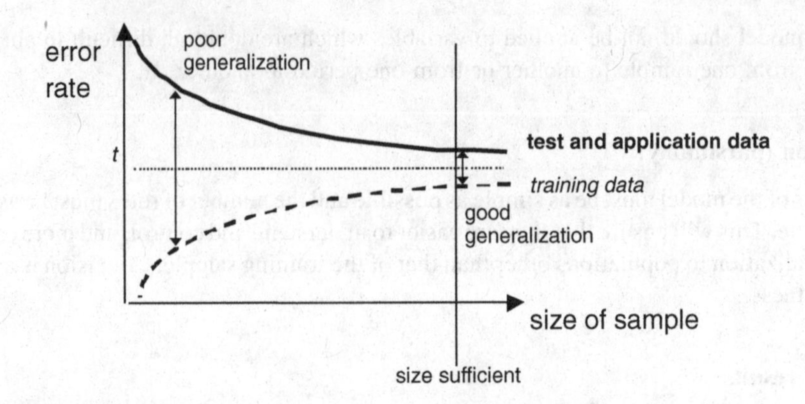

Figure 11.3 Error rates in training and testing, according to the size of the learning sample.

indefinitely with an increase in the size of the training sample, and the error rate in training does not rise indefinitely, but the training and test error rates generally converge towards the same limit t when the size of the training sample increases. This property, called *consistency* in learning theory, is not true of all learning processes, but it is found in those which concern us, namely the families of models in which the Vapnik–Chervonenkis dimension is finite (see the next section). The convergence is a mean convergence, and in practice the observed convergence is rarely perfect, because the error rates fluctuate from one sample to the next (sampling sensitivity) and the test error rate may even be lower than the learning error rate in a specific sample.

Figure 11.3 is an illustration of the above, and gives an idea of the minimum size of the training sample:

- below this size, the model resulting from the training is poorly generalizable to testing and application;
- above this size, there is no longer any significant decrease in the error rate in testing and application.

This critical size is generally greater than 1000 individuals, but depends on the complexity of the problem and the data, and, above all, on the dependent variable: in order to construct a sufficiently robust model (see above), it is advisable to have at least 300–500 individuals in each of the classes to be predicted.

Below this threshold, the population is too small for the selection of a test sample which is different from the training sample. The test has to be conducted in another way, using what is known as *cross-validation*. The population is split into, say, 10 random samples of equal size. The set of the first nine samples is used as the training sample, and the remaining tenth sample is used as the test sample. Thus a test error rate is found. The same operation is then repeated on every possible 9/10, using each remaining 1/10 as the test sample. Finally, the mean of the resulting 10 error rates is calculated, to estimate the error rate of the constructed model over the whole set of data.

TEST	TRAINING	TRAINING	TRAINING	TRAINING	TRAINING	TRAINING	TRAINING	TRAINING	TRAINING
TRAINING	TEST	TRAINING	TRAINING	TRAINING	TRAINING	TRAINING	TRAINING	TRAINING	TRAINING
TRAINING	TRAINING	TEST	TRAINING	TRAINING	TRAINING	TRAINING	TRAINING	TRAINING	TRAINING
TRAINING	TRAINING	TRAINING	TEST	TRAINING	TRAINING	TRAINING	TRAINING	TRAINING	TRAINING
TRAINING	TRAINING	TRAINING	TRAINING	TEST	TRAINING	TRAINING	TRAINING	TRAINING	TRAINING
TRAINING	TRAINING	TRAINING	TRAINING	TRAINING	TEST	TRAINING	TRAINING	TRAINING	TRAINING
TRAINING	TRAINING	TRAINING	TRAINING	TRAINING	TRAINING	TEST	TRAINING	TRAINING	TRAINING
TRAINING	TRAINING	TRAINING	TRAINING	TRAINING	TRAINING	TRAINING	TEST	TRAINING	TRAINING
TRAINING	TRAINING	TRAINING	TRAINING	TRAINING	TRAINING	TRAINING	TRAINING	TEST	TRAINING
TRAINING	TRAINING	TRAINING	TRAINING	TRAINING	TRAINING	TRAINING	TRAINING	TRAINING	TEST

This choice of 10 samples is most frequently used, in software packages and elsewhere, but it is also possible to have n samples, where the number is equal to the number of individuals in the population: at each step, a single individual is omitted (this is known as the 'leave-one-out' method). For reasons of computation time, this approach is only possible if n is not too large.

11.3.3 Vapnik's learning theory

This section is a more detailed examination of the matters discussed in the previous section, but it is not essential for understanding the rest of the book.

To evaluate the predictive quality of a model, we can measure the prediction error by means of various loss functions. The most widely used of these include:

- the quadratic function $L(y,f(x)) = (y - f(x))^2$ when y is continuous;
- the function $L(y,f(x)) = {}^1\!/_2\,|y - f(x)|$ when y is qualitative with two possible values, -1 and $+1$.

The risk (or real risk) is defined as the expectation (mean value) of the loss function, but since the joint probability distribution of x and y is not known, the risk can only be estimated. The commonest estimate is the empirical risk formula

$$\frac{1}{n}\sum_{i=1}^{n}(y_i - f(x_i))^2$$

or

$$\frac{1}{n}\sum_{i=1}^{n}\frac{1}{2}|y_i - f(x_i)|,$$

where n is the sample size. The latter formula includes the error rate mentioned in the previous section.

We know that the empirical risk measured on the training sample has an optimistic bias: it is generally lower than the real risk. The real risk is best estimated by measuring the empirical risk on another sample, called the test sample, and this generally results in curves similar to those of Figure 11.3 for the error rates.

The question arises of the convergence of the two curves towards a common value. This is because, if the two curves are close above a certain value of n, then the discriminating power of the model made to fit the n observations of the training sample will probably be generalized successfully to other samples.

On the theoretical level, Vladimir Vapnik[3] considered the convergence of the empirical risk for the training sample R_{emp} towards the risk R (to which the empirical risk for the test sample is assumed to be an approximation), and demonstrated two fundamental results concerning this convergence, one of which relates to the existence of a convergence while the other relates to the speed of convergence.

[3] Vapnik, V. (1995) *The Nature of Statistical Learning Theory*. New York: Springer.

Before stating these two theorems, we must define a quantity related to the model and called the *Vapnik–Chervonenkis* (VC) *dimension*. The VC dimension is measurement of the complexity of a model, which is actually defined for every family of functions $\mathbf{R}^p \to \mathbf{R}$ (and consequently for the $\{f(x) \geq 0$, yes or no$\}$ classification models associated with functions) for which it measures the separating power of the points of $\mathbf{R}^\mathbf{p}$. For example, linear discriminant analysis is associated with a linear function, namely the Fisher function, and the set of possible coefficients defines a family of functions. A frequently cited example is the family of functions $\{\sin(\alpha x), x \in \mathbf{R}\}$ for which the VC dimension is infinite. For further information, the reader should consult Section 19.5 of the book by Gilbert Saporta[4] or Section 7.9 of the book by Hastie, Tibshirani and Friedman.[5]

The importance of this concept is due to two findings by Vapnik:

- The empirical risk for the training sample R_{emp} of a model converges towards its risk R (the model is said to be consistent) if and only if its VC dimension is finite.

- If the VC dimension h of a model is finite, then, with a probability of error α, we obtain

$$R < R_{\text{emp}} + \sqrt{\frac{h(\log(2n/h)+1) - \log(\alpha/4)}{n}}. \tag{11.1}$$

The condition of finiteness set for the VC dimension, to ensure the convergence of R_{emp} towards R, is not trivial, as is shown by the preceding example.

The theoretical value of the bound (11.1) discovered by Vapnik is that it is universal: it can be applied to all models, without any particular assumption on the joint distribution of x and y. This upper bound is universal, as is the Berry–Esseen bound in the central limit theorem, for example. Like this bound, however, the Vapnik upper bound is far from optimal in specific cases where a better upper bound can be found. Indeed, the VC dimension of a model is equal to the number of parameters in some simple cases (linear models), but it is usually difficult to calculate and even to bound effectively, which limits the practical value of the bound (11.1). Support vector machines (SVMs) are among the first types of models for which the VC dimension could be calculated (Section 11.12.1). As with regularized regression, the models in this case are calculated by applying a constraint (such as the 'C' of ridge regression: see Section 11.7.2), and their VC dimensions decrease as the constraint increases, which explains their performance. In the case of SVMs, the VC dimension is a function of the inverse of the margin.

However, we must be aware that inequality (11.1) is only true with a given probability α, and that the upper limit tends towards infinity as α tends towards 0.

Inequality (11.1) has led to the structural risk minimization approach, in which nested models with increasing VC dimension $h_1 < h_2 < \ldots$ are considered, as is traditionally done with logistic or linear models nested by the addition of successive variables, or neural networks in which units are added to the hidden layer. When the VC dimension increases, the empirical risk generally decreases (on average), while the second term on the right-hand side of (11.1) increases. We then retain the model that minimizes the right-hand side of (11.1), in

[4] Saporta, G. (2006) *Probabilités, Analyse des Données et Statistique*, 2nd edn. Paris: Technip.

[5] Hastie, T., Tibshirani, R. and Friedman, J.H. (2009) *The Elements of Statistical Learning: Data Mining, Inference and Prediction*, 2nd edn. New York: Springer.

other words the one offering the best compromise between fit and robustness, and between bias and variance. The Vapnik theory provides a theoretical framework for this problem of seeking a fit while avoiding overfitting. When it can be used, the structural risk minimization approach is an alternative to other methods, such as those based on the Bayesian information criterion (Section 11.8.6).

One last point. It should be noted that, if the models we are interested in always have a finite VC dimension, this dimension can increase with the size of the training sample without causing an increase in the bound, as is shown by formula (11.1), because of the square root term, without even considering the reduction of the empirical risk. This square root term tends towards 0, and therefore the empirical risk converges towards the real risk as h/n tends towards 0. This corresponds to the well-known fact that the complexity of a model can increase as its training sample becomes larger.

11.3.4 Overfitting

When a link between the dependent variable and an independent variable appears in the training sample, and is therefore memorized in the model, even though it does not exist in the whole population to be modelled, we speak of the *overtraining*[6] or *overfitting* of the model. Overfitting is most likely to occur when the sample is too small with respect to the number of parameters of the model, and is particularly prevalent with some less robust modelling techniques such as decision trees and neural networks. This phenomenon is due to the fact that, when the model is excessively precise in the training phase, it starts to follow all the fluctuations in the training sample, as shown in Figures 11.6 and 11.7. Overfitting can be detected by testing the model on another sample, and can be limited by increasing the sample size or by simplifying the model (using fewer independent variables, fewer leaves on a decision tree, fewer units in the hidden layer of a perceptron, etc.).

The choice of the complexity of the model requires a judgement to be made between the quality of the fit to the training data and the capacity for subsequent generalization. This is what statisticians call the 'bias–variance dilemma': the bias measures the quality of the fit, and the variance measures the variability of the prediction for new cases. By seeking a bias which is too small, we increase the variance and compromise the durability of the model.

It is always helpful to try to understand the model by using specialist knowledge of the subject.

Not infrequently, overfitting occurs when the modelling base has a bias with respect to the dependent variable. An example of this is a model intended to predict the purchasing of a product. If the modelling base is too limited in time (i.e. the purchases are observed over a period which is too short) and in space (i.e. the purchases are observed in a geographical area which is too restricted), it may contain an abnormal proportion of customers who have benefited from promotions and whose tendency to buy the product is over-evaluated by comparison with the normal tendency of this population. The model will be distorted as a result of this, and may even be completely misleading if a strong promotion has targeted a population which does not usually buy very much. Furthermore, the resulting overfitting may not be noticed in the construction of the model, if the test sample has suffered from the same bias as the training sample. This is why it is essential to work on sufficiently large and

[6] Also called 'overadaptation' or 'overlearning'.

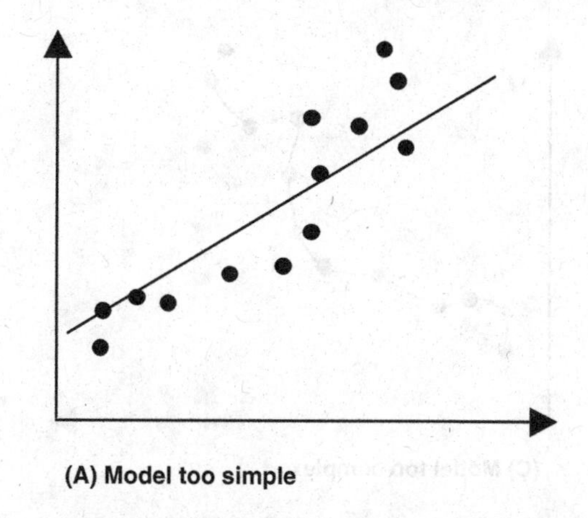

Figure 11.4 Underfitting of a regression model.

representative samples, covering a sufficiently long period of observation (of sales made, in this case). It is also preferable to find out about all the events which may have affected the population and may have an impact on the future models.

Overfitting also occurs when one of the independent variables is correlated by construction with the dependent variable. To take an extreme example, we might try to predict an unknown age by leaving among the independent variables a date of birth which would obviously be known only during the construction of the model, not during its application. It is essential not to rely on these independent variables whose presence is revealed by a model which is perfect in the training phase but useless in application. Clearly, this case of overfitting is less frequent, because it is more evident and easier to avoid.

Figures 11.4–11.6 show the fit to the data in the case of regression, for a model (A) which is too simple, a model (B) which is sufficiently complex, and a model (C) which is too complex. Similarly, Figure 11.7 shows the fit to the data for classification, for models A–C. These models correspond to the three abscissae (A), (B) and (C) in Figure 11.8, which shows

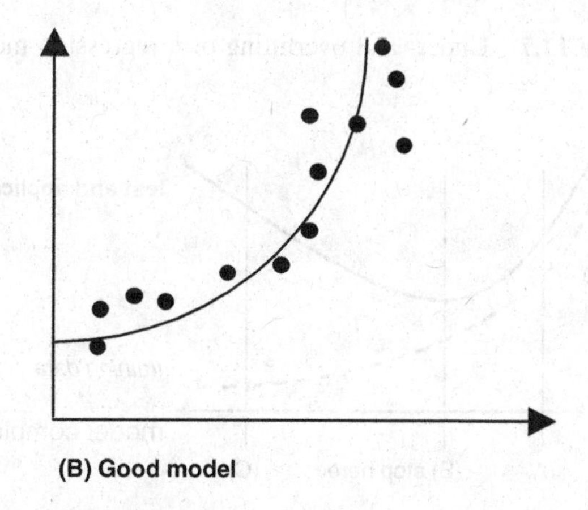

Figure 11.5 Good fitting of a regression model.

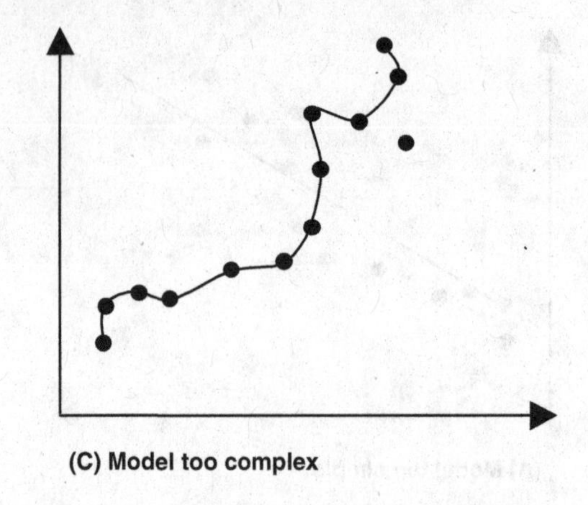

Figure 11.6 Overfitting of a regression model.

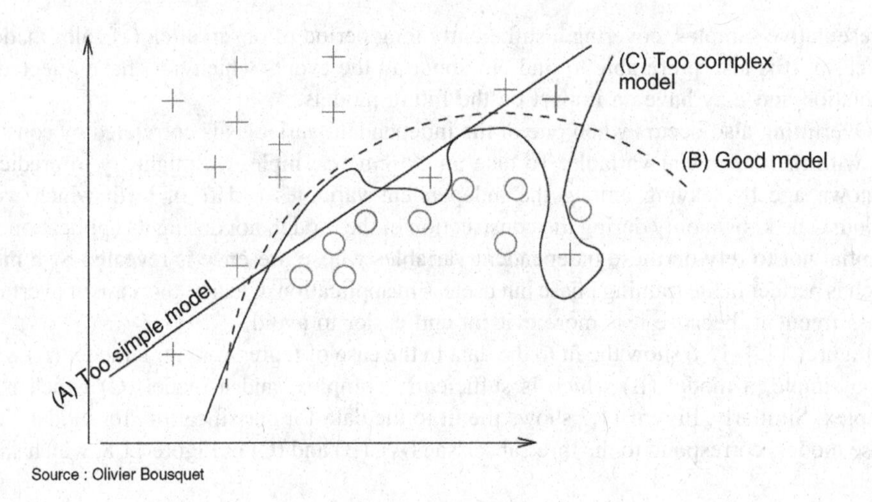

Figure 11.7 Under- and overfitting of a regression model.

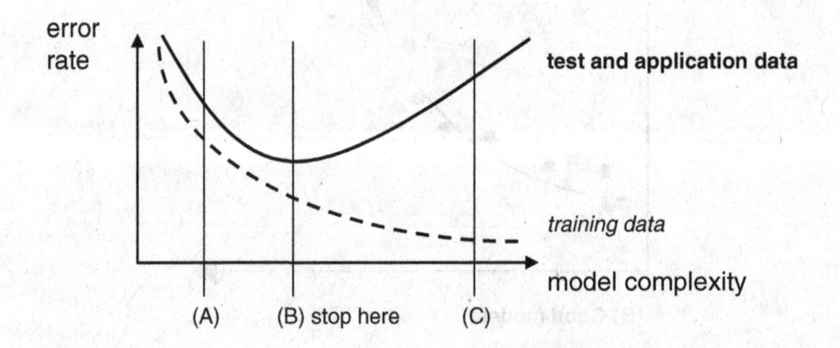

Figure 11.8 Error rates as a function of the complexity of the model.

the error rate as a function of the complexity of the model, in training and in testing. Comparing the error rates in training and in testing, we find that both rates decrease as the model begins to grow. The test error rate then rises again as the model not only learns the 'true' rules, which can be generalized, but also starts to be distorted in its adaptation to all the peculiarities of the training sample. At this point, corresponding to the abscissa (B) in Figure 11.8, the training should be halted.

The topic of overfitting will be considered again in the section on decision trees.

11.4 Classification by decision tree

The *decision tree* technique is one of the most intuitive and popular data mining methods, especially as it provides explicit rules for classification and copes well with heterogeneous data, missing data and non-linear effects. For applications concerned with database marketing, the only major competitor of the decision tree at present is logistic regression, which is preferred for risk prediction, owing to its greater robustness. It should be noted that decision trees are on the boundary between predictive and descriptive methods, since they create their classification by segmenting the population to which they are applied: thus they belong to the category of *supervised* divisive hierarchical methods.

11.4.1 Principle of the decision trees

The decision tree technique is used in classification to detect criteria for dividing the individuals of a population into n predetermined classes (in many cases, $n=2$). We start by choosing the variable which, by its categories, provides the best separation of the individuals in each class, thus providing sub-populations, called *nodes*, each containing the largest possible proportion of individuals in a single class; the same operation is then repeated on each new node obtained, until no further separation of the individuals is possible or desirable (according to criteria which depend on the type of tree). The construction is such that each of the terminal nodes (the *leaves*) mainly consists of the individuals of a single class. An individual is assigned to a leaf, and therefore to a certain class, with a reasonably high probability, when it conforms to all the rules for reaching this leaf. The set of rules for all the leaves forms the classification model (Figure 11.9).

11.4.2 Definitions – the first step in creating the tree

To construct a decision tree with the aim of dividing the individuals of a population into n classes, we must know how to choose the variable which best separates the individuals of each class: the precise criterion (C1) for the choice of the variable and the separation condition on this variable depends on the type of tree.

The number of possible separation conditions allowed by an independent variable depends on its type. A binary variable allows a single separation condition. For a continuous variable X having n separate values, there are $n - 1$ possible separation conditions on this variable. This is because, when the values $x_1, \ldots, x_n$ of X have been sorted in the order

$$x_1 \leq \ldots \leq x_n,$$

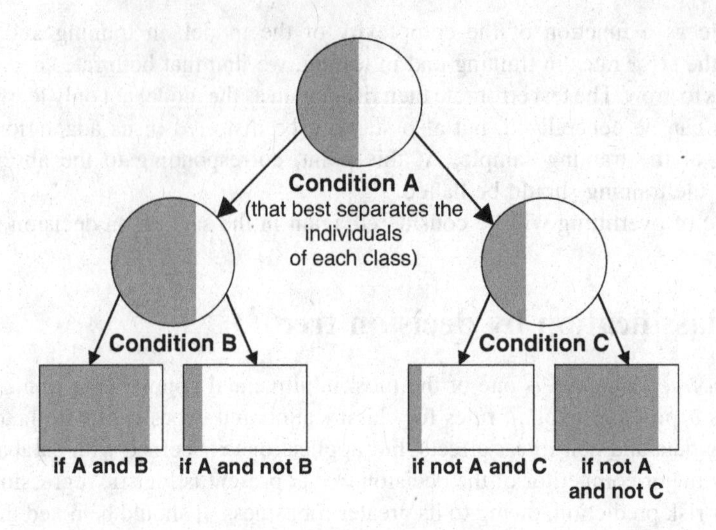

Figure 11.9 Decision tree.

the separation conditions are expressed in the form

$$X \leq \text{mean}\,(x_k, x_{k+1}).$$

A technique such as CART (see below) tests all the possibilities. If the independent variable X is qualitative, with a set E of n possible values $x_1, \ldots, x_n$, each separation condition on this variable will be in the form

$$X \in E', \quad \text{where } E' \subset E - \varnothing,$$

and we can see that there are $2^{n-1} - 1$ possible separation conditions on this variable (not $2^n - 1$, because the conditions $X \in E'$ and $X \in E - E'$ are equivalent). Here again, CART tests all the possibilities.

Here we can see that, when using data mining software, it is important to specify clearly that a variable which appears to be numeric (e.g. socioeconomic status) is actually qualitative. This is because the separation conditions on this variable will not take the same form.

When the best separation has been found, it is applied, and then the operation is repeated on each node to increase the discrimination, which gives rise to two or more *child nodes* for each node. Each child node in turn creates two or more nodes, and so on, until:

- the separation of the individuals cannot be repeated any further, either because there is only one individual left in each node, or because the individuals of a single node still all belong to the same class, or because they are all identical (in terms of the independent variables);
- or a certain criterion (C2) for stopping the deepening of the tree is met.

If each node of the tree has no more than two child nodes, the tree is said to be *binary*; but not all trees are binary. The first node of the tree is the *root*; the terminal nodes are the *leaves*. The path between the root and each leaf is the expression of a rule. For example: customers with an age less than x, a weight less than y and a height of at least z belong to class C in n% of cases. The percentage n is a membership score for the class C. The *density* of a node is the ratio of its number of individuals to the total number of the population; a common value for the minimum density of the leaves of a tree is 1–2%.

To see how to read a decision tree, let us return to the example of the prediction of survival following the sinking of the *Titanic*, as shown in the graphic produced by the IBM SPSS Decision Trees module (Figure 11.10). The root contains 2201 individuals, who are all the passengers on the liner. Of these, 711 (32.3%) will survive the shipwreck and 1490 (67.7%) will be drowned. The variable separating the survivors from the others most clearly is 'sex'. Among the 470 females (21.4% of the passengers), the survival rate is 73.2% (344 survivors), while for the 1731 males (78.6% of the passengers), the survival rate is only 21.2%

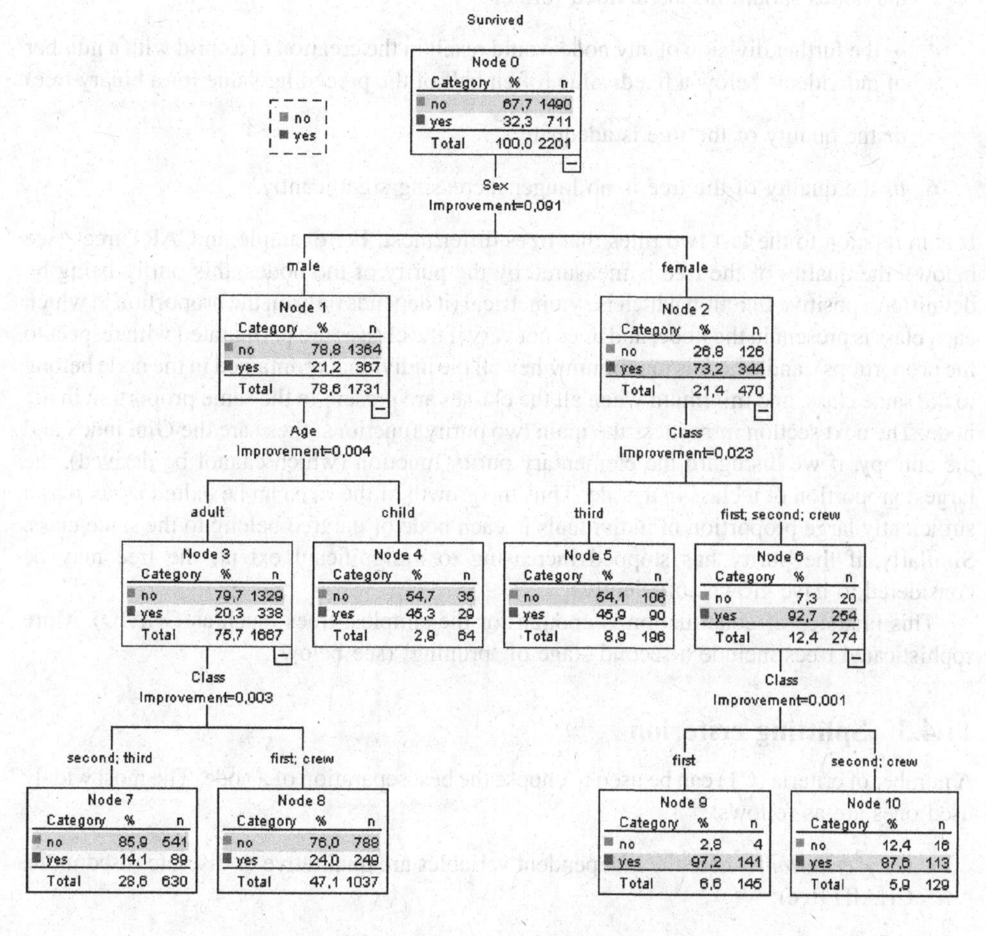

Figure 11.10 Decision tree for the *Titanic*.

(367 survivors). The best separation of the male population is provided by the binary variable 'age', the 64 boys having a survival rate of 45.3%, considerably lower than the 73.2% of the female population, but much greater than the adult male rate of 20.3%. As for those who were in second or third class, their survival rate is very low, at 14.1%. As for the females, if we leave aside the 196 unfortunates in third class, whose survival rate of 45.9% is very close to that of the boys, the others reach a survival rate of no less than 92.7%, and this rate rises to the exceptional level of 97.2% in first class. So there you have it. The fate of Leonardo DiCaprio and Kate Winslet was written in the statistics!

The stop criterion (C2), like the separation criterion (C1), depends on the type and parameter setting of the tree. (C2) often combines several rules, as follows:

1. the depth of the tree has reached a fixed limit
2. *or* the number of leaves (i.e. rules) has reached a fixed maximum
3. *or* the numbers contained in each node are less than a fixed value (often between 10 and 30, but it is preferable to choose between 75 and 100), below which it is considered that the nodes should not be divided further
4. *or* the further division of any node would result in the creation of a child with a number of individuals below a fixed value (often half of the preceding value for a binary tree)
5. *or* the quality of the tree is adequate
6. *or* the quality of the tree is no longer increasing significantly.

It is in relation to the last two rules that trees differ most. For example, in CART trees (see below), the quality of the tree is measured by the purity of the nodes, this purity being by definition a positive function which is symmetrical (it depends only on the proportion in which each class is present in the node, and does not vary if the classes are permutated with respect to the proportions), and which is maximum when all the individuals contained in the node belong to the same class, and minimum when all the classes are present in the same proportion in the node. The next section introduces the main two purity functions: these are the Gini index and the entropy, if we disregard the elementary purity function (which cannot be derived), the largest proportion of a class in a node. Thus the growth of the tree can be halted as soon as a sufficiently large proportion of individuals in each node of the tree belong to the same class. Similarly, if the purity has stopped increasing to a significant extent, the tree may be considered to have grown sufficiently.

This first step of construction is enough for the simplest trees (such as CHAID). More sophisticated trees include a second stage of 'pruning' (see below).

11.4.3 Splitting criterion

A number of criteria (C1) can be used to choose the best separation of a node. The most widely used ones are as follows:

- the χ^2 *criterion,* where the independent variables are qualitative or discrete (used in the CHAID tree);
- the *Gini criterion,* for all types of independent variables (used in the CART tree);

- the *Twoing criterion*, for any type of independent variable (used in the CART tree), if the dependent variable has $k \geq 3$ categories and we wish to convert the search for an optimal split on k categories to an optimal split on two super-categories composed of initial categories;
- the *ordered Twoing criterion,* if the dependent variable has $k \geq 3$ ordered categories, a criterion in which the two super-categories contain only adjacent categories from the initial categories;
- the *entropy,* or *information,* for all types of independent variables (used in the C4.5 and C5.0 trees).

The first criterion is well known and is detailed in the Appendix (see Section A.2.10). The choice of the independent variable and of the categories of this variable which are to separate a node into a number of child nodes is made so as to maximize the χ^2 of this variable cross-tabulated with the dependent variable (the class to be predicted).

The Gini index of a node is a purity function, calculated as follows:

$$\text{Gini (node)} = 1 - \sum_i f_i^2 = \sum_{i \neq j} f_i f_j$$

where f_i, $i = 1, \ldots, p$, are the relative frequencies in the node of the p classes to be predicted (the dependent variable). The more evenly distributed the classes are in a node, the higher the Gini index will be. As the purity of the node increases, the Gini index decreases. In the case of two classes, the Gini index ranges from 0 (pure node) to 0.5 (maximum mixing). With three classes, the index ranges from 0 to 2/3. The Gini index measures the probability that two individuals, picked at random with replacement from a node, belong to two different classes.

Remark. To simplify this description and the notation, I have adopted a simplification throughout this section which does not alter the essence of the explanations: it is assumed that the *a priori* probability of membership of a class is equal to the relative frequency of this class in the population. Decision tree software packages generally allow two other settings:

(i) the *a priori* probabilities are all equal;

(ii) the *a priori* probabilities are known by the user, who can enter them in a dialogue box.

It is this assumption of the equality of the *a priori* probability and the relative frequency that results in the same equality in each node, and enables us to write the above formula for the Gini index, which in a fully generalized form should be expressed as a function of the *a priori* probabilities, not the frequencies f_i. The same assumption also means that the probability that an individual belongs to a node is equal to the density of this node, in other words its number of individuals divided by the total number of the population.

Each separation into k child nodes (containing $n_1, n_2, \ldots, n_k$) should result in the greatest increase in purity and consequently the greatest decrease in the Gini index. In other words, we must minimize:

$$\text{Gini (separation)} = \sum_{i=1}^{k} \frac{n_i}{n} \text{Gini } (i\text{th node}).$$

An important property of the CART tree based on the Gini index is that the *impurity reduction* is always positive:

$$\text{Gini (parent node)} - \text{Gini (separation)} \geq 0.$$

The importance of a variable in a CART tree can be measured by calculating, for each node of the tree, the impurity reduction of the split created by the variable (when this variable has been selected for the split), and then adding up these impurity reductions for all the nodes of the tree. The IBM SPSS Decision Trees module provides this measurement.

This criterion enables us to take into account the costs C_{ij} of incorrect assignment of an individual of class j to class i. The Gini index of a node is then defined s

$$\text{Gini (node)} = \sum_{i,j} C_{ij} f_i f_j.$$

The simplest case, which is accepted by all software if no other setting is made, is the one in which $C_{ij} = 1$ if $i \neq j$ and $C_{ii} = 0$.

Note that, in contrast to discriminant analysis in which the costs of incorrect assignment do not alter the discriminating function, but only the decision threshold, these costs modify the decision tree itself.

Using the same principle of calculation, we can replace the Gini index with another purity function, the *entropy*, given by

$$\text{entropy (node)} = \sum_{i} f_i \log(f_i),$$

and we can then aim to minimize the entropy in the child nodes.

11.4.4 Distribution among nodes – the second step in creating the tree

When the tree has been constructed and the division criteria for each node have been established, each individual can be assigned to exactly one leaf, which is determined by the values of the independent variables for the individual. When this has been done, each leaf contains a certain proportion f_j of individuals of each class j. We can then deduce the class to which the leaf is assigned; this will define the class of all the individuals in the leaf.

The rule is that a leaf is assigned to a class j if the cost of assigning an individual of the leaf to class j is lower than the cost of assigning it to any other class. This cost C_j (still assuming that the *a priori* probabilities are identical to the proportions) is

$$C_j = \sum_{i=1}^{p} C_{ji} f_i.$$

Risk

Estimate	Std. Error
.217	.009

Growing Method : CRT
Dependent Variable:
Survived

Figure 11.11 Cost of a tree.

In the simplest case, where $C_{ij} = 1$ if $i \neq j$ and $C_{ii} = 0$, we find that

$$C_j = \sum_{i \neq j} f_i = 1 - f_j.$$

In this case, saying that class j of the leaf is the one that minimizes the cost C_j is equivalent to saying that it is the class that maximizes the proportion f_j of individuals in the leaf belonging to class j. Thus the class of the leaf is the one that is best represented in the leaf. More generally, if the *a priori* probabilities are different from the proportions, the class of the leaf is the most probable class in the leaf. Since all the individuals of the leaf are assigned to class j, the classification error rate of this leaf is $1 - f_j$.

Starting with the error rate of each leaf, the *error rate* of the tree, also called the *total cost* of the tree, or the *risk* of the tree, is calculated. This is the weighted sum of the error rates of the leaves, where the weighting is the probability of an individual's being assigned to the leaf. On the simplifying assumptions mentioned above concerning the *a priori* probabilities and the costs of incorrect assignment, the total cost of the tree is the proportion of individuals incorrectly classified by the tree.

In Figure 11.11, produced by IBM SPSS Decision Trees, the cost of the tree is 0.217, meaning that just over one individual in every five is incorrectly classified by the CART tree.

11.4.5 Pruning – the third step in creating the tree

Let us start with an example to show how useful pruning is. In a decision tree whose depth is very great, some nodes close to the leaves may contain very few individuals. The dependent variable may therefore appear to be correlated with all kinds of things. Thus, at the end of the training phase, income may appear to be correlated with the characteristics of Mr Smith, because Mr Smith is the only person with a high income in his small group in the training sample. When dealing with the whole population, the apparent correlation between these two variables will disappear, but it will already have been incorporated in the decision tree.

This example shows how necessary it is to shorten the branches of very deep trees, in order to avoid having very small nodes with no real statistical significance. There should be at least 20–30 individuals per node. Those branches that are too low should therefore be pruned before the error rate of the classification starts to rise. Pruning enables us to avoid the overfitting (overtraining) phenomenon discussed above and illustrated in Figure 11.12.

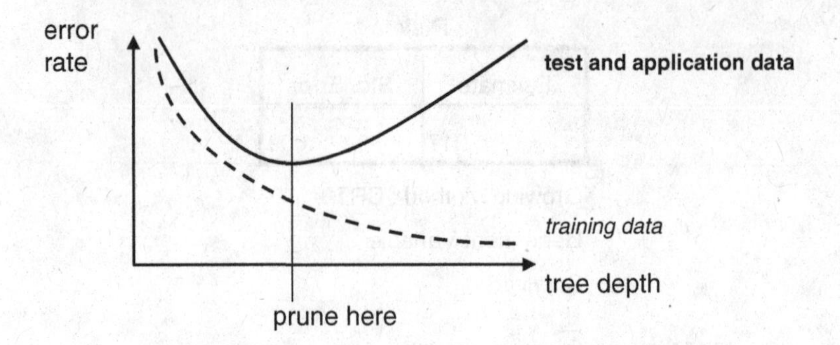

Figure 11.12 Error rate of a tree as a function of its depth.

A good algorithm should start by constructing a tree of maximum size (according to the above criterion (C2)), which it will then prune, after automatically detecting the optimal pruning threshold. This detection can be carried out in two ways.

If the size of the population is sufficient, a test sample, separate from the training sample, will have been created. It can be used to test each sub-tree of the maximum tree, and the sub-tree giving the lowest error rate in testing is then considered to be the best pruned tree.

If the population is not large enough for this, it will be necessary to use *cross-validation* (see Section 11.3.2) and combine the error rates found for all the possible sub-trees, again with the aim of choosing the best possible sub-tree.

11.4.6 A pitfall to avoid

There is a second cause of overfitting, where the dependent variable is present in a disguised form among the independent variables. During the training phase of a decision tree, the variables analysed must not include one that is *directly* correlated (by its nature or by construction) with the dependent variable. By way of example, if we wish to analyse the probability of taking out an insurance policy, the premiums for this contract should not be included among the variables analysed. Otherwise the first node will be as shown in Figure 11.13, and the tree will stop immediately. Now, the variable 'premiums' is not a cause but a consequence of taking out a policy. The tree has been overtrained with respect to the variable 'premiums'. It was partly in order to avoid this phenomenon that the advice was given in Section 2.3 to observe the dependent variable for a period of time not overlapping with the period of observation of the independent variables.

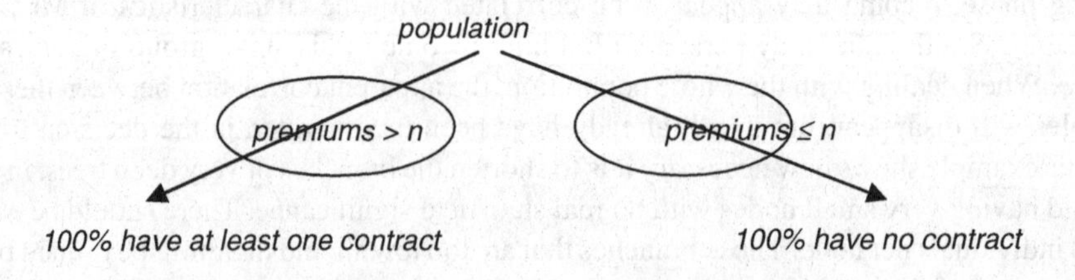

Figure 11.13 Decision tree with overfitting.

11.4.7 The CART, C5.0 and CHAID trees

The main decision tree algorithms are:

- CART (Classification And Regression Tree), for investigating all kinds of variables;
- C5.0 (developed by J.R. Quinlan), suitable for the investigation of all kinds of variables;
- CHAID (Chi-Square Automation Interaction Detection), initially reserved for the investigation of discrete and qualitative independent and dependent variables.

CART

The CART tree, invented in 1984 by the statisticians L. Breiman, J.H. Friedman, R.A. Olshen and C.J. Stone (Berkeley and Stanford Universities),[7] is one of the most effective and widely used decision trees. It is found, for example, in R (the *rpart* and *tree* functions), SAS Enterprise Miner, IBM SPSS Modeler and IBM SPSS Decision Trees, S-PLUS (TIBCO Software), Statistica (StatSoft), SPAD (Coheris-SPAD), and, of course, CART (Salford Systems). The name CART is registered, and only Salford Systems is allowed to use it, but this type of tree can be found under very similar names such as CRT or C&RT.

CART uses the Gini index to find the best separation of each node. In addition to this choice, the designers of CART have provided a number of technical solutions which yield two major benefits, namely generality and performance.

The generality is based primarily on the fact that the number of categories of the dependent variable can be finite or unlimited, and CART can be used for both classification and prediction: there is an appropriate node splitting criterion for each type of problem.

Its generality is also due to the fact that it can allow for the costs C_{ij} of incorrect assignment by incorporating them into the calculation of the Gini index (see Section 11.4.3 above).

Finally, the generality is enhanced by the capacity to process missing values by replacing each variable concerned with an equally splitting variable or an equally reducing variable. Equally splitting variables are those which provide approximately the same purity of the nodes as the original variable. Equally reducing variables are those which distribute the individuals in approximately the same way as the original variable. These variables can be used as 'surrogate' variables, but it is best to use equally reducing variables to maintain the consistency of the tree.

The performance of CART is primarily due to its pruning mechanism, which is more sophisticated than that of CHAID. A maximum tree is constructed to begin with, by continuing the node splitting process as far as possible. The algorithm then deduces a number of nested sub-trees by successive pruning operations, comparing the latter and then selecting the one with the lowest possible error rate, as measured by testing or cross-validation.

Another aspect of its performance is the absence of more or less arbitrarily fixed thresholds such as the χ^2 significance threshold of CHAID (see below). The determination of these thresholds, when necessary, is always difficult, because the best choice has to be made between a threshold providing a deep tree which lacks robustness because it depends too closely on the sample (overfitting) and a small tree with less predictive power.

---

[7] Breiman, L., Friedman, J.H., Olshen, R.A., and Stone, C.J. (1984) *Classification and Regression Trees*. Monterey, CA: Wadsworth & Brooks/Cole.

A final element of the performance of CART lies in its exhaustive search for all the possible splits, already mentioned in Section 11.4.2, which ensures that the optimal split is chosen. Clearly, this search can take a long time, especially when we are dealing with qualitative variables with a large number k of categories, since there will be $2^{k-1} - 1$ splits to be tested.

In its basic version, CART is binary. The drawback of this binary structure is that it produces trees which are 'narrow' but may be very deep, making the trees rather complex and difficult to read in some cases. Another possible problem with CART is that it is biased, with a tendency to favour those variables having the largest number of categories; because of this, it does not always have the greatest reliability. When there are qualitative variables with many categories, it may be preferable to use a tree that is both faster and unbiased, such as QUEST (Quick Unbiased Efficient Statistical Tree), invented in 1997 by Loh and Shih.[8] This tree operates with only one nominal qualitative dependent variable, but it is binary like CART, and has some features in common with the latter (treatment of missing values, etc.). It can be found in the *LohTools* package in R, available from the developers (see the website of their publisher, Springer).

Here is an illustration of the node splitting mechanism using the Gini criterion. A catalogue contains prices of articles and states whether or not they have been purchased. The aim is to find the decisive price for purchasing.

Article	Price	Purchase
1	125	N
2	100	N
3	70	N
4	120	N
5	95	Y
6	60	N
7	220	N
8	85	Y
9	75	N
10	90	Y

As mentioned above, the articles are initially classified by increasing value of the independent variable, i.e. the price. All the possible separation conditions are then tested, according to the value of the price.

Purchase	N	N	N	Y	Y	Y	N	N	N	N
Price	60	70	75	85	90	95	100	120	125	220

Threshold	55		65		72		80		87		92		97		110		122		172		230	
	≤	>	≤	>	≤	>	≤	>	≤	>	≤	>	≤	>	≤	>	≤	>	≤	>	≤	>
Y	0	3	0	3	0	3	0	3	1	2	2	1	3	0	3	0	3	0	3	0	3	0
N	0	7	1	6	2	5	3	4	3	4	3	4	3	4	4	3	5	2	6	1	7	0
Gini	0.420		0.400		0.375		0.343		0.417		0.400		0.300		0.343		0.375		0.400		0.420	

[8] Loh, W.-Y. and Shih, Y.-S. (1997) Split selection methods for classification trees. *Statistica Sinica*, 7, 815–840.

The optimal threshold is at the value of 97, because the Gini index of the separation is then

$$\frac{6}{10}(1-0.5^2-0.5^2)+\frac{4}{10}(1-0^2-1^2)=\frac{6}{10}\times 0.5=0.3.$$

At the threshold of 92, the Gini index would be equal to:

$$\frac{5}{10}(1-0.4^2-0.6^2)+\frac{5}{10}(1-0.2^2-0.8^2)=\frac{5}{10}\times 0.48+\frac{5}{10}\times 0.32=0.4.$$

It is worth re-examining the example of the operation of CHAID in Section 3.10. Compared with CHAID, we find (Figure 11.14) that the CART split is less balanced, with 81% of individuals in node 1 and 19 % in node 2. This is due to the ability of CART to detect very clear profiles rapidly, but it is possible, although not often implemented in the software, to adjust the CART splitting criterion so as to penalize unbalanced splits. To do this, we multiply the reduction in impurity by a coefficient depending on the proportion p_L of individuals sent to the left and the proportion p_R of individuals sent to the right. The new criterion to be maximized is:

$$(p_L p_R)^\alpha \cdot [\text{Gini (parent node)} - \text{Gini (separation)}],$$

where α is a non-negative integer. The left-hand term is maximized when $p_L = p_R = 0.5$. The value $\alpha = 0$ brings us back to the usual splitting criterion, while $\alpha = 1$ is the value most commonly used for penalizing unbalanced splits.

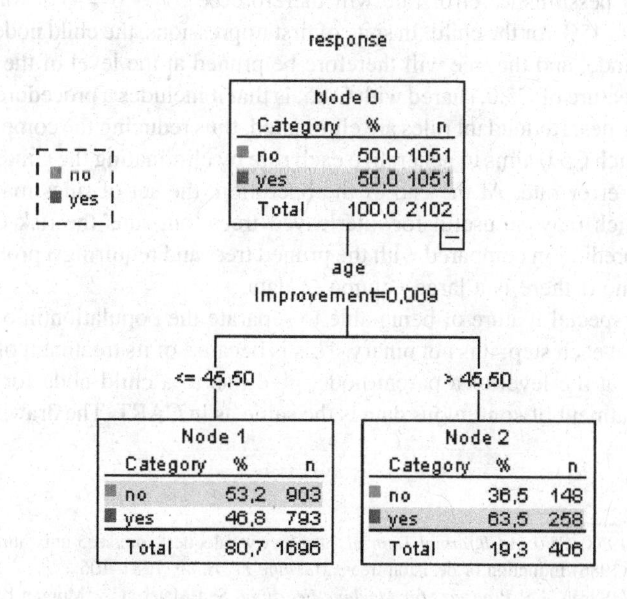

Figure 11.14 Split produced by CART.

C5.0

The C5.0 tree, the most recent (1998)[9] of those described here, was developed by the Australian researcher J. Ross Quinlan, as an improvement of his earlier trees ID3 (1986)[10] and C4.5 (1993).[11] Although less widely used than CART, it is implemented in SAS Enterprise Miner, IBM SPSS Modeler and R (*RWeka* package for C4.5). It is also marketed on Windows platforms under the name of See5, since the name C5.0 is reserved for Unix platforms.[12]

C5.0 works by aiming to maximize the information gain achieved by assigning each individual to a branch of the tree. It shares with CART its suitability for the investigation of all kinds of variables, its exhaustive search for all the possible splits and its device for optimizing the tree by the construction of a maximum tree followed by its pruning.

However, its pruning procedure is different from that of CART. Since the tree is constructed from a training sample, each node of the tree, corresponding to a set $\{C_i\}$ of conditions, is a sample of the set of all the individuals meeting these conditions. Having calculated the error rate of the node, we can determine the confidence interval Δ of this error rate t_ε by applying the conventional statistical formulae of Section A.2.6. At the fixed risk threshold, the maximum error rate that can be observed in this node when the tree is applied is therefore $t_\varepsilon + \Delta$. If the maximum error rate of the child node is greater than that of the parent node, the child node is deleted by pruning the tree at the level of the parent. The decision to prune or not to prune thus depends on both the error rate of the node and its confidence interval, in other words on the number of individuals in the node. Small nodes will tend to be pruned, even if they have a low error rate, as can be illustrated in a simple example.

Assume that we have a node of 1000 individuals with an error rate of 35%. This node has two child nodes, one of which has an error rate of 25% which is significantly lower than that of its parent. However, this node contains only 100 individuals out of the initial 1000. The formula for calculating the confidence interval (see the Appendix) shows that this interval has a width of about 6% for the parent node with 1000 individuals, as against 20% for the child node with 100 individuals. The 'pessimistic' error rate will therefore be $35 + 6 = 41\%$ for the parent, as against $25 + 20 = 45\%$ for the child. In spite of first impressions, the child node is less reliable than the parent node, and the tree will therefore be pruned at the level of the parent.

An original feature of C5.0, shared with C4.5, is that it includes a procedure for converting trees into sets of rules. Redundant rules are eliminated, thus reducing the complexity of the set of rules, after which C5.0 aims to generalize each rule by eliminating the conditions which do not decrease the error rate. At the end of the operation, the set of rules may be decreased significantly, which may be useful for interleaved trees, but runs the risk of reducing the accuracy of the prediction compared with the pruned tree, and requiring a prohibitive amount of processing time if there is a large volume of data.

C5.0 has the special feature of being able to separate the population into more than two sub-populations at each step: it is not binary. This is because of its treatment of the qualitative variables which, at the level of a parent node, give rise to a child node for each category. However, the treatment of continuous data is the same as in CART. The drawback of this tree

[9] Quinlan, J.R. (1998) *C5.0: An Informal Tutorial*. http://www.rulequest.com/see5-unix.html.

[10] Quinlan, J.R. (1986) Induction of decision trees. *Machine Learning*, 1, 81–106.

[11] Quinlan, J.R. (1993) *C4.5: Programs for Machine Learning*. San Mateo, CA: Morgan Kaufmann.

[12] Quinlan, J.R. (1998) Data mining tools See5 and C5.0. Technical report, RuleQuest Research.

is that the frequencies of the nodes decrease more rapidly, together with their statistical reliability and their generalizing capacity.

CHAID

This tree is older (the principle was stated in 1975 by J.A. Hartigan,[13] and the algorithm was devised in 1980 by G.V. Kass).[14] It is even described as being a product of the first decision tree, the AID tree (1963) of Morgan and Sonquist,[15] but the latter was based on the principle of analysis of variance for handling continuous dependent variables, while producing binary trees. CHAID is different.

It uses the χ^2 test to define the most significant variable of each node, so it can only be used with discrete or qualitative independent variables. Most software using CHAID has been designed to deal with continuous independent variables by discretizing them automatically, often into 10 classes, but sometimes into a number of classes that can be determined by the user. Unlike CART, it does not replace missing values with equally splitting or equally reducing values: it handles all the missing values as a single class, which it may merge with another class if appropriate (see step 3 below).

The χ^2 test is used in the successive steps of division of each node, steps 1–4 being the steps of merging the categories of the independent variables and step 5 being a node splitting step. These steps are carried out iteratively, on each child node after the parent node, until a stop condition is reached (Section 11.4.2). Of course, the frequency of the node to be split must be at least equal to the value specified when the tree parameters were set. Otherwise this node cannot be split, and the next steps will not be executed.

1. For each independent variable *X* having at least three categories, χ^2 is used to group the categories of *X* by cross-tabulating them with the *k* categories of the response variable (the dependent variable). We start by selecting the *admissible* pair of categories of *X* whose sub-table ($2 \times k$) is associated with the smallest χ^2 (the greatest associated probability). These are the two categories that differ the least on the response. If this χ^2 is not significant at the chosen threshold (having a probability greater than the stated threshold – see the Merging Categories field in Figure 11.15), the two categories are merged, and the result of this merger is considered to be a new composite category. Note that an *admissible* pair is an adjacent pair if *X* is ordinal or quantitative, or any pair if *X* is nominal.
2. Step 1 is repeated until all the pairs of categories (simple or composite) have a significant χ^2 (are significantly different on the response), or until there are no more than two categories. If one of these categories has a frequency below the minimum specified when the tree parameters were set, this category is merged with the category that is closest in terms of χ^2, even if this χ^2 was already significant. On each occasion, if a new merged category is made up of at least three initial categories with a sufficient frequency, it is possible to determine the binary split (among the initial categories)

[13] Hartigan, J.A. (1975) *Clustering Algorithms*. New York: John Wiley & Sons, Inc.

[14] Kass, G.V. (1980) An exploratory technique for investigating large quantities of categorical data. *Applied Statistics*, 29, 119–127.

[15] Morgan, J.N. and Sonquist, J.A. (1963). Problems in the analysis of survey data, and a proposal. *Journal of the American Statistical Association*, 58, 415–434.

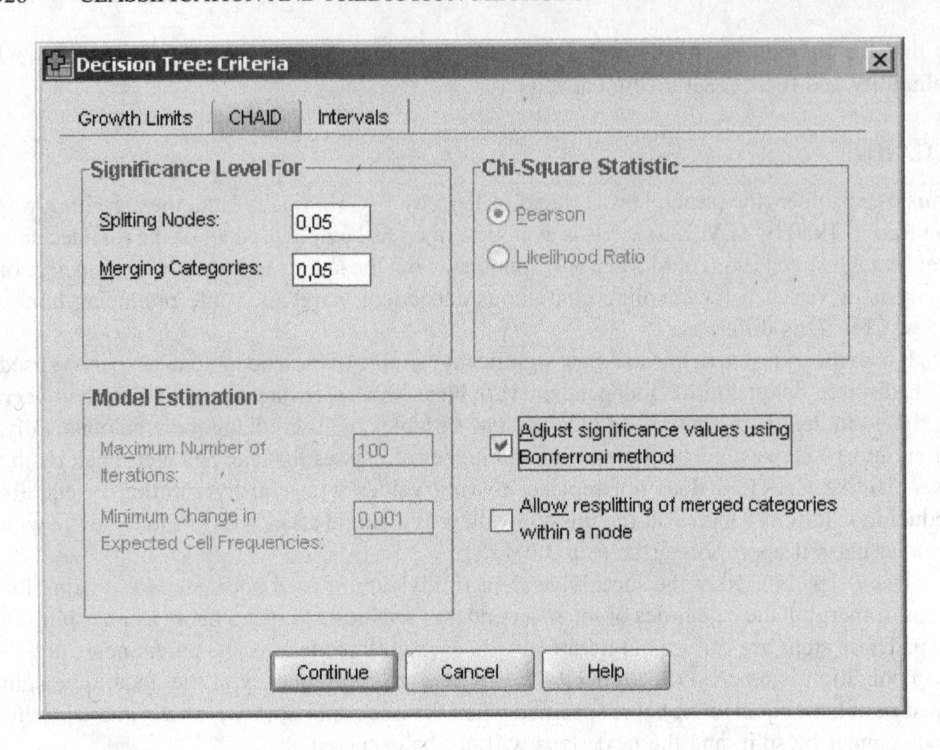

Figure 11.15 CHAID setting in AnswerTree.

having the greatest χ^2, and, if this value is significant, carry out the split. To do this, we must tick the box to 'Allow resplitting of merged categories' in the setting box shown in Figure 11.15. In practice, the effect of this setting is limited.

3. At the end of step 2, if the variable had six categories $\{a,b,c,d,e,f\}$, these would be, for example, grouped into three classes $\{a,d\}$, $\{b,c\}$ and $\{e,f\}$, or two classes $\{a,b,c,d\}$ and $\{e,f\}$. If the independent variable is nominal and has missing values, the set of missing values is considered to be a category that is treated in the same way as the others. However, if the variable is ordinal or quantitative, the missing values category is not included in the preceding merger processes. It is only after the end of these processes that CHAID attempts to merge it with another category, namely the one that is closest in terms of χ^2. It compares the probability of the χ^2 of the table produced by merging the category of the missing values with the one produced without the merger, and accepts the table for which the probability is lowest.

4. At the end of step 3, we have the probability associated with the χ^2 of the best table obtained. If necessary, this probability is multiplied by what is known as the Bonferroni correction (see Figure 11.15, where the 'Adjust significance values using Bonferroni method' box should be ticked). This coefficient is the number of possibilities for grouping the m categories of an independent variable into g groups ($1 \leq g \leq m$), and its multiplication by the probability associated with χ^2 prevents the over-evaluation of the significance of the multiple-category variables.

5. When the categories have been grouped optimally for each independent variable and the probability of the corresponding χ^2 has been calculated (and corrected by the Bonferroni adjustment if necessary), CHAID selects the variable for which the χ^2 is most significant, in other words the one for which the probability is lowest. If this probability is below the chosen threshold (in the Splitting Nodes field of Figure 11.15), we can divide the node into a number of child nodes equal to the number of categories of the variable after grouping. If this χ^2 does not reach the specified threshold, the node is not divided.

We can therefore see that:

- a reduction in the splitting threshold decreases the number of nodes in the tree, because it is difficult to have variables below the threshold (conversely, if the threshold equals 1, all the nodes are split);
- a reduction in the merging threshold decreases the number of categories found for each independent variable, since the pairs of categories are merged for longer before they fall below the threshold (conversely, if the threshold is 1, no pair of categories is merged).

Note also that, if a splitting threshold lower than the merging threshold is chosen, the categories of variables can be grouped without splitting the node: the probability of the χ^2 associated with each variable will have decreased following the groupings of categories, but not enough to be below the splitting threshold. It is therefore more logical to have a splitting threshold that is greater than or equal to the merging threshold.

Unlike AID and CART, CHAID is not binary, and therefore produces trees that tend to be wider rather than deeper. It does not have a pruning function: when the maximum tree has been constructed and the stop criteria are reached, the construction ceases. CHAID is still fairly widely used, and is found, for example, in SAS Enterprise Miner, IBM SPSS Modeler and IBM SPSS AnswerTree, Angoss KnowledgeSEEKER and Statistica (StatSoft).

Because of the way it is constructed, CHAID is useful for the discretization of continuous variables, as shown in Section 3.10 on the division of continuous variables into ranges. In this case, only steps 1–3 above are used.

11.4.8 Advantages of decision trees

First, the results are expressed as explicit conditions on the original variables (by contrast with neural networks). Because of this, the results are very easy for users to understand, the resulting model can easily be programmed by IT staff, and when the model is applied to new individuals the execution speed is greater, since the calculations consist of numeric comparisons ($X \leq n$) or inclusion tests ($X \in \{a,b,c,\ldots\}$), depending on whether X is a quantitative or qualitative variable.

Second, the decision tree method is non-parametric, meaning that the independent variables are not assumed to follow any particular probability distributions. These variables can be collinear. If they are not discriminating, the tree is not affected, because it does not need to select them. Moreover, the response of the dependent variable can be non-linear, or even non-monotonic, in terms of the independent variables. Interactions which may be present between a number of independent variables and the dependent variable will be detected by the tree. The tree is not modified by a monotonic transformation of the continuous independent variables. The data preparation and selection phase is greatly simplified by comparison with other techniques.

Third, trees are relatively unaffected by the presence of extreme individuals, which can be isolated in small nodes and do not affect the classification as a whole, in contrast to what happens in parametric and neural methods.

Fourth, trees can deal with missing data. For example, CHAID treats all the missing values of a variable as a category, which can either remain isolated or can be merged with another category. As for CART, I have already discussed its use of a surrogate variable which can replace a splitting variable which is not supplied for certain individuals.

Fifth, some trees such as CART and C5.0 allow variables of all types (continuous, discrete and qualitative) to be handled directly.

Finally, decision trees are therefore simple to use. Their computing times are quite reasonable, even during training.

11.4.9 Disadvantages of decision trees

First, the definition of the nodes at level $n + 1$ is very highly dependent on the definition at level n. Consider the following diagram:

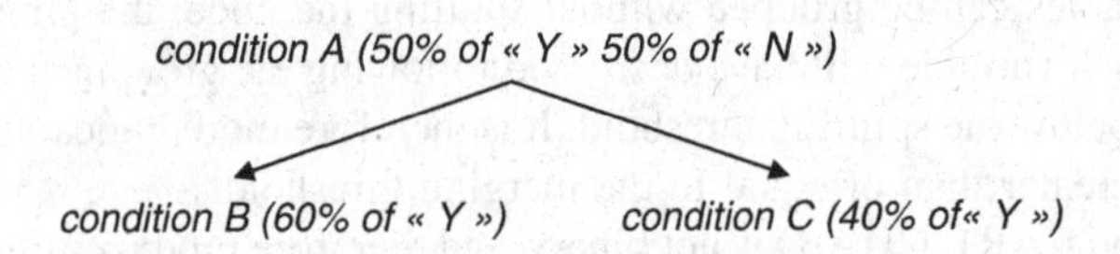

Condition B is valid for defining a group with a higher proportion of Y only if A is true. Otherwise, B may be false. In the definition of a class with a high proportion of Y,

$$(A_1) \text{ and } (A_2) \text{ and } (A_3) \ldots$$

condition (A_i) has no significance if taken alone, independently of the others. In an ideal classification, conditions (A_i) should all have the same weight, regardless of their order of appearance. Conversely, in classification by decision tree, the variables appearing in the first conditions have much more weight (they separate many more individuals) and affect the appearance of the other variables in the tree.

To sum up: the tree detects local, not global, optima. It evaluates all the independent variables sequentially, not simultaneously. The choice of a division for a node at a certain level of the tree is never revised subsequently. This may be troublesome, particularly as some trees (CART) make biased choices, giving preference to the variables having more categories.

Consequently, the modification of a single variable, if it is located near the top of the tree, may modify the whole tree. For example, imagine that an individual has all the characteristics of an individual of the group A, except for the value of a variable whose threshold it has slightly exceeded: it may be wrongly classified in group B, solely because the tree has tested this particular variable.

This lack of robustness of decision trees, which may sometimes be unacceptable, can often be overcome by resampling, in other words by constructing trees on a number of successive samples and aggregating them by a vote or a mean (see Section 11.15 on *bagging* and *boosting*), but this means losing the simplicity and readability of the model which are the advantages of decision trees.

Second, the training of a decision tree requires a sufficiently large number of individuals to provide at least 30–50 individuals per node, even in the lowest part of the tree (the leaves); otherwise the tree will soon 'overfit' on the sample. It is true that decision trees have at least one advantage over neural networks, in that the overfitting is easily seen and located by applying the rules of the tree to another sample.

Third, the form of the resulting models, ($X \leq n$) and ($X \in \{a,b,c\ldots\}$), leads to a definition of the rectangular areas of the variable space (see Section 11.16.1) which do not necessarily correspond to the distribution of the individuals. Thus individuals not having a rectangular distribution are difficult to classify with decision trees.

Some trees overcome this problem by replacing simple rules for dividing the nodes, of the form ($X \leq n$), with rules on a number of variables of the type ($aX + bY + \ldots \leq n$), which determine the divisions of the space as shown in Figure 11.16 and enable the classification to be at least as precise as if the tree had many more nodes.

A solution which can be used with any tree to detect rules of the $X \leq Y$ type is, if the possibility of such a rule is suspected, to create the variable X/Y, for which the tree will detect the rule $X/Y \leq 1$. This solution requires the creation of a number of additional variables.

Fourth, when a classification tree is used to explain a binary variable, a score function defined for any individual can be deduced from it. Decision trees are often used in this way in marketing to construct propensity scores used in business targeting. The 'score' of an individual then depends on the leaf to which the values of its predictors lead it, and it is equal to either 1 minus the classification error rate of the leaf when the leaf is assigned to the class to be predicted, or the classification error rate of the leaf when the leaf is assigned to the class not to be predicted. Returning to the example of the *Titanic* (Figure 11.10), the 'score' of a woman in first class is 0.972. The disadvantage of a score function constructed in this way, in comparison to a function constructed by logistic regression or discriminant analysis, is that the values of this score function are not uniformly distributed; this is because the number of different values of the score function cannot be greater than the number of leaves on the tree. A tree with 10 leaves can

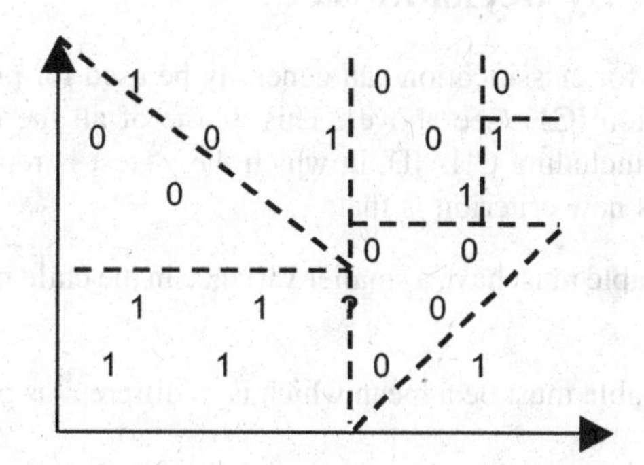

Figure 11.16 Regions of a tree with multi-variable nodes.

only provide 10 different score marks (and if we deepen the tree to obtain a sufficient number of nodes to smooth the distribution of the score values, we may cause overfitting of the tree).

But this disadvantage is not always too troublesome, especially for scores returned to their users in the form of several ranges, rather than a continuously varying quantity. In this case, the discontinuity of the score values can even be an advantage, because it provides natural cut-off points. Moreover, the distribution of the score values is often bimodal, since the tree favours the emergence of pure nodes. This leads to an overrepresentation of the extreme values (e.g. low- or high-risk customers) and an underrepresentation of the mean values (e.g. medium-risk customers), which is useful for a decision support tool which is required to provide clear-cut information.

Finally, there is a discontinuity of the response of the dependent variable as a function of the independent variables. A small change in the value of X may validate or invalidate a rule ($X \leq n$), and completely alter the prediction of the tree, especially if the rule ($X \leq n$) is close to the root of the tree.

The MARS (Multivariate Adaptive Regression Splines) algorithm devised by Friedman[16] overcomes this problem by abandoning the decision tree structure and using spline functions (which may be linear) instead. The MARS response surface is therefore continuous, not discontinuous as in decision trees. This behaviour is sometimes considered advantageous, especially in regression problems where a continuous response is expected.

On the other hand, like CART, of which it is a kind of generalization, MARS provides highly readable results in its own fashion; it selects the variables automatically, and can handle non-linearity, interactions and missing values. The readability of MARS is due to the fact that the model is written as a linear combination of functions of the form $\max(0, x - \textit{threshold})$, $\max(0, \textit{threshold} - x)$ or of products of these functions.

MARS is a little slower (depending on its settings) than CART, but less subject to overfitting, and is more robust overall.

MARS also segments the population, although not in the same way as a tree, which provides a degree of immunity to outliers (but less than that of a tree). The parameter setting is simple. We have seen that it has many advantages and would be worth implementing more commonly in software. It can be found in three R packages, namely *earth, mda* and *polspline*, as well as in the MARS software (a registered trademark, like CART) of Salford Systems and StatSoft's Statistica Data Miner.

11.5 Prediction by decision tree

Decision trees designed for classification can generally be used for prediction, by changing the node splitting criterion (C1) (see above). This is true of all the trees discussed above, except for QUEST, but including CHAID, in which the χ^2 test is replaced by Fisher's test.

The idea behind this new criterion is that:

- the dependent variable must have a smaller variance in the child nodes than in the parent node;
- the dependent variable must be a mean which is as different as possible from one child node to another.

---

[16] Friedman, J.H. (1991) Multivariate adaptive regression splines (with discussion). *Annals of Statistics*, 19(1), 1–141.

In other words, we must choose child nodes which minimize the intra-class variance and maximize the inter-class variance. The Fisher–Snedecor test (see the Appendix) can be applied to the ratio

intra-class variance/inter-class variance

and we can decide that a node should not be split unless the threshold of 20% is reached for Fisher's statistic, in other words when the Fisher distribution table indicates that the ratio of the variances has less than a 20% chance of being as high if the dependent variable is independent of the variable chosen to split the parent node.

The CHAID tree of Figure 11.17 divides 163 countries into five groups which are as different as possible in terms of GNP per citizen, the most discriminating criterion being energy consumption, and the group with the medium GNP being split again by life expectancy. This tree is a very simple way of focusing on one fifth of the countries which have a mean GNP 36 times greater than another fifth of the countries.

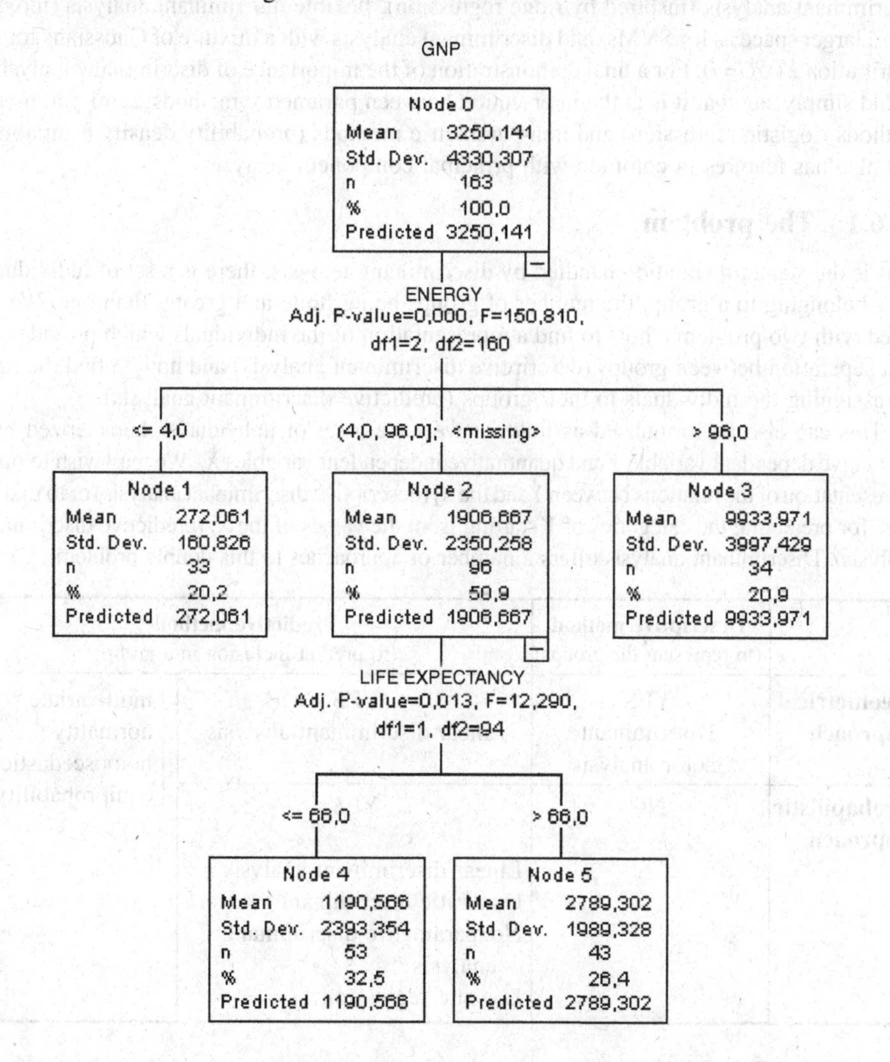

Figure 11.17 Regression tree.

11.6 Classification by discriminant analysis

Before the widespread use of logistic regression, Fisher's discriminant analysis was used for many years as the major classification method, in many fields ranging from biology, where Fisher's basic work was undertaken in 1936,[17] to credit scoring. This method is still preferred today by some central banks for scoring businesses. Initially limited to a framework which is discussed more fully below, this method is excellent within this framework, providing explicit, accurate and robust predictions, provided that the data have been prepared properly. This framework has been generalized to many different fields in recent decades. Thus one supplementary technique devised by Gilbert Saporta, known as the DISQUAL method, has extended the framework of discriminant analysis beyond quantitative independent variables so that it can also handle qualitative variables. More recently, the work of Hastie, Tibshirani and Friedman, described in their well-known book (cited above), has resulted in regularized discriminant analysis (inspired by ridge regression), flexible discriminant analysis (moving into a larger space as for SVMs) and discriminant analysis with a mixture of Gaussians for the distribution *P(x/G=i)*. For a final demonstration of the importance of discriminant analysis, I would simply say that it is at the intersection between parametric methods, semi-parametric methods (logistic regression) and non-parametric methods (probability density estimation), and also has features in common with principal component analysis.

11.6.1 The problem

This is the standard situation handled by discriminant analysis: there is a set of individuals, each belonging to a group, the number of groups being finite and greater than one. We are faced with two problems: how to find a representation of the individuals which provides the best separation between groups (descriptive discriminant analysis) and how to find the rules for assigning the individuals to their groups (predictive discriminant analysis).

This can also be formulated as follows: we have a set of individuals characterized by a qualitative dependent variable Y and quantitative independent variables X_i. We may wish to find a representation of the relations between Y and the X_i (descriptive discriminant analysis) or to find the rules for predicting the categories of Y starting from the values of the X_i (predictive discriminant analysis). Discriminant analysis offers a number of approaches to this double problem.

	Descriptive method (to represent the groups)	**Predictive method** (to predict inclusion in a group)	
Geometrical approach	YES Discriminant factor analysis	YES Linear discriminant analysis	↑ multivariate normality homoscedasticity equiprobability
Probabilistic approach	NO	YES Linear discriminant analysis Quadratic discriminant analysis Nonparametric discriminant analysis Logistic regression	

[17] Fisher, R.A. (1936) The use of multiple measurements in taxonomic problems. *Annals of Eugenics*, 7, 179–188.

11.6.2 Geometric descriptive discriminant analysis (discriminant factor analysis)

There is a qualitative response (dependent) variable Y with k categories, corresponding to k groups G_i whose frequencies are denoted n_i. The total frequency is n. On the other hand, there are p continuous independent variables X_j. Discriminant factor analysis is a matter of replacing the X_j with discriminant axes, in other words linear combinations of the X_j taking the most different possible values for individuals differing on the response variable. It will be seen that this mechanism entails an analysis into principal components of the cloud of the k centres of gravity of the classes (weighted by n_i/n). The number of axes is the minimum of $k-1$ and p. The historic example of discriminant analysis is that of Fisher's irises (three species and four variables: length and width of the petals and sepals).

The geometric descriptive approach can be illustrated in a simple way, as shown in Figure 11.18. In this example, we see that:

- the x axis clearly separates groups B and C, but not groups A and B;
- the y axis clearly separates groups A and B, but not groups B and C;
- while the z axis, the linear combination of x and y, clearly separates all three groups.

The straight line with equation $z=1$ separates the Bs and Cs, while the straight line with equation $z=-1$ separates the As and Bs: z is therefore a score function.

In mathematical terms, the n individuals form a cloud of n points in $\mathbf{R}^p$, formed by the k sub-clouds G_i to be differentiated. The inter-class ('between') variance is, by definition, the variance of the centres of gravity g_i (centroids) of the classes G_i, and the 'between' covariance matrix is $B=n^{-1}\sum n_i(g_i-g)(g_i-g)'$. The intra-class ('within') variance is, by definition, the weighted mean of the variances of the classes G_i, and the 'within' covariance matrix is $W=n^{-1}\sum n_iV_i$, calculated from the covariance matrix V_i of each class G_i. According to the Huygens theorem, $B+W=V$, the total covariance matrix.

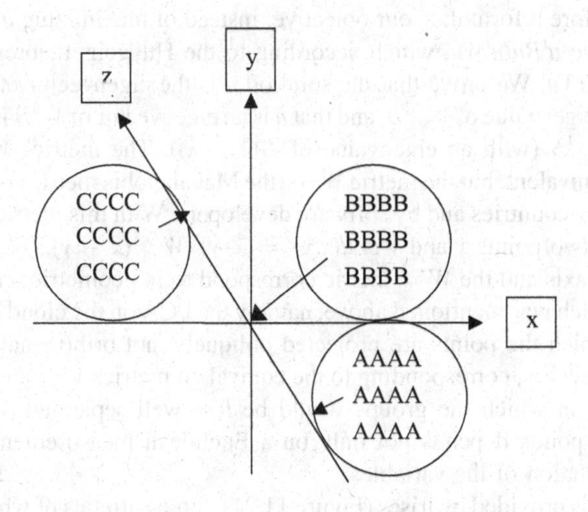

Figure 11.18 Discriminant factor analysis.

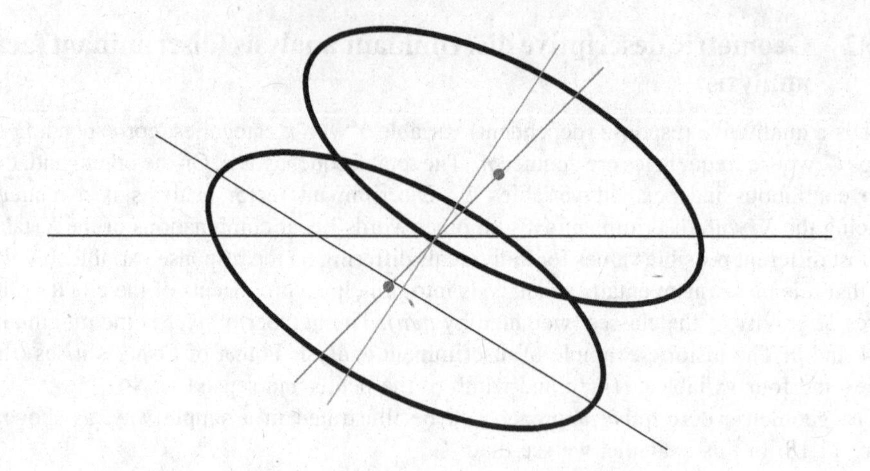

Figure 11.19 The double objective of discriminant analysis.

It is generally impossible to find an axis u which, in order to meet the objective of discriminant analysis, simultaneously

- maximizes the between-class variance on u: max $u'Bu$;
- and minimizes the within-class variance on u: min $u'Wu$.

This is clearly shown in Figure 11.19. If we are looking for the maximum between-class dispersion, we will choose an axis u parallel to the segment linking the centroids, while if we are looking for the minimum within-class dispersion we will choose an axis u perpendicular to the principal axis of the ellipses. We assume homoscedasticity, in other words the equality of all the covariance matrices V_i: this is the basic assumption of discriminant factor analysis.

We must therefore reformulate our objective: instead of maximizing $u'Bu$ or minimizing $u'Wu$, we maximize $u'Bu/u'Wu$, which according to the Huygens theorem is equivalent to maximizing $u'Bu/u'Vu$. We prove that the solution u is the eigenvector of $V^{-1}B$ associated with λ, the largest eigenvalue of $V^{-1}B$, and that u is an eigenvector of $V^{-1}B$ if and only if u is an eigenvector of $W^{-1}B$ (with an eigenvalue of $\lambda/(1-\lambda)$). The metrics V^{-1} and W^{-1} are therefore called equivalent, but the metric W^{-1} (the Mahalanobis metric) is used more widely in English-speaking countries and by software developers. With this metric, the square of the distance $d(x,y)$ of two points x and y is $d(x,y)^2 = (x-y)' W^{-1}(x-y)$.

What do the u axis and the W^{-1} metric correspond to in geometric terms? The u axis is that of the PCA which was mentioned above, namely the PCA on the cloud of centroids g_i, but it is an axis on which the points are projected obliquely, not orthogonally (Figure 11.20). Without this obliqueness, corresponding to the equivalent metrics V^{-1} and W^{-1}, this would be a simple PCA, in which the groups would be less well separated. In this metric, the separation of two points depends not only on a Euclidean measurement, but also on the variance and correlation of the variables.

An illustration is provided by irises (Figure 11.21), three varieties of which (*I. virginica*, *I. versicolor* and *I. setosa*) are (slightly) better separated by discriminating factor analysis than

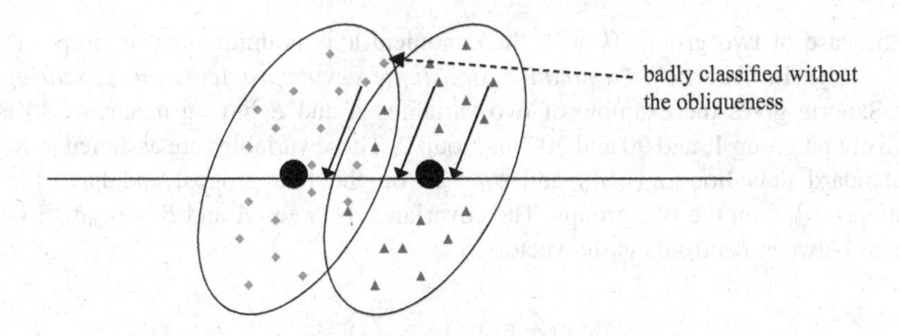

Figure 11.20 Effect of the metric W^{-1}.

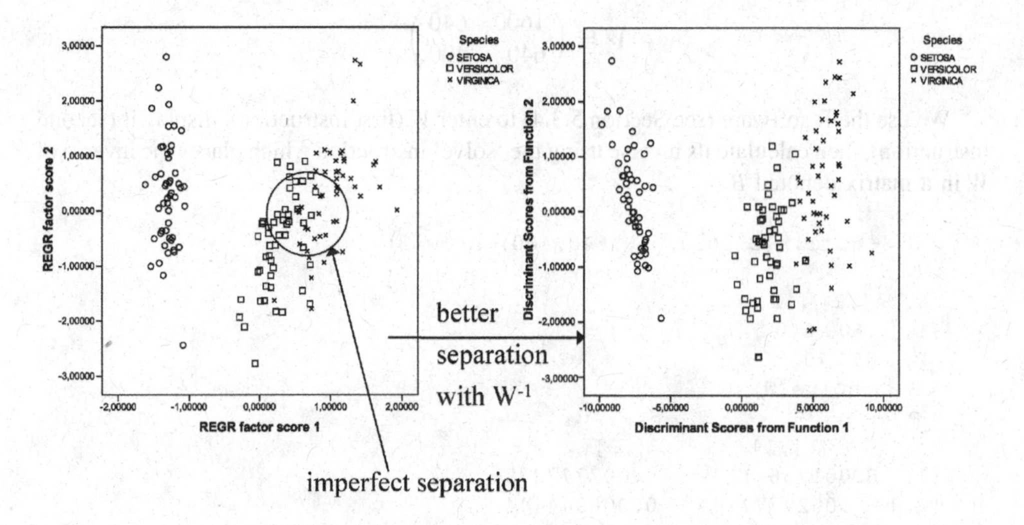

Figure 11.21 DFA on Fisher's irises.

by simple PCA (although the latter method is still worth considering, as mentioned by Jean-Paul Benzécri in Section 2.3.5 of his book *Histoire et Préhistoire de l'Analyse des Données*).[18]

The u axis is called the 'first canonical axis'. When the first canonical axis has been determined, we search for a second axis which is both the most discriminant and not correlated with the first. This procedure is repeated until the number of axes reaches the minimum of the two numbers $k - 1$ and p. As in PCA, the corresponding eigenvalues are related to their sum in order to evaluate their relative contributions. If the first eigenvalue represents a large percentage of the sum, this means that the k centroids lie roughly on a straight line; in this case, the first axis contains the essential information and is sufficient for a correct classification. In fact, it is not the canonical axes that are manipulated directly, but the *canonical variables*, which are the coordinates of the individual points on the canonical axes.

[18] Benzécri, J.-P. (1982) *Histoire et Préhistoire de l'Analyse des Données*, new edn. Paris: Dunod.

In the case of two groups ($k=2$), the canonical axis is unique and is proportional to $W^{-1}(g_1 - g_2)$. In his article *La notation statistique des emprunteurs ou ≪ scoring ≫*[19], Gilbert Saporta gives the example of two variables A and B having means of 40 and 90 respectively on group 1, and 90 and 100 on group 2. These variables are assumed to have the same standard deviations $\sigma_A = 40$ and $\sigma_B = 20$ on the two groups, and have the same correlation $r = 0.8$ on the two groups. The covariance between A and B is $r\sigma_A\sigma_B = 640$, the difference between centroids is the vector

$$g_1 - g_2 = \begin{pmatrix} -50 \\ -10 \end{pmatrix},$$

and the within-class covariance matrix is

$$W = \begin{pmatrix} 1600 & 640 \\ 640 & 400 \end{pmatrix}.$$

We use the R software (see Section 5.3.4) to enter W (first instruction), display it (second instruction), then calculate its inverse using the 'solve' instruction which places the inverse of W in a matrix denoted B:

```
> W <- matrix(c(1600,640,640,400),nrow=2)
> W
      [,1] [,2]
[1,] 1600 640
[2,] 640 400
> B <- solve(W)
> B
           [,1]          [,2]
[1,] 0.001736111    -0.002777778
[2,] -0.002777778    0.006944444
```

We then enter $(g_1 - g_2)$ in a column matrix denoted C, which we display, and then multiply (%*% operator) $B = W^{-1}$ by $C = (g_1 - g_2)$, which gives:

```
> C <- matrix(c(-50,-10),nrow=2)
> C
      [,1]
[1,] -50
[2,] -10
> B %*% C
         [,1]
[1,] -0.05902778
[2,]  0.06944444
>
```

[19] Saporta, G, (2002) La notation statistique des emprunteurs ou 'scoring'. www.eduscol.education.fr/D0015/ann_stat_6.pdf. This is an excellent eight-page summary of scoring, written for the Committee on Education in Mathematics (Report to the French Minister of National Education, under the direction of Jean-Pierre Kahane).

Pooled Within-Class Standardized Canonical Coefficients		
Variable	**Label**	**Can1**
nbproducts	number of products	0.4266757532
subscription1	subscription to other service 1	0.3416040633
nbchildren	number of children	-.3431234870
subscription2	subscription to other service 2	0.3885016925
changeconsum	change in consumption	0.2887758069
nbexits	number of exits with purchase	0.2736500676

Figure 11.22 Standardized coefficients of the canonical variable.

It follows from this that the canonical axis, which is proportional to $W^{-1}(g_1 - g_2)$, is also proportional to the vector (– 1 1.17647047) and that the canonical variable can be written $(x_1,x_2) \rightarrow -x_1 + 1.17647047x_2$. Clearly, in the case of two groups, this is unique.

This is how the SAS software outputs the coefficients of the canonical variable in the case of two groups of customers to be discriminated (see the syntax below). A first table (Figure 11.22) supplies the coefficients to be applied, not to the raw initial variables, but to the initial variables standardized (reduced centred) as a function of their means and standard deviations in the population. The second table (Figure 11.23) is applicable to non-reduced centred initial variables.

Thus the formula for the canonical variable starts with 0.4266757532(nbproducts – 8.93579)/4.354, or, in an equivalent way, with 0.0979952574(nbproducts – 8.93579). An SAS output (see Figure 11.24) displays the covariance matrix which contains the square of the standard deviation, and we can find in it the value 18.958 which is the square of 4.354.

Raw Canonical Coefficients		
Variable	**Label**	**Can1**
nbproducts	number of products	0.0979952574
subscription1	subscription to other service 1	0.0003287387
nbchildren	number of children	-.3019421367
subscription2	subscription to other service 2	0.0109642140
changeconsum	change in consumption	0.7603735609
nbexits	number of exits with purchase	0.0398690704

Figure 11.23 Raw coefficients of the canonical variable.

Pooled Within-Class Covariance Matrix, DF = 6383							
Variable	Label	nbproducts	subscription1	nbchildren	subscription2	changeconsum	nbexits
nbproducts	number of products	18.958	989.799	1.125	70.927	-0.010	7.590
subscription1	subscription to other service 1	989.799	1079802.363	58.693	4484.044	-10.186	2134.441
nbchildren	number of children	1.125	58.693	1.291	2.375	-0.063	1.052
subscription2	subscription to other service 2	70.927	4484.044	2.375	1255.541	-0.816	25.050
changeconsum	change in consumption	-0.010	-10.186	-0.063	-0.816	0.144	-0.017
nbexits	no. of exits with purchase	7.590	2134.441	1.052	25.050	-0.017	47.111

Figure 11.24 Intra-class covariance matrix.

Pooled Within Canonical Structure		
Variable	Label	Can1
nbproducts	number of products	0.669783
subscription1	subscription	0.539645
nbchildren	number of children	-0.211773
subscription2	subscription to other service 2	0.616690
changeconsum	change in consumption	0.302505
nbexits	number of exits with purchase	0.476046

Figure 11.25 Correlation of the canonical variable with the initial variables.

The first table is useful because it immediately emphasizes the most discriminating variables. In fact, a variable can be assigned a very low coefficient because it takes very high values itself. However, the coefficients implemented in the programs for calculating the canonical variable and applying it to prediction are the coefficients of the non-standardized variables (unless the variables are standardized before each application of the model).

Finally, SAS also supplies (Figure 11.25) the correlation coefficients of the initial variables with the canonical variable or variables.

11.6.3 Geometric predictive discriminant analysis

Each individual x is classed in the group G_i for which the distance to the centre g_i is minimal, this distance being calculated as shown above according to the W^{-1} metric:

$$d(x, g_i)^2 = (x-g_i)'W^{-1}(x-g_i) = x'W^{-1}x-2g'_iW^{-1}x+g'_iW^{-1}g_i.$$

As you can see, minimizing $d(x,g_i)^2$ is equivalent to maximizing $2g'_iW^{-1}x - g'_iW^{-1}g_i$, and $g'_iW^{-1}g_i = \alpha_i$ is a constant that does not depend on x. Thus for each of the k groups G_i we have a discriminant linear function found after inversion of the matrix W:

$$\alpha_i + \alpha_{i,1}X_1 + \alpha_{i,2}X_2 + \ldots + \alpha_{i,p}X_p,$$

and x is classed in the group for which the function is maximal.

In the example of Fisher's irises, the three discriminant functions are read from the output table of the program (IBM SPSS Statistics in this case), as shown in Figure 11.26. Thus the function associated with *I. setosa* is:

$$-86.308 + (2.354 \times \text{SepalLength}) + (2.359 \times \text{SepalWidth})$$

$$-(1.643 \times \text{PetalLength}) - (1.740 \times \text{PetalWidth}).$$

This shows how simply expressed a model calculated by discriminant analysis can be. As we will see subsequently, the credit scoring models used at the Banque de France are very similar in form.

In the specific case of two groups, which is commonly encountered, the descriptive aspect is simple (the discriminant axis links the two centroids) and what interests us is the predictive aspect. As a special case of what is shown above, x is classed in the group G_1 if

$$2g'_1W^{-1}x - g'_1W^{-1}g_1 > 2g'_2W^{-1}x - g'_2W^{-1}g_2,$$

a condition equivalent to

$$(g_1 - g_2)'W^{-1}x - \tfrac{1}{2}(g'_1W^{-1}g_1 - g'_2W^{-1}g_2) > 0.$$

Classification function coefficients

	Species		
	SETOSA	VERSICOLOR	VIRGINICA
SepalLength	2.354	1.570	1.245
SepalWidth	2.359	.707	.369
PetalLength	-1.643	.521	1.277
PetalWidth	-1.740	.643	2.108
(Constant)	-86.308	-72.853	-104.368

Fisher's linear discriminant functions

Figure 11.26 Linear discriminant functions for Fisher's irises.

The expression to the left of the inequality is, by definition, Fisher's score function $f(x)$ and x is classed in G_1 if $f(x) > 0$. This score function is obtained by finding the differences between the discriminant functions associated with each group, but it should be noted that we do not return exactly to the canonical variable shown above.

If, for example, we calculate the difference 0.47720 – 0.38455 for the variable nbproducts variable (Figure 11.27), we obtain a value of 0.09265 which is not equal to the coefficient 0.0979952574 of the canonical variable, but is equal to this coefficient multiplied by a quantity called the Mahalanobis distance, denoted by D below. The same is true of all the coefficients. Because of the normalization of the canonical variable, this must be multiplied by D to find the score function.

Since classification is based on the concept of distance from a point to a centroid, there is a naturally important quantity called the Mahalanobis D^2, which is the square of the distance between the two centroids:

$$D^2 = \mathrm{d}(g_1, g_2)^2 = (g_1 - g_2)'W^{-1}(g_1 - g_2).$$

The Mahalanobis D^2 measures the distance between the two groups to be discriminated, and thus it also measures the quality of the discrimination: a higher value means better discrimination. This is similar to the coefficient of determination R^2 of a regression, and can form the basis of a Fisher F test on the null hypothesis that all the centroids are equal (Figure 11.28). D^2 can be used in a stepwise regression.

The rule of classification of geometric discriminant analysis, which is a linear rule (which is why we speak of *linear discriminant analysis*), is therefore that we assign each individual to the group it is nearest to, using the W^{-1} metric to calculate the distance of the individual from the centroid of the group, in other words by carrying out an oblique projection of x on the discriminant axis. However, this rule should not be used if the two groups have different *a priori* probabilities or variances, as suggested by the example of Figure 11.29.

Linear Discriminant Function for response			
Variable	**Label**	**0**	**1**
Constant		-7.49693	-11.26600
nbproducts	number of products	0.38455	0.47720
subscription1	subscription to other service 1	-0.0001620	0.0001488
nbchildren	number of children	1.11604	0.83056
subscription2	subscription to other service 2	-0.00319	0.00717
changeconsum	change in consumption	8.42157	9.14047
nbexits	number of exits with purchase	0.05182	0.08951

Figure 11.27 Fisher's linear discriminant functions XXX.

Squared Distance to response		
From response	0	1
0	0	0.89390
1	0.89390	0

F Statistics, NDF=6, DDF=6378 for Squared Distance to response		
From response	0	1
0	0	133.48253
1	133.48253	0

Figure 11.28 Mahalanobis D^2 and associated Fisher test.

In this example, individual I is nearer to g_1 than to g_2, but the variance of the second class is such that it will be better to assign the individual I to it. In this case, we must use another method: we will now look at another form of discriminant analysis which is not descriptive but more general, which can deal with the above example by *quadratic discriminant analysis*, which is the optimal method when the two groups have normal distributions but different variances. We should say that this is a *theoretically* optimal method, because in practice it rarely provides better results than those of linear discriminant analysis, which is generally preferred. This practical weakness of quadratic discriminant analysis is due to the larger number of parameters to be estimated, meaning that the volumes of frequencies to be modelled must also be larger.

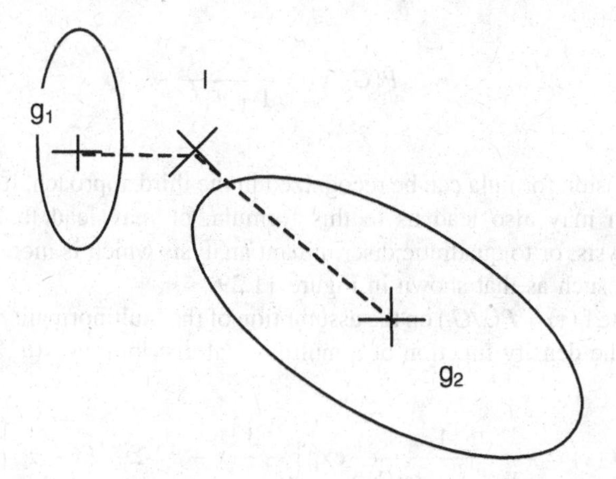

Figure 11.29 Limit of linear discriminant analysis.

11.6.4 Probabilistic discriminant analysis

This approach is also called *Bayesian*, because it was developed from Bayes' theorem:

$$P(G_i/x) = \frac{P(G_i)P(x/G_i)}{\sum_j P(G_j)P(x/G_j)},$$

where it should be noted that, for every $i \leq k$ (remember that k is the number of classes):

- $P(G_i/x)$ *is* the *a posteriori* probability of belonging to G_i, given x (where we know the characteristics of x, i.e. its 'personal file');

- $p_i = P(G_i)$ *is* the *a priori* probability of belonging to G_i (the proportion of G_i in the population);

- $f_i(x) = P(x/G_i)$ is the conditional density of the distribution of x, when its group G_i is known.

According to the *Bayesian classification rule*, x is classed in the group G_i where $P(G_i/x)$ is maximum.

We must therefore calculate $P(G_i/x)$. There are three ways of doing this:

- by initially calculating $P(x/G_i)$ by a parametric method (we assume the multinormality of x/G_i, possibly with the equality of the covariance matrices Σ_i, meaning that the number of parameters of the problem is finite);

- by initially estimating $P(x/G_i)$ by a non-parametric method (without any assumption as to the density function, which is written in the form $P(x/G_i)$ = frequency/volume and which we aim to estimate by the kernel method or the k-nearest-neighbour method), this method being restricted to large samples where there is no requirement for explicit formulae with tests of significance on the parameters;

- directly, by a semi-parametric approach in which $P(G_i/x)$ is written in the form

$$P(G_i/x) = \frac{e^{\alpha' x + \beta}}{1 + e^{\alpha' x + \beta}}.$$

The logistic regression formula can be recognized in the third approach. It can be shown that the first approach may also lead us to this formula, or may lead to linear (geometric) discriminant analysis, or to quadratic discriminant analysis which is more general and more suitable for cases such as that shown in Figure 11.29.

Let us calculate $f_i(x) = P(x/G_i)$ on the assumption of the multinormality of the distribution (Figure 11.30). The density function of a multinormal distribution $N(\mu_i, \Sigma_i)$ is written as

$$f_i(x) = \frac{1}{(2\pi)^{p/2}\sqrt{\det(\Sigma_i)}} \exp\left[-\frac{1}{2}(x - \mu_i)' \Sigma_i^{-1} (x - \mu_i) \right].$$

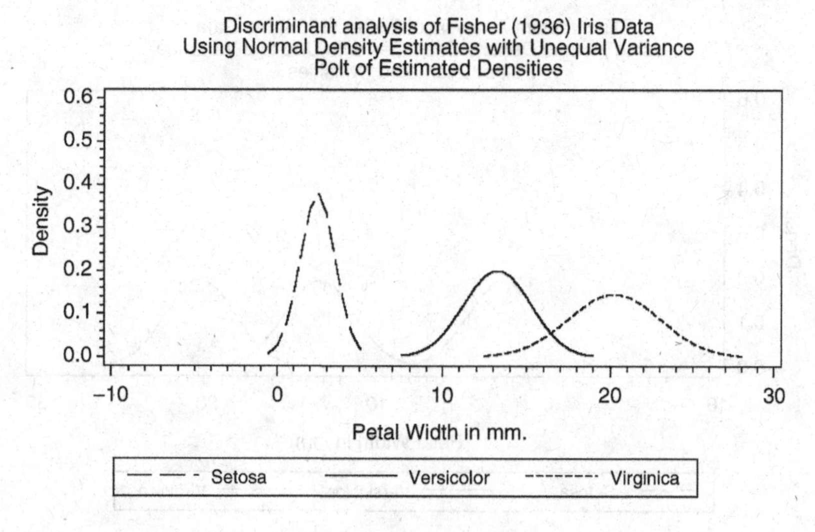

Figure 11.30 Multinormality without homoscedasticity (source: SAS).

Since we wish to maximize P(G_i/x), which according to Bayes is equivalent to maximizing $p_i f_i(x)$, we must find the i which maximizes

$$\left[\log(p_i) - \frac{1}{2}(x - \mu_i)'\Sigma_i^{-1}(x - \mu_i)\frac{1}{2}\log(\det(\Sigma_i))\right].$$

It can be seen that we obtain a quadratic rule in x. To simplify this rule, we must add the assumption of homoscedasticity: $\Sigma_1 = \Sigma_2 = \ldots = \Sigma_k = \Sigma$ (Figure 11.31). This shows that we classify x in the group G_i which maximizes

$$\left[\log(p_i) - \frac{1}{2}x'\Sigma^{-1}x - \frac{1}{2}\mu'_i\Sigma^{-1}\mu_i + x'\Sigma^{-1}\mu_i)\right].$$

Since $x'\Sigma^{-1}x$ is independent of i, this means that we are seeking the maximum of

$$\left[\log(p_i) - \frac{1}{2}\mu'_i\Sigma^{-1}\mu_i + x'\Sigma^{-1}\mu_i\right]$$

It can be seen that the assumption of homoscedasticity, added to the assumption of multinormality, converts a quadratic rule into a linear rule. Apart from the term $\log(p_i)$, we can recognize in this rule the Fisher discriminant function found by the geometric approach, and the *a priori* probability p_i only adds a constant to the discriminant function. Thus we find that, if we add the assumption of equiprobability to those of homoscedasticity and multinormality, the Bayesian rule (maximize $P(G_i/x)$) is equivalent to the geometric rule (maximize the Fisher function).

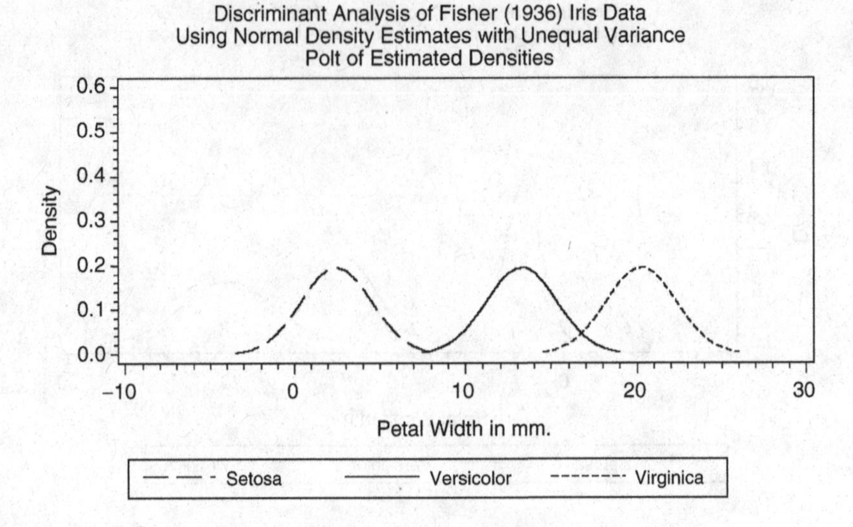

Figure 11.31 Multinormality with homoscedasticity (source: SAS).

By writing the Bayes rule for the case of two groups G_1 and G_2, we can easily deduce the formula

$$\frac{1}{P(G_1/x)} = 1 + \left(\frac{p_2}{p_1}\right) e^{-f(x)},$$

where $f(x) = (\mu_1 - \mu_2)'\sum^{-1} x - \frac{1}{2}(\mu'_1 \sum^{-1} \mu_1 - \mu'_2 \sum^{-1} \mu_2)$ is simply the Fisher score function. x is classed in G_1 if $P(G_1/x) > 0.5$, which is equivalent to $(p_2/p_1)e^{-f(x)} < 1$, or alternatively $f(x) > \log(p_2/p_1)$.

In the case of two groups, the Bayesian rule is therefore equivalent to the rule that the Fisher score function

$$f(x) > \log(p_2/p_1),$$

which generalizes the geometric rule $f(x) > 0$ when the *a priori* probabilities p_1 and p_2 are different. Additionally, the *a posteriori* probability $P(G_1/x)$ is written

$$P(G_1/x) = \frac{1}{1 + (p_2/p_1)e^{-f(x)}} = \frac{e^{f(x)}}{(p_2/p_1) + e^{f(x)}},$$

which is a generalization of the logistic function (see Section 11.8). However, it should be noted that the form

$$P(G_1/x) = \frac{1}{1 + e^{-f'(x)}} = \frac{e^{f'(x)}}{1 + e^{f'(x)}}$$

of the logistic model actually offers the same modelling power, since the term (p_2/p_1) which is not included in the denominator is incorporated in the linear function $f'(x)(\neq f(x) \text{ if } p_1 \neq p_2)$.

Summary

On the assumption of:	**the Bayesian rule is:**
multivariate normality	quadratic
multinormality + homoscedasticity	linear
	in the case of 2 groups, it is written $f(x) > \log(p_2/p_1)$, where $f(x)$ is the Fisher score function found by a geometrical procedure
multinormality + homoscedasticity + equiprobability	linear and equivalent to the geometric rule
	in the case of 2 groups, it is written $f(x) > 0$ and the *a posteriori* probability $P(G_1/x)$ is written in the logistic form

Note that, given the above assumptions of multinormality, homoscedasticity and equiprobability, linear discriminant analysis appears to be a special case of the logistic model in which the independent variables x act in a linear way. The logistic model in its specific form $P(G_i/x)$ may appear in other circumstances, even without the above constraining assumptions.

In addition to its greater generality, the Bayesian approach to discriminant analysis has the advantage of enabling the costs of incorrect classification to be included. Thus we use C_{ij} to denote the cost of classification in G_i instead of G_j, we assume that $C_{ii} = 0$, and the mean cost of classification in G_i is defined as the sum $\Sigma_j C_{ij} P(G_j/x)$. x is then classed in the group G_i which minimizes the cost. For example, in the case of two groups, x is classified in G_1 if $C_{12}P(G_2/x) < C_{21}P(G_1/x)$. We may thus decide to classify an x with a probability of $P(G_1/x) < P(G_2/x)$ in G_1 if the cost of classifying an element of G_1 in G_2 is very high.

11.6.5 Measurements of the quality of the model

In addition to the general quality measurements of a predictive model (see Section 11.16.5), the following measurements are applied specifically to discriminant analysis.

Wilks' lambda

This is the ratio $\Lambda = \det(W)/\det(V)$ of the determinant of the within-class covariance matrix to that of the total covariance matrix. It varies from 0 to 1. The lower the value, the better the model is. At the other extreme, if $\Lambda = 1$ all the centroids are equal. This indicator enables us to answer the two questions which arise concerning the independent variables of a model: do these variables discriminate between the classes? and can the classes be discriminated with fewer variables?

To answer the first question, we must combine a statistical test with Wilks' lambda so that we can reject the null hypothesis H_0 of the equality of the centroids. Once again, the number of observations is denoted n and the number of classes to be discriminated for these observations is denoted k. The lambda Λ of a model with p variables can be approximated (Rao, 1973)[20] by a Fisher distribution with (d_1,d_2) degrees of freedom, after subjecting Λ to a transformation of the form $(1–\Lambda)/\Lambda$ to generalize the F-ratio of linear regression and Fisher's discriminant analysis. The expression of d_1 and d_2 is rather complex, but is simplified in the case of two classes, where $d_1 = p$ and $d_2 = n - p - 1$ (as for the F-ratio of linear regression). In the case of three classes, $d_1 = 2p$ and $d_2 = 2n - 2p - 4$ (unless $p = 1$, in which case $d_2 = n - 3$). Another case where the expression is simplified is that in which $p = 1$, since 'testing that a variable can discriminate between the classes' is equivalent to 'testing the usefulness of the contribution of a first variable', a test which, as we will see subsequently, has the degrees of freedom $d_1 = k - 1$ and $d_2 = n - k$ (like the F-ratio of ANOVA). Note that the Wilks' lambda of a single variable is the quotient of its within-class variance divided by its total variance, and is therefore $1 - R^2$, where R^2 is the proportion of the variance of the variable explained by inclusion in one or other class to be discriminated: the variable becomes more discriminant as R^2 increases. In all cases, the Fisher statistic is associated with a probability which, if below a certain threshold (often fixed between 0.05 and 0.20) allows us to reject the hypothesis of the equality of the centroids and to accept that the variables are sufficient to discriminate the classes.

The second question relates to the number of variables required for discrimination, in other words the selection of the independent variables. Wilks' lambda can be used in a stepwise section, for example by using IBM SPSS Statistics or the SAS/STAT STEPDISC procedure as in the example below. Note that, in the STEPDISC procedure, the default selection threshold is set at 0.15, because this generally gives better results than a lower or higher threshold. SAS made this choice on the basis of the work by Costanza and Afifi[21] on linear discriminant analysis. We assume that $\Lambda_0 = 1$, and we start by finding the variable which leads to the lowest Wilks' lambda among all the models reduced to one variable, this value of lambda being denoted Λ_1. The selection process is then reiterated: when p variables have already been selected and have resulted in the lambda Λ_p, the lowest possible value for all models with p variables, we look for the $(p + 1)$th variable which minimizes Λ_{p+1}. Since we always need to find a stop criterion, we use the one which consists in finding that the quantity

$$\frac{n-k-p}{k-1}\left(\frac{\Lambda_p}{\Lambda_{p+1}}-1\right),$$

which measures the contribution of the $(p + 1)$th variable, follows a Fisher distribution with $(k - 1, n - k - p)$ degrees of freedom. This enables us to test the null hypothesis that the $(p + 1)$th variable does not improve the model. If no variable provides a sufficiently low

[20] Rao, C.R. (1973) *Linear Statistical Inference and Its Applications*. New York: John Wiley & Sons, Inc.

[21] Costanza, M.C. and Afifi, A.A. (1979) Comparison of stopping rules in forward stepwise discriminant analysis. *Journal of the American Statistical Association*, 74, 777–785.

Multivariate Statistics					
Statistic	Value	F Value	Num DF	Den DF	Pr > F
Wilks' Lambda	0.058628	1180.16	2	147	<.0001
Pillai's Trace	0.941372	1180.16	2	147	<.0001
Average Squared Canonical Correlation	0.470686				

Figure 11.32 *F* statistic of the model with one variable.

Λ_{p+1} to make the *F*-ratio above sufficiently large to ensure that the associated probability is below the chosen threshold (0.15 in STEPDISC), the addition of variables to the model must be halted. In the example of Fisher's irises, $\Lambda_0 = 1$ and $\Lambda_1 = 0.058628$, where the first variable is the petal length, giving a first *F*-ratio of

$$\frac{150-3}{2}\left(\frac{1}{\Lambda_1}-1\right) = 1180.16$$

and an associated probability less than 0.0001 (with degrees of freedom 2 and 147), as indicated in Figure 11.32, obtained as in the following figures by the STEPDISC procedure in SAS/STAT.

Since $\Lambda_2 = 0.036884$ when the sepal width is added, the second *F*-ratio is

$$\frac{150-3-1}{2}\left(\frac{\Lambda_1}{\Lambda_2}-1\right) = 43.035$$

and its associated probability is less than 0.0001 (with degrees of freedom 2 and 146). Note that the sepal width was not the second best variable in the first step of the selection (see Figure 11.33). This is because its lambda Λ_1 was 0.599217, giving an *F*-ratio equal to 49.16. In the first step, the best variable after the petal length is the petal width, with $\Lambda_1 = 0.071117$, markedly lower than the value for the sepal width, and only slightly higher than the value for the petal length. If the petal width was not selected in the second step, this is because the information that it contains is already largely contained in the petal

Statistics for Entry, DF = 2, 147				
Variable	**R-Square**	**F Value**	**Pr > F**	**Tolerance**
se_l	0.6187	119.26	<.0001	1.0000
se_w	0.4008	49.16	<.0001	1.0000
pe_l	0.9414	1180.16	<.0001	1.0000
pe_w	0.9289	960.01	<.0001	1.0000

Figure 11.33 Values of the *F* statistic of the four independent variables.

Stepwise Selection Summary									
Step	Number In	Entered	Partial R-Square	F Value	Pr > F	Wilks' Lambda	Pr < Lambda	Average Squared Canonical Correlation	Pr> ASCC
1	1	pe_l	0.9414	1180.16	<.0001	0.05862828	<.0001	0.47068586	<.0001
2	2	se_w	0.3709	43.04	<.0001	0.03688411	<.0001	0.55995394	<.0001
3	3	pe_w	0.3229	34.57	<.0001	0.02497554	<.0001	0.59495691	<.0001
4	4	se_l	0.0615	4.72	0.0103	0.02343863	<.0001	0.59594941	<.0001

Figure 11.34 Changes in Wilks' lambda in a stepwise procedure.

length, and the sepal width provides more new information, thus giving a better discrimination capacity.

We can continue by adding the petal width and then the sepal length, for which the corresponding lambdas Λ_3 and Λ_4 are low enough for the probabilities associated with the ratios Λ_2/Λ_3 and Λ_3/Λ_4 to be less than the threshold.

To sum up, in discriminant analysis, the variable selection process provides a sequence of decreasing lambdas: $\Lambda_1 > \Lambda_2 > \Lambda_3 \ldots > \Lambda_p$ (see Figure 11.34). The final value of Wilks' lambda Λ_p is a global performance indicator for the model, which will be better as Λ_p decreases. This sequence of Λ_i is associated with two series of Fisher statistics, one for each of the questions posed initially: whether the selected variables enable us to discriminate the classes, and whether the selected variables are all useful for discrimination. The answer to each question will be 'yes' if the probability associated with the Fisher test is below the specified threshold. For the selection of the first variable, but only for this, the two questions and the two tests are equivalent, and the F-ratios are the same.

In the case of Fisher's irises, we have seen that this first F-ratio is 1180.16. On the other hand, the F-ratio of the contribution of the second variable is 43.035 (with df 2 and 146; see Figure 11.35), whereas the F-ratio of the two-variable model is 307.105 (with df 4 and 292; see Figure 11.36). If these two statistics are different, their probabilities are both below the threshold and we can reject both null hypotheses, namely that (1) two variables cannot discriminate the species of iris, and (2) the addition of the second variable does not improve the discrimination.

Finally, $\Lambda_4 = 0.023439$, and the F-ratios associated with Λ_4 and Λ_3/Λ_4 respectively, which are 199.145 (see Figure 11.37) and 4.721 (see Figure 11.34), have probabilities less than the

Statistics for Entry, DF = 2,146				
Variable	Partial R-Square	F Value	Pr > F	Tolerance
se_l	0.3198	34.32	<.0001	0.2400
se_w	0.3709	43.04	<.0001	0.8164
pe_w	0.2533	24.77	<.0001	0.0729

Figure 11.35 Values of the F statistic for the contribution of the second variable.

Multivariate Statistics					
Statistic	Value	F Value	Num DF	Den DF	Pr > F
Wilks' Lambda	0.036884	307.10	4	292	<.0001
Pillai's Trace	1.119908	93.53	4	294	<.0001
Average Squared Canonical Correlation	0.559954				

Figure 11.36 *F* statistic of the model with two variables.

Multivariate Statistics					
Statistic	Value	F Value	Num DF	Den DF	Pr > F
Wilks' Lambda	0.023439	199.15	8	288	<.0001
Pillai's Trace	1.191899	53.47	8	290	<.0001
Average Squared Canonical Correlation	0.595949				

Figure 11.37 *F* statistic of the model with four variables.

threshold, enabling us to accept that the four variables can discriminate the three species of iris, and that four variables provide significantly better discrimination than three.

The coefficient of determination R^2, or the square of the canonical correlation

This coefficient R^2 is equal to the proportion of the variance of the response explained by its membership of one or other of the classes to be discriminated. In other words, R^2 is the ratio of the between-class variance to the total variance. This ratio is 98% in our example (Figure 11.38), a very high level which is found in scientific data but not in marketing or social science data. The aim of discriminant analysis is to maximize R^2.

The adjusted coefficient of determination R^2

R^2 is optimistic, because it increases with the number of variables. To reduce this bias, we can replace it with an adjusted R^2:

$$adjusted\,R^2 = 1 - \frac{(1-R^2)(n-1)}{n-p-1}.$$

	Canonical Correlation	Adjusted Canonical Correlation	Approximate Standard Error	Squared Canonical Correlation
1	0.984821	0.984508	0.002468	0.969872
2	0.471197	0.461445	0.063734	0.222027

Figure 11.38 Coefficient of determination.

11.6.6 Syntax of discriminant analysis in SAS

SAS/STAT offers a number of procedures, namely CANDISC for canonical (non-Bayesian) discriminant analysis, DISCRIM, the most comprehensive, offering geometrical and Bayesian methods, and STEPDISC, described above, which uses linear discriminant analysis for stepwise selection, not provided by either CANDISC or DISCRIM.

In the example below, the data in the data set DATA are analysed: the dependent variable is specified by CLASS, and the independent variables follow the instruction VAR. A first step, STEPDISC, is used to select the independent variables which are potentially most interesting (see Section 11.6.5). DISCRIM is then called for the first time, to carry out a parametric discriminant analysis (METHOD=normal), linear as with homoscedasticity (POOL=yes), with *a priori* probabilities of the classes to be discriminated proportional to their sizes (PRIORS proportional), with calculation of the canonical variables (CANONICAL) and testing by cross-validation (CROSSVALIDATE). Remember that there is only one canonical variable if the dependent variable can only take two values. The output data set OUT contains the input data, the value of the coefficients of the canonical variables (if the CANONICAL option has been chosen), the *a posteriori* probabilities of each input observation and the resulting assignment (predicted value of the dependent variable). Each observation is assigned to the class of the dependent variable which maximizes the *a posteriori* probability. The output data set OUTSTAT contains various statistics and the coefficients of the discriminant functions. OUTSTAT can then be used as the input of a new step, DISCRIM, in order to apply this discriminant function to a new data set (such as a test data set) specified in TESTDATA. The new scored data set is specified by TESTOUT.

```
PROC STEPDISC DATA=mytable.toscore;
CLASS response;
VAR var1 var2 ... varp; RUN;

PROC DISCRIM DATA=mytable.toscore METHOD=normal POOL=yes
     CROSSVALIDATE ALL CANONICAL OUT=mytable.scored
     OUTSTAT=mytable.ofstat;
CLASS response;
PRIORS proportional;
VAR var1 var2 ... varp; RUN;

PROC DISCRIM DATA=mytable.ofstat TESTDATA=mytable.test
TESTOUT=tout;
CLASS response;
VAR var1 var2 ... varp; RUN;
```

The start and end of the OUTSTAT data set contain the following observations, in particular the coefficients of the Fisher discriminant functions (lines 119–122):

The DISCRIM procedure, with the option ALL selected, displays a large number of outputs, some of which have been described above. These include:

- global within- and between-class covariance matrices, global within- and between-class correlation matrices, global and class means and variances, etc.;

Obs	response	_TYPE_	_NAME_	nbproducts	subscription 1	nbchildren	subscription 2	changeconsum	nbexits
1	.	N		6385.00	6385.00	6385.00	6385.00	6385.00	6385.00
2	0	N		5306.00	5306.00	5306.00	5306.00	5306.00	5306.00
3	1	N		1079.00	1079.00	1079.00	1079.00	1079.00	1079.00
4	.	MEAN		8.94	371.28	1.34	23.11	1.16	6.48
5	0	MEAN		8.47	281.68	1.38	19.62	1.14	5.96
6	1	MEAN		11.23	811.86	1.15	40.28	1.25	9.05
...	...	...	...	...	...	...	...	...	...
119	0	LINEAR	_LINEAR_	0.38	−0.00	1.12	−0.00	8.42	0.05
120	0	LINEAR	_CONST_	−7.50	−7.50	−7.50	−7.50	−7.50	−7.50
121	1	LINEAR	_LINEAR_	0.48	0.00	0.83	0.01	9.14	0.09
122	1	LINEAR	_CONST_	−11.27	−11.27	−11.27	−11.27	−11.27	−11.27

- Wilks' lambda, Pillai's trace, etc. (Figure 11.32);
- the canonical correlation coefficient, the adjusted coefficient, etc. (Figure 11.38);
- the linear discriminant functions (Figure 11.27) with the options METHOD=normal and POOL=yes;
- the coefficients of the canonical variables (Figure 11.23) with the option CANONICAL;
- the mean values of the canonical variables in each of the classes to be discriminated.

It also displays the confusion matrix (see Section 11.16.4), in other words the matrix having the real values of the dependent variable in the rows and the values predicted by the discriminant model in the columns, each cell containing the numbers of observations concerned. This matrix can reveal a classification error rate for each value of the dependent variable, with a global error rate. As the number of observations outside the diagonal of the matrix increases, the error rate rises and the model becomes less predictive. The DISCRIM procedure offers three different ways of calculating the confusion matrix.

The first method is always executed. In this method, each observation of the learning sample (the DATA table) is classified by means of the model constructed on this sample, in other words the model which is chosen and whose characteristics are output by the procedure. This is known as 'resubstitution classification' and it has an optimistic bias.

The second method corrects this bias by applying the model constructed on the learning sample to a distinct population, forming a validation sample. This validation sample is made up of the observations from the input table TESTDATA, and the specification of this kind of table results in the calculation of a confusion matrix on a validation sample.

The third approach also corrects the resubstitution bias, by what is known as the cross-validation method (see Section 11.3.2). This is done by using the CROSSVALIDATE option which causes each observation in the DATA table (i.e. the learning sample) to be classified using the model constructed on the other observations. The confusion matrix is then displayed, as above. An example is shown in Figure 11.39.

If we want to carry out a quadratic parametric discriminant analysis, we write

```
PROC DISCRIM DATA=mytable.toscore METHOD=normal POOL=no
```

and if we wish to carry out a non-parametric discriminant analysis (on the 10 nearest neighbours), we write

```
PROC DISCRIM DATA=mytable.toscore METHOD=npar k=10
```

11.6.7 Discriminant analysis on qualitative variables (DISQUAL Method)

Quantitative variables can be used to describe the financial operation of a business and construct a risk score by a linear discriminant analysis (Section 12.10). When attempts were made to apply this method to scores for lending to individuals, for use when a specialist consumer credit establishment wishes to examine the risk attached to a request from an individual, for example a customer buying furniture on credit in a department store, the problem arose of how to allow for the qualitative variables associated with individuals, such as sex, family status, socio-occupational category, etc. It was this kind of situation that Gilbert

The DISCRIM Procedure
Classification Summary for Calibration Data : WORK. APPRENT
Cross-validation Summary using Linear Discriminant Function

Number of Observations and Percent Classified into response			
From response	**0**	**1**	**Total**
0	3568 99.03	35 0.97	3603 100.00
1	219 96.05	9 3.95	228 100.00
Total	3787 98.85	44 1.15	3831 100.00
Priors	0.94049	0.05951	

Error Count Estimates for response			
	0	**1**	**Total**
Rate	0.0097	0.9605	0.0663
Priors	0.9405	0.0595	

Figure 11.39 Confusion matrix by cross-validation.

Saporta had in mind when he devised the DISQUAL method (DIScrimination on QUALitative variables) in 1975.[22]

In this method, we start with the qualitative variables, divide all the quantitative variables into classes (preferably with the same number of categories and similar frequencies), analyse the multiple correspondences in the indicator matrix of these variables (see Section 7.4), retrieve the (continuous) coordinates of the individuals on the most discriminating factor axes (the other axes represent 'statistical noise'), and then inject these coordinates at the input of a conventional linear discriminant analysis.

This gives us a Fisher score function for a linear combination of the factor axes. Now, since these axes are themselves linear combinations of the indicators of the categories of the initial (qualitative) variables, the Fisher function can be expressed as a linear combination of indicators of categories, which is equivalent to assigning a mark to each of these categories. It can be seen that the DISQUAL method not only has the useful property of processing qualitative variables and avoiding most of the disadvantages of discriminant analysis as listed in Section 11.6.9, but also provides its results in a very practical form: the coefficients (the 'marks') of two categories are comparable, because we are concerned here with indicators, instead of quantitative variables which may have widely differing magnitudes, and the coefficients can also be standardized to give a score between 0 and 100, for example.

DISQUAL has been implemented in SPAD for a long time, but can easily be programmed, as I have done in SAS language for an example of *credit scoring* in Section 12.9.

[22] Saporta, G. (1975) Liaisons entre plusieurs ensembles de variables et codage de données qualitatives. Doctoral thesis, University of Paris VI.

11.6.8 Advantages of discriminant analysis

1. It has a direct analytical solution (invert the W matrix).
2. Because of point 1, calculation is very fast. In this respect, discriminant analysis cannot be beaten.
3. It is optimal when the assumptions of homoscedasticity and multinormality are correct.
4. The coefficients of the linear combinations provide a result that is relatively explicit (but slightly less so than the odds ratios of logit regression).
5. The models that are produced are concise and easily programmed by IT personnel.
6. It is very good at detecting global phenomena (whereas decision trees detect local phenomena).
7. Unlike neural networks and decision trees, it needs far fewer cases for the construction of the model: a few hundred may be enough.
8. Many algorithms enable the variables explaining the response variable to be selected in a stepwise way.
9. It facilitates the integration of the classification error costs.
10. It is a method implemented in numerous software.

11.6.9 Disadvantages of discriminant analysis

Conventional linear discriminant analysis which does not make use of the improvement provided by DISQUAL has a number of drawbacks:

1. It only detects linear phenomena.
2. In principle, it can only be applied to continuous independent variables without missing values, even if discrete independent variables can be accepted. For discrete or qualitative variables, or in cases of non-linear phenomena, the DISQUAL method can be used.
3. It is sensitive to individuals outside the norm (outliers).
4. It theoretically requires the assumptions of multinormality, homoscedasticity and linear independence of the independent variables. Heteroscedasticity (non-homoscedasticity) may be due to the presence of outliers. As regards the existence of linear relations between the independent variables, or collinearity, this reduces the stability of the results and may lead to aberrations in the signs of the parameters. It should be noted that, if the last assumption is fundamental, failure to meet the requirements of multinormality and homoscedasticity does not necessarily invalidate discriminant analysis, but in this case the robustness of the discriminant analysis

will be greater if the classes to be discriminated have frequencies that are close and sufficiently large. If normal distributions of the variables cannot be achieved, we must at least ensure that their distributions are unimodal (with only one peak) and nearly symmetrical.

There is another drawback, from which even the DISQUAL method is not exempt: this is the absence of statistical tests of the significance of the coefficients, which are present in logistic regression and are very useful for the choice of independent variables and their division. However, we can find confidence intervals by bootstrapping (see Section 11.15.1), albeit that this complicates processing and increases execution times.

As a general rule, it is advisable to take the following measures in order to approach the constraining assumptions of discriminant analysis:

1. Standardize the variables as far as possible.

2. Carefully select the most discriminating variables.

3. Exclude any of these discriminating variables that are excessively intercorrelated.

4. Exclude extreme individuals or Winsorize (truncate) their extreme values.

5. If some heteroscedasticity persists, it is better to have classes of comparable sizes, even if it means sampling the population by retaining the whole of the smallest class and only a fraction of the largest class.

6. Work on homogeneous populations.

7. It is therefore preferable to segment the population in advance and construct a model by segment before synthesizing all these models (see Section 2.7).

11.7 Prediction by linear regression

There are two reasons for spending some time on the topic of linear regression in a book on data mining.[23] In the first place, linear regression forms the basis of all linear models and is universally applicable. Its modern variations – such as 'ridge' and 'lasso' regression – are very useful, especially in situations where the number of variables is greater than the number of observations, and in those more common situations where there is significant collinearity between the predictors. It will also be helpful to deal with linear regression before moving on to logistic regression and the generalized linear model, as this will enable us to introduce concepts which will be useful in logistic regression, such as the modelling of a conditional expectation and the concept of residuals. Certain analogies, such as that between the sum of squares of residuals in linear regression and deviance in logistic regression, will give us a better understanding of logistic regression.

[23] A more detailed introductory course can be found in Confais, J. and Le Guen, M. (2006) Premiers pas en régression linéaire avec SAS. *Modulad*, 35, 220–363.

11.7.1 Simple linear regression

Simple linear regression enables us to relate a continuous dependent variable Y to a continuous independent variable X. It is commonly assumed that the values $x_1, \ldots, x_n$ of X are controlled and not subject to measurement error, and the corresponding values $y_1, \ldots, y_n$ of Y are observed. The variable X can be time, and Y can be a quantity measured at different dates.[24] Y could also be the potential difference measured at the terminals of a resistor for different values of current strength X. It is assumed that X and Y are not independent, and that a knowledge of X enables us to improve our knowledge of Y. Of course, even if we know that $X = x$, this does not usually mean that the exact value of Y is known, but we assume that it will enable us to know the mean value $E(Y|X = x)$, the conditional expectation of Y given that $X = x$. More precisely, the basic postulate of linear regression is that $E(Y|X = x)$ is a linear function of x, which can be stated as

$$E(y_i) = \alpha + \beta x_i \text{ for every } i = 1, \ldots n,$$

or, in an equivalent way, as

$$y_i = \alpha + \beta x_i + \varepsilon_i, \text{ with } E(\varepsilon_i) = 0 \text{ for every } i = 1, \ldots n.$$

The term $(\alpha + \beta x_i)$ is the deterministic component of the model, (ε_i) is its stochastic component, and the values ε_i are called the 'errors'. The breakdown into a deterministic component and a stochastic component reflects the fact that individuals having the same value of x_i can have different responses Y (synchronic variation), or that one individual measured several times with the same value x_i can have different responses Y (diachronic variation), but in all cases the mean response of individuals having the value x_i or of the observations of an individual having the value x_i is completely determined. This is shown in Figure 11.40. This diagram also reveals the other hypotheses of the linear model, namely:

- the variance of the errors is the same for all values of X (homoscedasticity): $V(\varepsilon_i) = s^2$;
- the errors are linearly independent ($Cov(\varepsilon_i, \varepsilon_j) = 0 \; \forall i \neq j$);
- the errors are normally distributed ($\varepsilon_i \sim N(0, s^2)$).

Note also the essential condition that $n > 5$, or preferably $n > 15$.

[24] In this case, the existence of a trend (without seasonality) in the time series $x_1, \ldots, x_n$ can be tested in a more general context, without using the strong assumptions of linear regression. The non-parametric Mann–Kendall test can be used. This is applied to the statistic S of the same name, which is the sum of the signs of the differences $x_j - x_i$, where $j > i$. If S is divided by its maximum value, namely $n(n - 1)/2$ if there are no ties, we will recognize Kendall's tau (see the Section A.2.8). Under the null hypothesis that there is no tendency, and if $n > 10$, this statistic S is correctly approximated by a normal distribution with a mean of 0 and variance Var(S) which is a cubic function of n. If $n \leq 10$, an exact test must be used. By comparing the value of S with a tabulated value or one given by a quantile of the normal distribution, we can decide whether or not to reject the null hypothesis. Another test, the Sen test, can also be used to reject the null hypothesis, and also provides a robust estimator of the slope, which is the median of the slopes of all the pairs of points (i, x_i). In the Sen test, a confidence interval around the median slope is calculated, and this slope is considered to be significantly different from zero if its confidence interval does not contain 0, enabling us to reject the null hypothesis.

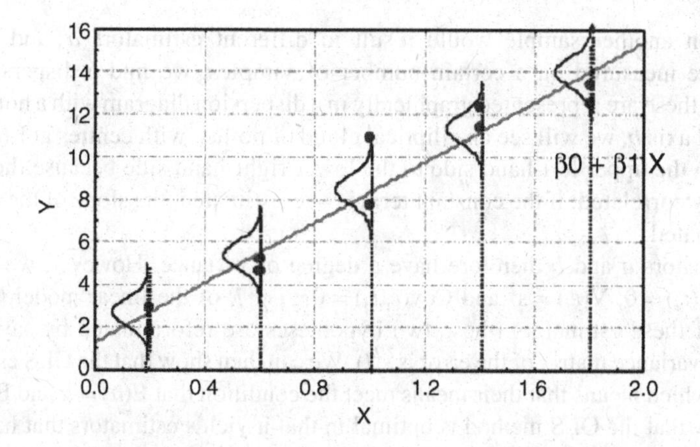

Figure 11.40 The hypotheses of linear regression.

Clearly, the last two hypotheses are transferred to the values of y_i: $\mathrm{Cov}(y_i,y_j) = 0$ and $y_i \sim N(\alpha + \beta x_i, s^2)$. The hypothesis of normality means that linear regression is included in the family of general linear models, whereas in generalized linear models (see Section 11.9.6) the conditional distribution of Y given $X = x_i$ is not necessarily normal.

To return to simple linear regression, we postulate the existence of a relation $\mathrm{E}(Y|X = x) = \alpha + \beta x$ which we will attempt to estimate on the basis of a sample. We therefore seek the estimators a and b of α and β. This is usually done with the ordinary least squares (OLS) method, which involves finding the coefficients a and b which minimize the differences

$$\sum_{i=1}^{n} (y_i - a - bx_i)^2.$$

We first find the estimator b of the slope,

$$b = \frac{\sum_i (x_i - \bar{x})(y_i - \bar{y})}{\sum_i (x_i - x)^2} = \frac{\mathrm{cov}(X, Y)}{\sigma_x^2},$$

with the usual notation for the means, $\bar{x} = \sum_i x_i / n$ and $\bar{y} = \sum_i y_i / n$. We then find the estimator a of the constant,

$$a = \bar{y} - b\bar{x}.$$

For every x_i, the predicted value of y_i is $a + bx_i$. We say that the straight line $\hat{Y} = a + bX$ fits the cloud of points. The errors ε_i are estimated by the *residuals* $y_i - (a + bx_i)$.

Note that we can never find the actual coefficients α and β, because:

- the linear model is often only an approximation to reality;
- we are only working on samples, not the whole population;
- measurement errors occur.

Estimation on another sample would result in different estimators a' and b'. If these estimators are measured on a certain number of samples, we find a dispersion of their values, and if these are represented graphically in a dispersion diagram with a horizontal axis a and vertical axis b, we will see an elliptical cloud of points, with centres at E(a) and E(b), inclined from the upper left-hand side to the lower right-hand side because the estimators are negatively correlated: if the constant term is overestimated, the slope of the straight line is underestimated.

The estimators a and b therefore have a degree of variance. However, we can use the hypotheses $E(\varepsilon_i) = 0$, $V(\varepsilon_i) = s^2$ and $Cov(\varepsilon_i, \varepsilon_j) = 0, \forall i \neq j$, of the linear model to refine our knowledge of these estimators (these two hypotheses are reformulated by saying that the variance—covariance matrix of the error is s^2I). We can then show that the OLS estimators are free of bias, which means that their means meet the condition that $E(a) = \alpha$ and $E(b) = \beta$. We can also show that the OLS method is optimal in that it yields estimators that have the least variance of all the unbiased linear estimators (the Gauss–Markov theorem). Such estimators are referred to as *best linear unbiased estimators* (BLUE). This property of minimal variance is useful because it guarantees the stability of the coefficients, the correct generalization of the model, and the reliability of the predictions. The BLUE property of OLS estimators explains the very widespread use of this method. However, an even lower variance is sometimes preferable to an absence of bias for the purpose of increasing the accuracy of predictions; this is discussed below and comes from the fact that the mean square error (MSE) of an estimator, i.e. the average of the square of the differences between the observed and the estimated values, is the sum of its variance and the square of its bias.

The supplementary hypothesis of normality $\varepsilon_i \sim N(0,s^2)$ signifies that the estimators a and b have a normal distribution. On this hypothesis, the maximum likelihood estimators coincide with the least squares estimators (see Section 11.8.5).

The variances V(a) and V(b) are related to the variance s^2 of the errors:

$$\sigma_a^2 = s^2\left[\frac{1}{n} + \frac{\bar{x}^2}{\sum_i (x_i - \bar{x})^2}\right] \text{ and } \sigma_b^2 = s^2\left[\frac{1}{\sum_i (x_i - \bar{x})^2}\right]$$

From this we can deduce the confidence intervals at the $100(1 - \alpha)\%$ level:

$$a \pm t_{\alpha/2,n-p-1} \cdot \sigma_a \text{ and } b \pm t_{\alpha/2,n-p-1} \cdot \sigma_b.$$

These formulae show that there are three ways of decreasing the variances:

- by increasing the size n of the sample,
- by increasing the range of the observed values of X (important: the model is only valid over this range),
- by reducing the variance s^2 of the errors in the sample.

Another way of reducing an excessive variance is to accept slightly biased estimators (see *ridge regression*, in Section 11.7.11).

The slope b of the regression line provides the direction (positive, negative or non-monotonic) of the relation between Y and X, but not its strength, which is given by the variance of the errors s^2, also known as the variance of the error ε. Like the coefficients α and β, s^2 is

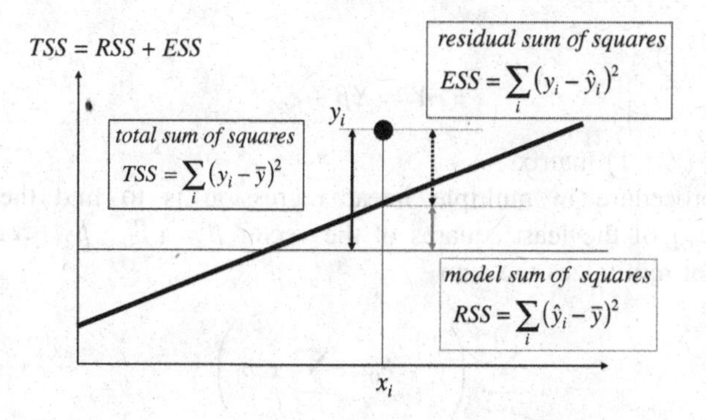

Figure 11.41 Analysis into sums of squares.

unknown, and we have to find an estimator of it. This is done by calculating the residual sum of squares,

$$ESS = \sum_{i=1}^{n} (y_i - a - bx_i)^2,$$

as shown in Figure 11.41, then dividing ESS by the number of degrees of freedom $n - p - 1$ (where $p = 1$, the number of independent variables, not counting the constant). The quantity

$$\frac{ESS}{n-2}$$

is an unbiased estimator of s^2. We recognize the mean square error and its square root is the root mean square error (RMSE).

11.7.2 Multiple linear regression and regularized regression

Simple linear regression can be generalized to the case of several independent variables X_i, giving *multiple linear regression*, and in this case we write

$$Y = \beta_0 + \beta_1 X_1 + \ldots + \beta_p X_p + \varepsilon,$$

adding an important supplementary hypothesis, namely the linear independence of the X_i. The above equation can be written in matrix form for the n observations:

$$\begin{pmatrix} y_1 \\ y_2 \\ \ldots \\ y_3 \end{pmatrix} = \begin{pmatrix} 1 & x_{11} & \ldots & x_{1p} \\ 1 & x_{21} & \ldots & x_{2p} \\ \ldots & \ldots & \ldots & \ldots \\ 1 & x_{n1} & \ldots & x_{np} \end{pmatrix} \begin{pmatrix} \beta_0 \\ \beta_1 \\ \ldots \\ \beta_p \end{pmatrix} + \begin{pmatrix} \varepsilon_1 \\ \varepsilon_2 \\ \ldots \\ \varepsilon_n \end{pmatrix}$$

or, more concisely, as

$$Y = X\beta + \varepsilon.$$

where X is a $n \times (p+1)$-matrix.

The usual procedure in multiple linear regression is to find the estimator $b = (b_0 \quad b_1 \quad \ldots \quad b_p)$ of the least squares of the vector $\beta = (\beta_0 \quad \beta_2 \quad \ldots \quad \beta_p)$, in other words the one that minimizes the sum

$$\sum_{i=1}^{n} \left(y_i - b_0 - \sum_{j=1}^{p} x_{ij} b_j \right)^2.$$

Using the matrix notation Y, X, A^t to denote the transposed matrix of any matrix A, and $\|A\|$ for the norm of A, the expression to be minimized is written:

$$\|Y - Xb\|^2 = (Y - Xb)^t (Y - Xb) = Y^t Y - 2b^t X^t Y + b^t X^t X b.$$

If this quantity is seen as a function of b, we must find b such that this quantity is minimal, which is done by differentiating it with respect to b,

$$\partial/\partial b(\|Y - Xb\|^2) = -2X^t Y + 2X^t X b,$$

and the condition

$$\partial/\partial b(\|Y - Xb\|^2) = 0$$

means that b must satisfy the least squares equation

$$b = (X^t X)^{-1} X^t Y. \tag{11.2}$$

This can easily be verified in the case where we have

$$X = \begin{pmatrix} 1 & x_1 \\ 1 & x_2 \end{pmatrix} \text{ and } Y = \begin{pmatrix} 1 & y_1 \\ 1 & y_2 \end{pmatrix}.$$

This is because

$$(X^t X)^{-1} X^t Y = X^{-1} Y = \frac{1}{x_2 - x_1} \begin{pmatrix} x_2 & -x_1 \\ -1 & -1 \end{pmatrix} \begin{pmatrix} y_1 \\ y_2 \end{pmatrix} = \frac{1}{x_2 - x_1} \begin{pmatrix} x_2 y_1 - x_1 y_2 \\ y_2 - y_1 \end{pmatrix}.$$

Now, it is clear that

$$\frac{y_2 - y_1}{x_2 - x_1}$$

is indeed the estimator of the slope, while

$$\frac{x_2 y_1 - x_1 y_2}{x_2 - x_1} = \frac{y_1 + y_2}{2} - \left(\frac{y_2 - y_1}{x_2 - x_1} \frac{x_1 - x_2}{2} \right)$$

is the estimator of the constant.

Since the vector of the estimated values is $\hat{Y} = Xb$, and $b = (X'X)^{-1}X'Y$, this vector is expressed as

$$\hat{Y} = X(X'X)^{-1}X'Y = HY,$$

where the $n \times n$ matrix $H = X(X'X)^{-1}X^t$ is called the 'hat matrix', because of the symbol appearing on top of the Y. The trace of this matrix, in other words the sum of its n diagonal elements h_i, is $p + 1$, since $Tr(H) = Tr((X^tX)^{-1}X^tX) = Tr(I_{p+1})$ and each of these is in the range from $1/n$ to 1 (inclusive). These terms are called 'leverages' or 'leverage values', because they are related to the effect of each observation i on the fit. The leverage in i is a measure of the distance between the values of X taken in i and the mean values over the n observations. The leverages depend only on X, not on Y, but their effect on the fit can easily be imagined, if we think of an observation very distant from the others: when its vector of Xs tends towards infinity, the fit will tend to be made to this observation only, and the residual will be very small. This is the leverage effect.

As in simple linear regression, the Gauss–Markov theorem holds, and the OLS estimators are the unbiased estimators which have minimal variance. Furthermore, if the errors are normally distributed, the estimators will also be normally distributed.

As in simple regression, the quantity

$$\hat{s}^2 = \frac{ESS}{n - p - 1}$$

is an unbiased estimator of the variance of the error s^2.

As for the variance of each residual $e_i = y_i - \hat{y}_i$, this is related to the variance of the error s^2 and to the leverage h_i:

$$\text{Var}(e_i) = s^2(1 - h_i).$$

It is therefore always less than the variance of the error s^2, and can even become very small if the leverage is very large. This is evidently due to the fact that the fit tends to be made only to the observation concerned. If the errors ε_i are normally distributed, the residuals e_i are too.

We can deduce from formula (11.2) that the variance–covariance matrix of the estimator b is

$$V(b) = s^2(X'X)^{-1}. \tag{11.3}$$

We can check this in our little example above, and find the variance σ_a^2 of the constant and σ_b^2 of the slope, as expressed above. For example, we have

$$\sigma_b^2 = \frac{s^2}{x_1^2 + x_2^2 - 2\mu} = \frac{s^2}{(x_1 - \mu)^2 + (x_2 - \mu)^2},$$

where $\mu = (x_1 + x_2)/2$.

The variances of the estimators (b_j), which, according to equality (11.3), are proportional to the diagonal terms of $(X'X)^{-1}$, become very large if the independent variables are collinear. In fact, the determinant of the matrix $(X'X)$ becomes close to 0 and this is difficult to invert. This leads to inaccuracies in the estimated parameters of the model and errors in the predictions, even if R^2 is large. We will return to these difficulties in Section 11.7.8.

These diagonal terms are also proportional to $1/(1 - R_j^2)$, where R_j is the multiple correlation coefficient of the jth variable with the $p - 1$ other independent variables. These ratios are those by which s^2 is multiplied to give the variances of the estimators (b_j) and this is why they are called 'variance inflation factors' (see Section 3.14). It is generally considered that they must be less than 10, and the mean of the p inflation factors is sometimes used as a global measurement of multicollinearity.

These diagonal terms are also related to the inverses of the eigenvalues of the matrix X^tX, some of which are close to 0 in cases of collinearity. This explains why condition indices are used to detect collinearity (see Section 3.14).

If collinearity is present and it is difficult to invert the matrix X^tX, one way of dealing with the collinearity is to add a constant k to the diagonal terms of X^tX to make it invertible without any numerical difficulty. This is the principle of 'ridge' regression as proposed by Hoerl and Kennard (1970).[25] The 'ridge' estimator is then

$$b_R = (X^tX + kI)^{-1}X^tY.$$

This is equivalent to finding the estimator b_R that minimizes

$$\sum_{i=1}^{n}\left(y_i - b_0 - \sum_{j=1}^{p} x_{i,j}b_j\right)^2 + k\sum_{j=1}^{p} b_j^2.$$

In vector terms, we can rewrite this expression as

$$\|Y - Xb\|^2 + k\|b\|^2.$$

This is equivalent to finding $b_R = (b_0\ b_1\ \ldots\ b_p)^t$ so as to minimize

$$\sum_{i=1}^{n}\left(y_i - b_0 - \sum_{j=1}^{p} x_{i,j}b_j\right)^2$$

subject to the condition $\sum_{j=1}^{p} b_j^2 \leq C^2$ (C is related to k). For this reason, we also describe this as the 'shrinkage method'. The greater the value of k (the smaller the value of C), the more the coefficients 'shrink'. The aim is to avoid unstable coefficients and very large coefficients which 'compensate' each other, sometimes at the cost of inconsistency of signs. The variance–covariance matrix of b_R is

$$V(b_R) = s^2(X^tX + kI)^{-1}(X^tX)(X^tX + kI)^{-1}.$$

It is interesting to note that a well-chosen value of k may not only allow us to invert $(X^tX + kI)$ without problems and to control the variance of the estimator, but also lead to a variance that is

[25] Hoerl, A.E. and Kennard, R.W. (1970) Ridge regression: biased estimation for nonorthogonal problems. *Technometrics*, 12, 55–67.

so greatly reduced that it compensates for the increase in bias ($E(b_R) \neq \beta$) in such a way that

$$E\left(\|b_R - \beta\|^2\right) \leq E\left(\|b - \beta\|^2\right)$$

The mean square error of the estimation of β by b_R is lower than that of the estimation by b. This is because the mean square error is the sum of the variance and the square of the bias of the estimator:

$$E(\|b_R - \beta\|^2) = V(b_R) + E(b - \beta)^2.$$

This parameter k can be seen as a 'cursor' enabling us to choose more bias and less variance (when k large) or less bias and more variance (when k is small), in a continuous way, thus optimizing the prediction accuracy. This fine-tuning is an advantage of shrinkage regression as compared with PLS regression and principal component regression.

In the case of simple linear regression, the unbiased estimator of the slope is

$$b = \frac{\sum_i (x_i - \bar{x})(y_i - \bar{y})}{\sum_i (x_i - \bar{x})^2},$$

and we can show that the 'ridge' estimator of parameter k is

$$b_R = \frac{\sum_i (x_i - \bar{x})(y_i - \bar{y})}{\sum_i (x_i - \bar{x})^2 + k}.$$

The mean square error of b_R reaches its minimum when $k = s^2/\beta^2$.

In practice, s^2 and β^2 are unknown, and we search for the optimal value of k by combining the following criteria:

- a variance inflation factor close to 1;
- a moderate increase in the RMSE;
- stable coefficients b_R from k onwards.

In order to apply this last criterion, we plot the values of the coefficient b_R for each independent variable as a function of k on the same graph, so that they are superimposed. This is known as a 'ridge trace'. Some examples will be discussed later on. Ridge regression is an effective method of regularization which is easy to apply and is widely used.

In 1996, Robert Tibshirani introduced a new method of regularisation (also known as 'penalization'), the lasso (least absolute shrinkage and selection operator).[26] In this linear method, the sum

$$\sum_{i=1}^{n} \left(y_i - b_0 - \sum_{j=1}^{p} x_{i,j}, b_j \right)^2$$

[26] Tibshirani, R. (1996) Regression shrinkage and selection via the lasso. *Journal of the Royal Statistical Society, Series B*, 58(1), 267–288.

is minimized subject to the condition $\sum_{j=1}^{p} |b_j| \leq t$, instead of $\sum_{j=1}^{p} b_j^2 \leq C^2$ as in ridge regression. In the same way, we aim to minimize

$$\sum_{i=1}^{n} \left(y_i - b_0 - \sum_{j=1}^{p} x_{i,j} b_j \right)^2 + k \sum_{j=1}^{p} |b_j|.$$

In 2005, Zou and Hastie proposed a compromise (the 'elastic net') between the lasso and ridge regression,[27] in which the penalization takes the form

$$\sum_{j=1}^{p} \left[\frac{1}{2}(1-\alpha) b_j^2 + \alpha |b_j| \right].$$

The lasso is present when $\alpha = 1$ and ridge regression is present when $\alpha = 0$.

Other kinds of penalization have recently been proposed by various researchers, in the form

$$\sum_{j=1}^{p} |b_j|^{\delta},$$

where δ is between 0 and 2, the values of δ lying between 0 and 1 being more suitable for cases where many of the coefficients to be estimated are small. The difficulty that arises in terms of computing is that of finding the simultaneous optimal values of k and δ quickly enough.[28] Each value of δ has a corresponding constraint domain: with two dimensions, this is a disc when $\delta = 2$, a lozenge when $\delta = 1$, and a 'star' when $\delta < 1$, this star contracting towards the axes as δ approaches 0. The arms of the star correspond to cases in which one of the coefficients to be estimated is small. If δ is less than 1, the constraint domain is concave and not convex, giving rise to problems of optimization as mentioned above. Various computing algorithms have been suggested, such as the LARS procedure of Efron and Hastie[29] (available in the *lars* package of R) and the GLMNET procedure based on 'coordinate descent', a much faster computing method (available in the *glmnet* package).

After multiple linear regression, these regularization methods have been applied in other contexts, as follows: logistic regression, the generalized linear model (Tibshirani, 1996;[30] Park & Hastie, 2007),[31] the Cox proportional hazards survival model (Tibshirani, 1997),[32] support vector machines (Hastie *et al.*, 2004),[33] quantile regression (Li & Zhu, 2007),[34] etc. In

[27] Zou, H. and Hastie, T. (2005) Regularization and variable selection via the elastic net. *Journal of the Royal Statistical Society, Series B*, 67(2), 301–320.

[28] See, for example, http://www-stat.stanford.edu/~jhf/talks/GPS_R.pdf and http://www-stat.stanford.edu/~jhf/ftp/GPSpaper.pdf, both presentations by Jerome H. Friedman.

[29] Efron, B., Hastie, T., Johnstone, I., and Tibshirani, R. (2004) Least angle regression (with discussion), *Annals of Statistics*, 32, 407–499.

[30] Tibshirani, Regression shrinkage and selection via the lasso, cited in n. 26.

[31] Park, M.Y. and Hastie, T. (2007) L_1-regularization path algorithm for generalized linear models. *Journal of the Royal Statistical Society, Series B*, 69(4), 659–677.

[32] Tibshirani, R. (1997) The lasso method for variable selection in the Cox model. *Statistics in Medicine*, 16, 385–395.

[33] Hastie, T., Rosset, S., Tibshirani, R. and Zhu, J. (2004) The entire regularization path for the support vector machine. *Journal of Machine Learning Research*, 5, 1391–1415.

[34] Li, Y. and Zhu, J. (2008) L_1-norm quantile regression. *Journal of Computational and Graphical Statistics*, 17, 163–165. See also Li, Y., Liu, Y. and Zhu, J. (2007). Quantile regression in reproducing kernel Hilbert spaces. *Journal of the American Statistical Association*, 102, 255–268.

	Sum of Squares	df	Mean Square	F	Significance
Regression	3267046.665	2	1633523.333	167.933	.000
Residual	116727.068	12	9727.256		
Total	3383773.733	14			

Figure 11.42 Illustration of the F test of a linear regression.

logistic regression, for example, it is not the square error but the log-likelihood that is penalized. The 'grouped lasso' was introduced in 2006 by Yuan and Lin[35] to include or exclude groups of variables; for example, such a group may be all the categories of a qualitative variable.

11.7.3 Tests in linear regression

In a linear regression on p independent variables, the number of degrees of freedom of the 'regression' sum of squares, RSS, is p. The equivalent of the ANOVA F-ratio is

$$F = \frac{RSS/p}{ESS/(n-p-1)}.$$

On the null hypothesis that all the regression coefficients are zero, F has a Fisher distribution with $(p, n - p - 1)$ degrees of freedom. In the following example of the regression of heating consumption on the external temperature and insulation thickness, the F test is considered significant because the probability associated with $F = 167.933$ is less than 5%. The table in Figure 11.42 also shows the estimator of the variance of the error, equal to 9727.56. This quantity is equal to the square of 98.627 which appears in Figure 11.47 ('standard error of the estimate').

The number of observations has a significant effect: a small sample (say, less than 20 observations) detects only the strong relations, but a large sample detects all the relations, even weak ones (H_0 is rejected even though RSS is small with respect to ESS).

In addition to the F-ratio, and the RMSE which must be as small as possible, another important quality indicator of the model is the coefficient of determination:

$$R^2 = \frac{RSS}{TSS} = 1 - \frac{ESS}{TSS}.$$

This represents the part explained by the regression to the sum of squares of the deviations from the mean. In the case of simple linear regression, R^2 is equal to r^2, the square of the Pearson correlation coefficient of X and Y. More generally, R^2 is the square of the Pearson coefficient of correlation of Y with its prediction $\hat{Y}$ (in other words, its projection on the regression line). R is called the coefficient of multiple correlation of Y with the variables X_i. In all cases, R^2 ranges from 0 to 1 and the fit improves as R^2 approaches 1. As this happens, the

[35] Yuan, M. and Lin, Y. (2006) Model selection and estimation in regression with grouped variables. *Journal of the Royal Statistical Society, Series B*, 68, 49–67.

cloud of points becomes closer to the regression line, and the equation for the regression line becomes a better predictor of the phenomenon measured. We say that there is a good fit, or that the regression line fits the cloud well. If $R^2 = 1$, the cloud is actually identical to the straight line, and the prediction by the regression straight line is perfect: the regression equation always yields the exact value that is to be predicted. In two dimensions, R^2 is the square of the coefficient of correlation between the dependent variable Y and the independent variable X (called the *regressor* or *predictor*), and is therefore related to the elongation of the cloud of points as shown in Figure A.7 in Appendix A.

Looking at the example (Figure 11.43) of heating consumption regressed on the insulation thickness, we find a degree of dispersion of the cloud around the regression line. This is indicated by a moderate R^2 of 0.22. Taking the example (Figure 11.44) of heating consumption regressed on the external temperature, we find a markedly greater concentration of the points around the regression line. The cloud is elongated along the straight line. This is indicated by a high R^2 of 0.76. In both cases, of course, the slope of the line is negative, because the consumption decreases when the external temperature rises or when the thickness of the insulation increases.

When the cloud deviates from the straight line, this may be due to measurement errors or other anomalies, but the usual cause is a lack of useful information for describing the data and predicting the phenomenon of interest. The work of modelling then consists of finding the variables that can be added to the model, in other words to the linear equation, to provide a better approximation to the cloud of points. Note, however, that we then move into a space with more dimensions, each new variable adding a new dimension.

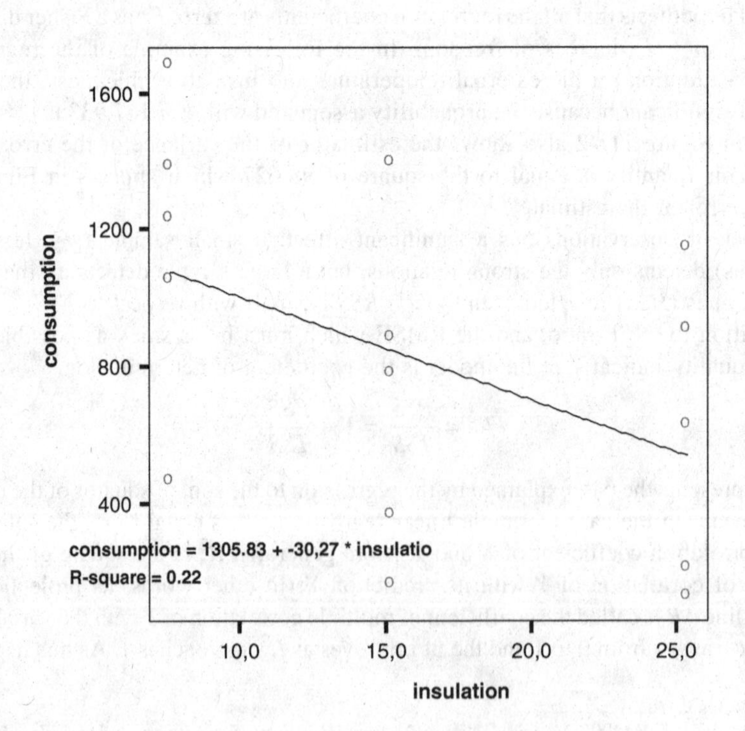

Figure 11.43 Regression of consumption on insulation.

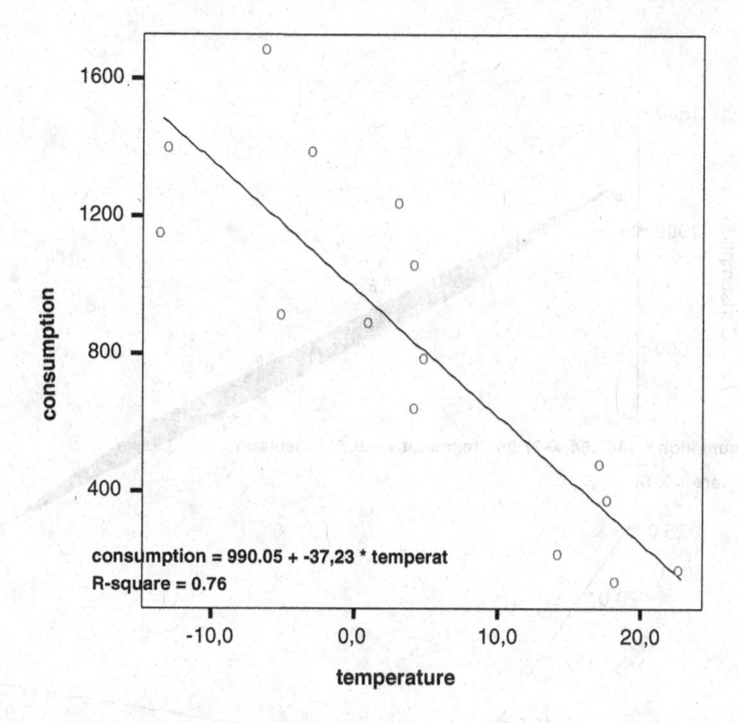

Figure 11.44 Regression of consumption on temperature.

Returning to our example of heating consumption, regressed simultaneously on the two variables of temperature and insulation thickness, we obtain the equation

$$\text{consumption} = 1467.64 - 37.06 \times \text{temperature} - 29.77 \times \text{insulation},$$

which, as Figure 11.45 shows, is much closer to the cloud of points than either of the straight lines, even the temperature line. Thus we have achieved our aim of finding regressors, in other words independent variables, which provide the best description of the phenomenon to be measured, in the sense that the heating consumption estimated by the resulting linear equation (the points on the plane) is very close to the actual consumption (the points in the cloud). This is indicated by a very high value of R^2, at 0.97, very close to the sum of the values of the R^2 of the simple regressions on insulation thickness and external temperature. This almost ideal situation is rather unusual. It is present here because the two regressors are hardly correlated at all (see Figure 11.49). Consequently, the information provided by each variable has hardly any redundancy in relation to the information provided by the other variable, and nearly all of their discriminant powers are added together.

On the other hand, if the two variables were highly correlated, we would find a markedly less satisfactory fit, as shown in Figure 11.46, where the cloud is clearly more dispersed around the plane. The R^2 that is displayed also shows that it does not exceed the R^2 of the first variable: the second variable adds virtually nothing to the regression or to the resulting model. And yet, the equation in Figure 11.46 shows that the sign of the 'temperature' variable remains negative: we shall see in Section 11.7.11 that collinearity can lead to an inversion of signs. For

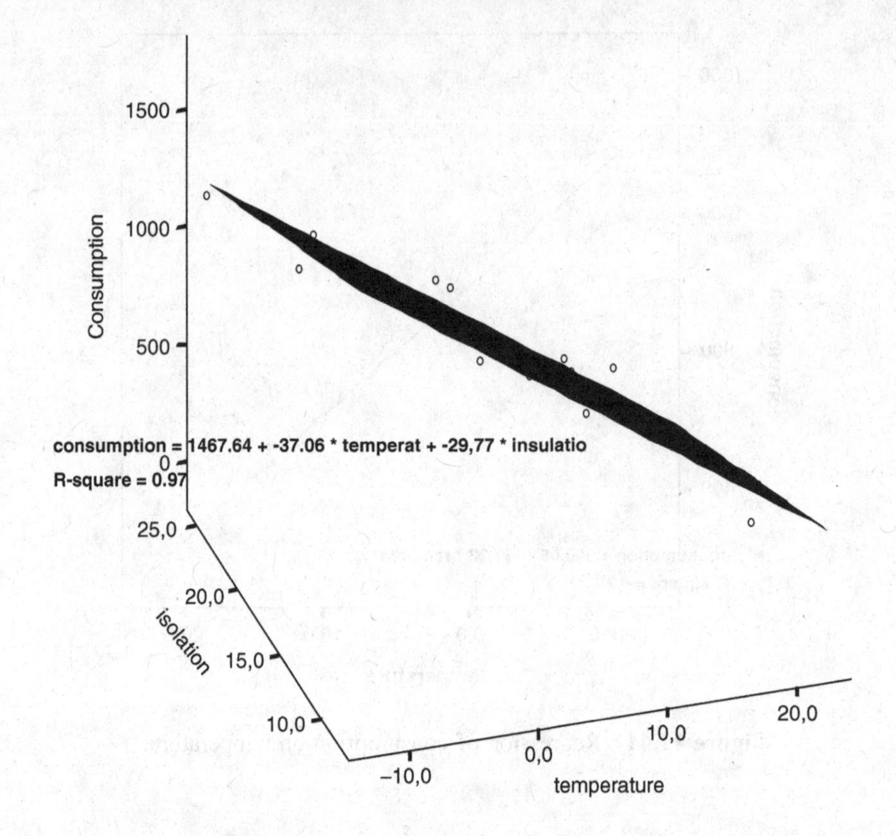

Figure 11.45 Regression of consumption on insulation and temperature.

the present example, the variable 'othervariable' is defined as the sum of the temperature and a random Gaussian term following an $N(0,1)$ distribution.

This is an extreme situation, but the opposite situation in which the variables are not correlated at all is rarely encountered, even though, as is well known, the R^2 of the regression generally increases with the number of variables. However, we should be wary of the excessive optimism of R^2, because this increases 'mechanically' with the number of variables, to a greater extent than the real fit of the linear model (a straight line with a regressor, a plane with two regressors, a 'hyperplane' with more than two regressors) to the cloud of points. It is therefore best to rely on the adjusted R^2, defined in the same way as in linear discriminant analysis:

$$R^2 adjusted = 1 - \frac{(1 - R^2)(n - 1)}{n - p - 1}$$

This adjusted R^2 is always smaller than R^2 (see Figure 11.47), even when $p = 1$, and may even be negative.

Leaving aside these global indicators, the contribution of each variable can be evaluated by dividing its coefficient by the standard deviation of the coefficient. When this quotient t is greater than $+2$ or less than -2, the coefficient is significantly different (at the 95% level)

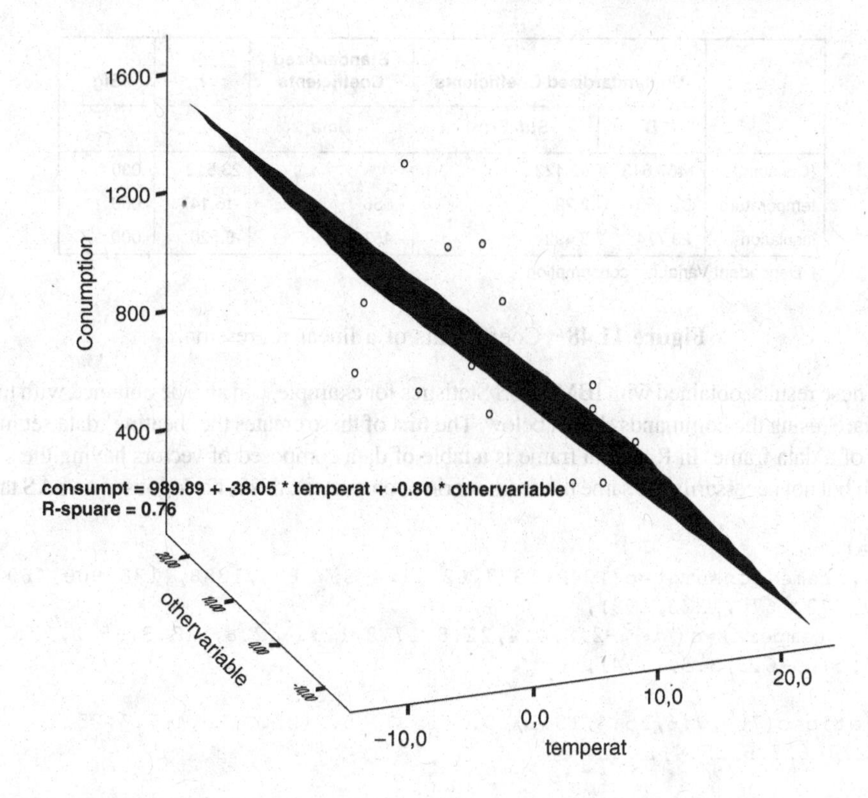

Figure 11.46 Regression of consumption on two highly correlated variables.

R	R square	Adjusted R square	Std Error of the Estimate	Durbin-Watson
.983	.966	.960	98.627	1.819

Figure 11.47 Quality indicators of a linear regression.

from 0. This is the case shown in Figure 11.48. If this were not the case, we would have to exclude the variable from the model. As the absolute value of *t* increases, the predictive power of the variable rises together with its contribution to the linear model. This is true, for example, of the temperature variable with respect to the insulation variable. Note that the value of *t* must not be confused with that of the coefficient of the variable: the coefficient only indicates the slope of the line, and a large coefficient can be associated with a small *t*, in other words a high standard deviation and a high uncertainty concerning the coefficient. The contribution of the variable is related to its *t* but not to its coefficient, which can also be multiplied or divided if the units of the variable are changed. As a special case, in simple linear regression the *p*-value of the Student test on the single predictor is equal to the *p*-value of the Fisher test, described above, on the mean squares.

	Unstandardized Coefficients		Standardized Coefficients	t	Sig
	B	Std. Error	Beta		
(Constant)	1467.643	62.422		23.512	.000
temperature	-37.060	2.295	-.866	-16.147	.000
insulation	-29.774	3.492	-.457	-8.526	.000

a Dependent Variable : consumption

Figure 11.48 Coefficients of a linear regression.

These results, obtained with IBM SPSS Statistics for example, can also be obtained with the R software, using the commands shown below. The first of these creates the 'heating' data set in the form of a data frame. In R, a data frame is a table of data composed of vectors having the same length but not necessarily the same mode (numeric or character); this is analogous to an SAS table.

```
> heating <-
data.frame(consumpt=c(1042,1377,622,154,357,874,1388,1138,900,460
,119,770,1670,1223,199),
  +    temperat=c(4.4,-2.8,4.4,22.8,17.8,1.1,-12.8,-13.3,-5.0,17.2,
18.3,5.0,-6.1,3.3,14.4),
  +
isolatio=c(7.6,7.6,25.4,15.2,15.2,15.2,15.2,25.4,25.4,7.6,25.4,
15. 2,7.6,7.6,25.4))
```

The *lm* function of R then executes a linear regression and creates the object 'heating.lm', whose *summary* function can be used to extract the main results and diagnostics of the regression.

```
> heating.lm <- lm(consumpt~temperat+insulatio,data=heating)
> summary(heating.lm)

Call:
lm(formula = consumpt ~ temperat + insulatio, data = heating)

Residuals:
     Min        1Q      Median         3Q         Max
-143.921   -63.025     1.599     52.781     202.574

Coefficients:
              Estimate   Std. Error    t value     Pr(>|t|)
(Intercept)   1467.643     62.422      23.512     2.09e-11 ***
temperat       -37.060      2.295     -16.147     1.67e-09 ***
insulatio      -29.774      3.492      -8.526     1.95e-06 ***
---
Signif. codes: 0 '***' 0.001 '**' 0.01 '*' 0.05 '.' 0.1 ' ' 1

Residual standard error: 98.63 on 12 degrees of freedom
Multiple R-squared: 0.9655, Adjusted R-squared: 0.9598
F-statistic: 167.9 on 2 and 12 DF, p-value: 1.685e-09
```

The plot function is used to extract the four graphs shown in Figure 11.50 from the object 'heating.lm'. It must be preceded by an instruction par(mfrow=c(n,m)), which displays $n \times m$ graphs on the same page, in n rows and m columns. The upper graphs enable us to check the hypotheses of homoscedasticity (the graph of residuals as a function of the predicted values) and normality (the graph of studentized residuals as a function of the quantiles of the normal distribution) of the residuals.

```
par(mfrow=c(2,2))
plot(heating.lm)
```

We can see that the sign of the coefficients is negative, meaning that the heating consumption decreases as the insulation thickness and external temperature increase. This check on the signs may appear to be trivial – and in this case it is – but it can allow us to detect any of the major anomalies which may always occur in databases (unexpected signs may also be due to interactions between the predictors).

More generally, when we have obtained a regression equation, we must examine the meaning of this equation and especially that of the constant. Sometimes an undesired negative constant appears in a linear regression. For example, if we are modelling a cost of a takeover, the constant may represent the fixed cost. In some cases, the constant may be negative although no interpretation with a negative value can be found. We must then check to see if this is due to the presence of aberrant points which should be excluded. We can also see if it is possible to add a variable (even one having only a moderate significance) to the model in order to change the sign of the constant. Or we can attempt to segment the population to replace the single regression equation with one equation for each segment (for each type of service, in the example of the takeover cost).

The resulting coefficients enable us to calculate the regression equation, which was our aim. We must also ensure that the hypotheses of the linear model have been confirmed.

Since this is a multiple linear regression, we must start by ensuring that there is no strong collinearity between the predictors. For this purpose, we calculate the Pearson linear regression coefficient as shown in Figure 11.49 (a more general test, valid when more than two predictors are present, is described in the next section). We have introduced the dependent variable (consumption) into this correlation calculation, enabling us to check not only that the two predictors 'insulation' and 'temperature' are practically uncorrelated, but also that these two predictors are reasonably strongly correlated (negatively, of course) with consumption. Note that the correlation coefficients -0.465 and -0.870 are close to the regression coefficients.

11.7.4 Tests on residuals

The three hypotheses on the residuals must then be checked: these are the normality of their distribution, the equality of their variance (homoscedasticity) and the absence of autocorrelation (no correlation between ε_i and ε_{i+1}). An example discussed below will show why these checks are so important and how they complement the analyses already carried out on the F-ratio, R^2, etc.

When the number of observations is not too large, graphic observation is a very useful way of making these checks. The graphics often show the *standardized* residuals r_i, which are the

Correlations

		consumption	temperature	insulation
Pearson Correlation	consumption	1.000	-.870	-.465
	temperature	-.870	1.000	.009
	insulation	-.465	.009	1.000
Sig. (1-tailed)	consumption		.000	.040
	temperature	.000		.488
	insulation	.040	.488	
N	consumption	15	15	15
	temperature	15	15	15
	insulation	15	15	15

Figure 11.49 Correlation coefficients.

residuals e_i standardized to allow for their variance $s^2(1 - h_i)$ (see above):

$$r_i = \frac{e_i}{\hat{s}\sqrt{1 - h_i}}.$$

In this formula, the variance of the error s^2 is replaced by its estimator,

$$\hat{s}^2 = \frac{ESS}{n - p - 1}.$$

We also define the studentized residuals which are calculated as shown above but with the usual estimator of s^2 replaced by the estimator found by making the adjustment without the ith observation, so that this estimator is independent of e_i.

If the standardized residual of an observation does not lie between -2 and $+2$, then it is significantly different from 0 at the 5% level. Remember that, under the hypotheses of the linear model, the residuals e_i follow a normal distribution $N(0, s^2(1 - h_i))$. Such a residual may indicate an aberrant observation, an outlier.

The normality of the distribution of the residuals is tested as mentioned in Section 3.6, using graphics, P-P plots and statistical tests such as the Kolmogorov–Smirnov test.

A visual tool for detecting heteroscedasticity (non-equality of variances) and autocorrelation of the residuals is the *residual plot*. The standardized residuals are represented as a function of the dependent variable Y (or of the predicted variable $\hat{Y}$). This diagram also allows us to check at a glance that no standardized residual exceeds the critical value of 2. In the preceding example (see Figure 11.50), one of the residuals is slightly greater than 2 and must be watched carefully. This point will be discussed further below.

To test the equality of the variance of the residuals, we can also use the Levene test, grouping the values of Y into classes. The residual plot shown on the right of Figure 11.52 is

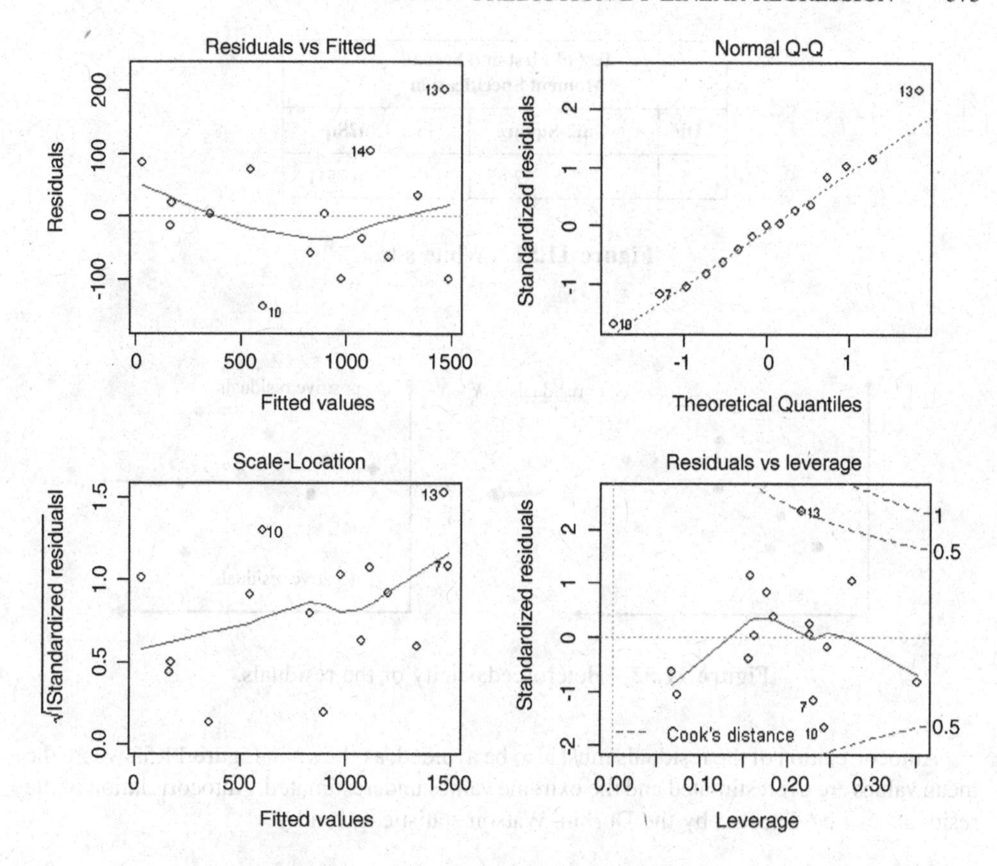

Figure 11.50 Diagrams of the residuals.

not acceptable, because the estimation of Y given X is accurate when Y is small, but inaccurate when Y is large. It is not uncommon for the variance to increase with X.

In SAS/STAT, there is an option of the linear regression procedure REG that can test the null hypothesis of the equality of the variance of the residuals. This is the SPEC option, which launches the White test:

```
PROC REG DATA=heating;
MODEL consumption = temperature insulation / SPEC;
```

This gives a result indicating that homoscedasticity is present when the probability (0.7541 in this case; see Figure 11.51) is greater than 0.05. The test is not very powerful for small samples (it is rare to have a probability of less than 0.05), but we can remember that a higher probability means a more satisfactory state of affairs.

When heteroscedasticity is present, it is advisable to replace the ordinary least squares with the *weighted least squares* (see Section 11.7.12 on robust regression), or to replace Y with its logarithm, square root or inverse (when the variance increases, as in Figure 11.52), or with its exponential or square (when the variance decreases). For example, we can regress log(Y) on X. Another cause of heteroscedasticity may be the lack of an essential variable in the model, as we shall see below, returning to the example of heating consumption.

Test of First and Second Moment Specification		
DF	**Chi2-Square**	**Pr > Chi2Sq**
5	2.65	0.7541

Figure 11.51 White's test.

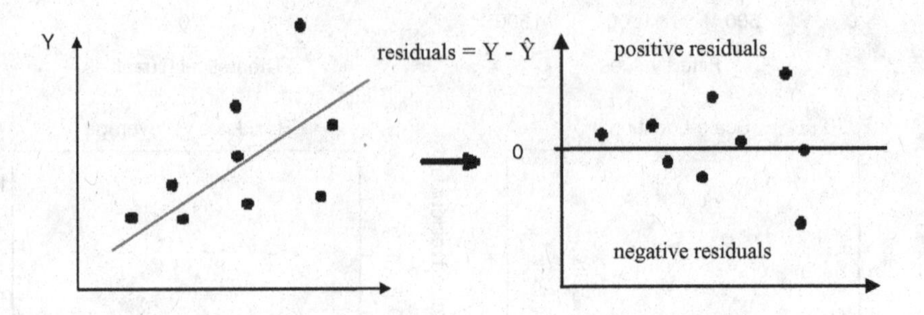

Figure 11.52 Heteroscedasticity of the residuals.

Autocorrelation of the residuals must also be avoided, as shown in Figure 11.53 where the mean values are overestimated and the extreme values underestimated. Autocorrelation of the residuals can be detected by the Durbin–Watson statistic, given by

$$\frac{\sum_{i=2}^{n} (\varepsilon_i - \varepsilon_{i-1})^2}{\sum_{i=1}^{n} \varepsilon_i^2}.$$

This ranges from 0 to 4, but must be close to 2 for the autocorrelation to be acceptable. It is less than 2 for positive correlations (because the successive values of the residuals are close to each other) and greater than 2 for negative correlations. As a general rule, the Durbin–Watson statistic should lie between 1.5 and 2.5. This test is valid even if the distribution of the

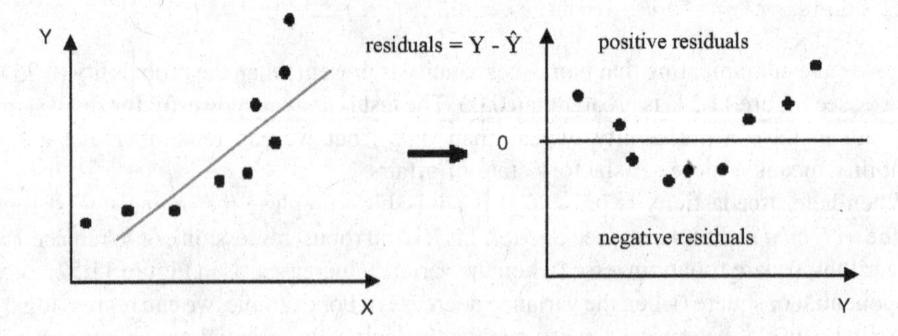

Figure 11.53 Autocorrelation of the residuals.

residuals is not normal. In the preceding example (Figure 11.47) we saw that the Durbin–Watson statistic is satisfactory (= 1.819). If autocorrelation is present, we can try to replace X with X^2 and regress Y on X^2, as in the example shown below of regression of *yb* on *x* according to Tomassone *et al.*[36]

11.7.5 The influence of observations

The last point to check concerning the quality of the prediction is the effect of outliers and aberrant observations.

We have already seen that a high standardized residual, especially one above the critical threshold of 2 in absolute terms, can be related to an aberrant observation. However, it is possible for an aberrant observation not to be associated with a large residual, and other criteria must be considered. Note that the converse is also true: in our example, observation 13, which has a standardized residual of 2.3, also has the highest value of Y, but this value is not actually outside the norm.

A criterion that is different from the standardized residual is the leverage introduced in Section 11.7.2. For each observation, this measures the difference between the observation and the mean, and the resulting effect of the observation on the fit. Since the leverages have a mean value of $(p + 1)/n$ (see above), we should not trust an observation if its leverage exceeds $2(p + 1)/n$. In our example, this threshold is 0.4. Figure 11.50 shows that no lever exceeds this threshold, even if one of them is close to it. Note that this is not observation 13. This demonstrates that the leverage and the residual are not completely equivalent criteria, but rather complementary. The *hatvalues* function in R can find the leverage values, in the same way as the INFLUENCE option of the SAS REG procedure (see below).

```
> lm <- lm(consumpt~temperat+insulatio,data=heating)
> hatvalues(lm)
```

1	2	3	4	5	6	7	8
0.15652928	0.18512746	0.17591508	0.24761044	0.16241773	0.07412309	0.23097134	0.35140711
9	10	11	12	13	14	15	
0.22686321	0.24432253	0.27596573	0.06770907	0.21699936	0.15726471	0.22677387	

Another way of measuring the effect of an individual is to remove it and then see if the regression coefficients remain within the confidence intervals of the initial coefficients. The *Cook's distance* of an observation measures the exact deviation of the vector of the coefficients with and without this observation, and it is generally best to be wary of Cook distances of more than 1. This value of 1 corresponds to the limit value of the Fisher–Snedecor distribution $F_{p,n-p-1}$ as n and p tend to infinity: the distribution is then concentrated towards this value. Given that p is always much smaller than infinity, we can refine the threshold of 1 by allowing for the real value of p and assuming that an observation has a very large effect when its Cook's distance exceeds the median value of $F_{p,n-p-1}$ (or $F_{p,\infty}$ if there are numerous observations). The Cook's distance is found for each observation when this is requested in IBM SPSS Statistics, or by using the R option in the REG procedure of SAS/STAT; this option also provides the real value to be predicted for each observation, the predicted value, the

[36] Tomassone, R., Lesquoy, É. and Millier, C. (1983) *La Régression: nouveaux regards sur une ancienne méthode statistique*. Paris: Masson.

standard error on the prediction of the mean, the residual, the standard error of the residual, the standardized residual which is the ratio of the two preceding quantities, and finally the position of the standardized residual in the range $[-2; +2]$ (Figure 11.54).

```
PROC REG DATA=heating;
MODEL consumption = temperature insulation / R INFLUENCE;
```

In the example of Figure 11.54, since $p=2$ and $n=15$, we examine the values of the distribution $F_{2,12}$, whose median is 0.73, as calculated using tables available on the Internet (http://geai.univ-brest.fr/~carpenti/statistiques/table1.php#fisher) or by using the R software:

```
> qf(0.5,2,12)
[1] 0.7347723
```

We can see that no observation has a Cook's distance exceeding the median 0.73 of $F_{2,12}$, the largest value being 0.498. If the number of observations tended towards infinity, the distribution of $F_{p,\infty}$ would be examined, and the median would then be 0.69, which is still higher than all the Cook's distances.

In most predictive models, there are a lot of observations and from six to eight variables, and the Cook's distance therefore has to be compared with the median of the distribution $F_{6,\infty}$, (i.e. 0.89) or $F_{8,\infty}$ (i.e. 0.92), these medians being close to the threshold of 1 which is normally recommended.

Note that some authors suggest a more prudent threshold of $4/(n-p-1)$, which in this case is 0.33; the thirteenth observation ($D=0.498$) exceeds this threshold and the tenth observation ($D=0.304$) is close to it.

<table>
<tr><th colspan="10">Output statistics</th></tr>
<tr><th>Obs.</th><th>Dependent Variable</th><th>Predicted Value</th><th>Std Error Mean Predict</th><th>Residual</th><th>Std Error Residual</th><th>Student Residual</th><th>-2-1 0 1 2</th><th>Cook's D</th></tr>
<tr><td>1</td><td>1042</td><td>1078</td><td>39.0205</td><td>-36.2933</td><td>90.580</td><td>-0.401</td><td>| | |</td><td>0.010</td></tr>
<tr><td>2</td><td>1377</td><td>1345</td><td>42.4356</td><td>31.8726</td><td>89.031</td><td>0.358</td><td>| | |</td><td>0.010</td></tr>
<tr><td>3</td><td>622.0000</td><td>548.3107</td><td>41.3663</td><td>73.6893</td><td>89.533</td><td>0.823</td><td>| |* |</td><td>0.048</td></tr>
<tr><td>4</td><td>154.0000</td><td>170.0991</td><td>49.0772</td><td>-16.0991</td><td>85.549</td><td>-0.188</td><td>| | |</td><td>0.004</td></tr>
<tr><td>5</td><td>355.4006</td><td>357.0000</td><td>39.7477</td><td>1.5994</td><td>90.263</td><td>0.0177</td><td>| | |</td><td>0.000</td></tr>
<tr><td>6</td><td>974.3076</td><td>874.0000</td><td>26.8517</td><td>-100.3076</td><td>94.901</td><td>-1.057</td><td>| **| |</td><td>0.030</td></tr>
<tr><td>7</td><td>1388</td><td>1489</td><td>47.3995</td><td>-101.4457</td><td>86.490</td><td>-1.173</td><td>| **| |</td><td>0.138</td></tr>
<tr><td>8</td><td>1138</td><td>1204</td><td>58.4656</td><td>-66.2779</td><td>79.429</td><td>-0.834</td><td>| *| |</td><td>0.126</td></tr>
<tr><td>9</td><td>896.6775</td><td>900.0000</td><td>46.9761</td><td>3.3225</td><td>86.721</td><td>0.0383</td><td>| | |</td><td>0.000</td></tr>
<tr><td>10</td><td>603.9215</td><td>460.0000</td><td>48.7503</td><td>-143.9215</td><td>85.736</td><td>-1.679</td><td>| ***| |</td><td>0.304</td></tr>
<tr><td>11</td><td>119.0000</td><td>33.1726</td><td>51.8111</td><td>85.8274</td><td>83.922</td><td>1.023</td><td>| |** |</td><td>0.133</td></tr>
<tr><td>12</td><td>829.7724</td><td>770.0000</td><td>25.6637</td><td>-59.7724</td><td>95.229</td><td>-0.628</td><td>| *| |</td><td>0.010</td></tr>
<tr><td>13</td><td>1670</td><td>1467</td><td>45.9435</td><td>202.5736</td><td>87.272</td><td>2.321</td><td>| |**** |</td><td>0.498</td></tr>
<tr><td>14</td><td>1223</td><td>1119</td><td>39.1121</td><td>103.9404</td><td>90.540</td><td>1.148</td><td>| |** |</td><td>0.082</td></tr>
<tr><td>15</td><td>177.7077</td><td>199.0000</td><td>46.9669</td><td>21.2923</td><td>86.726</td><td>0.246</td><td>| | |</td><td>0.006</td></tr>
</table>

Figure 11.54 Standardized residuals and Cook distance of the observations.

Output statistics								
						DFBETAS		
Obs.	**Residual**	**RStudent**	**Hat Diag H**	**Cov Ratio**	**DFFITS**	**Intercept**	**insolatio**	**temperat**
1	-36.2933	-0.3862	0.1565	1.4782	-0.1664	-0.1577	0.1260	0.0007
2	31.8726	0.3446	0.1851	1.5427	0.1643	0.1534	-0.1139	-0.0646
3	73.6893	0.8112	0.1759	1.3234	0.3748	-0.1703	0.2953	-0.0063
4	-16.0991	-0.1804	0.2476	1.7103	-0.1035	0.0072	-0.0135	-0.0883
5	1.5994	0.0170	0.1624	1.5499	0.0075	0.0015	-0.0006	0.0057
6	-100.3076	-1.0626	0.0741	1.0459	-0.3007	0.0331	-0.1611	0.0891
7	-101.4457	-1.1935	0.2310	1.1716	-0.6541	0.0369	-0.2692	0.5501
8	-66.2779	-0.8231	0.3514	1.6731	-0.6059	-0.3415	0.1272	0.4282
9	3.3225	0.0367	0.2269	1.6786	0.0199	-0.6915	0.0139	-0.0094
10	-143.9215	-1.8373	0.2443	0.7697	-1.0447	0.6390	-0.6915	-0.6263
11	85.8274	1.0248	0.2760	1.3639	0.6327	-0.2920	0.3947	0.3811
12	-59.7724	-0.6111	0.0677	1.2599	-0.1647	-0.0831	0.0195	-0.0064
13	202.5736	2.9939	0.2170	0.2774	1.5761	1.4015	-1.0069	-0.8320
14	103.9404	1.1650	0.1573	1.0867	0.5032	0.4811	-0.3801	-0.0345
15	21.2923	0.2357	0.2268	1.6539	0.1276	-0.0611	0.0880	0.0605

Figure 11.55 Residuals, leverages and DFBETAS of the observations.

When the influential observations have been detected, using the Cook's distance, the variables responsible for this influence can be identified more precisely, by means of the DFBETAS indicator defined for each variable. In SAS, this is obtained by the option INFLUENCE in the REG procedure; in R, it is found by the *dfbetas* function.

The DFBETAS indicator is the standardized difference between the estimated coefficient and the coefficient estimated without the observation. We need to examine any DFBETAS in excess of $2/\sqrt{n}$, a threshold which is equal to 0.52 in this case. The table obtained with SAS (Figure 11.55) shows values of DFBETAS which exceed this threshold for observations 10 and 13 (and also observation 7 for temperature), which were also those closest to the critical threshold of the Cook's distance and exceeded or approached the critical threshold of 2 in absolute terms for standardized residuals.

11.7.6 Example of linear regression

The example published by Tomassone et al. (cited in Section 11.7.4), which is similar to that of F. J. Anscombe,[37] clearly shows the importance of residual analysis, and of graphic

[37] Anscombe, F.J. (1973) Graphs in statistical analysis. *The American Statistician*, 27(1), 17–21.

x	ya	yb	yc	yd	xe	ye
7	5.535	0.113	7.399	3.864	13.72	5.654
8	9.942	3.77	8.546	4.942	13.72	7.072
9	4.249	7.426	8.468	7.504	13.72	8.491
10	8.656	8.792	9.616	8.581	13.72	9.909
12	10.74	12.69	10.69	12.22	13.72	9.909
13	15.14	12.89	10.61	8.842	13.72	9.909
14	13.94	14.25	10.53	9.919	13.72	11.33
14	9.45	16.55	11.75	15.86	13.72	11.33
15	7.12	15.62	11.68	13.97	13.72	12.75
17	13.69	17.21	12.75	19.09	13.72	12.75
18	18.10	16.28	13.89	17.20	13.72	12.75
19	11.29	17.65	12.59	12.33	13.72	14.16
19	21.37	14.21	15.04	19.76	13.72	15.58
20	15.69	15.58	13.74	16.38	13.72	15.58
21	18.98	14.65	14.88	18.95	13.72	17.00
23	1769	13.95	29.43	12.19	33.28	27.44

Figure 11.56 Example of linear regression, provided by Tomassone *et al.*

visualization in general. They construct a data set (Figure 11.56) which is the starting point for several regression analyses.

They studied the regression of *ya* on *x*, *yb* on *x*, *yc* on *x*, *yd* on *x*, and, finally, *ye* on *xe*. On each occasion, they found the same sums of squares, the same variance of the residuals RMSE^2, the same coefficient of variation (100 times RMSE divided by the mean of the dependent variable), the same *F*-ratio, the same R^2 and adjusted R^2, the same regression line, the same standard error of the coefficients, etc. (Figure 11.57).

However, the situations are very different, as shown by the dispersion diagrams in Figure 11.58, which show, reading from left to right and from top to bottom, *ya* cross-tabulated with *x*, *yb* cross-tabulated with *x*, *yc* cross-tabulated with *x*, *yd* cross-tabulated with *x*, and, finally, *ye* cross-tabulated with *xe*. Although the first cloud of points appears to fall within the scope of linear regression, this is clearly not the case with the clouds formed by *yb* and *yd* with *x* (on the right in the figure), which indicate a non-linear relationship. As for the clouds formed by *yc* with *x*, and *ye* with *xe,* each of these has one abnormal observation which artificially increases (*yc* with *x*) or decreases (*ye* with *xe*) the slope of the regression line. In the case of *yc* with *x*, the R^2 (equal to the square of the linear correlation coefficient of *x* and *yc* here) would be virtually equal to 1 without the abnormal observation, because the other observations are almost perfectly linearly related. Conversely, in the case of *ye* with *xe*, the abnormal observation causes a high correlation of 0.81 to appear. Although all the parameters calculated above are identical in the five regressions, we suspect that the analyses on the residuals will give very different results. This will certainly be instructive!

Analysis of variance					
Source	DF	Sum of Squares	Mean Square	F Value	Pr > F
Model	1	234.6	234.6	22.6	0.0003
Error	14	145.4	10.4		
Corrected Total	15	380.1			

Root MSE	3.22	R-Square	0.62
Dependent Mean	12.60	Adj R-Sq	0.59
Coeff Var	25.60		

Parameter Estimates					
Variable	DF	Parameter Estimate	Standard Error	t Value	Pr > |t|
Intercept	1	0.52	2.67	0.20	0.8476
x	1	0.81	0.17	4.75	0.0003

Figure 11.57 Statistics of the regressions provided by Tomassone *et al.*

For this purpose, a function introduced in SAS 9.1, ODS GRAPHICS, can be used to obtain the residual plot and several other graphic diagnostic elements,[38] simply by inserting the PROC REG syntax, which we have already considered, between the commands:

```
ODS HTML;
ODS GRAPHICS ON;
```

and

```
ODS GRAPHICS OFF;
ODS HTML CLOSE;
```

In order to be active, the ODS GRAPHICS instructions must always be associated with an HTML, PDF or RTF destination. The RTF destination enables the graph to be plotted in a Word document, while the HTML destination is used to display the graph in the SAS results window. Note that the SAS/GRAPH licence is essential for ODS GRAPHICS, even if an independent Java driver is used, rather than SAS/GRAPH procedures such as GPLOT or GCHART. The interested reader should consult the SAS reference[39] and *Reporting avec SAS* by Olivier Decourt,[40] which are full of useful information on ODS and ODS GRAPHICS. ODS GRAPHICS has been accompanied since SAS 9.1.3 by a new SAS language, GTL (*Graph Template Language*), for modifying the graphs produced by the procedures and also for creating

[38] And many other graphics for many procedures (more than 40 in SAS 9.2): CORR, FREQ, UNIVARIATE, BOXPLOT, NPAR1WAY, TTEST, CORRESP, PRINCOMP, FACTOR, CLUSTER, ANOVA, RSREG, ROBUSTREG, LOESS, LOGISTIC, GAM, GENMOD, GLM, MI, MIXED, ARIMA, LIFETEST, etc.

[39] Haworth, L.E., Zender C.L., Burlew M.M. (2009) *Output Delivery System: The Basics and Beyond*. SAS Publishing.

[40] Decourt, O. (2008) *Reporting avec SAS*. Paris: Dunod.

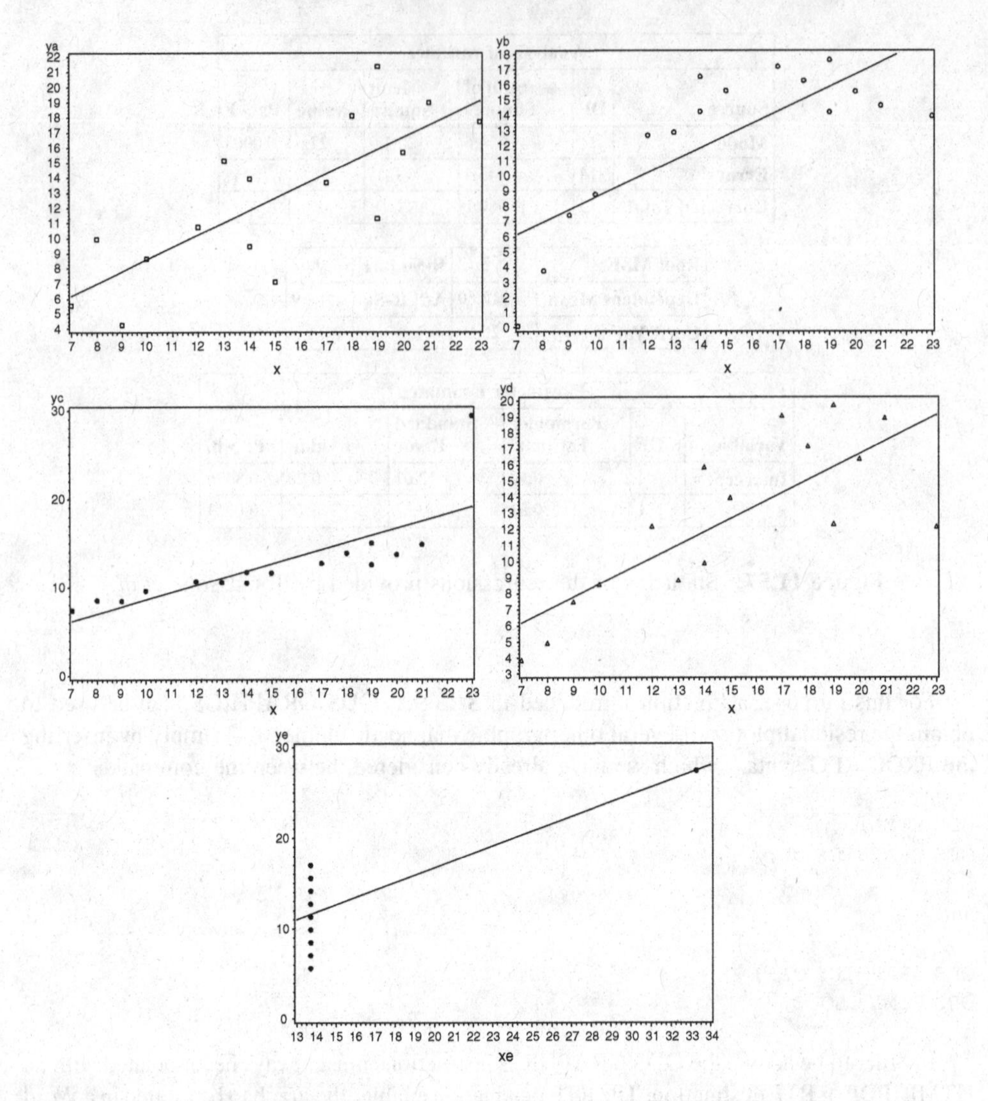

Figure 11.58 Dispersion diagrams of the regressions provided by Tomassone *et al.*

new graphs in a very flexible way, offering many more options than the standard ones provided in the procedures (an example of application will be given in Section 12.7).

In the REG procedure, the result is a finely crafted graph. An example is shown in Figure 11.59 for the regression of *ya* on *x*. The three basic hypotheses on the residuals (normality of distribution, equality of variance and absence of autocorrelation) are evidently satisfied. This can be confirmed by the calculation of the Durbin–Watson statistic.

Durbin–Watson D	2.538
Number of Observations	16
1st Order Autocorrelation	−0.277

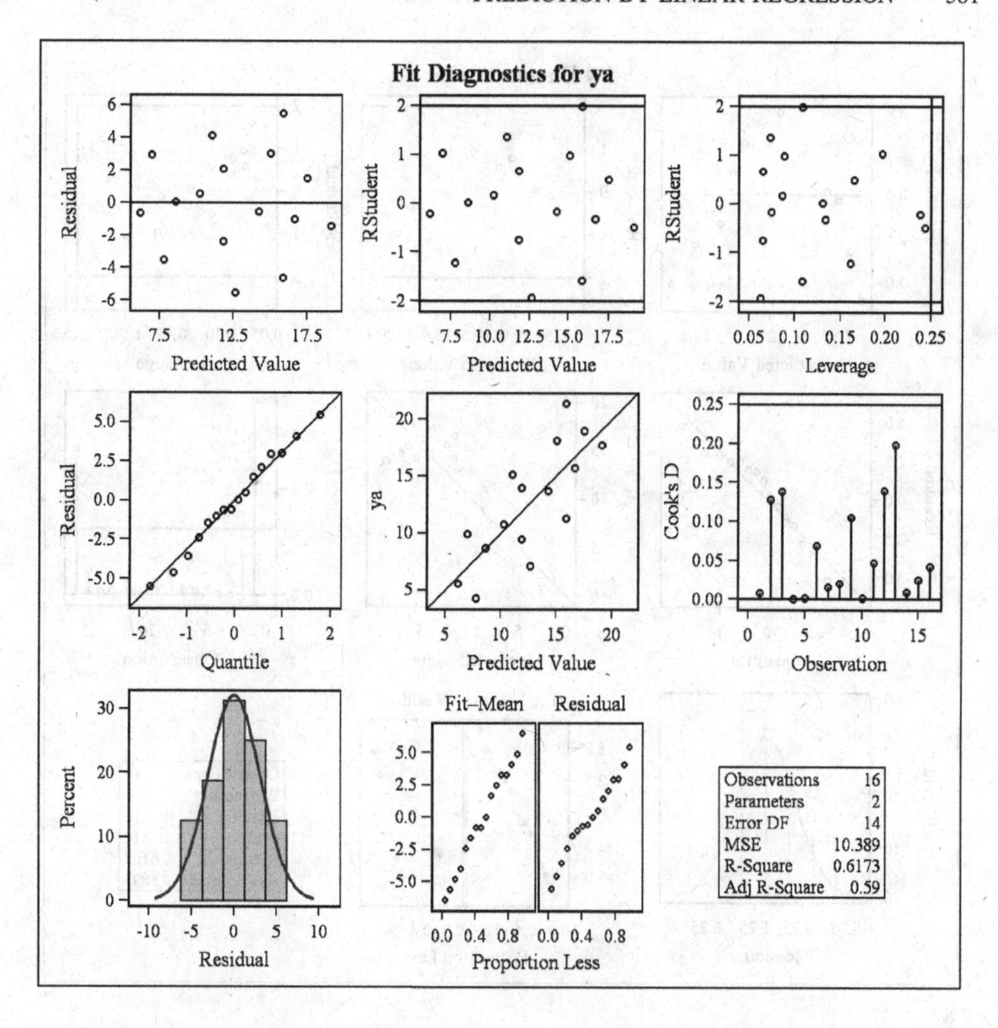

Figure 11.59 ODS GRAPHICS for the first regression of Tomassone *et al.*

This shows only a small negative autocorrelation.

For the regression of *yb* on *x*, the ODS graph shows a typical case of positive autcorrelation of the residuals (see the upper left-hand corner of Figure 11.60), confirmed by the Durbin–Watson statistic.

Durbin-Watson D	0.374
Number of Observations	16
1st Order Autocorrelation	0.595

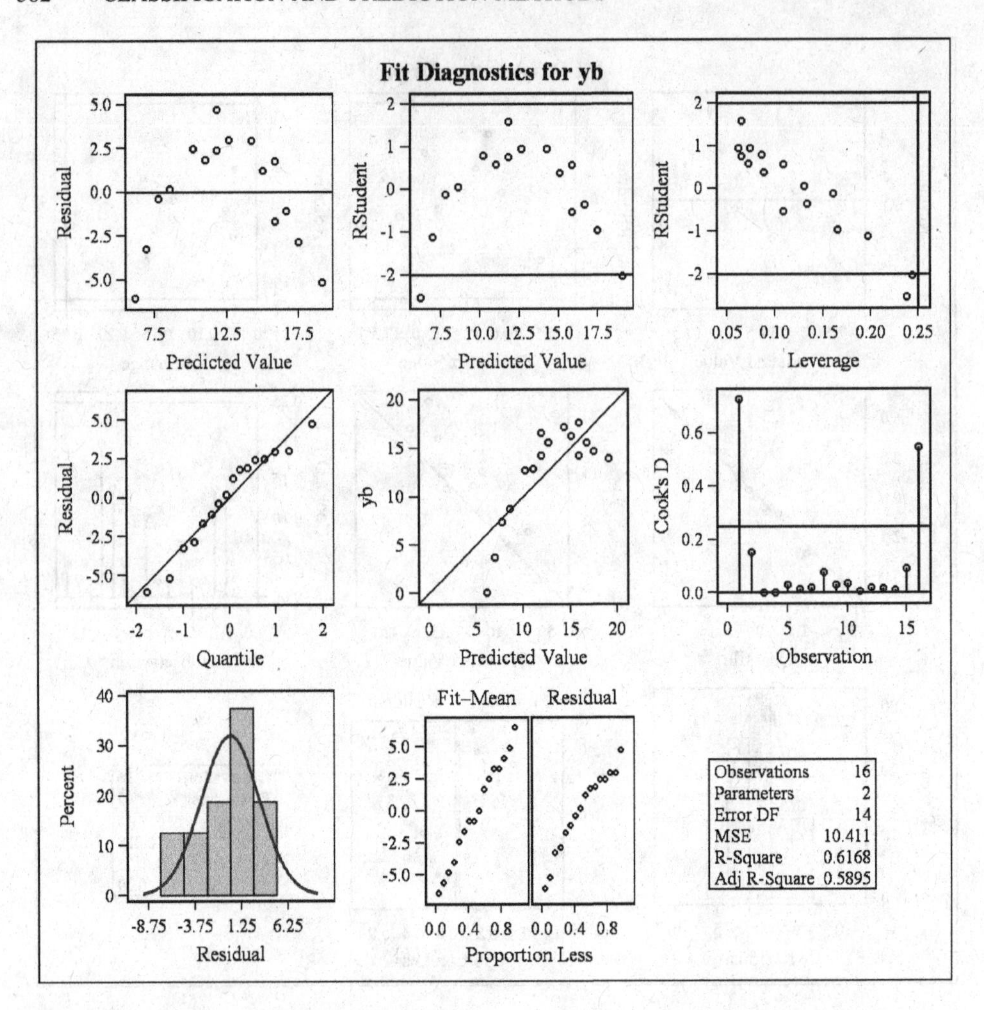

Figure 11.60 ODS GRAPHICS for the second regression of Tomassone *et al.*

It would be preferable to regress *yb* on x^2, or use the GLM procedure:

```
PROC GLM DATA= tomassone;
MODEL yb = x x*x;
RUN ;
```

In the regression of *yc* on *x*, the standardized residuals are all very close to 0, with the exception of the sixteenth observation. A notable element of the ODS graph (Figure 11.61) is the representation of the Cook's distance, which very clearly exceeds the limit value 1 for the sixteenth observation.

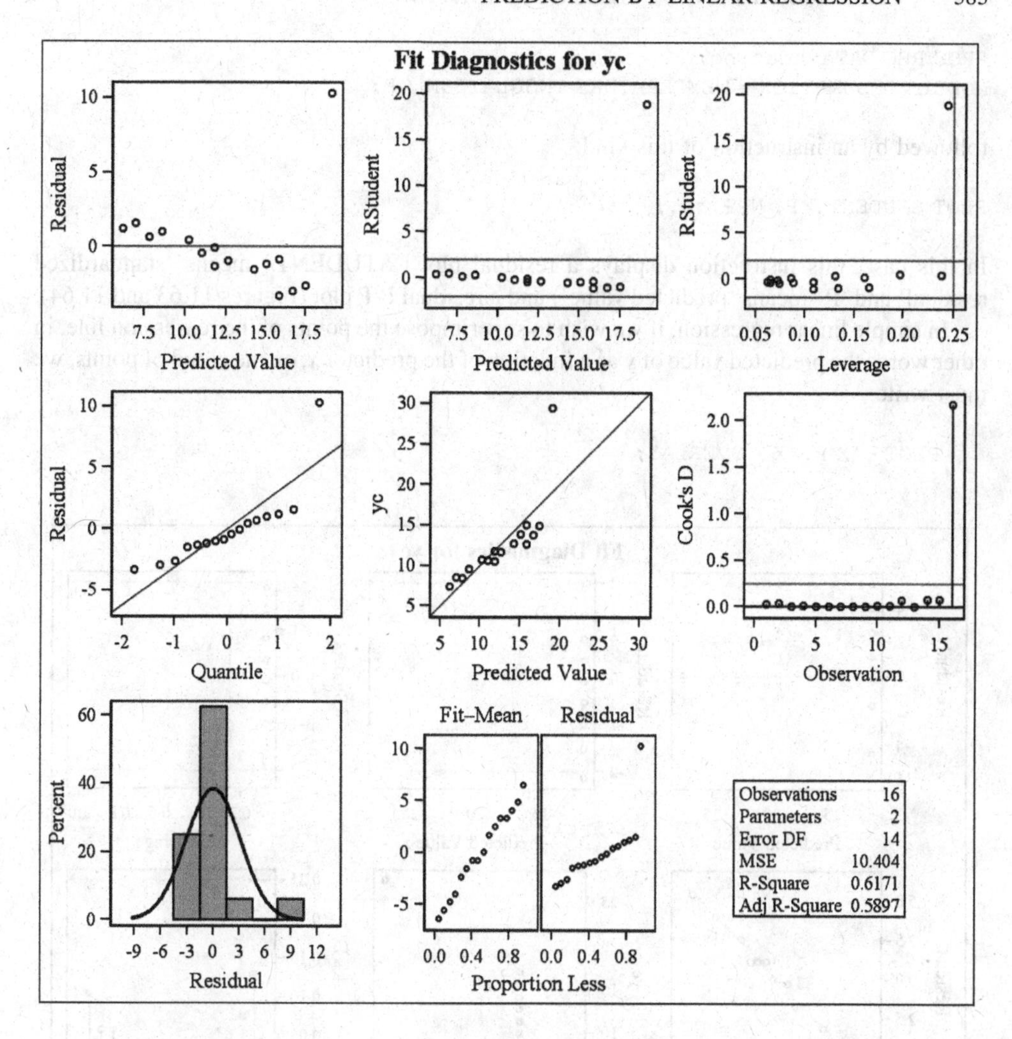

Figure 11.61 ODS GRAPHICS for the third regression of Tomassone *et al.*

We will skip the regression of *yd* on *x* and consider the regression of *ye* on *xe*, which is a limit case of heteroscedasticity of the residuals, as shown by the residual plot in Figure 11.62 (but even without this, there would be no doubt).

11.7.7 Further details of the SAS linear regression syntax

Even without ODS GRAPHICS, PROC REG in SAS/STAT can be used to obtain diagrams of residuals and many other graphic diagnostic elements. We can simply use the normal syntax:

```
PROC REG DATA=heating;
MODEL CONSUMPTION = TEMPERATURE INSULATION / R;
```

followed by an instruction of this kind:

```
PLOT STUDENT.*P. NPP.*R.;
```

In this case, this instruction displays a residual plot ('STUDENT.' means 'standardized residual' and 'P.' means 'predicted value') and a residual P-P plot (Figures 11.63 and 11.64).

In simple linear regression, if we wish to superimpose the points of the regression line, in other words the predicted value of y as a function of the predictor *x*, on the cloud of points, we must write:

```
PLOT Y*X P.*X / OVERLAY;
```

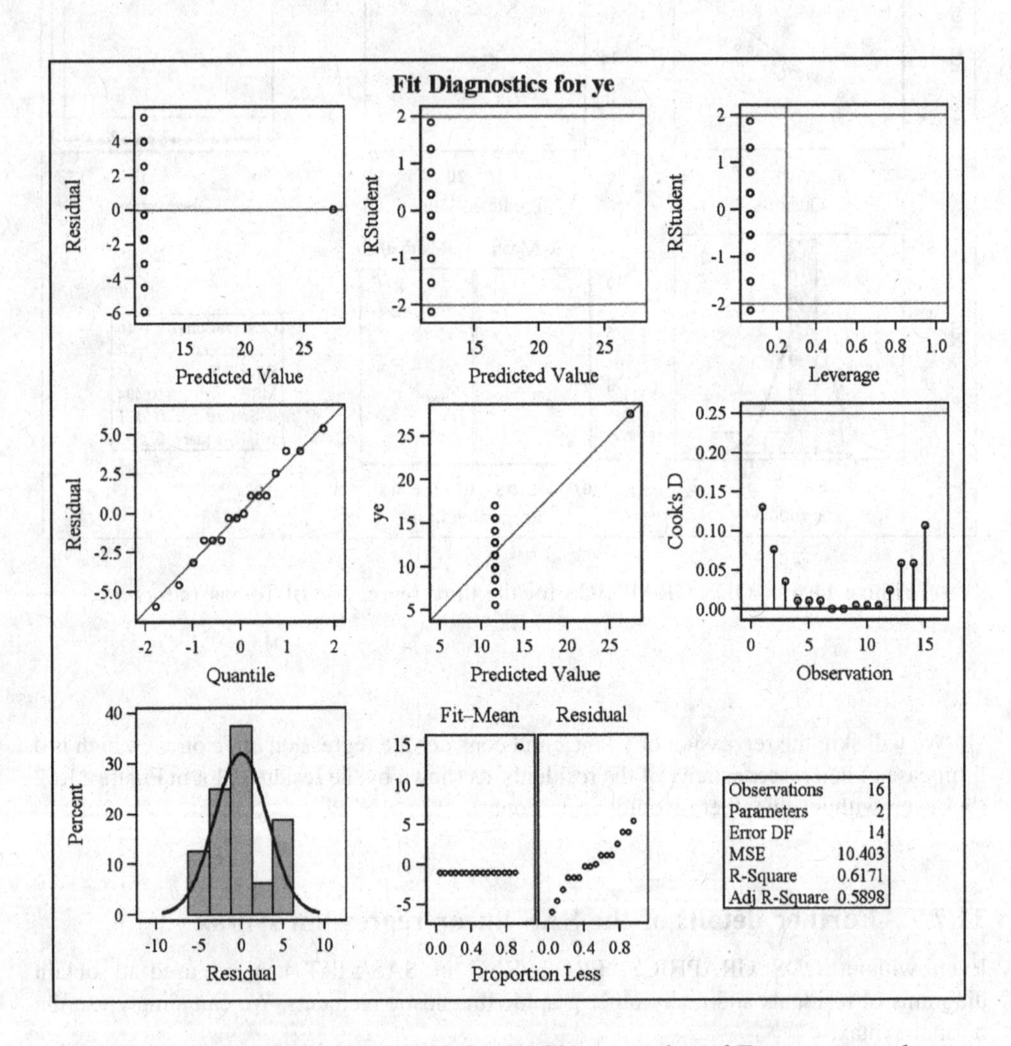

Figure 11.62 ODS GRAPHICS for the fifth regression of Tomassone *et al.*

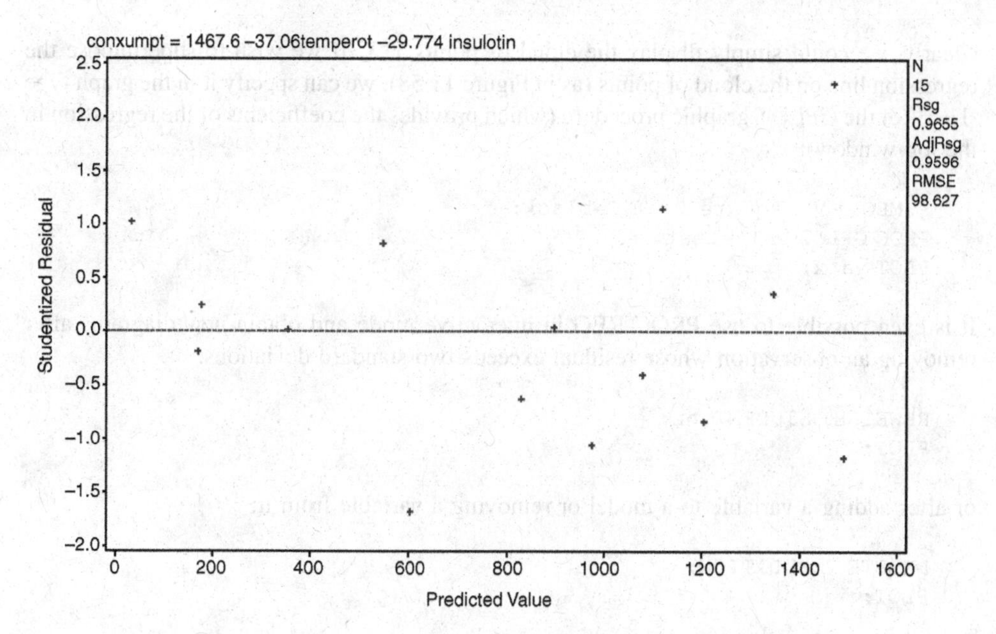

Figure 11.63 Residual plot output by PROC REG.

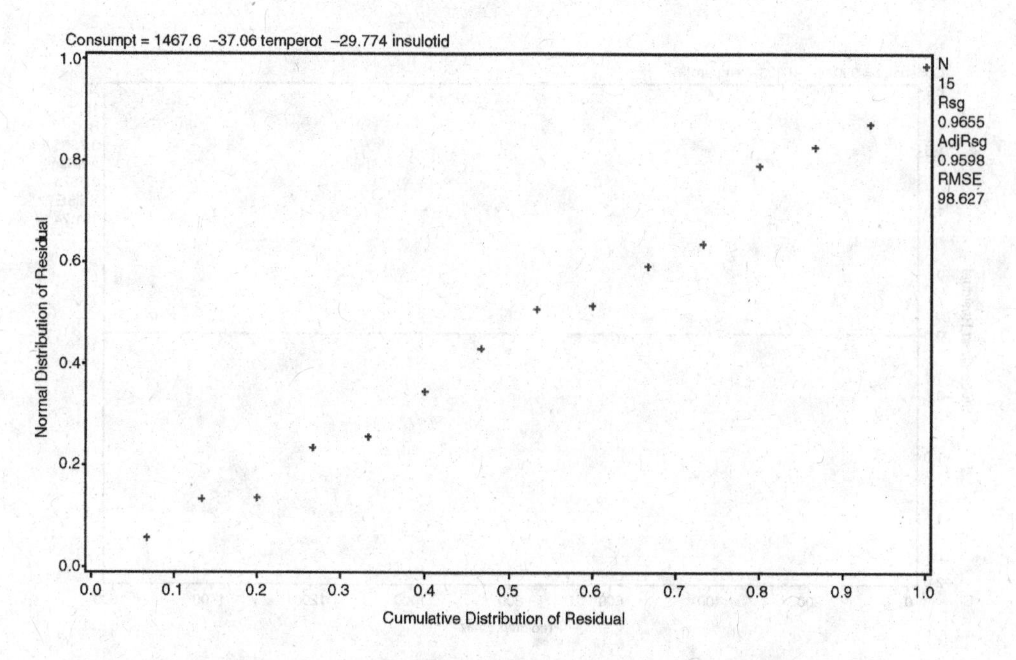

Figure 11.64 Residual P-P plot output by PROC REG.

Clearly, we could simply display the cloud of points Y^*X. If we wish to superimpose the regression line on the cloud of points (as in Figure 11.58), we can specify it in the graph type 'I = rl' of the GPLOT graphic procedure (which provides the coefficients of the regression in the log window):

```
SYMBOL1 V = square I =rl C=black;
PROC GPLOT DATA=test;
PLOT ya*x;
```

It is even possible to use PROC REG in interactive mode and obtain new diagrams after removing an observation whose residual exceeds two standard deviations:

```
REWEIGHT STUDENT.>2;
PLOT;
```

or after adding a variable to a model or removing a variable from it:

```
DELETE INSULATION;
PLOT;
```

The model is automatically recalculated each time and the diagram is redisplayed. Figure 11.65 shows the possible cost of removing a variable from the model. Without the variable 'insulation', R^2 has markedly decreased, and the residual plot also shows a tendency towards heteroscedasticity of the residuals. If such heteroscedasticity appears, we should try to establish whether a new variable could be added to the model. The heteroscedasticity can be quantified by the probability associated with the White test, which is now much less than the

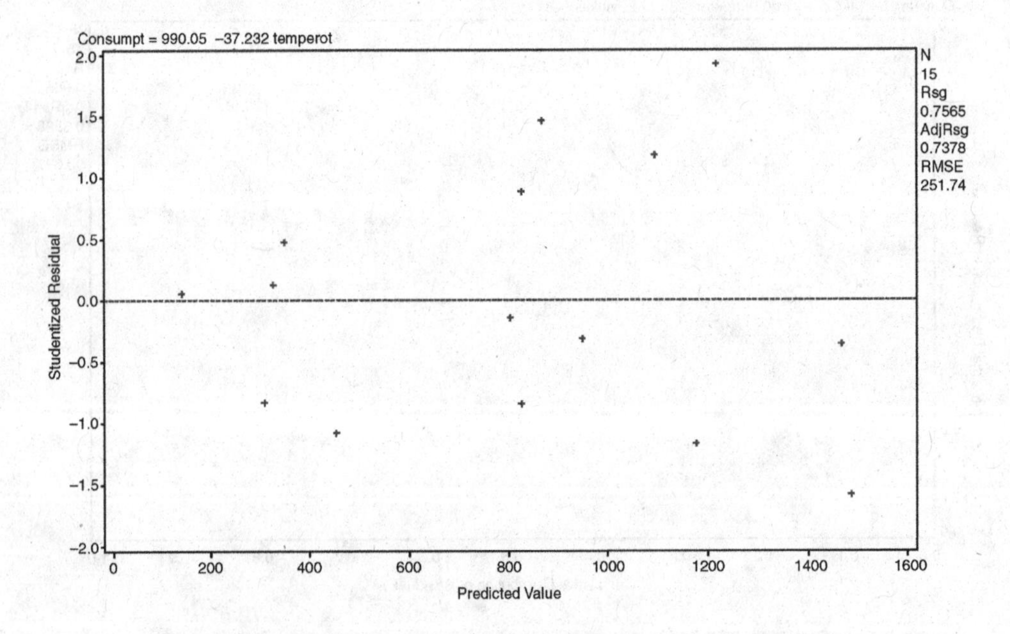

Figure 11.65 Residual plot after the exclusion of a predictor.

probability of 0.7541 found previously (see Figure 11.51) with the two predictors 'insulation' and 'temperature':

Test of First and Second Moment Specification		
DF	Chi-Square	Pr > Chi-Sq
2	3.61	0.1642

In interactive mode, the RUN instructions do not quit PROC REG, and another procedure, a DATA step, is required, or the instruction

```
QUIT;
```

11.7.8 Problems of collinearity in linear regression: an example using R

When a number of independent variables are (strongly) linearly correlated with each other, this has the following results:

- estimators, in other words regression coefficients, which are unstable because they are highly sensitive to even small variations of the variable to be regressed;
- regression coefficients which may be very large, if they 'compensate' each other;
- inverted signs on regression coefficients in some cases, making their interpretation incorrect and in any case counter-intuitive;
- high variances for the regression coefficients;
- reduced reliability of the predictions.

We will illustrate these phenomena with a simple, but striking, example provided by Bernadette Govaerts, included in her STAT2430 course at the Catholic University of Louvain (see the University's website).

There are four observations, on which a linear regression is to be carried out on two variables *X1* and *X2*, the variable to be regressed being first *Y*, and then *Z* in a second regression. The values of these variables are as follows:

$$X1 = \begin{pmatrix} 1 \\ 2 \\ 3 \\ 4 \end{pmatrix} \quad X2 = \begin{pmatrix} 1.01 \\ 1.99 \\ 3.01 \\ 3.99 \end{pmatrix} \quad Y = \begin{pmatrix} 16 \\ 34 \\ 44 \\ 46 \end{pmatrix} \quad Z = \begin{pmatrix} 17 \\ 34 \\ 44 \\ 46 \end{pmatrix}.$$

It would be hard to find more closely correlated variables than *X1* and *X2*.

These variables are entered into the R software in the following way, in the form of a data frame as before.

```
> colin <-
data.frame(X1=c(1,2,3,4),X2=c(1.01,1.99,3.01,3.99),Y=c(16,34,44,46 ),
Z=c(17,34,44,46))
```

The correlations are checked with the cor function. They are all positive and very close to 1.

```
> cor(colin)
           X1          X2          Y           Z
X1  1.0000000   0.9999677   0.9415545   0.9450493
X2  0.9999677   1.0000000   0.9415241   0.9450971
Y   0.9415545   0.9415241   1.0000000   0.9998956
Z   0.9450493   0.9450971   0.9998956   1.0000000
```

We then calculate the two linear regressions of *Y* on *X1* and *X2*, then *Z* on *X1* and *X2*.

```
> colin1.lm <- lm(Y~X1+X2,data=colin)
> colin2.lm <- lm(Z~X1+X2,data=colin)
```

The results of the first regression show that the coefficients are not significantly different from 0, that of *X1* being equal to 1 and that of *X2* being practically zero. However, there is a very high R^2 of 0.89.

```
> summary(colin1.lm)
Call:
lm(formula = Y ~ X1 + X2, data = colin)

Residuals:
 1   2   3   4
-4   4   4  -4

Coefficients:
              Estimate     Std. Error    t value    Pr(>|t|)
(Intercept)   1.000e+01    1.077e+01     0.928      0.524
X1            1.000e+01    4.454e+02     0.022      0.986
X2           -7.754e-12    4.472e+02    -1.73e-14   1.000

Residual standard error: 8 on 1 degrees of freedom
Multiple R-squared: 0.8865,      Adjusted R-squared: 0.6596
F-statistic: 3.906 on 2 and 1 DF, p-value: 0.3369
```

The results of the second regression should be very similar, given that the variable *Z* is equal to *Y*, except for the first observation, which is 17 instead of 16. The coefficients are still not significantly different from 0. However, they are completely different from those found for *Y*. The coefficient of *X1* has become negative, while that of *X2* is not zero at all.

```
> summary(colin2.lm)

Call:
lm(formula = Z ~ X1 + X2, data = colin)

Residuals:
    1      2      3      4
-3.75   3.75   3.75  -3.75
```

```
Coefficients:
                Estimate    Std. Error    t value    Pr(>|t|)
(Intercept)      10.88        10.10        1.077      0.476
X1               -2.75       417.60       -0.007      0.996
X2               12.50       419.26        0.030      0.981

Residual standard error: 7.5 on 1 degrees of freedom
Multiple R-squared: 0.8932, Adjusted R-squared: 0.6796
F-statistic: 4.182 on 2 and 1 DF, p-value: 0.3268
```

This example illustrates the instability of the coefficients which may even result in the inversion of the signs. Of course, it is not normal to have a negative sign for the coefficient of *X1*, which is positively correlated with *Z*. Note that R^2 is always very high, which clearly shows that this does not provide a guarantee of a good fit.

This also shows the usefulness of an explicit model as opposed to a 'black box' model, the archetype of which is the neural network. A fit can be found and result in a high global indicator of fit (R^2 in this case) which might satisfy us. However, it may happen, as in this case, that this fit cannot be relied on. This can be detected if we apply the model to another data set, but there is not always another comparable data set, or there may not be enough time, and in any case the detection of an unsatisfactory application of the model does not reveal the cause. However, when the coefficients and their standard errors are available, any difficulties are immediately revealed.

The instability of fit in the presence of closely correlated variables is very evident in the graphs, and R can generate an excellent 3D graphic display, as shown in Figures 11.66 and 11.67:

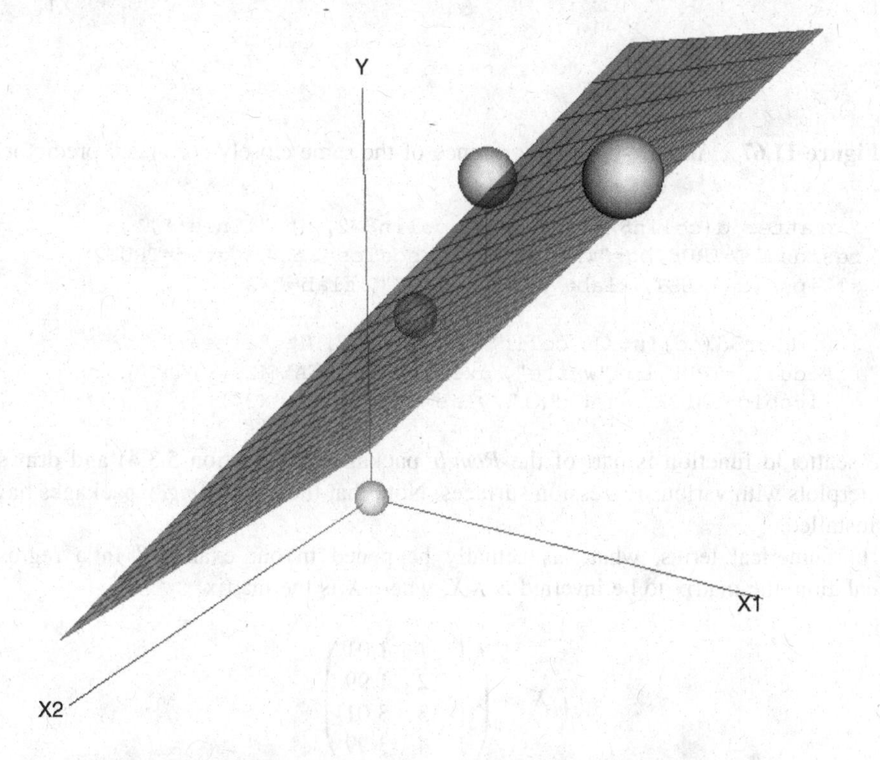

Figure 11.66 Fit in the presence of closely correlated predictors.

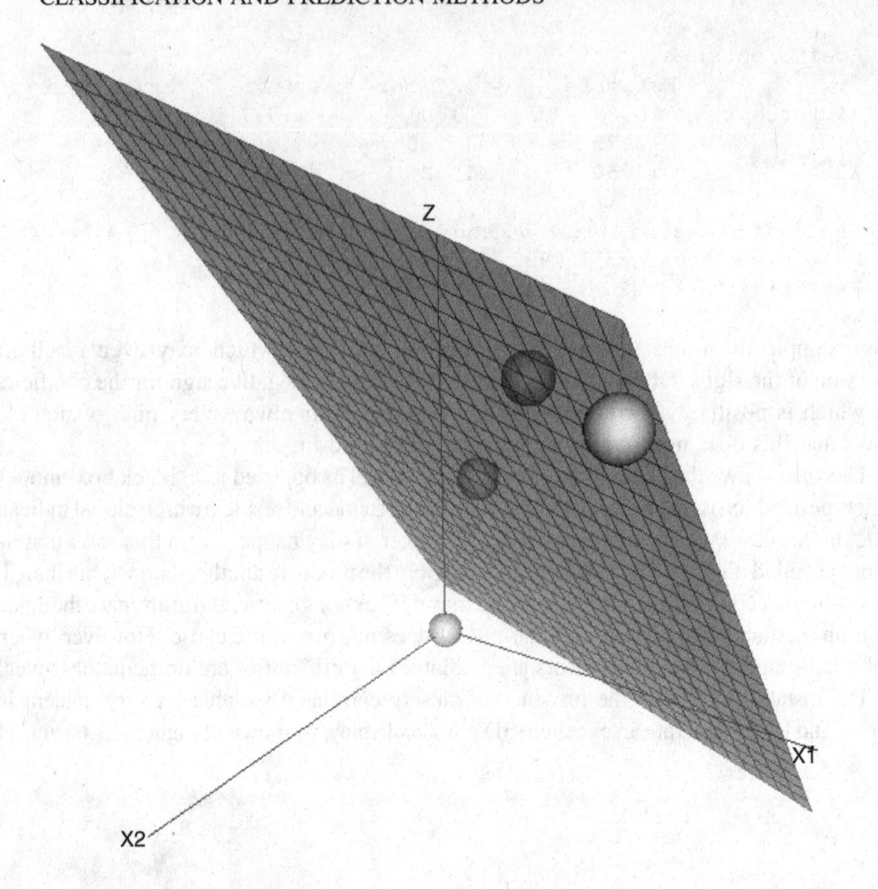

Figure 11.67 Another fit in the presence of the same closely correlated predictors.

```
> scatter3d(colin$X1, colin$Y, colin$X2, fit="linear",
residuals=TRUE,bg="white", axis.scales=FALSE, grid=TRUE,
ellipsoid=FALSE, xlab="X1",ylab="Y", zlab="X2")

> scatter3d(colin$X1, colin$Z, colin$X2, fit="linear",
residuals=TRUE,bg="white", axis.scales=FALSE, grid=TRUE,
ellipsoid=FALSE, xlab="X1",ylab="Z", zlab="X2")
```

The scatter3d function is part of the *Rcmdr* package (see Section 5.3.4) and draws 3D scatterplots with various regression surfaces. Note that the *rgl* and *mgcv* packages have to be installed.

In numerical terms, what has actually happened in our example? In a regression calculation, the matrix to be inverted is $X'X$, where X is the matrix

$$X = \begin{pmatrix} 1 & 1 & 1.01 \\ 1 & 2 & 1.99 \\ 1 & 3 & 3.01 \\ 1 & 4 & 3.99 \end{pmatrix}$$

We can input this matrix into R on a column by column basis, which is the default option:

```
> X <- matrix(c(1,1,1,1,1,2,3,4,1.01,1.99,3.01,3.99),nrow=4)
```

This is the matrix after input:

```
> X
     [,1]  [,2]   [,3]
[1,]   1     1    1.01
[2,]   1     2    1.99
[3,]   1     3    3.01
[4,]   1     4    3.99
```

The matrix X^tX is as follows:

```
t(X) %*% X
     [,1]        [,2]        [,3]
[1,]    4      10.00      10.0000
[2,]   10      30.00      29.9800
[3,]   10      29.98      29.9604
```

As for the inverse $(X^tX)^{-1}$, this is obtained thus:

```
> solve(t(X) %*% X)
          [,1]          [,2]          [,3]
[1,] 1.8125        30.625        -31.25
[2,] 30.6250     3100.250      -3112.50
[3,] -31.2500   -3112.500       3125.00
```

Clearly, the inversion of the matrix is problematic, with very high coefficients.

This is where the basic principle of ridge regression can be helpful (see Section 11.7.2). This principle consists in 'translating' the diagonal of the matrix X^tX to be inverted by a value k, which is equivalent to calculating an estimate under the constraint of the norm of the vector of the parameters.

The regression `lm.ridge` is available in the MASS package of R, which must be loaded in advance:

```
> library(MASS)
```

Its syntax is similar to that of linear regression, to which is added the specification of the translation to be performed on the matrix X^tX. We can test a range of values, which in this case are all the values of k, denoted lambda, in the range from 0 to 0.1, in steps of 0.001.

```
> ridge <- lm.ridge(Y~X1+X2,data=colin,lambda=seq(0,0.1,0.001))
```

We obtain a vector of parameters for each value of lambda. The first 20 and the last of these are displayed below, showing a good stability of the coefficients when lambda = 0.02.

```
> coef(ridge)
                          X1               X2
0.000    10.000000   10.000000    -4.948835e-12
0.001     9.959026    5.570771     4.445619e+00
0.002     9.959483    5.301762     4.714445e+00
0.003     9.961648    5.204298     4.811042e+00
0.004     9.964279    5.153741     4.860547e+00
0.005     9.967103    5.122655     4.890504e+00
0.006     9.970026    5.101509     4.910480e+00
0.007     9.973005    5.086121     4.924677e+00
0.008     9.976020    5.074365     4.935227e+00
0.009     9.979058    5.065048     4.943329e+00
0.010     9.982112    5.057448     4.949707e+00
0.011     9.985178    5.051102     4.954827e+00
0.012     9.988253    5.045699     4.959000e+00
0.013     9.991334    5.041025     4.962441e+00
0.014     9.994420    5.036924     4.965308e+00
0.015     9.997510    5.033284     4.967712e+00
0.016    10.000603    5.030017     4.969742e+00
0.017    10.003698    5.027060     4.971461e+00
0.018    10.006795    5.024360     4.972922e+00
0.019    10.009893    5.021878     4.974165e+00
0.020    10.012993    5.019580     4.975223e+00
...
0.100    10.259530    4.944714     4.951475e+00
```

This stability can be seen on the 'ridge trace' graph which shows the values of the parameter estimates as a function of the values of lambda.

```
> plot(ridge)
```

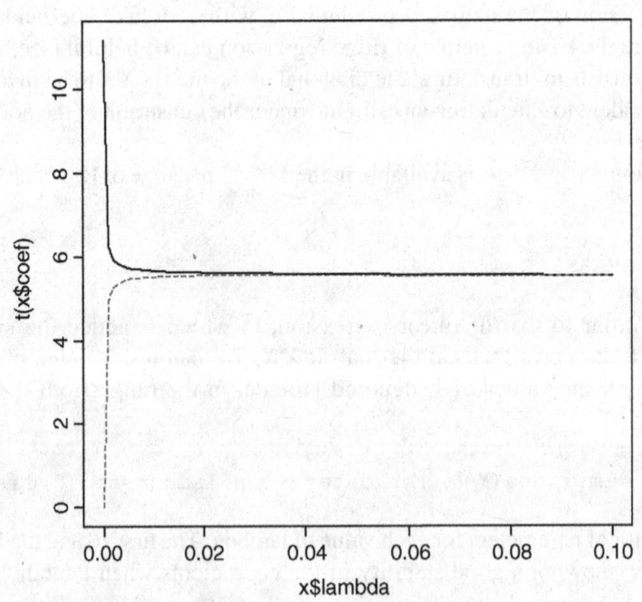

We then replace *Y* by *Z* and obtain similar results.

```
> ridge <- lm.ridge(Z~X1+X2,data=colin,lambda=seq(0,0.1,0.001))
> coef(ridge)
                           X1          X2
0.000     10.87500    -2.750000   12.500000
0.001     10.94508     3.981050    5.740919
0.002     10.95218     4.388398    5.330731
0.003     10.95668     4.534992    5.182336
0.004     10.96047     4.610285    5.105526
0.005     10.96397     4.655983    5.058430
0.006     10.96731     4.686572    5.026504
0.007     10.97057     4.708409    5.003365
0.008     10.97377     4.724725    4.985769
0.009     10.97693     4.737334    4.971895
0.010     10.98006     4.747334    4.960640
0.011     10.98318     4.755428    4.951300
0.012     10.98628     4.762088    4.943399
0.013     10.98937     4.767642    4.936610
0.014     10.99245     4.772324    4.930696
0.015     10.99552     4.776307    4.925484
0.016     10.99859     4.779722    4.920843
0.017     11.00165     4.782667    4.916673
0.018     11.00470     4.785222    4.912896
0.019     11.00776     4.787446    4.909452
0.020     11.01080     4.789389    4.906289
...
0.100     11.25059     4.780527    4.819236

> plot(ridge)
```

We find that the estimates of the coefficients are very similar for *Y* and *Z*, with values around 4.9 for the coefficients of *X1* and *X2*. In this case, this proximity of the values is quite compatible with the proximity of *Y* and *Z*.

We can use the value which seems appropriate for lambda, namely 0.02. We first create the diagonal matrix to be used for the translation of the matrix $X'X$ to be inverted. Note that ridge regression does not usually penalize the constant. In fact, if the constant were penalized, this would mean that the results of the ridge would depend on the origin of *Y*, and the translation of all the values of the dependent variable would not result in a similar translation of the predictions. Therefore it is not exactly the diagonal matrix [0.02 0.02 0.02], but rather the matrix [0 0.02 0.02] which is used to translate $X'X$.

```
> kI = diag(3)*0.02
> kI[1,1] <- 0
> kI

       [,1]  [,2]  [,3]
[1,]     0   0.00  0.00
[2,]     0   0.02  0.00
[3,]     0   0.00  0.02
```

We can check that the inverse matrix $(X^tX + kI)^{-1}$ does not include any more very high values:

```
> solve(t(X) %*% X + kI)
             [,1]          [,2]          [,3]
[1,]   1.504920281   -0.004969981   -0.4969981
[2,]  -0.004969981   24.752494931  -24.7505069
[3,]  -0.496998131  -24.750506938   24.9493062
```

Finally, we check that the products $(X^tX + kI)^{-1}X^tY$ and $(X^tX + kI)^{-1}X^tZ$ do actually provide us with the estimates of the vectors of coefficients, according to the formula given in Section 11.7.2:

```
> solve(t(X) %*% X + kI) %*% t(X) %*% Y
          [,1]
[1,]   10.000994
[2,]    5.049501
[3,]    4.950101
> solve(t(X) %*% X + kI) %*% t(X) %*% Z
          [,1]
[1,]   10.998976
[2,]    4.799014
[3,]    4.901396
```

11.7.9 Problems of collinearity in linear regression: diagnosis and solutions

Collinearity is usually measured with Pearson's linear correlation coefficient of the variables considered in pairs. If there are more than two predictors, however, we have seen that a more sophisticated and rigorous measurement is the variance inflation factor (VIF) and its inverse, the *tolerance* (see Section 3.14), which, for a variable X_i, is $1 - (R^2$ of the linear regression of X_i on the other variables). According to the usual criteria (see Section 3.14), the tolerance should be greater than 0.2, or at least 0.1. We can also look at the condition indices of the correlation matrix as mentioned in Section 3.14. These tests are requested in the REG procedure of SAS/STAT by adding the keywords TOL and COLLIN (and VIF for the variance inflation factor). Let us return to the previous example.

```
DATA COLIN;
INFILE DATALINES DELIMITER = ',' ;
INPUT X1 X2 Y Z ;
DATALINES;
1,1.01,16,17
2,1.99,34,34
3,3.01,44,44
4,3.99,46,46
RUN ;
```

```
PROC REG DATA = COLIN;
MODEL Y = X1 X2 / TOL VIF COLLIN ;
RUN ;
```

Very small values of tolerance and very large values of the VIF and condition index are a sign of collinearity (Figure 11.68).

If we return to the earlier example of the regression of energy consumption on external temperature and insulation thickness, we will obtain diametrically opposite results (Figure 11.69).

```
PROC REG DATA=heating;
MODEL consumption = temperature insulation / TOL VIF COLLIN;
RUN ;
```

In our example, the tolerance is much greater than 0.2, and the condition indices are therefore all less than 5. An absence of collinearity between the predictors is confirmed in this case, and supports the calculation of the Pearson correlation coefficients carried out above.

There are various remedies for the collinearity of a number of variables:

- the removal of the variables concerned (allowing R^2 to decrease a little in order to reduce the collinearity);
- the creation of a synthetic variable combining and replacing the variables concerned (for example, the ratio of two variables);
- transformation (logarithm, etc.) of the variables concerned (see Section 3.9);
- biased ridge regression as described in Section 11.7.2, or another regression with regularization (such as lasso regression, also described in Section 11.7.2);

Parameter Estimates

Variable	DF	Parameter Estimate	Standard Error	t Value	Pr > |t|	Tolerance	Variance Inflation
Intercept	1	10.00000	10.77033	0.93	0.5236		0
X1	1	10.00000	445.43911	0.02	0.9857	0.00006451	15501
X2	1	2.22045E-11	447.21360	0.00	1.0000	0.00006451	15501

Collinearity Diagnostics

Number	Eigenvalue	Condition Index	Proportion of variation		
			Intercept	X1	X2
1	2.88483	1.00000	0.01527	0.00000127	0.00000126
2	0.11516	5.00497	0.81505	0.00001507	0.00001465
3	0.00000536	733.75856	0.16967	0.99998	0.99998

Figure 11.68 Measures of collinearity – example 1.

Parameter Estimates							
Variable	**DF**	**Parameter Estimate**	**Standard Error**	**t Value**	**Pr > |t|**	**Tolerance**	**Variance Inflation**
Intercept	1	1467.64327	62.42223	23.51	<.0001		0
insulation	1	-29.77430	3.49207	-8.53	<.0001	0.99992	1.00008
temperature	1	-37.06030	2.29515	-16.15	<.0001	0.99992	1.00008

Collinearity Diagnostics					
			Proportion of variation		
Number	**Eigenvalue**	**Condition Index**	**Intercept**	**insulation**	**temperature**
1	2.14485	1.00000	0.03251	0.03274	0.06767
2	0.76635	1.67295	0.01592	0.02171	0.92453
3	0.08879	4.91488	0.95157	0.94555	0.00780

Figure 11.69 Measures of collinearity – example 2.

- principal component regression, which involves regressing the dependent variable not on the initial variables, but on their principal components in a PCA, and then moving from principal component regression coefficients to coefficients on the initial variables;

- PLS regression, described in one of the following sections.

Ridge, lasso, principal component and PLS regression will be compared on an example in Section 11.7.11.

The most important requirement is to make a good selection of the independent variables of a linear regression. As for logistic regression or discriminant analysis, there are stepwise selection procedures implemented in software such as R, SAS and IBM SPSS Statistics:

- In forward selection there is no variable in the model to begin with, and we add, one by one, those which have the largest F-ratio (largest sum of squares of the model), which must be significant with an associated probability equal to or below a certain fixed threshold: this threshold is called the threshold 'to enter' (in the SAS REG procedure, its default value is 0.5 and it is denoted SLE). The process is interrupted when there are no more variables external to the model which have a significant F-ratio at the fixed threshold.

- In backward selection, we start by entering all the variables into the model and then reject, one by one, those which have the smallest F-ratios (the smallest sum of squares of the model). The variables are rejected if their F-ratio is not significant, in other words as long as the associated probability is at least equal to a certain fixed threshold: this threshold is called the threshold 'to stay' (in the SAS REG procedure, its default value is 0.1 and it is denoted SLS). The procedure is interrupt when all the variables of the model have a significant F-ratio at the fixed threshold.

- In stepwise selection, there is no variable at the outset, and those having the largest F-ratios are added one by one, as in forward selection. However, at each step it is possible to remove one (and only one) of the variables if its F-ratio is no longer significant at the threshold 'to stay'. It is only after this operation that the next variable can be selected, if appropriate. The process is interrupted when there is no longer any variable external to the model with a significant F-ratio 'to enter' and when every variable in the model has a significant F-ratio 'to stay'.

- There are some cases in which the above methods do not provide the best possible selection of k variables (as in the example by Brenot, Cazes and Lacourly (1975)[41] mentioned in Section 4.5.2 of the course by Confais and Le Guen, cited in note 23 above). Another method, with a better performance than STEPWISE, is MAXR (Maximum R Improvement), implemented in the SAS software. This is also a stepwise method, which aims to maximize R^2 for each value of the number of independent variables. MAXR starts by choosing the variable that gives the greatest R^2, and then adds the one that causes the greatest increase in R^2. When this two-variable model has been established, all the possible permutations between one of the two variables of the model and an external variable are tested, the R^2 of the regression is calculated, and the permutation that is carried out is the one that provides the maximum increase in R^2. A third variable, which once again is the one that causes the greatest increase in R^2, is then added. This process of permutation of the variables is repeated until no further choice of variable increases R^2. The difference between the STEPWISE and MAXR methods is that all the possible permutations are evaluated in MAXR every time, but only the 'least good' variable is excluded in STEPWISE. Clearly, MAXR requires much more calculation.

Here is an example of the SAS syntax:

```
PROC REG DATA=heating;
MODEL consumption = temperature insulation / SELECTION = STEPWISE SLE =
0.05;
```

11.7.10 PLS regression

PLS regression finds a compromise between the two objectives of maximizing the explained variance of the predictors X_i (the principle of PCA) and maximizing the correlation between the X_i and the dependent variable Y (the principle of regression). To do this, we search for the linear combinations T_j of the X_i that maximize $\text{cov}^2(T_j,Y) = r^2(T_j,Y).\text{var}(T_j).\text{var}(Y)$. As in principal component regression, we perform a projection on linear combinations of the predictors that are not intercorrelated, but the PLS components differ from the principal components in that they optimize the correlation with the dependent variable at the same time as the variance of the predictors. In this sense, PLS regression is an extension of the finding that, in principal component regression, the components with the largest variance

[41] Brenot, J., Cazes, P. and Lacourly, N. (1975) Pratique de la régression: Qualité et protection. *Cahiers du BURO*, 23, 181.

are not necessarily the most predictive. A first conclusion to be drawn from this was that it might be best to select the principal components that were most correlated with the dependent variable, even if these were not the first in decreasing order of variance. PLS regression refines this idea by incorporating the constraint of correlation with the response in the determination of the components.

The advantage of PLS regression is that it can be used when there are a large number of variables showing collinearity, and even if the number of independent variables is greater than the number of observations. It has also been shown (de Jong, 1993)[42] that PLS on p components is always more predictive (the prediction is more correlated with the dependent variable) than regression on the first p principal components. We use cross-validation (see below) to determine a number p of components which are both small enough to avoid overfitting and large enough to explain most of the variance of the X_i (objective 1) and Y (objective 2). p is rarely higher than 3 or 4. The PLS regression algorithm is fast, because it consists of a sequence of simple regressions, without inversion or diagonalization of matrices. It is therefore efficient when large volumes of data are processed.

These properties make PLS regression particularly useful in chemistry, spectrometry, the petroleum industry, cosmetics, biology, medicine and the food industry. For example, in cosmetics, it allows the user to retain all the ingredients of a product, which represent a very large number of independent variables. In the food industry, it is used for sensory analysis, in which the classification of a product by a number of tasters (variable Y) is explained as a function of its physical, chemical and flavour properties (which may be up to several hundred in number). This method is starting to appear in statistical software, and is found in SAS, SPAD and R (*pls* package), in the specialist The Unscrambler product (Camo) and especially in the SIMCA (Soft Independent Modeling of Class Analogy) software which is the standard product in this field.

PLS regression was invented by Svante Wold (1983),[43] following the initial work of his father Herman Wold (1966)[44] on the PLS approach in structural equation models. This regression was followed by PSL2 regression, developed to predict a number of continuous dependent variables Y_j simultaneously, even if these variables are more numerous than the observations. This work is being continued by Michel Tenenhaus and others,[45] together with work on methods concerned with PLS logistic regression as described in Section 11.9.5, and the Cox-PLS model for survival data with independent variables that are strongly correlated or more numerous than the observations.

The PLS regression algorithm will now be outlined. The first step is to find a combination $T_1 = \Sigma_i \lambda_{1i} X_i$ of the X_i which maximizes both the variance of T_1 and the correlation between T_1 and Y. This is equivalent to maximizing the square of the covariance:

$$\mathrm{cov}^2(T_1, Y) = r^2(T_1, Y) \cdot \mathrm{var}(T_1) \cdot \mathrm{var}(Y).$$

[42] de Jong, S. (1993) PLS fits closer than PCR. *Journal of Chemometrics*, 7, 551–557.

[43] Wold, S., Martens, H. and Wold, H. (1983) The multivariate calibration problem in chemistry solved by the PLS method. In B. Kågström and A. Ruhe (eds), *Matrix Pencils*, Lecture Notes in Mathematics 973, pp. 286–293. Berlin: Springer.

[44] Wold, H. (1966) Estimation of principal component and related models by iterative least squares. In P.R. Krishnaiah (ed.), *Multivariate Analysis*, pp. 391–420. New York: Academic Press.

[45] See Tenenhaus, M. (1998) *La régression PLS: théorie et pratique*. Paris: Technip.

The solution is provided by the calculation of the covariance $\lambda_{1i} = \text{cov}(Y,X_i)$ followed by the normalization of the vector λ_{1i} so that $\|(\lambda_{11},\ldots, \lambda_{1p})\| = 1$. Thus we have

$$T_1 = \frac{1}{\sqrt{\sum_i \text{cov}(Y,X_i)^2}} \sum_i \text{cov}(Y,X_i) \cdot X_i.$$

The regression of Y on T_1 gives a residual Y_1,

$$Y = c_1 T_1 + Y_1,$$

while the regression of X_i on T_1 gives the residuals X_{1i},

$$X_i = c_{1i} T_1 + X_{1i}.$$

In a second step, the same operation is repeated, replacing Y with its residual Y_1 and replacing the X_i with their residuals X_{1i}. We obtain a combination $T_2 = \Sigma_i \lambda_{2i} X_{1i}$ with $\|(\lambda_{21},\ldots, \lambda_{2p})\| = 1$ and λ_{2i} proportional to $\text{cov}(Y_1,X_{1i})$. Then we regress Y_1 on T_2 and the X_{1i} on T_2, obtaining a residual Y_2 and residuals X_{2i}:

$$\begin{aligned} Y_1 &= c_2 T_2 + Y_2, \\ X_{1i} &= c_{2i} T_2 + X_{2i}. \end{aligned}$$

These operations are repeated until the number of components T_K gives a satisfactory result, which is checked by cross-validation as described below.

Finally, we can write

$$Y = c_1 T_1 + Y_1 = c_1 T_1 + c_2 T_2 + Y_2 = \ldots = \Sigma_j c_j T_j + \text{residual},$$

and this expression is replaced with an expression for the regression of Y as a function of the X_i in place of the T_j. The calculation of the PLS regression is then complete. Note that, by construction, the coefficients λ_{1i} of the first factor of the PLS regression have the same sign as that of the correlations between Y and the X_i. Therefore there is none of the kind of surprise that may occur with least squares linear regression. This property is not generalized to the set of coefficients of the X_i in the expression for Y when a number of factors are extracted. In fact, the coefficients λ_{2i} of T_2 with respect to the X_{1i} have the same sign as that of the correlations between Y_1 and the X_{1i}, but this sign is not necessarily that of the coefficients of T_2 with respect to the X_i.

Let us return to the way in which the choice of the number of components operates by cross-validation. At each step h, we wish to decide whether or not to retain the hth PLS component. This is done until we reach a step in which the PLS component is not retained.

For this purpose, we calculate the residual sum of squares (RESS_h), as in linear regression:

$$\text{RESS}_h = \sum_k \left(y_{(h-1),k} - \hat{y}_{(h-1),k} \right)^2$$

where $\hat{y}_{(h-1),k} = c_h t_{h,k}$ is the prediction of $y_{(h-1),k}$ calculated for each observation k.

The observations are then distributed into g groups, and the current step of the PLS algorithm is then executed g times on Y_{h-1} and the $X_{h-1,i}$, removing one group each time. We then calculate $PRESS_h$, the predicted residual sum of squares, which is similar to $RESS_h$ but avoids overfitting by replacing the prediction $\hat{y}_{(h-1),k}$ with the prediction $\hat{y}_{(h-1),-k}$ deduced from the analysis performed without the group containing the observation k. Thus we have

$$PRESS_h = \sum_k (y_{(h-1),k} - \hat{y}_{(h-1),-k})^2.$$

We retain the hth PLS component if $PRESS_h \leq \gamma \cdot RESS_{h-1}$. To define $RESS_0$, we postulate that it is equal to $\sum(y_i - \bar{y})^2$, where $\bar{y}$ is the mean of Y. The parameter γ ranges from 0 to 1, and we often specify that $\gamma = 0.95$ if $n < 100$, and $\gamma = 1$ if $n \geq 100$.

11.7.11 Handling regularized regression with SAS and R

Let us now compare ridge regression, principal component regression and PLS regression on a data set obtained from the book by Tomassone *et al.* cited in Section 11.7.4, on pine processionary caterpillars. We assume that the data set is available in an SAS table and is stored in the SAS XPORT format.

A conventional SAS7BDAT data set is converted to XPORT format as follows:

```
LIBNAME ridge 'C:\Users\Stéphane\Documents\Datamining\Data sets\
Data for linear regression' ;
LIBNAME To_R XPORT
'C:\Users\Stéphane\Documents\Datamining\Software
Datamining\R\data.xpt' ;

DATA To_R.data ;
SET ridge.caterpillars ;
RUN ;
```

This data set is then imported into R, using the *foreign* package (supplied with R) and the *Hmisc* package, which has to be installed, and which provides functions such as sasxport.get which can read the formatted values, the labels and the lengths of variables. *Hmisc* provides higher-performance imports than *foreign*.

```
> library(foreign) #Load the needed packages.
> library(Hmisc)

Package attachment: 'Hmisc'

    The following object(s) are masked from package:base

    format.pval,
    round.POSIXt,
    trunc.POSIXt,
    units

> caterpillars<-
sasxport.get("c:\\Users\\Stéphane\\Documents\\Datamining\\Software
Datamining\\R\\data.xpt")
```

Here are the data, displayed by the print command:

```
> print(caterpillars)
     x1  x2  x3  x4    x5  x6  x7   x8  x9 x10  x11         log
1  1200  22   1 4.0  14.8 1.0 1.1  5.9 1.4 1.4 2.37  0.86288996
2  1342  28   8 4.4  18.0 1.5 1.5  6.4 1.7 1.7 1.47  0.38526240
3  1231  28   5 2.4   7.8 1.3 1.6  4.3 1.5 1.4 1.13  0.12221763
4  1254  28  18 3.0   9.2 2.3 1.7  6.9 2.3 1.6 0.85 -0.16251893
5  1357  32   7 3.7  10.7 1.4 1.7  6.6 1.8 1.3 0.24 -1.42711636
6  1250  27   1 4.4  14.8 1.0 1.7  5.8 1.3 1.4 1.49  0.39877612
7  1422  37  22 3.0   8.1 2.7 1.9  8.3 2.5 2.0 0.30 -1.20397280
8  1309  46   7 5.7  19.6 1.5 1.3  7.8 1.8 1.6 0.07 -2.65926004
9  1127  24   2 3.5  12.6 1.0 1.7  4.9 1.5 2.0 3.00  1.09861229
10 1075  34   9 4.3  12.0 1.6 1.8  6.8 2.0 2.0 1.21  0.19062036
11 1166  24  17 5.5  16.7 2.4 1.5 11.5 2.9 1.7 0.38 -0.96758403
12 1182  41  32 5.4  21.6 3.3 1.4 11.3 2.8 2.0 0.70 -0.35667494
13 1179  15   0 3.2  10.5 1.0 1.7  4.0 1.1 1.6 2.64  0.97077892
14 1256  21   0 5.1  19.5 1.0 1.8  5.8 1.1 1.4 2.05  0.71783979
15 1251  26   2 4.2  16.4 1.1 1.7  6.2 1.3 1.8 1.75  0.55961579
16 1536  38  31 5.7  17.8 3.1 1.7 11.4 2.8 1.9 0.06 -2.81341072
17 1554  27  20 5.6  20.2 2.8 1.9  9.2 2.7 1.3 0.13 -2.04022083
18 1305  30   6 3.8  15.7 1.4 1.2  7.2 2.1 1.9 1.00  0.00000000
19 1316  34   8 3.1  11.4 1.5 1.8  5.0 1.6 2.0 0.41 -0.89159812
20 1427  39  19 4.6  15.2 2.4 1.6  9.1 2.4 1.9 0.72 -0.32850407
21 1575  20  32 5.2  18.9 3.0 1.7  9.4 2.5 1.8 0.67 -0.40047757
22 1397  26  16 4.2  14.8 2.2 1.6  7.7 2.2 1.8 0.12 -2.12026354
23 1377  29   4 5.3  19.8 1.2 1.8  6.8 1.6 1.9 0.97 -0.03045921
24 1574  24  23 5.2  17.8 2.4 1.8  7.8 2.2 2.0 0.07 -2.65926004
25 1396  45  13 4.7  15.2 1.7 1.6  7.8 2.1 1.4 0.10 -2.30258509
26 1393  27   5 4.7  18.3 1.2 1.7  7.5 1.7 2.0 0.68 -0.38566248
27 1433  23  18 6.5  21.0 2.7 1.8 13.7 2.7 1.3 0.13 -2.04022083
28 1349  24   1 2.7   5.8 1.0 1.7  3.6 1.3 1.8 0.20 -1.60943791
29 1208  23   2 3.5  11.5 1.1 1.7  5.4 1.3 2.0 1.09  0.08617770
30 1198  28  15 3.9  11.3 2.0 1.6  7.4 2.8 2.0 0.18 -1.71479843
31 1228  31   6 5.4  21.8 1.3 1.7  7.0 1.5 1.9 0.35 -1.04982212
32 1229  21  11 5.8  16.7 1.7 1.8 10.0 2.3 2.0 0.21 -1.56064775
33 1310  36  17 5.2  17.8 2.3 1.9 10.3 2.6 2.0 0.03 -3.50655790
```

We wish to predict the variable log using the functions x1, x2, x4 and x5. A correlation calculation shows that all the predictors are negatively correlated with the dependent variable log, and that x4 and x5 are strongly correlated.

```
> varselec <- subset(caterpillars,select=c(log,x1,x2,x4,x5))
> cor(varselec)
           log         x1         x2         x4         x5
log  1.0000000 -0.5336138 -0.4294398 -0.4252949 -0.2009383
x1  -0.5336138  1.0000000  0.1205209  0.3210528  0.2837739
x2  -0.4294398  0.1205209  1.0000000  0.1366877  0.1134163
x4  -0.4252949  0.3210528  0.1366877  1.0000000  0.9046552
x5  -0.2009383  0.2837739  0.1134163  0.9046552  1.0000000
```

We perform a linear regression and find that the coefficients of x4 and x5 are much larger than the others, and that the coefficient of x5 is inconsistent because it is positive. The risk of the inconsistency of signs in the presence of collinearity was pointed out previously and this has now occurred, since the coefficients of the collinear variables are very large and 'compensate' each other with opposite signs.

```
> lm <- lm(log~x1+x2+x4+x5,data=caterpillars)
> summary(lm)

Call:
lm(formula = log ~ x1 + x2 + x4 + x5, data = caterpillars)

Residuals:
Min      1Q Median       3Q       Max
-2.02086 -0.25012   0.09002    0.35179 1.71056

Coefficients:
              Estimate     Std.      t value      Pr(>|t|)
                           Error
(Intercept)   7.732144   1.488584     5.194      1.63e-05      ***
x1           -0.003924   0.001148    -3.419      0.001946      **
x2           -0.057343   0.019388    -2.958      0.006236      **
x4           -1.356138   0.319834    -4.240      0.000220      ***
x5            0.283058   0.076260     3.712      0.000905      ***

Signif. codes:   0 '***' 0.001   '**' 0.01   '*' 0.05   '.' 0.1 ' '

Residual standard error: 0.7906 on 28 degrees of freedom
Multiple R-squared: 0.6471, Adjusted R-squared: 0.5967
F-statistic: 12.83 on 4 and 28 DF, p-value: 4.677e-06
```

Logically, therefore, we attempt to carry out a ridge regression, as in Section 11.7.8, using the lm.ridge function.

```
> library(MASS)
> ridge <-
lm.ridge(log~x1+x2+x4+x5,data=caterpillars,lambda=seq(0,1,0.05))
> coef(ridge)
                            x1           x2          x4          x5
0.00  7.732144  -0.003923681  -0.05734270  -1.356138  0.2830582
0.05  7.716554  -0.003924173  -0.05732296  -1.336309  0.2782978
0.10  7.701162  -0.003924468  -0.05730150  -1.317100  0.2736848
0.15  7.685961  -0.003924575  -0.05727841  -1.298481  0.2692126
0.20  7.670941  -0.003924505  -0.05725376  -1.280427  0.2648748
0.25  7.656098  -0.003924265 -0.05722764  -1.262911  0.2606653
0.30  7.641423  -0.003923864  -0.05720012  -1.245910  0.2565785
0.35  7.626912  -0.003923310  -0.05717127  -1.229403  0.2526092
0.40  7.612557  -0.003922611  -0.05714116  -1.213367  0.2487522
0.45  7.598353  -0.003921772  -0.05710984  -1.197782  0.2450030
0.50  7.584296  -0.003920801  -0.05707738  -1.182631  0.2413569
0.55  7.570380  -0.003919704  -0.05704383  -1.167895  0.2378098
0.60  7.556600  -0.003918487  -0.05700925  -1.153558  0.2343577
```

```
0.65  7.542952  -0.003917154  -0.05697367  -1.139603  0.2309969
0.70  7.529433  -0.003915712  -0.05693715  -1.126015  0.2277236
0.75  7.516037  -0.003914165  -0.05689974  -1.112781  0.2245346
0.80  7.502760  -0.003912518  -0.05686147  -1.099886  0.2214266
0.85  7.489601  -0.003910776  -0.05682238  -1.087318  0.2183965
0.90  7.476554  -0.003908942  -0.05678252  -1.075065  0.2154414
0.95  7.463616  -0.003907021  -0.05674191  -1.063114  0.2125586
1.00  7.450785  -0.003905017  -0.05670059  -1.051455  0.2097454
> plot(ridge)
```

We find that the coefficients only change very slightly when the parameter of the ridge regression varies between 0 and 1 (Figure 11.70), unlike the case in the preceding section. We can check that the lambda parameter must be more than 50 if we want the coefficients to converge towards consistent values, particularly if they are to be negative for x5. This will be easily understood if we remember that a matrix $X'X$ is translated from a diagonal *kI*: this translation must be greater as $X'X$ increases. According to the order of magnitude of the regressors, we can therefore obtain *k* parameters which are very different and not comparable from one data set to another or from one situation to another. Furthermore, we do not obtain comparable ridge coefficients, since they depend on the order of magnitude of the regressors. For these reasons, it is often preferable to apply ridge regression to standardized variables, in other words those which are centred and reduced. This is not done by the *ridge* function in R, but is provided by the RIDGE option of the SAS REG procedure.

Here is the SAS syntax for ridge regression:

```
PROC REG DATA = ridge.caterpillars RIDGE = 0 to 1 BY 0.05
OUTVIF OUTEST = coeff_ridge ;
MODEL log = X1 X2 X4 X5 ;
PLOT / RIDGEPLOT ;
RUN ;
QUIT;
```

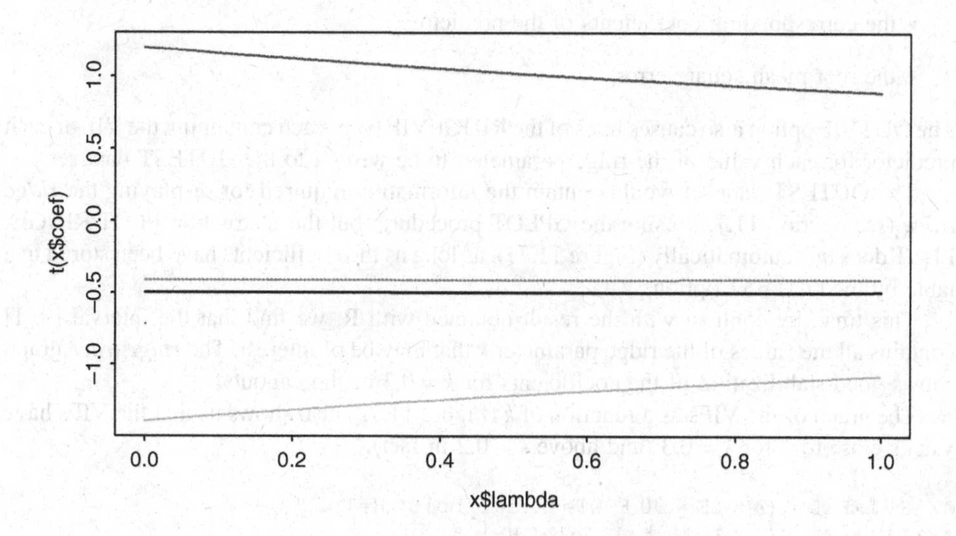

Figure 11.70 Ridge trace with R.

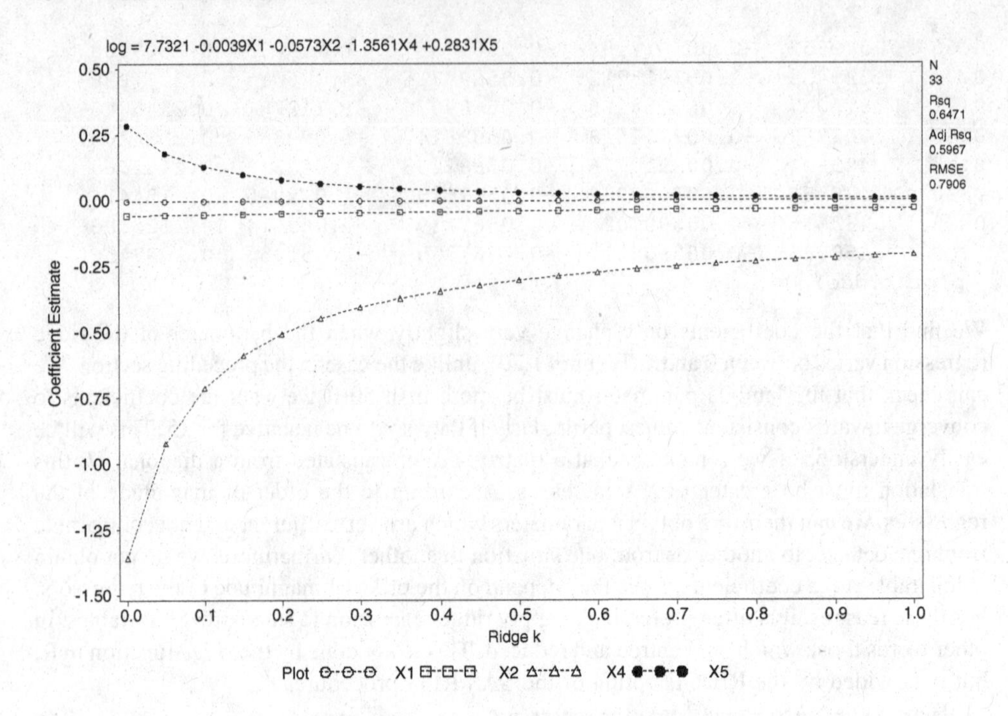

Figure 11.71 Ridge trace with SAS.

The RIDGE option is followed by the range of values which we wish to assign to the ridge parameter. The instruction OUTEST = <file_name> is used to write lines of the RIDGE type to the specified data set, these lines containing:

- each value of the ridge parameter;
- the corresponding coefficients of the predictors;
- the root mean square error.

The OUTVIF option also causes lines of the RIDGEVIF type, each containing the VIF of each predictor for each value of the ridge parameter, to be written to the OUTEST data set.

The OUTEST data set would contain the information required for displaying the *ridge trace* (see Section 11.7.8) using the GPLOT procedure, but the instruction PLOT/RIDGE-PLOT does this automatically (Figure 11.71), as long as the coefficients have been stored in a table by the OUTEST option.

This time, by contrast with the result obtained with R, we find that the interval [0, 1] contains all the values of the ridge parameter k that may be of interest. The *ridge trace* graph shows good stabilization of the coefficients for $k = 0.3$ or thereabouts.

The graph of the VIFs as a function of k (Figure 11.72) also shows us that the VIFs have values close to 1 for $k = 0.3$ (and above $k = 0.2$ in fact).

```
AXIS1 LABEL = (ANGLE = 90 FONT='Arial/bold' H=1.5 "VIF") ;
AXIS2 LABEL = (FONT='Arial/bold' H=1.5) ;
PROC GPLOT DATA = coeff_ridge ;
WHERE _TYPE_ = "RIDGEVIF" ;
```

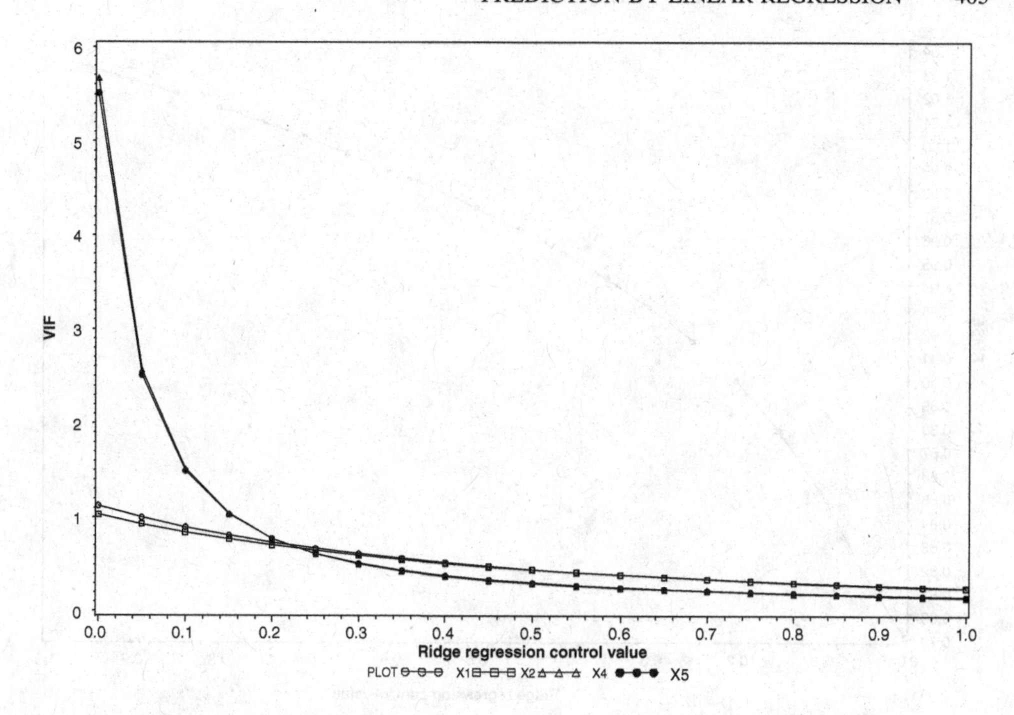

Figure 11.72 Value of VIFs as a function of the ridge parameter.

```
PLOT (X1 X2 X4 X5) * _RIDGE_ _ / OVERLAY LEGEND VAXIS = axis1
HAXIS=axis2 ;
RUN ;
QUIT ;
```

Finally, the graph of RMSE as a function of k (Figure 11.73) shows that RMSE still increases significantly for $k > 0.3$, showing the usefulness of stopping at this value of k to avoid unnecessarily increasing the RMSE.

```
AXIS1 LABEL = (ANGLE = 90 FONT='Arial/bold' H=1.5 "RMSE") ;
AXIS2 LABEL = (FONT='Arial/bold' H=1.5) ;
PROC GPLOT DATA = coeff_ridge ;
WHERE _TYPE_ = "RIDGE" ;
PLOT _RMSE_ _ * _RIDGE_ / OVERLAY VAXIS = axis1 HAXIS=axis2 ;
RUN ;
QUIT ;
```

Although the ridge function of the MASS package in R does not carry out ridge regression on standardized data and does not provide the above information, it is still possible to write a simple function in R to obtain it. This is an interesting exercise, using the formulae shown in earlier sections. I have not incorporated the cross-validation which is executed by the ridge function in R but not by the SAS RIDGE option.

```
ridges <- function (X,Y,lambda)
{
```

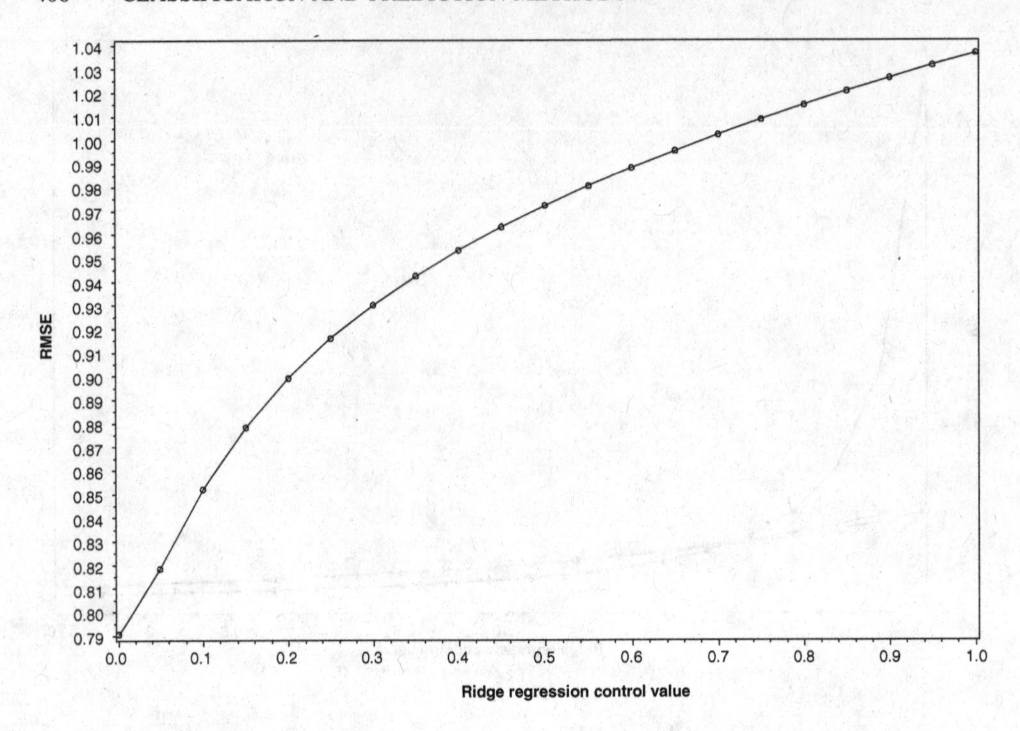

Figure 11.73 Value of RMSE as a function of the ridge parameter.

```
n <- length(Y)
X.std <- cbind(1,scale(X)/sqrt(n-1))
# standardization of variables
p <- dim(X.std)[[2]] # p = number of columns of X
xscale <- sqrt(apply(X,2,var)*(n-1))
xmeans <- apply(X,2,mean) # mean of each variable
if(sum(lambda==0)==0) lambda <- c(0,lambda)
# if necessary, add 0 to the values of the ridge parameter
r <- length(lambda) # number of values of parameter
VIF <- matrix(NA,r,p-1)
beta <- matrix(NA,r,p)
RMSE <- matrix(NA,r,1)
for(i in 1:r)
{
k <- lambda[i]
kI <- diag(p)*k
# note: the intercept is not penalized
kI[1,1] <- 0
X.k <- t(X.std) %*% X.std + kI
inv <- solve(t(X.k)) # solve(x) = inverse of x
beta[i,] <- inv %*% t(X.std) %*% Y
# ridge coefficients on standardized data
```

Here we apply the formula

$$b_R = (X^tX + kI)^{-1}X^tY.$$

```
# calculation of VIF
VIF[i,] <- diag(inv%*%t(X.std)%*%X.std%*%inv)[-1]
```

The VIF is deduced from the formula

$$V(b_R) = s^2(X^tX + kI)^{-1}(X^tX)(X^tX + kI)^{-1}.$$

```
# calculation of RMSE
E <- Y - X.std%*%cbind(beta[i,]) # vector containing the residuals of the
observations
RMSE[i] <- sqrt(sum(E^2)/(n-p)) # sum of squared residuals = RMSE
```

Note that the RMSE is the same regardless of whether it is calculated on the standardized data or not.

```
} # end of loop on lambda
# return to non standardized data
beta.orig <- t(t(beta[,2:p])/xscale)
intercept <- beta[,1]-apply(t(t(beta.orig)*xmeans),1,sum)
beta.orig <- cbind(intercept,beta.orig)
# summary of results
result = cbind(lambda,RMSE,beta.orig,VIF)
colnames(result) <-
c("lambda","RMSE","intercept",colnames(X),paste("VIF",colnames(X)))
return(result)
} # end of function
```

Before calling this function, we must write the predictors in the form of a matrix X and the dependent variable in the form of a matrix (vector) Y:

```
> X <- as.matrix(subset(caterpillars,select=c(x1,x2,x4,x5)))
> Y <- as.matrix(subset(caterpillars,select=log))
```

The function is then called with these matrices entered as parameters. We also specify that we wish to test the ridge parameters in the range from 0 to 1, at intervals of 0.05.

The table output by our *ridges* function contains the same information as the OUTEST data set of the REG procedure in SAS, the VIFs being placed on the same row as the coefficients of the predictors. The values calculated by R are exactly the same as those calculated in SAS.

Before going on to discuss principal component regression, I will show how to move from ridge coefficients on the predictors to ridge coefficients on their principal components in SAS, and then point out an interesting feature of this.

```
> ridges(X,Y,lambda=seq(0,1,0.05))
      lambda      RMSE  intercept           x1          x2         x4          x5   VIF x1    VIF x2    VIF x4    VIF x5
 [1,]   0.00 0.7906457  7.732144 -0.003923681 -0.05734270 -1.3561376 0.283058218 1.1226491 1.0264271 5.6720928 5.5110555
 [2,]   0.05 0.8185528  7.292476 -0.003872789 -0.05611179 -0.9222523 0.178514258 0.9970852 0.9264669 2.5644138 2.5083573
 [3,]   0.10 0.8520473  6.941476 -0.003762661 -0.05439221 -0.7123137 0.127482111 0.8941393 0.8411502 1.5161179 1.4934876
 [4,]   0.15 0.8783931  6.637951 -0.003639039 -0.05259143 -0.5885637 0.097174800 0.8072301 0.7673305 1.0354242 1.0268062
 [5,]   0.20 0.8991023  6.366848 -0.003515143 -0.05082609 -0.5069205 0.077075464 0.7328510 0.7029336 0.7729488 0.7710813
 [6,]   0.25 0.9159576  6.120608 -0.003395569 -0.04913543 -0.4489423 0.062767912 0.6685869 0.6463876 0.6122958 0.6139278
 [7,]   0.30 0.9301538  5.894636 -0.003281930 -0.04753189 -0.4055623 0.052070070 0.6126304 0.5964509 0.5056715 0.5091733
 [8,]   0.35 0.9424531  5.685779 -0.003174656 -0.04601725 -0.3718094 0.043777999 0.5635795 0.5521239 0.4304881 0.4349792
 [9,]   0.40 0.9533492  5.491703 -0.003073668 -0.04458881 -0.3447337 0.037171982 0.5203243 0.5125914 0.3749255 0.3799056
[10,]   0.45 0.9631716  5.310588 -0.002978676 -0.04324198 -0.3224758 0.031794678 0.4819733 0.4771825 0.3322991 0.3374745
[11,]   0.50 0.9721467  5.140965 -0.002889300 -0.04197150 -0.3038070 0.027341146 0.4478023 0.4453404 0.2985897 0.3037848
[12,]   0.55 0.9804350  4.981621 -0.002805140 -0.04077204 -0.2878837 0.023599998 0.4172176 0.4165996 0.2712570 0.2763665
[13,]   0.60 0.9881534  4.831532 -0.002725802 -0.03963841 -0.2741075 0.020419919 0.3897281 0.3905684 0.2486279 0.2535894
[14,]   0.65 0.9953894  4.689825 -0.002650914 -0.03856575 -0.2620431 0.017689645 0.3649249 0.3669155 0.2295610 0.2343391
[15,]   0.70 1.0022095  4.555745 -0.002580129 -0.03754954 -0.2513657 0.015325470 0.3424648 0.3453586 0.2132541 0.2178303
[16,]   0.75 1.0086661  4.428636 -0.002513130 -0.03658563 -0.2418289 0.013263167 0.3220582 0.3256563 0.1991285 0.2034953
[17,]   0.80 1.0148008  4.307920 -0.002449626 -0.03567020 -0.2332418 0.011452615 0.3034592 0.3076012 0.1867570 0.1909137
[18,]   0.85 1.0206473  4.193088 -0.002389353 -0.03479977 -0.2254546 0.009854104 0.2864581 0.2910140 0.1758179 0.1797684
[19,]   0.90 1.0262333  4.083687 -0.002332073 -0.03397116 -0.2183482 0.008435762 0.2708746 0.2757394 0.1660646 0.1698154
[20,]   0.95 1.0315823  3.979314 -0.002277566 -0.03318146 -0.2118263 0.007171705 0.2565536 0.2616418 0.1573049 0.1608641
[21,]   1.00 1.0367140  3.879605 -0.002225634 -0.03242802 -0.2058106 0.006040691 0.2433608 0.2486032 0.1493867 0.1527633
```

First of all, we must calculate a ridge regression on the standardized variables in order to obtain ridge coefficients that can be used with the coefficients of the variables on the principal components. These coefficients, produced by the PRINCOMP procedure, are those of the *standardized* variables. This is done as follows:

```
PROC STANDARD DATA= ridge.caterpillars OUT=reduced MEAN=0 STD=1 ;
VAR x1 x2 x4 x5 ;
RUN ;

PROC REG DATA = reduced RIDGE = 0 to 1 BY 0.05 OUTEST =
coeff_ridge_std ;
MODEL log = x1 x2 x4 x5 ;
PLOT / RIDGEPLOT ;
RUN ;
QUIT ;
```

We obtain the ridge coefficients:

```
DATA coeff_ridge_pc (DROP = _model_ _type_ _depvar_ _
pcomit_) ; SET coeff_ridge_std ;
WHERE _TYPE_ = "RIDGE" ;
RUN ;
```

The coefficients of the variables on the principal components are obtained as the output of a PCA executed by the PRINCOMP procedure.

```
PROC PRINCOMP DATA = reduced OUT = pcr OUTSTAT = coef ;
VAR x1 x2 x4 x5 ;
RUN ;
```

The SCORE procedure multiplies the ridge regression coefficients on standardized variables (DATA set) by the coefficients of the principal components on the variables X_i (SCORE data set). For each line of the first data set (DATA), in other words each value of the ridge parameter, this procedure will calculate the scalar product of the vector (x1 x2 x3 x4) of the first data set by each vector (x1 x2 x3 x4) of the second, for each line of the _SCORE_ type, therefore each principal component of this second data set. For each line, the result will be placed in a variable whose name is given by the _NAME_ variable of the SCORE data set.

```
PROC SCORE DATA=coeff_ridge_pc SCORE=coef OUT=coeff_ridge_pc2 ;
VAR x1 x2 x4 x5 ;
RUN ;
DATA coeff_ridge_pc2 ;
  SET coeff_ridge_pc2 ;
  prin1 = prin1*SQRT(2.1141) ;
  prin2 = prin2*SQRT(0.9758) ;
  prin3 = prin3*SQRT(0.8158) ;
  prin4 = prin4*SQRT(0.0943) ;
RUN ;
```

The first data set (DATA) contains the ridge regression coefficients on standardized variables:

Obs	_RIDGE_	_RMSE_	Intercept	X1	X2	X4	X5	log
1	0.00	0.79065	−0.81328	−0.50630	−0.41881	−1.41143	1.21787	−1
2	0.05	0.81855	−0.81328	−0.49973	−0.40982	−0.95985	0.76807	−1
3	0.10	0.85205	−0.81328	−0.48552	−0.39726	−0.74135	0.54850	−1
4	0.15	0.87839	−0.81328	−0.46957	−0.38411	−0.61256	0.41810	−1
5	0.20	0.89910	−0.81328	−0.45358	−0.37121	−0.52759	0.33162	−1
6	0.25	0.91596	−0.81328	−0.43815	−0.35887	−0.46725	0.27006	−1
7	0.30	0.93015	−0.81328	−0.42349	−0.34715	−0.42210	0.22403	−1
8	0.35	0.94245	−0.81328	−0.40965	−0.33609	−0.38697	0.18836	−1
9	0.40	0.95335	−0.81328	−0.39662	−0.32566	−0.35879	0.15993	−1
10	0.45	0.96317	−0.81328	−0.38436	−0.31582	−0.33562	0.13680	−1
11	0.50	0.97215	−0.81328	−0.37283	−0.30654	−0.31619	0.11764	−1
12	0.55	0.98044	−0.81328	−0.36197	−0.29778	−0.29962	0.10154	−1
13	0.60	0.98815	−0.81328	−0.35173	−0.28950	−0.28528	0.08786	−1
14	0.65	0.99539	−0.81328	−0.34207	−0.28167	−0.27273	0.07611	−1
15	0.70	1.00221	−0.81328	−0.33293	−0.27425	−0.26161	0.06594	−1
16	0.75	1.00867	−0.81328	−0.32429	−0.26721	−0.25169	0.05707	−1
17	0.80	1.01480	−0.81328	−0.31609	−0.26052	−0.24275	0.04928	−1
18	0.85	1.02065	−0.81328	−0.30831	−0.25416	−0.23465	0.04240	−1
19	0.90	1.02623	−0.81328	−0.30092	−0.24811	−0.22725	0.03630	−1
20	0.95	1.03158	−0.81328	−0.29389	−0.24234	−0.22046	0.03086	−1
21	1.00	1.03671	−0.81328	−0.28719	−0.23684	−0.21420	0.02599	−1

The second data set (SCORE) contains the coefficients of the principal components on the variables X_i, in the observations of the SCORE type. The MEAN and STD observations have values of 0 and 1 respectively, because the PCA was conducted on standardized variables. This changes nothing in the SCORE coefficients, but if the MEAN and STD observations contained the mean and standard deviation of each variable X_i, the values of the X_i would have been standardized before the calculation of the scalar product.

Obs	_TYPE_	_NAME_	X1	X2	X4	X5
1	MEAN		0.0000	0.0000	0.0000	0.0000
2	STD		1.0000	1.0000	1.0000	1.0000
3	N		33.0000	33.0000	33.0000	33.0000
4	CORR	X1	1.0000	0.1205	0.3211	0.2838
5	CORR	X2	0.1205	1.0000	0.1367	0.1134
6	CORR	X4	0.3211	0.1367	1.0000	0.9047
7	CORR	X5	0.2838	0.1134	0.9047	1.0000
8	EIGENVAL		2.1141	0.9758	0.8158	0.0943
9	SCORE	Prin1	0.3697	0.1846	0.6484	0.6394
10	SCORE	Prin2	0.2156	0.9359	−0.1793	−0.2131
11	SCORE	Prin3	0.9033	−0.2995	−0.1980	−0.2350
12	SCORE	Prin4	0.0312	0.0157	−0.7129	0.7003

Now, to express the coordinates of the variables $X_1, \ldots, X_n$ on the principal components Prin1, we have to multiply the previous coefficients 0.3697, ..., 0.6394 by the square root 1.4540 of the eigenvalue 2.1141 of Prin1 (see section 7.1.1).

The first line of the data set OUT corresponds to the ridge parameter 0.00 and contains the coefficients −0.50630, −0.41881, ... of Y on $X_1, X_2, \ldots$, so that the scalar product

$$(-0.50630 \times 0.3697 \times 1.4540) - (0.41881 \times 0.1846 \times 1.4540) - (1.41143 \times 0.6484 \times 1.4540) + (1.21787 \times 0.6394 \times 1.4540) = -(0.4009 \times 1.4540) = -0.5829$$

is equal to the coefficient of Y on Prin1. We do the same calculation for Prin2, Prin3 and Prin4. The set of ridge regression coefficients on the principal components is as follows:

RIDGE	Prin1	Prin2	Prin3	Prin4
0.00	−0.58289	−0.50150	−0.30594	0.56406
0.05	−0.56943	−0.47705	−0.28827	0.36856
0.10	−0.55657	−0.45488	−0.27253	0.27370
0.15	−0.54428	−0.43468	−0.25842	0.21767
0.20	−0.53252	−0.41619	−0.24571	0.18069
0.25	−0.52125	−0.39921	−0.23418	0.15444
0.30	−0.51046	−0.38357	−0.22369	0.13486
0.35	−0.50010	−0.36910	−0.21409	0.11968
0.40	−0.49015	−0.35569	−0.20529	0.10757
0.45	−0.48060	−0.34321	−0.19718	0.09769
0.50	−0.47140	−0.33159	−0.18969	0.08947

Finally, the GPLOT procedure is used to display the absolute values of the ridge coefficients on principal components, as a function of their ridge parameter value.

```
DATA coeff_ridge_pc3 ;
SET coeff_ridge_pc2 ;
prin1 = ABS(prin1) ;
prin2 = ABS(prin2) ;
prin3 = ABS(prin3) ;
prin4 = ABS(prin4) ;
RUN ;

GOPTIONS RESET=all;
SYMBOL1 v=CIRCLE i=JOIN c=BLACK w=2 h=1.5 l=2 ;
SYMBOL2 v=SQUARE i=JOIN c=BLACK w=2 h=1.5 l=2 ;
SYMBOL3 v=TRIANGLE i=JOIN c=BLACK w=2 h=1.5 l=2 ;
SYMBOL4 v=DOT i=JOIN c=BLACK w=2 h=1.5 l=2 ;
AXIS1 LABEL = (ANGLE = 90 FONT='Arial/bold' H=1.5
"Coefficients") ;
AXIS2 LABEL = (FONT='Arial/bold' H=1.5) ;
PROC GPLOT DATA = coeff_ridge_pc3 ;
PLOT (prin1 - prin4) * _ridge_ / OVERLAY LEGEND VAXIS = axis1
HAXIS=axis2 ;
RUN ;
QUIT ;
```

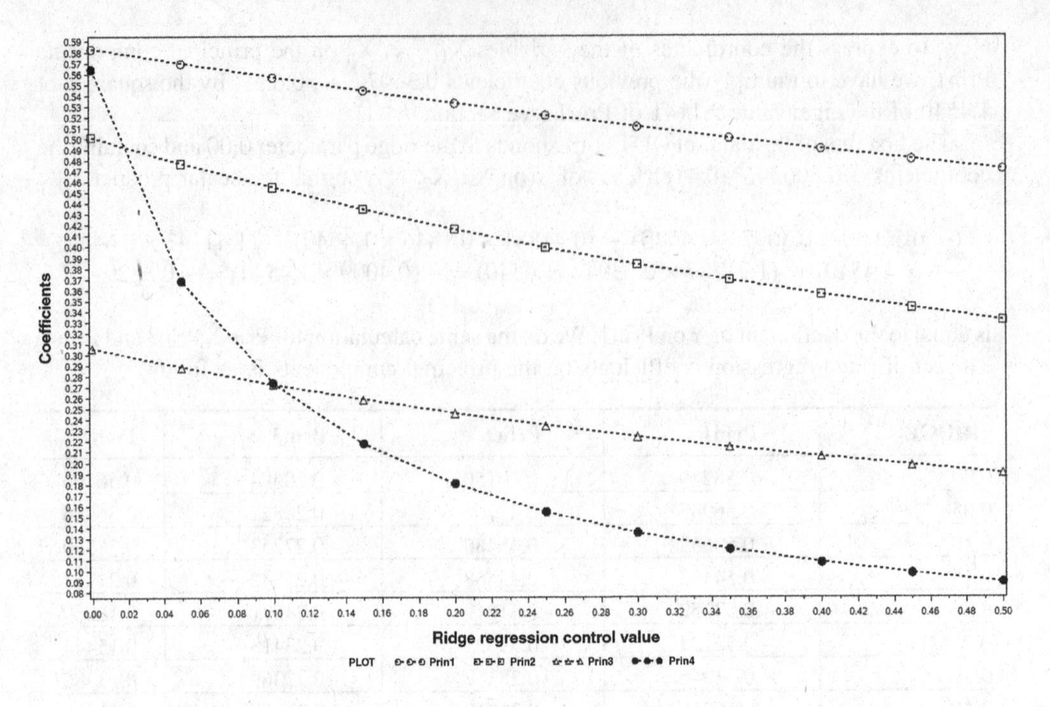

Figure 11.74 Variation of the ridge coefficients of the principal components.

Figure 11.74 and the preceding table of coefficients reveal a remarkable property of ridge regression:[46] it progressively decreases the absolute values of the coefficients of all the principal components. This is not necessarily the case with the coefficients of the predictors, which may increase in absolute value, as shown by the example in Section 11.7.8.

Moreover, the coefficient of a principal component is decreased by the ridge regression to a greater extent if the variance of this component is smaller: the coefficient of the first principal component decreases moderately, whereas the coefficient of the last principal component falls abruptly. We say that ridge regression penalizes the principal components with low variance.

Figure 11.74 is at least as informative as the conventional 'ridge plot', and the phenomenon which it reveals is sometimes much more apparent. It also shows a consistency between ridge regression and the regression which will be carried out subsequently by eliminating, one by one, the principal components that are less useful to the regression: as we will see, Prin3 is eliminated first, and this is the component whose ridge coefficient has the lowest absolute value for $k = 0$, i.e. for least squares regression.

Unlike ridge regression, principal component regression (PCR) does not progressively decrease the coefficients of the principal components, but deletes them one after another, since it eliminates the principal components one after another, starting with those having the smallest variance and finishing with the first principal component. The discrete and discontinuous nature of this process causes significant, even abrupt, variations in the coefficients of the predictors, which contrast with the progressive variation of the coefficients in ridge

[46] See Section 3.4.3 of Hastie, T., Tibshirani, R. and Friedman, J. (2001) *The Elements of Statistical Learning: Data Mining, Inference and Prediction.* New York: Springer.

regression. For PCR, the analogue of the lambda parameter of ridge regression is the number of principal components that are retained. This is the regularization parameter, which is varied to a greater or lesser degree according to the collinearity of the predictors. Its minimum value (lambda = 0, or number of principal components equal to the number of predictors) corresponds to conventional linear regression, and its value increases in the presence of collinearity and when there are abnormally large regression coefficients to be corrected.

Let us return to our example. Here is the PCOMIT option in the SAS REG procedure which launches the PCR. An advantage of this option is that it enables the regression to be carried out once only on a set of possible values of the number of principal components. For example, the following syntax calculates the regression with the omission of 0, 1, 2 and then 3 components, starting with the components with the smallest variance.

```
PROC REG DATA = ridge.caterpillars PCOMIT = 0 1 2 3 OUTEST =
coeff_pcr ;
MODEL log = x1 x2 x4 x5 ;
RUN ;
QUIT ;
```

In R, the syntax is as follows:

```
library(pls)
pcr= pcr(log~x1+x2+x4+x5,data=caterpillars,ncomp=2)
```

Only one value of the number of components can be specified. This number can also be chosen by cross-validation.

Returning to SAS, we can represent the values of the coefficients as a function of the number of components omitted (Figure 11.75). These are contained in the data set written at the output of the procedure by means of the instruction OUTEST = <file_name>, and they are extracted by selecting the records of the IPC (incomplete principal components) type.

```
GOPTIONS RESET=all ;
SYMBOL1 v=CIRCLE i=JOIN c=BLACK w=2 h=1.5 l=2 ;
SYMBOL2 v=SQUARE i=JOIN c=BLACK w=2 h=1.5 l=2 ;
SYMBOL3 v=TRIANGLE i=JOIN c=BLACK w=2 h=1.5 l=2 ;
SYMBOL4 v=DOT i=JOIN c=BLACK w=2 h=1.5 l=2 ;
AXIS1 LABEL = (ANGLE = 90 FONT='Arial/bold' H=1.5
"Coefficients") ;
AXIS3 LABEL = (ANGLE = 90 FONT='Arial/bold' H=1.5 "RMSE") ;
AXIS2 LABEL = (FONT='Arial/bold' H=1.5) ;
PROC GPLOT DATA = coeff_pcr ;
WHERE _TYPE_ = "IPC" ;
PLOT (x1 x2 x4 x5) * _PCOMIT__ / OVERLAY LEGEND VAXIS = axis1
HAXIS=axis2 ;
PLOT _RMSE__ * _PCOMIT__ / VAXIS = axis3 HAXIS=axis2;
RUN ;
QUIT ;
```

We can also represent the RMSE as a function of the number of components dropped (Figure 11.76).

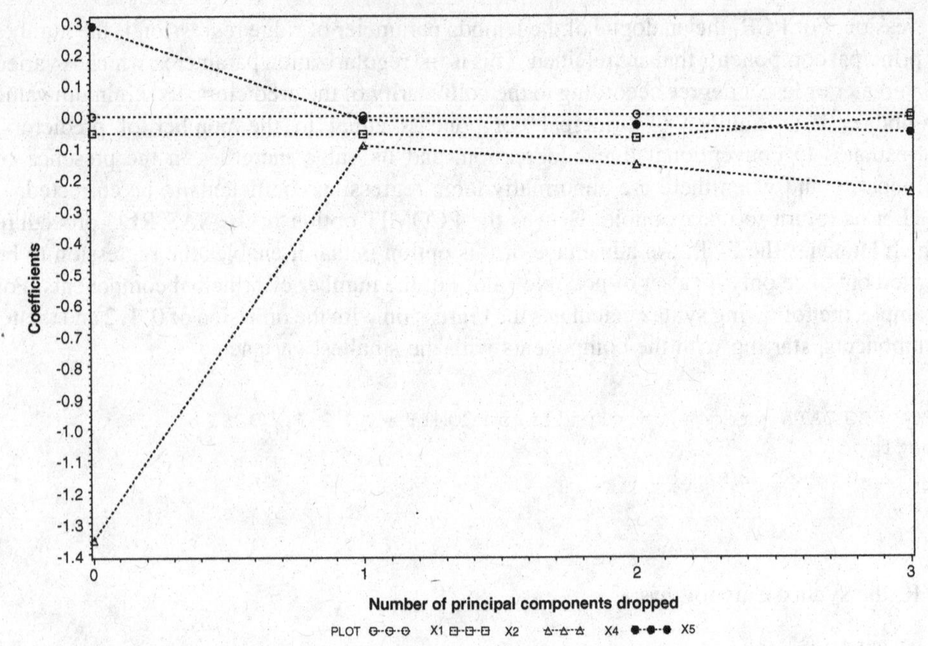

Figure 11.75 Variation of the PCR coefficients according to the number of principal components dropped.

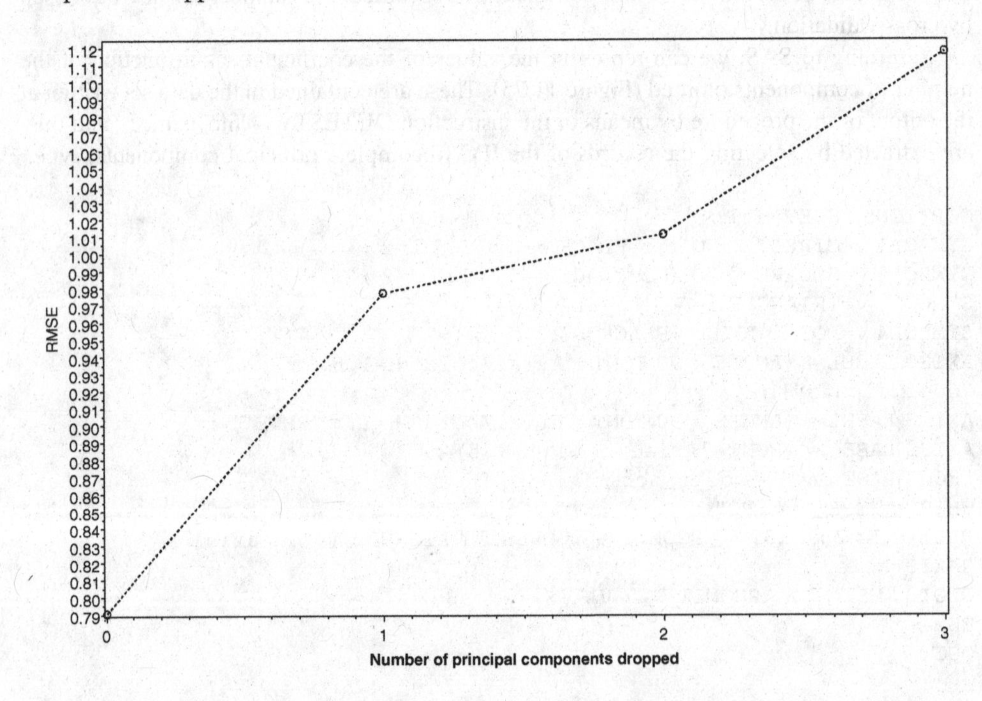

Figure 11.76 Variation of the RMSE according to the number of principal components dropped.

We can see that we must exclude at least one principal component in order to stabilize the coefficients (x4 and x5, which are closely correlated, have strong coefficients which 'compensate' each other). The removal of a principal component causes a marked increase in RMSE, but this increases less when a second component is dropped, then rises steeply again if a third one is dropped. A good choice is to retain two or three principal components.

Another type of principal component regression can be carried out, based on the principle that the principal components most closely related to the dependent variable are not necessarily those with the largest variance.

A PCA is carried out, followed by linear regression with global selection of the principal components, in order to find, for each possible value p of the number of variables in the model, where p varies over a chosen range, the best combinations of p principal components with respect to a fixed criterion, for example R^2. This search, which is global rather than stepwise, is carried out by using the leaps and bounds algorithm (Section 3.13).

```
PROC PRINCOMP DATA = ridge.caterpillars OUT = pcr OUTSTAT = coef ;
VAR X1 X2 X4 X5 ;
RUN ;

PROC REG DATA = pcr ;
MODEL log = prin1-prin4 / SELECTION = rsquare CP AIC BIC RMSE ;
RUN ;
QUIT ;
```

The SELECTION = rsquare option detects the best subsets of k variables with respect to R^2, and displays these subsets. If nothing else is specified, SAS only considers the subsets of 1, 2, ..., p variables, where p is the total number of predictors. The minimum number of subsets searched for can be set by the option START = n (default value 1), while the maximum number of subsets searched for can be set by the option STOP = n (default value p). In this way we can specify a range. Also, if we only wish to retain the best k subsets of each size, we can specify the option BEST = k. The default value of BEST is p if $p > 10$. If $p \leq 10$, all the subsets are selected by default. In large data sets, a small value of BEST considerably reduces the computing time.

Two criteria other than R^2 can be optimized by this option, namely the adjusted R^2 (SELECTION = adjrsq) and the Mallows C_p (SELECTION = cp). But R^2 is most widely used, although it has the drawback of always favouring the most complex model, unlike the Mallows C_p. However, it is possible to display other criteria, as we have done. Thus our syntax displays the CP, RMSE, AIC and BIC. The last two of these are defined in Section 11.8.6 in the context of logistic regression, but their definition is the same for linear regression. The C_p statistic, introduced by Mallows (1973),[47] resembles the (more general) AIC, because it is a measure of the sum of squares of the errors, penalized by the number of predictors:

$$C_p(q) = (n - p - 1)\frac{ESS(q)}{ESS(p)} - n + 2(q + 1),$$

where $ESS(q)$ is the sum of residual squares for q predictors, and $q \leq p$.

[47] Mallows, C.L. (1973). Some comments on C_p. *Technometrics*, 15(4), 661–675.

For C_p, the Akaike information criterion (AIC) and the Bayesian information criterion (BIC), we seek the lowest possible value, preferably close to $p+1$ for C_p (the latter can be less than $p+1$, or even negative).

Number in model	R-square	C(p)	AIC	BIC	Root MSE	Variables in model
1	0.2192	32.9454	9.2779	8.9721	1.11765	Prin1
1	0.2052	34.0582	9.8654	9.5043	1.12764	Prin4
1	0.1623	37.4646	11.6016	11.0806	1.15770	Prin2
1	0.0604	45.5467	15.3886	14.5381	1.22607	Prin3
2	0.4244	18.6653	1.2157	1.3977	0.97547	Prin1 Prin4
2	0.3815	22.0717	3.5898	3.4102	1.01120	Prin1 Prin2
2	0.3675	23.1845	4.3298	4.0407	1.02260	Prin2 Prin4
2	0.2796	30.1539	8.6212	7.7283	1.09130	Prin1 Prin3
2	0.2656	31.2666	9.2575	8.2796	1.10187	Prin3 Prin4
2	0.2227	34.6730	11.1324	9.9106	1.13362	Prin2 Prin3
3	0.5867	7.7916	−7.7128	−5.6621	0.84074	Prin1 Prin2 Prin4
3	0.4848	15.8738	−0.4424	−0.0577	0.93865	Prin1 Prin3 Prin4
3	0.4419	19.2801	2.1992	2.0314	0.97698	Prin1 Prin2 Prin3
3	0.4279	20.3929	3.0182	2.6850	0.98918	Prin2 Prin3 Prin4
4	0.6471	5.0000	−10.9258	−7.2038	0.79065	Prin1 Prin2 Prin3 Prin4

The table displayed by SAS shows that the best two-variable model is not the one with prin1 and prin2, which would be selected by the PCOMIT option, but the model with prin1 and prin4. Furthermore, the best three-variable model contains prin1, prin2 and prin4, and does not contain prin3. To optimize R^2 or another of the indicators, therefore, we must not select the principal components in the order of the eigenvalues.

We can now restart the procedure to find the RMSE and the coefficients associated with the best combinations of variables.

```
PROC REG DATA = pcr OUTEST = coeff_pcr ;
MODEL log = prin1-prin4 / SELECTION = rsquare CP AIC BIC RMSE BEST = 1 ;
RUN ;
QUIT ;
```

Number in model	R-square	C(p)	AIC	BIC	Root MSE	Variables in model
1	0.2192	32.9454	9.2779	8.9721	1.11765	Prin1
2	0.4244	18.6653	1.2157	1.3977	0.97547	Prin1 Prin4
3	0.5867	7.7916	−7.7128	−5.6621	0.84074	Prin1 Prin2 Prin4
4	0.6471	5.0000	−10.9258	−7.2038	0.79065	Prin1 Prin2 Prin3 Prin4

```
GOPTIONS RESET=all ;
SYMBOL1 v=CIRCLE i=JOIN c=BLACK ;
SYMBOL2 v=SQUARE i=JOIN c=BLACK ;
SYMBOL3 v=TRIANGLE i=JOIN c=BLACK ;
SYMBOL4 v=DOT i=JOIN c=BLACK ;
PROC GPLOT DATA = coeff_pcr ;
PLOT _RMSE_ * _IN_ ;
RUN ;
QUIT ;
```

Looking at Figure 11.77, three appears to be a good number of components to retain, because the RMSE increases sharply with one or two principal components.

The expression for the principal component regression coefficients is then used, with the expression for the principal components as a function of the standardized initial predictors, to deduce the expression for the coefficients of regression on the initial predictors, as a function of the number of principal components retained.

dim	X1	X2	X4	X5
1	– .001148501	– 0.010134	– 0.24975	– 0.05958
2	– .000704437	– 0.006174	– 1.50800	0.23941
3	– .001552613	– 0.071232	– 1.42056	0.26456
4	– .003923681	– 0.057343	– 1.35614	0.28306

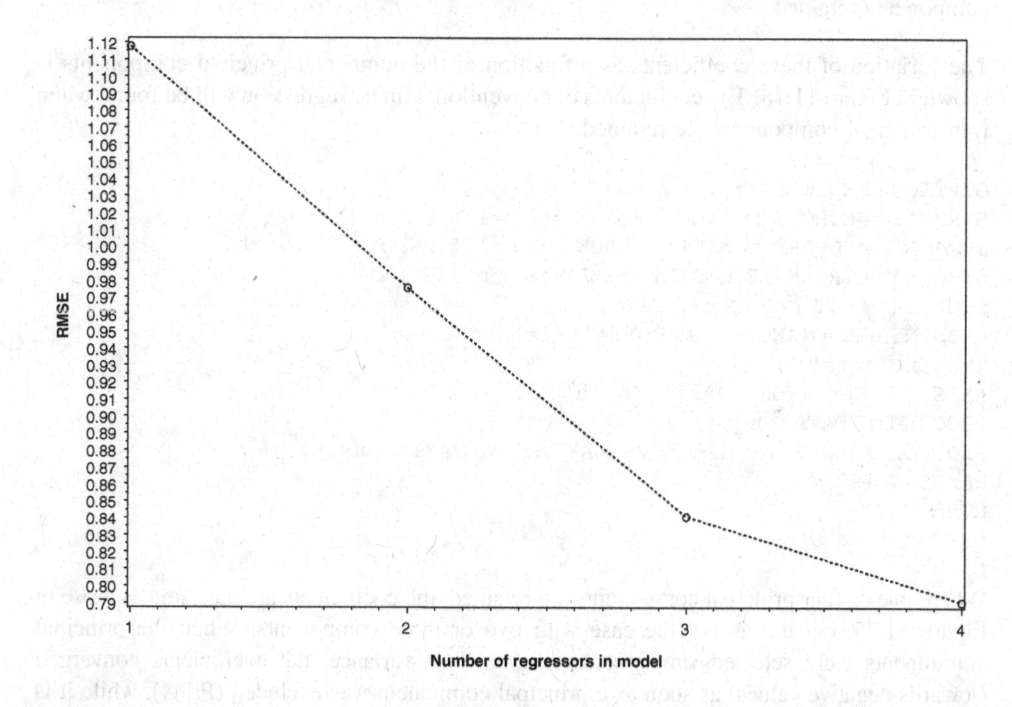

Figure 11.77 Variation of the RMSE according to the number of principal components retained.

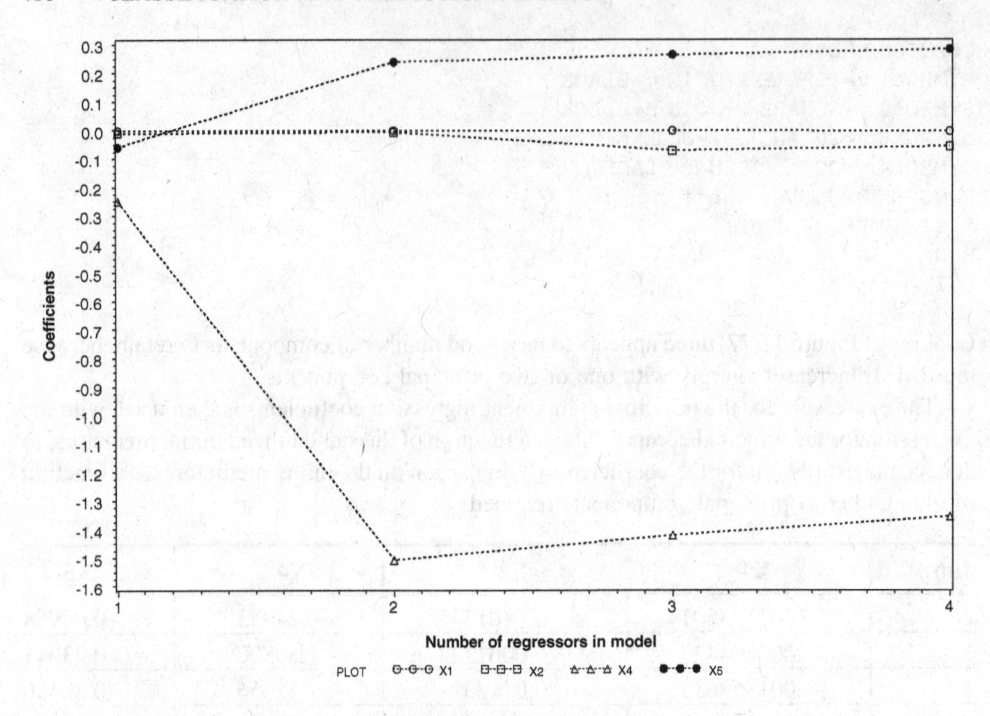

Figure 11.78 Variation of the PCR coefficients according to the number of principal components retained.

The variation of these coefficients as a function of the number of principal components is shown in Figure 11.78. The coefficients of conventional linear regression will be found when four principal components are retained.

```
GOPTIONS RESET=all;
SYMBOL1 v=CIRCLE i=JOIN c=BLACK w=2 h=1.5 l=2 ;
SYMBOL2 v=SQUARE i=JOIN c=BLACK w=2 h=1.5 l=2 ;
SYMBOL3 v=TRIANGLE i=JOIN c=BLACK w=2 h=1.5 l=2 ;
SYMBOL4 v=DOT i=JOIN c=BLACK w=2 h=1.5 l=2 ;
AXIS1 LABEL = (ANGLE = 90 FONT='Arial/bold' H=1.5
"Coefficients") ;
AXIS2 LABEL = (FONT='Arial/bold' H=1.5) ;
PROC GPLOT DATA = t ;
PLOT (X1 X2 X4 X5) * dim / OVERLAY LEGEND VAXIS = axis1
HAXIS=axis2 ;
RUN ;
QUIT ;
```

Where one or four principal components are retained, the coefficients are the same as those in Figure 11.75, but this is not the case with two or three components. When the principal components were selected simply according to their variance, the coefficients converged (towards negative values) as soon as a principal component was excluded (Prin4), while it is necessary to wait until there is only one principal component left (Prin1) when these components have been selected with a view to their relationship with the dependent variable. It is the fourth principal component, prin4, that interferes with the convergence of the coefficients: when it was excluded and the other three were kept, the coefficients converged (first regression), which is not the case when the third component is excluded (second regression). This result is consistent with Figure 11.74. This discrepancy between results of both regressions corresponds to the fact that Prin4 curve intersects Prin2 and Prin3 curves in Figure 11.74.

This example shows the difficulty of choosing between optimization of the regression, based on the RMSE for example, and fast continuous convergence of the coefficients. These difficulties occur with principal component regression but not with ridge regression, which may be preferred for this reason.

We will now take a look at PLS regression, which has some similarities with the regression we have just carried out. As I mentioned above, the aim of PLS regression is to explain, by an appropriate choice of the number of components extracted, the greater part of the variance of both the X_i (objective 1) and the dependent variable (objective 2).

In the R software, PLS regression requires the use of the *pls* package, and is carried out as follows:

```
library(pls)
simpls= mvr(log~x1+x2+x4+x5,data=caterpillars, ncomp=4,
method="simpls")
```

We will use the SIMPLS method of de Jong (1993).[48] The calculations for this method are fast, and in this case, with only one dependent variable, it is equivalent to the NIPALS algorithm forming the basis of PLS regression. Other methods are available, and it is also possible to request cross-validation to find the optimal number of components.

```
simpls= mvr(log~x1+x2+x4+x5,data=caterpillars, validation="CV",
method="simpls")
```

In SAS, PLS regression is carried out with the PLS procedure, which has a very simple syntax. In the first example below, I have added the option NFAC = 4 to request the extraction of four components. By default, NFAC is min(15,*p*,*n*), where *p* is the number of predictors and *n* is the number of observations. It is useful to start by requesting the maximum number of components, to obtain Figure 11.79, which displays the percentage variance of the predictors (on the left) and of the dependent variable (on the right) as a function of the number of components extracted. We can see that the fourth component has hardly any effect on the variance of the dependent variable (which changes from 62.3% to 64.7%), but has more of an effect on the variance of the predictors(rising from 82.5% to 100%).

I have specified the choice of the SIMPLS method as for R. With only one dependent variable, the different methods give the same results.

```
PROC PLS DATA = ridge.caterpillars METHOD = simpls NFAC = 4 ;
MODEL log = X1 X2 X4 X5 ;
RUN ;
```

[48] de Jong, S. (1993) SIMPLS: an alternative approach to partial least squares regression. *Chemometrics and Intelligent Laboratory Systems*, 18, 251–263.

Percent Variation Accounted for by Partial Least Squares Factors				
Number of Extracted Factors	Model Effects		Dependent Variables	
	Current	Total	Current	Total
1	47.9283	47.9283	40.5213	40.5213
2	27.6749	75.6032	9.4396	49.9609
3	6.9465	82.5497	12.3992	62.3601
4	17.4503	100.0000	2.3480	64.7081

Figure 11.79 Variance accounted for, as a function of the number of components.

To determine the optimal number of components, we will use the variation in the value of PRESS found by cross-validation, as mentioned in Section 11.7.10. One option which is widely used for this purpose is 'CV =one'. It can find the value of PRESS, which, as shown here, reaches its minimum when the four components are extracted. Of course, we may sometimes need fewer components, or even a single component, to find the minimum PRESS. However, we can see that the PRESS decreases more between two and three principal components than between three and four, suggesting that it may be sufficient to extract three components.

```
PROC PLS DATA = ridge.caterpillars METHOD = simpls CV = one ;
MODEL log = X1 X2 X4 X5 ;
RUN ;
```

Cross Validation for the Number of Extracted Factors	
Number of Extracted Factors	**Root Mean PRESS**
0	1.03125
1	0.836625
2	0.79649
3	0.717582
4	0.679275

Minimum root mean PRESS	0.6793
Minimizing number of factors	4

However, the procedure has extracted four components, the number being determined automatically by cross-validation. Since this is equivalent to performing the conventional linear regression which has already been done, we will restart the procedure and specify three components, which appears to be a useful number in terms of both the PRESS and the resulting variance.

```
PROC PLS DATA = ridge.caterpillars METHOD = simpls NFAC = 3 ;
MODEL log = X1 X2 X4 X5 / SOLUTION ;
RUN ;
```

The SOLUTION option on the MODEL line displays the PLS coefficients on both the standardized predictors and the raw predictors.

Parameter Estimates for Centered and Scaled Data

	log
Intercept	0.0000000000
X1	–.5250356913
X2	–.2429739212
X4	–.9601628308
X5	0.8423477018

Parameter Estimates

	log
Intercept	8.447241975
X1	–0.005065509
X2	–0.041416195
X4	–1.148520786
X5	0.243732672

These coefficients are closer to the coefficients found by regression on the principal components 1, 2 and 4 than to those for components 1 to 3 found by PCOMIT (see Figure 11.84).

As before, it is interesting to see how the coefficients vary as a function of the number of components extracted. As the PLS procedure only finds the coefficients for the number of components specified or determined by cross-validation, this procedure has to be run for every possible number of components extracted. This is done with a macro that records the calculated coefficients in an ODS data set each time, concatenates them in a single data set, and then prints them. The parameters of this macro are the data set name, the variables and the maximum number of components to be extracted.

```
%MACRO pls (data,n,y,x) ;

%DO i = 1 %TO &n ;

ODS OUTPUT ParameterEstimates = pls&i ;

PROC PLS DATA = &data METHOD = simpls NFAC = &i ;
MODEL &y = &x / SOLUTION ;
RUN ;

DATA pls&i ;
   SET pls&i ;
   nbfact = &i ;
RUN ;
```

```
%IF &i = 1 %THEN %DO ;
DATA pls ;
SET pls&i ;
RUN ;
%END ;
%ELSE %DO ;
PROC APPEND BASE = pls
DATA = pls&i ; RUN ;
%END ;

%END ;

PROC TRANSPOSE DATA = pls OUT=t (DROP= _name_) ;
VAR &y ;
BY nbfact ;
ID RowName ;
RUN ;

PROC PRINT DATA = t ;
RUN ;

%MEND ;

%pls (ridge.caterpillars,4,log,X1 X2 X4 X5) ;
```

This is what is produced when the macro is run:

nbfact	Intercept	X1	X2	X4	X5
1	6.257883169	– 0.003021586	– 0.042962193	– 0.298577984	– 0.034123901
2	7.824476462	– 0.004530997	– 0.070126309	– 0.281904679	0.040173384
3	8.447241975	– 0.005065509	– 0.041416195	– 1.148520786	0.243732672
4	7.732143743	– 0.003923681	– 0.057342697	– 1.356137608	0.283058218

```
GOPTIONS RESET=all;
SYMBOL1 v=CIRCLE i=JOIN c=BLACK w=2 h=1.5 l=2 ;
SYMBOL2 v=SQUARE i=JOIN c=BLACK w=2 h=1.5 l=2 ;
SYMBOL3 v=TRIANGLE i=JOIN c=BLACK w=2 h=1.5 l=2 ;
SYMBOL4 v=DOT i=JOIN c=BLACK w=2 h=1.5 l=2 ;
AXIS1 LABEL = (ANGLE = 90 FONT='Arial/bold' H=1.5 "Coefficients") ;
AXIS2 LABEL = (FONT='Arial/bold' H=1.5) ;
PROC GPLOT DATA = t ;
PLOT (x1 x2 x4 x5) * nbfact / OVERLAY LEGEND VAXIS = axis1
HAXIS=axis2 ;
RUN ;
QUIT ;
```

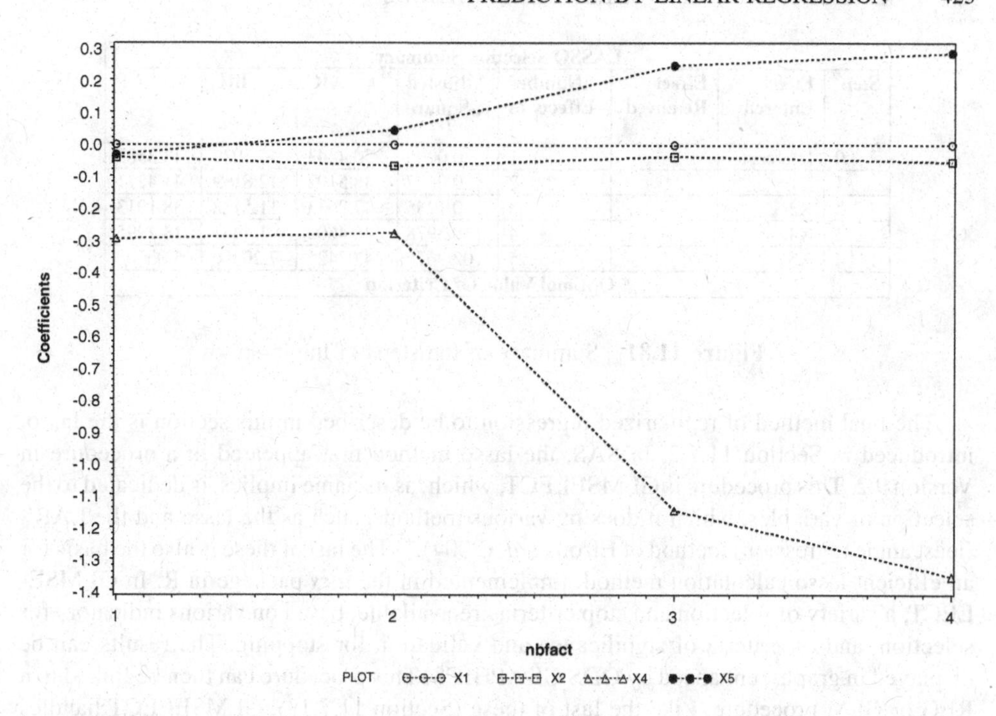

Figure 11.80 Variation of the PLS coefficients according to the number of components retained.

Figure 11.80 shows that the coefficients with three PLS components are close to those found by regression on the principal components prin1, prin2 and prin4. It also shows that, unlike this regression whose coefficients are only stabilized with a single principal component, the coefficients of the PLS regression are stabilized with two components, suggesting that this solution should be tested. PLS regression on two components also yields a lower RMSE (0.89) than those of regressions on two principal components (in accordance with the de Jong theorem mentioned above). No principal component regression provides such a low RMSE, except for the regression on prin1, prin2 and prin4 (RMSE = 0.84), and of course the regression on all the principal components (RMSE = 0.79). Now, these two regressions do not have stabilized coefficients. So far, only PLS regression on two principal components and ridge regression with $k = 0.3$ have a relatively low RMSE (0.93) combined with stabilized coefficients.

With two PLS components, we obtain coefficients close to those of ridge regression, with a slightly lower RMSE (more precisely, the square root of PRESS). PLS regression on two components therefore appears to be a possible alternative to ridge regression in this case, but we have to bear in mind the rather abrupt variation of the coefficients and the difficulty of determining the correct number of components to be retained. It is a difficult choice.

The results shown above demonstrate that, in PLS regression, the optimization of the variance of the predictors tends to lead to the optimization of the correlation with the dependent variable, so that PLS regression becomes less similar to least squares regression and more similar to ridge regression and principal component regression, although it is closer to the variant of PCR in which the principal components are chosen not in decreasing order of variance, but as a function of their discriminant power.

LASSO Selection Summary							
Step	**Effect Entered**	**Effect Removed**	**Number Effects In**	**Adjusted R-Square**	**AIC**	**BIC**	**CP**
0	Intercept		1	0.0000	50.4443	15.5939	48.3383
1	X1		2	0.0837	48.5102	12.8199	41.4221
2	X2		3	0.1235	47.9633	11.5128	38.1913
3	X4		4	0.3978	36.4602	1.4441	18.2995
4	X5		5	0.5967*	24.0742*	-7.2038*	5.0000*
*** Optimal Value Of Criterion**							

Figure 11.81 Summary of the steps of the lasso.

The final method of regularized regression to be described in this section is the lasso, introduced in Section 11.7.2. In SAS, the lasso method first appeared in a procedure in Version 9.2. This procedure is GLMSELECT, which, as its name implies, is dedicated to the selection of variables, which it does by various methods, such as the lasso and the LARS (least angle regression) method of Efron *et al.* (2004).[49] The last of these is also the basis for an efficient lasso calculation method, implemented in the *lars* package in R. In GLMSELECT, a variety of selection and stop criteria are available, based on various indicators for selection, and on criteria of significance and validation for stopping. The results can be displayed in graphs, enhanced by ODS GRAPHICS. This procedure can then be linked to a REG or GLM procedure. Like the last of these (Section 11.7.13), GLMSELECT handles qualitative predictors.

The following syntax applies a lasso regression to the preceding variables, taking the Mallows C_p as the predictor selection criterion and as the stop criterion. This means that the predictor selected at each step is the one which minimizes the Mallows C_p, and the selection process is interrupted when this criterion increases again. This stop criterion is not always strict enough, because it is sometimes better to avoid a supplementary variable that does little to decrease the Mallows C_p (or the AIC, BIC, etc.). This is the case here. The instruction STATS = . . . displays the criteria specified in the selection summary table. The instruction DETAILS = all provides a detailed display of the selection process, with the values of the indicators, for each step in the process (Figure 11.81). Finally, with ODS GRAPHICS, the instruction PLOTS = all displays a set of graphs describing the variation of the selection quality indicators.

```
ODS GRAPHICS ON ;

ODS OUTPUT ParameterEstimates = lasso ;

PROC GLMSELECT DATA = ridge.caterpillars PLOTS=all ;
MODEL log = x1 x2 x4 x5 / selection=lasso (CHOOSE=cp STOP=cp)
STATS=(rsquare adjrsq bic aic cp) DETAILS=all ; RUN ;

ODS GRAPHICS OFF;
```

[49] Efron, B., Hastie, T., Johnstone, I., and Tibshirani, R. (2004) Least angle regression (with discussion), *Annals of Statistics*, 32, 407–499.

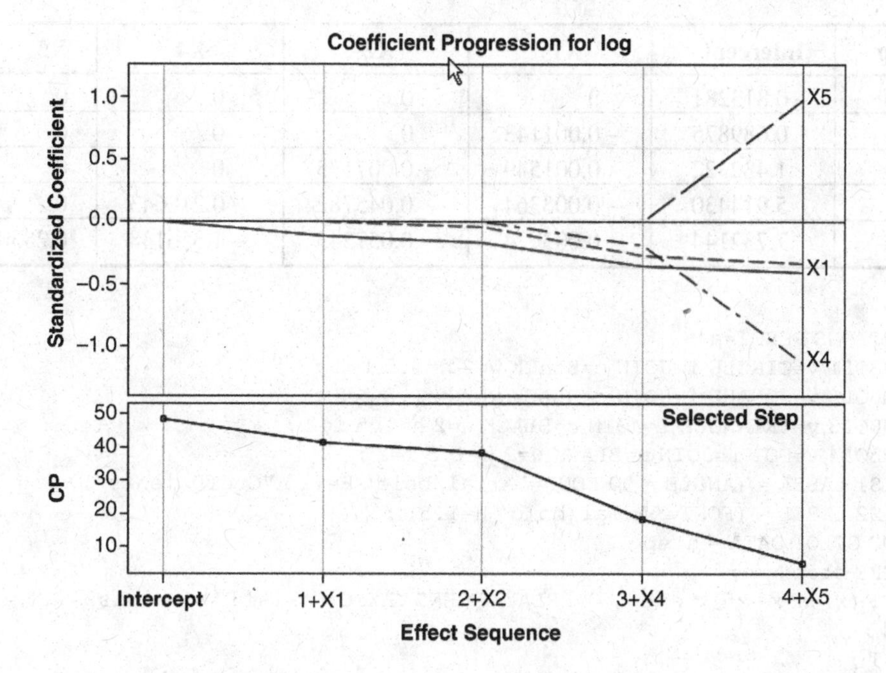

Figure 11.82 Variation of the standardized lasso coefficients and the Mallows Cp.

One of the graphs (Figure 11.82) shows the values of the coefficients of the standardized predictors as a function of the selection step. None of the graphs produced by SAS shows the coefficients of the raw predictors, but we can do this with the GPLOT procedure, after retrieving the values of these coefficients into an ODS data set specified before the procedure. This will allow us to compare these coefficients with those calculated by the other regression methods used in this section.

```
PROC TRANSPOSE DATA = lasso OUT=step (DROP= _name_) ;
WHERE Step GE 0 ;
VAR Estimate ;
BY Step ;
ID Effect ;
RUN ;

DATA step ;
SET step ;
IF x1 = . THEN x1 = 0 ;
IF x2 = . THEN x2 = 0 ;
IF x4 = . THEN x4 = 0 ;
IF x5 = . THEN x5 = 0 ;
RUN ;

PROC PRINT DATA = step ;
RUN ;
```

Step	Intercept	X1	X2	X4	X5
0	– 0.813281	0	0	0	0
1	0.689875	– 0.001143	0	0	0
2	1.432322	– 0.001549	– 0.007175	0	0
3	5.914430	– 0.003364	– 0.045782	– 0.218643	0
4	7.732144	– 0.003924	– 0.057343	– 1.356138	0.283058

```
GOPTIONS RESET=all;
SYMBOL1 v=CIRCLE i=JOIN c=BLACK w=2 h=1.5 l=2 ;
SYMBOL2 v=SQUARE i=JOIN c=BLACK w=2 h=1.5 l=2 ;
SYMBOL3 v=TRIANGLE i=JOIN c=BLACK w=2 h=1.5 l=2 ;
SYMBOL4 v=DOT i=JOIN c=BLACK w=2 h=1.5 l=2 ;
AXIS1 LABEL = (ANGLE = 90 FONT='Arial/bold' H=1.5 "Coefficients") ;
AXIS2 LABEL = (FONT='Arial/bold' H=1.5) ;
PROC GPLOT DATA = step ;
WHERE step > 0 ;
PLOT (x1 x2 x4 x5) * step / OVERLAY LEGEND VAXIS = axis1 HAXIS=axis2 VREF = 0;
RUN ;
QUIT ;
```

Figure 11.83 shows that the coefficients are stable up to the third step, which suggests that we should retain three predictors, a choice confirmed by the graph of the Mallows C_p, which shows a significant drop between the second and third steps. The information displayed by the DETAILS = all option (not reproduced here) shows that the RMSE of the lasso with three

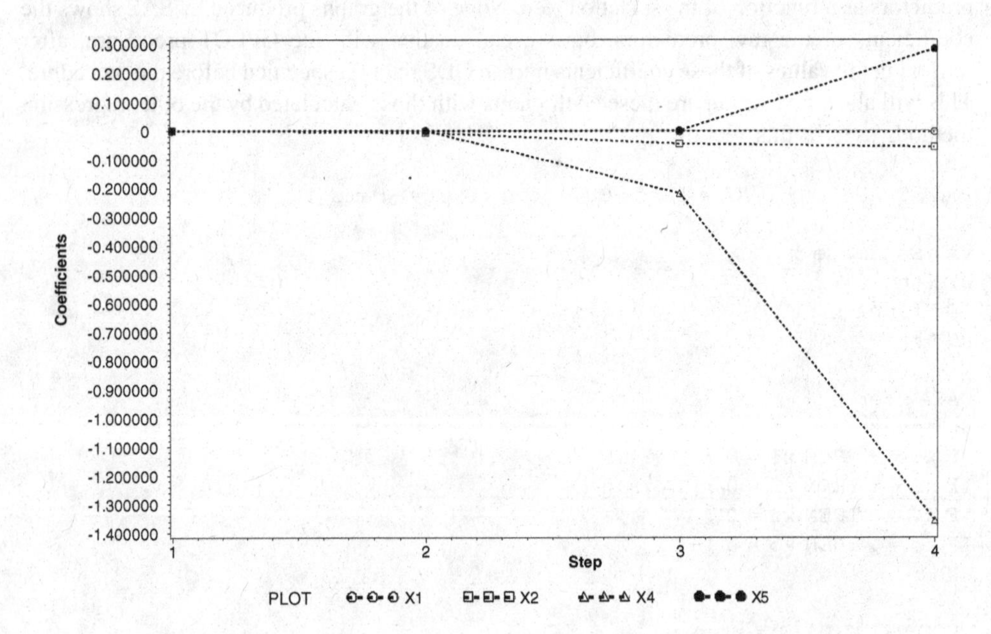

Figure 11.83 Variation of the lasso coefficients.

variables is 0.96611. We can see that this is greater than the value for ridge regression with $k = 0.3$ (RMSE = 0.93015) and for PLS regression on two components (RMSE = 0.89246).

More generally, the graph of the coefficients reveals two facts:

- Unlike the 'ridge plot', the horizontal axis is not a continuous scale, because each value of this axis corresponds to a step, not to a value of the lasso parameter, which changes continuously.
- Each change of step is characterized by the elimination of a coefficient, whereas ridge coefficients can tend towards 0, but are not eliminated in succession.

This can be understood if we remember (see Section 11.7.2) that these coefficients are found under the constraint $\sum_{j=1}^{p} |b_j| \leq t$, instead of the constraint $\sum_{j=1}^{p} b_j^2 \leq C^2$ used for ridge regression. Clearly, if t is sufficiently small, some coefficients can have no other solutions than $b_j = 0$, because other solutions would result in a sum of more than t. However, we have $b_j^2 < b_j$ for $b_j < 1$, which always permit non-zero solutions, even if they are very small, to the quadratic constraint. Because of this characteristic, lasso regression is not only a method of regularization, like ridge regression, but also a method of selecting variables.

In the SAS GLMSELECT procedure, the lasso is implemented solely as a method of selecting variables, and this is why the only values of the parameter t which are considered (though the value of t is not stated) are those which coincide with the elimination of the coefficient of a variable. The intermediate values are disregarded in order to produce the graph and the outputs.

We may feel that it is a pity to eliminate the coefficient of a predictor entirely and thus lose the information that it might provide. We may also regret that the GLMSELECT procedure cannot make the lasso parameter vary continuously, which would allow us to slightly allow for the variable x5, and possibly to bring the RMSE closer to the value for ridge regression.

Where our present problem is concerned, the lasso method does not perform as well as ridge regression and PLS regression on two components, which are still the best methods because they combine a stability of coefficients with a reasonably low RMSE.

Here is a summary of the results:

We can see that the coefficients of PLS regression on three components are close to those of regression on principal components 1, 2 and 4, and also that those of PLS regression on two components are close to those of ridge regression (Figure 11.84).

	x1	x2	x4	x5	RMSE
linear regression	– 0.00392	– 0.05734	– 1.35614	0.28306	0.79065
ridge regression (k = 0.3)	– 0.00328	– 0.04753	– 0.40556	0.05207	0.93015
lasso regression (on 3 variables)	– 0.00336	– 0.04578	– 0.21864	0	0.96611
PCR on components 1, 2, 4	– 0.00155	– 0.07123	– 1.42056	0.26456	0.84074
PCR on components 1, 2, 3	– 0.00437	– 0.06130	– 0.09789	– 0.01593	0.97698
PLS regression on 2 components	– 0.00453	– 0.07013	– 0.28190	0.04017	0.89246
PLS regression on 3 components	– 0.00507	– 0.04142	– 1.14852	0.24373	0.84710

Figure 11.84 Comparison of linear, ridge, lasso, PCR and PLS regressions.

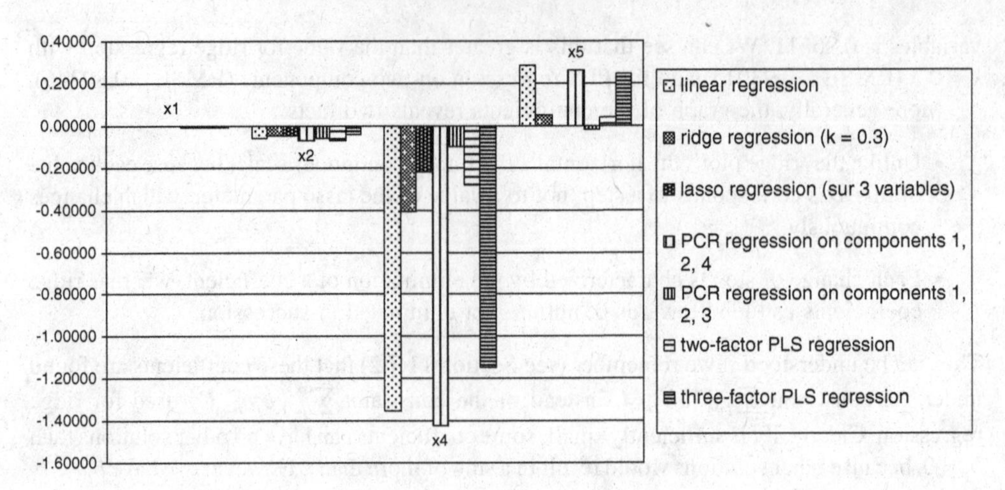

Figure 11.85 Comparison of coefficients in different regression methods.

As for ridge regression, we can express the lasso coefficients as a function of the principal components, instead of the predictors. This leads us to Figure 11.86, which shows a behaviour of the lasso which is intermediate between that of ridge regression and principal component regression: the coefficients are not eliminated one after another as in PCR, but they are decreased less progressively than in ridge regression.

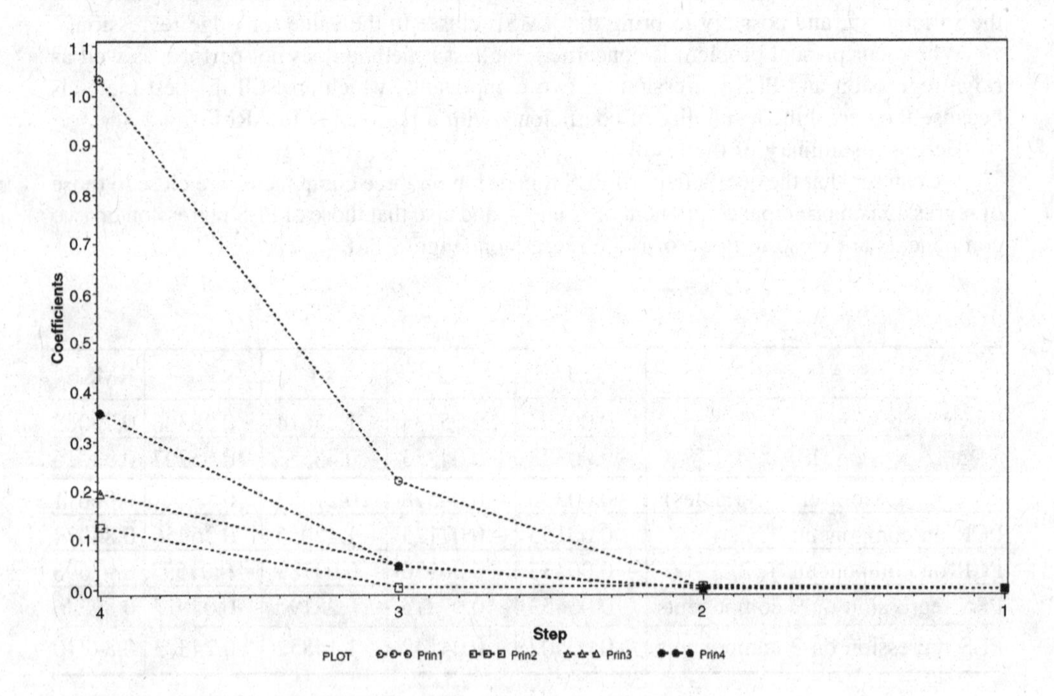

Figure 11.86 Variation of the lasso coefficients of the principal components.

In the lasso, it is the coefficients of some of the predictors, not those of the principal components, that are eliminated. The effect of this selection is therefore less abrupt than in PCR, because it has less effect on the coefficients of the other predictors than the elimination of the coefficient of a principal component. However, we then lose all the information that was contributed by the eliminated predictors.

Before leaving this topic, it is worth looking at the *lars* package mentioned previously, which implements the lasso in R. The lars function is called by entering the predictors and the dependent variable in matrix form, and specifying that we wish to calculate the lasso (other variants, including LARS regression, are available).

```
> library(lars)

> X <- as.matrix(subset(caterpillars,select=c(x1,x2,x4,x5)))
> Y <- as.matrix(subset(caterpillars,select=log))

lasso= lars(X,Y,type="lasso")
```

We obtain the results given above.

```
> summary(lasso)
LARS/LASSO
Call: lars(x = X, y = Y, type = "lasso")
  Df    Rss     Cp
0  1  49.596  48.338
1  2  44.022  41.422
2  3  40.752  38.191
3  4  27.067  18.299
4  5  17.503   5.000
> coef (lasso)
            x1            x2            x4           x5
[1,]   0.000000000   0.000000000   0.0000000   0.0000000
[2,]  -0.001142795   0.000000000   0.0000000   0.0000000
[3,]  -0.001548896  -0.007174823   0.0000000   0.0000000
[4,]  -0.003364430  -0.045782227  -0.2186428   0.0000000
[5,]  -0.003923681  -0.057342697  -1.3561376   0.2830582
```

The example used in this section has enabled us to compare the different methods of regularization, namely ridge, lasso, PLS and PCR. Depending on the choice of the regularization parameter, the regression coefficients that are selected may vary significantly. The values of this parameter are continuous in ridge and lasso regression, but discrete in PLS and PCR. The choice of this parameter is therefore more progressive for ridge and lasso regression; it is made easier by the ‘ridge plot’ which plots the variation of the coefficients as a function of the regularization parameter. The lasso is less progressive, because it eliminates some of the coefficients, and consequently some of the predictors.

With PLS and PCR, the choice of the parameter, in other words the number of components, is less simple and obvious, because the variation of the coefficients is discontinuous (very much so in some cases) and their convergence may be slow. Sometimes, convergence does not take place until the number of components is reduced to one. This means a loss of information and a worsening of the RMSE which is a criterion of fit. In all cases, the RMSE in PLS regression is smaller than the value in principal component regression, in accordance with the de Jong theorem.

An interesting discussion of these methods can be found in Section 3.6 of the book by Hastie, Tibshirani and Friedman (2009), already mentioned in Section 11.3.3 above, which cites a comparative study which finds that ridge regression has the advantage in minimizing the prediction error.

11.7.12 Robust regression

Robust regression algorithms are intended to provide predictions which are less sensitive to outliers and are valid when the residuals of the observations do not have a normal distribution. They are also widely used in statistical software, such as SAS, R, S-PLUS and STATA. They operate by replacing the least sums (or means, which is equivalent) of squares with one of the following expressions (this is not an exhaustive list):

- the sum of the absolute values $\Sigma_i |y_i - a - bx_i|$, a form of regression called L_1, which has been known since 1757 but has the drawback of not having a formula which yields the coefficients and standard errors;
- the 'least winsored squares', with replacement of the values exceeding the extreme percentiles Q_x and Q_{100-x} (for example, Q_1 and Q_{99}) with Q_x and Q_{100-x} themselves;
- the 'least trimmed squares', with elimination of the values exceeding the extreme centiles;
- the median least squares;
- the weighted least squares (weighted by the inverse of the variance of the dependent variable, to compensate for the heteroscedasticity, assuming for example $p_i = s^2/s_i^2$ in the neighbourhood of a point x_i);
- locally weighted least squares on neighbours (LOESS: LOcal regrESSion).

Let us take a closer look at LOESS regression (Cleveland, 1979),[50] also called LOWESS (LOcally WEighted Smoothing Scatter). We choose a smoothing parameter, which represents a certain percentage of the set of points. In the set of x (vector of predictors), a certain number of neighbourhoods are determined by the value of the smoothing parameter. The smaller this percentage, the more numerous the neighbourhoods will be. Note that these neighbourhoods generally overlap, and, even in the case of a parameter equal to 1, there are several neighbourhoods (three in our example below). But the smoothing parameter s determines the number n of points to be taken into account in each local regression: n is the smallest integer less than or equal to (total frequency $\times$ s). If there are 16 observations and $s = 0.6$, then $n = 9$. In each neighbourhood, we weight each of the n points according to a decreasing function of its distance from the centre of the neighbourhood. We then perform a linear (or if necessary quadratic) regression in each neighbourhood, with allowance for the weightings. This principle is similar to that of the moving means method.

[50] Cleveland, W. (1979). Robust locally weighted regression and smoothing scatterplots. *Journal of the American Statistical Association*, 74(368), 829–836.

In SAS, the regression is linear by default, but it is possible to request quadratic regression by adding the option DEGREE=2 to the MODEL instruction (see the syntax below). It is not a global function that is determined, but a fit of polynomials of degree 1 or 2, which are more or less numerous according to the value of the smoothing parameter (Figure 11.87). This modelling by polynomial fitting is both the strength and the weakness of LOESS regression. Its strength lies in its capacity to model complex situations. Its weakness is seen in the absence of a mathematical function to describe the regression in analytical terms (and enable it to be programmed), in the need for a rather large number of observations, and in the greater complexity of calculation, which may be significant for large data volumes. LOESS regression is implemented in SAS, R and S-PLUS. In SAS it is implemented by the special LOESS procedure:

```
PROC LOESS DATA = test;
MODEL Y = X / SMOOTH = .2 TO 1 BY .2 ;
ODS OUTPUT outputstatistics = testloess;
RUN;

PROC SORT DATA = testloess; BY SmoothingParameter X; RUN;
SYMBOL1 c=black v=dot;
SYMBOL2 c=black v=point i=join l=1;
PROC GPLOT DATA = testloess;
BY SmoothingParameter;
PLOT DepVar*X = 1 Pred*X = 2 / OVERLAY;
RUN;
QUIT;
```

The LOESS procedure can be used to test several values of the smoothing parameter at once, and to record the results in a table specified by the command ODS OUTPUT

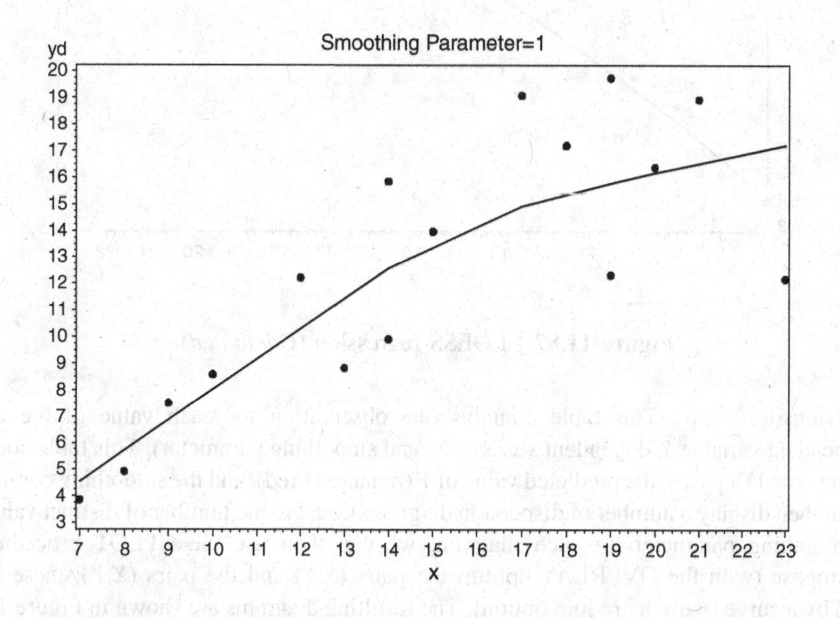

Figure 11.87 LOESS regression: effect of the smoothing parameter.

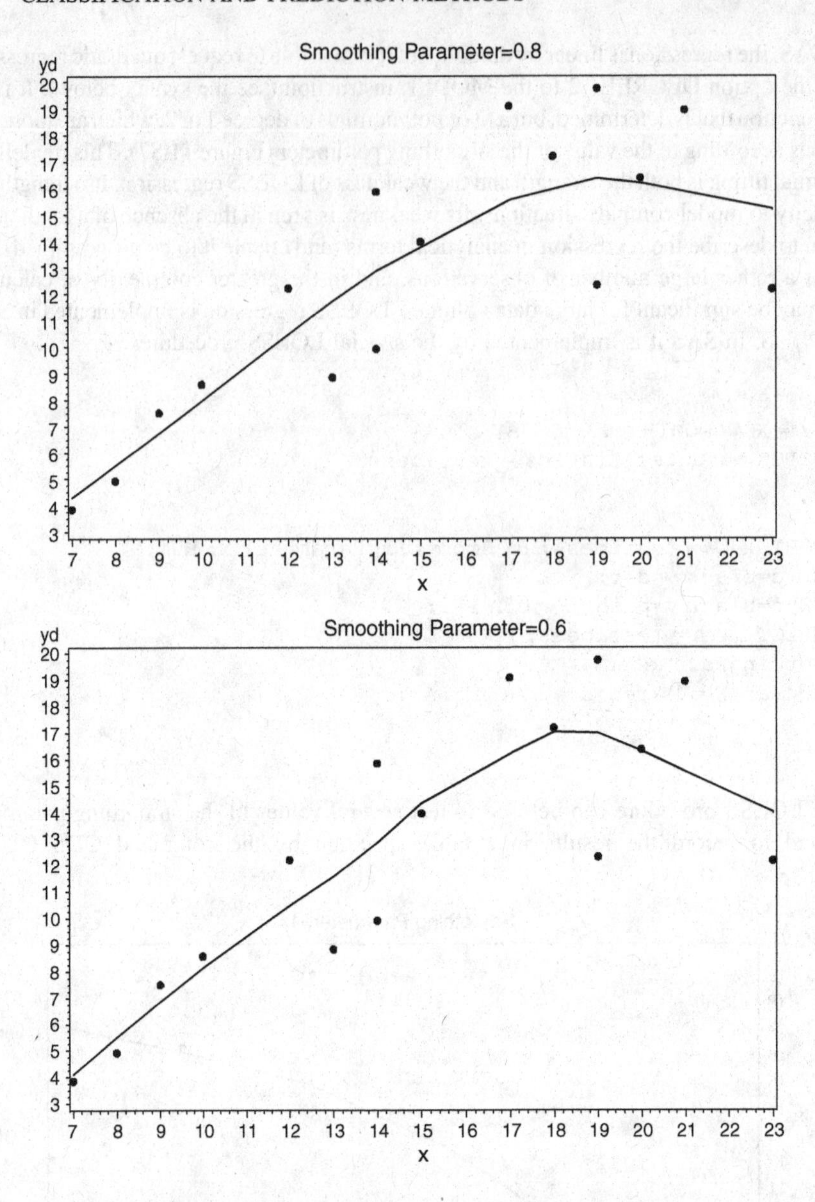

Figure 11.87 LOESS regression (*Continued*).

outputstatistics = ... This table contains one observation for each value of the triplet (independent variable *Y*, dependent variable *X*, and smoothing parameter). This table contains *X*, *Y* (renamed DepVar), the predicted value of *Y* (renamed Pred), and the smoothing parameter. We can then display a number of dispersion diagrams equal to the number of distinct values of the smoothing parameter. In each diagram, we can then use the GPLOT procedure to superimpose (with the OVERLAY option) the pairs (X,Y) and the pairs $(X,\hat{Y})$, these being linked by a curve (with the i=join option). The resulting diagrams are shown in Figure 11.87, which applies, by way of example, to the fourth (*yd*) of Tomassone *et al.* (see Section 11.7.6).

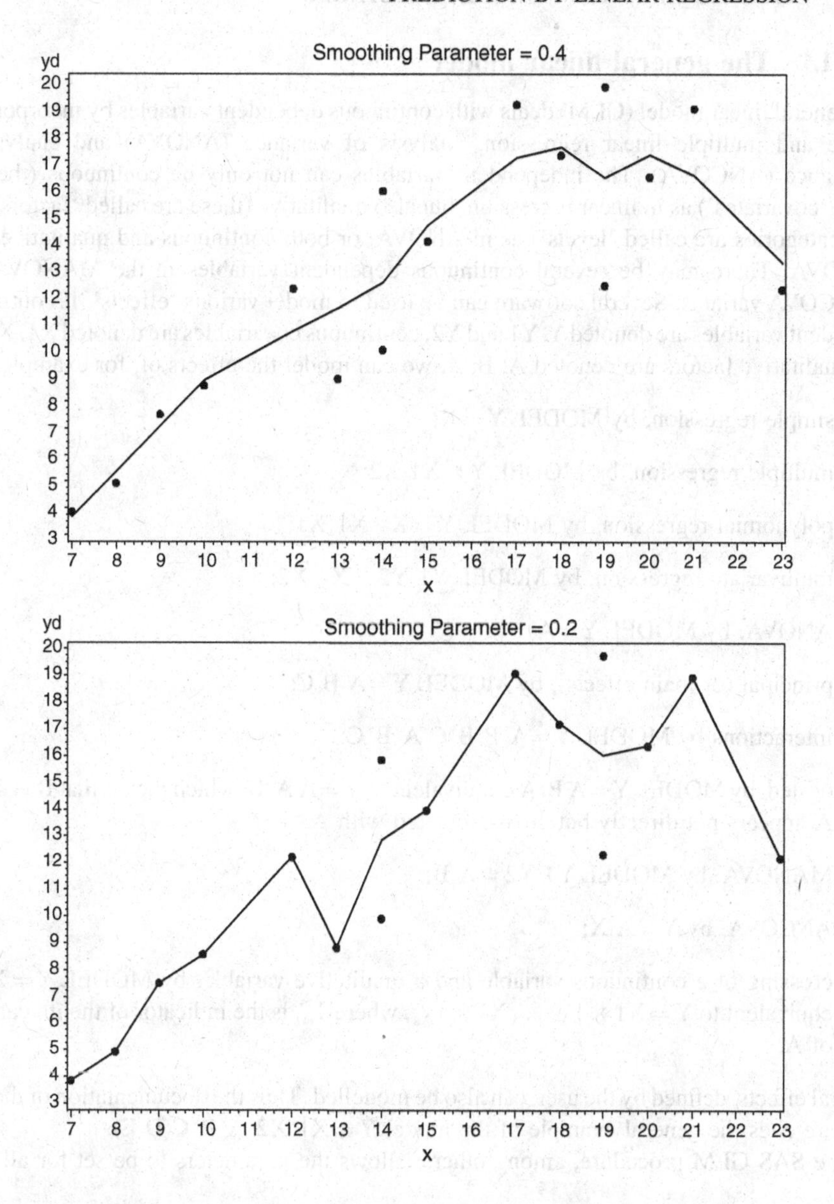

Figure 11.87 LOESS regression (*Continuation and Conclusion*).

The LOESS procedure also offers an option for specifying a quality-of-fit criterion, such as the AIC (Section 11.8.5), and for automatically obtaining the value of the smoothing parameter which optimizes the criterion (by means of the option SELECT = <criterion> in the MODEL instruction). In fact, there are variants of these criteria for avoiding overfitting (if the smoothing parameter is too small) which would result from the use of the conventional criteria. If two values of the smoothing parameter are optimal, the largest value is chosen. In our example, with the modified AIC, the optimal smoothing parameter is 0.78. Referring to the LOESS curves described above, and to Figure 11.5, the choice of a value close to 0.78 appears entirely justified.

11.7.13 The general linear model

The general linear model (GLM) deals with continuous dependent variables by incorporating simple and multiple linear regression, analysis of variance (ANOVA) and analysis of covariance (ANCOVA). The independent variables can not only be continuous (they are called 'covariates') as in linear regression, but also qualitative (these are called 'factors', and their categories are called 'levels') as in ANOVA, or both continuous and qualitative as in ANCOVA. There may be several continuous dependent variables in the MANOVA and MANCOVA variants. Several software can be used to model various 'effects'. If continuous dependent variables are denoted Y, Y1 and Y2, continuous covariables are denoted X1, X2,. . ., and qualitative factors are denoted A, B. . ., we can model the effects of, for example:

- simple regression, by MODEL Y = X;
- multiple regression, by MODEL Y = X1 X2;
- polynomial regression, by MODEL Y = X1 X1*X1;
- multivariate regression, by MODEL Y1 Y2 = X1 X2;
- ANOVA, by MODEL Y = A;
- principal (or main effects), by MODEL Y = A B C;
- interactions by MODEL Y = A*B B*C A*B*C;
- nested, by MODEL Y = A B(A), equivalent to Y = A A*B, which means that B nested in A appears not directly but cross-tabulated with A;
- MANOVA, by MODEL Y1 Y2 = A B;
- ANCOVA, by Y = A X;
- crossing of a continuous variable and a qualitative variable, by MODEL Y = X1*A, equivalent to $Y = X1 \times 1_{A1} \ldots X1 \times 1_{Ak}$, where 1_{Ai} is the indicator of the *i*th category of A.

General effects, defined by the user, can also be modelled. Thus the documentation of the SAS software cites the general example of the model Y = X1*X2*A*B*C(D E).

The SAS GLM procedure, among others, allows the parameters to be set for all these effects.

The GLM can handle models with fixed, random or mixed effects. It also handles models with repeated measures. Factors and covariables can have *fixed effects*. They are controlled by the experimenter and all their values must be taken into account. Since the aim is to quantify their effect on the dependent variable, this analysis is similar to a regression analysis (predictive purpose).

Factors and covariables can also have *random effects*, only a sample of the values of these being known. This is not a matter of quantifying their effect, but simply the proportion of the variance of the dependent variable that they account for. This analysis is similar to a correlation analysis (descriptive purpose). It is for example performed to compare the effect of two packages of a product on the purchases of consumers in a number of stores, or to compare two treatments given to patients in a number of hospitals. If the aim is not to predict the result of a treatment as a function of the hospital, but simply to avoid the bias due to the place where treatment is given, the 'hospital' variable is stated to be a random effect.

Mixed-effects models are mixtures of fixed and random effects models.

In the SAS software, the GLM and MIXED procedures take fixed, random and mixed effects into account. Here is an example of the syntax:

```
PROC GLM DATA=example ;
CLASS department degree ;
MODEL salary = department degree ;
RANDOM department ;
LSMEANS degree ;
RUN ;
```

In this example, we wish to account for an employee's salary on the basis of his degree (fixed effect) while allowing for the department to which he belongs (random effect).

The syntax of the MIXED procedure is very similar to that of the GLM procedure, and the estimates it provides are the same. However, the standard errors associated with these estimates are larger in the MIXED procedure, which takes into account not only the variance of the error (mean squared residuals) like the GLM procedure, but also the variance of the random effect. The MIXED procedure is preferable for the analysis of random effects, even if the calculations are more complex and lengthy for large data volumes. Other reasons for preferring the MIXED procedure are indicated below, in the context of repeated measures models. In both procedures, the LSMEANS instruction launches the calculation of the mean least squares for each category of the fixed effect (or fixed effects if appropriate).

```
proc MIXED data=example ;
CLASS department degree ;
MODEL salary = degree ;
RANDOM department ;
LSMEANS degree ;
RUN ;
```

The measurements $y_1, y_2, \ldots, y_k$ of the dependent variable Y on a number of individuals can be correlated (longitudinal data) in the case of a single individual observed k times (for example, before and after a medical treatment) or k individuals sharing a common characteristic (same family, same segment). We therefore have to abandon the assumption of linear regression that there is no correlation between the measures for a number of individuals. In this case, we must use a repeated measures model. As we will see, there is a model of this type for logistic regression (Section 11.9.2), because Y can be continuous or discrete. A repeated measures model can cope with both fixed and random effects simultaneously. A repeated measures model can handle the following effects:

- 'within-subject effects', which can vary for a single individual, such as the effect of time or treatment (comparison of a patient before and after treatment);
- 'between-subject effects', constant for the observations of a single individual, such as the effect of the patient's characteristics, age, sex, blood group, etc. (comparison of the patient with others);
- 'within-subject-by-between-subject effects', such as the interactions between the treatment and the patient's characteristics.

In the SAS software, GLM and MIXED are used to execute the above analyses, with the following differences:

- the GLM procedure is based on least squares and the MIXED procedure is based on the maximum likelihood;
- the GLM procedure does not handle observations with missing values;
- the MIXED procedure handles qualitative and continuous within-subject effects, whereas GLM handles only qualitative within-subject effects;
- the MIXED procedure automatically determines within-subject-by-between-subject effects, and automatically distinguishes within-subject effects from between-subject effects, unlike GLM, in which they have to be distinguished explicitly;
- the MIXED procedure requires an explicit specification of the within-subject variance–covariance matrix, but offers a choice of different matrix forms.

In the example below, sex is a between-subject effect and age is a within-subject effect; a measurement is made four times for each individual, at four different ages. GLM considers age to be a qualitative variable, with four specified levels: 8, 10, 12 and 14 years. If unspecified, they are considered to be 1, 2, 3 and 4.

```
PROC GLM DATA=glm_example ;
CLASS sex ;
MODEL y1-y4 = sex ;
REPEATED age 4 (8 10 12 14) ;
RUN ;
```

The equivalent syntax with the MIXED procedure is as follows:

```
PROC MIXED DATA= mixed_example ;
CLASS sex age individual ;
MODEL y = sex|age ;
REPEATED / TYPE=cs SUB=individual ;
RUN ;
```

We find that the between-subject and within-subject effects are not explicitly distinguished, but that a within-subject variance–covariance matrix has been specified (see Section 11.9.2 for details of this). The variable 'individual' has been specified, because the input data set structure is not the same for both procedures.

For GLM, the values of age are not specified, because they are given in the syntax of the procedure:

```
individual    sex  y1    y2     y3     y4
1              F  100   115    130    145
2              M  110   130    150    170
...
```

For the MIXED procedure:

```
individual   sex    y    age
1             F    100     8
1             F    115    10
1             F    130    12
1             F    145    14
2             H    110     8
```

11.8 Classification by logistic regression

Although its roots lie in the early history of data analysis,[51] logistic regression was introduced into software more recently than linear discriminant analysis, possibly because of its greater complexity of calculation, and has therefore only recently become a regularly used tool for most statisticians. Because of its many qualities, logistic regression is now tending to take the place of its rival in many classification problems, especially scoring. This development is being aided by the continual improvement and generalization of logistic regression in the context of the generalized linear model, which I will describe briefly. Daniel L. McFadden was responsible for some of these enhancements, for which he won the Nobel Prize for Economics in 2000. Logistic regression is becoming universal, because it can handle dependent variables with two values (without making such restrictive assumptions as those of discriminant analysis), with $k \geq 3$ *ordered* values (which discriminant analysis cannot tackle), or with $k \geq 3$ nominal values, and the independent variables can be quantitative or qualitative. Moreover, its results are highly explicit, especially in its *logit* version using the odds ratios which are very popular in medicine and epidemiology. Finally, logistic regression is one of the most reliable classification methods, and this reliability is easy to monitor using a number of statistical indicators. Linear discriminant analysis continues to be very useful in the case of multinormal and homoscedastic continuous independent variables, since it provides a direct solution, whereas logistic regression only offers an approximation to this. However, the generality, interpretability and robustness of logistic regression are three major strong points of this fundamental technique. I might add that the logistic distribution and its S-shaped distribution function are encountered in many fields, including demographics, epidemiology (in relation to the spread of an epidemic), psychology (progression of learning), technology (spread of a new technology) and marketing (sales of a new product).

11.8.1 Principles of binary logistic regression

Binary logistic regression deals with one binary dependent ('target') variable $Y = 0$ or 1, and p independent variables X_j which may be continuous, binary or qualitative (for which the indicators lead to the case of a binary variable). In the following notation the p variables X_j are grouped in a vector $X = (X_1, X_2, \ldots, X_p)$. As we will see below, qualitative dependent variables

[51] In 1838, Pierre-François Verhulst wrote about a 'logistic equation', which he introduced to model the growth of a human population while taking the available resources into account, following the ideas of Thomas Malthus. See Verhulst, P.-F. (1838) Notice sur la loi que la population poursuit dans son accroissement. *Correspondance Mathématique et Physique*, 10, 113–121.

with $k > 2$ categories are treated with what is known as polytomous logistic regression, in which the k categories can be ordered (ordinal logistic regression) or not (multinomial logistic regression = nominal logistic regression). Here, if $p = 1$ the logistic regression is *simple*; if $p > 1$, it is *multiple*.

In any regression problem, the aim is to write the conditional expectation of the dependent variable Y as a linear combination of regressors X. We consider the expectation because it is only on average, as calculated for each p-tuplet of values of the p regressors, that the dependent variable is linear with respect to the regressors: the dependent variable may fluctuate about its mean. The aim of logistic regression is therefore the same as in all regressions, namely to model the conditional expectation E(Y/X=x). We wish to know the mean of Y for any value of X. For a value Y equal to 0 or 1 (Bernoulli distribution), this mean is the probability that $Y = 1$. Thus we have

$$\mathrm{E}(Y/X = \mathrm{x}) = \mathrm{Prob}(Y = 1/X = \mathrm{x}).$$

In linear regression, we aim to draw a hyperplane through the middle of the cloud of points $(x_1, x_2, \ldots, x_p, y)$, so that the set of the mean values of Y for all values of X is approximated by this hyperplane, which has the equation

$$\mathrm{E}(Y/X = x) = \beta_0 + \beta_1 X_1 + \beta_2 X_2 + \ldots + \beta_p X_p.$$

Clearly, this approximation is not appropriate if $Y = 0$ or 1, since the term $\beta_0 + \beta_1 X_1 \ldots + \beta_p X_p$ is then unbounded, whereas Prob($Y = 1/X = x$) is bounded in the interval [0,1].

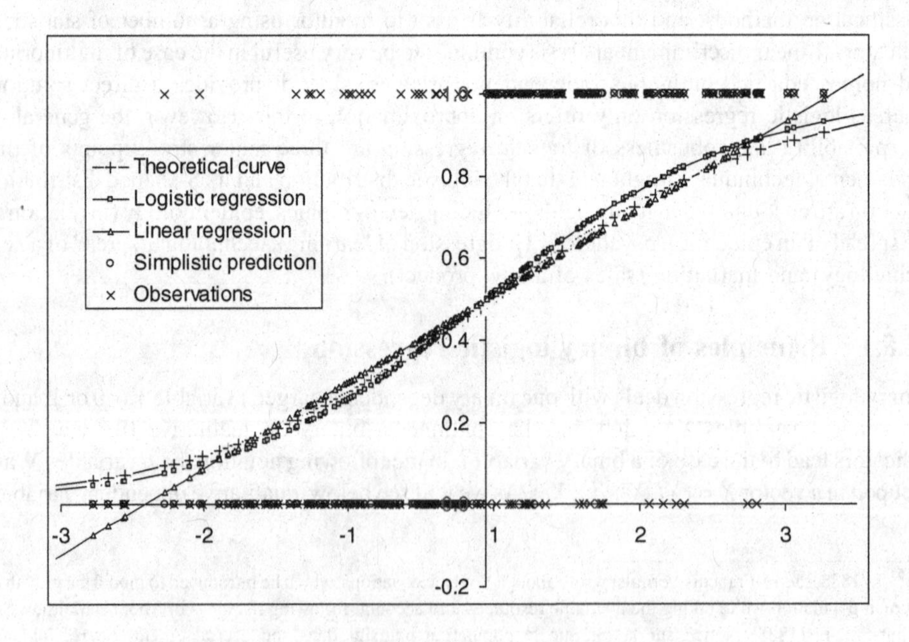

Figure 11.88 Comparison of linear and logistic regression.

In fact, in the favourable situation (in which we are interested) where X can discriminate the values of Y, the behaviour of the cloud of points (x,y) is typically as shown in Figure 11.88 (where $p=1$): when x is small, then $y=0$ as a rule, and when x is large, $y=1$ in most cases. To simplify, this is a classic case of a multinormal homoscedastic variable x: $x \approx N(0,1)$ over the set of points such that $y=0$, and $x \approx N(1,1)$ over the set of points such that $y=1$. We will subsequently consider the extreme case of 'complete separation' where there is an x_0 such that $y=0$ for every $x \leq x_0$ and $y=1$ for every $x > x_0$: paradoxically, logistic regression cannot find a solution to this apparently simple problem.

Let us return to the case of Figure 11.88: evidently, the values $\text{Prob}(Y=1/X=x)$ as x varies follow the theoretical curve represented by the crosses '+' in the diagram. It is an S-shaped curve, not a straight line. More precisely, with the assumptions accepted here on the distribution of x, the Bayes formula (see Section 11.6.4) shows that the curve of the function $x \rightarrow \text{Prob}(Y=1/X=x)$ is given by $f_{N(1,1)}(x)/[f_{N(1,1)}(x)+f_{N(0,1)}(x)]$, where $f_{N(\mu,\sigma)}$ is the density function of the distribution $N(\mu,\sigma)$. With the SAS software, this is found by the syntax pdf('normal',x,1,1)/(pdf('normal',x,0,1) + pdf('normal',x,1,1)). Note that S-shaped curves are common in medicine and biology, fields which saw the birth of logistic regression.

In practice, if the points are highly dispersed, in other words if there are very few points for a given value x of X, it will not be possible to calculate $\text{Prob}(Y=1/X=x)$ directly, and we will have to group the values of X in brackets to estimate the probability that $Y=1$ given x by the proportion of the $Y=1$ given x. This is what is done in the classic example in Chapter 1 of the well-known treatise by Hosmer and Lemeshow.[52]

Patient no.	Age	CHD
1	20	0
2	23	0
3	24	0
4	25	0
5	25	1
...	...	...
97	64	0
98	64	1
99	65	1
100	69	1

This example relates to coronary heart disease (CHD). Hosmer and Lemeshow studied 100 patients whose ages ranged from 20 to 69 years, and noted the occurrence (CHD = 1) or non-occurrence (CHD = 0) of the disease in them.

Given the small number of cases, we cannot calculate $\pi(x) = \text{Prob}(\text{CHD}=1/\text{age}=x)$ and we must create groupings by age classes:

[52] Hosmer, D.W. and Lemeshow, S. (1989) *Applied Logistic Regression*. New York: John Wiley & Sons, Inc.; 2nd edn, 2000.

Age class	No. of patients	CHD = 0	CHD = 1	Proportion of CHD = 1
20–29	10	9	1	0.10
30–34	15	13	2	0.13
35–39	12	9	3	0.25
40–44	15	10	5	0.33
45–49	13	7	6	0.46
50–54	8	3	5	0.63
55–59	17	4	13	0.76
60–69	10	2	8	0.80
TOTAL	100	57	43	0.43

Note that this procedure of grouping in classes is very common in scoring. It enables us to plot Figure 11.89: another S-curve! This shape of curve has already been encountered in relation to neural networks: it is the sigmoid or logistic curve. Following the expression of this curve, we can write $\pi(x) = \text{Prob}(Y=1/X=x)$ in the form

$$\pi(x) = \frac{e^{\beta_0 + \sum_j \beta_j x_j}}{1 + e^{\beta_0 + \sum_j \beta_j x_j}},$$

an equation equivalent to

$$\log\left(\frac{\pi(x)}{1-\pi(x)}\right) = \beta_0 + \beta_1 x_1 + \ldots + \beta_p x_p.$$

The function $f(p) = \log(p/(1-p))$ is called the *logit*. It is a special case of the *link functions* found in logistic regression and in the generalized linear models that I will discuss later. In this

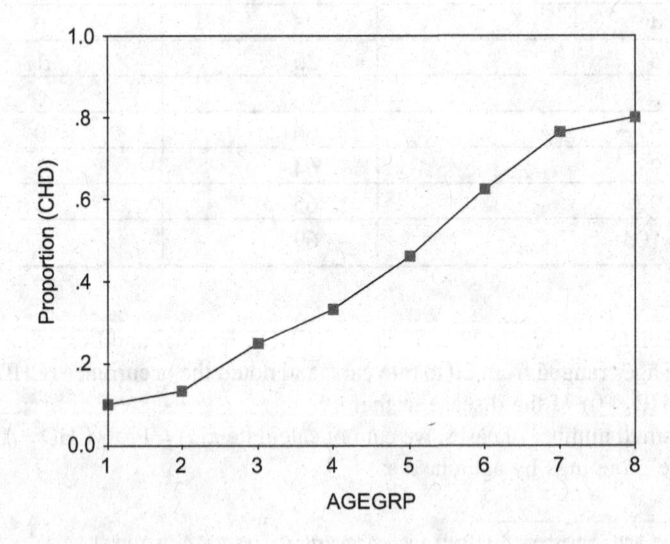

Figure 11.89 Proportion of CHD by age group.

Table 11.1 Logistic regressions.

Model	**Link function**	**Transfer function**
Logit	$\log(\mu/(1-\mu))$	$\frac{e^t}{1+e^t}$
Probit (normit)	Inverse function of the distribution function of the reduced centred normal distribution	$s(t) = \int_{-\infty}^{t} \frac{e^{-z^2/2}}{\sqrt{2\pi}} dz$
Log-log (complementary)	$\log[-\log(1-\mu)]$	$1 - e^{-e^t}$

kind of model, it is not the expectation $\pi(x) := E(Y/X=x)$ that is written as a linear combination of the independent variables, but $f(\pi(x))$, where f is the link function. In the most widespread form of logistic regression, the logit of the conditional expectation is modelled as a linear combination of the independent variables. This format is compatible with the Bayesian rule of discriminant analysis and the calculation of the *a posteriori* probability based on this in the case of a normal distribution of X/Y with equality of the variances and equality of the *a priori* probabilities (see Section 11.6.4). But there are other S-curves as well.

11.8.2 Logit, probit and log-log logistic regressions

Although the *logit* variant of logistic regression appears to be the natural form, if only because of its link with discriminant analysis (see Section 11.6.4), there are other variants, listed in Table 11.1 together with their link and *transfer functions* (the inverse of the link).

The *log-log* (also called *gombit* with reference to Gompertz) has an S-curve which is not symmetrical but very close to that of the logit when t is small and the probability is less than 0.1 (see Figure 11.90). It is widely used in epidemiology and toxicology, for example in the calculation of the probability that a person will be infected as a function of his age, or that an insect will be killed as a function of the dose of insecticide.

The log-log model is also found in survival analysis. Prentice and Gloeckler[53] have shown that if the survival data (collected for a set of individuals at discrete time intervals) follow the Cox proportional hazards regression model (see Section A.2.14), then their likelihood is that of a binary logistic model with a log-log link function, in which each individual i contributes for k_i terms which correspond to k_i independent observations. In this case, the survival model and the log-log logistic model are equivalent. We can carry out a logistic regression, treating each time interval for each individual as an observation: for each observation, there is a binary response indicating whether or not the individual has died (or, more generally, whether or not a certain event has occurred) in this time interval.[54]

Because of its link with the normal distribution, the *probit* is sometimes called *normit*. It is much less popular than the *logit* at present. Although its transfer function has an S-shaped plot

[53] Prentice, R.L. and Gloeckler, L.A. (1978). Regression analysis of grouped survival data with applications to breast cancer data. *Biometrics*, 34, pp. 57–67.

[54] For further details, see Nakache, J.-P. and Confais, J. (2003) *Statistique Explicative Appliquée*. Paris: Technip.

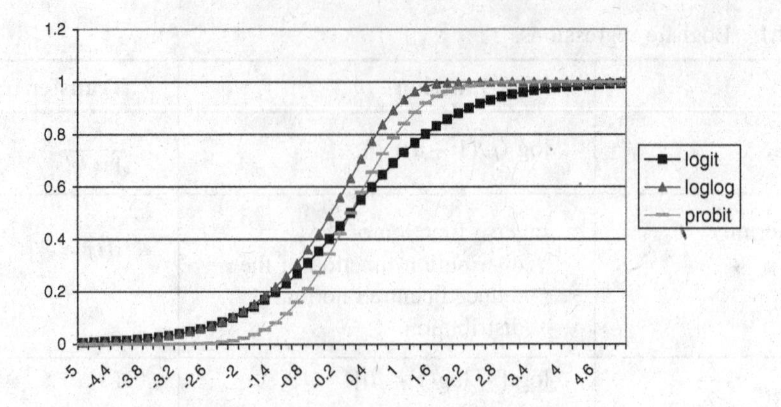

Figure 11.90 Comparison of S-curves.

resembling that of the logit, the probit has fewer benefits that the logit, especially in the interpretation of coefficients, as we shall see in due course. Moreover, if we compare the curves, we will see that the probit curve falls faster towards probabilities close to 0, and rises faster towards probabilities close to 1: the probit is therefore not recommended when numerous cases have a high or low probability, in other words when the distribution is long-tailed (as in a risk score with a high concentration of risk and a large population free from risk). In reality, the logit corresponds to the behaviour of a larger number of physical and chemical phenomena, which explains its popularity. In any case, the probit generally leads to the same classification as the logit, because it supplies coefficients which are approximately proportional to those of the logit:

$$\text{coefficients (logit)} \approx \frac{\pi}{\sqrt{3}} \text{ coefficients (probit)}.$$

So we may as well benefit from the flexibility and simplicity of calculation of the logit. It is worth noting that, when the probabilities to be predicted are close to 0.5, a good estimate is obtained by multiplying the probit coefficients not by 1.81 (the value shown above), but by 1.6, which is virtually the ratio of the probability densities in 0 of the standardized normal distribution $\left(e^{-z^2/2}/\sqrt{2\pi}\right)$ and logistic distribution of parameter 1 $(e^z/(1+e^z)^2)$.[55] Why are we introducing these probability densities? Because the two transfer functions can be written as a distribution function. The probit transfer function is the distribution function of the standardized normal distribution (see Table 11.1). The logit transfer function can be written

$$\frac{e^t}{1+e^t} = \int_{-\infty}^{t} \frac{e^z}{(1+e^z)^2} dz.$$

[55] Davidson, R. and McKinnon, J.G. (1993) *Estimation and Inference in Econometrics*. New York: Oxford University Press.

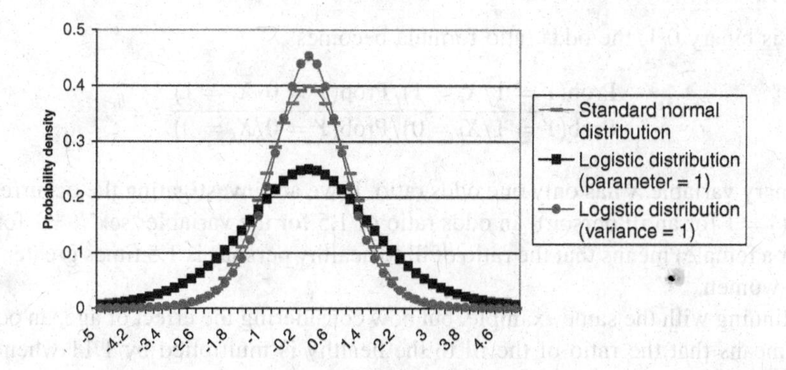

Figure 11.91 Density functions of logistic and normal distributions.

Thus the logit transfer function takes the form of the distribution function of the logistic distribution with parameter 1. You can see that the two transfer functions are similar. In fact, the two distributions – normal and logistic – are themselves similar, with the same symmetrical 'bell curve' shape; the logistic distribution is slightly more pointed than the normal distribution, with a kurtosis of 4.2 instead of 3. The resemblance is more striking when we consider the logistic distribution with parameter $\sqrt{3/\pi^2}$, which has a variance of 1 like the reduced normal distribution, but is more pointed (Figure 11.91). The logistic distribution of parameter 1 is more flattened than the reduced normal distribution, and consequently the logit transfer function decreases more slowly than the probit function towards probabilities close to 0, and increases more slowly towards probabilities close to 1. However, it has a standard deviation of $\pi/\sqrt{3}$, which explains why the logit coefficients have this proportionality factor with the probit coefficients (see above).

11.8.3 Odds ratios

The odds ratio of an independent variable measures the variation of the ratio of the probabilities of the occurrence of the event $Y=1$ against $Y=0$ (like the 'odds' in gambling[56]) when X_i changes from x to $x+1$. In this case, logit($\pi(x)$) increases by the coefficient β_i of X_i, and the odds $\pi(x)/[1-\pi(x)]$ are multiplied by $\exp(\beta_i)$. This is written

$$OR = \frac{\pi(x+1)/[1-\pi(x+1)]}{\pi(x)/[1-\pi(x)]} = e^{\beta_i}.$$

Watch out for a misleading (but sometimes deliberate) simplification: the odds ratio is different from the relative risk $\pi(x+1)/\pi(x)$, unless $\pi(x)$ is small (as in the detection of a rare phenomenon).

[56] If the probability of winning is 0.8, the odds are $0.8/0.2=4$, because there are four times as many chances of winning as losing. If winning is certain, the odds are infinite.

If X_i is binary 0/1, the odds ratio formula becomes

$$OR = \frac{\text{Prob}(Y=1/X_i=1)/\text{Prob}(Y=0/X_i=1)}{\text{Prob}(Y=1/X_i=0)/\text{Prob}(Y=0/X_i=0)} = e^{\beta_i}.$$

A binary variable X has only one odds ratio. If we are investigating the occurrence of a disease ($Y=1$ for an ill person), an odds ratio of 1.5 for the variable 'sex' (= 1 for a male and 0 for a female) means that the ratio of ill to healthy persons is 1.5 times greater for men than for women.

Continuing with the same example, but now considering the effect of age, an odds ratio of 1.11 means that the ratio of the ill to the healthy is multiplied by 1.11 whenever the age increases by a year. The ratio will be multiplied by 10 after 9 years. We can guess that it is not always relevant to compare age 61 and 60, 60 and 59, etc., with the same odds ratio, because the progress of morbidity is not necessarily the same over the whole of life. There is also a risk of being derailed by a lack of robustness of the model, due to a lack of data (see the CHD example above). The odds ratios of continuous variables therefore give rise to two difficulties, namely the failure to take non-linearity into account and the lack of robustness. When one of these problems threatens to arise, we can divide the continuous variable into classes.

As for qualitative variables, these have a number of odds ratios equal to the number of categories minus one, because one of the categories is used as a reference and its coefficient is generally set to 0 (this is the most common and convenient arrangement, but this coefficient can also be the opposite of the sum of all the other coefficients). Imagine that we are comparing the probability $\pi(x)$ of occurrence of the disease in cities, towns and the countryside. If we choose 'countryside' as the reference category, we might obtain the results shown in Table 11.2. This table shows that, when we move from the reference category 'countryside' to the 'town' category, the proportion $\pi(x)/[1-\pi(x)]$ of ill persons with respect to healthy persons is multiplied by the exponential 0.573 of the difference of the coefficients B associated with the 'town' category ($B = -0.557$) and with the reference category ($B = 0$). In other words, the proportion of ill persons is almost twice as small in a town as it is in the countryside.

In general terms, an odds ratio less than 1 (a B coefficient less than 0) indicates a negative effect of the independent variable on the dependent variable, and an odds ratio of greater than 1 (a B coefficient greater than 0) indicates a positive effect.

Software packages often suggest the last category as the reference category by default, but sometimes they offer the user a choice (see the example of SAS syntax below). In this case, we

Table 11.2 Example of odds ratios.

							95% CI for Exp(B)	
	B	S.E.	Wald	DF	Sig.	Exp(B)	Lower	Upper
Country			36.671	2	0.000			
Small town	– 0.557	0.136	16.784	1	0.000	0.573	0.438	0.748
City	0.288	0.143	4.057	1	0.044	1.334	1.008	1.765
Constant	– 1.256	0.236	28.363	1	0.000	0.285		

may choose the most frequent category for reference, but it may be preferable to choose a reference category which is extreme with respect to the target (being weaker or more hazardous, for example), to ensure that the *B* coefficients of the categories of the variable all have the same sign. There may be a gain in readability, especially when the number of categories exceeds three.

If all the independent variables are qualitative, the set of reference categories is represented by the constant β_0: a 'mean' individual whose categories are all reference categories has a probability of $\pi(x) = \text{Prob}(Y=1/X=x) = \exp(\beta_0)/[1 + \exp(\beta_0)]$. Clearly, therefore, it will be useful to choose the most frequent category of all the variables for reference: there is less risk of incompatibility of the various reference categories, and there will be more chance of the existence of the 'average' individual.

11.8.4 Illustration of division into categories

To demonstrate the usefulness of dividing a continuous variable into classes in the construction of a predictive model, I have compared the performance of the models constructed where the same four independent variables are:

- left in their initial continuous form;
- divided into classes and considered as ordinal variables;
- divided into classes and considered as nominal variables.

I have used the area under the ROC curve to compare the performances; I will show later that this area is a good global performance indicator, and that the model becomes more precise as the area approaches 1.

We can see (Figure 11.92) that the division into nominal categories gives the best performance. This is because, instead of having a single odds ratio, which would assume that the probabilities vary identically between 0 and 1, 1 and 2, 2 and 3, etc., there are as many odds ratios as there are categories, allowing us to take non-linear, and even non-monotonic, responses into account. Division into ordinal categories means that we consider the number of variables as a discrete variable 1, 2, and so on, with which a single odds ratio is associated.

Area Under the Curve

				Asymptotic 95% Confidence Interval	
Test Result Variable(s)	Area	Std. Error[a]	Asymptotic Sig.[b]	Lower Bound	Upper Bound
Ordinal variables	,834	,008	,000	,818	,850
Nominal variables	,836	,008	,000	,820	,852
Scale variables	,820	,010	,000	,801	,839

a. Under the nonparametric assumption

b. Null hypothesis: true area = 0.5

Figure 11.92 Effect of the division of a continuous variable into classes.

By comparison with division into nominal categories, we lose the modelling power of multiple odds ratios, but we gain in robustness by comparison with the original continuous variable, which explains the very acceptable performance of this model. However, we may wonder whether the slightly better performance of nominal division as compared with ordinal division justifies the complexity of the multiple odds ratios.

11.8.5 Estimating the parameters

The parameters to be estimated in a logit logistic model are the coefficients β_i of the linear combination expressing the logit of the probability Prob(Y=1/X=x). Logistic regression, and more generally the generalized linear model, differs from the simple linear model in that the parameters are estimated not by the least squares method, but by the maximum likelihood method.[57] We must examine this method more closely.

Invented by the leading statistician Sir Ronald Fisher, it involves estimating a parameter β of the distribution of a random variable X in view of a certain number of independent observations, by writing a likelihood function which is a function of β, of which the maximum is to be found.

If the distribution is discrete, the likelihood function is written as follows, by definition:

$$L(\beta, x^1, x^2, \ldots, x^n) = \text{Prob}_\beta(X = x^1) \times \text{Prob}_\beta(X = x^2) \times \ldots \times \text{Prob}_\beta(X = x^n).$$

If the distribution is continuous with a density f_β, the likelihood function is written as follows, by definition:

$$L(\beta, x^1, x^2, \ldots, x^n) = f_\beta(x^1) \times f_\beta(x^2) \times \ldots \times f_\beta(x^n).$$

The value of β which maximizes $L(\beta, x^1, x^2, \ldots, x^n)$ is the value which maximizes the probability of the observed results. The situation is the inverse of that in which the parameter of the distribution of X is known and the probability of observing the results x^i is calculated: here, we have observed the x^i and we look for the parameter that maximizes the probability of observing them. $L(\beta, x^1, x^2, \ldots, x^n)$ is a density function if we see it as a function of $(x^1, x^2, \ldots, x^n)$ and a likelihood function if we see it as a function of β. The observations must be independent if we are to be able to write the likelihood in $(x^1, x^2, \ldots, x^n)$ as the product of the likelihoods of each observation x^i. Note that this method can be generalized to the case of truncated data, whose conditional likelihood is calculated by dividing the density function by Prob($X > s$), where s is the truncation threshold.

We can search for a maximum by searching for a value for which the first derivative is cancelled and the second derivative is negative (assuming that these derivatives exist). Most

[57] But it is also possible to estimate the coefficients by the maximum likelihood method in linear regression. If the residuals follow a uniform distribution, the maximum likelihood estimators are those which minimize the maximum of the $|e_i|$. If the residuals follow a normal distribution, the maximum likelihood estimators are those which minimize the sum of squares of the e_i. The estimators of the coefficients are then the same as those found by the least squares method. This can seen if we know that the y_i follow a normal distribution like the residuals, and that the mean of this distribution is $b_0 + x_{i1}b_1 + x_{i2}b_2 + \ldots + x_{ip}b_p$ and its variance is s^2, and write the density function of the y_i. However, the estimator of the variance of the residuals is not the same.

The log-likelihood of a linear model is $-(n/2)\log(ESS/n)$, where n is the number of observations. The AIC, defined below for the logistic model, is therefore valid for the linear model: $n\log(ESS/n) + 2(p + 1)$, where p is the number of regressors in the model. The choice of model may be made in such a way as to minimize this quantity.

frequently, we aim to maximize the logarithm of the likelihood, which is an equivalent problem but simpler to solve, since the logarithm converts products into sums. This search for a maximum is a problem of optimization which is handled by algorithms such as Newton–Raphson.

The estimation of the maximum likelihood enables us to estimate the parameter λ of a Poisson distribution, the parameter α of an exponential distribution, the parameters μ and σ of a normal distribution, etc. As a general rule, the maximum likelihood estimator may exist and be unique, or may not be unique, or may not exist.

In a regression problem, the laws are not simple but conditional, and one of the following functions must be maximized:

$$\text{Prob}_\beta(Y = y^1/X = x^1) \times \text{Prob}_\beta(Y = y^2/X = x^2) \times \ldots \times \text{Prob}_\beta(Y = y^n/X = x^n)$$

or

$$f_\beta(y^1/X = x^1) \times f_\beta(y^2/X = x^2) \times \ldots \times f_\beta(y^n/X = x^n).$$

f_β is a conditional density function.

In binary logistic regression, we observe the data $[(x^1,y^1), (x^2,y^2), \ldots, (x^n,y^n)]$ in which every y^i is 0 or 1, and x^i is the vector of variables accounting for the ith observation. If $y^i = 1$, the probability of obtaining (x^i,y^i) will be, by definition, $\text{Prob}(Y=1/X=x^i) = \pi(x^i)$, where $\pi(x^i)$ is expressed using the same notation as in Section 11.8.1 but without, as yet, replacing $\pi(x^i)$ with its value in the logistic model. If $y^i = 0$, the probability of obtaining (x^i,y^i) will be $\text{Prob}(Y=0/X=x^i) = 1-\pi(x^i)$. We can combine these two cases by writing that the probability of obtaining (x^i,y^i) is

$$\pi(x^i)^{y^i}(1 - \pi(x^i))^{1-y^i},$$

for any value of y^i. Now, in order to continue the calculation and apply the above formula, we must introduce a fundamental assumption for logistic regression, namely that the observations (x^i,y^i) are independent. This assumption can only be dispensed with in a generalization of logistic regression introduced in Section 11.9.2, known as logistic regression with correlated data. This assumption of independence enables us to write the likelihood function as a product of the probabilities:

$$\prod_{i=1}^{n} \pi(x^i)^{y^i}(1 - \pi(x^i))^{1-y^i}.$$

If we now replace $\pi(x^i)$ with its expression in the logistic model, the likelihood function appears as a function of the coefficient vectors $(\beta_0, \beta_1, \ldots, \beta_p)$:

$$\prod_{i=1}^{n} \left(\frac{e^{\beta_0 + \sum_j \beta_j x_j^i}}{1 + e^{\beta_0 + \sum_j \beta_j x_j^i}} \right)^{y^i} \left(1 - \frac{e^{\beta_0 + \sum_j \beta_j x_j^i}}{1 + e^{\beta_0 + \sum_j \beta_j x_j^i}} \right)^{1-y^i}.$$

It is this function $L(\beta_0, \beta_1, ..., \beta_p)$ of the coefficients β_j that is to be maximized: we need to find the coefficients such that $L(\beta_0, \beta_1, ..., \beta_p)$ is as close as possible to 1, as this will mean that the model provides the best possible fit to the observed data. In fact, the likelihood cannot be 1, and the model cannot perfectly fit to the observed data, except in a single case – when the model contains as many coefficients as there are separate observations (x^i,y^i). Such a model is said to be *saturated*. By way of analogy, a simple linear model is saturated when the cloud of points is reduced to two points, and the straight line is a perfect fit to the cloud formed by the two points.

The determination of the best logistic model is therefore based on a search for the coefficients that maximize the likelihood. This is a problem which does not have an analytical solution, that is to say a solution expressed directly from the initial data, like the discriminant functions of discriminant analysis which are found by inverting the covariance matrix. In this case, the optimal solution $(\hat{\beta}_0, \hat{\beta}_1, ..., \hat{\beta}_p)$ will be found by an iterative numerical method, the most widely used of which are the Newton–Raphson and Fisher algorithms. This lack of an analytical solution may be the main drawback of logistic regression, making it harder for a software developer to program, and requiring more computing time than discriminant analysis, and may make it impossible to achieve a reliable solution in some exceptional cases. The algorithm does not converge if the groups are completely separated, whereas discriminant analysis is still effective in this case. Even in a commonplace example such as that of Section 11.8.7, a single point can determine the convergence of the algorithm.

$(\hat{\beta}_0, \hat{\beta}_1, ..., \hat{\beta}_p)$ denotes the estimate of the coefficients by maximization of the likelihood, leading to an estimate $\hat{\pi}(x^i)$of the conditional probability $\hat{\pi}(x^i)$ of having $Y=1$ given that $x=x^i$. This is the value predicted by the logistic model. It is worth noting that

$$\sum_{i=1}^{n} y_i = \sum_{i=1}^{n} \hat{\pi}(x_i).$$

In other words, the sum of the observed values of Y is equal to the sum of the predicted values.

As in linear regression, we never find the true coefficients β_j (except for a saturated model) and the estimators $\hat{\beta}_j$ have a certain level of variance. An estimator will obviously be more reliable if its variance is lower, and the variables X_j for which this variance is low will be preferred. More precisely, the Student test (see the Appendix, Section A.2.5) suggests that we should specify an analogue of the squared Student's t, which is called the Wald statistic of a variable X_j, which is $(\hat{\beta}_j/\text{standard deviation }(\hat{\beta}_j))^2$. According to the null hypothesis H_0 that $\hat{\beta}_j=0$, the ratio $(\hat{\beta}_j/\text{standard deviation }(\hat{\beta}_j))^2$ follows a standard normal distribution. The Wald statistic can be used to test the significance of the estimated coefficient $\hat{\beta}_j$ and the contribution of the variable X_j. This contribution is real only if the Wald statistic is greater than 4, or more precisely if it is greater than 3.84 (= 1.96^2). In fact, saying that $(\hat{\beta}_j/\text{standard deviation }(\hat{\beta}_j))^2 > 3.84$ is equivalent to saying that $|\hat{\beta}_j/\text{standard deviation }(\hat{\beta}_j)| > 1.96$ and is separated by more than 1.96 standard deviations from the mean, because $(\hat{\beta}_j/\text{standard deviation }(\hat{\beta}_j))$ follows a standard normal distribution. This leads to a p-value of <5% associated with the null hypothesis H_0, and $\hat{\beta}_j/\text{standard deviation }(\hat{\beta}_j)$ is in the H_0 rejection region at the 95% threshold. We therefore deduce that $\hat{\beta}_j$ is significantly different from 0. In this case, the confidence interval at 95% of the odds ratio $\exp(\hat{\beta}_j)$ does not contain 1. This odds ratio must be significantly greater than 1 (positive effect of the variable) or significantly less than 1 (negative effect) for the estimator of the coefficient to be significantly different from zero. This is certainly the case in the example given below for the variable AGE, which makes a real contribution to the prediction of coronary heart disease in the Hosmer and Lemeshow example, because the Wald indicator is 21.254 > 4 and the confidence interval of the odds ratio [1.066, 1.171] does not contain 1.

Variables in the equation

							95% CI for Exp(B)	
	B	S.E.	Wald	DF	Sig.	Exp(B)	Lower	Upper
AGE	.111	.024	21.254	1	.000	1.117	1.066	1.171
Constant	– 5.309	1.134	21.935	1	.000	.005		

Care should be taken, because an odds ratio may be long way from 1 but may still have such a wide confidence interval that it contains the value 1, meaning that the variable is not particularly relevant. We will see that this happens when collinear variables are present, and the absence of collinearity is an essential assumption that must be checked very carefully.

The Wald criterion can be used in stepwise regression.

11.8.6 Deviance and quality measurement in a model

Let n be the total number of individuals (or observations), k the number of degrees of freedom of a given adjusted model (each quantitative variable counting for 1 and each qualitative variable with m categories counting for m–1 in the calculation of k), $L(\beta_k)$ the likelihood of this model, $L(\beta_0)$ the likelihood of the model reduced to the constant, and $L(\beta_{max})$ the likelihood of the *saturated model.* $L(\beta_{max})$ is the maximum likelihood with which the likelihood $L(\beta_k)$ of any model is compared. This leads to the concept of *deviance*:

$$D(\beta_k) = -2[\log L(\beta_k) - \log L(\beta_{max})] = \log[L(\beta_{max})/L(\beta_k)]^2.$$

This is similar to the residual sum of squares (RSS) in linear regression,[58] and it is calculated as a sum over the set of the n individuals. It is equal to the sum of squares of the residuals of the individual deviances. In the common case of a dependent 0/1 variable, the likelihood of the saturated model is 1, and therefore

$$D(\beta_k) = -2\log\text{ (likelihood of the adjusted model)},$$

[58] The deviance formula (11.4) applied to a linear regression, where y is the real value and $\pi(x)$ is the conditional expectation $E(Y/X = x)$, shows that the deviance and the RSS are actually equal. In fact, since the likelihood is between 0 and 1, the value 1 corresponding to a perfect fit, we return to the scale of values of the sum of squares, ranging from 0 to $+\infty$ (infinity), the value 0 corresponding to a perfect fit, taking firstly the logarithm of likelihood (log(1) = 0), then twice this logarithm, since the expression

$$2.\log L(\beta_0, \beta_1, \ldots, \beta_p) = \log L(\beta_0, \beta_1, \ldots, \beta_p)^2$$

leads to a square, finally changing the sign of the expression to produce a value of –2.log(likelihood) between 0 and $+\infty$.

Model Fit Statistics		
Criterion	**Intercept Only**	**Intercept and Covariates**
-2 log L	6196.452	4543.570

Testing Global Null Hypothesis: BETA=0			
Test	**Chi-Square**	**DF**	**Pr > ChiSq**
Likelihood Ratio	1652.8817	13	<.0001
Score	3497.7522	13	<.0001
Wald	1542.3349	13	<.0001

Figure 11.93 The deviance criterion.

i.e., according to the likelihood formula described in the preceding section,

$$D(\beta_k) = -2\sum_{i=1}^{n}\left[y^j \log\left(\pi(x^i)\right) + (1-y^j)\log\left(1-\pi(x^j)\right)\right]. \qquad (11.4)$$

This deviance can be seen in the SAS output shown in Figure 11.93, in the 'intercept and covariates' column, where it is 4543.57 in the example.

The residual of the deviance of the ith observation is

$$\pm\sqrt{2[y^j\log(\pi(x^i)) + (1-y^j)\log(1-\pi(x^j))]},$$

the sign of the expression being positive if the observed value y^i is greater than the predicted value $\pi(x^i)$, and negative otherwise. Note that, as in linear regression, it is useful to examine the observations in which the residual of the deviance exceeds a certain threshold, around 2 in absolute terms. To do this, as we will see subsequently in our discussion of the syntax of logistic regression, we can write the residual of the deviance of each individual in an output table.

The aim of logistic regression is to maximize the likelihood $L(\beta_k)$ of the adjusted model or the log-likelihood log $[L(\beta_k)]$, which is equivalent to minimizing the deviance $D(\beta_k)$.

For a given set of variables with k degrees of freedom (number of quantitative variables or categories of qualitative variables), we look for the coefficients which maximize the likelihood $L(\beta_k)$, as indicated in the preceding section. When these k coefficients have been found and if they are significantly different from 0 (those which are not significantly different from 0 beeing removed), we need to know if we could improve the model by adding l degrees of freedom. To do this, we calculate the difference between the deviances,

$$\begin{aligned} D(\beta_k) - D(\beta_{k+1}) &= -2[\log L(\beta_k) - \log L(\beta_{\max})] + 2[\log L(\beta_{k+l}) - \log L(\beta_{\max})] \\ &= -2[\log L(\beta_k) - \log L(\beta_{k+l})], \end{aligned}$$

which is positive. We can then make use of a basic finding: on the hypothesis H_0 of the nullity of all the l new coefficients, the difference between the deviances follows a χ^2 distribution with l degrees of freedom. Therefore we will not add the l new degrees of freedom (quantitative variables or categories of qualitative variables) unless the difference between the deviances is greater than the critical threshold of the χ^2 with l degrees of freedom. Note that, even in this case, some coefficients may be zero, so that we generally prefer to add only one variable at a time, thus setting $l = 1$ (quantitative variable) or $l = m - 1$ (qualitative variable with m categories).

In particular, the null hypothesis that the model fit the best to the data can be reformulated to state that the p parameters of the model (including the constant) are all significantly different from 0, and that the supplementary parameters $p+1, p+2, \ldots, n$ required to reach the number of parameters of the saturated model are all zero. Under this null hypothesis, the deviance of the adjusted model follows a χ^2 distribution with $n-p$ degrees of freedom. We should note that it is a relative criterion and not an absolute criterion of goodness-of-fit which will be given by other criteria we shall see later on (Hosmer-Lemeshow, AUC...)

The concept of likelihood is therefore essential for comparing a logistic regression model with a sub-model, in other words for comparing two nested models. It is the most commonly used concept and can be found in any stepwise regression method, whether forward, backward or mixed (see Section 3.13).

If we are uncertain about adding a quantitative variable to a model with k variables, a good criterion for the decision, before comparing the two models with an ROC curve, is to compare the value $D(\beta_k) - D(\beta_{k+1})$ with the theoretical value of the χ^2 with one degree of freedom at the 5% threshold: if $D(\beta_k) - D(\beta_{k+1}) > 3.84$, we can add the variable. Similarly, a qualitative variable with m categories should cause a decrease in deviance according to a χ^2 distribution with m–1 degrees of freedom. What we need to know, therefore, is not whether the likelihood has increased when a variable was added – because it always increases – but whether it has increased enough.

As a special case of the above, the difference

$$D(\beta_0) - D(\beta_k) = -2[\log L(\beta_0) - \log L(\beta_k)]$$

follows a χ^2 distribution with k degrees of freedom under the hypothesis H_0 of the nullity of all the coefficients $\beta_1, \beta_2, \ldots, \beta_k$. H_0 is rejected if the difference $D(\beta_0) - D(\beta_k)$ exceeds the critical threshold of the χ^2 with k degrees of freedom. Thus, in Figure 11.93, we find that the difference $6196.452 - 4543.570 = 1652.882$ is indeed greater than the critical threshold of the χ^2 with 13 degrees of freedom.

As this difference increases, the model becomes better. In other words, we would like the $-2 \log [L(\beta_k)]$ (equal to 4543.570 in this case) to be as small as possible.

This requirement is found in an equivalent form in two other well-established criteria:

- the *Akaike information criterion* (AIC), $-2 \log [L(\beta_k)] + 2(k + 1)$;
- the *Bayesian information criterion* (BIC), $-2 \log [L(\beta_k)] + (k + 1) \log(n)$, often called the *Schwarz criterion.*

The difference between these criteria is that BIC penalizes complex models more than AIC (if $n > \exp(2) \approx 7.4$). These two criteria enable us to compare two models, the better one being that for which AIC and BIC are lower. In our example (Figure 11.94), we have

$$k = 13, n = 135\,782, \text{AIC} = 4543.570 + 28 = 4571.570 \text{ and}$$
$$\text{BIC} = 4543.570 + 14.\log(135\,782) = 4709.033.$$

When n is small, BIC may lead to the selection of an over-simple model, whereas AIC will lead to a better compromise between precision and robustness (this is known as the 'bias–variance' dilemma). However, the BIC is preferable when n is large and we can show

Model Fit Statistics		
Criterion	**Intercept Only**	**Intercept and Covariates**
AIC	6198.452	4571.570
SC	6208.270	4709.033

Figure 11.94 The AIC and BIC.

(see Section 19.4.1.3 of the book by Saporta, cited in Section 11.3.3 above) that the probability that the BIC will choose the best model tends towards 1 when n tends towards infinity, which is false for the AIC. The concept of a 'good' model can be defined thus: when data have been generated using a model, the 'good' model, out of all those tested, is the one which has produced the chosen data. In other words, the BIC identifies the model which has the greatest probability of having generated the observed data: the model which minimizes the BIC is the one that maximizes the *a posteriori* probability of the model conditionally given the observed data, the *a priori* probabilities of all the models being considered equal (see Section 7.7 of the 2009 book by Hastie *et al.*, also cited in Section 11.3.3 above). It is this expression of the *a posteriori* probability by means of the Bayes formula that gives the name 'Bayesian' to the BIC.

These two criteria are used to compare non-nested models for which it is not possible to use the previous approach based on the difference $D(\beta_k) - D(\beta_{k+l})$. We can also use these criteria with nested models, in a stepwise variable selection process, representing the curve of the AIC and BIC as a function of the number of selected variables. A minimum in these curves, especially in the BIC curve if n is large, can indicate the correct number of variables to retain. An example of the use of this method is given in Section 2.21.4 of my book *Étude de cas en statistique décisionnelle.*[59]

As demonstrated by some researchers, where the selection of variables is concerned, these criteria, based on the concept of likelihood, appear preferable to the other common criterion which is that of the Wald statistic (see the previous section), at least in two cases – when the number of observations is small, or when certain coefficients β_i are large. In the latter case, Hauck and Donner,[60] as well as Jennings[61] (see Section 1.3 of the book by Hosmer and Lemeshow cited in the References), have underlined the lack of power of the Wald test, which may thus fail to reject the hypothesis H_0 of the nullity of the coefficient, even when the latter is significantly different from zero.

If the Wald test is too conservative, this is because it has a weakness compared with the likelihood tests: while these only apply to the estimation of the parameter itself according to the maximum likelihood, the Wald test also applies to an estimate of the standard deviation of the estimator of the parameter.

It has a further weakness. It does not necessarily give the same result regardless of whether it is applied to a variable V, to its logarithm $\log(V)$ or to another transformation of V; this is because there is no simple relation between the standard deviation of V and that of its transforms.

[59] Tufféry, S. (2009) *Étude de Cas en Statistique Décisionnelle*. Paris: Technip.

[60] Hauck, W.W. and Donner, A. (1977). 'Wald's Test as applied to hypotheses in logit analysis', *Journal of the American Statistical Association*, 72, 851–853.

[61] Jennings, D.E. (1986). 'Judging inference adequacy in logistic regression', *Journal of the American Statistical Association*, 81, 471–476.

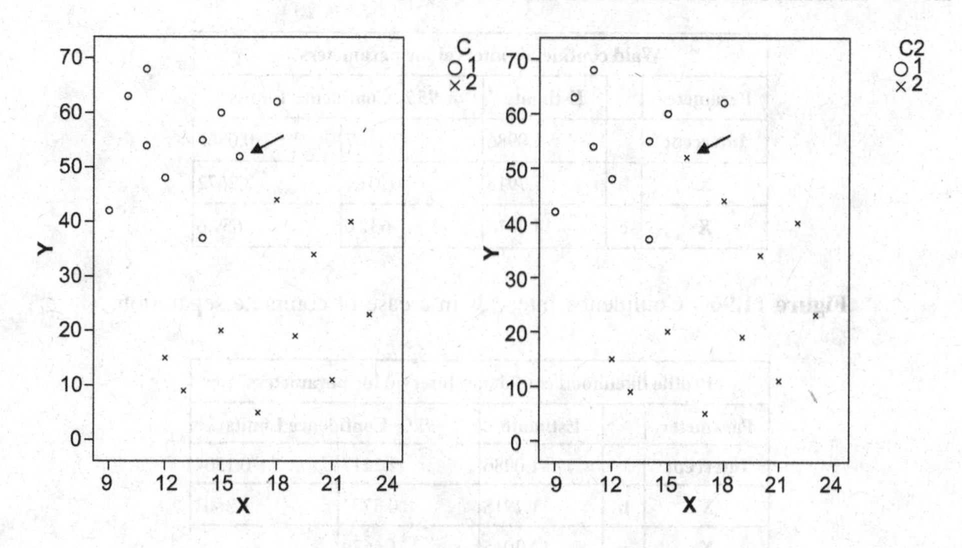

Figure 11.95 Complete separation in logistic regression.

11.8.7 Complete separation in logistic regression

I have said that logistic regression does not converge when separation is complete. In the second graph of Figure 11.95, a single point has changed its group and is sufficient to ensure the absence of complete separation, owing to which the logistic regression will indeed converge. The problem posed by complete separation will be understood if we remember that the exponential of each coefficient β of the model is the ratio of the odds $\pi(x)/(1 - \pi(x))$ of the corresponding variable. Now, separation means that $\pi(x)$ is 1 for certain values of x, giving us infinite odds $\pi(x)/(1 - \pi(x))$ and a coefficient β which should become infinite.

This problem of complete separation is less theoretical than it appears, and is not uncommon in small populations or when a descending ('backward') stepwise regression requires the manipulation of numerous variables in the first iterations. Suppose, for example, that we wish to predict a variable Y based on an independent variable X which has three categories a, b and c, their contingency table being as follows:

	$Y=0$	$Y=1$	TOTAL
$X=a$	15	5	20
$X=b$	5	10	15
$X=c$	0	5	5
TOTAL	20	20	40

It seems easy to predict that $Y=0$ if $X=a$, and $Y=1$ otherwise. However, the logistic regression does not converge in this case, because of the probability $\pi(c)=1$!

A remedy for this kind of situation may be to regroup the categories of the independent variable concerned.

With the SAS LOGISTIC procedure, we can also use the options CLPARM=PL and CLODDS=PL. Used in cases of separation, where the Wald confidence intervals of the coefficients and of their odds ratios are overwhelmingly large, these options enable us to

Wald confidence interval for parameters				
Parameter		**Estimate**	**95% Confidence Limits**	
Intercept		-1.0986	-2.1107	-0.0865
X	b	1.7918	0.3164	3.2672
X	c	14.3015	-631.0	659.6

Figure 11.96 Confidence intervals in a case of complete separation.

Profile likelihood confidence interval for parameters				
Parameter		**Estimate**	**95% Confidence Limits**	
Intercept		-1.0986	-2.2217	-0.1504
X	b	1.7918	0.3732	3.3602
X	c	14.3015	1.6679	.

Figure 11.97 Confidence intervals with the CLPARM=PL option in a case of complete separation.

obtain at least an upper or a lower bound for these intervals (CLPARM for the coefficients and CLODDS for the odds ratios). If we cannot be certain of the value of a coefficient, we can at least be sure of one of its bounds, which provides useful information in some cases. In the above example, the estimated coefficients are not aberrant (Figure 11.96), because the chosen reference category is a, and the coefficient of b is therefore positive and that of c is positive and greater than that of b. This is entirely consistent with the proportions of values $Y=0$ and $Y=1$ (the modelled value of Y) observed for the values a, b and c of X, but the Wald confidence interval of the coefficient of c is very large and, in particular, contains 0.

On the other hand, the CLPARM=PL option provides the confidence interval estimates shown in Figure 11.97, and, in particular, gives us a lower bound greater than 0 for the coefficient of c.

I wish to thank Jean-Pierre Nakache for drawing my attention to this possibility.

11.8.8 Statistical tests in logistic regression

Without embarking on the description of the general tests on predictive models formed by the confusion matrix, the ROC curve and the Gini index, which will be discussed later, we can list the tests that are specific to logistic regression. The first four have already been presented:

- the χ^2 test on the Wald indicators, which must be greater than 3.84;
- the 95% confidence intervals of the odds ratios must not contain {1};
- the value of $-2 \log L(\beta_k)$ must be as small as possible, or a χ^2 test is carried out on the modification of the -2 log-likelihood when a coefficient β_k is removed (null hypothesis: $\beta_k=0$);
- the AIC and BIC must be as small as possible;

- the Cox–Snell R^2 and Nagelkerke's adjusted R^2;
- the Hosmer–Lemeshow test on the comparison of the theoretical and the observed proportions;
- the normalized deviance test and the normalized Pearson's χ^2 test;
- concordance tests (related to the area under the ROC curve and the Gini index).

The Cox–Snell R^2 is an equivalent of the R^2 of linear regression, defined on the basis of the likelihoods by

$$R^2 = 1 - \left[\frac{L(\beta_0)}{L(\beta_k)}\right]^{2/n},$$

which therefore may not exceed (for a saturated model)

$$R^2_{\max} = 1 - [L(\beta_0)]^{2/n}.$$

Nagelkerke's R^2 ('max-rescaled R-square') is the quotient $R^2/R^2_{\max}$, which varies between 0 and 1.

R-Square	0.0121	**Max-rescaled R-Square**	0.2712

The equivalence of the Cox–Snell R^2 with the $R^2 = ESS/TSS$ of linear regression is shown as follows. It will be remembered (Section 11.8.5) that the log-likelihood $\log(L)$ of a linear model is expressed as a function of the number n of observations and of the sum ESS of the residual squares:

$$-(n/2)\log(ESS/n).$$

For the model reduced to the constant, $ESS = TSS$, the total sum of squares. The result is found by successive transformations:

$$R^2 = 1 - \left(\frac{L_0}{L_k}\right)^{2/n} = 1 - \left(\frac{e^{\log L_0}}{e^{\log L_k}}\right)^{2/n} = 1 - \left(e^{\log L_0} - e^{\log L_k}\right)^{2/n} = 1 - \left(e^{-\frac{n}{2}\log\frac{TSS}{n} + \frac{n}{2}\log\frac{ESS}{n}}\right)^{2/n}$$

$$= 1 - \left(e^{\frac{n}{2}\left(\log\frac{ESS}{n} - \log\frac{TSS}{n}\right)}\right)^{2/n} = 1 - e^{\log\left(\frac{ESS}{TSS}\right)} = 1 - \frac{ESS}{TSS} = \frac{RSS}{TSS}.$$

The Hosmer–Lemeshow test is an absolute criterion of goodness-of-fit carried out by distributing the observations among $g = 10$ groups in increasing order of probability (calculated according to the model), in other words in deciles of the *a posteriori* probability. We calculate the χ^2 of the $g \times 2$ table of the observed and expected frequencies for the modelled event (CHD = 1 in this case), which is compared with the distribution of χ^2 with $g - 2$ degrees of freedom. The loss of one degree of freedom, so that we have $g - 2$ instead of $g - 1$, corresponds to the fact that the grouping into 10 groups is not specified in advance but depends on the estimated parameters, which thus 'consume' one degree of freedom. The null hypothesis to be tested is that the observed frequencies are the expected ones, these being calculated by multiplying the calculated probabilities by the frequencies in each decile. If the χ^2 is large (i.e. the associated probability is small), the null hypothesis must be rejected and we must conclude that the model is not a good fit to the data. However, the model is a very good fit to the data in the example of CHD (Figure 11.98). The Hosmer–Lemeshow test is intuitive and easy to carry out and interpret. However, with small frequencies (a few tens or hundreds of observations) this test lacks power, meaning that the null hypothesis may be false (the fit of the model is poor) but not rejected (the calculated probability exceeds 0.05). In spite of all this, when the observations are grouped this test limits the problems of sparse data mentioned below, which can lead to poor estimation of the expected frequencies and a misleading comparison with the observed frequencies. A second drawback of the Hosmer–Lemeshow test is its dependence on the way in which the observations are grouped, with the result that two different software may yield rather different χ^2 probabilities and conclusions.

Before we discuss the next test, which is the normalized deviance and normalized Pearson χ^2 test, we need to introduce a variant of the concept of deviance described above. Unlike that deviance, which is calculated as a sum over the set of individuals, the variant which we will now be talking about is the deviance D' which is calculated as a sum over the set of m

Contingency table for Hosmer and Lemeshow test

		CHD = 0		CHD = 1		
		Observed	Expected	Observed	Expected	Total
Step 1	1	9	9.213	1	.787	10
	2	9	8.657	1	1.343	10
	3	8	8.095	2	1.905	10
	4	8	8.037	3	2.963	11
	5	7	6.947	4	4.053	11
	6	5	5.322	5	4.678	10
	7	5	4.200	5	5.800	10
	8	3	3.736	10	9.264	13
	9	2	2.134	8	7.866	10
	10	1	.661	4	4.339	5

Hosmer-Lemeshow test

Step	Chi square	DF	Signif.
1	.890	8	.999

Figure 11.98 The Hosmer–Lemeshow test.

sub-populations defined by the independent variables, whose number m is equal to the number of possible cross-tabulations of all the categories of the independent variables. This number is less than or equal to the size n of the population, and the presence of continuous variables evidently tends to reduce the difference between n and m.

As before, the null hypothesis that the model fits the best to the data can be reformulated to state that the p parameters of the model (including the constant) are sufficient for the fit, and that the supplementary parameters $p + 1, p + 2, \ldots, m$ required to reach the number of parameters of the saturated model are all zero. In simple terms, the number m of parameters of the saturated model in this form of deviance is not equal to the number of individuals, but is equal to the number of sub-populations. Under this null hypothesis, the deviance D' of the adjusted model follows a χ^2 distribution with $m - p$ degrees of freedom. The deviance D' is smaller than the deviance D as m is smaller than n, but the number of degrees of freedom also decreases. It is only when $n = m$ that the two deviances D and D' are equal.

When m is large, as is the case when continuous independent variables are present, the m sub-populations have small frequencies. In this case, D', which approaches D, ceases to be useful as a measure of fit, and the deviance D is preferred.

When m is small, because the independent variables are all qualitative and do not have too many categories, the deviance D' is small with respect to D, and is more useful than the deviance D because the frequency n then largely exceeds the real number of parameters of the saturated model, of which m is a much more accurate measurement. Under the null hypothesis that a model with p parameters fits the best to the data, I have said that the deviance D' of the adjusted model follows a χ^2 distribution with $m - p$ degrees of freedom. If the associated probability (significance) is greater than 0.05, the null hypothesis cannot be rejected at the 95% confidence level, and we can accept the hypothesis that the model fits the best to the data. Figure 11.99 shows an example of this. This is an output of the SAS/STAT PROC LOGISTIC, in which the calculation of the deviance D' is launched by the options AGGREGATE and SCALE=NONE (see Section 11.9.6 for more details).

In this example, where $m = 207$ and $p = 10$, the normalized deviance, which by definition is equal to the deviance $D' = 182.451$ divided by 197, the number of degrees of freedom, is close to 1, indicating that there is no overdispersion (see Section 11.9.6). Any overdispersion would have no effect on the value of the model parameters, but their variances would be underestimated and the Wald statistics would be overestimated, meaning that the fit of the model might be less good than it appeared. To be absolutely precise, we should say that the normalized deviance as an estimate of overdispersion improves as the frequencies of each sub-population decrease; in other words, it is better when m is smaller. In our example, the value of

Deviance and Pearson Goodness-of-Fit Statistics

Criterion	Value	DF	Value/DF	Pr > ChiSq
Deviance	182.451	197	0.9261	0.764
Pearson	183.147	197	0.9297	0.752

Figure 11.99 Deviance and Pearson's χ^2.

Association of Predicted Probabilities and Observed Responses			
Percent Concordant	81.4	**Somers' D**	0.765
Percent Discordant	4.9	**Gamma**	0.885
Percent Tied	13.6	**Tau-a**	0.005
Pairs	62787552	**c**	0.882

Figure 11.100 Tests of concordance.

m is not very small, but its closeness to the normalized Pearson χ^2 which is equal to 183.147 is a good sign of the quality of the estimate provided by the normalized deviance.

Tests of concordance are not a feature of logistic regression, but SAS has implemented them in its LOGISTIC procedure. This is how they are calculated (Figure 11.100). We assume that the dependent variable *Y* takes the values 0 and 1, and n_1 (or n_2 respectively) denotes the number of observations in which $Y = 0$ (or $Y = 1$), and $n = n_1 + n_2$ is the total number of observations. We are interested in the $t = n_1 n_2$ pairs formed by an observation in which $Y = 1$ and an observation in which $Y = 0$. In this case, $t = 135\,318 \times 464 = 62\,787\,552$. Concordance is present among these *t* pairs if the estimated probability that $Y = 1$ is greater when $Y = 1$ than when $Y = 0$. We then use n_c to denote the number of concordant pairs (81.4% of the pairs in this case), n_d to denote the number of discordant pairs, and $t - n_c - n_d$ to denote the number of tied pairs. We then calculate:

- Somers' $D = (n_c - n_d)/t$;
- gamma $= (n_c - n_d)/(n_c + n_d)$;
- tau-a $= 2\,(n_c - n_d)/n(n-1)$;
- $c = (n_c + 0.5[t - n_c - n_d])/t$.

The closer these indices are to 1, the better the model is. Note that the quantity c is simply the area under the ROC curve which will be discussed later on (Section 11.16.5), and Somers' *D* is simply the Gini index.

11.8.9 Effect of division into categories and choice of the reference category

For a qualitative or discretized variable, we sometimes find that one or two categories have a weak effect on the model, with a Wald index that is too low (less than 3.84) and a confidence interval of the odds ratio containing {1}. They may be worth retaining if the other categories are useful. Note that the constant is almost always retained in the model, and no tests of significance are applied to it. However, if the Wald indicator of the constant shows that it is not significant, the constant can be excluded, in which case the probability of $Y = 1$ for an individual having all the reference categories is 0.5.

To resolve this problem, we could try to group the categories in question with other categories whose coefficients have the same sign. Another solution which may be effective is to change the reference category. This solution shows that a Wald criterion greater than 3.84 should not always be taken at face value when judging the quality of a model. In fact, it is strictly identical regardless of the reference category. The benefit of changing the reference category, and thus reducing the standard deviations of the coefficients, is that we have more significant odds ratios which can be analysed more legitimately.

To illustrate this subtle point concerning the odds ratios of qualitative variables, we will take the example of two odds ratios OR(*A* v. *C*) and OR(*B* v. *C*) which are significant, while OR(*A* v. *B*) = OR(*A* v. *C*)/OR(*B* v. *C*) is not significant. We consider a variable *X* with three categories *A*, *B* and *C*, for which logistic regression using *B* as the reference variable gives a result in which OR(*A* v. *B*) and OR(*C* v. *B*) are significant:

		OR	Lower boundary of CI	Upper boundary of CI	Wald
X	*A* v. *B*	0.804	0.650	0.994	4.0644
X	*C* v. *B*	0.698	0.544	0.894	8.0934

Now let us change the reference variable: we will choose *C*. As we can see below, OR(*A* v. *C*) = OR(*A* v. *B*)/OR(*C* v. *B*) = 0.804/0.698 = 1.152.

		OR	Lower boundary of CI	Upper boundary of CI	Wald
X	A v. C	1.152	0.915	1.452	1.4501
X	B v. C	1.434	1.119	1.837	8.0935

However, OR(A v. C) is not significant, because its confidence interval contains 1. This example shows that a Wald criterion greater than 3.84 is not always enough in itself to draw a conclusion as to the correct division of a variable into a number of categories or the necessity of redividing it.

11.8.10 Effect of collinearity

We will now consider the damage that can be done to the predictive power of a variable VAR1, which is initially satisfactory, by the introduction of a variable VAR2 which is closely correlated with it (Pearson's correlation coefficient = 0.89):

		B	S.E.	Wald	DF	Sig.	Exp(B)	95% CI for Exp(B)	
								Lower	Upper
Step 1[a]	VAR1	.064	.010	43.988	1	.000	1.066	1.046	1.087
	Constant	− 3.537	.160	486.531	1	.000	.029		

[a]Variable(s) entered on step 1: VAR1.

Variables in the Equation

		B	S.E.	Wald	DF	Sig.	Exp(B)	95% CI for Exp(B)	
								Lower	Upper
Step 2[b]	VAR1	−.034	.017	3.851	1	.050	.967	.934	1.000
	VAR2	.111	.018	39.372	1	.000	1.118	1.079	1.157
	Constant	−4.138	.220	352.609	1	.000	.016		

[b]Variable(s) entered at step 2: VAR2.

We find that the confidence interval of the odds ratio of VAR1 contains {1} after the introduction of VAR2, and that VAR1 can no longer be considered as correctly predictive: its Wald indicator has been divided by more than 11 and the uncertainty about its coefficient increases accordingly. If the variable VAR2 were even more closely correlated with VAR1, the decrease in the Wald indicator would be even greater.

We even find that the sign of the coefficient of VAR1 is reversed. This entirely destroys the readability of the variable, and therefore of the model, because we no longer know if a variable has a positive or negative effect. The result is not necessarily completely false, and we can try to explain it by the effect of the interaction between the variables, but this is not what is usually meant by readability.

Finally, experience shows that the gain in the performance of the model due to the addition of such a variable VAR2 is limited, even in the training sample. And if we measure the discriminant power of the model on a validation sample, especially if this is taken from another parent population (from observations made at another date, for example), the discriminant power may be greatly decreased by the addition of a closely correlated variable.

We must therefore carry out the tests of collinearity mentioned in Section 3.14, and we must be aware that, unless we have extremely numerous and complex data (to which a preliminary automatic clustering should be applied in any case), it is rarely justifiable to have more than 10 variables in a model.

11.8.11 The effect of sampling on *logit* regression

Logistic regression with the *logit* link function involves finding $\pi(x) = \text{Prob}(Y=1/X=x)$ in the form

$$\log\left(\frac{\pi(x)}{1-\pi(x)}\right) = \beta_0 + \beta_1 x_1 + \ldots + \beta_p x_p,$$

with coefficients maximizing the likelihood. If we carry out a sampling E independent of X, we can show that the probability $\pi_E(x) = \text{Prob}(Y=1/X=x,\ X \in E)$ meets this condition:

$$\log\left(\frac{\pi_E(x)}{1-\pi_E(x)}\right) = \beta'_0 + \beta_1 x_1 + \ldots + \beta_p x_p,$$

where $\beta'_0 = \beta_0 +$ a constant equal to $\log(p_{1,E}/p_{0,E}) - \log(p_1/p_0)$, with

- p_i = proportion of cases Y=i in the total population (*a priori* probability);
- $p_{i,E}$ = proportion of cases Y=i in the sample E.

Therefore, if E is independent of X, the same score function can be used to decide if $Y=1$ in E and in the whole population, by changing the decision threshold only. In particular, if $p_{1,E}/p_{0,E} = p_1/p_0$, then $\beta'_0 = \beta_0$ and the threshold is unchanged.

This means that a score calculated on a sub-population E can be applied to a sub-population E' if the distribution of the independent variables is the same in E and E', even if the event to be predicted is rarer (or more frequent) in E'. This attractive property is very useful in scoring, but beware: it is only valid for the *logit* model, not the *probit* model.

If the calculation of Prob(Y=1/X=x, $X \in E$) is applied to the $X \in E'$ and the same acceptance threshold Prob(Y=1/X=x, $X \in E$) $> s_0$ is specified, there will be the same percentage of acceptances in E' (because the probability is expressed as a function of the independent variables X which have the same distributions in E and E'), but the frequency of the event will be lower in the acceptances of E' (where the event is assumed to be rarer), because their probability Prob(Y=1/X=x, $X \in E'$) is less than Prob(Y=1/X=x, $X \in E$) (independence of E and X). This is the reason for replacing β_0 with β'_0.

If E is dependent on X, such that some categories of the independent variables, and not only of the dependent variable, are under- or overrepresented in E, there is no such simple result.

11.8.12 The syntax of logistic regression in SAS Software

The LOGISTIC procedure can be used to apply a logistic regression to a binary dependent variable, even if this is polytomous (see below). This dependent variable is named immediately after the instruction MODEL, before the '=' sign, the independent variables being listed after the '=' sign. The qualitative variables must be cited on the CLASS line, the keyword REF being used to indicate the reference category for each qualitative variable, including the dependent variable (although it is simpler to indicate the category modelled by the keyword EVENT='x' for the latter). Otherwise, the last category is taken as the reference by default, but it is also possible to specify that the reference category is to be the first, by entering REF=FIRST after the '/'.

The instruction PARAM=REF indicates that the coefficient of the reference category is 0, so that the coefficient β of another category represents the effect of this category with respect to that of the reference category, and the corresponding odds ratio is $\exp(\beta - \beta_{\text{ref}}) = \exp(\beta)$. If nothing is specified, or if PARAM=EFFECT is specified, the coefficient of the reference category will be the opposite of the sum of the coefficients of the other categories, in such a way that the coefficient β of another category would represent the effect of this category with respect to the mean effect of all the categories, and the odds ratio of this category with respect to the reference category will not be $\exp(\beta)$, but $\exp(\beta - \beta_{\text{ref}}) = \exp(\beta + \sum$ coefficients of all the categories $\neq$ reference). We can see that the default setting PARAM=EFFECT makes it difficult to interpret $\exp(\beta)$, and the setting PARAM=REF is generally preferred. It may also be preferable to use the PARAM=GLM setting which causes the explicit display of the reference category and its null coefficient. By contrast with PARAM=REF, the setting PARAM=GLM does not allow us to force the reference category, but we can then use a SAS format to place the category that we want to be the reference category in the final position by alphabetical order.

The different values of the PARAM coding do not modify the value of the odds ratios, which are displayed by SAS thus:

VARQUALIi MOD1 vs VARQUALIi MODREF o1
VARQUALIi MOD2 vs VARQUALIi MODREF o2
...

What does change is the link between the odds ratios o1, o2, ... and the logistic regression coefficients. Consider the example of a dependent variable 'occurrence of a disease = yes' and an independent variable which is the age group, with three categories, from the lowest group T1 to the highest group T3. We find that the risk of falling ill increases with age, such that, with PARAM=EFFECT, the coefficients associated with the age groups are − 0.1098 (T1), − 0.0120 (T2), the last category T3 being the reference category and having the coefficient 0.1098 + 0.0120 = 0.1218. With PARAM=REF, we have the coefficients − 0.2315 and − 0.1337, the coefficient of T3 being zero. The odds ratio of T1 with respect to T3 is exp(− 0.2315) = 0.7933, but if it is calculated from the coefficients determined by the coding EFFECT, it is exp(− 0.1098 − 0.1098 − 0.0120) = 0.7933. In this case, the coefficient − 0.1098 of T1 is less negative than in the first case, because it represents the (beneficial) effect of the group T1, not with respect to the worst case (T3), but with respect to the medium case.

Let us return to the syntax of the LOGISTIC procedure. We specify the type of stepwise selection[62] (SELECTION), the level of significance for the selection of a variable (SLE), and the maximum number of iterations (MAXITER). We can use the keywords RSQUARE and LACKFIT to specify that we wish to calculate the R^2 or perform the Hosmer–Lemeshow test. For the specification of interactions between the variables, see Section 3.12. The CTABLE option, described more fully in Section 11.16.6, produces a classification table, in other words a set of confusion matrices presented in lines for a whole set of possible thresholds for the score function deduced from the *a posteriori* probability. The threshold is the probability level above which the event is considered to be predicted, and below which the non-event is considered to be predicted. The PPROB option following CTABLE sets the interval of the threshold.

I have not mentioned the LINK=<function> option, because the LINK=LOGIT link function is implemented by default, but it is also possible to specify LINK=PROBIT, LINK=CLOGLOG or LINK=GLOGIT for the generalized logit used when the dependent variable is nominal with $k > 2$ categories (see below).

The input data are found in the DATA table, and the command P=proba writes a variable called 'proba', containing the predicted probability for the dependent category of the dependent variable, to the output data set OUT. The command RESDEV=deviance writes a variable 'deviance' containing the residual of the deviance of each observation. Optionally, the points on the ROC curve can be stored in a data set specified by OUTROC, and the points are stored for the various steps _STEP_ of the regression in the case of stepwise regression. There is a better option: from SAS 9 onward, the model itself can be stored, in a special SAS format, in the OUTMODEL data set. This data set can be used at the input of a procedure in which the recorded model is simply applied to a new DATA data set, using the SCORE command.

[62] In this procedure, the Wald test is used for the stepwise selection.

Here is the SAS syntax:

```
PROC LOGISTIC DATA=mytable.toscore OUTMODEL=my.model;
/* determination of the model */
CLASS varquali1 (REF='A1') varquali2 (REF='A2') ... /
PARAM=ref;
MODEL target (EVENT='1') = varquali1 varquali2 ... varquanti1
var_quanti2 ...
/ SELECTION=forward SLE=.05 MAXITER=25 OUTROC=roc RSQUARE
LACKFIT CTABLE PPROB=(0 TO 1 BY 0.1);
OUTPUT OUT=mytable.scored P=proba RESDEV=deviance;
RUN;

SYMBOL1 i=join v=none c=blue;
/* tracing of the ROC curve */
PROC GPLOT DATA=roc;
WHERE _step_ in (1 7);
TITLE 'ROC curve';
PLOT _sensit_*_1mspec_=1 / VAXIS=0 to 1 by .1 CFRAME=ligr;
RUN;

PROC LOGISTIC INMODEL=my.model;
/* application of the model */
SCORE DATA=othertable.toscore;
RUN;
```

11.8.13 An example of modelling by logistic regression

Let us return to the unfortunate example of the *Titanic* which has already served to illustrate the phenomena of interaction (in Section 3.12) and decision trees (in Section 11.4.2). We therefore model the category '1' (survival) of the dependent variable 'survived', which is specified in the SAS syntax by taking '0' (drowned) as the reference category of the dependent variable. In this way we obtain the basic model, with the three variables SEX, CLASS and AGE (see Section 3.12):

```
PROC LOGISTIC DATA=titanic;
MODEL survived (ref='0') = class age sex
/SELECTION=stepwise RSQUARE;
```

We find that the variables selected by the 'stepwise' process are, in sequence, SEX, CLASS and then AGE. No variable is eliminated from the process.

Summary of the sequential selection

Step	**Effect**		**DF**	**Number in**	**Chi 2 of the score**	**Wald chi 2**	**Pr > Chi 2**
	Entered	**Excluded**					
1	**SEX**		1	1	456.8742		<.0001
2	**CLASS**		1	2	41.9271		<.0001
3	**AGE**		1	3	17.0481		<.0001

The adjusted R^2 ('Max-rescaled R-Square') changes from 0.2502 to 0.2727 and then 0.2811. The deviance $-2 \log L$ changes from 2769.457 (model reduced to the constant) to 2334.988 (model with SEX), then 2291.456 (model with SEX and CLASS), and finally 2274.902 (model with SEX, CLASS and AGE). The difference between two successive deviances is therefore always greater than the threshold of 3.84 corresponding to χ^2 with 1 degree of freedom, even if this difference between deviances decreases as the variables are entered.

Model fit statistics		
Criterion	**Coordinate at the origin only**	**Coordinate at the origin and covariables**
AIC	2771.457	2282.902
SC	2777.153	2305.689
– 2 log L	2769.457	2274.902

R-Square	0.2012	**Max-rescaled R-Square**	0.2811

Looking at the coefficients, we find that

- the coefficient of SEX is negative, because fewer males (SEX = 1) than females (SEX = 0) survive the shipwreck;
- the coefficient of AGE is negative, because fewer adults (AGE = 1) than children (AGE = 0) survive the shipwreck;
- the coefficient of CLASS is negative, because there is less survival in third class than in second class, and less in second class than in first class.

Analysis of the estimates of maximum likelihood					
Parameter	**DF**	**Estimate**	**Std. error**	**Wald chi 2**	**Pr > Chi 2**
Intercept	1	2.6096	0.2936	79.0083	<.0001
CLASS	1	– 0.3290	0.0465	50.0996	<.0001
AGE	1	– 1.0062	0.2456	16.7769	<.0001
SEX	1	– 2.6141	0.1333	384.6431	<.0001

Overall, these signs are consistent with the survival rates calculated by cross-tabulating each independent variable with the dependent variable, using the FREQ procedure:

```
PROC FREQ DATA=titanic;
TABLES (class age sex)*survived;
```

Table of CLASS by SURVIVED

CLASS	SURVIVED		
Frequency Percentage Percent in row Percent in col.	0	1	Total
0	673 30.58 76.05 45.17	212 9.63 23.95 29.82	885 40.21 .
1	122 5.54 37.54 8.19	203 9.22 62.46 28.55	325 14.77
2	167 7.59 58.60 11.21	118 5.36 41.40 16.60	285 12.95
3	528 23.99 74.79 35.44	178 8.09 25.21 25.04	706 32.08
Total	1490 67.70	711 32.30	2201 100.00

Table of AGE by SURVIVED

AGE	SURVIVED		
Frequency Percentage Percent in row Percent in col.	0	1	Total
0	52 2.36 47.71 3.49	57 2.59 52.29 8.02	109 4.95
1	1438 65.33 68.74 96.51	654 29.71 31.26 91.98	2092 95.05
Total	1490 67.70	711 32.30	2201 100.00

Table of SEX by SURVIVED

SEX	SURVIVED		
Frequency Percentage Percent in row Percent in col.	0	1	Total
0	126 5.72 26.81 8.46	344 15.63 73.19 48.38	470 21.35
1	1364 61.97 78.80 91.54	367 16.67 21.20 51.62	1731 78.65
Total	1490 67.70	711 32.30	2201 100.00

The model has just one fault: the risk of the crew (CLASS = 0) is not accounted for well, because it is predicted by the model as being less than the risk of first class, whereas it is actually at the risk level of third class. This is a case of a non-monotonic response as a function of an input variable, which is 'class' in this example.

But the Wald indicators of the variables are all greater than 4, although that of AGE, the last variable entered into the model, is much lower than that of SEX, the first variable.

The performance of the model – its capacity to distinguish the survivors from the victims – is given by the indicator 'c' (this is the area under the ROC curve), which is equal to 0.742.

Association of the predicted probabilities and the observed responses

Percent Concordant	65.9	**Somers' D**	0.483
Percent Discordant	17.5	**Gamma**	0.579
Percent Tied	16.6	**Tau-a**	0.212
Pairs	1059390	**c**	0.742

Therefore the model is satisfactory, but we may still ask if we should not allow for the risk to the crew (CLASS = 0), which is higher than the level predicted by the model, and if we should not also allow for the interactions, such as the interaction between class and sex mentioned in Section 3.12.

The simplest way to examine the nature of the non-monotonic risk as a function of CLASS is to consider the variable CLASS as qualitative (nominal) instead of quantitative (discrete). To do this, we can simply add a row for CLASS (as chance would have it, the same name as the variable):

```
PROC LOGISTIC DATA=titanic;
CLASS class / PARAM = ref;
MODEL survived (ref='0') = class age sex
/SELECTION=stepwise RSQUARE;
```

We find that the introduction of the variable CLASS into the model in nominal form causes an increase in the adjusted R^2 (from 0.2727 to 0.3042) and a decrease in the deviance (from 2291.456 to 2228.913), which ensures the better fit of the model. The complete model, with the age, has an R^2 of 0.3135 and a deviance of 2210.061, which are better than those of the previous model.

Model fit statistics

Criterion	Coordinate at the origin only	Coordinate at the origin and covariables
AIC	2771.457	2222.061
SC	2777.153	2256.241
−2 Log L	2769.457	2210.061

R-Square	0.2244	Max-rescaled R-Square	0.3135

Type 3 analysis of effects

Effect	DF	Wald Chi 2	Pr > Chi 2
CLASS	3	108.2432	<.0001
AGE	1	18.9236	<.0001
SEX	1	297.0678	<.0001

The CLASS instruction, in other words the presence of qualitative variables, displays a table showing the 'type 3 analysis of effects'. This analysis is performed for each variable by comparing the sub-model excluding the variable with the model including this variable and the others, in order to test the null hypothesis that this variable has no effect in the model if the other variables are included. This test can be based on the log-likelihood or on the Wald χ^2. The type 3 analysis on the Wald χ^2 is calculated more rapidly than that on the log-likelihood, but it is less reliable if the sample is small (see Section 11.8.6).

As shown by the 'Analysis of the estimates' table, for a binary variable, the Wald χ^2 of a variable and the corresponding probability coincide with those of the category which is not the reference category. In this case, testing the significance of the variable is equivalent to testing the significance of the coefficient of the category which is not the reference category, the null hypothesis then being that this coefficient is 0.

For qualitative variables which, like CLASS, have more than two categories, the sum of the Wald χ^2 of the various categories is not necessarily equal to the Wald χ^2 of the variable; it may be smaller or larger.

Analysis of the estimates of maximum likelihood

Parameter		DF	Estimate	Std. error	Wald Chi 2	Pr > Chi 2
Intercept		1	1.3276	0.2480	28.6490	<.0001
CLASS	0	1	0.9201	0.1486	38.3441	<.0001
CLASS	1	1	1.7778	0.1716	107.3697	<.0001
CLASS	2	1	0.7597	0.1764	18.5558	<.0001
AGE		1	– 1.0615	0.2440	18.9236	<.0001
SEX		1	– 2.4201	0.1404	297.0678	<.0001

Let us look at the coefficients. For AGE and SEX, the comments are the same as previously. As regards CLASS, we have three coefficients, corresponding to categories 0, 1 and 2 (the default reference category is the last of these, which is third class). The coefficient of category 2 is positive, and that of category 1 is even higher, corresponding to their higher survival rates than those of third class passengers. However, the coefficient of category 0 is inconsistent, because it is positive instead of being slightly negative, as it should be in view of the slightly lower survival rate for the crew (23.95%) compared with third class (25.21%). Moreover, the coefficient of category is greater than that of category 2, even though the survival rate is higher in second class (41.40%). The coefficients of these two categories must be reconsidered.

Using the variable CLASS in the nominal form appears to be useful in terms of performance, but at least one category of CLASS poses a problem which means that it must be eliminated. How can we do this without returning to the initial discrete variable? What we need to do is to create 'indicator' variables, as in the following syntax.

```
DATA titanic;
SET titanic;
class0 = (class = 0);
class1 = (class = 1);
class2 = (class = 2);
class3 = (class = 3);
RUN;
```

We then run the following syntax:

```
PROC LOGISTIC DATA=titanic;
MODEL survived (ref='0') = class0 class1 class2 class3 age sex /SELECTION=
stepwise RSQUARE;
```

This time, the variables enter the model in the following order: SEX, CLASS1, CLASS3 and AGE. Age is still the least discriminating variable. The area under the ROC curve has increased from 0.742 to 0.750.

Association of the predicted probabilities and the observed responses

Percent Concordant	64.9	**Somers' D**	0.500
Percent Discordant	14.8	**Gamma**	0.628
Percent Tied	20.3	**Tau-a**	0.219
Pairs	1059390	**c**	0.750

The coefficients all have Wald values greater than 4, and their signs are consistent: those of SEX, AGE and CLASS3 are negative, and that of CLASS1 is positive, because the survival rate is much higher in first class.

Analysis of the estimates of maximum likelihood

Parameter	DF	Estimate	Std. error	Wald Chi 2	Pr > Chi 2
Intercept	1	2.1448	0.2766	60.1435	<.0001
class1	1	0.9022	0.1498	36.2820	<.0001
class3	1	– 0.8634	0.1352	40.7638	<.0001
AGE	1	– 1.0315	0.2412	18.2893	<.0001
SEX	1	– 2.3813	0.1338	316.9023	<.0001

We have not yet considered the interactions between variables. We have seen in Section 3.12 that the survival rate decreased with the class number for females, whereas the survival rate for males was slightly higher in third class than in second class. It was also pointed out that the testing of all possible interactions by the LOGISTIC procedure is carried out by writing the following syntax, with the vertical bars '|'.

```
PROC LOGISTIC DATA=titanic;
MODEL survived (ref='0') = class | age | sex
/SELECTION=stepwise RSQUARE;
```

With this syntax, we obtain the model described in Section 3.12, which has the advantage of allowing for the interaction CLASS*SEX by assigning a positive coefficient to it, corresponding to the slightly higher survival rate for males in third class. However, this model has the drawback of not allowing for the special feature of class 0 (the crew) and the resulting non-monotonic form for the survival rate as a function of the class number.

We must therefore run the following syntax:

```
PROC LOGISTIC DATA=titanic;
MODEL survived (ref='0') = class0 | class1 | class2 | class3 | age | sex
/SELECTION=stepwise RSQUARE;
```

The variables are included in the model as follows:

Summary of the sequential selection

Step	Effect Entered	Effect Excluded	DF	Number in	Chi 2 of the score	Wald Chi 2	Pr > Chi 2
1	SEX		1	1	456.8742		<.0001
2	class1		1	2	73.9363		<.0001
3	class3		1	3	34.0874		<.0001
4	class3*SEX		1	4	53.5742		<.0001
5	AGE		1	5	18.8664		<.0001
6	class3*AGE		1	6	25.1637		<.0001
7		class3*AGE	1	5		0.0048	0.9449

The specific interaction between third class and the male sex appears very precisely in the form of the variable CLASS3*SEX. The coefficients of the model are as follows:

Analysis of the estimates of maximum likelihood

Parameter	DF	Estimate	Std. error	Wald Chi 2	Pr > Chi 2
Intercept	1	3.1208	0.3278	90.6106	<.0001
class1	1	0.7760	0.1632	22.6144	<.0001
class3	1	− 2.4654	0.2834	75.6842	<.0001
AGE	1	− 0.9709	0.2278	18.1663	<.0001
SEX	1	− 3.5167	0.2453	205.5231	<.0001
class3*SEX	1	2.1465	0.3088	48.3280	<.0001

We can see that the positive sign of the CLASS3*SEX interaction is compensated, to avoid giving an excessively high survival rate to males in third class, by coefficients which are more negative than previously for CLASS3 (− 2.4654 instead of − 0.8634) and SEX (− 3.5167 instead of − 2.3813).

Association of the predicted probabilities and the observed responses

Percent Concordant	65.0	Somers' D	0.503
Percent Discordant	14.7	Gamma	0.631
Percent Tied	20.3	Tau-a	0.220
Pairs	1059390	c	0.752

The area under the ROC curve has now increased to 0.752.

We have thus obtained a model without deficiencies which performs better than the previous ones. However, we may wonder if it would not be appropriate to include interactions in the model without including the variables which make up the interaction. It might be possible to reveal some interactions which do not appear at present because the variables are not sufficiently discriminating when taken in isolation (main effect). To enable a variable to be manifested in an interaction without itself appearing directly, which is not the default option of the LOGISTIC procedure, we must use the 'HIERARCHY = none' option:

```
PROC LOGISTIC DATA=titanic;
MODEL survived (ref='0') = class0 | class1 | class2 | class3 | age | sex
/ HIERARCHY = none /SELECTION=stepwise RSQUARE;
```

The area under the ROC curve leaps from 0.752 to 0.768, but one of the Wald indicators is close to 4 and another is not very high either, and the interaction of order 3 which is included in the model does not simplify it and may give rise to the problem of an excessively low frequency at the intersection of three conditions.

Association of the predicted probabilities and the observed responses

Percent Concordant	68.4	**Somers' D**	0.537
Percent Discordant	14.8	**Gamma**	0.645
Percent Tied	16.8	**Tau-a**	0.235
Pairs	1059390	**c**	0.768

Analysis of the estimates of maximum likelihood

Parameter	DF	Estimate	Std. error	Wald Chi 2	Pr > Chi 2
Intercept	1	2.9375	0.2940	99.8583	<.0001
class1	1	0.5393	0.1747	9.5325	0.0020
class3	1	– 3.1011	0.3270	89.9148	<.0001
class2*AGE	1	– 1.1329	0.2541	19.8720	<.0001
class3*SEX	1	– 0.8268	0.3550	5.4233	0.0199
AGE*SEX	1	– 4.1915	0.2867	213.6821	<.0001
class3*AGE*SEX	1	3.5410	0.4513	61.5748	<.0001

We therefore rerun the preceding syntax with the addition of the symbol @2, which specifies that only the interactions of order 2 or below are tested.

```
PROC LOGISTIC DATA=titanic;
MODEL survived (ref='0') = class0 | class1 | class2 | class3 | age | sex @2
/ HIERARCHY = none /SELECTION=stepwise RSQUARE;
```

This gives us an area under the ROC curve as great as that of the preceding model, Wald indicators which are all greater than 16, and a simpler model without a third-order interaction.

Association of the predicted probabilities and the observed responses

Percent Concordant	68.4	**Somers' D**	0.535
Percent Discordant	14.8	**Gamma**	0.644
Percent Tied	16.8	**Tau-a**	0.234
Pairs	1059390	**c**	0.768

Analysis of the estimates of maximum likelihood

Parameter	DF	Estimate	Std. error	Wald Chi 2	Pr > Chi 2
Intercept	1	3.4837	0.3014	133.6350	<.0001
class0	1	– 0.6878	0.1685	16.6631	<.0001
class3	1	– 3.6473	0.3337	119.4584	<.0001
class2*AGE	1	– 1.7510	0.2681	42.6509	<.0001
SEX	1	– 2.8045	0.3987	49.4888	<.0001
class3*SEX	1	2.4913	0.3485	51.1143	<.0001
AGE*SEX	1	– 1.2523	0.2793	20.1003	<.0001

In addition to the CLASS3*SEX interaction which was found previously, a second interaction appears: this is CLASS2*AGE, corresponding to the fact that it is only for adults that the survival rate decreases in second class, whereas for children the survival rate in second class is 100% (24 out of 24 children saved), as shown by the contingency table produced by the following syntax.

Table 1 of CLASS by SURVIVED

Test for AGE=0

CLASS	SURVIVED		
Frequency **Percentage** **Percent in row** **Percent in col.**	**0**	**1**	**Total**
0	0 0.00 . 0.00	0 0.00 . 0.00	0 0.00
1	0 0.00 0.00 0.00	6 5.50 100.00 10.53	6 5.50
2	0 0.00 0.00 0.00	24 22.02 100.00 42.11	24 22.02
3	52 47.71 65.82 100.00	27 24.77 34.18 47.37	79 72.48
Total	52 47.71	57 52.29	109 100.00

```
PROC FREQ DATA=titanic;
TABLES age*class*survived;
```

Table 2 of CLASS by SURVIVED

Test for AGE=1

CLASS	**SURVIVED**		
Frequency **Percentage** **Percent in row** **Percent in col.**	**0**	**1**	**Total**
0	673 32.17 76.05 46.80	212 10.13 23.95 32.42	885 42.30
1	122 5.83 38.24 8.48	197 9.42 61.76 30.12	319 15.25
2	167 7.98 63.98 11.61	94 4.49 36.02 14.37	261 12.48
3	476 22.75 75.92 33.10	151 7.22 24.08 23.09	627 29.97
Total	1438 68.74	654 31.26	2092 100.00

The adjusted $R^2 = 0.3582$ is markedly higher than before, and the deviance $= 2117.451$ is markedly lower, showing that the model is even better adjusted. In reality, in a situation like this, the statistician will check with experts in the field to ensure that the rule he has discovered (the survival rate of children is the same in first and second class) is plausible and can be explained, and can therefore be legitimately modelled. Without this precaution, there would be a risk of overfitting by modelling a special case which is not a general rule.

Model fit statistics

Criterion	**Coordinate at the origin only**	**Coordinate at the origin and covariables**
AIC	2771.457	2131.451
SC	2777.153	2171.327
– 2 log L	2769.457	2117.451

R-Square	0.2564	**Max-rescaled R-Square**	0.3582

We therefore consider that we have obtained a model which fits well enough to the data, with a robustness attested by the Wald indicators, and a better performance, as measured by the area under the ROC curve, than that of the other models tested. Strictly speaking, we should measure the area under the ROC curve on the test sample, rather than on the training sample, but if we were to do this the conclusion would still be the same in this case.

This is how we make this more rigorous measurement. A variable TARGET is created in the data set analysed, this variable being equal to the variable SURVIVED for two thirds of the randomly chosen observations, and being missing for the other third of observations. We then model TARGET, instead of SURVIVED as before. Thus, the observations for which TARGET is missing will not be included in the construction of the logistic model, but the LOGISTIC procedure can calculate the *a posteriori* probability of these at the output of the model, provided that the independent variables of the model are not missing (which is always the case here). This probability is written to an output data set which also contains the entered independent variable, namely SURVIVED, and the area under the ROC curve is calculated for this third of the observations. Thus these observations form a true test sample, which has not been used for the training of the model but can be used to test it.

The SAS syntax is as follows.

```
DATA titanic;
SET titanic;
IF ranuni(0) < 0.66 THEN target = survived;
RUN;

PROC LOGISTIC DATA=titanic;
MODEL target (ref='0') = class0 | class1 | class2 | class3 | age
| sex @2
/ HIERARCHY = none /SELECTION=stepwise RSQUARE;
OUTPUT OUT=model PREDICTED=proba ;
RUN;
%AUC(model,survived,proba);
```

The macro %AUC which is used to calculate the area under the ROC curve, which is applied to the Mann–Whitney test, is the one described in Section 11.16.5, with the addition of a single WHERE condition to avoid selecting the training sample:

```
PROC NPAR1WAY WILCOXON DATA=table CORRECT=no;
WHERE target = .;
CLASS &target;
VAR &score;
RUN;
```

11.8.14 Logistic regression with R

The results of the preceding section can also be obtained using the R software, especially the *glm* function of the *stats* package which is automatically loaded when R is started. We start by changing the default directory, before reading the 'titanic.txt' file, which is done very simply because it is a text file with separators and without a header for the name of the variables.

```
> setwd("C:/Documents and Settings/tuffery/My Documents/Data Mining/
Data sets/Data for logistic regression/Titanic")

> titanic <- read.table("titanic.txt", header = FALSE, sep = ";",
quote="\"",dec=",",,col.names=c("class","age","sex","survived"))
```

We 'attach' the data frame 'titanic', enabling us to state the names of the variables subsequently without specifying the data frame to which they belong.

```
> attach(titanic)
```

Several contingency tables can be produced, but this cannot be done with a single command as it can in the FREQ procedure of SAS or SPSS.

```
> table (class,survived)
        survived
class      0     1
    0    673   212
    1    122   203
    2    167   118
    3    528   178
> prop.table(table(class,survived),1)
     survived
class          0          1
    0  0.7604520  0.2395480
    1  0.3753846  0.6246154
    2  0.5859649  0.4140351
    3  0.7478754  0.2521246
```

The logistic regression (logit) is simply calculated by the *glm* function, in which the link function 'logit' is specified as a parameter. If not specified otherwise, all the predictors are considered to be quantitative variables. The *summary* function retrieves the results of the SAS LOGISTIC procedure, with the coefficients, the deviance, and the AIC. The glm function additionally displays the quartiles of the residual of the deviance; we note here that this hardly ever exceeds 2 in absolute terms, in line with what is required.

```
> logit <-
glm(survived~class+age+sex,data=titanic,family=binomial(link =
"logit"))
> summary(logit)

Call:
glm(formula = survived ~ class + age + sex, family =
binomial(link = "logit"),
    data = titanic)

Deviance Residuals:
    Min        1Q     Median        3Q       Max
-1.8989   -0.7879    -0.5877    0.7022    2.0615
```

```
Coefficients:
             Estimate Std. Error  z value  Pr(>|z|)
(Intercept)   2.60985    0.29360    8.889   < 2e-16 ***
class        -0.32904    0.04648   -7.079  1.45e-12 ***
age          -1.00627    0.24565   -4.096  4.20e-05 ***
sex          -2.61420    0.13329  -19.613   < 2e-16 ***
---
Signif. codes: 0 '***' 0.001 '**' 0.01 '*' 0.05 '.' 0.1 ' ' 1

(Dispersion parameter for binomial family taken to be 1)

    Null deviance: 2769.5  on 2200  degrees of freedom
Residual deviance: 2274.9  on 2197  degrees of freedom
AIC: 2282.9

Number of Fisher Scoring iterations: 4
```

If we wish to consider the predictors as qualitative variables as before, we can use the *factor* function, applied to all the rows and each of columns 1 to 4. The effect of this function can be seen by requesting a 'summary' of the file before and after:

```
> summary(titanic)
     class            age               sex              survived
 Min.   :0.000   Min.   :0.0000   Min.   :0.0000   Min.   :0.0000
 1st Qu.:0.000   1st Qu.:1.0000   1st Qu.:1.0000   1st Qu.:0.0000
 Median :1.000   Median :1.0000   Median :1.0000   Median :0.0000
 Mean   :1.369   Mean   :0.9505   Mean   :0.7865   Mean   :0.3230
 3rd Qu.:3.000   3rd Qu.:1.0000   3rd Qu.:1.0000   3rd Qu.:1.0000
 Max.   :3.000   Max.   :1.0000   Max.   :1.0000   Max.   :1.0000
> for (i in 1:4) titanic[,i] <- factor(titanic[,i])
> summary(titanic)
 class    age      sex      survived
 0:885   0: 109   0: 470   0:1490
 1:325   1:2092   1:1731   1: 711
 2:285
 3:706
```

The *glm* function can be restarted without having anything else to specify. However, we will find that it does not choose the same reference category as SAS for the CLASS variable, because it chooses category 0. We can change this reference category in order to facilitate a comparison with the SAS results. This is done outside the glm function, either on the definition of the factors, by the levels option, or as follows (note that we must specify the file, in spite of the previous 'attach'):

```
> titanic$class <- relevel(titanic$class,ref="3")
>
summary(glm(survived~class+age+sex,data=titanic,family=binomial
(link = "logit")))
```

```
Call:
glm(formula = survived ~ class + age + sex, family =
binomial(link = "logit"),
     data = titanic)

Deviance Residuals:
    Min          1Q      Median        3Q        Max
-2.0812     -0.7149     -0.6656    0.6858     2.1278

Coefficients:
              Estimate Std. Error  z value   Pr(>|z|)
(Intercept)    1.3276     0.2480     5.352   8.67e-08 ***
class0         0.9201     0.1486     6.192   5.93e-10 ***
class1         1.7778     0.1716    10.362    < 2e-16 ***
class2         0.7597     0.1764     4.308   1.65e-05 ***
age1          -1.0615     0.2440    -4.350   1.36e-05 ***
sex1          -2.4201     0.1404   -17.236    < 2e-16 ***
---
Signif. codes: 0 '***' 0.001 '**' 0.01 '*' 0.05 '.' 0.1 ' ' 1

(Dispersion parameter for binomial family taken to be 1)

    Null deviance: 2769.5  on 2200  degrees of freedom
Residual deviance: 2210.1  on 2195  degrees of freedom
AIC: 2222.1

Number of Fisher Scoring iterations: 4
```

This gives us the same coefficients as those estimated previously by the SAS LOGISTIC procedure.

11.8.15 Advantages of logistic regression

1. It can handle discrete, qualitative or continuous independent variables.
2. It can handle an ordinal or nominal dependent variable.
3. The conditions for using logistic regression are less restrictive than those of linear discriminant analysis (no assumption of multinormality or homoscedasticity of the independent variables).
4. It provides models that are often very accurate.
5. It generally works well with small samples, perhaps even better than linear discriminant analysis (and therefore quadratic discriminant analysis) which requires the estimation of a larger number of parameters (the means of each group and the intraclass covariance matrix).
6. The models that are produced are concise and easily programmed by IT personnel.

7. It can handle non-monotonic responses.
8. It can allow for interactions between independent variables.
9. It directly models a probability.
10. It supplies confidence intervals for the results.
11. Many statistical tests, such as tests of significance of coefficients, are available. They are asymptotic and even exact.
12. It allows stepwise selection of variables.
13. It detects global phenomena (whereas decision trees only detect local phenomena).
14. It is implemented in numerous software.

11.8.16 Advantages of the logit model compared with probit

1. The coefficients of the logit are easily interpreted in terms of odds ratios.
2. Independent sampling of the independent variables only changes the constant of the logit (see Section 11.8.11).

11.8.17 Disadvantages of logistic regression

1. The explanatory variables must be linearly independent (non-collinearity).
2. It is a numeric approximation, with the following consequences:
   a. It is an iterative calculation which is much slower than the direct calculation of linear discriminant analysis.
   b. It is less precise than discriminant analysis when the assumption of normality of the latter is true (there is an asymptotic rise in the error rate of 30–50% in logistic regression, as shown by Efron).[63]
   c. Logistic regression does not always converge towards an optimal solution. In particular, it does not work if the groups are completely separated.
3. It does not handle the missing values of continuous variables (unless they are divided into classes and the missing values are grouped in a special class).
4. It is sensitive to extreme values of continuous variables (unless they are divided into classes).

[63] Efron, B. (1975) The efficiency of logistic regression compared to discriminant analysis. *Journal of the American Statistical Association*, 70, 892–898.

11.9 Developments in logistic regression

11.9.1 Logistic regression on individuals with different weights

Our whole description of logistic regression up to this point has been concerned with individual data, where each observation, in other words each line in the data set, relates to a different individual or object. This situation is by far the most common, especially in scoring. However, it is unavoidable when at least one of the independent variables is continuous, because it is most unlikely that two individuals will have the same value for this variable. On the other hand, if all the independent variables are qualitative, it is possible and even logical to group the individuals. Suppose that we wish to model the type of leisure pursuit (outdoor or otherwise) followed by an individual as a function of his/her sex, age group (five values) and rural or urban place of residence. There are only 20 possible intersections of all the categories of the independent variables, and we can very easily replace the initial data set, containing one line per individual, showing the type of leisure, with a data set of 20 lines, showing the combination of the independent variables (all different), the number of individuals for each combination, and among these the number of individuals who have outdoor leisure pursuits. Therefore, instead of modelling a 0/1 Bernoulli variable, we must model a binomial variable and its proportion y_i/n_i (the number of 'tails' found, out of the total number of 'heads' and 'tails'). Conventionally, we say that y_i is the number of events and n_i is the number of trials. In this case, y_i is the number of individuals in group i having outdoor leisure pursuits, and n_i is the total number of individuals in this group. Note that, in contrast to the correlated data model described in the next section, the individuals of a single group are independent, even if they have features in common. This situation can be dealt with by the SAS/STAT LOGISTIC procedure, and by the GENMOD procedure for which a command in the following form is shown (a more detailed example of this procedure is given below):

```
MODEL y/n = var1 var2 ... vark / DIST=bin LINK=logit
```

The PROBIT procedure of the IBM SPSS Regression module can also carry out this calculation, with the probit or logit link function.

11.9.2 Logistic regression with correlated data

Here, the situation is more or less the converse of that described above: instead of having one observation covering a number of individuals, each individual is represented by a number of observations. This happens if the individual is subjected to several different processes or is observed on several occasions over time. Such data are known as 'longitudinal'. A similar situation arises where a single group is represented by a number of individuals having features in common. In this case, unlike the situation in the previous section, the individuals in a single group are not independent (but may have their own individual characteristics); this may be useful, for example, in the epidemiological study of a contagious disease which may be transmitted between members of a single family.

The common feature of these different situations is that the observations to be modelled are not all independent, and therefore they negate a fundamental assumption of the basic models, used in the calculation of likelihood: the non-correlation of the observations.

I will now discuss the modelling of correlated data in the context of logistic regression and the generalized estimating equations (GEE) method of Liang and Zeger[64] for the analysis of repeated measurements, bearing in mind that this approach is generalized to the case of a discrete non-binary dependent variable (following a Poisson distribution, for example), and that there are also other approaches based on mixed models which take into account random effects which are specific to each individual. A discussion of these methods, illustrated by a specific application, can be found in an article by Guéguen, Zins and Nakache.[65] Mixed models are recommended when there are a large number of observations per individual. In these models, we simultaneously model the mean and the variance of the dependent variable, and we must define the variance–covariance matrix, whereas in the method which I will now describe we simply have to define a working correlation matrix, as you will shortly see.

For example, the observations can be presented as follows:

ID (family)	Individual	Target	Seniority	Housing	Income	Age	Height	Weight	...
1	1	1	5	house	2500	35	175	65	
1	2	0	5	house	2500	8	130	32	
1	3	1	5	house	2500	6	110	28	
2	1	0	10	apartment	2500	55	165	55	
...	...	...	...	...	...	...	...	...	

We can see that the same family (same 'ID') can be present several times, with some of the data being identical each time (seniority, income, housing) and some being different (age, height, weight). SAS/STAT can model this kind of situation as follows:

```
PROC GENMOD DATA=my.table;
CLASS id individual;
MODEL target = seniority housing income age height weight
/ DIST=bin TYPE3 WALD;
REPEATED SUBJECT=id
/ WITHINSUBJECT=individual TYPE=exch COVB CORRW;
OUTPUT OUT=my.model PRED=proba ;
RUN;
```

The command SUBJECT=id defines the variable 'id' (this could be a set of variables) as the identifier of the observations of a single group, in which the values of the dependent variable are intercorrelated, whereas these values are independent for the observations of two separate groups. As a general rule, an 'id' can correspond to one individual observed n times, or n individuals belonging to the same segment, the same establishment, the same family, etc. TYPE indicates the structure of the correlation matrix $M_{jk} = [\text{corr}(y_{ij}, y_{ik})]_{jk}$, a matrix in which y_{ij} is the value of the dependent variable for the jth observation on the ith individual.

[64] Liang K.-Y. and Zeger S.L. (1986) Longitudinal data analysis using generalized linear models. *Biometrika*, 73, 13–22.

[65] Guéguen, A. Zins, M. and Nakache, J.P. (2000),Utilisation des modèles marginaux et des modèles mixtes dans l'analyse de données longitudinales concernant mariage et consommation d'alcool des femmes de la cohorte Gazel. *Revue de Statistique Appliquée*, XLVIII(3), 57–73.

The TYPE is chosen according to the matrix structure which appears likely in view of the problem in question. Here, we have chosen the 'exchangeable' type, which assumes that corr $(y_{ij},y_{ik}) = \alpha$ if $j \neq k$. The possible types are:

- 'exchangeable': corr$(y_{ij},y_{ik}) = \alpha$ if $j \neq k$, denoted EXCH or CS;
- 'independent': corr$(y_{ij},y_{ik}) = 0$ if $j \neq k$, denoted IND, the default type in GENMOD;
- 'm-dependent': corr$(y_{ij},y_{i,j+t}) = 0$ if $t > m$ and α_t if t is between 1 and m, denoted MDEP (m);
- 'unstructured': corr$(y_{ij},y_{ik}) = \alpha_{jk}$ if $j \neq k$, denoted UNSTR;
- 'autoregressive': corr$(y_{ij},y_{i,j+t}) = \alpha^t$ if $t = 0, 1, \ldots, n_i - j$, denoted AR;
- 'fixed': user-defined in the form

```
TYPE = USER (  1.0   0.4   0.6   0.7
               0.4   1.0   0.9   0.5
               0.6   0.9   1.0   0.3
               0.7   0.5   0.3   1.0)
```

Clearly, in all cases corr$(y_{ij},y_{ik}) = 1$ if $j = k$.

The WITHINSUBJECT=individual command defines the variable 'individual' as identifying the different observations (the repeated measurements) of a single group. A group must not have two observations with the same value of the variable defined by WITHINSUBJECT=individual. It is this variable that defines the repetition of a group several times (it is often a 'time' variable in medical trials, where the progress of a disease is followed with allowance for the patient's characteristics and the treatment given). Two points are worth noting about this variable: first, it is not necessarily one of the independent variables; and second, its categories are not necessarily sorted in the data set, and are not necessarily all present for all the individuals (one patient may miss a medical examination, for example).

If all the categories are present and well sorted, the WITHINSUBJECT=individual command is superfluous for the use of the correlation matrix, because the variables y_{ij} will be in the same order for all i. Otherwise, this command is necessary; for example, if the second measurement were missing for an individual i, the third measurement would be treated as the second, and y_{i3} would be treated as y_{i2}.

If the categories are correctly sorted but not all present, we add the 'sorted' option to the WITHINSUBJECT=individual command to stop GENMOD sorting the observations again.

The GENMOD procedure calculates logistic regression coefficients for the set of variables to the right of the 'model target=' command, regardless of whether or not they are identical for all the observations having the same 'id' (in some repeated measures models, all the independent variables are identical for a given 'id'). We use the logit link function if the distribution of the dependent variable is binomial ('dist=bin'), but there are also other functions. You should be aware that GENMOD is much slower than LOGISTIC, and the computation times can be very long for large data volumes.

11.9.3 Ordinal logistic regression

Here we assume that the variable Y to be predicted takes m ordered values, denoted $1, 2, \ldots, m$. The simplest and most widely used model, implemented in IBM SPSS Regression and in the SAS LOGISTIC procedure for example, is the *equal (or parallel) slopes model*: it is assumed that the logit of the cumulative probabilities is written

$$\text{logit}(\text{Prob}(Y \leq r/X = x)) = \alpha_r + \sum_i \beta_i x_i, \quad \text{for } 1 \leq r < m,$$

where only the constant is dependent on r. The probabilities are written

$$\text{Prob}(Y \leq r/X = x) = \frac{\exp(\alpha_r + \sum_i \beta_i x_i)}{1 + \exp(\alpha_r + \sum_i \beta_i x_i)}, \quad r = 1, \ldots, m-1.$$

We speak of a 'proportional odds model', because the odds ratios for a fixed r are all proportional with respect to each other, the ratio being independent of r:

$$\frac{\text{Prob}(Y \leq r/X = x)/\text{Prob}(Y > r/X = x)}{\text{Prob}(Y \leq r/X = x')/\text{Prob}(Y > r/X = x')} = \frac{\exp(\alpha_r + \sum_i \beta_i x_i)}{\exp(\alpha_r + \sum_i \beta_i x'_i)} = \exp\left(\sum_i \beta_i (x_i - x'_i)\right).$$

This model is often preferred to more complex models, which quickly become too complex when there is an increase in the number of independent variables or the number of categories of the dependent variable. In common applications, the dependent variable may be a degree of seriousness of an illness or a customer satisfaction level.

In ordinal logistic regression, we can use the following link functions:

- logit;
- probit, whose coefficients are less easy to interpret than those of the logit;
- log-log, given by $\log[-\log(1-\mu)]$ and used when high values of the dependent variable are more likely (e.g., marks of 3 to 5 out of 5 in a satisfaction survey);
- cauchit, given by $\tan[\pi(\mu-0.5)]$ and used when the extreme values of the dependent variable are more likely (e.g., a mark of 5/5 in a satisfaction survey).

The cauchit link function is not very widely used; it is available in IBM SPSS Regression and R, but not in SAS.

11.9.4 Multinomial logistic regression

Here, we assume that the variable Y to be predicted takes m non-ordered values. In this case, the logit function $\text{Prob}(Y = 1/X = x)$ can be replaced with the *generalized logit*

$$\log\left(\frac{\text{Prob}(Y = j/X = x)}{\text{Prob}(Y = m/X = x)}\right),$$

calculated with respect to a reference category ($Y = m$), and it is this expression, defined for $1 \le j \le m$, that is written as a linear combination of the independent variables, whose coefficients $(\beta_{ij})_i$ vary for each j. The probabilities are written:

$$\text{Prob}(Y = j/X = x) = \frac{\exp(\alpha_j + \sum_k \beta_{jk} x_k)}{1 + \sum_{i=1}^{m-1} \exp(\alpha_i + \sum_k \beta_{ik} x_k)}, j = 1, \ldots, m-1;$$

$$\text{Prob}(Y = m/X = x) = \frac{1}{1 + \sum_{i=1}^{m-1} \exp(\alpha_i + \sum_k \beta_{ik} x_k)}.$$

Here, we cannot reasonably assume that the coefficients β_{jk} are independent of j. Since Version 8.2, SAS/STAT has incorporated this algorithm in its LOGISTIC procedure, but a more general algorithm, providing other functions as well as the generalized logit, is available in CATMOD.

It may be helpful to summarize the treatment of logistic regression in different SAS procedures.

Logistic regression	**LOGISTIC**	**GENMOD**	**CATMOD**
dichotomous	yes	yes	yes (avoid quantitative X_i)
ordinal polytomous	equal slopes model, with cumulative logit	yes, with cumulative logit, cumulative probit or cumulative log-log	yes, with adjacent, cumulated or 'mean' logit
nominal polytomous	yes, with generalized logit	no	yes, with generalized logit or 'identity'
repeated measures	no	yes (GEE method)	yes (WLS method)
conditional model	yes (since SAS 9)	no	no
overdispersion control	yes	yes	no
stepwise selection of the variables	yes	no	no
exact tests of the parameters	yes	no	no
ROC curve	yes	no	no

11.9.5 PLS logistic regression

PLS logistic regression was developed from 2000 onwards by Michel Tenenhaus (see his book cited in Section 11.7.10 above). The algorithm for finding the components is similar to that of ordinary PLS regression (see Section 11.7.10), but there is a method offering the same performance which is easier to apply: this is logistic regression on PLS components. We start with a PLS regression of the indicator of Y (or the indicators of Y, if Y has more than two categories) on the independent variables X_i and, if necessary, on their interactions $X_i{}^*X_j$. We obtain k PLS components ($k = 1$, if appropriate), and then perform a logistic regression of Y on these PLS components.

PLS logistic regression is useful when we wish to retain numerous closely correlated independent variables in an analysis, possibly with missing values, as may be the case in

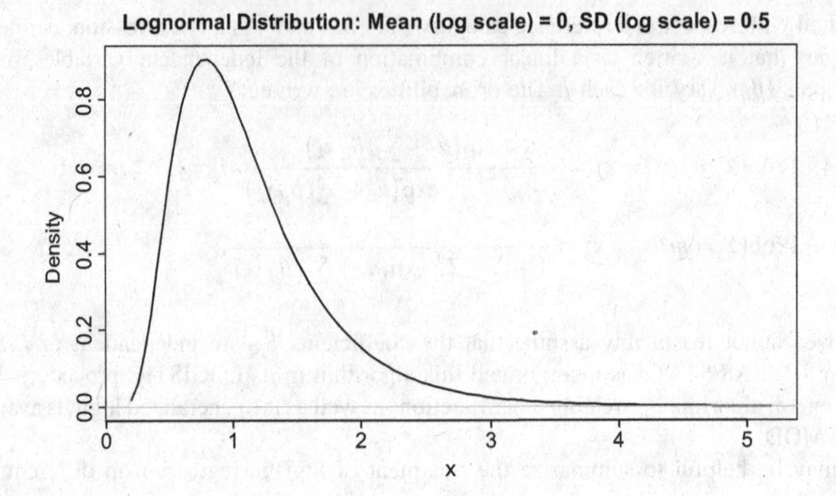

Figure 11.101 Density of the lognormal distribution.

sensory analysis and satisfaction surveys. In more common situations, this method does not necessarily provide any significant benefit compared with ordinary logistic regression.

11.9.6 The generalized linear model

The *generalized linear model* incorporates in a wide-ranging generalization the general linear model (linear regression, analysis of variance and analysis of covariance), logistic regression, Poisson regression, the log-linear model, and other models. It was developed by John Nelder and R. W. M. Wedderburn (1972).[66] This type of model is widely used in non-life insurance for modelling the frequencies (Poisson distribution) and annual costs of losses (gamma and lognormal distribution; see Figures 11.101 and 11.102). These models are available in Statistica, R, S-PLUS, SAS and IBM SPSS Statistics. In SAS/STAT, we use the GENMOD procedure, in which we use DIST to specify the distribution of $Y/X=x$ and LINK to specify the link function. For example, given that a variable Y has a lognormal distribution if log(Y) has a normal distribution, we model log(y) as a function of the x_i with the parameters DIST = normal and LINK = id. For a conventional logistic regression, the parameters are DIST = binomial and LINK = logit.

By contrast with the general linear model, the variable Y to be predicted is not necessarily continuous. Its mathematical expectation is expressed in the form

$$g(\mathrm{E}(Y/X = x) = \beta_0 + \sum_i \beta_i X_i,$$

where g is a differentiable monotonic link function. In this case, the general linear model is doubly generalized, because $g \neq 1$. The variables X_i can be initial variables of the model, interactions $X_k = X_i{}^*X_j$ or powers $(X_i)^P$ introduced to allow for non-linear effects.

[66] Nelder, J.A and Wedderburn, R.W (1972) Generalized linear models. *Journal of the Royal Statistical Society, Series A*, 135(3), 370–384.

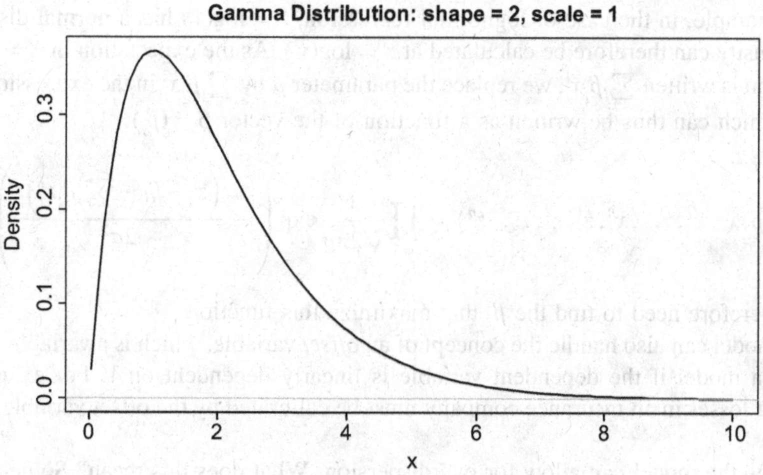

Figure 11.102 Density of the gamma distribution.

The main models are as follows:

Distribution of Y/X=x	Problem	Link function (examples)
Normal (continuous)	Regression	$g(\mu) = \mu$
Gamma (continuous positive)	Regression on asymmetric quantitative dependent variable (e.g. service life of equipment, cost, amount of a loss)	$g(\mu) = -1/\mu$ (or in some cases $g(\mu) = \log(\mu)$)
Lognormal (continuous positive)	Regression on asymmetric quantitative variable (preceding examples, number of words in a sentence)	$g(\mu) = \mu$
Bernoulli (discrete: yes/no)	Logistic regression	logit $g(\mu) = \log(\mu/1-\mu)$, probit, log-log $g(\mu) = \log(-\log(1-\mu))$
Poisson (discrete: counting)	Modelling a rare phenomenon (number of losses in motor insurance) or frequency of a contingency table (log-linear model)	$g(\mu) = \log(\mu)$
Multinomial (categories not ordered)	Logistic regression on nominal dependent variable	Generalized logit
Multinomial (ordered categories)	Logistic regression on ordinal dependent variable	Cumulative, adjacent logit

The estimation of the model is carried out by the maximum likelihood method as for logistic regression. It is evaluated by calculating the deviances of the log-likelihoods and by the χ^2 test.

For example, in the case of lognormal regression, $Z = \log(Y)$ has a normal distribution whose density can therefore be calculated at $z^i = \log(y^i)$. As the expectation in $X = x^i$ of this distribution is written $\sum_j \beta_j x_j^i$, we replace the parameter μ by $\sum_j \beta_j x_j^i$ in the expression for the density which can thus be written as a function of the vector $\beta = (\beta_j)$:

$$L(\beta, x^1, x^2, \ldots, x^n, z^1, z^2, \ldots, z^n) = \prod_{i=1}^{n} \frac{1}{\sqrt{2\pi}\sigma} \exp\left(- \frac{\left(z^i - \beta_0 - \sum_j \beta_j x_j^i\right)^2}{2\sigma^2} \right).$$

We therefore need to find the β_j that maximize this function.

The model can also handle the concept of an *offset* variable, which is a variable V used to calibrate a model if the dependent variable is linearly dependent on V. For example, the number of losses in an insurance company must be calibrated by the offset variable 'number of policies'.

Finally, the model can allow for overdispersion. What does this mean? Sometimes, for example if there are repeated observations, the variance of the dependent variable, assumed to be of the binomial or Poisson type, which is theoretically fixed by the model ($V(\mu) = n \cdot \mu(1-\mu)$ and $V(\mu) = \mu$, respectively), is multiplied by a scale factor. If this parameter is greater than 1, overdispersion is considered to be present. It may also be less than 1, in which case we speak of underdispersion, but this is less common. Overdispersion indicates a poor fit of the model to the data. The dispersion parameter is currently estimated by one of two quantities:

- the scaled deviance, in other words the deviance calculated over the set of the m sub-populations defined by the independent variables (see Section 11.8.8) and divided by its degrees of freedom;
- the scaled Pearson's χ^2, in other words the Pearson's χ^2 of the model divided by its degrees of freedom.

The number of degrees of freedom common to these two statistics is $m(k-1) - p$, where k is the number of categories of the dependent variable, p is the number of parameters (continuous variables or categories of qualitative variables, without omitting the constant), and m is the number of sub-populations defined by the independent variables.

Like the Hosmer–Lemeshow test, these two statistics compare the observed frequencies with those predicted by the model, and therefore measure the quality of fit of the model to the data. They are also both estimators of the same parameter, and should therefore have similar values. If this is not the case, it is generally because the frequencies of the m sub-populations are too small, which may well occur when there are many variables, especially when there are continuous variables or variables having many categories. In this case we speak of *sparse data*. In this common case, Pearson's χ^2 approaches $m(k\text{-}1)$, the deviance depends less and less on the observed frequencies, and the two estimates of the dispersion parameter are false and may even suggest underdispersion; in any case they no longer depend on the quality of fit of the model. The data will certainly be sparse for Pearson's χ^2 if a large number of sub-populations have less than five observations each.

Conversely, when the sub-populations are large enough, these two statistics follow a χ^2 distribution with $m(k-1) - p$ degrees of freedom, and it is possible to test the null hypothesis

Deviance and Pearson Goodness-of-Fit Statistics				
Criterion	**Value**	**DF**	**Value/DF**	**Pr > ChiSq**
Deviance	826.3152	392	2.1079	<.0001
Pearson	1102.9418	392	2.8136	<.0001

Number of unique profiles: 402

Note: The covariance matrix has been multiplied by the heterogeneity factor (Deviance / DF) 2.10795.

Figure 11.103 Overdispersion and deviance and Pearson's χ^2.

that the model fits the data perfectly. An associated probability less than 0.05 will indicate the presence of a dispersion problem, as in Figure 11.103.

The SAS GENMOD and LOGISTIC procedures, together with NOMREG in IBM SPSS Statistics, calculate the deviance and scaled χ^2, carry out the associated test and therefore enable any overdispersion to be detected. GENMOD calculates them automatically when the dependent variable is of the 'events/trials' type as in Section 11.9.1, while LOGISTIC calculates them when the SCALE=NONE (or SCALE=N) option is selected. If the dependent variable is a 0/1 Bernoulli variable, SAS has to be prompted to determine the sub-populations defined by the independent variables, which is done by adding the AGGREGATE option (and, for GENMOD, by adding the list of independent variables, which is unnecessary for LOGISTIC). Also, when SCALE=DEVIANCE or SCALE=PEARSON is selected, these procedures estimate the overdispersion parameter using the specified statistic, and divide the Wald statistics by this parameter (see the note on Figure 11.103).

Overdispersion may be harmful. It has no effect on the value of the model coefficients, but their variances will be underestimated and the Wald statistics will be overestimated, meaning that the fit of the model may be less good than it appears. We must therefore be cautious, especially if the Wald statistics are not much above 4, because the criterion may not really be satisfied even if it appears to be.

If overdispersion occurs, we may need to exclude any outliers, find a more appropriate link function, transform the independent variables or add an important independent variable which may have been omitted from the model.

11.9.7 Poisson regression

Poisson regression is used to model rare events such as the number of losses in motor insurance. In this section, I will describe a small example of motor insurance claims (taken from the SAS help data sets), enabling us to compare the treatment of Poisson regression in SAS and IBM SPSS Statistics. For the basic functionality, the estimated parameters are identical, but there are differences in the interpretation aids (the type 1 and 3 analyses offered by SAS). Some advanced functions are present only in SAS, such as repeated measures modelling or using the SCALE option to allow for overdispersion.

In the data sets below, the first column is the number of insured persons, the second is the number of losses, the third is the type of car (three levels) and the fourth is age (two levels).

n	nbloss	car	age
500	42	small	1
1200	37	medium	1
100	1	large	1
400	101	small	2
500	73	medium	2
300	14	large	2

We assume that the number of losses (nbloss) follows a Poisson distribution and that its mean μ/(age, car) is linked by the link function 'log' to:

- a linear combination of the 'age' and 'car' variables, for which the coefficients are to be found;
- and an offset variable (in this case, the Napierian logarithm of the number 'n' of policies) having a constant coefficient equal to 1.

We therefore have $\log(\mu) =$ linear predictor (age, car) $+ \log(n)$. This is a Poisson regression. In SAS, the Poisson regression is obtained as follows:

```
LN = LOG(N) ;
PROC GENMOD DATA=sasuser.insurance ;
    CLASS car age;
    MODEL nbloss = car age /   DIST = POISSON
                               LINK = LOG
                               OFFSET = LN
                               TYPE1 TYPE3;
OUTPUT OUT=output
    RESRAW = RES RESCHI = ZRE STDRESCHI = ADJ
    RESDEV = DEV PRED = PRE;
RUN;
```

GENMOD can be used to specify any distribution and any link function, provided that they can be defined by programming. The offset variable is not 'n' itself but its Napierian logarithm LN created by the syntax LN = LOG(n). The MODEL instruction enables us to allow for various effects, which are the same as those of the GLM procedure (see Section 10.7.7). A certain number of results can be written to the output data set specified by OUTPUT (Figure 11.108); I have only mentioned the ones handled by IBM SPSS Statistics, but there are others such as the linear predictor XBETA for which the inverse of the link function (i.e. the exponential in this case) gives the predicted value RESRAW, or the lower bound LOWER and the upper bound UPPER of the confidence interval of the predicted value.

The 'SCALE' row always appears in the parameters of the model (Figure 11.104), even if there is no overdispersion and even if the scale factor is fixed at 1, as it is here. Note that the 'scaled deviance' is 1.41 in this case, and is therefore not drastically increased (Figure 11.105). However, if we wanted to avoid it and thus avoid over-optimistic standard deviations on the estimates, we could add the command SCALE = DEVIANCE on the MODEL line in the SAS code. The effect of this would be to estimate the scale factor SCALE by the square root of the scaled deviance (the square root is replaced by the inverse function in gamma regression).

Analysis Of Parameter Estimates								
Parameter		**DF**	**Estimate**	**Standard Error**	**Wald 95% Confidence Limits**		**Chi-Square**	**Pr > ChiSq**
Intercept		1	-1.3168	0.0903	-1.4937	-1.1398	212.73	<.0001
car	large	1	-1.7643	0.2724	-2.2981	-1.2304	41.96	<.0001
car	medium	1	-0.6928	0.1282	-0.9441	-0.4414	29.18	<.0001
car	small	0	0.0000	0.0000	0.0000	0.0000		
age	1	1	-1.3199	0.1359	-1.5863	-1.0536	94.34	<.0001
age	2	0	0.0000	0.0000	0.0000	0.0000		
Scale		0	1.0000	1.0000	1.0000			

Figure 11.104 Parameters of a Poisson regression using the SAS GENMOD procedure.

Criteria For Assessing Goodness Of Fit			
Criterion	**DF**	**Value**	**Value/DF**
Deviance	2	2.8207	1.4103
Scaled Deviance	2	2.8207	1.4103
Pearson Chi-Square	2	2.8416	1.4208
Scaled Pearson X2	2	2.8416	1.4208
Log Likelihood		837.4533	

Figure 11.105 Deviance and Pearson's χ^2 with the SAS GENMOD procedure.

We can also replace the scaled deviance with the scaled Pearson's χ^2, using the SCALE = PEARSON command. This method of controlling overdispersion is not provided in IBM SPSS Statistics.

SAS also supplies the likelihood deviances (type 1 and type 3) which are often used for testing the contributions of the factors. In type 1 analysis (Figure 11.106), each value in the 'deviance' column is the deviance of the model containing the variable of the line and of the preceding ones. For example, the model containing the constant, the type of car and the age has a deviance of 2.8207, while the model containing only the constant and the type of car has

LR Statistics For Type 1 Analysis				
Source	**Deviance**	**DF**	**Chi-Square**	**Pr > ChiSq**
Intercept	175.1536			
car	107.4620	2	67.69	<.0001
age	2.8207	1	104.64	<.0001

Figure 11.106 Type 1 analysis with the SAS GENMOD procedure.

LR Statistics For Type 3 Analysis			
Source	**DF**	**Chi-Square**	**Pr > ChiSq**
car	2	72.82	<.0001
age	1	104.64	<.0001

Figure 11.107 Type 3 analysis with the SAS GENMOD procedure.

a deviance of 107.462, showing the significance of the contribution of the age variable, since the difference between the deviances, which is compared to the distribution of the χ^2, is 104.64. In type 1 analysis, the order in which the variables enter the model is taken into consideration. If we want to disregard this order, we use type 3 analysis. The χ^2 shown in Figure 11.107 is the difference between the deviances of the models with and without the variable considered on the line. The difference between the deviances of the models with (1) the constant, the car type and the age, and (2) the constant and the age, is therefore 72.82, showing the contribution of this variable regardless of the order in which it appears.

Where IBM SPSS Statistics is concerned, the first thing to note is that qualitative factors cannot be introduced in the GENLOG procedure, and they must therefore be recoded as numeric variables. The log-linear approach also means that we must use the dependent variable (nbloss) to weight the cells defined by the factors. This is done using the WEIGHT command. Note also that the qualitative variable 'car' is replaced with the numeric variable 'carnum' (this transformation is obligatory, as mentioned above). Additionally, the offset variable, which is specified by the /CSTRUCTURE command, is 'n', instead of being its logarithm as it would be in the SAS software.

```
WEIGHT BY nbloss.
GENLOG
age carnum /CSTRUCTURE = n
/MODEL = POISSON
/PRINT = ESTIM
/PLOT = NONE
/CRITERIA = CIN(95) ITERATE(20) CONVERGE(.001) DELTA(.5)
/DESIGN age carnum
/SAVE = RESID ZRESID ADJRESID DEV PRED.
WEIGHT OFF.
```

Obs	n	nbloss	car	age	pre	low	up	zre	dev	Res	adj
1	500	42	small	1	35.799	27.6490	46.351	1.03642	1.00847	6.20110	1.68547
2	1200	37	medium	1	42.975	33.5520	55.043	-0.91138	-0.93383	-5.97456	-1.62471
3	100	1	large	1	1.227	0.6992	2.152	-0.20455	-0.21139	-0.22654	-0.21572
4	400	101	small	2	107.201	89.8159	127.951	-0.59892	-0.60484	-6.20110	-1.68547
5	500	73	medium	2	67.025	54.1187	83.010	0.72977	0.71931	5.97456	1.62471
6	300	14	large	2	13.773	8.2997	22.857	0.06104	0.06088	0.22654	0.21572

Figure 11.108 A data set output by the SAS GENMOD procedure.

Goodness of fit tests [a,b]

	Value	df	Sig.
Likelihood ratio	2.821	2	.244
Pearson Chi-Square	2.842	2	.242

a Model: Poisson
b Design: Constant + age + carnum

Parameter Estimate [b,c]

					95% confidence interval	
Parameter	Estimate	Std. error	Z	Sig.	Lower bound	Upper bound
Constant	-1.317	.0903	-14.5854	.000	-1.4937	-1.1398
[age = 1]	-1.320	.1359	-9.7128	.000	-1.5863	-1.0536
[age = 2]	0[a]					
[carnum = 1]	-1.764	.272	-6.478	.000	-2.298	-1.230
[carnum = 2]	-.693	.128	-5.402	.000	-.944	-.441
[carnum = 3]	0[a]					

a This parameter is set to zero because it is redundant.
b Model: Poisson
c Design: Constant + age + carnum

Figure 11.109 Output of the GENLOG procedure in IBM SPSS Statistics.

There are fewer outputs (Figure 11.109) than in the SAS software, and the absence of type 1 and type 3 analyses is regrettable. Of course, the results RESID ZRESID ADJRESID DEV PRED are identical to those supplied by SAS.

11.9.8 The generalized additive model

The *generalized additive model* of Hastie and Tibshirani (1990)[67] is even more general than the generalized linear model, because in this case we write:

$$g(E(Y/X = x)) = \beta_0 + \sum_i f_i(x),$$

still using a link function g, but replacing the product of the coefficients β_i with a general function f_i (such as a spline function). The model is called additive because of the summation on the i. This modelling is powerful, but, as with neural networks, we must be careful about

[67] Hastie, T. and Tibshirani, R. (1990) *Generalized Additive Models*. New York: Chapman & Hall.

overfitting and the interpretability of the results. It is fully implemented in the R software (the *mgcv* package) and in the SAS GAM procedure.

11.10 Bayesian methods

Bayes' theorem is one of the most important results in probability theory. It concerns the inversion of probabilities and relates, for two events A and B, the conditional probability of A given B to the conditional probability of B given A:

$$P(A/B) = P(B/A)\frac{P(A)}{P(B)}.$$

Now, if $\{A_1, A_2, \ldots, A_n\}$ is a complete system of events, in other words such that the intersection of two events is always empty and the set of events is equal to the whole universe of possible outcomes (we say that the system is a partition of Ω), then we can write

$$P(B) = \sum_{k=1}^{n} P(B/A_k)P(A_k),$$

and, for every i,

$$P(A_i/B) = P(B/A_i)\frac{P(A_i)}{\sum_{k=1}^{n} P(B/A_k)P(A_k)}.$$

This form, called Bayes' second formula, is very widely used. This fundamental theorem is true for both discrete and continuous probability distributions. In the latter case, the sum is replaced by an integral.

Bayes' theorem is the foundation of Bayesian statistics. In this context, we are concerned with a probability model characterizing the behaviour of observations $X_1, \ldots, X_n$, conditionally on a parameter θ, and the *a posteriori* probability of the parameter conditionally on ('updated by') the observations, $P(\theta \mid X_1, \ldots, X_n)$, is expressed as being proportional to the product of the *a priori* probability $P(\theta)$ of the parameter and of the likelihood $f_x(\theta) = P(X_1, \ldots, X_n \mid \theta)$ of θ for the observations.

11.10.1 The naive Bayesian classifier

As its name suggests, this is a classification method developed from Bayes' theorem. It has already been used in Section 11.6.4 as an inversion formula for calculating the probability $P(G_i/x)$ of an individual's belonging to a group G_i conditionally on its characteristics x based on the probability $P(x/G_i)$ of having these characteristics conditionally on its belonging to G_i. We also need to know the *a priori* probability of G_i. According to Bayes' theorem,

$$P(G_i/x) = \frac{P(G_i)P(x/G_i)}{\sum_j P(G_j)P(x/G_j)},$$

where the denominator is equal to the probability $P(x)$. We can reformulate this property in terms normally used for classification problems. The binary variable to be explained is Y, we attempt to explain $Y=1$, and the explanatory variables are $X_1, \ldots, X_n$. Thus we have:

$$P(Y=1/X_1,\ldots,X_p)=\frac{P(X_1,\ldots,X_p/Y=1)P(Y=1)}{P(X_1,\ldots,X_p)}$$

$$=\frac{P(X_1,\ldots,X_p/Y=1)P(Y=1)}{P(X_1,\ldots,X_p/Y=1)P(Y=1)+P(X_1,\ldots,X_p/Y=0)P(Y=0)}.$$

Assuming 'naively' that the variables are independent (conditionally on Y), we can write

$$P(X_1,\ldots,X_p/Y=k)=\prod_{i=1}^{p}P(X_i/Y=k),\quad k=0,1,$$

which leads to the formula used by the naive Bayesian classifier:

$$P(Y=1/X_1,\ldots,X_p)=\frac{P(Y=1)\prod_{i=1}^{p}P(X_i/Y=1)}{P(Y=1)\prod_{i=1}^{p}P(X_i/Y=1)+P(Y=0)\prod_{i=1}^{p}P(X_i/Y=0)}$$

The desired probability $P\,(Y=1/X_1,\ldots,X_p)$ is therefore calculated simply on the basis of the estimates of the probabilities $P\,(X_i/Y=1)$ supplied by the cross-tabulations of the independent variables X_i with the dependent variable Y.

To illustrate the operation of this classifier, let us take the example of *credit scoring* which will be described more fully at the end of Chapter 12. We have a first variable X_1 to account for the risk of default, which is the mean balance in a current account. If an individual has the category 'CA < 0 euros', the table below shows that

- Prob(AC < 0 euros | Credit=Bad) = 45%,
- Prob(AC < 0 euros | Credit=Good) = 19.86%.

The table also shows that

- Prob(Credit=Bad) = 30%,
- Prob(Credit=Good) = 70%.

This information can be used to calculate the probability of default for a recipient of credit, given characteristics such as the mean balance in his current account.

Table of Accounts by Credit

Accounts	Credit		
FREQUENCY **Percent** **Row Pct** **Col Pct**	**Good**	**Bad**	**Total**
CA < 0 euros	139 13.90 50.73 19.86	135 13.50 49.27 45.00	274 27.40
CA [0-200 euros[	164 16.40 60.97 23.43	105 10.50 39.03 35.00	269 26.90
CA >= 200 euros	49 4.90 77.78 7.00	14 1.40 22.22 4.67	63 6.30
No checking account	348 34.80 88.32 49.71	46 4.60 11.68 15.33	394 39.40
Total	700 70.00	300 30.00	1000 100.00

The naive Bayesian classifier is fairly simple to program. Here is an example, in the form of an SAS macro. The parameters are:

- table, the name of the data set used to calculate the naive Bayesian classifier;
- appli, the name of the data set to which the naive Bayesian classifier is applied (it is possible to have appli = table);
- variablesX, the list of independent variables;
- variableY, the dependent variable (binary);
- ref, the (reference) category to be predicted of the dependent variable;
- nref, the other category of the dependent variable.

```
%MACRO Bayes (table, appli, variablesX, variableY, ref, nref ) ;
```

We start by retrieving in an SQL procedure the mean value of the indicator (&variableY = &ref), i.e. the proportion of individuals such that (&variableY = &ref), in other words $P(Y = ref)$.

```
PROC SQL NOPRINT ;
  SELECT MEAN(&variableY = &ref)
    INTO : probaEvent
  FROM &table ;
QUIT ;

%PUT Proba(&variableY = &ref) = &probaEvent ;
```

To display the list of independent variables in the SAS LOG window:

```
%PUT Variables : &variablesX ;
```

We must then count the number of words in the parameter '&variablesX', that is to say the number of independent variables:

```
%LET nb_var = %EVAL(%SYSFUNC(COUNTC(%SYSFUNC(COMPBL
(&variablesX)),' ')) +1);
```

The number of variables is displayed in the LOG window:

```
%PUT Number of independent variables = &nb_var ;
```

A 'loop' is then created to process all the independent variables entered as parameters. On each iteration of the loop, we obtain $P(X_i/Y = \text{ref})$ and $P(X_i/Y = \text{nref})$ at the output of a FREQ procedure.

```
%DO i = 1 %TO &nb var ;
```

We start processing the *i*th variable and the %SCAN function is used to extract the '&i'th word of &variablesX.

```
%LET varX = %SCAN(&variablesX,&i);

PROC FREQ DATA = &table ;
TABLES &varX * &variableY / OUTPCT OUT=pct&i (KEEP = pct_col
&varX &variableY) ;
RUN ;

PROC SORT DATA = pct&i ; BY &varX ; RUN ;

PROC TRANSPOSE DATA = pct&i OUT=pctr&i (DROP = _name_ _label_)
PREFIX=pct ;
BY &varX ; ID &variableY ; VAR pct_col ;
RUN ;

%IF &i = 1 %THEN %DO ;
DATA temp_table ;
SET &appli ;
probaref = 1 ; probnref = 1 ;
RUN ;
PROC SORT DATA=temp_table OUT=temp_table ; BY &varX ; RUN ;
```

```
%END ;
%ELSE %DO ;
PROC SORT DATA=bayes%EVAL(&i-1) OUT=temp_table
(DROP = pct&ref pct&nref) ; BY &varX ; RUN ;
%END ;
```

We calculate $\prod_{i=1}^{i} P(X_j/Y = 1)$ recursively from $\prod_{j=1}^{i-1} P(X_j/Y = 1)$ and from $P(X_i/Y = 1)$, the first term being contained in 'temp_table' and the second in 'pctr&i':

```
DATA bayes&i ;
MERGE temp_table (IN=a) pctr&i (IN=b);
BY &varX ;
probaref = (pct&ref / 100) * probaref ;
probnref = (pct&nref / 100) * probnref ;
RUN ;
```

We reach the end of the loop for the set of independent variables:

```
%END ;

DATA predic_bayes (DROP = pct&ref pct&nref) ;
SET bayes%EVAL(&i-1) ;
probacas = SUM ( probaref * &probaEvent , probnref * (1 - &probaEvent)) ;
probayes = (probaref * &probaEvent) / probacas ;
RUN ;
```

And this is the end of the macro:

```
%MEND Bayes ;
```

This macro is called in the following way in the credit scoring case described in Chapter 12. I have used the same independent variables as those chosen for the logistic model. Category 2 (or 1) of the Credit (dependent) variable is 'Bad' (or 'Good'), meaning that the applicant for credit has defaulted in the past.

```
%Bayes (train_score, valid_score, Accounts Credit_history
Credit_duration Age Savings Guarantees Other_credits
Status_residence,
Credit, 2 , 1 ) ;
```

The output of the macro is a data set 'predic_bayes', calculated from the data set &appli ('valid_score' in this case), in which the variable 'probayes' is the calculated probability $P(\text{Credit=Bad}/X_1, \ldots, X_p)$.

In this example, in terms of the area under the ROC curve (see Section 11.16.5), the prediction quality of the Bayesian classifier is close to that of logistic regression. However, the division into deciles of the probability calculated by the logistic model shows a regular progression of the default rate, decile by decile, which is less marked in the 'probayes' probability.

This suggests that the naive Bayesian classifier requires a rather larger volume of data than other methods, without which some of the probabilities $P(X_i/\text{Credit=Bad})$ may be very small and non-significant, thus distorting the model. If only one of the probabilities $P(X_i/\text{Credit=Bad})$ is zero, the final probability will also be zero. The problem here is that this probability

P(Credit=Bad/$X_1,\ldots,X_p$) is calculated as a *product* of probabilities. This makes it less robust. It should still be noted that, if one of the probabilities P(X/Credit=Bad) is zero, the logistic regression (in contrast to linear discriminant analysis) will not be able to estimate the final probability either: this is the phenomenon of complete separation (Section 11.8.7).

To sum up, the naive Bayesian classifier can be useful, if only for providing a reference base for the performance of other models. In spite of the 'naivety' of its assumptions, it can sometimes perform surprisingly well.

11.10.2 Bayesian networks

The use of Bayes' theorem in classification methods is not restricted to the naive Bayesian classifier. The example below shows another application, relating to a score for consumer credit approval used on the Internet, which is detailed in Chapter 12 of Naïm *et al.*, *Réseaux Bayésiens* (2007).[68]

We construct a Bayesian network, in other words a directed acyclic graph, relating the dependent variable to the (discrete) independent variables.[69,70] Such a network represents the dependences (and independences) between the variables, where an arrow between a variable A and a variable B (we say that A is the parent of B) indicates that the conditional probability $P(B|A)$ is different from $P(B)$. B depends on A. A Bayesian network is also defined by the set of conditional probabilities. It therefore consists of two elements: a qualitative element which is the representation of the dependences, and a quantitative element, which is the measurement of these dependences.

In the naive Bayesian classifier, we calculate the probabilities $P(X_i|Y=1)$, and the dependent variable is the parent of each of the independent variables, but an independent variable is never the parent of another. This can be seen in the Bayesian network, where all the independent variables are at the same level and are never interconnected by a link. There is a more general Bayesian network, called the *tree augmented naive Bayes classifier*,[71] in which each independent variable can have a parent other than the dependent variable. An arc between an independent variable X_1 and an independent variable X_2 (conditionally on the dependent variable Y) signifies that the conditional probability $P(X_2|X_1,Y)$ is different from the conditional probability $P(X_2|Y)$: it depends on X_1. The effect of Y on X_2 depends on the value of X_1. In the same way, the probability $P(X_2|X_1,Z)$ is different from the product of $(X_1|Z)$ and $P(X_2|Z)$. An independent variable can thus have age or bank balance as its parent. This situation is more complex than that of the naive Bayesian classifier, but it is common and may be more effective.

The advantage of Bayesian networks is that they enable knowledge to be represented in graphic form (the expression 'white box' is sometimes used, as opposed to the 'black box' of the neural network); they are easy to use and modify; and they can be used for drawing inferences, in other words calculating the conditional probability of a set of variables of the network after the other variables have been observed. The dependences between variables can

[68] Naïm, P., Wuillemin, P.-H., Leray, P., Pourret, O. and Becker, A. (2007) *Réseaux Bayésiens*, 3rd edn. Éditions Eyrolles.

[69] Guéguen, A., Zins, M. and Nakache, J.-P. (1996) Utilisation des réseaux probabilistes en analyse discriminante sur variables qualitatives. *Revue de Statistique Appliquée*, 44(1), 55–75.

[70] Judea Pearl. *Probabilistic Reasoning in Intelligent Systems: Networks of Plausible Inference*. Morgan Kaufmann, 1988.

[71] Friedman, N., Geiger, D. and Goldszmidt, M. (1997) Bayesian network classifiers. *Machine Learning*, 29(2–3), 131–163.

even be used to draw inferences when certain variables cannot be observed. A deduction can be made from incomplete data, as in the example shown below. The first users of Bayesian networks in scoring problems were a group of researchers at Fair and Isaac.[72] Other examples can be found in Chapter 12 of the book by Thomas, Edelman and Crook,[73] and in an article by Baesens *et al.*[74]

In our example, a tree augmented naive Bayesian network is constructed by creating links between the dependent variable and a certain number of independent variables. The network enables us to calculate the conditional probabilities of the variables with respect to each other. If no category is specified for any independent variable, the network supplies the *a priori* probability of each category of the dependent variable, such as the probability of defaulting in the case of a credit risk score. We can decide that credit should be approved (or refused) if the conditional probability of the category 'good payer' of the dependent variable is above (or below) a certain score threshold s_2 (or s_1).

When the first question has been put to the customer applying for credit, his response determines the category of one of the independent variables and supplies the conditional probabilities of the categories of the dependent variable. The first variable is usually age, which is the parent of several other variables which it influences – family status, number of children, income, etc. We then calculate, for each of the categories X_{ij} of each of the independent variables X_i not yet used, the conditional probability of the corresponding 'good payer' category, in other words the probability calculated conditionally on the response Q_1 to the first question and to the category X_{ij}. We also calculate the probability $P(X_{ij} \mid Q_1)$ of X_{ij} conditionally on Q_1 (in fact, we estimate these probabilities by calculating their proportions). We are interested in the categories X_{ij} such that $P(\text{'good payer'} \mid Q_1, X_{ij})$ is less than s_1 or greater than s_2, because we assume that if this probability is less than s_1 (or greater than s_2) the credit application can be interrupted and the application rejected (or accepted). The advantage of a decision based on a partial questionnaire is its greater rapidity. The drawback is that the probability calculated from a partial number of responses is not necessarily equal to the probability calculated over all the responses: a very good response may 'rescue' an initially poor file, and vice versa. However, the authors of the book cited above estimate that the error rate due to the incomplete nature of the questionnaire is 5%, given that, in their study, 65% of the questionnaires were incomplete (8.5 questions asked on average, out of a possible 14).

This is how the second question is chosen, given the response Q_1 to the first question. For each variable X_i, we calculate the sum P_i of the probabilities $P(X_{ij} \mid Q_1)$ for the set of categories X_{ij} such that $P(\text{'good payer'} \mid Q_1, X_{ij})$ is less than s_1 or greater than s_2. The variable (and therefore the question) that is chosen is the one that maximizes P_i, because it is the one that maximizes the probability of being able to end the questionnaire. The choice of an *n*th question given the responses to the preceding questions is made in the same way, and the procedure is interrupted as soon as $P(\text{'good payer'} \mid Q_1, Q_2 \ldots)$ is less than s_1 or greater than s_2.

This scoring method, known as the adaptive questionnaire, is no more precise than a conventional method, but it can be used to minimize the number of questions put to the

[72] Chang, K.C., Fund, R., Lucas, A., Oliver, R. and Shikaloff, N. (2000) Bayesian networks applied to credit scoring. *IMA Journal of Management Mathematics*, 11(1), 1–18.

[73] Thomas, L.C., Edelman, D.B. and Crook, J.N. (2002) *Credit Scoring and Its Applications*. Philadelphia: Society for Industrial and Applied Mathematics.

[74] Baesens, B., Egmont-Petersen, M., Castelo, R. and Vanthienen, J. (2002) Learning Bayesian network classifiers for credit scoring using Markov chain Monte Carlo search. In *The Sixteenth International Conference on Pattern Recognition (ICPR '02)*, Vol. 3, pp. 30049. Washington, DC: IEEE Computer Society.

customer, which is important for a credit application on the Internet, where the customer may very easily abandon his purchase if the process appears to be taking too long. At the same time, it makes it harder for competitors to use trial and error methods to discover the scoring algorithm that is used on-line. However, the presence of numerous incomplete questionnaires in the scoring base can make it difficult to develop future scores.

11.11 Classification and prediction by neural networks

Supervised learning neural networks, especially the multilayer perceptron and the radial basis function network, are used for both classification and prediction.

Classification can take place in two ways. We must bear in mind that the units of the output layer can take continuous values between 0 and 1. We can create one unit for each class to be predicted in the output layer, and assume that the class of an individual is the one for which the unit outputs the highest value (closest to 1). We may decide to be more restrictive: we could assign a class to an individual only if the corresponding unit outputs a value above a given threshold of acceptance (e.g., 0.5), and if the other units all output values below a given threshold of rejection (e.g., 0.5). If these conditions are not fulfilled (e.g., if the values are 0.8 and 0.6), the class of the individual is considered to be unpredictable. This type of network learns by associating each individual of the learning set with unit output values of 1 for the unit corresponding to the (known) class of the individual and 0 for the other units.

The second method can really only be used if there are two classes to be predicted. In this case there is only one unit in the output layer, and the value of 0 for this unit corresponds to one class and the value of 1 to the other class. During learning, the unit takes only the values 0 and 1, but in the application phase any of the values in the range from 0 to 1 can be taken. In the test phase, using a sample which is different from the learning sample, we therefore choose a threshold value which separates the classes. This can be done by looking on the ROC curve or in the set of confusion matrices for a compromise between the sensitivity of the network (its capacity to detect events) and its specificity (its capacity to avoid detecting false events).

11.11.1 Advantages of neural networks

Neural networks have certain special advantages. Firstly, they are good at allowing for non-linear relations and complex interactions between variables, at least if the necessary investment is made in terms of the number of units in the hidden layer(s). The downside of this complexity and the possibly large number of units in the network is the risk of overtraining and the fact that it is not easy to extract the subset of the most relevant variables from the set of all the potential predictive variables.

A second advantage is that the neural network method is *non-parametric*, meaning that the independent variables are not assumed to follow any particular probability distributions.

There is also a third advantage. Some networks have a greater resistance to defective data than other methods (the multilayer perceptron is better than the RBF in this respect). If an input variable of a multilayer perceptron is too noisy, the weight of the corresponding unit will fall to zero.

Finally, neural networks can model a wide range of problems, including clustering, classification, prediction, time series (economic forecasting), optical character recognition (for signatures) and automatic reading of handwriting on envelopes (for postal sorting)[75] and cheques (seven-layer neural networks used by AT&T), linguistic analysis (text mining), speech recognition and synthesis, face recognition, object recognition based on their shape or their signal in military and industrial fields (scanners of video images at underground stations to automatically detect overcrowding, detection of malfunctions in machines by vibration analysis), automatic pilots in aircraft, machine control in an industrial production process, signal processing in medicine (diagnosis of a cardiac signal, estimation of the size of a tumour), weather forecasting, and chess (a neural network based learning module enabled the Deep Blue computer to beat the world champion Garry Kasparov in 1997).

11.11.2 Disadvantages of neural networks

The use of neural networks has a number of serious drawbacks, already mentioned in my discussion of neural clustering:

- Convergence towards the best global solution is not always certain.
- The considerable risk of overfitting, if the number of cases is too small with respect to the number of units (see Section 8.4).
- The impossibility of handling an excessively large number of variables.
- The non-explicit nature of the results, which is unacceptable for some applications such as medical diagnosis or automatic pilot systems (if there is a hidden layer, an *a posteriori* analysis is required to discover the weights of the different variables used in the score calculation).
- The difficulty of using the networks correctly, because the parameters (number of hidden layers, number of units, learning rate, momentum, etc.) are numerous and hard to control.
- Neural networks are only naturally applicable to continuous variables in the range [0,1], and therefore the number of units has to be multiplied for qualitative variables.

As I have mentioned in Chapter 8 on neural networks, a network does not always converge towards a correct solution, because it considers that it has found a correct solution when the error function, which is to be minimized by the adjustment of the weights, reaches a minimum and ceases to diminish with further adjustment of the weights. However, this minimum may be only a local rather than a global one. We therefore need to prevent the network from being 'trapped' in a local minimum. To do this, we start the learning with a high weight adjustment rate, and then decrease this rate during the learning, as the network approaches a correct solution. We also control the *momentum* of the network in order to limit the oscillation of the network about the solutions; the momentum is what impels the weights to continue to change in the direction of their development.

[75] Automatic recognition of handwritten post codes is one of the classic applications of neural networks, because of its economic implications, the huge databases available (in the US Postal Service, for example) and the interesting problems that it raises: before attempting to decipher each character, taking its size, orientation and style into account, it is first necessary to separate them from each other, which is harder than it may appear.

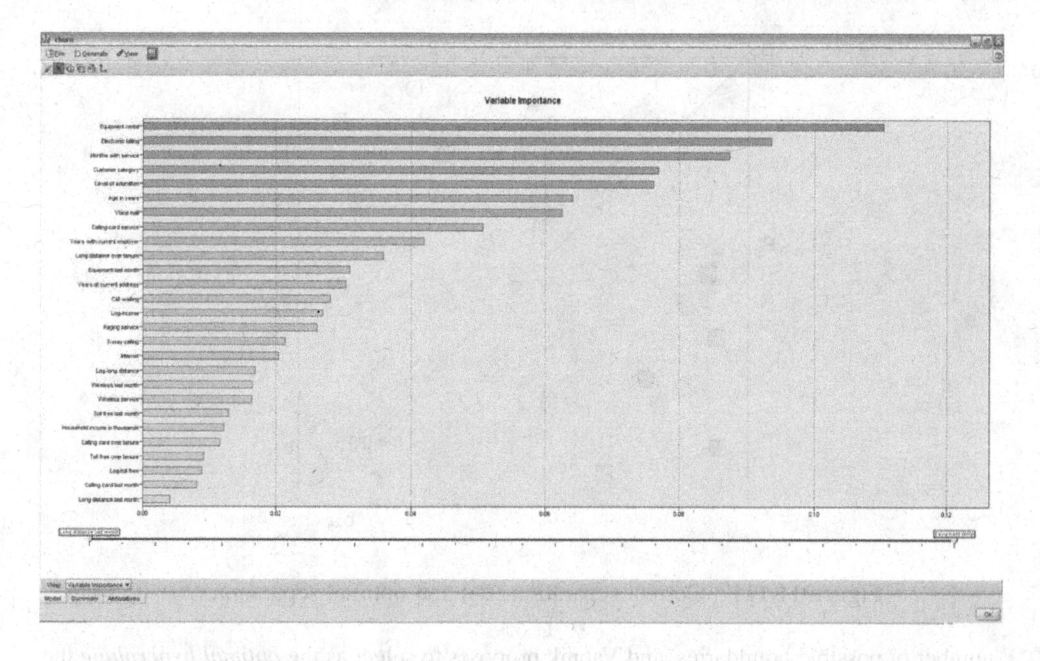

Figure 11.110 Result of the learning of a neural network in IBM SPSS Modeler.

On the subject of non-explicit results, we may recall (see Section 8.4) that a network with n input units, a single hidden layer, m units in the hidden layer and one output unit has $nm + m$ weights. Thus the number of parameters of a neural network can soon become large. Furthermore, different sets of weights can lead to similar predictions. So the set of weights of a network cannot be used to understand the results supplied by this network; this is quite different from the case of decision trees and their leaves. However, most software packages can classify the model variables in order of relevance.

By way of example (Figure 11.110), after learning, IBM SPSS Modeler iterates on each of the input variables and calculates the performance of the network in the absence of an input signal on this variable. This calculation gives a weight to each variable and thus enables them to be classified in decreasing order of effect on the accuracy.

To sum up, we only use neural networks if we have a large enough learning sample and if conventional methods are found to be unsatisfactory, for example because of highly non-linear relationships between the variables.

11.12 Classification by support vector machines

11.12.1 Introduction to SVMs

The separable case

Based on the work of Vladimir Vapnik from 1995 onwards, this recent classification method is mainly concerned with the case of linearly separated observations, in other words those which can be cut off from each other by a linear boundary. In this case, however, there are an infinite

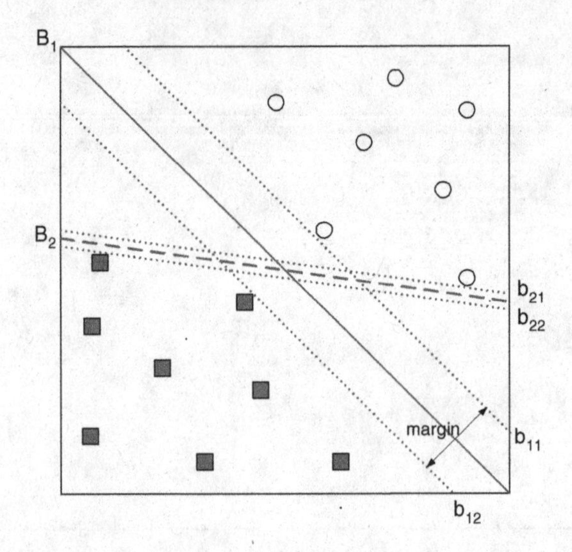

Figure 11.111 Correct separation (B_2) and optimal separation (B_1).

number of possible boundaries, and Vapnik proposes to select as the *optimal hyperplane* the one that maximizes the width of the margin between the observations. This hyperplane is assumed to guarantee not only the fit of the model but also its robustness.

Finding this optimal hyperplane is a matter of finding a hyperplane with the equation $a{\cdot}x + b = 0$ ($a{\cdot}x$ is the scalar product of a and x) which meets the following two conditions:

- it gives good separation of the groups A and B to be discriminated (goodness-of-fit of the model), in the sense that the function defined by $f(x) = a{\cdot}x + b$ is positive if and only if $x \in A$, and $f(x) \leq 0$ if and only if $x \in B$;
- it is as distant as possible from all the observations (robustness of the model), given that distance from an observation x to the hyperplane is $|a{\cdot}x + b|/\|a\|$.

The margin, which by definition is the width of the space between observations, is $2/\|a\|$ and is to be maximized. Because of this constraint, the term 'wide-margin separator' is used.

Figure 11.111, taken from a presentation by Tan, Steinbach, Kumar, and Eick,[76] shows how this second condition appears visually: the hyperplane B_1 is better than hyperplane B_2, because it maximizes the margin. Both hyperplanes satisfy the first condition.

Given the points (x_i,y_i), where $y_i = 1$ if x_i is in A and $y_i = -1$ if x_i is in B, finding the optimal hyperplane $a{\cdot}x + b = 0$ is a matter of finding a pair (a,b) which meets two conditions:

- for every i, $y_i(a{\cdot}x_i + b) \geq 1$ (correct separation);
- $\frac{1}{2}\,\|a\|^2$ is minimal (maximum margin).

This is a problem of optimization subject to constraints. We can see the analogy with regularized regression (Section 11.7.2) in which we also seek a solution subject to a constraint, which in this case is that the margin is to be maximized. This criterion of a wide margin to

[76] http://www2.cs.uh.edu/~ceick/DM/dm_ibl.ppt#605,16,Support Vector Machines

ensure correct generalization can be appreciated intuitively: in a sample other than the learning sample, the points will not all fall outside the margin, and some of them may therefore be incorrectly classified, but this risk evidently decreases as the width of the margin increases.

The solution of this problem provides an expression

$$a = \sum_i \alpha_i y_i x_i$$

and therefore

$$f(x) = \sum_i \alpha_i y_i (x \cdot x_i) + b, \quad \alpha_i \geq 0,$$

in which the sign of *f(x)* indicates the class to which observation is to be assigned. What is remarkable about this expression, which has the appearance of an ordinary score function, is that we can show that the only non-zero coefficients α_i are those which correspond to the points x_i that are exactly on the boundaries of the margin: these are the support points or *support vectors* (Figure 11.112). In other words, the optimal hyperplane depends only on the support points, namely the closest points. This is different from the situation in linear discriminant analysis, where the remote points also have an effect on the solution. This is generally considered to be a favourable aspect of support vector machines (SVMs) in terms of robustness, because remote points may be aberrant or at any rate harmful to a good capacity for generalization. However, this is not true in all cases, such as the heteroscedastic multinormal case where linear discriminant analysis is the best method. This is because the points on the boundary of a class are not necessarily the best representatives of this class for modelling.

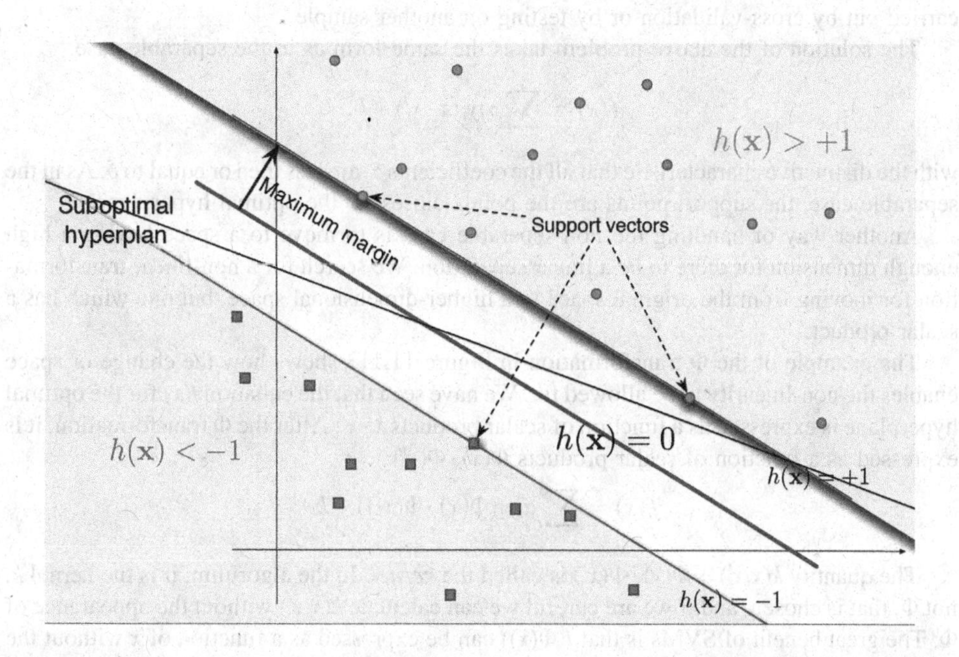

Figure 11.112 Optimal hyperplane and support vectors of an SVM (source: Antoine Cornuéjols).

Another interesting finding by Vapnik shows that the generalization capacity of an SVM increases as the number of support points decreases.

The non-separable case

In practice, if the two populations to be discriminated are not perfectly separated but overlap, a term measuring the classification error must be added to each of the two conditions shown above, but this does not alter the principle of the optimization problem.

This term is defined for each observation x_i on the wrong side of the boundary, by measuring the distance separating it from the boundary of the margin on the side of its class. This distance is then normalized by dividing it by the half-margin $1/\|a\|$, giving a term i, called the 'slack variable'. An 'error' in the model is an observation for which $\xi_i > 1$. The sum of all the ξ_i represents the set of classification errors. The previous two constraints for finding the optimal hyperplane thus become:

- for every i, $y_i(a \cdot x_i + b) \geq 1 - \xi_i$;
- $^1/_2 \|a\|^2 + \delta \Sigma_i \xi_i$ is minimal.

If $\xi_i > 0$, the condition $y_i(a \cdot x_i + b) \geq 1 - \xi_i$ is a relaxation of the initial condition $y_i(a \cdot x_i + b) \geq 1$. This occurs not only for incorrectly classified observations ($\xi_i > 1$), but also for those within the margin ($\xi_i \in]0, 1]$).

The quantity δ is a parameter which penalizes errors and controls the adaptation of the model to the errors. As this increases and the sensitivity to errors rises, the adaptation also becomes greater. We should therefore make a good choice of δ to reach a good compromise between fit and robustness, in other words the generalization capacity. This choice of δ can be carried out by cross-validation or by testing on another sample.

The solution of the above problem takes the same form as in the separable case:

$$f(x) = \sum\nolimits_i \alpha_i y_i (x \cdot x_i) + b,$$

with the distinctive characteristic that all the coefficients α_i are less then or equal to δ. As in the separable case, the support points are the points closest to the optimal hyperplane.

Another way of handling the non-separable case is to move to a space having a high enough dimension for there to be a linear separation. We search for a non-linear transformation for moving from the original space to a higher-dimensional space, but one which has a scalar product.

The example of the Φ transformation in Figure 11.113 shows how the change of space enables the non-linearity to be allowed for. We have seen that the equation $f(x)$ for the optimal hyperplane is expressed as a function of scalar products $x \cdot x'$. After the Φ transformation, it is expressed as a function of scalar products $\Phi(x) \cdot \Phi(x')$:

$$f(x) = \sum\nolimits_i \alpha_i y_i (\Phi(x) \cdot \Phi(x_i)) + b.$$

The quantity $k(x,x') = \Phi(x) \cdot \Phi(x')$ is called the *kernel.* In the algorithm, it is the kernel k, not Φ, that is chosen, and if we are careful we can calculate $k(x,x')$ without the appearance of Φ. The great benefit of SVMs is that $f(\Phi(x))$ can be expressed as a function of x without the explicit intervention of Φ. In this way, the calculations are done in the original space, and are therefore simpler and faster. This is why we speak of a 'kernel machine'.

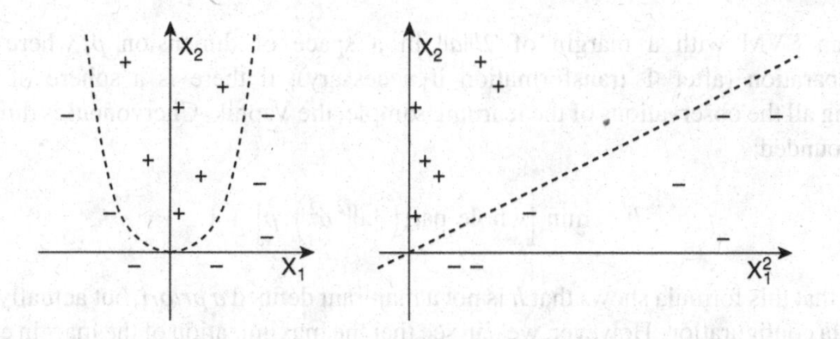

Figure 11.113 Example of transformation in an SVM.

Here are some examples of kernels:

- linear: $k(x,x') = x \cdot x'$;
- polynomial: $k(x,x') = (x \cdot x')^d$;
- Gaussian (RBF): $k(x,x') = \exp\left(\|x - x'\|^2 / 2\sigma^2\right)$, one of the most widely used;
- sigmoid: $k(x,x') = \tanh\ \{\kappa(x \cdot x') + \theta\}$, where κ is the *gain* and θ is the *threshold.*

Taking the example of the second-degree polynomial kernel and the function

$$x = (x_1, x_2) \to \Phi(x) = (x_1^2, \sqrt{2}x_1x_2, x_2^2),$$

we see that the scalar product $\Phi(x) \cdot \Phi(x') = (x_1x'_1 + x_2x'_2)^2 = (x \cdot x')^2$ is expressed in the arrival space without the appearance of Φ.

The computation time varies depending on the choice of kernel, but this choice also enables us to model various problems with a quality of results which certainly makes SVMs an important technique for the future and one that is already in fashion today. After the SVMs for classification (SVCs) introduced by Corinna Cortes and Vladimir Vapnik[77] in 1995, a variant appeared in 1996 for the case in which the independent variable is continuous, in the form of the support vector regression (SVR) of Harris Drucker, Chris Burges, Linda Kaufman, Alex Smola and Vladimir Vapnik.[78]

The Vapnik–Chervonenkis dimension of an SVM

As mentioned previously in the context of Vapnik's learning theory (Section 11.3.3), SVMs are one of the first types of model for which an explicit expression of the Vapnik–Chervonenkis dimension has been found. Remember that this dimension measures the complexity of the model and is used to evaluate the generalization capacity of the model and its robustness, by limiting the difference between the (unknown) theoretical risk of the model and the empirical risk calculated on the learning sample. The risk function can be the error rate, for example.

---

[77] Cortes, C. and Vapnik, V. (1995) Support-vector networks, *Machine Learning*, 20, 1–25.

[78] Drucker, H., Burges, C., Kaufman, L., Smola, A. and Vapnik, V. (1997) Support vector regression machines. In M.C. Mozer, M.I. Jordan and T. Petsche (eds), *Advances in Neural Information Processing Systems 9, Proc. NIPS '96*, pp. 155–161. Cambrige, MA: MIT Press.

For an SVM with a margin of $2/\|a\|$ in a space of dimension p where there is linear separation (after Φ transformation if necessary), if there is a sphere of radius ρ containing all the observations of the learning sample, the Vapnik–Chervonenkis dimension h is thus bounded:

$$h \leq \min\left[\text{whole part}\left(\|\mathrm{a}\|^2\rho^2\right), p\right] + 1.$$

Note that this formula shows that h is not a majorant defined *a priori*, but actually depends on the data configuration. However, we can see that the maximization of the margin entails the minimization of the Vapnik–Chervonenkis dimension, and that the two criteria, the special and the general, are consistent. We can also see that it is the case where the data are not linearly separated in the original space, and where p may become large, that the maximization of the margin plays a part in ensuring good generalization. In the case of linearly separable data, good generalization is ensured by the term p.

11.12.2 Example

Very few commercial software packages implement SVMs, although IBM SPSS Modeler contains this feature (Figure 11.114). However, there is also plenty of freeware: this includes software dedicated to SVMs and also more comprehensive software packages including an SVM function. Examples of this are Weka, which implements the standard sequential minimal optimization algorithm, and R, whose *kernlab* package contains the ksvm function which implements a number of SVM algorithms. Among the dedicated software, SVM^{light} is

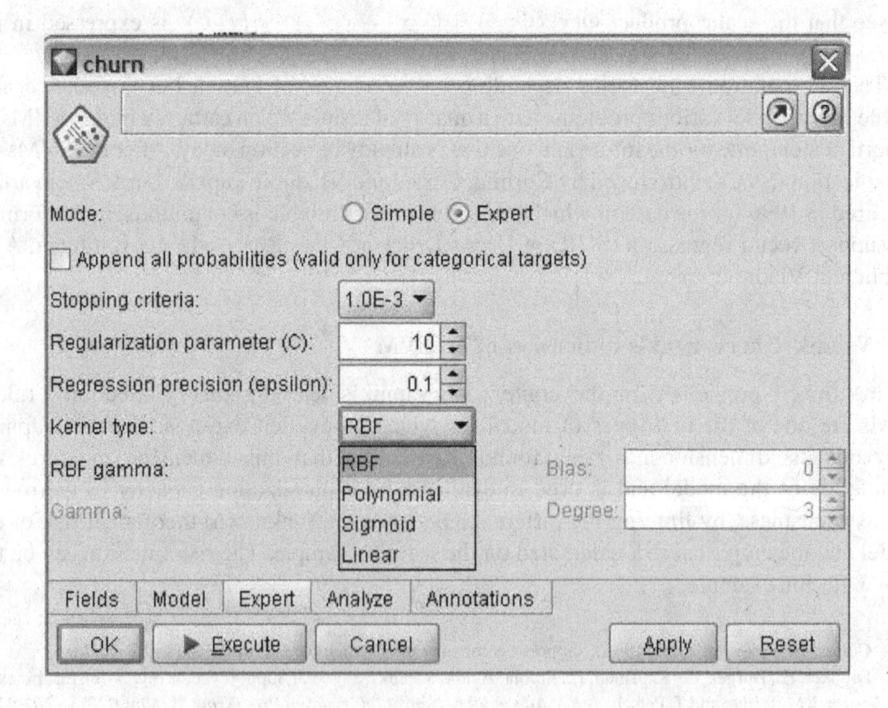

Figure 11.114 SVM in IBM SPSS® Modeler.

one of the most popular, as it has an optimized algorithm allowing it to work with large volumes. Its executable code is also available for Linux, Windows, Cygwin and Solaris, and its C++ source is of course equally available. We can mention mySVM and SVMTorch. All of these implement methods for clustering and regression, and the kernels mentioned above (linear, polynomial, Gaussian and sigmoid). A much more comprehensive list can be found at www.support-vector-machines.org/SVM_soft.html, a portal for SVM software and libraries of routines.

By way of an example, this is how the Rattle graphic interface in R uses the ksvm function. We can choose SVM from a list of modelling techniques in the 'Model' tab of Rattle (Figure 11.115). After running the algorithm, we can select the resulting syntax from the Log window (Figure 11.116), and then paste it into the R console. The command is sent and the result shown in Figure 11.117 is obtained.

If we wish to evaluate the performance of the SVM model and compare it with that of other models, we return to the Rattle interface and choose the Evaluate tab. Thus, in Figure 11.118, we have ticked the three models to be evaluated: 'Decision Tree', 'SVM' and 'Regression', the decision tree and the logistic regression having been constructed previously. We have also ticked 'ROC' to obtain the ROC curve of these models (see Section 11.16.5). The ROC curves of the three models are displayed in the graphic window of the R console (Figure 11.119). We find that SVMs provide the best model, followed by logistic regression; the tree model (with two nodes only) has the poorest performance.

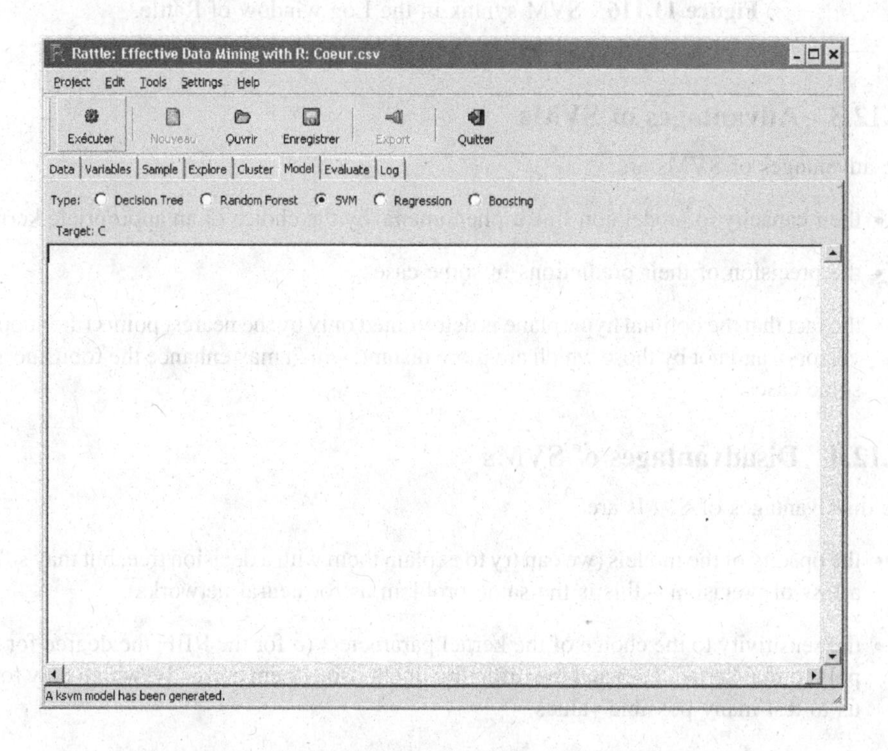

Figure 11.115 SVM in the modelling window of Rattle.

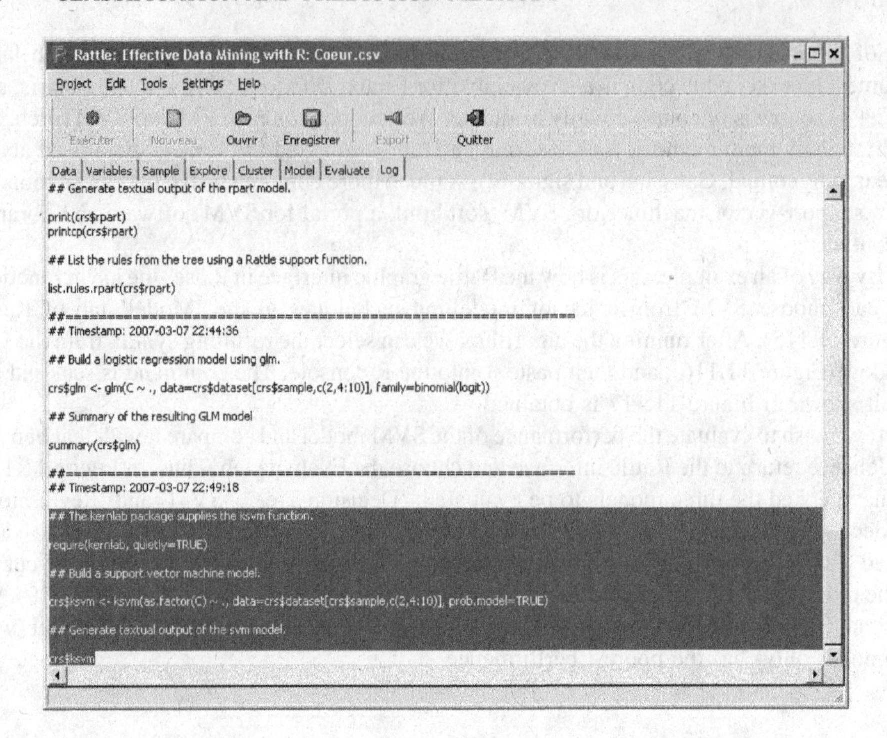

Figure 11.116 SVM syntax in the Log window of Rattle.

11.12.3 Advantages of SVMs

The advantages of SVMs are:

- their capacity to model non-linear phenomena, by the choice of an appropriate kernel;
- the precision of their predictions in some cases;
- the fact that the optimal hyperplane is determined only by the nearest points (the support vectors), and not by those which are more distant, which may enhance the robustness in some cases.

11.12.4 Disadvantages of SVMs

The disadvantages of SVMs are:

- the opacity of the models (we can try to explain them with a decision tree, but may suffer a loss of precision – this is the same problem as for neural networks);
- the sensitivity to the choice of the kernel parameters (σ for the RBF, the degree for the polynomial kernel, etc.) and the difficulty of choosing them correctly, which may force us to test many possible values;
- the computation time which is sometimes lengthy;

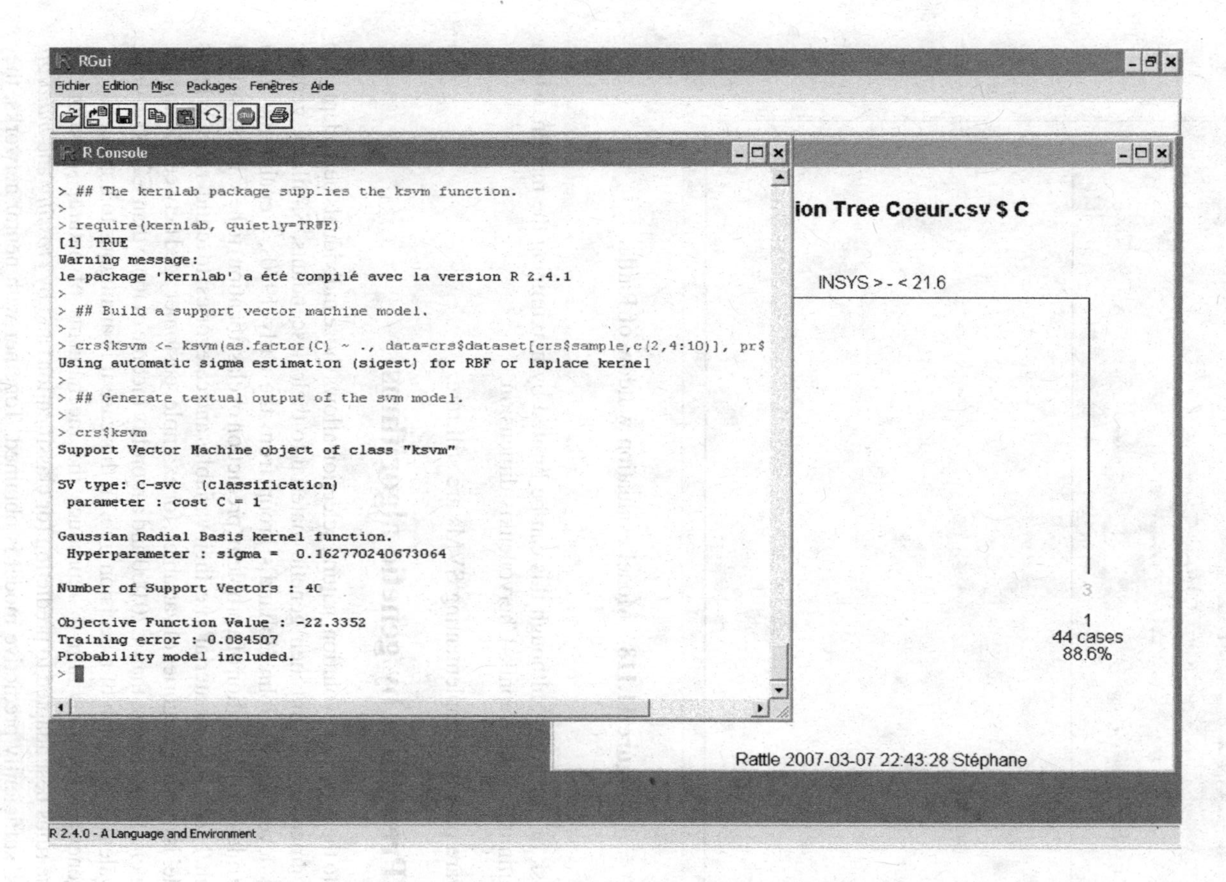

Figure 11.117 Execution of the ksvm function in the R console.

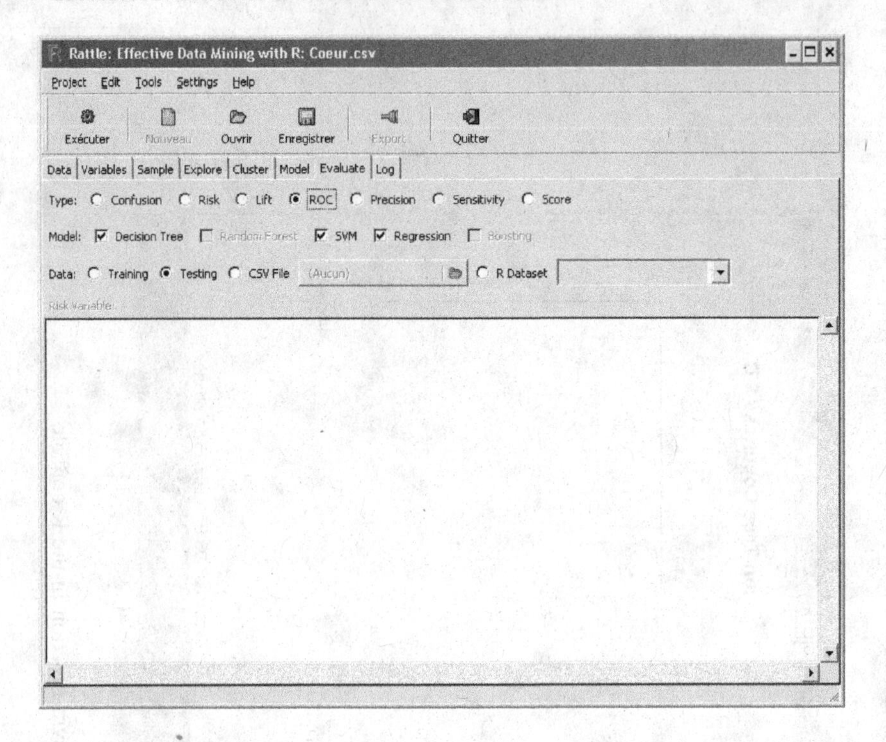

Figure 11.118 Model evaluation window of Rattle.

- the risk of overfitting, although this can be limited by maximizing the margin, which also minimizes the Vapnik–Chervonenkis dimension;
- software packages implementing SVMs are still rare.

11.13 Prediction by genetic algorithms

According to the theory of evolution, natural selection allows those individuals best adapted to their environment to transmit their genetic material to their descendants. Similarly, genetic algorithms, developed by John Holland's group from the early 1970s,[79] enable the most appropriate rules for the solution of a problem (prediction or classification) to be selected so that they transmit their 'genetic material' (i.e. their variables and categories) to 'child' rules. What is called a 'rule' is a set of categories of variables, for example 'customer aged between 36 and 50, having financial assets of less than €20 000 and a monthly income of more than €2000'. A rule is the equivalent of a branch of a decision tree. In this case it is analogous to a *gene*.

Thus genetic algorithms aim to reproduce the mechanisms of natural selection, by *selecting* the rules best adapted to prediction (or classification) and by *crossing* and *mutating* them until a sufficiently predictive model is obtained. Together with neural networks, they form the second type of algorithm which mimics natural mechanisms to explain phenomena which are not necessarily natural.

[79] Holland, J. H. (1975) *Adaptation in Natural and Artificial Systems.* Ann Arbor: University of Michigan Press.

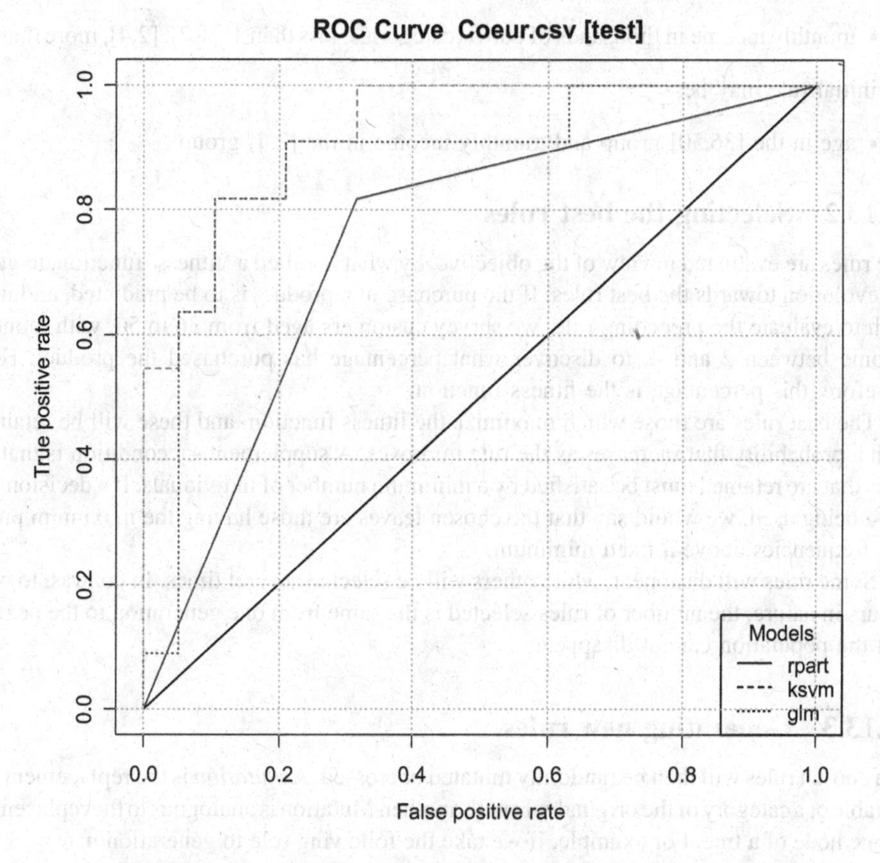

Figure 11.119 ROC curves of SVM, logit and CART models.

The execution of a genetic algorithm includes three steps:

1. random generation of initial rules;

2. selection of the best rules;

3. generation of new rules by mutation or crossing, step 3 looping back to step 2 until the execution of the algorithm stops.

11.13.1 Random generation of initial rules

The first rules are generated, the only constraint being that they must all be distinct. Each rule contains a random number of variables, chosen at random, each having a randomly chosen category.

Suppose that the variables are:

- age in years (categories: [18,35], [36,50], [51,65], more than 65);

- financial assets in thousands of euros (categories: less than 10, [10,20], [20,60], more than 60);

- monthly income in thousands of euros (categories: less than 1, [1,2], [2,4], more than 4).

An initial rule may be:

- age in the [36,50] group and monthly income in the [2,4] group.

11.13.2 Selecting the best rules

The rules are evaluated in view of the objective, by what is called a 'fitness' function, to guide the evolution towards the best rules. If the purchase of a product is to be predicted, and if we wish to evaluate the preceding rule, we survey customers aged from 36 to 50, with monthly income between 2 and 4, to discover what percentage has purchased the product. Here, therefore, this percentage is the fitness function.

The best rules are those which maximize the fitness function, and these will be retained, with a probability that increases as the rule improves. A supplementary condition is that the rules that are retained must be satisfied by a minimum number of individuals. If a decision tree were being used, we would say that the chosen leaves are those having the maximum purity and frequencies above a fixed minimum.

Some rules will disappear, while others will be selected several times. In contrast to what occurs in nature, the number of rules selected is the same from one generation to the next, so that the population cannot disappear.

11.13.3 Generating new rules

The chosen rules will then be randomly mutated or crossed. A *mutation* is the replacement of a variable or a category of the original rule with another. Mutation is analogous to the replacement of one node of a tree. For example, if we take the following rule to generation n:

- age $\in$ [36,50] and monthly income $\in$ [2,4],

a mutation can give rise to the following rule (the 'child' rule) in generation $n + 1$:

- age $\in$ [36,50] and financial assets $\in$ [10,20].

A *crossing* of two rules (which must be distinct from each other) is the exchange of some of their variables or categories to produce two new rules. Crossing is analogous to exchanging the places of two sub-trees. For example, if we take the following rules to generation n:

- age >65 and financial assets >60,
- age $\in$ [18,35] and financial assets $\in$ [10,20] and monthly income $\in$ [2,4],

the crossing can produce the following 'child' rules in generation $n + 1$:

- financial assets >60 and monthly income $\in$ [2,4],
- financial assets $\in$ [10,20] and monthly income $\in$ [2,4] and age $>$ 65.

As in nature, crossing is much more common than mutation, which is accidental, and has negative rather than positive consequences in most cases. In genetic algorithms, however, even if the mutations do not increase the value of the fitness function, it is better to have one mutation in each generation, which makes it possible to reintroduce useful conditions which disappeared by chance, and to avoid the premature convergence of the algorithm towards a local optimum.

The 'child' rules that are retained for evaluation are those that are distinct from the 'parent' rules, distinct from each other and satisfied by a minimum number of individuals. After evaluation, some of the 'child' rules are retained to become the new 'parent' rules and continue the algorithm.

11.13.4 End of the algorithm

The algorithm ends when one of the following two conditions is met: either a previously specified number of iterations has been reached, or, starting from a generation of rank n, the rules of generations $n, n-1$ and $n-2$ are (almost) identical. The number of iterations varies between several tens and several hundreds.

11.13.5 Applications of genetic algorithms

Genetic algorithms are used for solving optimization problems such as the travelling salesman problem. They are also used to improve the performance of other prediction tools such as neural networks. For the crucial question of calculating the weights of the nodes of a neural network, genetic algorithms provide a solution. This is to represent all the weights of the network with a gene, and then, starting from several possible sets of weights (i.e. genes), to select, cross and mutate the best genes from generation to generation, until an optimal set of weights is achieved. Genetic algorithms are also useful in situations in which many local optima are present, using mutation to avoid premature convergence towards one of these.

Genetic algorithms, and more generally biomimetic algorithms, have also been used since the mid-1990s in clustering problems, where they have several advantages: graphic visualization, the possibility of parallelizing the calculations, and the ease of combining them in hybrid approaches with other methods, such as k-means. In these biomimetic algorithms, each data element is represented by an artificial animal. A concept of similarity is defined between the data, and we study the movement of the animals in a group ('swarm intelligence') given that an animal has only a local perception but tends to move towards similar animals.[80] Similarly, the movement and grouping of ants ('artificial ants') have been studied, as well as the way in which they transport objects. The ants are more likely to collect the objects if they are dissimilar to neighbouring objects; then they move at random, and deposit the object in an area with a probability that increases with the number of similar objects already present in this area. The first algorithm describing the carrying of objects by ants was proposed by Lumer and Faieta.[81] It was subsequently improved, notably by combining it with the k-means algorithm, to profit from the best features of both algorithms:[82] the ant algorithm provides an initial

---

[80] These studies are used in the film industry to create realistic behaviour in synthetic images showing moving animals.

[81] Lumer, E.D. and Faieta, B. (1994) Diversity and adaptation in populations of clustering ants. In *Proceedings of the Third International Conference on Simulation of Adaptive Behaviour*, pp. 501–508. Cambridge, MA: MIT Press.

[82] Monmarché, N., Slimane, M. and Venturini, G. (1999) On improving clustering in numerical databases with artificial ants. In *5th European Conference on Artificial Life (ECAL'99)*, Lecture Notes in Artificial Intelligence, 1674, pp. 626–635. Berlin: Springer-Verlag.

classification, without the more or less arbitrary choice of the number of initial classes and centres that is found in the *k*-means algorithm. The centres of this initial classification are then used as the initial centres of the *k*-means algorithm, which reallocates some of the objects to more appropriate classes. The ant algorithm is then used again, and so on. In this case the contribution of the *k*-means method is useful because the ant algorithm does not always manage to classify all the objects (when the algorithm stops, some of the ants are still carrying objects), and it may take some time to reclassify an object that was incorrectly classified before. The *k*-means algorithm accelerates the convergence of Lumer and Faieta's artificial ant algorithm.

Some of these biomimetic algorithms provide results as good as those of single-linkage hierarchic ascendant classification, but with faster calculation and handling larger data volumes.[83] One advantage of these algorithms is that they can be run in parallel, thus increasing their speed. The interested reader will find a survey of this research in H. Azzag et al. (2004).[84]

11.13.6 Disadvantages of genetic algorithms

This type of algorithm is generally rather slow. Its complexity increases exponentially as a function of the number of rules used, because each rule in each generation must be evaluated, and there may be several thousand rules. It can only be used on rather small volumes of data. Furthermore, it is quite tricky to adjust. It is not yet widely included in software, with the notable exception of Version 9 of SAS which incorporates it in its SAS/OR module (the GA procedure).

11.14 Improving the performance of a predictive model

Even if the performance of a predictive model depends much more on the field studied (health, insurance, banking, etc.), the problems encountered (studies of risk, propensity, etc.) and the available data than on the method of modelling used, it is nearly always possible to build a more precise, and especially a more robust model, either by partitioning models after pre-segmentation, or by aggregating a number of models built by the same method applied to a number of samples, or by combining a number of models built by different methods applied to the same sample, or by combining these possibilities.

When pre-segmentation (or pre-clustering), mentioned in Section 2.5, is used, the modelling step is preceded by a step of clustering the population, then building a specific model for each of the clusters, before making a synthesis of these. Surprisingly, perhaps, adding as many variables as possible does not improve a model, but usually detracts from it. This makes it preferable to segment the population before modelling it, so as to be able to work with homogenous groups which require fewer variables to describe them.

The aggregation of models, to be described in the next section, is a way of applying the same method of modelling for a reasonably large number of times to slightly different

---

[83] Azzag, N., Monmarché, N., Slimane, M., Venturini, G. and Guinot, C. (2003) Anttree: a new model for clustering with artificial ants. In *IEEE Congress on Evolutionary Computation*, Canberra, Australia.

[84] Azzag, H., Picarougne, F., Guinot, C. and Venturini, G. (2004): Un survol des algorithmes biomimétiques pour la classification. *Revue des Nouvelles Technologies de l'Information*, RNTI-C-1, pp. 13-24. Cépaduès Éditions, France.

samples obtained from the same original population, before making a synthesis of the resulting models.

The third technique, the combination of models, applies a number of different modelling methods to the same population, for example a discriminant analysis and a decision tree, before making a synthesis of these. The simplest method of making the synthesis is to calculate for each individual the arithmetic or geometric mean of the scores resulting from the different models, but we can also use the prediction obtained from a model as an independent variable for the second model, or cross-tabulate the scores in a more complex way, for example by assigning weightings to the models.

By calculating the mean of the scores obtained from two classification methods, we can use the best parts of each method. This mean clearly has a meaning when the two initial scores are closely correlated (if this is not the case, there is a problem). Of course, there are differences, due for example to cases which are better detected by the decision tree than by discriminant analysis, or cases of incorrect classification by the tree when it has been misled by a threshold effect (when there is an individual at the limit of the splitting threshold of a node). As it would be risky to interpret the mean of two completely opposing scores, it will be preferable not to assign a final overall score to the individuals involved, but to give them a mean score.

It can be seen why we can expect a better prediction quality from the synthesis of two models: a customer having a maximum synthetic score is a customer who has obtained the best score by two different methods, on variables which may also be distinct. By taking the mean of a score found by a decision and one found by discriminant analysis, we have stabilized the fluctuations and the prediction of these two models, and obtained a better performing model, as is shown by the superimposition of the lift curves (Figure 11.120).

We can represent the aggregation and combination of models in a table (see Table 11.3).

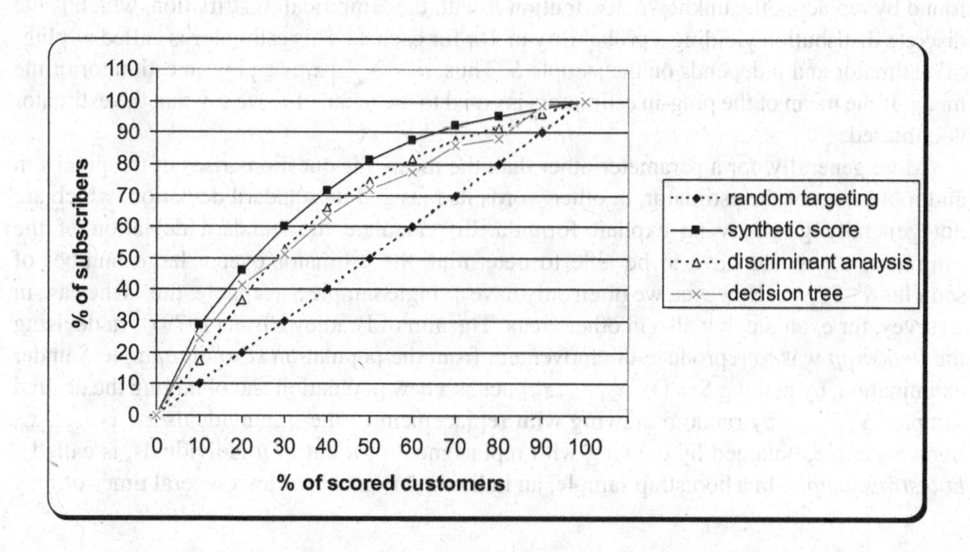

Figure 11.120 Superimposition of the lift curves of different models.

Table 11.3 Methods of improving performance.

Use:		On:	
		The same sample	**Different samples**
What:	**The same method**	Single model	Aggregation of models
	Different methods	Combination of models	Mixture*

*This may be a sequence of bootstrap samples, as in bagging (see Section 11.15.2), to which a decision tree and a neural network are applied each time.

11.15 Bootstrapping and ensemble methods

Before describing model ensemble methods more commonly called ensemble methods, I must recapitulate some aspects of resampling, which is the application of Monte Carlo simulation methods to statistics, and which includes a number of methods, of which the jackknife (used for cross-validation) and the bootstrap described below.

11.15.1 Bootstrapping

A classic problem encountered in statistics, and more generally in data mining, is that of the estimation of a statistical parameter. Such a parameter is defined for a global population Ω, and is a function of the statistical distribution F defined on Ω. This parameter can be the mean of F. Now, the global population and the distribution F are generally unknown, especially as the population (such as a set of customers) may be continually changing, or there may be errors of measurement, input, etc. When we are working on a data set, therefore, it is nearly always a sample $S = \{x_1, x_2, \ldots, x_n\}$ taken from the unknown global population, and we try to approximate the parameter with an estimator defined on the sample S, this estimator being found by replacing the unknown distribution F with the 'empirical' distribution, which is the discrete distribution yielding a probability of $1/n$ for each x_i. This estimator is called a 'plug-in' estimator and it depends on the sample S. Thus, $n^{-1}\sum_{i=1}^{n} x_i$ is a plug-in estimator of the mean. If the mean of the plug-in estimators is equal to the mean of F, we say that the estimator is unbiased.

More generally, for a parameter other than the mean, the question arises of the precision and robustness of the estimator, in other words its bias and its standard deviation, which are not generally given by an explicit formula. To calculate the standard deviation of the estimator, we would have to be able to determine the estimator over a large number of samples $S', S'', \ldots$. However, we often only have a single sample S available; this is the case in a survey, for example, but also in other areas. The aim of Bradley Efron (1979)[85] in devising the *bootstrap* was to reproduce the movement from the population Ω to the sample S under examination, by making $S = \{x_1, x_2, \ldots, x_n\}$ act as a new population and obtaining the desired samples $S', S'', \ldots$ by random drawing with replacement of the n individuals $x_1, x_2, \ldots, x_n$. Such a sample, obtained by drawing with replacement of n out of n individuals, is called a *bootstrap sample*. In a bootstrap sample, an individual x_i may be drawn several times or may

[85] Efron, B. (1979) Bootstrap methods: another look at the jackknife. *Annals of Statistics*, 7(1), 1–26.

not be drawn at all. The probability that a given individual x_i will be drawn is $1 - (1 - 1/n)^n$, which tends towards 0.632 as n tends towards $+\infty$.

When a certain number B (generally, $B \geq 100$) of bootstrap samples S^* has been drawn and the plug-in estimator Θ^* of the sample S^* has been calculated for each of them, we obtain a distribution of the bootstrap plug-in estimators Θ^*, from which we deduce a standard deviation of the plug-in estimator Θ of the sample S. We can also deduce confidence intervals from the quantiles of the distribution: we specify a fairly large B, for example $B = 1000$ (the minimum according to Efron) and examine the 25th weakest value $Q_{2.5}$ and the 25th strongest value $Q_{97.5}$ of the bootstrap estimator in order to gain some idea of the 95% confidence interval $[Q_{2.5}, Q_{97.5}]$ of the estimator.

As for the bias, its bootstrap approximation is equal to the difference between the mean of the bootstrap estimators Θ^* and the estimator Θ calculated on S.

The principle of the bootstrap can be summarized as follows: the sample S is made to act as the global population Ω, and the bootstrap sample S^* is made to act as the sample S, given that the estimator Θ^* behaves with respect to Θ as Θ with respect to the desired parameter over the global population, and the knowledge of Θ^* (distribution, variance, bias) contributes to the knowledge of Θ.

In classification and scoring problems, the parameters to be estimated may be:

- the error rate (or correct classification rate) or other measure of the performance of the score model (area under the ROC curve, Gini index, etc.);
- the coefficients of the score function;
- the predictions (*a posteriori* probabilities of belonging to each class to be predicted).

As the global population on which the model is to be built is unknown, the above parameters can only be estimated. We start by constructing B bootstrap samples from the initial sample, and then build a model on each bootstrap sample. We obtain B classification models. The bootstrap on the error rate or the area under the ROC curve enables us to obtain confidence intervals of these performance indicators of the model. This situation is illustrated in Figure 11.121. Note that the mean error rate on the bootstrap samples is an estimate biased by optimism, since these error rates are calculated by resubstitution on the individuals which have been used for training the model. In a variant shown in Figure 11.122, the errors are calculated only on the individuals not included in the bootstrap sample: we speak of an out-of-bag estimate. Since this estimate is biased by pessimism, Efron and Tibshirani have suggested that the optimistic bias of resubstitution estimation and the pessimistic bias of out-of-bag estimation could be rectified simultaneously by the 'magic formula' of the .632-bootstrap:

$$\text{Estimate}_{.632} = 0.368 \times \text{estimate (resubstitution)} + 0.632 \times \text{estimate(bootstrap-oob)}.$$

This formula allows for the probability of 0.632 of the selection of each individual in one of the various bootstrap samples (in other words, an individual belongs on average to $0.632B$ bootstrap samples[86]), which causes the excessive fluctuation of the out-of-bag estimator. It can be applied to a performance indicator such as the area under the ROC curve. As this

[86] Conversely, a sample contains $0.632n$ different individuals on average.

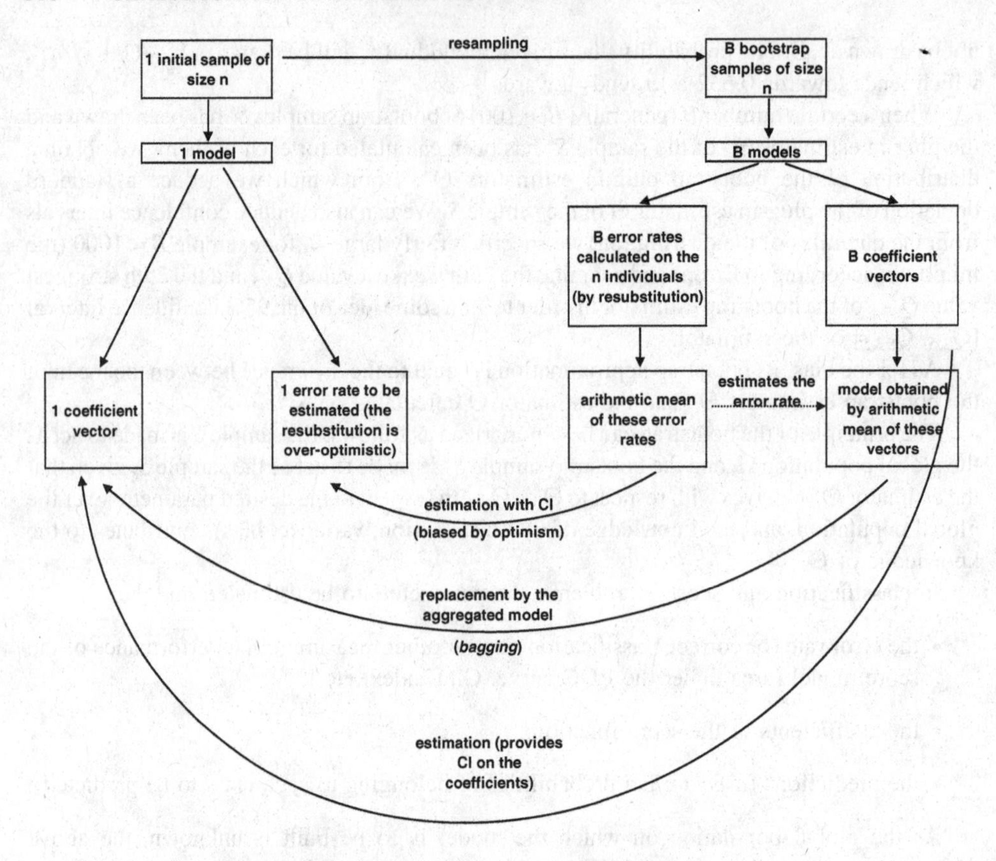

Figure 11.121 Bootstrap and bagging resampling.

estimate is itself over-optimistic in some cases with a high level of overfitting, the same authors proposed a more elaborate variant in 1997, called the '.632 + bootstrap'.[87]

The bootstrap on the coefficients of the classification function is used, for example, with linear discriminant analysis to find confidence intervals of the coefficients in order to assess the actual contribution of each independent variable.

11.15.2 Bagging

Bootstrapping on predictions has been known since the work of Leo Breiman (1996) under the name of *bagging*, or 'Bootstrap AGGregatING'.[88] It involves the construction of a family of models on *m* bootstrap samples, followed by the aggregation of the predictions of each model. This aggregation is carried out by voting (classification) or averaging (regression). Averaging can also be used with logistic regression, by calculating the averages of the a posteriori probabilities supplied by the model; in this case, the average probabilities are

[87] Efron, B. and Tibshirani, R. (1997) Improvements on cross-validation: The .632 + bootstrap method. *Journal of the American Statistical Association*, 92, 548–560.

[88] Breiman, L. (1996) Bagging predictors. *Machine Learning*, 26(2), 123–140.

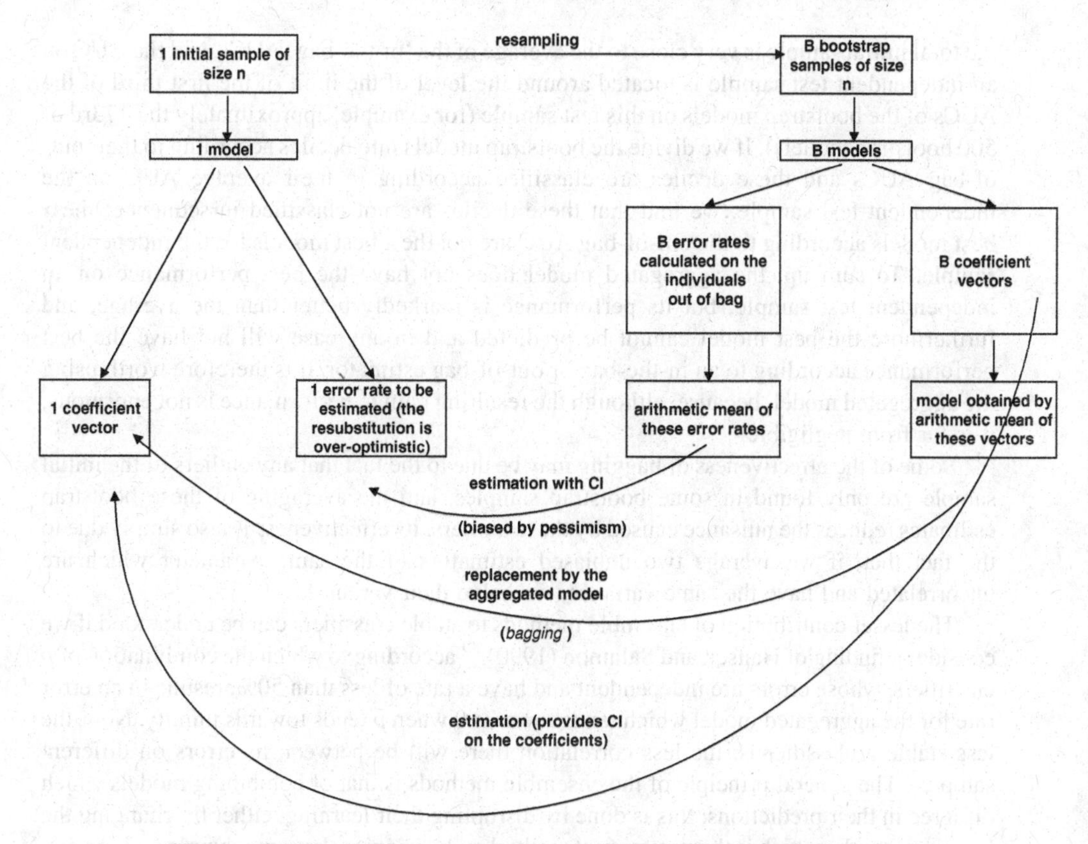

Figure 11.122 Bootstrap resampling with out-of-bag estimation.

approximated by the probabilities supplied by the 'average' model, in other words the model whose coefficients are the average coefficients of the different models for each variable (or for each category).

Bagging can reduce the variance, but not the bias, of a model, and it is particularly useful for remedying the lack of robustness of unstable classifiers such as decision trees and neural networks. In the case of trees, however, it should be pointed out that the aggregation of a number of trees destroys the simple structure of the decision tree so that we lose the main advantage of these trees, which is their readability. In linear discriminant analysis and logistic regression, however, bagging does not increase the complexity of the model compared with the base model, because the result is simply the 'average' model. However, bagging is less useful for stable classifiers: the variance of the model may decrease, but to a lesser degree. However, it is still helpful, because it improves the generalization of the model, especially when there are only a small number of individuals to be modelled. This is because, in the distribution of bootstrap estimators, models having a maximum 'out-of-bag' AUC (area), which are the natural candidates for good generalization, are often not those whose area is maximal in a new sample which is independent of the initial sample (for instance, a sample created at a later date). In other words, the out-of-bag AUCs and the AUCs measured in an independent sample are weakly correlated (the correlation coefficient can be less than 0.1). As for the aggregated model, in the examples that we have tested its AUC on the total initial sample is very close to the average of the 'in-the-bag' AUCs, and the AUC on an independent test sample is located around the level of the limit of the first third of the AUCs of the bootstrap models on this test sample (for example, approximately the 173rd of 500 bootstrap models). If we divide the bootstrap models into deciles according to their out-of-bag AUCs and these deciles are classified according to their average AUC on the independent test sample, we find that these deciles are not classified in sequence: the x best models according to the out-of-bag AUC are not the x best models for the independent sample. To sum up, the aggregated model does not have the best performance on an independent test sample, but its performance is markedly better than the average, and furthermore the best model cannot be predicted and in any case will not have the best performance according to an in-the-bag or out-of-bag estimator. It is therefore worth using the aggregated model, because, although the resulting gain in performance is not enormous, it is far from negligible.

Some of the effectiveness of bagging may be due to the fact that any outliers of the initial sample are only found in some bootstrap samples, and the averaging of these bootstrap estimates reduces the nuisance caused by these outliers. Its effectiveness is also simply due to the fact that, if we average two unbiased estimators of the same parameter which are uncorrelated and have the same variance, we halve their variance.

The lesser contribution of ensemble methods to stable classifiers can be understood if we consider a finding of Hansen and Salamon (1990),[89] according to which the combination of p classifiers, whose errors are independent and have a rate of less than 50%, results in an error rate for the aggregated model which tends towards 0 when p tends towards infinity. Now, the less stable a classifier is, the less correlation there will be between its errors on different samples. The general principle of the ensemble methods is that of combining models which disagree in their predictions; this is done by disrupting their learning, either by changing the learning sample, which is the commonest method, or by keeping the same sample and varying the learning parameters, which may be done with neural networks.

With this aim, Opitz and Maclin (1999)[90] propose the combination of neural networks built on the same sample, by varying only the parameters and the topology of each base network.

There is a special kind of bagging algorithm which is applied to decision trees with the introduction of a different random selection of candidate independent variables, taken from the set of all the independent variables, at each split of a node. At each split, therefore, what is tested is not the set of independent variables, but simply a sample of variables, chosen at random. This can be used to ensure that the same variables (the ones having the highest individual discriminant power) do not appear every time; it also decreases the correlation between the successive trees and thus decreases the variance of the aggregated model.

This random selection of variables was first suggested by Ho[91] and Dietterich,[92] independently of bagging. Breiman[93] subsequently devised a way of combining this

[89] Hansen, L. and Salamon, P. (1990) Neural network ensembles. *IEEE Transactions on Pattern Analysis and Machine Intelligence*, 12, 993–1001.

[90] Opitz, D. and Maclin, R. (1999) Popular ensemble methods: an empirical study. *Journal of Artificial Intelligence Research*, 11, 169–198.

[91] Ho, T. K. (1995) Random decision forests. In M. Kavanaugh and P. Storms (eds), *Proc. Third International Conference on Document Analysis and Recognition*, Vol. 1, pp. 278–282. New York: IEEE Computer Society Press.

[92] Dietterich, T. (2000) An experimental comparison of three methods for constructing ensembles of decision trees: bagging, boosting and randomization. *Machine Learning*, 40(2), 139–157.

[93] Breiman, L. (2001) Random forests. *Machine Learning*, 45, 5–32.

randomization of the variable selection with the randomization of the learning sample. He was responsible for developing this double randomization, known as the 'random forests' method. Random forests have since become extremely popular because of their benefits: they provide performance often comparable to that of boosting (see below), but are simpler to use, faster to calculate and more resistant to noise in the data (a weak point in boosting, as we shall see). However, we should not let this apparent simplicity conceal a subtle feature of the algorithm, namely that a new random sample of variables is taken at each node of each tree, not simply once for each tree. There is a parameter that can be 'tuned' to obtain the best results: this is the number of variables selected at each split (see Section 11.15.4). Of course, the number of iterations must also be chosen. It will generally be lower than in boosting, because the performance of random forests becomes stable in less time. On the other hand, boosting sometimes achieves a better final level of performance.[94]

Unlike simple bagging, random forests can be used successfully on trees limited to two leaves ('stumps'), without causing the appearance of trees which use the same variables so frequently that they become overcorrelated (see Section 11.15.4).

This double randomization mechanism is particularly beneficial for decision trees, but of course it can be applied to other base classifiers. Breiman has also proposed (*loc. cit.*) the application of this mechanism to boosting, in other words the combination of the learning sample weighting mechanism (see below) with the random variable selection mechanism. The initial results of this procedure showed that it could be beneficial for data sets of a certain size.

11.15.3 Boosting

A new approach to the combination of models came from machine learning, in the form of the *boosting* method devised in 1996 by Yoav Freund and Robert E. Schapire. The first polynomial time algorithm was described by Schapire in 1989,[95] and in 1990 Freund devised an improved, but still imperfect, algorithm.[96] Following research which commenced in 1995, they published their founder paper in 1997,[97] describing the AdaBoost.M1 algorithm, subsequently renamed Discrete AdaBoost in a paper by Friedman *et al.*[98] A later paper by Schapire and Singer[99] reformulated the Discrete AdaBoost algorithm and proposed the Real AdaBoost algorithm (reformulated by Friedman *et al.*), in which the model does not predict a class $Y \in \{-1; +1\}$, but the probability $P(Y=1|x)$. The papers by Freund and Schapire, and then Schapire and Singer, also proposed generalizations of AdaBoost to the case of a dependent variable with more than two categories; the AdaBoost.MH algorithm of the latter authors is worth mentioning.

In boosting, the same classification algorithm, such as a decision tree, is applied successively to versions of the initial training sample which are modified at each step to

[94] See Section 15.2 of Hastie *et al.*, *The Elements of Statistical Learning*, cited in Section 11.3.3 above.

[95] Schapire, R.E. (1990) The strength of weak learnability. *Machine Learning*, 5(2):197–227.

[96] Freund, Y. (1995) Boosting a weak learning algorithm by majority. *Information and Computation*, 121 (2):256–285.

[97] Freund, Y. and Schapire, R.E. (1997) A decision-theoretic generalization of online learning and an application to boosting. *Journal of Computer and System Sciences*, 55(1), 119–139.

[98] Friedman, J., Hastie, T. and Tibshirani, R. (2000) Additive logistic regression: a statistical view of boosting (with discussion), *Annals of Statistics*, 28, 337–407.

[99] Schapire, R.E. and Singer, Y. (1999) Improved boosting algorithms using confidence rated predictions. *Machine Learning*, 37(3), 297–336.

allow for classification errors of the preceding step, and the classifiers (which may be weak) constructed in this way are then combined to produce a stronger classifier.

The modifications made at each step are generally an overweighting of the observations incorrectly classified at the preceding step and an underweighting of the others, but a variant called 'arcing' (short for 'adaptative resampling and combining'), also devised by Freund and Schapire (initially under the name of boosting), introduces a random factor by drawing a training sample with replacement in each iteration, in the initial training set, with a greater probability of drawing for the observations which were incorrectly classified in the preceding iteration. This is therefore a probability proportional to size (pps) sampling method. This random sampling introduces greater diversity and greater independence into the resulting ensemble models. It avoids the determinism which causes the same models (the same trees, for example) to recur after a number of interval iterations. However, it reduces the percentage of observations contributing to the training of each classifier, which may be less than 20%, as against the 63.2% which would be drawn in a bootstrap with equal probabilities (Breiman 1996,[100] Section 5.1).

This sampling resembles bagging in some ways, but bagging only reduces the variance of the classifiers, whereas arcing, and boosting in general, reduces both the variance and the bias, as is shown by the examples of application to trees with two leaves ('stumps') or other depth 1 trees as base classifiers (see below). These trees typically have a low variance but a high bias.

This reduction of the bias has been quantified in a theorem of Freund and Schapire (1997),[101] later generalized by Schapire and Singer (1999),[102] which establishes a limit on the training error rate of the boosted classifier.

If each classifier f_m has an error rate $\varepsilon(f_m)$ on the training sample (with a frequency N), this error rate being weighted for each observation, so that we have

$$\varepsilon(f_m) = \frac{1}{N}\sum_{i=1}^{N} p_i \cdot 1_{[f_m(x_i)\neq y_i]},$$

then the error rate of the boosted classifier F on the training sample decreases exponentially with the number of steps M. More precisely, given that

$$\varepsilon(F) = \frac{1}{N}\sum_{i=1}^{N} 1_{[F(x_i)\neq y_i]},$$

we find that

$$\varepsilon(F) \leq \prod_{m=1}^{M} 2\sqrt{\varepsilon(f_m)(1-\varepsilon(f_m))}.$$

A good illustration of the boosting mechanism is provided by Schapire in his lecture which can be found at http://videolectures.net/mlss05us_schapire_b/ and in the book *Boosting: Foundations and Algorithms*, by Robert E. Schapire and Yoav Freund, to be published by MIT Press.

[100] Breiman, L. (1996) Bias, variance, and arcing classifiers. Internal report 460, Department of Statistics, University of California, Berkeley.

[101] See n. 95 above.

[102] See n. 97 above.

In step 1, all the observations have the same weight and a classifier h_1 is determined:

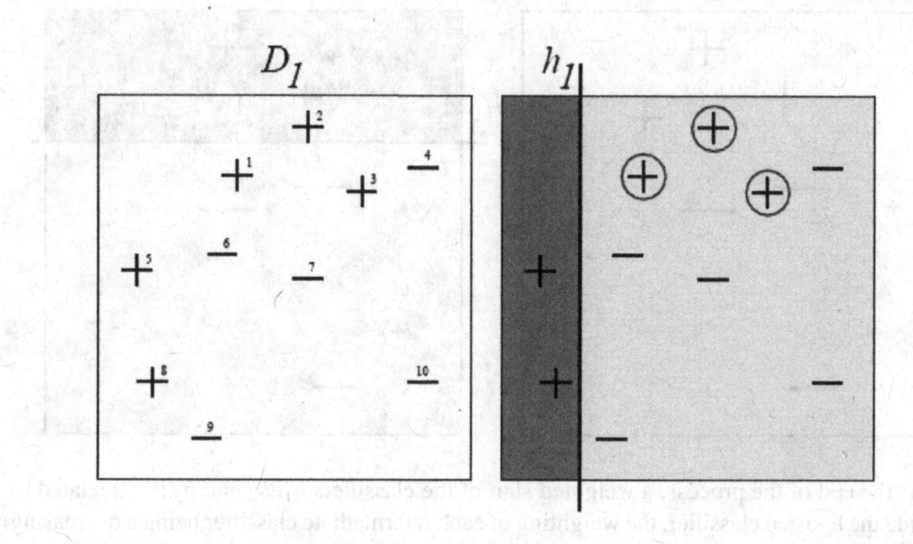

In step 2, the observations incorrectly classified by h_1 (circled in the diagram) are overweighted and a classifier h_2 is determined:

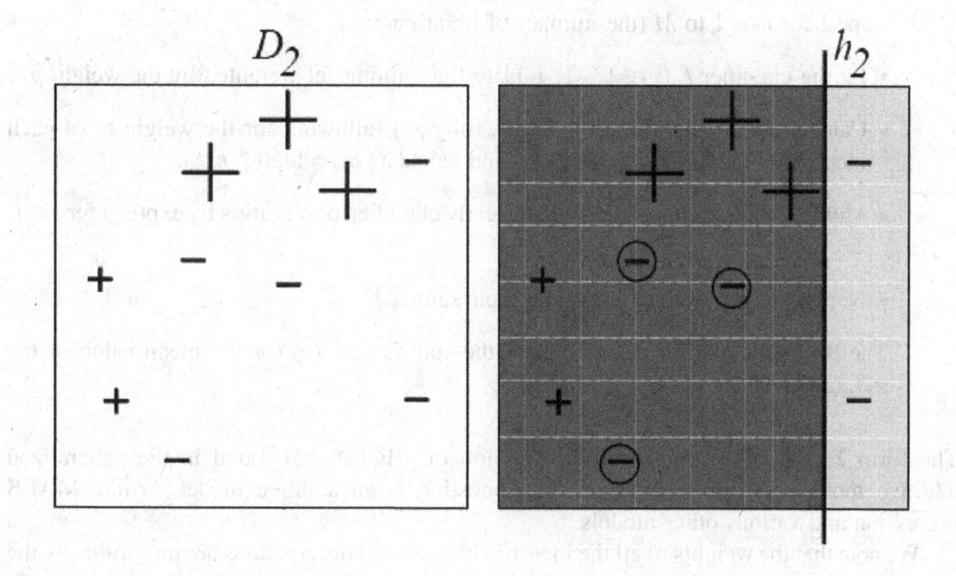

In step 3, the observations incorrectly classified by h_2 are overweighted and a classifier h_3 is determined:

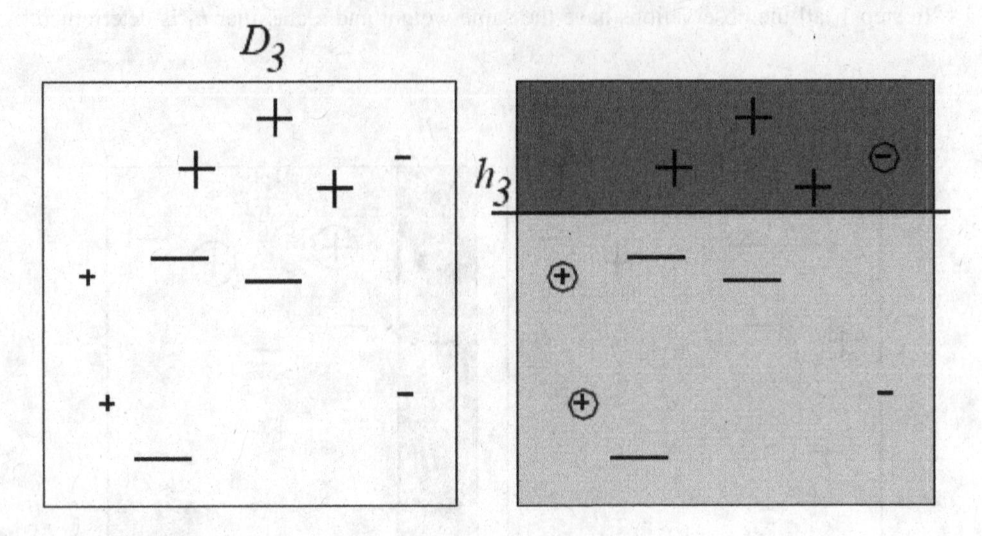

At the end of the process, a weighted sum of the classifiers h_1, h_2 and h_3 is calculated to provide the boosted classifier, the weighting of each intermediate classifier being a decreasing function of its error rate (Figure 11.123).

More precisely, the Discrete AdaBoost algorithm is as follows:

1. Initialize the weights of the N observations of the training sample: $p_i = 1/N$, $i = 1, 2, \ldots, N$
2. Repeat for $m = 1$ to M (the number of iterations):
   - Fit the classifier $f_m(x) \in \{-1, +1\}$ to the training set weighted by the weight p_i
   - Calculate the weighted error rate ε_m of $f_m(x)$ (allowing for the weight p_i of each incorrectly classified observation) and calculate $\alpha_m = \log((1-\varepsilon_m)/\varepsilon_m)$
   - Multiply the weight p_i of each incorrectly classified observation by $\exp(\alpha_m)$ for $i = 1, 2, \ldots, N$
   - Normalize the weights p_i so that their sum is 1
3. The boosted classifier is the sign of the sum $\Sigma_m \alpha_m f_m(x)$ (or the mean value of the $\alpha_m f_m(x)$).

The form $\Sigma_m \alpha_m f_m(x)$ resembles the expression of $g(\mathrm{E}(Y/X{=}x))$ found in the generalized additive model (Section 11.9.8): in fact, boosting is an additive model,[103] like MARS regression and various other models.

We note that the weights of all the incorrectly classified observations are multiplied by the same value, by contrast with the arc-x4 described below, and that this multiplier decreases as the error rate increases. Thus we prevent a model which is a poor fit in one iteration from having an inappropriate importance in the development of the weightings.

[103] See Section 10.2 of Hastie *et al.*, *The Elements of Statistical Learning*, cited in Section 11.3.3 above.

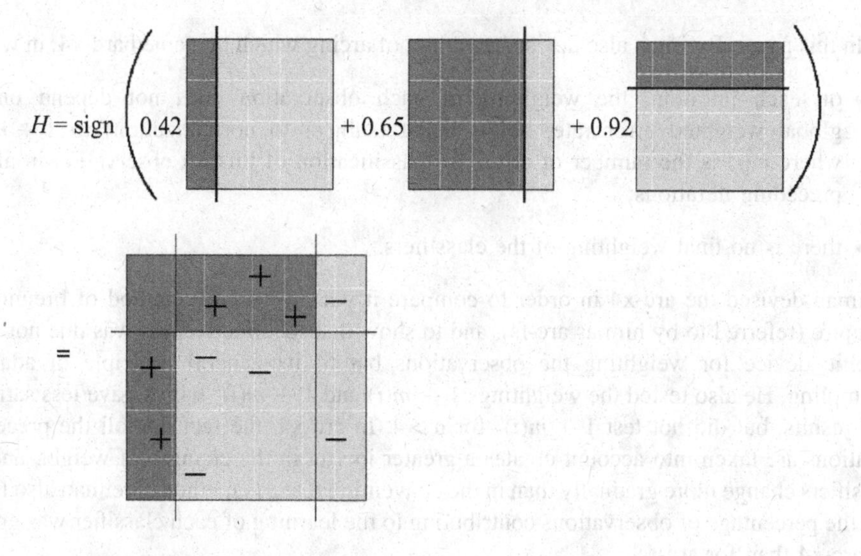

Figure 11.123 Illustration of the boosting mechanism.

The arcing algorithm is similar, except for the pps random sampling:

1. Initialize the weights of the N observations of the training sample: $p_i = 1/N$, $i = 1, 2, \ldots, N$
2. Repeat for m = 1 to M:
   - In the training sample, draw N observations with replacement, each according to the probability p_i
   - Fit the classifier $f_m(x) \in \{-1, +1\}$ to the bootstrap sample drawn in this way
   - On the initial training sample:
     - Calculate the weighted error rate ε_m of $f_m(x)$ and calculate $\alpha_m = \log((1 - \varepsilon_m)/\varepsilon_m)$
     - Multiply the weight p_i of each incorrectly classified observation by $\exp(\alpha_m)$ for $i = 1, 2, \ldots, N$
     - Normalize the weights p_i so that their sum is 1
3. The boosted classifier is the sign of the sum $\Sigma_m \alpha_m f_m(x)$ (or the mean value of the $\alpha_m f_m(x)$).

In his previously cited paper of 1996, Breiman made an improvement to this algorithm. When the error rate ε_m reaches a value of at least 0.5 in an iteration, the weights $p_i = 1/N$ are reinitialized, instead of the algorithm being stopped as Freund and Schapire had originally proposed. They are also reinitialized when $\varepsilon_m = 0$ and $\log((1 - \varepsilon_m)/\varepsilon_m)$ is therefore undefined. This device is not applicable to the algorithm for boosting by reweighting, since in this case, with no resampling, the reinitialization of the weights would cause the same model to recur every time.

In this paper, Breiman also devised a variant of arcing which he named arc-x4, in which:

- on each iteration, the weighting of each observation does not depend on the global weighted error rate, but is equal (subject to normalization) to $1 + m(i)^4$, where $m(i)$ is the number of errors of classification of this ith observation in all the preceding iterations;
- there is no final weighting of the classifiers.

Breiman devised the arc-x4 in order to compare it with the arcing method of Freund and Schapire (referred to by him as arc-fs), and to show that its effectiveness was due not to its specific device for weighting the observations but to its general principle of adaptive resampling. He also tested the weightings $1 + m(i)$ and $1 + m(i)^2$ which gave less satisfactory results, but did not test $1 + m(i)^n$ for $n > 4$. In arc-x4, the fact that all the preceding iterations are taken into account creates a greater inertia in the changes of weight, and the classifiers change more gradually than in the conventional arc-fs method. Breiman also found that the percentage of observations contributing to the learning of each classifier was greater for arc-x4 than for arc-fs.

Here is now the Real AdaBoost algorithm:

1. Initialize the weights of the N observations:
   - $p_i = 1/N$, $i = 1, 2, \ldots, N$
2. Repeat for $m = 1$ to M
   - Calculate the probability $p_m(x) = P(Y = 1|x)$ on the training sample weighted by the weights p_i
   - Calculate $f_m(x) = ½ \log(p_m(x)/(1 - p_m(x))$
   - Multiply the weight p_i of each observation (x_i, y_i) by $\exp(-y_i.f_m(x_i))$ for $i = 1, 2, \ldots, N$
   - Normalize the weights p_i so that their sum is 1
3. The boosted classifier is the sign of the sum $\Sigma_m f_m(x)$.

Bagging and boosting are both ensemble algorithms, the essential difference being that bagging is a purely random process while boosting is an adaptive process. This also means that a bagging algorithm can be parallelized, but a boosting algorithm cannot. In addition to this computational aspect, it is generally accepted that the convergence of boosting towards an optimal performance (minimum error rate or maximum AUC) tends to be slower than in bagging. This phenomenon is particularly marked when the boundary between the classes to be predicted is poorly delimited, and Adaboost loses time by reweighting the classification errors of observations which are close to the boundary.

The paper by Friedman *et al.* cited above proposes a simple idea for significantly reducing the amount of computation and the time required. The authors find that, in the course of the iterations, an increasingly large proportion of the training sample is correctly classified and is therefore given very low weights. If we decide to carry out the training process on each iteration only on observations representing 90–99% of the total weight, the number of observations used in the training will decrease considerably and the computation will be faster. Of course, the adjusted model is then applied to all of the training sample in order to recalculate the weights of all the observations and in order to include them in the rest of the iterations.

The main characteristic of boosting is therefore that it concentrates its action on individuals that are difficult to model and that behave in ways that are harder to predict. Consequently, an iteration in which the outliers are poorly classified is followed by an iteration which classifies them well but does not classify the rest of the population so well. The next iteration re-establishes the balance, but in turn it will classify some of the outliers poorly. Thus balancing takes place during the iterations, in which, in contrast to bagging, the models are only locally optimal (for certain observations), instead of being globally optimal. It is their aggregation that is globally optimal, as is clearly shown in the example provided by Schapire (above).

However, this mechanism can have drawbacks in certain circumstances. Firstly, the variance of the boosted model can increase if the base classifier is stable (e.g. linear discriminant analysis, logistic regression on specified variables, etc.). This is because the same observations are often poorly classified in this case, even with a modified sample. The weight of these observations will increase rapidly, together with the error rate in training.

The variance can also increase with noisy data. Where noise is present, the error of the boosted model increases with the number of aggregated models. The sensitivity of boosting to noise is due to the adaptive device which puts the emphasis on observations that are difficult to classify (by weighting them more or by choosing them more often), which by their nature are more likely to be noisy observations or outliers. The sensitivity of AdaBoost to noise has been pointed out by Opitz and Maclin (1999) and Dietterich (2000), both cited in Section 11.15.2. As this sensitivity is due to an exponential weighting of the errors, this has led some researchers to propose other weighting formulae. Examples are the Friedman paper cited at the beginning of this section, with Gentle AdaBoost, and Freund (1999) with BrownBoost.[104] In Gentle AdaBoost, the function $f_m(x) = \frac{1}{2}\log(p_m(x)/(1-p_m(x))$ of Real AdaBoost is replaced with a bounded function. The results of Gentle AdaBoost with noisy data are generally much better than those of Discrete AdaBoost or Real AdaBoost.

Thus, although boosting generally exhibits better performance than bagging, with good resistance to overfitting, it may be less effective in some cases. Note the generally beneficial effect of the weighting of each base classifier by α_m in the final aggregation: the weight of each classifier decreases as its error rate rises, and therefore this weighting limits the risk of overfitting. In practice, we have found that the gain to be expected from weighting in the final aggregation is limited, especially if there is a sufficient number of observations in each leaf to ensure the robustness of the trees and reduce the risk of overfitting.

Boosting is currently used when the base model is a decision tree, although it can be applied to other base models, at least those that are unstable. It frequently improves the robustness of the models (by decreasing their variance), which also occurs with bagging, but, unlike bagging, it can also improve their precision (by reducing their bias).

[104] Freund Y. (1999). An adaptive version of the boost by majority algorithm. In *Proceedings of the Workshop on Computational Learning Theory*. Morgan Kaufmann.

The performance of boosting appears to be improved when the trees obtained in the successive iterations are as independent as possible. This can be done, as in arcing, by replacing the weighting mechanism by random sampling with replacement (pps). Another principle, similar to that used in random forests, is that of increasing the choice of variables (in random forests, a random element is added to the choice of variables): the initial selection of the most discriminating variables would therefore not necessarily be beneficial. For the same reason, a non-binary tree such as CHAID may provide a small gain by comparison with CART, even though this is the tree that is nevertheless most commonly used in boosting, ahead of C4.5.

There are certain criteria that must be specified for the implementation of a boosting method:

- the use of bootstrap samples or of the complete initial sample;
- the method of calculation of the adaptation error (how do we calculate an error rate when the classifier outputs a continuous prediction?);
- the method of calculating the weightings (should they be bounded?);
- the method of final aggregation of the base models.

With a tree, non-trivial results can be obtained even with a base tree having only a stump, although better results are clearly achieved with a more complex tree and if the recommended number of leaves of the base tree is in the range from 4 to 8, or equal to the square root of the number of independent variables. The book by Hastie, Tibshirani and Friedman (2009, section 10.1) gives the example of a stump used on a simulated test set (based on Gaussian sums), for which the test error rate is 46% (hardly better than a random prediction) but decreases to 12.2% after 400 boosting iterations, whereas a conventional decision tree has an error rate of 26% in this case. Like random forests, and unlike simple bagging, boosting is effective on stumps. This is another difference between these two ensemble methods. The difference in performance between boosting and bagging is naturally maximal for depth 1 trees, because with bagging it is often the same variable, or at least a very small number of variables, that is chosen for the single split of the .tree (with a variable division if necessary). Boosting, especially in its arcing variant, introduces a greater variety of selection. In this configuration, the difference in performance between boosting and random forests on one hand, and bagging on the other hand can be very great.

Many algorithms based on the boosting principle have been published, including Discrete AdaBoost, Arcing, Real AdaBoost and Gentle AdaBoost which I have already mentioned, LogitBoost (see the Friedman *et al.* paper cited at the beginning of this section), and more recently Friedman's gradient boosting[105] and many others. They do not always differ much in performance on real data, as opposed to simulated data sets: see, for example, Section 11.7 of Friedman *et al.* The main feature of all these studies is the remarkable resistance of boosting to overfitting.

11.15.4 Some applications

Other researchers than those mentioned above have compared the performance of these methods on real or simulated data sets. Opitz and Maclin tested the bagging and boosting of

[105] Friedman, J. (2002) Stochastic gradient boosting. *Computational Statistics & Data Analysis*, 38, 367–378.

neural networks and decision trees on 23 data sets.[106] Their paper is interesting in several ways, particularly because neural networks are mentioned less frequently in the literature on ensemble methods. There are at least two reasons for this: firstly, the longer processing times of neural networks are a disadvantage when working on multiple samples; secondly, the performance of neural networks is rather sensitive to parameters which are not always easy to choose (from this point of view, CART is ideal). However, neural networks are useful because of their wide range of application and their performance which may be better than that of decision trees.

As for the number of models to be aggregated, these authors consider that it depends on the method:

- about 10 are sufficient for neural network bagging and boosting and decision tree bagging;
- for decision tree boosting, 25 models are required to achieve the greater part of the error reduction.

These numbers are smaller than those mentioned by Hastie *et al.*, but have been determined using real data sets instead of simulated ones. For their part, Bauer and Kohavi made a detailed study, on 12 data sets, of bagging and boosting with decision trees, but not with neural networks.[107] They also investigated the naive Bayesian classifier, which is very stable, and found that bagging or boosting could not be expected to really reduce the variance of this classifier (although boosting decreased its bias).

As a general rule, I have said that these ensemble methods are not particularly beneficial for stable classifiers. As well as the naive Bayesian classifier, there is always the logit model. This is particularly true for boosting and even arcing, because of their unwelcome tendency to overweight the same observations in a stable model at all times (see above). Conversely, random forests and their double randomization introduce a beneficial variability into models. Bagging lies between these groups, with a fast performance (less than twenty iterations) resulting in improved performance by comparison with the base model, although the improvement is not so great as that provided by random forests. We can see this in the tests conducted on the data set put on line by The Insurance Company (TIC) Benchmark[108] and currently used in competitions or benchmarking[109]. With this data set, a propensity score for taking out a caravan insurance policy was modelled. The models were constructed on a training sample of 5822 customers and the areas under the ROC curve were measured on a validation sample of 4000 customers. Figure 11.124 compares the area under the ROC curve (AUC) of models produced by bagging, arcing and random forests, as a function of the number of bootstrap samples. The base classifier is a logit model produced by stepwise selection (at the 1% threshold) on a set of 17 variables (preselected from a larger set of 59 variables). If we had used a base model without selection of variables, the aggregated models would have been so similar that the overall performance would not have improved. The performance curve for arcing on CART trees is also shown for comparison (see the diagram below).

[106] Opitz, D. and Maclin, R. (1999). Popular ensemble methods: an empirical study. *Journal of Artificial Intelligence Research*, 11, 169–198.

[107] Bauer, E. and R. Kohavi (1999) An empirical comparison of voting classification algorithms: Bagging, boosting, and variants. *Machine Learning* 36 (1–2), 105–139.

[108] http://www.liacs.nl/~putten/library/cc2000/

[109] See S. Tufféry (2009), *Étude de cas en statistique décisionnelle*, Édition Technip, Section 2.23.

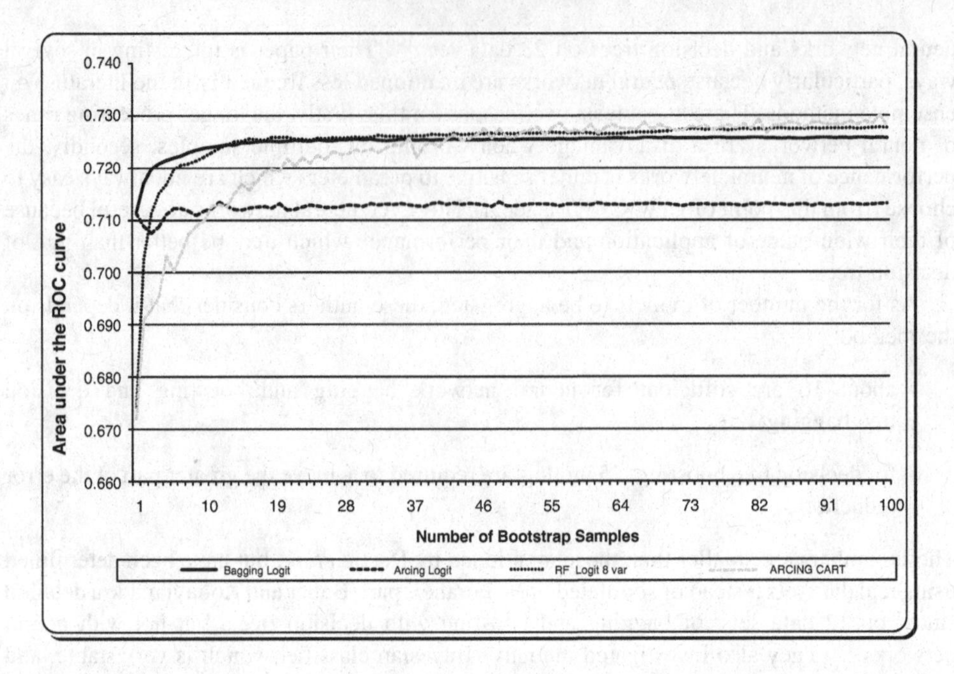

Figure 11.124 AUC of models obtained by bagging, arcing and random forests of logit base classifier, as a function of the number of bootstrap samples.

The values of AUC shown in this diagram and the next one are the mean values found for each model over thirty simulations.

We can see that logistic regression does not benefit from arcing at all, but it clearly does benefit from an aggregation of random forests produced by drawing a new random sample from 8 independent variables on each iteration (the reason for the number 8 is given below). In this case, each model is produced by stepwise selection from a set of 8 variables chosen from the 17 possible variables. This introduces variability into the variables used in each of the aggregated models, resulting in variability in the logit models and better performance by the aggregation. However, we can see in the diagram that, although the arcing of CHAID trees converges more slowly, it finally provides a slightly better performance than that achieved by aggregation of logistic models. The diagram below shows that, more generally, a weak classifier such as a decision tree gains more from bagging and boosting than a stable classifier, and this gain may be considerable. However, the initial performance of the decision tree is poorer, and the aggregation of models does not take it far above or take it even below the level of logistic regression. In this case, the areas under the ROC curve are approximately 0.680 for the initial decision tree, 0.714 for the initial logit model, 0.728 for random forests of logit models and arcing of CART trees, and 0.734 for the arcing of CHAID trees. From 0.714 to 0.734, the gain is 2.8%: not negligible, but not very large either.

In the credit scoring data set ("German credit data") used in Section 12.7, the areas under the ROC curve for the test sample are approximately 0.695 for the initial decision tree, 0.762 for the initial logit model (see Section 12.8), and 0.768 for random forests of logit models, but only 0.745 for arcing with CART or CHAID trees.

We should note the interesting performance of random forests on logit models, especially since the coefficients resulting aggregated model are the mean coefficients of the elementary logit models, making it as concise, readable and easily understood and used as any logistic model.

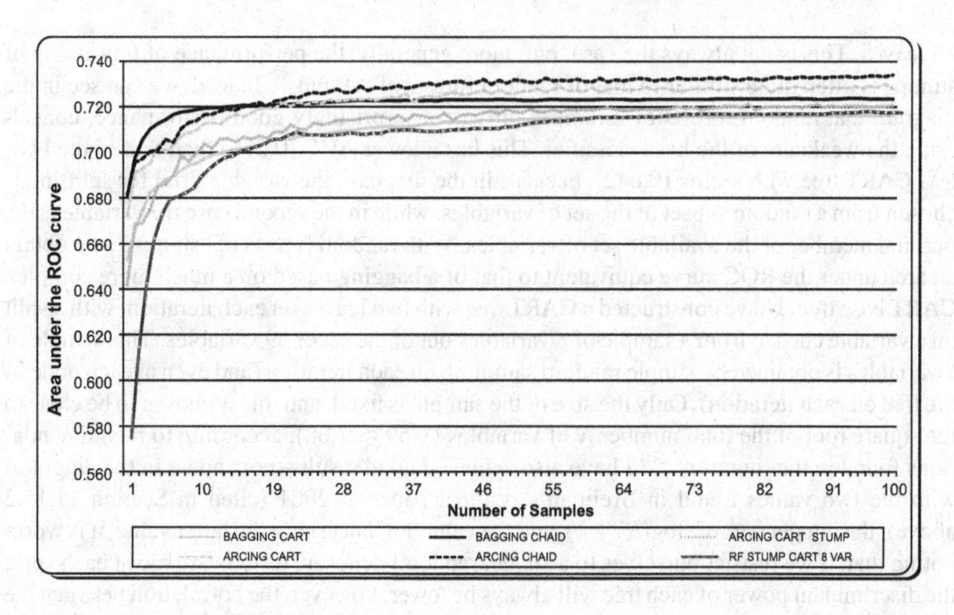

Figure 11.125 AUC of models obtained by bagging, arcing and random forests as a function of the number of bootstrap samples.

Figure 11.125 shows the area under the ROC curve for models created from the same (TIC) insurance data set by bagging and arcing of a base classifier which is a CHAID or a CART tree. In all cases, the trees constructed on each iteration have no more than 15 leaves, each with at least 100 individuals, and the depth of the tree is limited to 3 for CHAID and 6 for CART. The CART trees are binary and the CHAID trees have a maximum of five children at each node. These values were chosen because they led to good results.

The noise level of the data is sufficiently low, and the population is sufficiently large, to avoid the introduction of fluctuations or supplementary noise. This case is therefore favourable to arcing, and we find that the performance with this method is better than with bagging. However, the convergence is slower, and takes place after about fifty iterations for arcing, compared with thirty for bagging with CHAID or forty for bagging with CART. We also find that bagging with CHAID is better than bagging with CART; this was also the case with some other data sets (such as the "German credit data") but not with all. I have not shown them in this diagram, but the performance of arcing without final weighting of the classifiers is worse than that of conventional arcing with weighting, but by only a small amount (4/1000 of the area under the ROC curve). This is also a common finding. Another general phenomenon is the convergence of the performance of arcing towards a level which has little dependence on the base CART or CHAID tree, by contrast with bagging.

We can say that arcing is superior to bagging here, but this may not be the case with less numerous or noisier data. Starting with a classifier (CART or CHAID) for which the area under the ROC curve is about 0.680, we find that arcing gives us an AUC of about 0.730, a gain of 7.3%. The gain provided by bagging is 6.3% for CHAID and 5.4% for CART.

I have also tested arcing with two-leaf CART trees ("stumps") which, as mentioned above, provides very satisfactory results despite the simplicity of the base classifier. In fact, the performance in this case is very similar to that of bagging with CART trees with (a maximum of)

15 leaves. This is not always the case, but, more generally, the performance of this arcing of stumps is often quite similar to that of random forests of "stumps". Indeed, we can see in the diagram that random forests of "stumps" provide a surprisingly good performance, considering the weakness of the base classifier. This has a lower AUC (0.576) than that of the two-leaf CART tree with arcing (0.642), because in the first case the variable used for splitting is chosen from a random subset of the set of variables, while in the second case the variable is the optimal member of the available set of variables. With random forests of "stumps", we obtain an area under the ROC curve equivalent to that of a bagging based on a much more complex CART base tree. I have constructed a CART tree with two leaves on each iteration, with a split on a variable chosen from a sample of 8 variables out of the set of 59 variables. This sample of 8 variables is obtained by simple random sampling on each iteration (and even at each node of the tree on each iteration). Only the size of the sample is fixed, and this is chosen to be close to the square root of the total number *N* of variables ($\sqrt{59} \approx 7.68$), according to recommendations found in the literature.[110] I have also achieved good results (not shown in the diagram) with the two values found in Breiman's original paper of 2001 (cited in Section 11.15.2 above): the integer part of $(\log(N) + 1)$, and the value 1. Concerning the latter value, it is worth noting that, if we restrict ourselves to a single, randomly chosen, tested variable at each split, the discriminant power of each tree will always be lower; however, the correlation between the trees decreases markedly, which is advantageous for the final ensemble model whose variance decreases, as observed by Breiman. Breiman also proposed a method for limiting the number of variables selected at each node without unduly decreasing the performance of each tree: this was to take linear combinations of these variables (the coefficients of these combinations were, of course, random). He carried out conclusive tests with three variables selected each time and with 2 to 8 linear combinations of these three variables. Note, however, that it is important not to use an excessively small number of variables if the proportion of truly discriminant variables is low. This is because there would be a low probability of selecting a discriminant variable in such a case, and the performance would inevitably be poor, despite the weak correlation between the trees.

In our data, random forests with random drawing of a single variable provides truly amazing performance levels (not shown in the diagram), since we start with an AUC of 0.540 and reach 0.680 in about a hundred iterations, i.e. a gain of 26%! With other data, we can achieve even greater gains (although their practical usefulness is limited). In our example, with a choice among eight variables, we only need about twenty iterations to make the performance of random forests of CART trees with two leaves match that of bagging with more complex CART trees. This provides a demonstration of the claims made in Section 11.15.2. For these random forests, I have shown only the results found with a CART tree, because it appears to be more suited to simple binary splitting than the CHAID tree. However, I have also obtained satisfactory results with CHAID.

11.15.5 Conclusion

Whatever method is chosen - bagging, random forests or boosting - these ensemble methods can often make a marked improvement in the quality of predictions, especially in terms of robustness. The other side of the coin is the loss of readability of the decision trees, in certain cases (for decision trees and neural networks), the need to store all the models so that they can be combined and, in general, the large amount of computation time required, which may

[110] See Section 15.3 of Hastie *et al.*, *The Elements of Statistical Learning*, cited in Section 11.3.3 above.

become troublesome when the number of iterations exceeds several hundred. However, it should be noted that the loss of readability does not affect models such as discriminant analysis or logistic regression models, because in these cases the coefficients of the final aggregated model are the mean coefficients of the elementary models. In cases where the aggregation of models improves the discriminating capacity of the basic model, as in certain cases of random forests of logit models (see above), aggregation can therefore be of considerable interest. More generally, the benefits of model aggregation are such that these techniques have been the subject of many theoretical studies, and are beginning to appear in commercial software, while also being available in the R software. The latter software

Table 11.4 Comparison of bagging and boosting.

BAGGING	BOOSTING
Characteristics	
Bagging is a random mechanism	Boosting is an adaptive mechanism and is generally deterministic (except for arcing)
On each iteration, learning takes place on a different bootstrap sample	Generally (except in arcing), learning takes place on the whole initial sample on each iteration
On each iteration, the resulting model must perform well over all the observations	On each iteration, the resulting model must perform well on certain observations; a model performing well on certain outliers will perform less well on other observations
In the final aggregation, all the models have the same weight	In the final aggregation, the models are generally weighted according to their error rate
Advantages and disadvantages	
A method for reducing variance by averaging models	Can reduce the variance and bias of the base classifier
	But the variance can increase with a stable base classifier
Loss of readability if the base classifier is a decision tree	Loss of readability if the base classifier is a decision tree
Ineffective on stumps (unless double randomization is provided as in random forests)	Effective on stumps
Faster convergence	Slower convergence
The algorithm can be parallelized	A sequential algorithm, which cannot be parallelized
No overfitting: better than boosting in the presence of noise	Risk of overfitting, but better than bagging overall on non-noisy data (arcing is less sensitive to noise)
Bagging is effective more often than boosting...	... but when boosting is effective, it is better than bagging

provides the boosting functions adaboost and logitboost in the *boost* package, adaboost.M1 in the *adabag* package, and the *ada* package which implements Discrete AdaBoost, Real AdaBoost, LogitBoost and Gentle AdaBoost. Gradient boosting is available in the *gbm* and *mboost* packages. For CART tree bagging, R offers the bagging function of the *ipred* (improved predictive models) and *adabag* packages, which is used with the rpart function that implements CART. Finally, Breiman's random forests are provided in the randomForest function of the package with the same name.

The main characteristics of bagging and boosting are summarized and compared in Table 11.4.

11.16 Using classification and prediction methods

11.16.1 Choosing the modelling methods

In view of what has already been said in this chapter on modelling methods, we must keep these facts in mind when making a choice:

1. Linear regression deals with continuous variables, discriminant analysis deals with nominal dependent variables and continuous independent variables, DISQUAL discriminant analysis deals with nominal dependent variables and qualitative independent variables, logistic regression deals with qualitative dependent variables (nominal or ordinal) and continuous or qualitative independent variables, neural networks deal with continuous variables on [0,1] (and transform the rest), some decision trees (CHAID) natively handle discrete qualitative variables (and transform the rest), other trees (CART, C5.0) can also handle continuous variables, and MARS deals with binary or continuous dependent variables and all types of independent variable.

2. If we want a *precise* model, we will prefer linear regression, discriminant analysis and logistic regression, possibly MARS, and perhaps SVMs and neural networks, taking care to avoid overfitting (by making sure that there is no more than one hidden layer, and not having too many units in the hidden layer).

3. For *robustness*, we should avoid decision trees and be wary of neural networks, and prefer a robust regression to least squares regression if necessary.

4. For *conciseness* of the model, we should prefer linear regression, discriminant analysis and logistic regression, and to a certain extent MARS and decision trees, provided that the trees do not have too many leaves.

5. For *readability* of the rules, we should prefer decision trees and avoid neural networks and SVMs. Logistic regression, DISQUAL discriminant analysis, linear regression and MARS also provide easily interpreted models.

6. If there are *few data*, avoid decision trees and neural networks.

7. If we have data with *missing values*, we can try using a tree, MARS, PLS regression or logistic regression, coding the missing values as a special class.

8. *Extreme values* (outliers) of continuous variables do not affect decision trees and are not too much of a problem for MARS, logistic regression and DISQUAL if the

continuous variables are divided into classes and the extremes are placed in one or two classes.

9. If the independent variables are very numerous or *highly correlated*, decision trees, PLS regression and regularized regression (ridge, lasso) are appropriate.

10. If we have *large data volumes*, it is best to avoid neural networks, SVMs, and to a lesser extent logistic regression, if we wish to reduce the computation time.

11. Neural networks are more useful when the structure of the data is not clear. When the structure is evident, it is best to make use of this with other types of model.

12. The choice of method may also be guided by the topography of the classes to be discriminated. This is because inductive classification methods divide the independent variable space into regions, each associated with one of the classes to be predicted. Any new individual who falls into one of these regions is classified accordingly. The shape of these regions depends on the method used. For example, if there are linear structures in the data, they will not be detected by decision trees. The behaviour of the regions delimited by some classification methods are shown in Figures 11.126–11.128, together with their effect on the classification of an individual denoted '?'.

13. If we are seeking a method for direct application to the data, without having to prepare them (by normalization, discretization, transformation or selection of variables), to homogenize them (when they are of different types), or to adjust fine parameters – in other words, if we are looking for an 'off the shelf' method – we should consider decision trees, MARS, bagging and boosting.

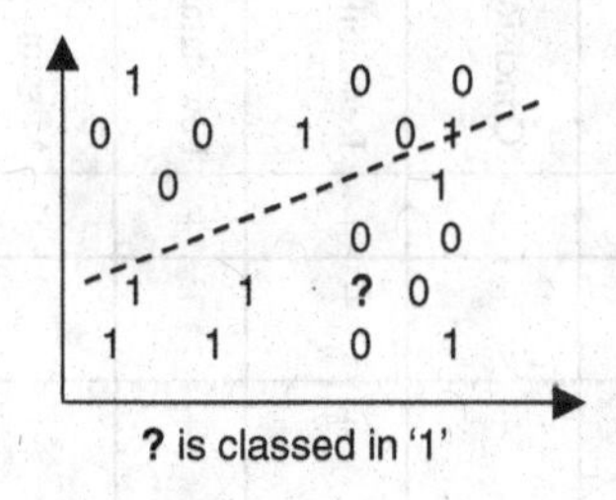

Figure 11.126 Regions in a discriminant analysis.

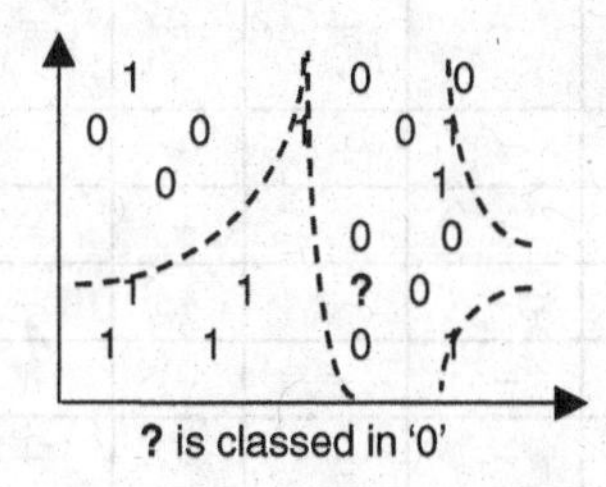

Figure 11.127 Regions in a neural network.

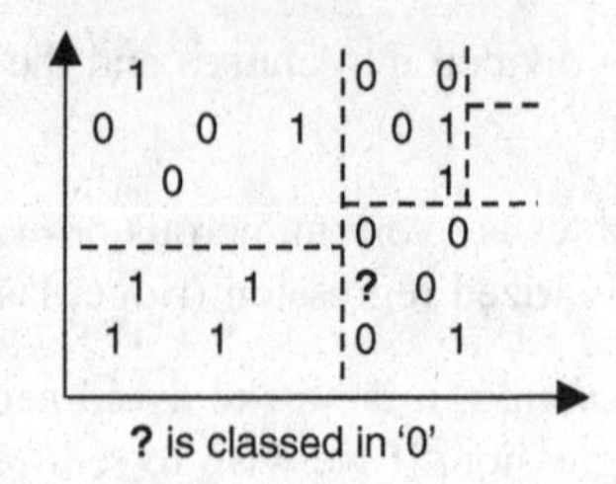

Figure 11.128 Regions in a decision tree.

The table below summarizes the advantages and disadvantages of each classification method according to the main evaluation criteria. The symbol '=' denotes a generally acceptable behaviour of the method or a behaviour which may be good or bad depending on the circumstances; thus, in the context of neural networks, the perceptron is not as badly affected as the RBF network by the presence of correlated variables. The robustness of an SVM depends on its parameter setting. Boosting is generally robust, but not with noisy data.

Table 11.5 Advantages ans disadvantages of classification methods.

	Precision	Robustness	Concision	Readability	Few data	Missing values	Outliers	Correlated variables	Computation speed	Off-the-shelf
Linear regression	+	+	+	+	+	−	−	−	+	−
Regularized regression	+	+	+	+	+	−	−	+	+	−
Linear discriminant analysis	+	+	+	+	+	−	−	−	+	−
DISQUAL analysis	+	+	+	+	+	=	=	+	+	−
Logistic regression	+	+	+	+	+	=	=	−	=	−
Decision trees	=	−	=	+	−	+	+	+	+	+
Neural networks	+	−	−	−	−	−	=	=	−	−
SVM	+	=	−	−	−	−	=	=	−	−
Bagging (of trees)	+	=	−	−	−	+	+	+	−	+
Boosting (of trees)	+	=	−	−	−	+	+	+	−	+
MARS	+	=	+	+	−	+	=	+	=	+

11.16.2 The training phase of a model

The learning phase begins with the construction of the learning sample. This is a sample taken from the population under study, from which the predictive model is built. It must therefore contain enough individuals to provide a statistically reliable support for the discovery of rules and the building of a model (see Section 11.3.2). It must also be sufficiently representative for the model built from this sample to be generalized successfully to the whole population, with predictions that remain reliable.

This does not mean that this sample is necessarily an exact reproduction, on a smaller scale, of the whole population, in other words the result of simple random sampling. On the one hand, this situation cannot always be achieved, because there may be selection biases in propensity studies (see below) just as there are in risk studies (see Section 11.16.3). On the other hand, the learning sample may sometimes be *deliberately biased*, to facilitate learning, especially in a neural network or a decision tree, so that it contains the same proportion of each of the classes to be predicted, even if this is not the case with the whole population. If we wish to classify individuals into two categories, 'yes' and 'no', the sample will contain 50% 'yes' and 50% 'no': the sample is stratified on the dependent variable. The need for such an adjustment of the sample is evident in the case of a CART decision tree (see Section 11.4.7). If we analyse the results of a direct marketing campaign in which the rate of return is 3%, then obviously we cannot construct a CART tree capable of classifying 3% of purchasers and 97% of non-purchasers. Since this tree uses a division criterion based on purity, division may well become impossible from the root of the tree onwards: how can we divide a node which is already 97% pure? The tree will say that there are no purchasers, with a quite acceptable error rate of 3%.

This sample adjustment may also be recommended in the case of a linear discriminant analysis carried out with independent variables which do not have the same variance in the different classes to be discriminated: the heteroscedasticity is considered to be less serious when the frequencies are of the same size. For logistic regression, the question of sample adjustment is debated by statisticians, but does not appear to be essential.

Now let us consider how we define the class of each individual – in other words, how we define the dependent variable. This definition is crucial and decisive for the results. There are three ways of tackling this problem, which I will illustrate with the example of a propensity score for predicting customer response to a marketing campaign for a product which is not new and has been marketed before. If it were a new product, we would have to carry out a preliminary propensity survey on a panel of customers, or use some known results for a similar product, or purchase a ready-made model from a scoring specialist.

The first approach is simplest, most obvious, and least satisfactory. The 'good' (i.e. 'receptive') customers are defined as those who have bought the product (and not returned it to the vendor), and the 'bad' (i.e. 'resistant') customers are those who have not purchased it. The drawback of this approach is that it makes the results entirely dependent on the marketing campaigns and targeting carried out previously for this product. The model is completely dependent on the past, and can only reproduce the targeting rules of earlier campaigns. If customers aged over 50 have been omitted, out of prejudice, then some of these may have bought the product from a more enterprising competitor. However, the model will insist that customers aged over 50 have a low propensity to buy. This may well be false: the over-50s might have been the keenest purchasers of the product, if it had been offered to them.

The second approach is the opposite of the first. It defines 'good' and 'bad' customers only among those who were targeted in a previous marketing campaign. This neutralizes the targeting bias. In our example, the condition 'aged over 50' will no longer be discriminating for separating the 'good' from the 'bad', because all the individuals in the sample will be aged under 50. However, this approach also has some deficiencies. First of all, we have to be able to identify the previously targeted customers, but the IT system may not allow this (if this is the case, the system needs to be adapted for this purpose). Then the previous campaigns must be recent enough for the data on the customers at the time of these campaigns to be still available: we need to know what the customer's equipment was and how he used it at the time of the campaign, not at the present time. The campaigns must have been on a large enough scale to enable us to construct a learning sample: if fewer than 500 products were sold, we cannot do anything. The conditions in which the previous campaigns were carried out must be comparable to those of future campaigns. The results will be distorted if a previous campaign was accompanied by an exceptional promotion, especially if some customers benefited more than others: these customers will obviously distort the model if we cannot detect and isolate them. Finally, the previous campaigns must have covered a substantial proportion of the customer base. In our example, the propensity score for customers over 50 years of age will not be artificially low, but it may not be very reliable, if these customers have never been targeted. Thus there may be risks involved in training a model on previously targeted customers only.

We may consider that the situation relating to propensity is not symmetrical: a non-targeted non-purchasing customer does not necessarily have a low propensity, while a non-targeted purchaser has a high propensity (even higher than a targeted purchaser, we may be tempted to think). It is therefore tempting to keep all the purchasers, even if non-targeted, in the sample of customers with propensity. This also has the advantage of enlarging and diversifying the sample. However, this extension to non-targeted purchasers may introduce a significant bias. Suppose that the targeted customers are mostly home-owners: if we add non-targeted purchasers to the 'with propensity' sample, we will add tenants to the sample and this may give the erroneous impression that tenants have a greater propensity than owner-occupiers.

The third method of constructing the learning sample lies between the first two. We can use it if we are unwilling or unable to use the history of earlier targeted campaigns, while not resorting to the simplistic 'good' and 'bad' definitions. In this approach, we define the 'bad' customers as the ones, out of those who have not purchased the product, who meet another criterion. This criterion is not the fact of having been targeted and having had the product offered to them, but a criterion determined by common sense (constraints due to legislation, age, etc.), by marketing intuition, or by segmenting the population and finding the dominant criteria in the segments (not too small) which only contain customers who do not have the product. We can assume that the customers who meet this criterion will be resistant to the product. Although this approach avoids the drawbacks of the first two, it suffers from other disadvantages. In the first place, if we fix a 'criterion' for 'bad' in a more or less arbitrary way, we cannot use the variables relating to this criterion in the predictive model, as this would evidently lead to overfitting. There is also a risk that the criterion of 'bad' will have a small membership, in other words only the customers meeting this criterion will resemble them. In the phase of application to the whole population, the model will be unable to classify the customers without propensity, if they do not resemble the 'bad' customers artificially defined in the learning phase.

So which of these methods should we choose? If we have sufficiently complete, comprehensive and representative records of marketing campaigns similar to the one for which propensity is to be modelled, the second method is preferable. Otherwise, the third is better, but we must recognize that many users simply opt for the first method.

11.16.3 Reject inference

The question that arises in the propensity score example in the previous section is that of the learning of a model based on a non-random sample. This also arises in the medical and insurance fields. It is posed here because marketing policies and targeting are such that some customers always receive fewer offers than the rest. Therefore they will take out fewer contracts and bias the propensity models. This problem of selection bias arises in a more radical way in the development of risk models, especially in credit scoring, because the customer may be refused credit, which is much worse than not having it offered.

The process of taking rejected files into account in credit scoring (*reject inference*) poses a problem because the lending organization does not allow them to exist and reveal themselves as 'good' or 'bad', so that the dependent variable is not known. Now, these customers have not been rejected by chance. Consequently, it is incorrect to apply a model built on a population of accepted applicants to this population. Furthermore, the files classed as 'no action' because of rejection by customers, who may for example have found better offers among the competition, must be added to those rejected by the organization: this is another source of bias. At the present time, despite numerous attempts,[111] no fully satisfactory statistical solution has been found for the problem of reject inference. This is evidently more crucial where the proportion of rejections is higher: it depends on the type of credit, the type of borrower and the type of lender. For example, it is a greater problem for personal credit to individuals from specialist establishments than for property lending by retail banks.

Various different approaches can be tried. The first three of these are rather elementary and the next two are statistical.

First method: ignore the existence of the 'rejects' and simply model those 'accepted'. This is a fairly commonplace approach in the field of risk, but less so in propensity studies (it is the second approach described in the previous section).

Second method: treat all the rejected files as 'bad'. This is a similar stance to that of considering every non-purchaser as 'bad' in terms of propensity, regardless of whether or not he has received a marketing offer (this is the first approach described in the previous section).

Third method: assume that the 'rejects' have been rejected randomly and assign them at random to the 'good' or 'bad' category while keeping the same proportion of these two categories as in the 'accepted' population.

To explain the *fourth method* ('augmentation', Hsia, 1978),[112] one of the most widely used, we must break down the probability $P(b|x)$ of a file having a profile x being 'good', by writing it as follows:

$$P(b|x) = P(b|x,a)P(a|x) + P(b|x,r)P(r|x),$$

[111] See, for example, Crook, J. and Banasik, J. (2002) Does reject inference really improve the performance of application scoring models? Working Paper 02/3, Credit Research Centre, University of Edinburgh.

[112] Hsia, D. C. (1978) Credit scoring and the Equal Credit Opportunity Act. *Hastings Law Journal*, 30, 371–448.

where $P(r|x)$ ($P(a|x)$) is the probability that a file with features x will be rejected (accepted). In this method, we assume that the unknown probability $P(b|x,r)$ conforms to the equality

$$P(b|x,r) = P(b|x,a).$$

This equality means that the risk of a file depends only on its own characteristics and not on whether it has been accepted or rejected. We could therefore carry out the score training on the accepted files to which the rejected files would be added, after classifying each rejected individual as 'good' or 'bad' according to the accepted files which resemble him, using for example the *k*-nearest neighbours (see Section 11.2) to bring him towards the nearest neighbouring profile. We can see that an isolated accepted file among rejected files will have a large weight in the inferred model, because all these rejected files will be given the value of the dependent variable of the accepted file. More precisely, each accepted file is entered into the model with a weight inversely proportional to the probability $P(a|x)$ that it had of being accepted. The augmentation method is therefore normally applied in two stages: first we determine an acceptance model to calculate the probability $P(a|x)$ of each file, and then we weight each accepted file by $1/P(a|x)$ and build the final model on the accepted files which have been weighted in this way. This weighting enables us to 'compensate' to some degree for the absence of the rejected individuals.

Fifth method ('iterative reclassification', Joanes, 1994):[113] the score training initially takes place on 'good' and 'bad' accepted files. The resulting score is then applied to the 'rejects' to classify them as 'good' or 'bad'. A score is then recalculated by adding the rejected files predicted as 'good' ('bad') by the preceding model to the 'good' ('bad') accepted files. This new score is then applied to the 'rejects' to classify them as 'good' or 'bad', and the reiteration continues until the resulting scores become stable.

Sixth method: define *a priori* criteria based on the available data, such that we can assert that a rejected file is good or bad. For example, we can say that not all the non-targeted non-purchasing customers have a low propensity, but only those who meet certain supplementary conditions (e.g. age, income, etc.)

Seventh method (if allowed by the subject and the law): complete the knowledge of the 'rejects' using external data (e.g. credit bureaus), taking care to ensure that the internal and external definitions of the dependent variable match.

Eighth method (the most reliable): accept some of the files which should have been rejected. Very few establishments will risk this, but the resulting losses might be less than the benefit gained from a more reliable score.

11.16.4 The test phase of a model

However well constructed the training sample may be, it obviously cannot tell us about the model's capacity to be generalized to the whole population. One cannot be both referee and player; the same sample cannot be used for both the development and the validation of a model, because the result of using a sample to validate a model built on the same sample is always too optimistic. We must therefore have a separate sample for testing each model and selecting the best of the models built during the training phase.

[113] Joanes, D.N. (1993–4) Reject inference applied to logistic regression for credit scoring. *IMA Journal of Mathematics Applied in Business and Industry*, 5, 35–43.

This sample is also taken by drawing at random, which is not necessarily simple but may require stratification on the dependent variable. The test sample is generally half the size of the training sample. If it is really impossible to take a test sample, because of an insufficient number of individuals in each class (less than 500), we can perform a cross-validation (see Section 11.3.2). It is sometimes considered that cross-validation is sufficient to decide between a number of models belonging to the same family, whereas a choice between models of different types must be made with a test sample.

The following indicators are clearly a better measure of the quality of the model when they are calculated on a test sample than when they are calculated on the training sample.

For a prediction model, the quality is evaluated by the statistical indicators described previously in the context of linear regression (Section 11.7), particularly the coefficient of determination R^2 and the mean square error RMSE.

For a classification model, the performance can be stated in a *confusion matrix* of the following form:

		Observed		
Predicted		Yes	No	Total
	Yes	250	150	400
	No	50	550	600
	Total	300	700	1000

This enables us to measure the *error rate,* or incorrect classification rate, which in our example is $(50 + 150)/1000 = 20\%$. In this case, a random classification would result in the following table:

		Observed		
Predicted		Yes	No	Total
	Yes	90	210	300
	No	210	490	700
	Total	300	700	1000

with a correct classification rate of $(90 + 490)/1000 = 58\%$, as compared with the 80% of the example. Rather than perform a random classification, it is even better to assign all the individuals to the largest class, in other words the one with 700 individuals, which yields a correct classification rate of 70%.

In this example, if we send a mailing to the predicted purchasers only, we send only 400, instead of 1000, and we only lose 50 purchasers out of 300. We can thus make considerable savings in our mailshots. Some of the 150 non-purchasers who are predicted to be purchasers are score errors, and some are prospects who should be followed up because of their 'purchaser' profile. In this example, we find that there are far fewer purchasers predicted as non-purchasers (50 out of 600) than non-purchasers predicted as purchasers (150 out of 400). Such an asymmetry is not unusual. In the present case, the score is rather optimistic, and this is preferable in a marketing development context where it is better to send a mailing with no result than to lose business. In the field of risk (financial, and especially medical), it is generally preferable to be over-pessimistic, and to class an individual as 'at risk' even where he is not, rather than to fail to detect an individual who really is at risk.

To repeat: the confusion matrix calculated on the test data supplies an error rate that is a better measure of the quality of the model than the error rate in training. As mentioned in Section 11.3.4, the error rate in training decreases constantly with the complexity of the model, while the error rate in testing eventually increases if the complexity of the model increases. The increase in the complexity of the model must be halted at this point. What is important is to have the lowest error rate in testing, not in training.

Thus the test phase enables us to select the best model out of all those considered during the training phase. But this is not all: In completely explicit models such as decision trees, this phase can be used not only to adjust the complexity, in other words the depth of the tree, to the correct level globally, but also to check the validity of each rule one by one, ensuring that the prediction made in training (for each leaf of the tree) is confirmed during testing.

To evaluate the performance of score models, we shall now look at some indicators that are more elaborate and useful than the error rate: these are the ROC curve, the lift curve, and the measurements of area associated with them.

11.16.5 The ROC curve, the lift curve and the Gini index

Faced with the multiplicity of modelling methods, each with its own statistical quality indicators (such as the Wilks lambda for discriminant analysis and the log-likelihood for logistic regression), statisticians have attempted to find universal criteria for the performance of a model. The best-developed and most widely used criteria are described below. They are applied to models for classification into two classes. For three or more classes, there appears to be no simple generalization of the curves described here.

ROC curve

The discriminating power of a score model can be visualized with a curve called the receiver operating characteristic (ROC) curve, a term that originated in signal processing. It represents (on the Y axis) the proportion of events (such as the appearance of a risk) detected as such because their score is greater than s, as a function (on the X axis) of the proportion of false events, in other words non-events detected as events because their score is greater than s (e.g. 'not at risk' detected as 'at risk'), when the score separation threshold s is varied. More specifically, two functions of s are defined:

- the *sensitivity* $\alpha(s)$, the probability of correctly detecting an event at the threshold $s = \text{Prob}(\text{score}(x) \geq s \mid x = \text{event})$,
- the *specificity* $\beta(s)$, the probability of correctly detecting a non-event at the threshold $s = \text{Prob}(\text{score}(x) < s \mid x = \text{non-event})$,

and we can say that the proportion of false events among the non-events is $1 - \beta(s) = \text{Prob}(\text{score}(x) \geq s \mid x = \text{non-event})$. Therefore the ROC shows $\alpha(s)$ as a function of $1 - \beta(s)$,

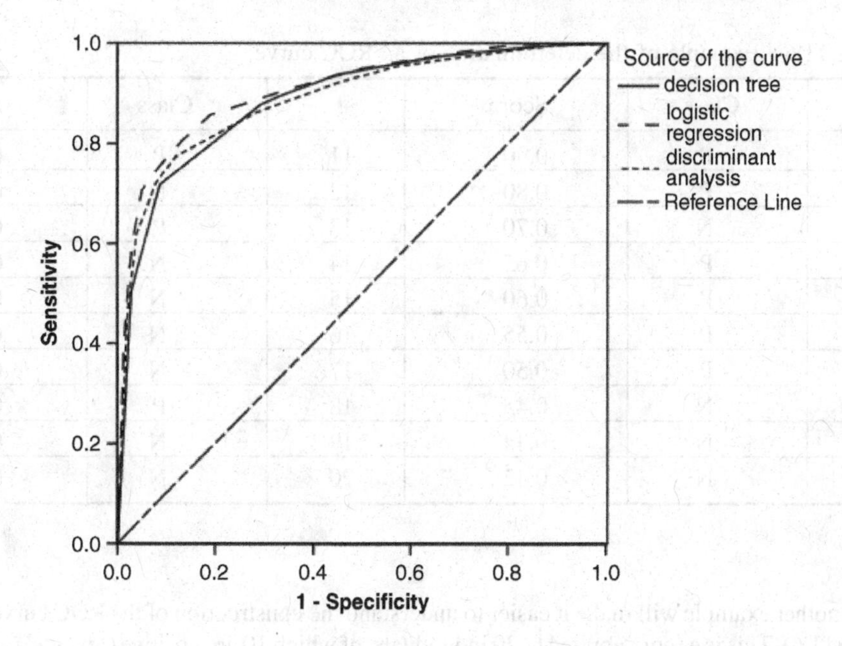

Figure 11.129 ROC curve.

for values of s ranging from the maximum (where all individuals are considered to be non-events, and thus $\alpha(s) = 1 - \beta(s) = 0$) to the minimum (where all individuals are considered to be events, and thus $\alpha(s) = 1 - \beta(s) = 1$).

If this curve coincides with the diagonal, the model performs no better than a random model, as is the case with the density curves shown in Figure 3.14, when the variable Z is replaced with the score. The more closely the ROC approaches the top left-hand corner of the square in Figure 11.129, the better the model is, because it captures the largest possible number of true events with the fewest possible false events. The ROC curve of a perfect model is composed of the two segments linking the points (0, 0) to (0, 1) and (0, 1) to (1, 1). This case corresponds to completely separate density curves in Figure 3.12. The model is improved by having large values on the vertical axis associated with small values on the horizontal axis, in other words a large area between the ROC curve and the horizontal axis. The convexity of this curve is equivalent to the property of the increase of the probability of the event conditionally on the score, as a function of this score function. The ROC curve is invariant for any increasing monotonic transformation of the score function, an interesting property when we realize that the score function is frequently normalized, for example in order to bring its values into the range from 0 to 100.

For example, the ROC of a logistic regression in Figure 11.129 passes through the point $x = 0.3$ and $y = 0.9$. This point corresponds to a threshold s which is such that, if we consider all individuals having a score greater than s as 'events', then 30% of the non-events have been detected incorrectly (30% of the non-events have a score greater than s) and 90% of the true events have been correctly detected (90% of the events have a score greater than s). Warning: 0.3 does not correspond to 30% of the total population, and the ROC curve must not be confused with the lift curve defined below.

Table 11.6 Example of the determination of an ROC curve.

#	Class	Score	#	Class	Score
1	P	0.90	11	P	0.40
2	P	0.80	12	N	0.39
3	N	0.70	13	P	0.38
4	P	0.65	14	N	0.37
5	P	0.60	15	N	0.35
6	P	0.55	16	N	0.30
7	P	0.50	17	N	0.25
8	N	0.45	18	P	0.20
9	N	0.44	19	N	0.15
10	N	0.42	20	N	0.10

Another example will make it easier to understand the construction of the ROC curve (see Table 11.6). This is a score applied to 20 individuals, of which 10 are positive ('events') and 10 are negative ('non-events'). The score must be higher for the positives. If we set a threshold at 0.9, such that every individual with a score of at least 0.9 is considered positive, we have 10% of true positives (1 in 10) and 0% of false positives (see Figure 11.130). Similarly, if we set a threshold at 0.8, such that every individual with a score of at least 0.8 is considered positive, we have 20% of true positives and 0% of false positives. Now, if we set a threshold at 0.7, such that every individual with a score of at least 0.7 is considered positive, we still have 20% of true positives but also 10% of false positives, because the third individual is negative and has a score of 0.7. And so on.

Each point on the ROC curve corresponds to the confusion matrix defined by a certain threshold value. To plot the ROC curve, therefore, we must examine the set of confusion matrices defined for the set of possible values of the threshold.

The link between the confusion matrix and the ROC curve is illustrated in Figure 11.131 with the example of heart disease discussed in Section 10.8.1 on logistic regression. At the threshold of 0.5, the sensitivity of the model is 27/43 and its specificity is 45/57.

Classification table[a]

		Predicted		
		CHD		
Observed		0	1	Percentage correct
CHD	0	45	12	78.9
	1	16	27	62.8
Global percentage				72.0

[a] The cut-off value is .500

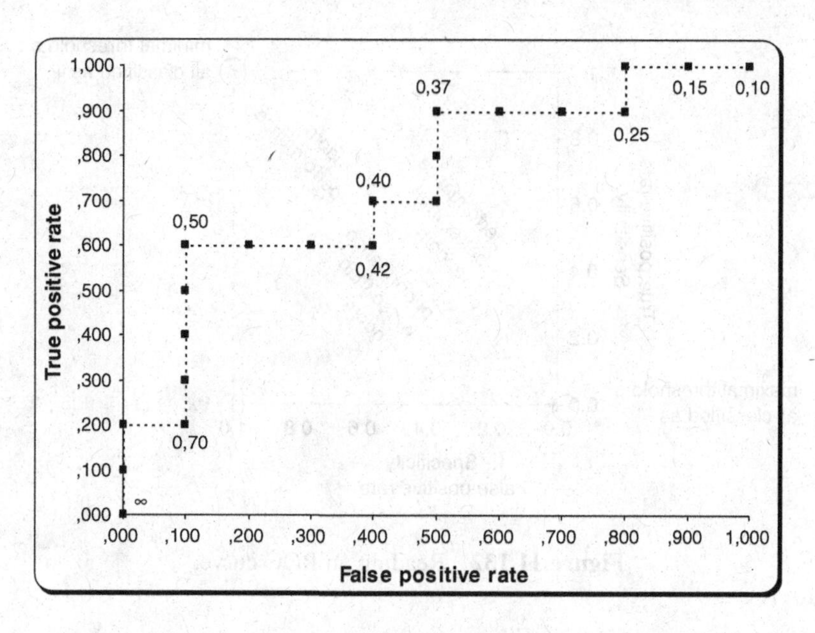

Figure 11.130 Example of the determination of an ROC curve.

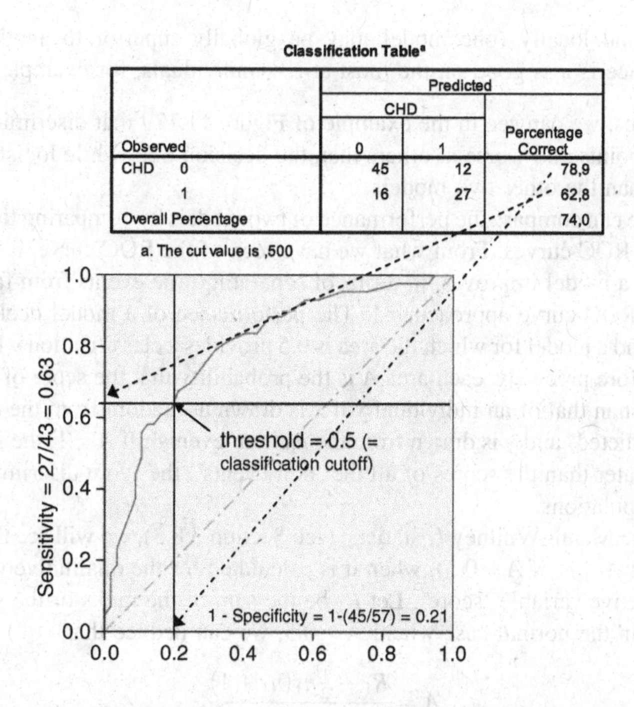

Classification Table[a]

Observed		Predicted CHD 0	Predicted CHD 1	Percentage Correct
CHD	0	45	12	78,9
	1	16	27	62,8
Overall Percentage				74,0

a. The cut value is ,500

Figure 11.131 The link between the confusion matrix and the ROC curve.

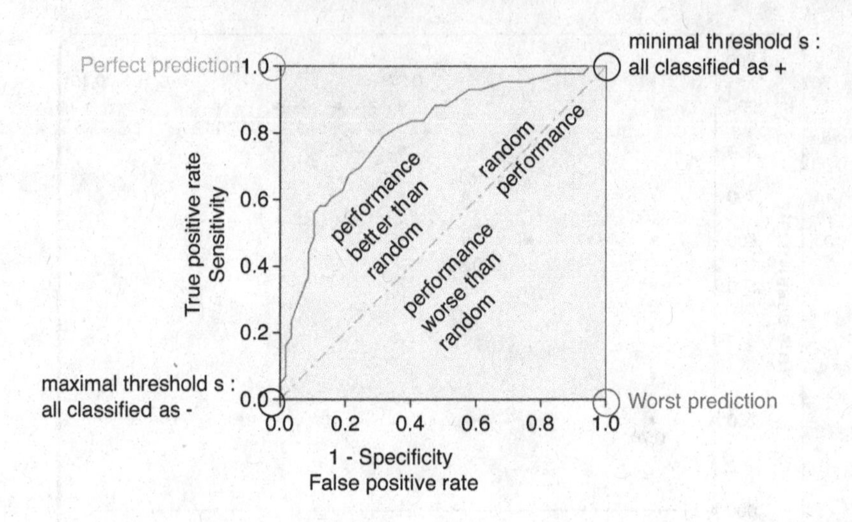

Figure 11.132 Reading an ROC curve.

Figure 11.132 shows how to read the ROC curve of a model. The ROC curve enables us to compare models:

- of different types (even when their performance indicators are not directly comparable);
- globally and locally (one model may be globally superior to another, even if its performance is less good on the most at-risk individuals, for example).

In local terms, we can see in the example of Figure 11.129 that discriminant analysis is better at some points and worse at others than the decision tree, while logistic regression is always better than the other two models.

Globally, we can compare the performance of two models by comparing the areas A under their respective ROC curves. From what we have seen of the ROC curve, it is clear that the performance of a model improves, in terms of separating true events from false ones, as its area under the ROC curve approaches 1. The performance of a model declines as its area approaches 0, and a model for which the area is 0.5 provides a classification which is no better than random. More precisely, each area A is the probability that the score of an individual x will be greater than that of an individual y if x is drawn at random from the n_1 'events' (the group to be predicted) and y is drawn from the n_0 'non-events'. If $A = 1$, the scores of all the 'events' are greater than the scores of all the 'non-events': the score discriminates perfectly between the populations.

Recalling the Mann–Whitney U statistic (see Section 3.8.3), we will see that this is equal to $(1–A)n_1n_0$ (or An_1n_0, if $A < 0.5$), when it is calculated for the qualitative variable 'target' and the quantitative variable 'score'. Let R_1 be the sum of the ranks of the scores of the n_1 'events'; then, in the normal case where $A \geq 0.5$, we can deduce that

$$A = \frac{R_1 - \frac{1}{2}n_1(n_1+1)}{n_1 n_0}.$$

In fact, if $A \geq 0.5$, the score of an individual in the group to be predicted is more likely to be greater than the score of an individual in the other group than vice versa. Therefore, $U_1 < U_0$ (in the notation of Section 3.8.3) and $U = \min(U_0, U_1) = U_1$. Now,

$$U_1 = n_1 n_0 + \frac{n_1(n_1+1)}{2} - R_1,$$

and therefore

$$\frac{U}{n_1 n_0} = 1 + \frac{1}{n_1 n_0}\left[\frac{n_1(n_1+1)}{2} - R_1\right]$$

and

$$A = 1 - \frac{U}{n_1 n_0}$$

takes the value mentioned above.

SAS provides a simple method of programming this calculation of the area under the ROC curve. The syntax below creates a macro with the name AUC, for which the code starts with '%macro AUC' and ends with '%mend AUC'. Its parameters are the scored data set DATA, the dependent variable TARGET and the score SCORE. As with every SAS macro, these parameters are preceded by the symbol '&' in the code. When it reads this symbol, the SAS macro compiler knows that it has to replace the parameter with the value specified when the macro was called. This call is made, as many times as required, by the instruction containing the name of the macro (preceded by %) and the values of the parameters, which can vary with each call of the macro:

```
%AUC(model,survived,proba);
```

When it finds this instruction, SAS calls the macro and replaces '&data' with 'model', '&target' with 'survived' and '&score' with 'proba' in the macro code. Thus the macro is transformed into a program which can be analysed and executed by the ordinary SAS compiler, instead of the macro language compiler.

In this macro, the NPAR1WAY procedure calculates the rank sums R_0 and R_1, and the following DATA step calculates the Mann–Whitney statistic according to the formula shown in Section 3.8.3 and reproduced above, before using this to deduce the area under the AUC, displayed by the PRINT procedure. As in Section 3.8.4, we can use the SAS ODS instruction, which, when placed before the NPAR1WAY procedure, causes the results (the rank sums) to be sent to the specified data set (wilcoxon).

```
%macro AUC(data,target,score);
ODS OUTPUT WilcoxonScores = wilcoxon;
PROC NPAR1WAY WILCOXON DATA=&data CORRECT=no;
CLASS &target;
VAR &score;
RUN;

DATA auc;
SET wilcoxon;
n0 = N; R0 = SumOfScores ;
n1 = lag(N); R1 = lag(SumOfScores) ;
```

```
U1 = (n1*n0) + (n1*(n1+1)/2) - R1;
U0 = (n1*n0) + (n0*(n0+1)/2) - R0;
U = min(U1,U0);
AUC = 1- (U/(n1*n0));
RUN;

PROC PRINT DATA=auc (KEEP = AUC) NOOBS;
TITLE "Area under the ROC curve";
WHERE AUC > .;
RUN;

%mend AUC;
```

To make the syntax clearer, let us look at the content of the AUC data set:

Obs	Class	N	SumOfScores	n0	R0	n1	R1	U1	U0	U	AUC
1	1	711	1038858.0	711	1038858	.	.	.	.	.	.
2	0	1490	1384443.0	1490	1384443	711	1038858	273648	785742	273648	0.74169

We can calculate the confidence interval of A, and a test can be used to reject the null hypothesis that $A = 0.5$. In the preceding example we see that the three models perform significantly better than a random classification (Figure 11.133), and that discriminant analysis is globally slightly better than the decision tree, which was not easily visible in Figure 11.129.

Lift curve

A variant of the ROC curve is the *lift curve*, which has long been used in marketing. It is also used in econometrics, under the name of the Lorenz curve or power curve. It represents the proportion $\alpha(s)$ of events detected (the sensitivity) as a function of the proportion of the individuals selected (those with a score greater than s). We can see that the lift curve has the same ordinate $\alpha(s)$ as the ROC curve, but a different abscissa. Since

Test Result Variable(s)	Area	Std. error[a]	Asymptotic sig.[b]	Asymptotic 95% Confidence Interval	
				Lower bound	Upper bound
discriminant analysis	.889	.008	.000	.873	.904
logistic regression	.906	.007	.000	.892	.921
decision trees	.887	.008	.000	.872	.902

a. Under the nonparametric hypothesis

b. Null hypothesis: true area = 0.5

Figure 11.133 Area under the ROC curve.

$\alpha(s) > 1 - \beta(s)$, this abscissa will be greater than the abscissa $1 - \beta(s)$ of the ROC curve, because it is the proportion of all the individuals having a score greater than s, instead of the proportion of all the 'non-event' individuals having a score greater than s. Since the lift curve generally has a larger abscissa than the ROC curve for a given ordinate, the lift curve lies below the ROC curve.

This abscissa, the proportion of individuals having a score greater than s, is composed of the sum of two terms:

- the proportion of 'event' individuals with a score greater than s, given by

$$\text{Prob}(x = \text{event}) \times \text{Prob}(\text{score}(x) > s/x = \text{event}) = p\alpha(s);$$

- the proportion of 'non-event' individuals with a score of greater than s, given by

$$\text{Prob}(x = \text{non-event}) \times \text{Prob}(\text{score}(x) > s/x = \text{non-event}) = (1-p)(1-\beta(s)).$$

Here p denotes the *a priori* probability of the event in the whole population (corresponding, for example, to the proportion of purchasers or at-risk customers), $\alpha(s)$ is the sensitivity at the threshold s and $\beta(s)$ is the specificity as defined above.

We therefore deduce that the area under the lift curve is:

$$\begin{aligned}\text{AUL} &= \int \text{ordinate d}\{\text{abscissa}\}\\ &= \int \alpha \text{d}\{p \cdot \alpha + (1-p)(1-\beta\} = p\{\int \alpha \text{d}\alpha\} + (1-p)\{\int \alpha \text{d}(1-\beta)\}\\ &= p/2 + (1-p)(\text{area under the ROC curve}).\end{aligned}$$

We can conclude again that the area under the ROC curve (AUC) is always greater than the area under the lift curve (AUL), provided that AUC > 0.5, in other words that the ROC curve is above the diagonal. We find that:

$$\text{AUC} - \text{AUL} = \text{AUC} - p/2 - (1-p)\text{AUC} = p \cdot \text{AUC} - p/2 = p(\text{AUC} - 0.5).$$

This formula also shows that:

- if AUC = 1 (score providing perfect separation), then AUL = $p/2 + (1 - p) = 1 - p/2$ (as shown by the plot of the lift curve in Figure 11.134);
- if AUC = 0.5 (random prediction), then AUL = $p/2 + 1/2 - p/2 = 0.5$;
- if the probability p is very small, the areas under the two curves are very similar;
- in all cases, measuring the area under the lift curve is equivalent to measuring the area under the ROC curve for the purposes of deciding whether one model is better than another.

So why should we prefer the area under the ROC as the universal measure of performance? There are three reasons for this:

- the area under the ROC curve can be simply interpreted as a probability;
- it can be deduced directly from the Mann–Whitney statistic and is suitable for non-parametric statistical testing;

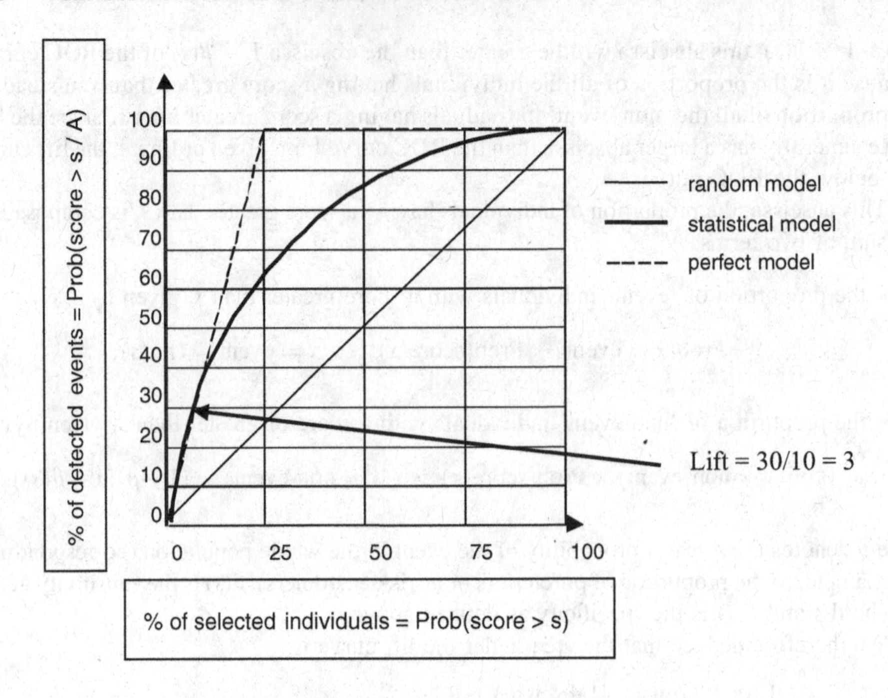

Figure 11.134 Lift curve.

- as a performance indicator of a model, it has an absolute significance, which, unlike the area under the lift curve, does not depend on the *a priori* probability of the event (even the shape of the lift curve depends on this probability).

The lift curve is plotted by classifying the individuals in decreasing order of score, grouping them into percentiles for example, determining the percentage of events in each percentile, then plotting the cumulative curve of these percentages, so that a point with coordinates (n,m) on the curve signifies that the n% of individuals having the highest score contain m% of the events. The lift curve of a random model is the straight line with the equation $y = x$, because n% of the events are attained by targeting n% of the individuals. Conversely, a perfect model would be one in which all the events were selected first of all. The corresponding lift curve is shown in Figure 11.134: this is a straight line touching the ordinate of 100% at an abscissa equal to the global percentage p of events, extended by a horizontal line with the equation $y = 100$. The area under this ideal lift curve is $1 - p/2$. The model improves as its lift curve approaches the ideal curve.

The example in this figure shows that we reach 30% of the events (purchasers, for example) by targeting 10% of the individuals. This means that we have been able to find a threshold s such that 10% of the individuals have a score greater than s and 30% of the events have a score greater than s.

By definition, the lift at n% is the ratio m/n such that (n,m) is on the lift curve; in other words, such that n% of the individuals contain m% of the events. Thus, in the example above, we have a lift of 3 at 10%. This corresponds to a response rate multiplied by 3 with respect to a random model.

Note that some authors call the lift curve the 'response curve', and reserve the term 'lift curve' for the curve which has the same abscissa and has an ordinate which is no longer the percentage of purchasers but is directly the lift. In our example, the abscissa 10 corresponds to the ordinate 3 (a lift of 30/10 at 10%) and the abscissa 50 corresponds to the ordinate 1.7 (a lift of 85/50 at 50%). This is a decreasing curve.

In the field of propensity to apply for consumer credit, it is sometimes considered that a good score model will find 60% of purchasers with 30% of the customers, and that a very good score model will find 90% of purchasers with 30% of the customers (giving a lift at 30% of 2 to 3). In a study of *churn* (in mobile telephone systems), a good score model will find 35–50% of *churners* among 10% of the customers (lift at 10% between 3.5 and 5).

The Gini index

We have seen that the area under the lift curve has no absolute significance, because it depends on the probability p of the event. However, we know that this curve must tend towards the ideal lift curve. This suggest another measure of the performance of a predictive model: the ratio

$$\frac{\text{surface between the real lift curve and the diagonal}}{\text{surface between the ideal lift curve and the diagonal}}$$

which must obviously be as close as possible to 1. Given that the numerator is AUL $-$ 0.5 $= p/2 + (1 - p)\text{AUC} - 1/2$ and that the denominator is $1 - p/2 - 1/2 = (1 - p)/2$, we can easily deduce that this ratio is equal to

$$2.\text{AUC} - 1$$

and that it therefore has the useful property, like the AUC, of being independent of the probability p. It is called the Gini index or coefficient (or 'accuracy ratio'), and is identical to Somers' D statistic. This index is twice the area between the ROC curve and the diagonal.

11.16.6 The classification table of a model

The classification table of a model is the series of confusion matrices obtained for a whole set of possible thresholds of the score function. The threshold is the value of the score function above which the event is considered to be predicted, and below which the non-event is considered to be predicted.

For a logistic model, the score function is given by the *a posteriori* probability at the output of the model. The SAS LOGISTIC procedure has a useful option called CTABLE (see Section 11.8.12), which produces a classification table in which the series of confusion matrices is shown in rows, with one row for each possible threshold. The PPROB option following CTABLE sets the interval of the threshold.

The example used to illustrate the operation of CTABLE is the coronary heart disease (CHD) study of Hosmer and Lemeshow which we have examined previously. The event to be predicted is the onset of CHD, corresponding to category 1 of the dependent variable. Setting a threshold at 0.5 means that we consider any individual whose probability exceeds 0.5 to be ill. The coefficients of the confusion matrix at the 0.5 threshold can be read on the row associated with the probability level of 0.5 (in bold type in Figure 11.135). We find 27 and 45 in the

Classification table										
Prob. level	Correct		Incorrect		Percentage					
	Event	Non-event	Event	Non-event	Correct	Sensi-tivity	Speci-ficity	False POS	False NEG
0.000	43	0	57	0	43.0	100.0	0.0	57.0	.
0.100	42	6	51	1	48.0	97.7	10.5	54.8	14.3
0.200	39	24	33	4	63.0	90.7	42.1	45.8	14.3
0.300	36	32	25	7	68.0	83.7	56.1	41.0	17.9
0.400	32	41	16	11	73.0	74.4	71.9	33.3	21.2
0.500	**27**	**45**	**12**	**16**	**72.0**	**62.8**	**78.9**	**30.8**	**26.2**
0.600	25	50	7	18	75.0	58.1	87.7	21.9	26.5
0.700	19	51	6	24	70.0	44.2	89.5	24.0	32.0
0.800	7	55	2	36	62.0	16.3	96.5	22.2	39.6
0.900	1	57	0	42	58.0	2.3	100.0	0.0	42.4
1.000	0	57	0	43	57.0	0.0	100.0	.	43.0

Figure 11.135 Classification table.

'Correct' columns: these are the numbers of correctly predicted individuals, which are on the diagonal of the confusion matrix. The 12 individuals who are not ill but are predicted to be ill are in the 'Incorrect' column (because they are incorrectly predicted) and in the 'Event' column (because the illness, which is the event, has been predicted). The same applies to the 16 individuals who are ill but were predicted not to be. The confusion matrix at the 0.5 threshold is therefore

Predicted → Observed ↓	0	1	Total
0	45	12	57
1	16	27	43
total	61	39	100

The meaning of the classification table columns under the 'Percentage' heading is as follows:

- The 'Correct' column contains the percentage of correct predictions, in this case (45 + 27)/100 = 72%.
- The 'Sensitivity' column (defined in the previous section) contains the percentage of events predicted as such; in this case it is 27 (predicted events) divided by 43 (observed events), i.e. 62.8%.
- The 'Specificity' column (see the previous section) contains the percentage of non-events predicted as such; in this case it is 45 (predicted non-events) divided by 57 (observed non-events), i.e. 78.9%.
- Symmetrically, the 'False POS' column contains the proportion of incorrectly predicted events (wrongly predicted as ill) within the set of predicted events. In this case, 12

individuals were wrongly predicted to be ill, out of 39 individuals predicted to be ill. Hence the percentage is 12/39 = 30.8%.

- Finally, the 'False NEG' column contains the proportion of incorrectly predicted non-events (wrongly predicted as healthy) within the set of predicted non-events. In this case, 16 individuals were wrongly predicted to be healthy, out of 61 individuals predicted to be healthy. Hence the percentage is 16/61 = 26.2%.

This classification table can be useful for setting a score threshold, which can be chosen so as to minimize the global error rate, or the proportion of false positives or false negatives, depending on the aim of the study.

11.16.7 The validation phase of a model

Because each of the tested models is optimized on a test sample, it is useful, for the comparison of different optimized models, to have a third sample to enable us to decide between the models with the lowest possible bias. This phase enables us to confirm the test results and to predict the results that will be obtained for the population as a whole. However, it can be omitted if the total frequency in each class is not large enough to allow three samples to be taken. If some observations (e.g. outliers) have been excluded from the training sample, they will be reincorporated in the validation sample to check that the behaviour of the model is consistent in all cases. If possible, it is advisable to validate the models on an 'out-of-time' sample, in other words a picture of the modelled population taken on a different date from the training and test samples.

11.16.8 The application phase of a model

Unlike the previous three phases, which are only executed during the building or revision of a model, the application of a finished model to the population takes place:

- either automatically, on the scale of the whole population, whenever new values of the independent variables enable us to predict, using the model, a new estimate of the dependent variable (often once per month or once per quarter);

- or manually, for a given individual, when his file is under examination (e.g. a customer whose risk score is calculated before he is sold a piece of furniture on credit).

12

An application of data mining: scoring

To provide a specific illustration of the contribution made by the application of data mining methods in the business world, this chapter will describe the use of an important branch of data mining in the finance industry – the calculation of scores. It is based on the business application of the classification techniques examined in the last chapter. What is it that banks and financial institutions wish to predict? This is the first thing to be considered. Because of its many years of use and universal application, *scoring* can be seen as the archetypal business application of data mining. Our examination of this topic will also add depth to the knowledge imparted in Chapter 1, on the use of data mining, and in Chapter 2, on ways of implementing a data mining project. The end of this chapter will be a practical example of *credit scoring* using two of the main classification methods described in Chapter 11, namely logistic regression and the DISQUAL version of linear discriminant analysis.

12.1 The different types of score

The main types of score used in banking are:

- propensity scores;
- risk (behaviour) scores;
- application scores;
- recovery scores;
- attrition scores.

These are defined as follows.

Data Mining and Statistics for Decision Making, First Edition. Stéphane Tufféry.

A *propensity score* measures the probability that a customer will be interested in a given product or service. It is calculated for an individual or a household that has been a bank customer for several months, based on data describing the operation of their accounts and banking products during this period, as well as their sociodemographic characteristics.

A *risk (behaviour) score* measures the probability that a customer with a current account, bank card, arranged overdraft or credit agreement will be involved in a payment or repayment problem. It is calculated for an individual who has been a bank customer for several months, based on his sociodemographic characteristics and data describing the operation of his accounts and banking products during this period. It is therefore a *behavioural* risk score.

Cross-tabulating these first two scores gives what is sometimes called a *pre-acceptance score*, as it provides a credit customer base with pre-accepted credit offers: if the customer responds favourably, and subject to legal checks (to ensure that he is not recorded in the incident files), the credit is granted. This score is illustrated by the diagram below, in which we can see that quadrant I will be targeted.

	+	I	III
Propensity	–	II	IV
		–	+
		Risk	

An *application score* (or *acceptance score*) is a risk score calculated for a new customer or one who has had few interactions with the bank. There are no historical data (or not enough of them) for this new customer, and the risk is calculated in (virtually) real time, at the moment of the customer's application to the bank, based on declarative data (such as socio-occupational data) supplied by the customer, matched with geodemographic data describing the standard of living and consumption habits in the district where the customer lives. A granting score can also be calculated for a known customer, if elements specific to the application are incorporated in the calculation.

A *recovery score* is an evaluation of the amount that can be recovered on an account or loan in dispute, and can suggest the most effective action for recovery, while avoiding disproportionate action in respect of loyal, profitable customers who do not pose any real risk.

An *attrition score* measures the probability that a customer will leave the bank. It is calculated for an individual who has been a bank customer for several months at least, based on data describing the operation of his accounts and banking products during this period, his relations with the bank, and his sociodemographic characteristics. Attrition is harder to calculate for a customer than for a product, for two reasons. The first is that there are several ways of ending the relationship: the customer might reduce his holdings, or might retain his holdings but decrease his cash flows. The second reason is that the score is less reliable when used to detect the departure of the customer far enough in advance, and it often only becomes very reliable at a moment so close to the customer's departure that there is no way of forestalling this.

12.2 Using propensity scores and risk scores

By using propensity scores it is possible to improve the rate of return on marketing campaigns (mailshots, telesales, etc.) – in other words, we can:

- reach more receptive customers without increasing the number of customers contacted, or
- reach the same number of receptive customers while decreasing the number of customers contacted,
- ensure that customers are not pestered with poorly targeted campaigns.

Using propensity scores enables sales staff to make more sales with fewer appointments.

By using risk scores in the field of credit, we can improve:

- the efficiency of the examination of credit applications (by rapidly processing files with high or low scores, and concentrating on the medium scoring files);
- the possibilities of delegation (more feasible for high-scoring files), particularly the use of young sales staff, who can be authorized to sell credit on the basis of the score in some cases, which was previously not always possible;
- the feeling of security among sales staff, at least for the less experienced ones;
- the satisfaction of customers, who can receive a quicker response to their applications for credit;
- the distribution circuit (pre-accepted credit can be distributed by direct marketing);
- the uniformity of decisions among different distribution channels (bank branches, call centres, etc.);
- the uniformity of the decisions of an individual operator (without an objective tool to assist in decision making, the analysis of a credit file can vary according to whether the previous file was very good or very bad);
- the adaptation of pricing to the risk incurred;
- the allocation of assets, according to the recommendations of the Basel Committee which has substituted the McDonough ratio[1] (or Basel II ratio) for the Cooke ratio.

Even among the 'medium' files, the score provides a detailed view that facilitates decision making; the single shade of grey is transformed into a whole spectrum of different tones. By way of example, in scores used in housing finance, the differences in risk between the highest and lowest scoring files may range from 1 to 50,[2] whereas the differences in debt ratios of

[1] Since the early 1990s, every bank has been required to have capital assets equal to not less than 8% of its weighted risks. This 8% ratio is called the solvency ratio. In the Cooke version of the solvency ratio (1988), the risks correspond to the amount of credit (market risks were to be added in 1996), which must be weighted according to the quality of the debtor. This weighting is very approximate, as the 8% applies to the whole amount of credit advanced to a business, 20% in the case of a bank, and 0% in the case of a government, regardless of the nature of the business, the bank and the government. The reform undertaken in 2004 by the Basel Committee did not alter the solvency ratio, rechristened the *McDonough ratio*, but increased the range of risks taken into account (including operational risks) and refined the method of weighting the risk, particularly by allowing the use of systems ('internal ratings') for classifying borrowers on the basis of the probability of default predicted in the various types of portfolio of each bank, including sovereign, bank, corporate, retail bank (for private and business customers), equity, securitization, and others. A move towards Basel III started in 2010 with a strengthening of the capital requirements planned by the Basel Committee for the period from 2013 to 2019.

[2] Michel, B. (1998) Les vertus du score. *Banque & Stratégie*, no. 154.

the files examined vary from 1 to 2, and the differences in rates of personal contribution vary from 1 to 5.

Other benefits of the use of risk scores include:

- a reduction in arrears (in spite of the increased efficiency);
- the limitation of the risk of over-indebtedness;
- the winning of a cohort of low-risk customers previously excluded from targeted sales because of their riskiness, and for whom scoring has enabled the cost of credit (allowing for interest rates and management costs) to be precisely adapted, or guarantees specified, according to the risk incurred;
- quantification, for control purposes, of the risk of the operation in future years;
- and, more generally, the fine adjustment of the business's policy.

Indeed, by adjusting the acceptance threshold on the score scale, we can emphasize either an increase in sales or the reduction of risk: we can increase the acceptance rate with a constant non-payment rate, or decrease the non-payment rate while keeping the acceptance rate constant. These results can be achieved by accepting low-risk customers who are currently rejected, and by rejecting high-risk customers who are currently accepted.

In the area of methods of payment, scoring can be used to automate payment decisions relating to banking transactions (cheques issued, direct debits paid, etc.) which reduce the funds in a customer's current account beyond the threshold of his authorized credit limit: we can decide whether to allow or refuse any given transaction. Even if scoring is not used to fully automate the payment decision process, it can be used to draw up a hierarchy of risks involved in each overdraft transaction and to display a list of transactions on the account manager's computer screen, starting with the most critical items and ending with the least risky. Scoring allows the manager to concentrate on the essential risk elements, thereby saving time. If he gains just a quarter of an hour per day, the bank gains 3.6% in productivity: 27 people can do the same work as 28 would have done ($28 \times 1/4$ hour = 7 hours = one working day).

12.3 Methodology

This section is complementary to Chapter 2 in some ways, while focusing on certain particularities of scoring. I shall therefore follow the plan of Chapter 2, which will enable us to maintain the correct logical and chronological order.

12.3.1 Determining the objectives

We must decide on the type of score required by the business, according to the objectives and the available technology. For example, the creation of a behavioural score will require the use of historical data covering at least one year; if we only have six months' worth of historical data, it will be better to buy a generic score; if we have even fewer historical data, we had better use a granting score.

We must also decide how the score will be used: will it be a decision-making tool ('operational score', see below) or a targeting tool for direct marketing ('strategic score')?

Even if the underlying models are identical in both cases, the output of the score will depend on the use made of it, as well as the computer resources available for implementing it.

We must draw up a schedule and milestones for the project. We must provide for a test phase on part of the system, and choose the deployment date so as to avoid conflict with any unrelated major marketing campaigns already in the pipeline.

We must also decide whether to develop the scoring in-house, with the assistance of a specialist firm if necessary, or to subcontract the score calculations to a service provider (see below). In the latter case, the provider will probably have to sign a confidentiality clause to meet legal requirements.

We can then specify the perimeter of the customer base to be scored, together with the 'at risk/no risk' customers (for a risk score), or 'purchasers/non-purchasers' (for a propensity score) or those 'leaving/not leaving' (for an attrition score). The definition of the perimeter must specify whether or not any 'dormant' customers (those carrying out few, or no, banking transactions, but with non-zero balances) should be included. If they are scored, it is best to separate them from the rest. They should be processed with a special scoring model, in the absence of data on their banking transactions, which will rely more heavily on their sociodemographic characteristics. The choice of target also depends on the marketing policy of the business and the markets which it wishes to develop or gain. We must also decide whether each statistical individual to be scored will be a person, a couple (if appropriate) or a household, including dependent children.

12.3.2 Data inventory and preparation

The data are collected and then analysed and prepared before the creation of the scoring model (or models if the population to be scored has been pre-segmented). In the collection of data, some will give rise to more questions than others. This is particularly true of income. The bank may know of three types of income for a customer. These are the income he has declared, the income measured via the credit entries in his accounts, and those which can be estimated using geodemographic databases. The first type of information is not provided by all customers; and it may not be accurate, or up to date. The third type is known for all customers, as well as for prospects, but is only an approximation subject to the limits of accuracy of geodemographic information. The first and third types of income information have the advantage of representing the total income of the customer, even if he uses more than one bank. The second type of data, relating to the income held in the customer's accounts at a bank, is fragmented for the bank observing it. The information does not match the total household income, which for INSEE, the French Statistical Office, 'is understood as the total net monetary income received by all members of the household; it includes income from work, from property and from transfers'. This distinction between the three components of income cannot always be maintained on the basis of the computer data available to the bank; in some cases the data only inform the bank that a movement originating outside the bank has credited the current account of a customer, although the bank does not know, from the description of the movement, whether it is due to a wage payment, a rent, a maintenance payment, or some other payment.

12.3.3 Creating the analysis base

For risk score learning, we need a sample of customers who have been accepted by the bank and whose files have been found to be good or bad, but it is preferable to have a sample of customers refused by the bank as well. This is because their absence from the learning sample would mean that they could not be scored, even though their risk profile is probably far from neutral. The procedure for taking rejected customers into account is a form of 'reject inference' (see Section 11.16.3). It is not a simple matter, because, even if the files of these rejected customers are available and comprehensive (in fact they are not always retained, or even completed in full), their credit variable is inherently unknown. The aim of reject inference is to predict what the credit variable (good or bad file) would have been if the credit had been granted, so that the rejected customers can be incorporated in the construction of the model.

To ensure that the model is reliable and has a good generalization capacity, the training sample should, if possible, contain at least:

- 500–1000 credit files confirmed as 'good';
- 500–1000 credit files confirmed as 'bad';
- ideally, 500–1000 rejected credit files.

In the worst case, the number of files may be reduced to 200–300, while ensuring that only a small number of variables are introduced into the model. If there is more than one segment in the population, and a score model is developed for each segment, then the numbers given must of course relate to each segment.

Obviously, we must define in advance what we mean by a 'good' and 'bad' file. A 'bad' file often contains at least two or three (monthly) instalments unpaid and still outstanding, even if the file has not led to legal action or was closed without losses. A 'good' file records no unpaid instalments; in no circumstances must there be two or more consecutive unpaid instalments outstanding. Sometimes a neutral area is defined, corresponding to a single missed payment.

'Good' and 'bad' as described in Section 11.16.2 must also be defined for learning purposes in a propensity score model. In this type of score, allowance must be made for preferential pricing granted to some types of customer, such as employees of the business and certain partners. Even if these customers are ultimately scored, they are generally excluded from the score model training sample because their atypical behaviour may affect the modelling. Customers who have benefited from a free period are also excluded from the learning phase, because these customers more often cancel their contract at the end of this period.

If the quantity of one of the file types is insufficient, we can try to resample by bootstrapping (see Section 11.15.1) the files concerned, for example the 'bad' files, in order to bring their number closer to that of the 'good' ones. Even though the representativeness of the 'bad' files is not improved by this, the 'good' ones are still as representative as before, which would not be the case if their number had to be reduced in order more closely to match the number of 'bad' files. This bootstrap resampling can be carried out as follows. The number of non-payments, which is assumed to be small (less than 500, say), is denoted d. B bootstrap samples of d individuals are then taken from the 'bad' files, B being large enough to ensure that $Bd > 500$, and Bd individuals are drawn from the 'good' files by a simple random procedure. In this way we will obtain a 50–50 sample of 'good' and 'bad', with adequate representation of the 'bad'.

12.3.4 Developing a predictive model

The predictive techniques most commonly used for scoring are logistic regression, discriminant analysis and decision trees; sometimes support vector machines or naive Bayesian classifiers are used. However, neural networks are rarely used, because their advantages cannot make up for their drawbacks, particularly their opaque nature, the difficulty of setting their parameters correctly and the lack of appropriate statistical tests. They are especially prone to overfitting, particularly if the number of neurons in the hidden layer or layers is large; on the other hand, if the number of neurons is low, the goodness of fit of the network generally tends to approach that of logistic regression, so there is no obvious reason to use a neural network.

12.3.5 Using the score

If we wish to use a score at a point of sale, we must be able to retrieve the information in a simple and immediately comprehensible form. Consultation with end users often leads to the division of scores into three classes: low, medium and high.

To choose the boundaries of these three groups, in the case of a propensity score for example, the response rates of the scored customers are shown as a function of their score marks. This representation also enables us to verify the score, by comparing the response rates of the customers included in the modelling, the customers excluded from the modelling (atypical customers, privileged customers, employees, etc.), and all the scored customers (Figure 12.1).

We also check that the response rate is an increasing function of the score, even for customers who have not been taken into account in the development of the model.

Where the score division is concerned, we find two natural thresholds in the example of Figure 12.1: the first is between the scores of 6 and 7, and the second is between the scores of 8 and 9. Thus we have customers with low propensity, whose score is between 1 and 6, and whose

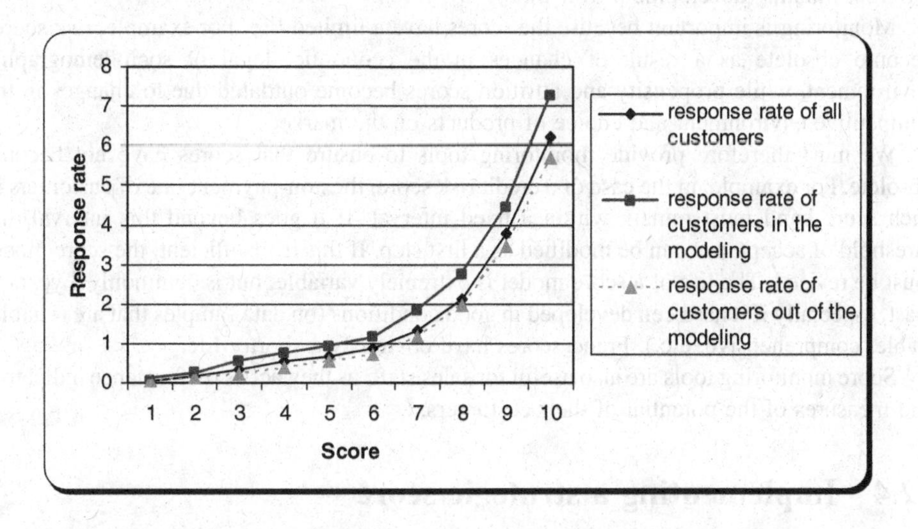

Figure 12.1 Response rate as a function of the score points.

response rate is less than 1%. Then we have customers with medium propensity, with scores of 7 or 8, and a response rate ranging from 1% to 3%. Finally, there are customers with high propensity, whose score is between 9 and 10, with a response rate of more than 3%.

12.3.6 Deploying the score

The score is used after it has been presented to its future users for their approval, if necessary, and after a full-scale test period. The deployment of an operational score (see below) takes more time, because it has to be retrievable at the workstations of the sales staff.

12.3.7 Monitoring the available tools

The use and quality of the score are monitored. This monitoring is carried out on:

- the number of customers by value or score band;
- the number of customers moving from one value of the score to another between two calculation dates, output in the form of a transition matrix;
- the response rate and the amounts taken up by score value or band;
- the forcing rate ('forcing', in risk scoring, means that credit has been granted to a customer whose poor score, if taken in isolation, does not justify acceptance of the file);
- the non-payment or dispute rate for each score value or band (risk);
- the reasons for refusal to purchase (propensity) or sell (risk).

We then compare the results for the different customer segments, the different products, the different sales channels (branches, call centres, Internet), the different regions and the different months (development over time).

Monitoring is important because the scores have a limited life. For example, risk scores become obsolete as a result of changes in the economic, legal or sociodemographic environment, while propensity and attrition scores become outdated due to changes in the competitive environment and choice of products on the market.

We must therefore provide monitoring tools to ensure that scores have not become obsolete. For example, in the case of a credit risk score, the non-payment rate of customers in each score band must remain within a fixed interval. If it goes beyond this interval, the threshold of acceptance can be modified as a first step. If this is insufficient, the score model must be revised. The life of a score model is extremely variable, but is commonly 5 years at least, especially if it has been developed in good conditions (on data samples that are reliable, stable, comprehensive, etc.). Fraud scores have obviously a shorter life.

Score monitoring tools are also useful for sales staff, as they act as performance indicators and measures of the potential of their customers.

12.4 Implementing a strategic score

A strategic score is a score used in the proactive targeting of customers for more or less centralized marketing operations. Generally, therefore, it is a propensity score which incorporates a risk behaviour score, in other words a pre-acceptance score (see Section 12.1), to which is added the calculation of the amount (of a loan, investment, standing order, etc.) that can be offered to the customer. This amount is calculated on the basis of the customer's income, outgoings, commitments and scores. It is difficult to evaluate the actual disposable income and irreducible financial commitments of the customer; these must be found by analysing the database, not by means of customer declarations, because a strategic score is calculated for all customers, even those who have never contacted a customer service representative and have never needed to supply this information. It may be useful to have a score model for each distribution channel.

If we wish to conduct a preliminary test on a number of branches, it will be preferable to choose branches that are fairly well differentiated in terms of the type of housing and customers. We may wish to conduct this test because it is our first venture into this field and we need to train our customer service representatives, or because we have no feedback from previous campaigns, or simply in order to refine the scoring system (especially the recommended acceptance and rejection thresholds).

The statistical determination of the profile of customers who have or have not responded during the test will be useful for generalizing the score. It will be useful to compare the resulting response rates with the response rates of a control sample, taking care to ensure that we are comparing comparable operations (same product, same price and same marketing presentation). The control sample will consist of customers taken randomly from two populations. On the one hand, there are customers who have been scored, but whose scores have not been passed on to the sales staff: all the score values, including the least favourable ones, will be represented, ensuring that customers with low scores have been given them correctly, and that information is obtained on customers with a sufficient variety of segments and scores to enrich the forthcoming score calculation. On the other hand, there are customers chosen not because of their score, but because they belong to a conventional target group for marketing. By not divulging the score, we can evaluate the 'placebo' effect on sales staff who may sell better to customers whom they consider to be well scored. This comparison will demonstrate the relevance of data mining (by comparing the response rates of the control sample as a function of the score points and as a function of membership of the conventional target group), as well as the relevance of the use of data mining (by comparing the response rates of the main population with those of the control sample).

12.5 Implementing an operational score

An operational score is a score of risk behaviour (for customers) or credit approval (for customers and prospects), designed as a decision support system for customer service representatives, to suggest whether or not to grant credit, and to propose a maximum amount to be lent or a guarantee requirement (surety, mortgage, etc.). It can also be used to automate payment decisions for accounts which have exceeded their credit limit (see above). The 'risk' element of a strategic score (see the preceding section) can act as an operational score. The difference between the two types of score, strategic and operational, is that the latter is preferably used in a 'reactive' way (in response to a customer's approach) in the course of a contact (in a branch, on the telephone or on the business's website), while the former is preferably used in a 'proactive' way (the bank takes the initiative) and in direct marketing.

Table 12.1 Decision table for property loans.

Risk	Risk score	Percentage of personal contribution			
		<10%	10–20%	20–50%	>50%
−−	between 75 and 100	to be examined	€150 000	€200 000 preferential rate	€300 000 preferential rate
−	between 50 and 75	to be examined	€100 000 mortgage	€150 000 mortgage	€200 000
+	between 25 and 50	to be examined	€100 000 mortgage + guarantee	€150 000 mortgage + guarantee	€150 000 mortgage
++	between 0 and 25	to be examined	to be examined	to be examined	to be examined

The score result can be output in the form of 'traffic lights' (green, amber or red) or as a decision table, produced by cross-tabulating the score value and another variable. Table 12.1 is an (imaginary) example relating to the approval of a property loan.

Although the application score is calculated in real time, the risk behaviour score is generally calculated monthly, with daily updates for customers whose financial resources have decreased (e.g. if there is a new loan or direct debit on their account) or for whom a payment incident has been recorded, when the score must immediately be downgraded.

12.6 Scoring solutions used in a business

12.6.1 In-house or outsourced?

There exist consultancy firms that provide services in scoring strategy, development and monitoring. Evidently, these bureaus can offer immediate access to the know-how and experience of specialists in scoring.

On the other hand, in-house scoring has the benefit of developing skills in the business, so that in a few years it will be possible:

- to build up a thorough knowledge of the business's data (which may have other positive effects for marketing and management);
- to develop synergy and a scoring culture among all the personnel involved in the operation;
- to develop greater reactivity whenever a new study is required;
- and to develop a range of resources for scoring, clustering, product association searching, and the like, for various requirements and users.

As a general rule, the desirability of outsourcing data mining operations varies from one business to another. It will be less advantageous for those that have more data to mine, and those that gain a greater competitive advantage by doing this themselves, such as financial institutions, insurance companies, retail businesses, and telephone companies. However, banks may outsource some of their activities which are concerned less with the historic customer data and more with the granting scores used to estimate the risk of non-payment of applicants for credit who are unknown to the bank.

This kind of outsourcing in the banking system is common practice in the USA, where it is facilitated by the fact that is permissible to generate 'positive' files on borrowers, in other words files containing information on a person's income and loans (past and present). The use of 'positive' files is permitted in most European countries, including the United Kingdom, Ireland, Germany, Austria, the Netherlands, Belgium, Switzerland, Norway and Italy. In France, however, it is only permissible at present to use 'negative' files, which record the details of over-indebted borrowers. Some people fear that the existence of positive files 'promotes the development of aggressive credit marketing'. Indeed, this does appear to be the case in some countries where positive files are used and where the credit risks have paradoxically increased.

In the USA there are service providers, known as *credit bureaus*,[3] which create vast databases with the aid of information supplied by lending institutions or obtained from legal announcements, yearbooks, etc. They work for credit institutions that can ask them for information about an applicant for credit, and can obtain a list of all existing loans, accompanied by a risk score if required. This information can be provided in real time, for example to a motor showroom where a vehicle is about to be sold on credit. A credit institution which refuses to grant credit must send the applicant, free of charge, a letter explaining the reasons for rejection, and stating the name of the credit bureau that has been used. If the reasons given are incorrect, the customer can ask the credit bureau to modify the data in question.

The three leading American credit bureaus are Equifax, Experian and TransUnion. They are beginning to establish a presence in countries outside the USA, including those where positive files are prohibited, and where an institution's knowledge of credit granted by its competitors is based solely on each customer's declaration and cannot be verified until the borrower is a non-payer and recorded in a negative file. Credit bureaus are most widespread in North America, South Africa, the UK, Italy and Spain.

In these countries, and in an increasing number of others, a credit institution can therefore obtain the score rating of a person who applies for credit (or, in particular, a credit card). This rating is provided by a credit bureau, which has calculated the score by applying a model (the *scorecard*) to the data it holds on this person. The most widespread score models in the USA have been developed, not by the credit bureaus themselves, but by 'model producers', the most famous of which is Fair, Isaac and Company (see Section 12.10). This firm supplies the three credit bureaus, via a software package, with score models known by the name of FICO scores (although the credit bureaus have developed other models, which are less widely used and may be less expensive, such as Experian's ScoreX). As each credit bureau calculates its scores on its own data, the FICO scores for a given individual may differ from one credit bureau to another. However, the FICO scores are regularly adjusted to ensure that a rating represents the same risk regardless of which credit bureau calculates it.

These score models are called generic models, or generic scores, because they have not been specifically determined on the basis of data from any one institution, but on the data

[3] Sometimes spelt 'credit bureaux'.

collected in numerous institutions and organizations. The performance of a generic score may be slightly less perfect than that of a personalized score, but it is still very satisfactory and highly stable over time, even over 10 years, because it is inherently less sensitive to changes in the scored population. There are also some generic scores adapted to each type of portfolio or product.

Thus there are three stages in the US system: the score producers (1) which supply score models to credit bureaus (2), which apply them to their mega-databases to rate individuals whose ratings are then notified to the credit institutions (3), which may match these generic scores with personalized scores or internal criteria relating to the additional knowledge of the individuals that they may have, to additional granting rules (debt to income ratio, personal contribution, etc.) or to the allowance for the specific features of the type of credit applied for (since the FICO scores rate an individual, not any specific application).

Score producers may also work directly for businesses, to which they supply generic or personalized scores.

A scorecard such as that shown in Table 12.2 can be used to calculate a number of points for a customer, to form the score, which, if high, signifies that the risk associated with the customer is low. The development of this type of scorecard is examined below in a detailed example of credit scoring.

Table 12.2 Scorecard.

Data	Value	Points
Age	18–25	10
	35–45	20
	45–60	25
	>60	15
Marital status	married/widowed	20
	cohabiting	10
	divorced	5
	unmarried	15
Number of children	0	15
	1	20
	2	25
	3	20
	>3	15
Number of overdraft days per month	<10	20
	10–20	10
	21–25	5
	>25	0
Occupation	employee/executive/retired	10
	other	0

FICO scores are mostly based on the following data:

- the person's payment history and the presence or absence of late payments (35% of the score);
- the current use of loans taken out and especially the proportion of credit facilities used (30% of the score);
- the length of the credit history (15% of the score);
- the types of credit in use (10% of the score);
- the number of recent credit applications (10% of the score).

FICO scores do not take age, occupation or seniority at work into account, although these data are legally usable. They range over a scale which runs, not from 0 to 100, but from 300 to 850, with a median at 723. They are generally used in the following way, but there may be variations between institutions, with specific acceptance thresholds. Below a certain number of points, about 540, credit is refused. Just above this level, but below 640, are the notorious 'subprime' loans which triggered the financial crisis that began in 2007 and led to the crash of autumn 2008. These are loans granted in spite of statistical predictions, disregarding the low ratings which suggested a high rate of non-payment. The best credit is 'prime' (a score of 680 or above). Depending on the rating, the credit institution may reject the application, raise the interest rate, make additional checks, or demand additional guarantees.

12.6.2 Generic or personalized score

As we have just seen, the business can choose between a generic and a customized score. The advantages of a generic score are those of all off-the-shelf solutions: it can be obtained quickly (in about a month); it is cheaper in the short term; and it requires little technical and human investment (less data history, for example).

There are three drawbacks of generic scores. First, a generic score is less precise than a made-to-measure score, obtained after an in-depth statistic analysis of each variable conducted in partnership with experts in the business. Secondly, it does not allow for feedback from previous operations in the business, new or specialized products of the business, local competition, etc. Thirdly, it is more suitable for use in English-speaking countries where positive risk files exist and where financial institutions exchange their data, allowing credit bureaus to work with very large databases and develop generic scores adapted to each type of institution.

12.6.3 Summary of the possible solutions

The various possible solutions are summarized in Figure 12.2 and their advantages and disadvantages shown in Table 12.3.

12.7 An example of credit scoring (data preparation)

In this section, we will use two of the methods described in this book on a public data set, in order to construct a credit scoring tool which can be used for granting consumer credit in a bank or a specialist credit institution.

We will use qualitative or discrete independent variables for the most part, with some continuous variables. When the continuous variables have been discretized, we can use all the

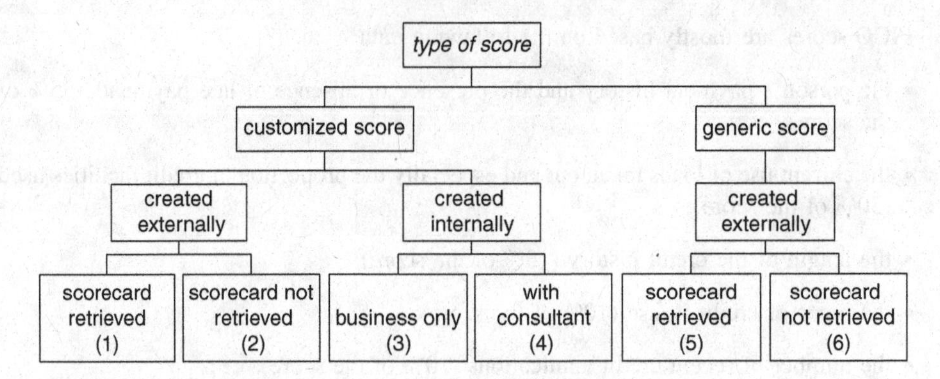

Figure 12.2 Possible solutions for developing a score.

Table 12.3 Advantages and disadvantages of the different solutions.

	Performance of the score	Technology transfer	Short-term life of the score	Long-term life of the score	Speed of production
(1)	+	−	+	−	−
(2)	+	−	− (for one-time use)	−	−
(3)	+	−	+	+	−
(4)	+	+	+	+	−
(5)	−	−	+	−	+ + (30 days)
(6)	−	−	− (for one-time use)	−	+

variables in a multiple correspondence analysis (Section 7.4), which will help us to explore the data and ensure their consistency. We will then go on to the modelling phase, calculating two score models, using:

- binary logistic regression (Section 11.8);
- the DISQUAL method of linear discriminant analysis (Section 11.6.7).

We will use these to construct two scorecards and then compare their performance. After choosing one of these scorecards, for appropriate reasons, we will draw up the rules for granting credit based on this scorecard. The naive Bayesian classifier has also been tested on this data set, but the results are rather less satisfactory (Section 11.10.1).

The data set which is used is well known under the name of 'German credit data' and is available on the Internet.[4] It contains 1000 consumer credit files, of which 300 are classed as defaulting. It is actually a stratified sample adjusted on the dependent variable, the real non-payment rate being 5%.

[4] At http://www.stat.uni-muenchen.de/service/datenarchiv/kredit/kredit.html. For a description of the data, see: http://www.stat.uni-muenchen.de/service/datenarchiv/kredit/kreditvar.html.

This data set was originally provided by Hans Hofman, of the University of Hamburg, and there are several variants, with different transformations of some of the variables. It was used by Fahrmeir and Tutz (1994).[5] It consists of the 'Credit' variable (the dependent variable), with its 700 'Good' categories (no non-payments) and its 300 'Bad' categories (non-payments present), and 19 independent variables (one of which, relating to the applicant's place of origin, has been omitted):

- three continuous numeric variables – credit duration in months, amount of credit in euros and age in years;
- seven discretized numeric variables – mean current account balance, savings outstanding, number of loans already recorded at the bank, affordability ratio, seniority at work, length of time at the address, number of dependents;
- nine qualitative variables: purpose of the credit, applicant's repayment history, other borrowings (outside the bank), valuables owned by the applicant, his guarantees, his marital status, his residential status, his type of employment and the existence of a telephone number.

Some variables relate to the credit itself (duration, amount, purpose), others to the financial profile of the borrower (amounts outstanding, other borrowings, etc.) and some to the personal profile of the borrower (age, marital status, etc.). All these data are of course measured at the time of the credit application and not subsequently. I have added a variable, ID, which is a customer number from 1 to 1000.

Given that only three variables are not divided into classes, we shall start by discretizing them, enabling them to be used together with the others, with the same methods. This will provide greater simplicity and readability.

The SAS syntax below will provide an illustration of the distributions of the three continuous variables, for the 700 'Good' files on one hand, and for the 300 'Bad' files on the other. These graphs will also enable us to detect immediately any anomalies and links between the continuous variable and the dependent variable.

```
PROC UNIVARIATE DATA=test ;
VAR Credit_duration Credit_amount Age ;
CLASS Credit ;
HISTOGRAM Credit_duration Credit_amount Age
/ NORMAL (MU=EST SIGMA=EST COLOR=BLACK L=2) ;
INSET MEAN MEDIAN CV NMISS ;
FORMAT Credit CREDIT. ;
RUN ;
```

The credit duration has predictable peaks at 12, 24, 36, 48 and 60 months (Figure 12.3). The high proportion of longer-term credit among the defaulters is clearly visible.

Very small amounts of credit are rare: the minimum is €250, and 95% are above €700 (Figure 12.4). A maximum frequency is reached at about €1200, after which the trend is

[5] Fahrmeir, L. and Tutz, G. (1994) *Multivariate Statistical Modelling Based on Generalized Linear Models*. New York: Springer.

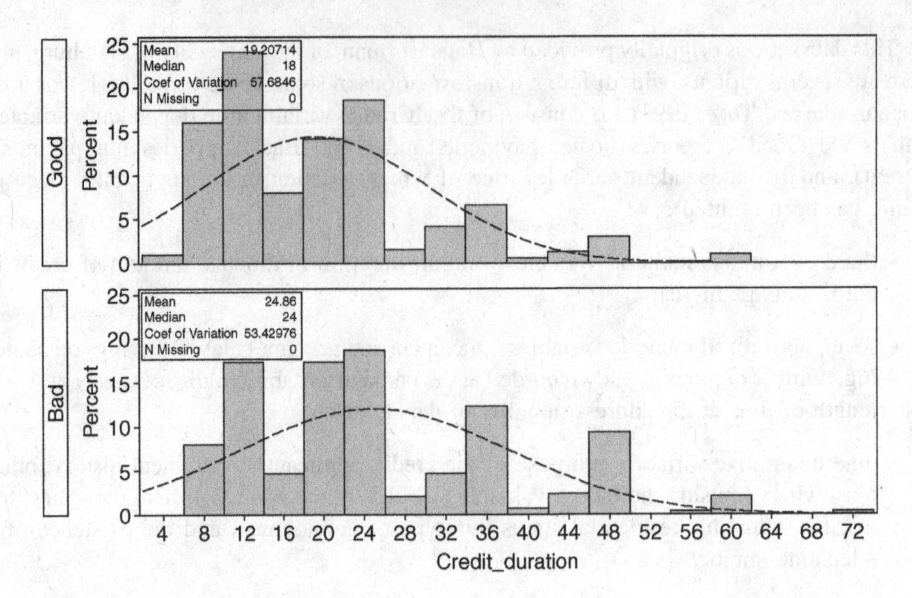

Figure 12.3 Distribution of the duration of credit (for good customers and others).

entirely downwards. The larger proportion of higher amounts among the non-payments is evident.

The ages of the borrowers range from 19 to 75 (Figure 12.5). The distribution for the non-payment-free borrowings is much more uniform than for the others. The non-payments mainly relate to borrowers who were under 40 when the credit was granted.

Each of the three continuous variables that we have examined has a significant link with the dependent variable; the next step is to discretize them. Let us take the example of age. We start by dividing it into deciles, like the other two variables.

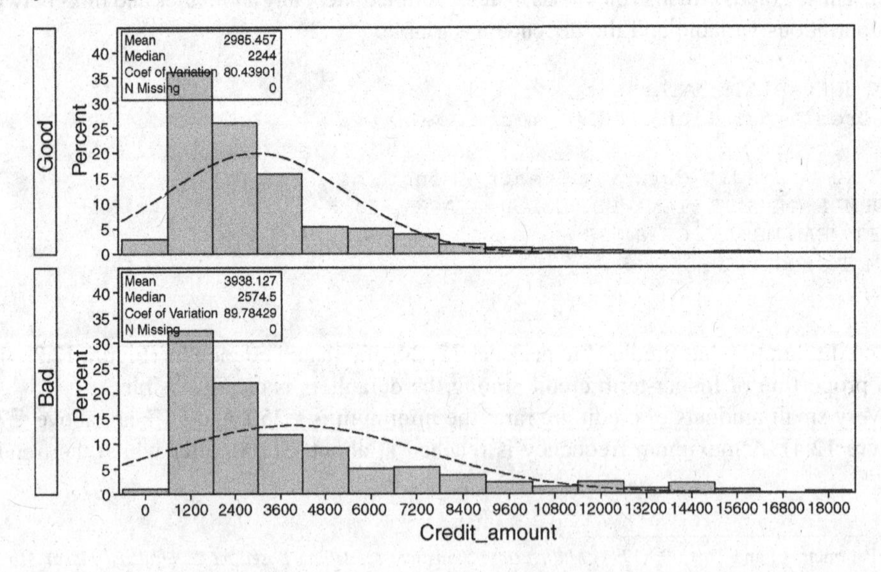

Figure 12.4 Distribution of the amount of credit (for good customers and others).

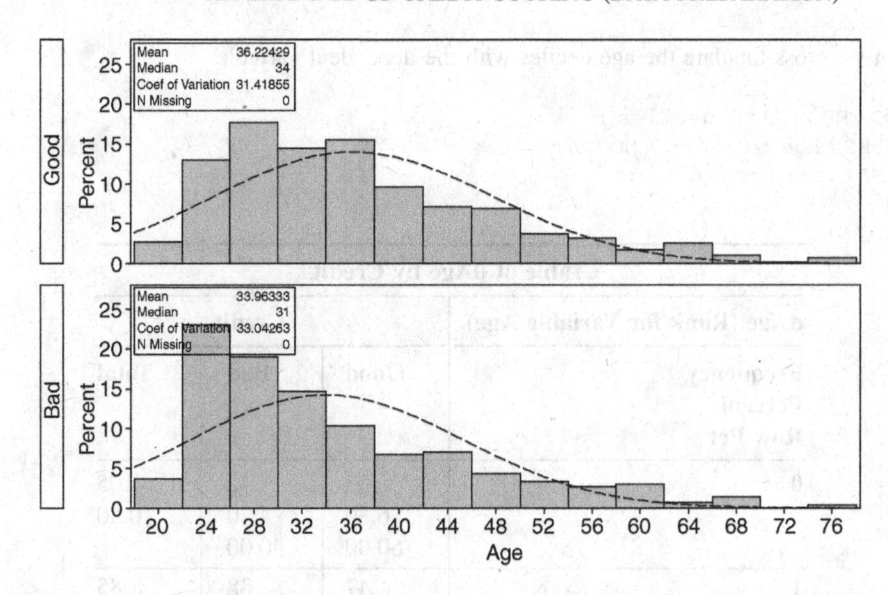

Figure 12.5 Distribution of the applicant's age (for good customers and others).

```
PROC RANK DATA = test GROUPS = 10 OUT = deciles ;
VAR Credit_duration Credit_amount Age ;
RANKS dCredit_duration dCredit_amount dAge ;
RUN ;
```

We then display the thresholds of the deciles.

```
PROC MEANS DATA = deciles MIN MAX ;
CLASS dAge ;
VAR Age ;
RUN ;
```

Analysis Variable: Age

Rank for Variable Age	N Obs	Minimum	Maximum
0	105	19.0000000	23.0000000
1	85	24.0000000	25.0000000
2	101	26.0000000	27.0000000
3	120	28.0000000	30.0000000
4	105	31.0000000	33.0000000
5	72	34.0000000	35.0000000
6	113	36.0000000	39.0000000
7	98	40.0000000	44.0000000
8	105	45.0000000	52.0000000
9	96	53.0000000	75.0000000

Then we cross-tabulate the age deciles with the dependent variable.

```
PROC FREQ DATA = deciles ;
TABLES dAge * Credit / NOCOL ;
FORMAT Credit Credit. ;
RUN ;
```

Table of dAge by Credit

dAge (Rank for Variable Age)	Credit		
Frequency Percent Row Pct	Good	Bad	Total
0	63 6.30 60.00	42 4.20 40.00	105 10.50
1	47 4.70 55.29	38 3.80 44.71	85 8.50
2	74 7.40 73.27	27 2.70 26.73	101 10.10
3	79 7.90 65.83	41 4.10 34.17	120 12.00
4	72 7.20 68.57	33 3.30 31.43	105 10.50
5	55 5.50 76.39	17 1.70 23.61	72 7.20
6	89 8.90 78.76	24 2.40 21.24	113 11.30
7	70 7.00 71.43	28 2.80 28.57	98 9.80
8	84 8.40 80.00	21 2.10 20.00	105 10.50
9	67 6.70 69.79	29 2.90 30.21	96 9.60
Total	700 70.00	300 30.00	1000 100.00

The contingency table shows that the first two age deciles correspond to a non-payment rate markedly higher than that of the other deciles. Thus there is a threshold at 25 years. No other threshold is clearly evident, as the non-payment rate subsequently fluctuates between 20% and a little over 30%. We can therefore decide on how to divide 'age' into two bands.

This procedure is repeated for the duration and the amount of credit. I have omitted it here for reasons of space. I will simply say that the credit duration has a threshold at 15 months, and then at 36 months (inclusive). It also has a threshold at 8 months, which is nevertheless unusable because of the size too small of this band: 94 files. The non-payment rate certainly low of 10.64% has a 95% confidence interval which leads it to an upper bound of 16.87%, close to the default rate of the second decile. As for the amount of credit, this shows a threshold at €4716, between the 7th and 8th decile, but the threshold is not very clear, and could be between the 6th and 7th decile. The examination of vingtiles helps us here to decide, with a clearer cut between the 14th and 15th vingtiles, which corresponds to €3972, which we round off to €4000. We have therefore divided the variable into twenty bands of equal size and then measured the non-payment rates in each band. Here are the SAS formats corresponding to these divisions:

```
PROC FORMAT ;
  value DURATION
    0-15 = '<= 15 months'
    16-36 = '16-36 months'
    37-high = ' > 36 months' ;
  value AMOUNT
    0-<4000 = '< 4000 euros'
    4000-high = '> 4000 euros' ;
  value AGE
    0-25 = '<= 25 years'
    26-high = '> 25 years' ;
RUN ;
```

It is worth noting that a discretization method based on the principle of minimizing the entropy, as implemented in the IBM SPSS Data Preparation module (see Chapter 5 on software), yields a single threshold at 15 months for the duration, a threshold at €4000 for the amount, and considers that the age variable is too weakly linked to dependent variable to be discretized.

The formats of the other variables, available on the Internet, are as follows:

```
PROC FORMAT ;
  value ACCOUNTS
    1 = 'CA < 0 euros'
    2 = 'CA [0-200 euros['
    3 = 'CA >= 200 euros'
    4 = 'No checking account' ;
  value $HISTORY
    'A30' = 'Current non-payment at another bank'
    'A31' = 'Previous non-payments'
    'A32' = 'Current credits without delay till now'
    'A33' = 'Previous credits without delay
    'A34' = 'No credit at any time' ;
  value $PURPOSE
    'A40' = 'New vehicle'
    'A41' = '2nd hand vehicle'
```

```
    'A42' = 'Furniture'
    'A43' = 'Video HIFI'
    'A44' = 'Household appliances'
    'A45' = 'Improvements'
    'A46' = 'Education'
    'A47' = 'Holidays'
    'A48' = 'Training'
    'A49' = 'Business'
    'A50' = 'Others' ;
  value SAVINGS
    0 = 'No savings'
    1= '< 100 euros'
    2 = '[100-500 euros['
    3 = '[500-1000 euros['
    4 = '>= 1000 euros' ;
  value EMPLOYMENT
    1 = 'Unemployed'
    2 = 'Empl < 1 year'
    3 = 'Empl [1-4[ years'
    4 = 'Empl [4-7[ years'
    5 = 'Empl >= 7 years' ;
  value $STATUS
    'A91' = 'Man: divorced/separated'
    'A92' = 'Woman: divorced/separated/married'
    'A93' = 'Man: unmarried'
    'A94' = 'Man: married/widowed'
    'A95' = 'Woman: unmarried' ;
  value $GUARANTEE
    'A101' = 'No guarantee'
    'A102' = 'Co-borrower'
    'A103' = 'Guarantor' ;
value RESID
    1 = 'Res < 1 year'
    2 = 'Res [1-4[ years'
    3 = 'Res [4-7[ years'
    4 = 'Res >= 7 years' ;
  value DEPENDENTS
    1 = '0-2'
    2 = '>=3' ;
  value NBCRED
    1 = '1 credit '
    2 = '2 or 3 credits'
    3 = '4 or 5 credits'
    4 = '>= 6 credits' ;
  value INSTALL_RATE
    1 = '< 20'
    2 = '[20-25['
    3 = '[25-35['
    4 = '>= 35' ;
  value $ASSETS
    'A121' = 'Property'
    'A122' = 'Life insurance'
    'A123' = 'Vehicle or other'
```

```
    'A124' = 'No known assets' ;
  value $CREDIT
    'A141' = 'Other banks'
    'A142' = 'Credit institutions'
    'A143' = 'No credit' ;
  value $HOUSING
    'A151' = 'Tenant'
    'A152' = 'Owner'
    'A153' = 'Rent-free accommodation' ;
  value $JOB
    'A171' = 'Unemployed'
    'A172' = 'Unskilled'
    'A173' = 'Skilled employee / worker'
    'A174' = 'Executive' ;
  value $TELEPHONE
    'A191' = 'No Tel'
    'A192' = 'With Tel' ;
  value CREDIT
    1 = 'GOOD'
    2 = 'BAD' ;
RUN ;
```

As all the variables are now available in the form of classes, we can perform a multiple correspondence analysis. We use the CORRESP procedure with the BINARY option to create an indicator matrix. This enables us to collect in the output data set OUT the records of _OBS_ type containing the factor coordinates Dim1 and Dim2 of each individual. The _name_ variable of this data set contains an observation number. It is not usually directly linked to any ID of the input data set, but in this case, as the data set to be scored has an ID that we have created as equal to the observation number, this ID can be matched with the _name_ variable to add the factor coordinates to the input data set to be scored. This then enables us to represent the factor plane of the individuals, distinguishing them according to the value of the dependent variable (which the OUT data set does not contain). This variable has also been specified as 'supplementary' in the MCA, so that it can be represented in the variable plane, but without making it participate in the factor axis calculations.

```
PROC CORRESP DATA=test BINARY DIMENS=2 OUT=output NOROW=PRINT ;
TABLES Accounts Credit_duration Credit_amount Savings Seniority_
employment Installment_rate Seniority_residence Age Nb_credits
Number_of_dependents Credit_history Purpose_credit Marital_status
Guarantees Assets Other_credits Status_residence Job Telephone Credit ;
SUPPLEMENTARY Credit ;
FORMAT Accounts ACCOUNTS. Credit_duration DURATION. Credit_amount
AMOUNT. Savings SAVINGS. Seniority_employment EMPLOYMENT.
Installment_rate INSTALL_RATE. Seniority_residence RESID. Age AGE.
Nb_credits NBCRED. Number_of_dependents DEPENDENTS. Credit_history
$HISTORY. Purpose_credit $PURPOSE. Marital_status $STATUS. Guarantees
$GUARANTEE. Assets $ASSETS. Other_credits $CREDIT. Status_residence
$HOUSING. Job $JOB. Telephone $TELEPHONE. Credit CREDIT. ;
RUN ;

DATA coord ;
SET output;
WHERE _type_ = 'OBS' ;
```

```
id = INPUT(STRIP(_name_),8.) ;
KEEP id Dim1 Dim2 ;
RUN ;
```

We have to carry out the minor task of transforming the _name_ variable, which is not numeric but alphanumeric, and in which the observation number is not left-justified. The STRIP function left-justifies it and removes the terminal spaces, and the INPUT function transforms it into a numeric variable.

```
PROC SORT DATA = test_score ; BY id ; RUN ;
PROC SORT DATA = coord ; BY id ; RUN ;

DATA test_acm ;
MERGE test (IN=a) coord (IN=b) ;
BY id ;
IF a ;
RUN ;
```

After linking with the initial data set, the stated graphic representation can be obtained, using the GPLOT procedure for high definition display. The individuals are differentiated as a function of the 'Credit' variable by adding the instruction '= Credit' to the PLOT line.

```
GOPTIONS RESET=all;
SYMBOL1 v=CIRCLE c=BLACK;
SYMBOL2 v=DOT c=BLACK;
PROC GPLOT DATA = test_acm ;
WHERE _type_ = 'OBS' ;
PLOT Dim2*Dim1 = Credit ;
FORMAT Credit CREDIT. ;
RUN ;
QUIT ;
```

The key to Figure 12.6 shows that the at-risk individuals are represented by solid dots, while good payers are represented by circles. It can be seen that the at-risk individuals are concentrated towards the top of the factor plane, and more precisely in the upper left-hand area. They are much rarer in the lower part.

We shall now display the factor plane of the variables. We will not use the customary GPLOT procedure, as it gives a poor picture of a large number of categories, with too much superimposition. Instead, we will use the recent GTL language and ODS GRAPHICS, mentioned above, which appeared with Version 9 of SAS. These will give us the elegant graphic shown in Figure 12.7. The SAS 9.2 syntax, which differs in several ways from that of Version 9.1.3, is shown below.

```
PROC TEMPLATE ;
DEFINE STATGRAPH example.mca ;
BEGINGRAPH ;
LAYOUT OVERLAY / XAXISOPTS=(LABEL="Axis 1" GRIDDISPLAY=ON)
YAXISOPTS=(LABEL="Axis 2" GRIDDISPLAY=ON) ;
SCATTERPLOT X=Dim1 Y=Dim2 / DATALABEL=_name_ GROUP=_type_ ;
ENDLAYOUT ;
ENDGRAPH ;
```

```
END ;
RUN ;

ODS HTML ;
ODS GRAPHICS ON ;

DATA _NULL_ ;
SET output;
WHERE _type_ IN ('VAR' 'SUPVAR') AND (SqCos1 > 0.02 OR SqCos2 > 0.02) ;
FILE PRINT ODS=(TEMPLATE="example.mca") ;
PUT _ODS_ ;
RUN ;
ODS GRAPHICS OFF ;
ODS HTML CLOSE ;
```

Note that we only display the categories that are sufficiently well represented on at least one of the factor axes and whose positioning in the plane can therefore be commented on legitimately. We exclude the categories for which none of the squared cosines is greater than 0.02 (see Chapter 7 on factor analysis). We will see subsequently, in the contingency tables for the independent variables and the dependent variable, that the categories which are not represented are also those which are least frequent.

The principal plane shows the coherence of the categories with each other. For example, the categories '<= 25 years', 'tenant' and 'empl < 1 year' are close together, in the upper left-hand area. The categories 'property' and 'owner' are also fairly close together. Oddly, they are close to the 'non-qualified' type of employment, whereas we would have expected to see the 'executive' category at this position. However, the 'unemployed' category of employment type and seniority of employment are in the same part of the plane. They are also fairly close to the categories of 'rent-free accommodation' and 'no known assets'. It is very surprising that the last two are near the 'executive' category. Possibly we should be wary of the 'employment type' variable. The category '>= 35%' is very close to the 'Good' category, which is remarkable,

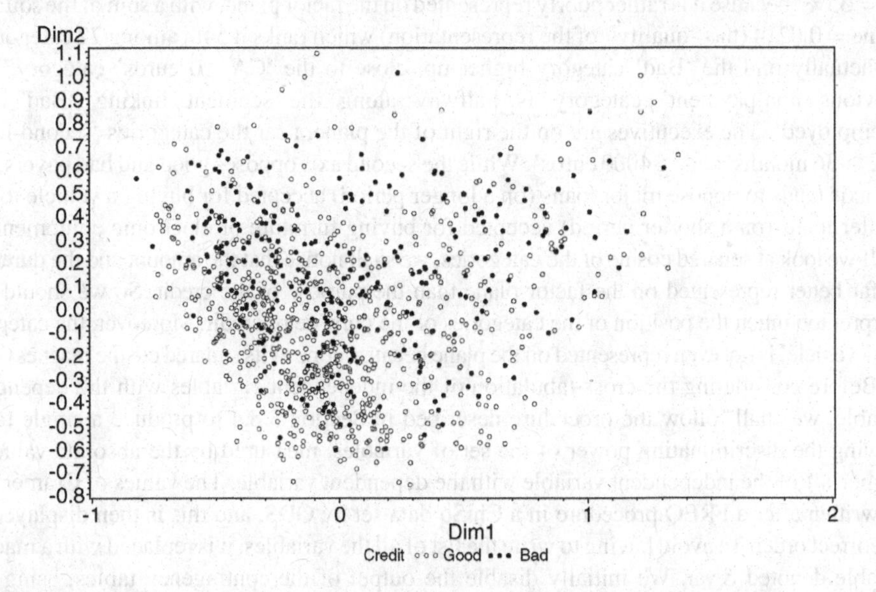

Figure 12.6 Multiple correspondence analysis on the credit data set (individuals).

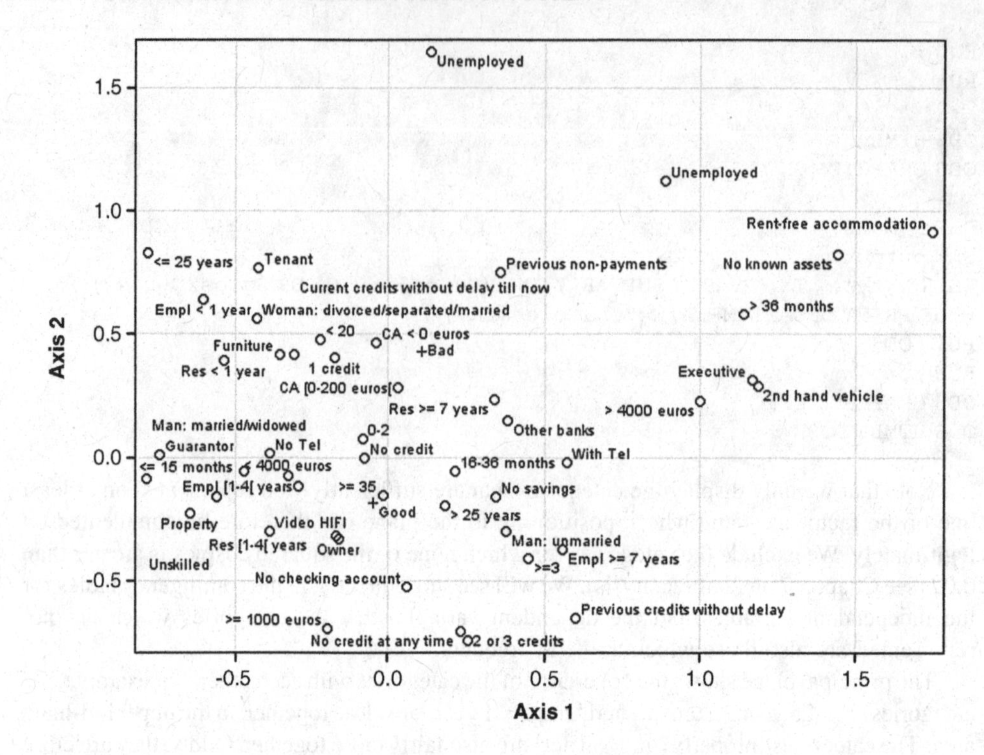

Figure 12.7 Multiple correspondence analysis on the credit data set (variables).

because we expect it to be linked to a risk of non-payment, as it corresponds to the highest indebtedness. However, there are some possible explanations for this, such as the fact that higher indebtedness may be due to a property loan. This form of credit is only granted to the most solvent customers. But we should not interpret anyway too much the position of this category "$>=35\%$" because it is rather poorly represented on the factor plane, with a sum of the squared cosine $=0.0214$ (the "quality" of the representation) which ranks it 54th among 72 categories. We actually find the 'Bad' category higher up, close to the 'CA <0 euros' category. The 'previous non-payment' category is halfway along the segment linking 'Bad' and 'unemployed'. The executives are on the right of the plane, near the categories 'second-hand car', '>36 months' and '>4000 euros'. While the second axis opposes good and bad payers, the first axis tends to oppose major loans (on a longer period) accepted for buying a vehicle to the smaller loans (on a shorter period) accepted for buying furniture or electronic equipment.

If we look at squared cosine of the categories, we shall notice that the amount and the duration are far better represented on the factor plane than the purpose of the credit. So we should not interpret too much the position of the categories of the purpose of credit. Moreover, the category 'New vehicle' is not even represented on the plane because none of its squared cosine reaches 0.02.

Before considering the cross-tabulation of the independent variables with the dependent variable, we shall follow the procedure described in Section 3.8.4 to produce a single table showing the discriminating power of the set of variables, measured by the absolute value of Cramér's $V$of the independent variable with the dependent variable. The values of Cramér's V are written after a FREQ procedure in a ChiSq data set by ODS, and this is then displayed in the correct order. To avoid having to write the list of all the variables, it is replaced with a macro-variable denoted &var. We initially disable the output of the contingency tables, using the NOPRINT instruction. We use the SCAN(c,n) function to extract the *n*th word of the chain c.

```
ODS OUTPUT ChiSq = ChiSq ;
PROC FREQ DATA=test ;
TABLES (&var) * credit / CHISQ NOCOL NOPRINT ;
FORMAT Accounts ACCOUNTS. Credit_duration DURATION.
Credit_amount AMOUNT. Savings SAVINGS. Seniority_employment EMPLOYMENT.
Installment_rate INSTALL_RATE. Seniority_residence RESID. Age AGE.
Nb_credits NBCRED. Number_of_dependents DEPENDENTS. Credit_history
$HISTORY. Purpose_credit $PURPOSE. Marital_status $STATUS. Guarantees
$GUARANTEE. Assets $ASSETS. Other_credits $CREDIT. Status_residence
$HOUSING. Job $JOB. Telephone $TELEPHONE.
Credit CREDIT. ;
RUN ;

DATA ChiSq ;
SET ChiSq ;
WHERE Statistic CONTAINS "Cramer" ;
abs_V_Cramer = ABS(Value) ;
Variable = SCAN (Table,2) ;
KEEP Variable Value abs_V_Cramer ;
RUN ;

PROC SORT DATA = ChiSq ; BY DESCENDING abs_V_Cramer ; RUN ;

PROC PRINT DATA = ChiSq ; RUN ;
```

Here is the list of variables, arranged in decreasing order of discriminating power:

Obs	**Value**	**abs_V_Cramer**	**Variable**
1	0.3517	0.35174	Accounts
2	0.2484	0.24838	Credit_history
3	0.2050	0.20499	Credit_duration
4	0.1900	0.19000	Savings
5	0.1735	0.17354	Purpose_credit
6	0.1581	0.15809	Credit_amount
7	0.1540	0.15401	Assets
8	0.1355	0.13553	Seniority_employment
9	0.1349	0.13491	Status_residence
10	−0.1279	0.12794	Age
11	0.1133	0.11331	Other_credits
12	0.0980	0.09801	Marital_status
13	0.0815	0.08152	Guarantees
14	0.0740	0.07401	Installment_rate
15	0.0517	0.05168	Nb_credits
16	0.0434	0.04342	Job
17	−0.0365	0.03647	Telephone
18	0.0274	0.02737	Seniority_residence
19	−0.0030	0.00301	Number_of_dependents

A decrease in Cramér's V can be seen after the 12th variable, with another decrease after the 14th variable. Variables after the 15th have V values less than 0.05, and consequently these are not retained for use in the score model.

For example, the type of employment is found to have little predictive value for the risk of non-payment, and not in an intuitive way, since executives are most at risk (perhaps related to their position on the MCA graphic) with a non-payment rate of 34.46%.

Table of Job by Credit

Job	Credit		
Frequency **Percent** **Row Pct**	**Good**	**Bad**	**Total**
Unemployed	15 1.50 68.18	7 0.70 31.82	22 2.20
Unskilled	144 14.40 72.00	56 5.60 28.00	200 20.00
Skilled employee/worker	444 44.40 70.48	186 18.60 29.52	630 63.00
Executive	97 9.70 65.54	51 5.10 34.46	148 14.80
Total	700 70.00	300 30.00	1000 100.00

The length of time at the current address also has a surprising effect on the risk of non-payment, since those who have been at the address for the shortest time are least at risk.

Table of Seniority_residence by Credit

Seniority_residence	Credit		
Frequency **Percent** **Row Pct**	**Good**	**Bad**	**Total**
Res <1 year	94 9.40 72.31	36 3.60 27.69	130 13.00
Res [1–4[years	211 21.10 68.51	97 9.70 31.49	308 30.80
Res [4–7[years	106 10.60 71.14	43 4.30 28.86	149 14.90

Res >= 7 years	289 28.90 69.98	124 12.40 30.02	413 41.30
Total	700 70.00	300 30.00	1000 100.00

I have not reproduced the table for the possession of a telephone number. As might be expected, the absence of a telephone number is a risk factor, but a very weak one, as indicated by the low value of its Cramér's *V*. This variable is sometimes examined, but it is not often of much use, especially as it fails to distinguish between landlines, mobiles and work numbers.

Similarly, I have omitted the table for the number of dependents, as Cramér's *V* for this variable is practically zero and there is practically no link with the dependent variable.

Before going on to examine the modelling based on the independent variables which were not eliminated at this stage, we should make sure that there are no excessively strong links between these variables. As the variables are all qualitative or discrete, Cramér's *V* is an appropriate measure of the links. The previous syntax, based on the output of the contingency tables in an ODS data set which is subsequently formatted, will now be slightly modified. In this case, the FREQ procedure cross-tabulates the set of variables with itself, and we must of course exclude any trivial pairs (variables cross-tabulated with themselves) and symmetrical pairs, which is done by the IF SCAN... instruction.

```
ODS OUTPUT ChiSq = ChiSq ;

PROC FREQ DATA=test ;
TABLES (&var) * (&var) / CHISQ NOCOL NOPRINT ;
FORMAT Accounts ACCOUNTS. Credit_duration DURATION. Credit_amount
AMOUNT. Savings SAVINGS. Seniority_employment EMPLOYMENT.
Installment_rate INSTALL_RATE. Seniority_residence RESID. Age AGE.
Nb_credits NBCRED. Number_of_dependents DEPENDENTS. Credit_history
$HISTORY. Purpose_credit $PURPOSE. Marital_status $STATUS. Guarantees
$GUARANTEE. Assets $ASSETS. Other_credits $CREDIT. Status_residence
$HOUSING. Job $JOB. Telephone $TELEPHONE.
Credit CREDIT. ;
RUN ;

DATA ChiSq ;
SET ChiSq ;
WHERE Statistic CONTAINS "Cramer" AND value NE 1 ;
abs_V_Cramer = ABS(Value) ;
Variable = SCAN(Table,2)!!" x "!!SCAN(Table,3) ;
KEEP Variable Value abs_V_Cramer ;
IF SCAN(Table,2) < SCAN(Table,3) ;
RUN ;

PROC SORT DATA = ChiSq ; BY DESCENDING abs_V_Cramer ;

PROC PRINT DATA = ChiSq ; RUN ;
```

Obs	Value	abs_V_Cramer	Variable
1	0.5532	0.55318	Assets × Status_residence
2	0.5066	0.50660	Credit_duration × Credit_amount
3	0.4257	0.42569	Job × Telephone
4	0.3782	0.37822	Credit_history × Nb_credits
5	0.3416	0.34158	Credit_amount × Purpose_credit
6	0.3113	0.31131	Job × Seniority_employment
7	0.3096	0.30965	Age × Status_residence
...	...	...	...

The first pairs displayed are very closely linked and values of Cramér's V above 0.4 in absolute terms are considered to be problematic.

Table of Assets by Status_residence

Assets	Status_residence			
Frequency Percent Row Pct	Tenant	Owner	Rent-free accommodation	Total
Property	55 5.50 19.50	226 22.60 80.14	1 0.10 0.35	282 28.20
Life insurance	46 4.60 19.83	184 18.40 79.31	2 0.20 0.86	232 23.20
Car or other	60 6.00 18.07	271 27.10 81.63	1 0.10 0.30	332 33.20
No known assets	18 1.80 11.69	32 3.20 20.78	104 10.40 67.53	154 15.40
Total	179 17.90	713 71.30	108 10.80	1000 100.00

The contingency table clearly shows that the most valuable asset is related to the status at residence by only one category for each variable, by the link between the 'no known assets' category and the 'rent-free accommodation' category. This does not justify the exclusion of one of the two variables at this stage, but we must bear this link in mind in case it becomes relevant.

The second most closely linked pair is the duration and amount of credit. The strength of the link is due to the fact that 95% of the loans for periods of more than 15 months are for amounts below €4000, while 83% of the loans for periods of more than 36 months are for amounts above €4000. It is unlikely that both variables will appear simultaneously in a model, and the duration (Cramér's $V = 0.20$) is appreciably more discriminating than the amount (Cramér's $V = 0.16$). We may even wonder if the discriminating power of the amount is due to its link with the duration, because the link with the risk of non-payment is more evident for the duration than for the amount, since there is no difference between the orders of magnitude of the monthly instalments on these amounts, which are much smaller than the amount of a property loan.

Table of Credit_duration by Credit_amount

Credit_duration	Credit_amount		
Frequency Percent Row Pct	<4000 euros	>4000 euros	Total
<= 15 months	408 40.80 94.66	23 2.30 5.34	431 43.10
16–36 months	331 33.10 68.67	151 15.10 31.33	482 48.20
>36 months	15 1.50 17.24	72 7.20 82.76	87 8.70
Total	754 75.40	246 24.60	1000 100.00

The cross-tabulation with the dependent variable supports the view that the duration is preferable. It offers a more balanced division of the population, because no category exceeds half of the credit applications. One of the categories (≤15 months) is also markedly less risky than the average, whereas another (>36 months) is markedly more risky. This enables us to distribute the files more satisfactorily between higher and lower risk. We therefore exclude the credit amount from the list of variables to be tested in the model.

Table of Credit_duration by Credit

Credit_duration	Credit		
Frequency Percent Row Pct	Good	Bad	Total
<=15 months	342 34.20 79.35	89 8.90 20.65	431 43.10
16–36 months	316 31.60 65.56	166 16.60 34.44	482 48.20
>36 months	42 4.20 48.28	45 4.50 51.72	87 8.70
Total	700 70.00	300 30.00	1000 100.00

Table of Credit_amount by Credit

Credit_amount	Credit		
Frequency Percent Row Pct	**Good**	**Bad**	**Total**
<4000 euros	559 55.90 74.14	195 19.50 25.86	754 75.40
>4000 euros	141 14.10 57.32	105 10.50 42.68	246 24.60
Total	700 70.00	300 30.00	1000 100.00

The third pair of strongly linked variables has disappeared with the exclusion of the 'telephone' variable, as has the fourth pair with the removal of the 'number of credits' and the fifth pair with the removal of the 'amount of credit'. The pairs shown below have acceptable links.

Let us examine the other variables, starting with those most strongly linked to the dependent variable. To obtain the contingency tables shown below, we rerun the FREQ procedure without the NOPRINT option.

For the average balance in the current account, a negative balance clearly increases the risk of non-payment. A balance of more than €200 reduces the non-payment rate by more than half.

Table of Accounts by Credit

Accounts	Credit		
Frequency Percent Row Pct	**Good**	**Bad**	**Total**
CA < 0 euros	139 13.90 50.73	135 13.50 49.27	274 27.40
CA [0–200 euros]	164 16.40 60.97	105 10.50 39.03	269 26.90
CA >= 200 euros	49 4.90 77.78	14 1.40 22.22	63 6.30
No checking account	348 34.80 88.32	46 4.60 11.68	394 39.40
Total	700 70.00	300 30.00	1000 100.00

Very logically, the non-payment rates are linked to the applicant's repayment history. There is a clearly marked gradation from the applicant who has never had any credit to the applicant who has had outstanding payments at the bank or has other outstanding payments at other banks. Two intermediate classes have non-payment rates very similar to each other, and very close to the average: those who have already repaid their loans in the past (31.82% of non-payments) and those who have a loan being repaid without any delay at present (31.89% of non-payments). Similar non-payment rates, a similar significance for the business, and one of the classes which represents less than 9% of the files: these are three reasons for grouping these two classes together for modelling. This is also programmed in the FORMAT procedure:

```
value $HISTORY
"A30" = "Current non-payment at another bank"
"A31" = "Previous non-payments"
"A32","A33" = "Credits without delay"
"A34" = "No credit at any time" ;
```

Table of Credit_history by Credit

Credit_history	Credit		
Frequency Percent Row Pct	Good	Bad	Total
Current non-payment at another bank	15 1.50 37.50	25 2.50 62.50	40 4.00
Previous non-payments	21 2.10 42.86	28 2.80 57.14	49 4.90
Current credits without delay till now	361 36.10 68.11	169 16.90 31.89	530 53.00
Previous credits without delay	60 6.00 68.18	28 2.80 31.82	88 8.80
No credit at any time	243 24.30 82.94	50 5.00 17.06	293 29.30
Total	700 70.00	300 30.00	1000 100.00

Similarly, for savings, there is a low non-payment rate for those having no savings product (which is different from having a savings product with a zero balance), which simply means that their savings are held at another bank. The non-payment rate then rises sharply, taking similar values under €100 and between €100 and €500. These categories are therefore combined. We may hesitate to combine the two bands between €500 and €1000 and over €1000, because their non-payment rates are somewhat different, but the small size of these two bands is a good reason for combining them. We could even consider grouping them together with the first category.

```
value SAVINGS
  0 = 'no savings'
  1-2 = '< 500 euros'
  3-4 = '> 500 euros' ;
```

Table of Savings by Credit

Savings	Credit		
Frequency Percent Row Pct	Good	Bad	Total
No savings	151 15.10 82.51	32 3.20 17.49	183 18.30
<100 euros	386 38.60 64.01	217 21.70 35.99	603 60.30
[100–500 euros]	69 6.90 66.99	34 3.40 33.01	103 10.30
[500–1000 euros]	52 5.20 82.54	11 1.10 17.46	63 6.30
>=1000 euros	42 4.20 87.50	6 0.60 12.50	48 4.80
Total	700 70.00	300 30.00	1000 100.00

For the purpose of the credit applied for, the categories are too numerous, and some of them are much too small. We must therefore create some combinations. It is logical, and in accordance with the non-payment rates, to group the purposes 'furniture', 'household appliances' and 'improvements' together. We will also group 'training' with 'education', in spite of a very high non-payment rate for 'training', which in fact means nothing because it is based on a single non-payment. In this grouping, therefore, we have taken a logical approach. We might also consider grouping this category with 'business'. We find that the non-payment rates are doubled between second-hand cars and new cars. Note that the purpose is always specified, and the credit granted is not a revolving credit, for example.

```
value $PURPOSE
  'A40' = "New vehicle"
  'A50' = "Others"
  'A41' = "2nd hand vehicle"
  'A42', 'A44', 'A45' = "Internal fittings"
  'A43' = "Video HIFI"
  'A46', 'A48' = "Education"
  'A49' = "Business"
  'A47' = "Holidays" ;
```

Table of Purpose_credit by Credit

Purpose_credit	Credit		
Frequency Percent Row Pct	Good	Bad	Total
New vehicle	145 14.50 61.97	89 8.90 38.03	234 23.40
2nd hand vehicle	93 9.30 80.87	22 2.20 19.13	115 11.50
Furniture	123 12.30 67.96	58 5.80 32.04	181 18.10
Video HIFI	218 21.80 77.86	62 6.20 22.14	280 28.00
Household appliances	8 0.80 66.67	4 0.40 33.33	12 1.20
Improvements	14 1.40 63.64	8 0.80 36.36	22 2.20
Education	28 2.80 56.00	22 2.20 44.00	50 5.00
Training	8 0.80 88.89	1 0.10 11.11	9 0.90
Business	63 6.30 64.95	34 3.40 35.05	97 9.70
Total	700 70.00	300 30.00	1000 100.00

The link between the asset of highest value possessed and the non-payment risk is logical. The absence of known assets doubles the non-payment rate as compared with the possession of a property asset. An applicant in the latter class has a better financial position, especially if he has repaid his property loan. Even if this is not the case, he is more likely than the average customer to ensure that his instalments are paid on time (out of fear of repossession, maybe). In any case, if a property loan has been granted to him, he must have had a degree of financial security. The two intermediate categories of 'life insurance' and 'vehicle or other' have equal non-payment rates, and are therefore grouped together into the 'non-property asset' category.

```
value $ASSETS
'A121' = "Property"
'A122','A123' = "Other assets"
'A124' = "No known assets" ;
```

Table of Assets by Credit

Assets	Credit		
Frequency Percent Row Pct	**Good**	**Bad**	**Total**
Property	222 22.20 78.72	60 6.00 21.28	282 28.20
Life insurance	161 16.10 69.40	71 7.10 30.60	232 23.20
Vehicle or other	230 23.00 69.28	102 10.20 30.72	332 33.20
No known assets	87 8.70 56.49	67 6.70 43.51	154 15.40
Total	700 70.00	300 30.00	1000 100.00

Seniority in employment follows a logical general trend, since an applicant with greater seniority is less at risk. Some of the detailed results are rather surprising, and may simply be due to a number of non-payments which is too small to yield reliable non-payment rates. However, there may be other explanations. For instance, an unemployed person is slightly less at risk than a person who has worked for less than a year. This may be due to the fact that the 'unemployed' category includes not only job seekers, but also people such as retired persons who do not need to work. As for the category of people who have been in their present employment for more than 7 years, this presents slightly more risk than the category of 4–7 years. This may be due to the presence of older employees who are more likely to be made redundant. Whatever the reasons, the first two and the last two categories will be grouped together to avoid any inconsistencies and to ensure that the categories are large enough.

```
value EMPLOYMENT
1-2 = 'Unemployed or lt 1 year'
3 = 'Empl [1-4[ years'
4-5 = 'Empl >= 4 years';
```

Table of Seniority_employment by Credit

Seniority_employment	Credit		
Frequency Percent Row Pct	Good	Bad	Total
Unemployed	39 3.90 62.90	23 2.30 37.10	62 6.20
<1 year in employment	102 10.20 59.30	70 7.00 40.70	172 17.20
Employed [1–4]years	235 23.50 69.32	104 10.40 30.68	339 33.90
Employed [4–7]years	135 13.50 77.59	39 3.90 22.41	174 17.40
Employed >=7 years	189 18.90 74.70	64 6.40 25.30	253 25.30
Total	700 70.00	300 30.00	1000 100.00

For residential status, the grouping of the categories follows the same logic as before.

```
value $HOUSING
'A151','A153' = 'Not Owner'
'A152' = 'Owner' ;
```

Table of Status_residence by Credit

Status_residence	Credit		
Frequency Percent Row Pct	Good	Bad	Total
Tenant	109 10.90 60.89	70 7.00 39.11	179 17.90
Owner-occupier	527 52.70 73.91	186 18.60 26.09	713 71.30
Rent-free accommodation	64 6.40 59.26	44 4.40 40.74	108 10.80
Total	700 70.00	300 30.00	1000 100.00

For age, no changes need to be made, and it will be tested as it stands.

Table of Age by Credit

Age	Credit		
Frequency **Percent** **Row Pct**	**Good**	**Bad**	**Total**
<=25 years	110 11.00 57.89	80 8.00 42.11	190 19.00
>25 years	590 59.00 72.84	220 22.00 27.16	810 81.00
Total	700 70.00	300 30.00	1000 100.00

Where loans from other institutions are concerned, it is logical that these should entail a higher risk of non-payment, because the customer is more indebted and events taking place at the other institutions cannot be monitored or controlled. The nature of the banking institution – conventional or specialist – makes no difference in the non-payment rate, and the first two categories are combined.

```
value $CREDIT
'A141','A142' = "Other banks or institutions"
'A143' = "No credit" ;
```

Table of Other_credits by Credit

Other_credits	Credit		
Frequency **Percent** **Row Pct**	**Good**	**Bad**	**Total**
Other banks	82 8.20 58.99	57 5.70 41.01	139 13.90
Credit institutions	28 2.80 59.57	19 1.90 40.43	47 4.70
No credit	590 59.00 72.48	224 22.40 27.52	814 81.40
Total	700 70.00	300 30.00	1000 100.00

Turning to marital status, Cramér's V with the dependent variable goes below the threshold of 0.1, and the decrease in discriminating power is evident. The difference between the non-payment rates of the different categories diminishes. The two categories having the most similar non-payment rates will be grouped together, but it is unlikely that the variable will play a major part in the prediction. Counterintuitively, the weakness of this variable is often encountered in scoring.

It is a pity, however, that the 'divorced/separated woman' category has been grouped with 'married woman' for which the non-payment rate is probably lower. The model may suffer from this.

We may also note the rather surprising absence of an 'unmarried woman' category from the sample. The score model will therefore be developed without this variable, and will be unable to score a credit application made by an unmarried woman, if this occurs. To avoid this unforeseen situation, it is preferable to add an *a priori* rule on unmarried women, which can be validated by credit analysts. This rule will be that the 'unmarried woman' category will be combined with one of the existing categories, either 'woman: divorced/separated/married' or 'man: unmarried'.

```
value $STATUS
'A91' = "Man: divorced/separated"
'A92' = "Woman: divorced/separated/married"
'A93', 'A94' = "Man: unmarried/married/widowed"
'A95' = "Woman: unmarried" ;
```

Table of Marital_status by Credit

Marital_status	**Credit**		
Frequency **Percent** **Row Pct**	**Good**	**Bad**	**Total**
Man: divorced/separated	30 3.00 60.00	20 2.00 40.00	50 5.00
Woman: divorced/separated/married	201 20.10 64.84	109 10.90 35.16	310 31.00
Man: unmarried	402 40.20 73.36	146 14.60 26.64	548 54.80
Man: married/widowed	67 6.70 72.83	25 2.50 27.17	92 9.20
Total	700 70.00	300 30.00	1000 100.00

The existence of a guarantor tends to decrease the non-payment rate, by contrast with the effect of a co-borrower, but this situation is fairly uncommon. However, we can test this variable, which may provide further useful information for the individuals concerned.

Table of Guarantees by Credit

Guarantees	Credit		
Frequency Percent Row Pct	Good	Bad	Total
No guarantee	635 63.50 70.01	272 27.20 29.99	907 90.70
Co-borrower	23 2.30 56.10	18 1.80 43.90	41 4.10
Guarantor	42 4.20 80.77	10 1.00 19.23	52 5.20
Total	700 70.00	300 30.00	1000 100.00

Inspection of the instalment rate table confirms a fact that is well known to credit scoring experts: this variable, which appears to be popular among credit analysts, is actually a poor predictor of the risk of non-payment. The non-payment rates are so close that this variable cannot be of any use at all, and we remove it from the selection.

Table of Installment_rate by Credit

Installment_rate	Credit		
Frequency Percent Row Pct	Good	Bad	Total
<20	102 10.20 75.00	34 3.40 25.00	136 13.60
[20–25]	169 16.90 73.16	62 6.20 26.84	231 23.10
[25–35]	112 11.20 71.34	45 4.50 28.66	157 15.70
>=35	317 31.70 66.60	159 15.90 33.40	476 47.60
Total	700 70.00	300 30.00	1000 100.00

When the categories have been regrouped, we can recalculate the contingency tables with the credit variable, to obtain a precise measurement of the effect of the changes on the non-payment rates. This is also the way to measure the changes in Cramér's *V* following the regrouping. These tables are not reproduced here for reasons of space. We find that Cramér's *V* for savings changes from 0.190 to 0.188, for the purpose of the credit it changes from 0.174 to 0.161 (more regroupings), and for seniority in employment it changes from 0.136 to 0.133. The other *V* values vary imperceptibly or not at all.

We can use all 12 selected variables for modelling in an SAS macro-variable. This is more convenient and will avoid the need to specify them all in the syntax.

```
%let varselec =
Accounts Credit_history Credit_duration Savings Seniority_employment
Age Purpose_credit Marital_status Guarantees Assets
Other_credits Status_residence ;
```

Regardless of the modelling method, we have seen that the population to be modelled had to be divided in a random way into two samples, namely one for training and one for validation. Here we will do this by simple random sampling, which gives a satisfactory result. For the details of stratified random sample, see Section 2.16.2 of my book *Étude de Cas en Statistique Décisionnelle*.[6]

```
%LET seed = 123;

DATA train valid ;
SET test ;
IF RANUNI (&seed) < 0.66
  THEN DO ; sample = "T" ; OUTPUT train ; END ;
  ELSE DO ; sample = "V" ; OUTPUT valid ; END ;
RUN ;

PROC FREQ DATA = train; TABLE Credit; FORMAT Credit CREDIT.; RUN;
PROC FREQ DATA = valid ; TABLE Credit; FORMAT Credit CREDIT.; RUN;
```

Here is the distribution of the Good and Bad files in the training sample:

Credit	Frequency	Percent	Cumulative frequency	Cumulative percent
Good	451	70.03	451	70.03
Bad	193	29.97	644	100.00

and in the validation sample:

Credit	Frequency	Percent	Cumulated frequency	Cumulative percent
Good	249	69.94	249	69.94
Bad	107	30.06	356	100.00

These distributions are very close to those of the total population.

The performance of the models is evaluated by calculating the area under the ROC curve for the test sample, using the SAS %AUC macro described in Section 11.16.5.

[6] Tufféry, S. (2009) *Étude de Cas en Statistique Décisionnelle*. Paris: Technip.

12.8 An example of credit scoring (modelling by logistic regression)

The first modelling method used in this case study is logistic regression, which is the standard method for this kind of situation, for the reasons mentioned previously. I will not repeat the detailed description of the principles of logistic regression and its implementation in the SAS LOGISTIC procedure, as these were covered in Chapter 11.

Let us assume that we are modelling the 'Bad' category of the 'Credit' variable, in other words the existence of non-payments, using the variables preselected on the basis of the initial statistical tests. We carry out a stepwise selection, with the usual 5% thresholds. The FORMAT instruction takes into account the completed discretization and regrouping of categories, without the need for any physical modification of the input data set.

```
PROC LOGISTIC DATA=train ;
CLASS &varselec / PARAM=glm;
MODEL Credit (ref='Good') = &varselec / SELECTION=stepwise
SLE=0.05
SLS=0.05 ;
SCORE DATA=train OUT=train_score;
SCORE DATA=valid OUT=valid_score;
FORMAT Accounts ACCOUNTS. Credit_history $HISTORY. Credit_duration
DURATION. Age AGE. Purpose_credit $PURPOSE. Savings SAVINGS.
Seniority_employment EMPLOYMENT. Marital_status $STATUS. Guarantees
$GUARANTEE. Assets $ASSETS. Other_credits $CREDIT. Status_residence
$HOUSING. Credit CREDIT.;
RUN ;
```

We obtain a model with eight variables. Note that the first five variables included in the model are the five most closely linked to the dependent variable in terms of Cramér's *V*. Age appears in tenth place according to Cramér's *V*, but it is usually included in this type of scoring because it is less correlated with the other variables and contributes a different kind of intrinsic information which is independent of the banking relationship. The presence of credit at other institutions appears in eleventh position, but is included here because it provides external information which is less correlated with the other information of internal origin.

Summary of Stepwise Selection

Step	Effect		DF	Number	Score	Wald	Pr > ChiSq
	Entered	Removed		In	Chi-Square	Chi-Square	
1	Accounts		3	1	89.1537		<.0001
2	Credit_duration		2	2	32.6682		<.0001
3	Purpose_credit		5	3	31.1379		<.0001
4	Credit_history		3	4	15.6917		0.0013
5	Savings		2	5	10.2912		0.0058
6	Age		1	6	7.2765		0.0070
7	Guarantees		2	7	7.2307		0.0269
8	Other_credits		1	8	5.7753		0.0163

A type 3 analysis of effects has already been described in the *Titanic* case study (Section 11.8.13). It is performed for each variable by comparing the sub-model excluding the variable with the model including this variable and the others, in order to test the null hypothesis that this variable has no effect in the model if the other variables are included.

Type 3 analysis of effects

Effect	DF	Wald Chi-Square	Pr > ChiSq
Accounts	3	48.8490	<.0001
Credit_history	3	13.6473	0.0034
Credit_duration	2	39.8004	<.0001
Savings	2	11.7869	0.0028
Age	1	7.6924	0.0055
Purpose_credit	5	34.2489	<.0001
Guarantees	2	7.6865	0.0214
Other_credits	1	5.7013	0.0170

We must then carefully read the table of parameters of the logistic model. It is particularly important to identify Wald statistics below 3.84 (the 5% significance threshold) and coefficients that are inconsistent with the actual non-payment rates.

There is no problem with the 'accounts', 'credit duration', 'age' and 'other credits' variables.

For credit history, the 'Credits without delay' category has an abnormally high coefficient because it is hardly lower than that of the 'Previous non-payments' category. Moreover, the Wald statistic for this coefficient is too low.

For savings, the coefficient of 0.8267 for the '>500 euros' category is inconsistent with the fact that this category presents the least risk. Also, the Wald statistic is too low. This category will therefore be grouped with the 'no savings' category, for which the non-payment rate is practically the same.

In the case of guarantees, the 'co-borrower' category has a negative sign which is inconsistent with the fact that it presents a higher risk than the 'no guarantee' category. These two categories will be grouped together.

For the purpose of credit, the coefficients are completely inconsistent and have little significance. We could group 'business' and 'education' together as their non-payment rates are similar. However, the 'internal fittings' and 'video Hi-Fi' categories are problematic: they have different risk levels, but should be grouped together. This can be done by creating a category called 'Home':

```
PROC FORMAT;
value $PURPOSE
 'A40' = "New vehicle"
 'A50' = "Others"
 'A41' = "2nd hand vehicle"
 'A42','A43','A44','A45' = "Home"
 'A46','A48','A49' = "Education-Business"
 'A47' = "Holidays" ;
RUN ;
```

Table of Purpose_credit by Credit

Purpose_credit	Credit		
Frequency Percent Row Pct	Good	Bad	Total
New vehicle	145 14.50 61.97	89 8.90 38.03	234 23.40
2nd hand vehicle	93 9.30 80.87	22 2.20 19.13	115 11.50
Home	363 36.30 73.33	132 13.20 26.67	495 49.50
Education-Business	99 9.90 63.46	57 5.70 36.54	156 15.60
Total	700 70.00	300 30.00	1000 100.00

When we run a test with this grouping, we find that the 'New vehicle' coefficient is much higher than that for 'Education-Business', which does not fit the real situation, even if this grouping slightly improves the performance. We can see that it is difficult to achieve satisfactory groupings, and we will try to dispense with this variable subsequently. A further reason is that seven variables are quite enough in view of the relatively limited number of observations for training the model.

Analysis of Maximum Likelihood Estimates

Parameter		DF	Estimate	Standard Error	Wald Chi-Square	Pr > ChiSq
Intercept		1	−3.5215	0.6479	29.5442	<.0001
Accounts	**CA < 0 euros**	1	1.9927	0.2913	46.7810	<.0001
Accounts	**CA >= 200 euros**	1	1.0816	0.4466	5.8662	0.0154
Accounts	**CA [0–200 euros]**	1	1.4617	0.2835	26.5899	<.0001
Accounts	**No checking account**	0	0	.	.	.
Credit_history	**Credits without delay**	1	−0.1701	0.4633	0.1348	0.7135
Credit_history	**Current non-payment at another bank**	1	0.8270	0.6314	1.7153	0.1903
Credit_history	**No credit at any time**	1	−0.8344	0.4924	2.8715	0.0902
Credit_history	**Previous non-payments**	0	0	.	.	.
Credit_duration	**>36 months**	1	1.9666	0.3836	26.2768	<.0001
Credit_duration	**16–36 months**	1	1.3391	0.2396	31.2309	<.0001

Credit_duration	<= 15 months	0	0	.	.	.
Savings	<500 euros	1	1.0611	0.3097	11.7404	0.0006
Savings	>500 euros	1	0.8267	0.4400	3.5300	0.0603
Savings	No savings	0	0	.	.	.
Age	<= 25 years	1	0.7323	0.2640	7.6924	0.0055
Age	>25 years	0	0	.	.	.
Purpose_credit	2nd hand vehicle	1	−0.7367	0.4544	2.6281	0.1050
Purpose_credit	Business	1	0.1404	0.3914	0.1287	0.7198
Purpose_credit	Education	1	0.8946	0.4545	3.8734	0.0491
Purpose_credit	Internal fittings	1	0.2388	0.3160	0.5709	0.4499
Purpose_credit	New vehicle	1	1.3640	0.3108	19.2645	<.0001
Purpose_credit	Video HIFI	0	0	.	.	.
Guarantees	Co-borrower	1	−0.0619	0.5057	0.0150	0.9026
Guarantees	Guarantor	1	−1.6185	0.5838	7.6854	0.0056
Guarantees	No guarantee	0	0	.	.	.
Other_credits	No credit	1	−0.6159	0.2579	5.7013	0.0170
Other_credits	Other banks or institutions	0	0	.	.	.

The %AUC macro calculates the area under the ROC curve, which is 0.835 for the training sample and 0.742 for the validation sample.

```
%AUC(train_score,Credit,P_Bad);
%AUC(valid_score,Credit,P_Bad);
```

We may find that a backward stepwise selection would produce exactly the same model in this case, which would support our confidence in the choice of variables.

We will group the categories as shown below, before restarting the logistic regression on the 'accounts', 'credit history', 'credit duration', 'savings', 'age', 'guarantees' and 'other credits' variables, leaving out the purpose of the credit for the reasons stated above.

```
PROC FORMAT ;
value $HISTORY
  'A30','A31' = "Credits with non-payments"
  'A32','A33' = "Credits without delay"
  'A34' = "No credit at any time";
value SAVINGS
  0 = "No savings or > 500 euros"
  1-2 = "< 500 euros"
  3-4 = "No savings or > 500 euros" ;
value $GUARANTEE
  'A101', 'A102' = "No guarantor"
  'A103' = "Guarantor" ;
RUN ;
```

Here is the resulting model:

Analysis of Maximum Likelihood Estimates

Parameter		DF	Estimate	Standard error	Wald Chi-Square	Pr > ChiSq
Intercept		1	−3.1995	0.3967	65.0626	<.0001
Accounts	**CA < 0 euros**	1	2.0129	0.2730	54.3578	<.0001
Accounts	**CA >= 200 euros**	1	1.0772	0.4254	6.4109	0.0113
Accounts	**CA [0–200 euros]**	1	1.5001	0.2690	31.1067	<.0001
Accounts	**No checking account**	0	0	.	.	.
Credit_history	**Credits with non-payments**	1	1.0794	0.3710	8.4629	0.0036
Credit_history	**Credits without delay**	1	0.4519	0.2385	3.5888	0.0582
Credit_history	**No credit at any time**	0	0	.	.	.
Credit_duration	**>36 months**	1	1.4424	0.3479	17.1937	<.0001
Credit_duration	**16–36 months**	1	1.0232	0.2197	21.6955	<.0001
Credit_duration	**<=15 months**	0	0	.	.	.
Age	**<=25 years**	1	0.6288	0.2454	6.5675	0.0104
Age	**>25 years**	0	0	.	.	.
Savings	**<500 euros**	1	0.6415	0.2366	7.3501	0.0067
Savings	**No savings or >500 euros**	0	0	.	.	.
Guarantees	**Guarantor**	1	−1.7210	0.5598	9.4522	0.0021
Guarantees	**No Guarantor**	0	0	.	.	.
Other_credits	**No credit**	1	−0.5359	0.2439	4.8276	0.0280
Other_credits	**Other banks or institutions**	0	0	.	.	.

The area under the ROC curve is 0.806 in the training sample and 0.762 in the validation sample, showing a noticeable convergence of the two areas by comparison with the first model, demonstrating the greater robustness of the new model.

Only 'Credits without delay' has a Wald statistic slightly below the critical threshold of 3.84. However, its coefficient is consistent. I have tested the grouping of this category with the 'No credit at any time' category, but this led to a reduction in performance (AUC = 0.750 in validation). So we shall retain the model above in its existing form. We may recall (see Section 11.8.6) that the Wald test sometimes lacks power when the number of observations is rather limited, with the result that it may fail to detect the significance of a coefficient.

When the logistic model has been chosen, with its variables and their division into categories, we no longer need to use a training sample and a validation sample, and it is preferable to adjust the coefficients in the table above by re-evaluating them against the whole population. This enables us to find the coefficients with the smallest possible bias, and this becomes more useful if the population is not very large, as in this case. In the interests of brevity, we shall assume that this has already been done.

We can then transform the above table of coefficients into a scorecard, following the principle described in Section 2.27 of my book *Étude de Cas en Statistique Décisionnelle* cited in Section 12.7 above.

The transcription of a model into scorecard form is common practice in credit scoring. In the case of a logistic regression on qualitative or discretized variables, producing one coefficient per category of each variable, we simply need to do the following:

- substitute the logit (linear combination of the indicators of the categories) for the probability EXP(logit)/{1 + EXP(logit)} as the score value;
- then normalize the logit so that it lies between 0 and 100 (or, in many cases, 1000, in order to limit the effects of rounding).

In this normalization of the logit, the logistic regression coefficients are replaced with new coefficients, called 'numbers of points', each associated with a category. For example, instead of assigning the coefficient 0.6288 to the category '<=25 years', we assign 8 points to it. The number of points assigned to each category is determined in such a way that each individual has a total number of points in the range from 0 to 100, these two bounds being achievable, at least in theory. This number of points is the score of the individual.

This number of points is perfectly linearly correlated (Pearson correlation = 1) with the logit, but not with the 'true' logistic score, in other words the probability EXP(logit)/ {1 + EXP(logit)}. However, it is perfectly correlated with the logistic score in terms of ranks (Spearman correlation = 1), and its discriminating power is exactly the same, since the ranks, and therefore the classification of the individuals, are retained by the increasing function EXP(x)/{1 + EXP(x)}. The area under the ROC curve for the scorecard is therefore equal to that for the logistic score (allowing for rounding).

This calculation of the score as a sum of points is explained in Section 4.8.3 of Nakache and Confais (2003).[7]

The coefficient of the model associated with category i of variable j is denoted $c(j,i)$. For each variable j, we look for the smallest coefficient $c(j,i)$, denoted min(j), and the largest coefficient $c(j,k)$, and we then calculate Deltamax(j), the difference between them, as the largest difference between the two coefficients of a single variable. We then calculate Total_weight, the sum of all the Deltamax for j. Finally, a number of points is assigned to each category i of the variable j:

$$N(j,i) = 100 \times \frac{c(j,i) - \min(j)}{\text{Total\_weight}}$$

To sum up, we find that:

$$N(j,i) = 100 \times \frac{c(j,i) - \min_k(c(j,k))}{\sum_l [\max_m c(l,m) - \min_m c(l,m)]}$$

This calculation can be carried out simply with SAS (see my book cited above) or with an Excel type spreadsheet. In the present case, we obtain the following grid:

[7] Nakache, J.-P. and Confais, J. (2003) *Statistique Explicative Appliquée*. Paris: Technip.

Variable	**ClassVal0**	**nbpoints**
Age	>25 years	0
Age	<=25 years	8
Other_credits	No credit	0
Other_credits	Other banks or institutions	7
Accounts	No checking account	0
Accounts	CA >= 200 euros	13
Accounts	CA [0–200 euros]	19
Accounts	CA < 0 euros	25
Credit_duration	<= 15 months	0
Credit_duration	16–36 months	13
Credit_duration	>36 months	18
Savings	no savings or >500 euros	0
Savings	<500 euros	8
Guarantees	Guarantor	0
Guarantees	No guarantor	21
Credit_history	No credit at any time	0
Credit_history	Credits without delay	6
Credit_history	Credits with non-payments	13

The advantage of this scorecard is its readability. I have highlighted the category with the maximum weight for each variable. The sum of the highlighted numbers is 100. Credit analysts who are not statisticians can easily understand and comment on this scorecard. This will facilitate their use of the scoring tool and they will be able to compare the numbers of points with their professional understanding. It may also suggest that a given division is inappropriate or that a particular variable has too much weight, and may lead the statistician to re-examine his model.

In our example, the personal criteria (age) or external criteria (credits at other institutions) have less weight. On the other hand, an overdrawn current account is a major risk factor, which is common knowledge, as is the absence of a guarantor. Rather surprisingly, the duration of the credit has more weight than the presence of non-payments.

By way of example, let us calculate the number of points of two applicants. A young person aged under 25, applying for credit for the first time at the institution and having no other credit, with no non-payments, with an account having a slightly positive balance (but less than €200), with a small amount of savings (less than €500), without a guarantor, applying for credit for 36 months, will have a score of

$$8 + 0 + 19 + 13 + 8 + 21 + 0 = 69 \text{ points.}$$

An applicant aged over 25, with credits at competing institutions, without non-payments, with an account having an average balance of more than €200, with more than €500 in savings, without a guarantor, applying for credit for 12 months, will have a score of:

$$0 + 7 + 13 + 0 + 0 + 21 + 0 = 41 \text{ points.}$$

The decision made as to their risk will be considered below.

The implementation of a scorecard in an IT system is very simple. In SAS, the implementation is as follows:

```
DATA test_score ;
SET train_score valid_score ;
nbpoints = SUM (
(age<=25)*8 , (other_credits NE "A143")*7 , (accounts=3)*13 ,
(accounts=2)*19 , (accounts=1)*25 , (credit_duration >= 16 AND
credit_duration <= 36)*13 , (credit_duration > 36)*18 ,
(savings = 1 OR savings = 2)*8 , (guarantees NE "A103")*21 ,
(credit_history in ("A32" "A33"))*6 ,
(credit_history in ("A30" "A31"))*13
) ;
RUN ;
```

We can thus assign a number of points to any credit application. But this calculation cannot be considered to be a decision support tool, because it does not allow the credit analyst to reach an opinion immediately on the credit application that he is examining. A number of points is not enough to tell him whether he should accept or reject the application.

The last step in the construction of the scoring tool is to divide the numbers of points into bands. Three score bands are generally created:

- the least risky, for which a few checks need to be made and the customer must be asked for the minimum requisite documents;
- an intermediate band, for which the file must be examined rather more closely and a standard risk analysis must be conducted;
- the most risky, for which the application is either rejected, or at least sent to the line manager for a more thorough examination of the file.

I shall now show how this division is carried out.

The first step is to use the RANK procedure to create a data set containing the variable 'nbpoints' and its division into deciles, 'dnbpoints'. We then cross-tabulate the deciles with the dependent variable, using the FREQ procedure.

```
PROC RANK DATA = test_score GROUPS = 10 OUT = deciles_score ;
VAR nbpoints ;
RANKS dnbpoints ;
RUN ;

PROC FREQ DATA = deciles_score ;
TABLES dnbpoints * Credit / NOCOL NOPERCENT ;
```

```
FORMAT Credit CREDIT.;
RUN ;
```

The contingency table shows a phenomenon which is well known in scoring: the non-payment rates (more generally, the proportion of occurrence of the event to be predicted) does not increase linearly. Thus we see that this rate is very low and gradually increasing in the first three deciles, then jumps to 15.83% in the fourth decile, and then to 27.55% in the next decile. The non-payment rate then goes through a region of moderate increase, up to the last two deciles in which the rate increases strongly. The grouping of these bands is therefore self-evident, and we can make it match the thresholds of the numbers of points calculated by the MEANS procedure.

Table of dnbpoints by Credit

dnbpoints (Rank for Variable nbpoints)	**Credit**		
Frequency Row Pct	**Good**	**Bad**	**Total**
0	99 95.19	5 4.81	104
1	89 93.68	6 6.32	95
2	100 93.46	7 6.54	107
3	101 84.17	19 15.83	120
4	71 72.45	27 27.55	98
5	60 64.52	33 35.48	93
6	48 59.26	33 40.74	81
7	60 57.69	44 42.31	104
8	38 41.30	54 58.70	92
9	34 32.08	72 67.92	106
Total	700	300	1000

```
PROC MEANS DATA = deciles_score MIN MAX ;
CLASS dnbpoints ;
VAR nbpoints ;
RUN ;
```

Analysis Variable: nbpoints			
Rank for Variable nbpoints	**N Obs**	**Minimum**	**Maximum**
0	104	6.0000000	29.0000000
1	95	33.0000000	37.0000000
2	107	39.0000000	42.0000000
3	120	43.0000000	48.0000000
4	98	49.0000000	54.0000000
5	93	55.0000000	60.0000000
6	81	61.0000000	65.0000000
7	104	66.0000000	69.0000000
8	92	70.0000000	74.0000000
9	106	75.0000000	95.0000000

The thresholds of the numbers of points are 48 and 69, highlighted in the table above. We create a format corresponding to the bands defined in this way, and we can apply it to a table cross-tabulating the number of points and the occurrence of non-payment. We can see that the minimum (maximum) number of points is 6 (95) in the modelling sample; however, it could be smaller (greater) in another sample.

```
PROC FORMAT ;
VALUE nbpoints
0-48 = 'low risk'
49-69 = 'medium risk'
70-high = 'high risk' ;
RUN ;

PROC FREQ DATA = test_score ;
TABLES nbpoints * Credit / NOCOL ;
FORMAT Credit CREDIT. nbpoints nbpoints. ;
RUN ;
```

We have established a band of 42.6% of the credit applications for which the risk is very low, because the non-payment rate is 8.69%, far below the average rate of 30% (remember that these rates are not real, because the sampling has been stratified on the dependent variable). We then have a band of 37.6% of credit applications for which the non-payment rate is slightly above average. Finally, we have a very high risk band in which almost two-thirds of the files show non-payment. They represent about 20% of the applications which must be rejected or at least closely examined.

At 69 points, the young applicant described above presents a medium risk, but on the borderline of high risk. The other applicant has a low risk. Both would have moved into a higher risk band if they had applied for longer-term credit.

Table of nbpoints by Credit

nbpoints	Credit		
Frequency Percent Row Pct	Good	Bad	Total
low risk	389 38.90 91.31	37 3.70 8.69	426 42.60
medium risk	239 23.90 63.56	137 13.70 36.44	376 37.60
high risk	72 7.20 36.36	126 12.60 63.64	198 19.80
Total	700 70.00	300 30.00	1000 100.00

This gives us a practical tool. I shall show another way of obtaining it in the next section.

12.9 An example of credit scoring (modelling by DISQUAL discriminant analysis)

The second modelling method used for our credit scoring example is Saporta's DISQUAL linear discriminant analysis, described in Section 11.6.7. We will test this method because its principles are interesting and different from those of logistic regression, and also because it was invented to overcome some of the problems of credit scoring, in which it is very successful. It is particularly useful in this case, because most of the variables are qualitative.

The first step of the DISQUAL procedure is a multiple correspondence analysis, enabling us to find the factor coordinates of the individuals. We use the same syntax as before, with the BINARY option which creates a complete binary table with one row per individual. We will restrict ourselves to the variables selected in the data exploration, a list of which is given by the &varselec macro variable (see Section 12.7). This time, however, we use the DIMENS=21 instruction because we wish to extract 21 axes, in other words all the axes, because, as you may remember (Section 7.4.1), the total number of axes is the difference between the number of categories (33) and the number of variables (12). One of the outputs of the MCA shows that the total inertia is 33/12–1 = 1.75 and that the first eight axes account for 51% of this inertia. The SOURCE option is used to add to the output data set a variable _VAR_ containing the name of the variable on each row of the _VAR_type (otherwise we only have the name of the category, in the variable _NAME_). This will be useful later on.

```
PROC CORRESP DATA=test_score BINARY DIMENS=21 OUT=output
NOROW=PRINT SOURCE ;
TABLES &varselec ;
FORMAT Accounts ACCOUNTS. Credit_history $HISTORY. Credit_duration
DURATION. Age AGE. Purpose_credit $PURPOSE. Savings SAVINGS.
Seniority_employment EMPLOYMENT. Marital_status $STATUS. Guarantees
$GUARANTEE. Assets $ASSETS. Other_credits $CREDIT. Status_residence
$HOUSING. Credit CREDIT.;
RUN ;

DATA coord ;
SET output ;
WHERE _type_ = 'OBS' ;
id = INPUT(STRIP(_name_),8.) ;
KEEP id Dim1-Dim21 ;
RUN ;

PROC SORT DATA = test_score ; BY id ; RUN ;
PROC SORT DATA = coord ; BY id ; RUN ;

DATA test_acm ;
MERGE test_score (IN=a) coord (IN=b) ;
BY id ;
IF a ;
RUN ;
```

After saving the factor coordinates of the individuals in the test_acm data set at the same time as the initial variables (by merging with the initial data set), we can move on to the second step of the DISQUAL method, namely the linear discriminant analysis on the factor coordinates. We start by selecting the variables with the SAS STEPDISC procedure, adapted for this operation as shown in Section 11.6.6. We wish to know which factor axes to keep, out of the 21 produced by the MCA. In a DISQUAL analysis, these axes will not necessarily be the first in terms of eigenvalues, because there is no reason why the axes accounting for most of the inertia in the cloud of independent variables should be those which are most closely linked to the dependent variable.

```
PROC STEPDISC DATA= test_acm ;
CLASS credit ;
VAR Dim1-Dim21 ;
RUN ;
```

The summary table shows that 10 axes are selected, because they meet the (default) threshold of 15% for the F test applied to Wilks' lambda. However, we shall restrict ourselves to the first eight axes selected, which satisfy the F test at the 5% threshold. Eight is a convenient number of axes, and avoids a long search for axes with low inertia which only weakly differentiate the individuals.

Stepwise Selection Summary

Step	Number in	Entered	Removed	Partial R-square	F value	Pr > F	Wilks' lambda	Pr < Lambda	Average Squared Canonical Correlation	Pr > ASCC
1	**1**	Dim2		0.1235	140.68	<.0001	0.87645066	<.0001	0.12354934	<.0001
2	**2**	Dim1		0.0433	45.09	<.0001	0.83852569	<.0001	0.16147431	<.0001
3	**3**	Dim5		0.0296	30.39	<.0001	0.81369748	<.0001	0.18630252	<.0001
4	**4**	Dim3		0.0232	23.60	<.0001	0.79484596	<.0001	0.20515404	<.0001
5	**5**	Dim9		0.0188	19.06	<.0001	0.77989449	<.0001	0.22010551	<.0001
6	**6**	Dim15		0.0161	16.24	<.0001	0.76734394	<.0001	0.23265606	<.0001
7	**7**	Dim11		0.0073	7.25	0.0072	0.76177807	<.0001	0.23822193	<.0001
8	**8**	Dim4		0.0041	4.04	0.0447	0.75868553	<.0001	0.24131447	<.0001
9	**9**	Dim12		0.0030	2.96	0.0858	0.75642611	<.0001	0.24357389	<.0001
10	**10**	Dim16		0.0023	2.27	0.1324	0.75469550	<.0001	0.24530450	<.0001

We then perform a conventional linear discriminant analysis on the factor axes selected in this way, namely those of rank 2, 1, 5, 3, 9, 15, 11 and 4. We use the standard DISCRIM procedure (Section 11.6.6).

```
PROC DISCRIM DATA=test_acm (WHERE=(sample="T")) METHOD=normal
POOL=yes CROSSVALIDATE ALL CANONICAL OUT=scores
OUTSTAT=statdescr TESTDATA=test_acm (WHERE=(sample="V"))
TESTOUT=testout ;
CLASS Credit ;
PRIORS prop ;
VAR Dim2 Dim1 Dim5 Dim3 Dim9 Dim15 Dim11 Dim4 ;
RUN ;
```

Using the %AUC macro, we find an area under the ROC curve of 0.823 for the learning sample and 0.777 for the test sample, indicating a degree of overfitting.

We now progressively simplify the model, removing the axes one by one:

Removal of axis	AUC in learning	AUC in testing
Dim4	0.822	0.776
Dim11	0.812	0.780
Dim15	0.794	0.792
Dim9	0.783	0.785
Dim3	0.772	0.767

We find that the AUCs in training and validation converge towards a very similar value, a sign that the performance is free of optimistic bias in both case, and therefore has a good chance of being obtained subsequently with other samples. The optimum is achieved with a model with five variables, namely Dim2, Dim1, Dim5, Dim3 and Dim9.

We rerun the linear discriminant analysis on the five variables and retrieve the coefficients of the Fisher discriminant function in an ODS file.

```
ODS OUTPUT LinearDiscFunc = Fisher ;
PROC DISCRIM DATA=test_acm (WHERE=(sample="T")) METHOD=normal
POOL=yes CROSSVALIDATE
ALL CANONICAL OUT=scores OUTSTAT=statdescr
TESTDATA=test_acm (WHERE=  (sample="V")) TESTOUT=testout ;
CLASS credit ;
PRIORS prop ;
VAR Dim2 Dim1 Dim5 Dim3 Dim9 ;
RUN ;
```

Linear Discriminant Function for Credit

Variable	Bad	Good
Constant	−1.50516	−0.42748
Dim2	1.65305	−0.91802
Dim1	0.95066	−0.35603
Dim5	−1.10068	0.53319
Dim3	0.75686	−0.17947
Dim9	−0.81894	0.36289

The difference between the two columns gives us a score function on the factor axes. We find that the coefficients of this function are not an increasing function of the eigenvalue, and that an axis such as the ninth has a coefficient that is not negligible. It is impossible to know in advance which axes will be selected and what their coefficients will be in the score function. However, we should certainly avoid the ones with the lowest eigenvalue.

We should also note the absence of standard deviations for the coefficients, which is not the case with logistic regression. They are not provided by linear discriminant analysis, and we must carry out multiple evaluations of the coefficients on bootstrap samples in order to obtain an estimate of their standard deviations.

```
DATA Fisher ;
SET Fisher ;
SCORE = Bad - Good ;
RUN ;
```

Variable	Bad	Good	SCORE
Constant	−1.50516	−0.42748	−1.07768
Dim2	1.65305	−0.91802	2.57107
Dim1	0.95066	−0.35603	1.30669
Dim5	−1.10068	0.53319	−1.63387
Dim3	0.75686	−0.17947	0.93633
Dim9	−0.81894	0.36289	−1.18184

In the next step, we move from the coefficients on the factor axes to the coefficients on the categories of the initial variables, using the definition of the axes as linear combinations of the indicators of the categories. We start by transforming the data set of the coefficients of the Fisher score function.

```
PROC TRANSPOSE DATA=Fisher OUT=t NAME=_TYPE_ ;
VAR SCORE ;
RUN ;
```

TYPE	COL1	COL2	COL3	COL4	COL5	COL6
SCORE	−1.07768	2.57107	1.30669	−1.63387	0.93633	−1.18184

```
DATA coeff_score ;
SET t ;
RENAME COL1 = Constant COL2 = Dim2 COL3 = Dim1 COL4 = Dim5 COL5 = Dim3
COL6 = Dim9 ;
_NAME_ = "Estimate" ;
RUN ;
```

TYPE	Constant	Dim2	Dim1	Dim5	Dim3	Dim9	_NAME_
SCORE	−1.07768	2.57107	1.30669	−1.63387	0.93633	−1.18184	Estimate

The above 'coeff_score' data set can then be applied by the SCORE procedure to the 'output' data set produced by the MCA, which contains, in particular, the coordinates Dim1, ..., Dim21 of each category on the factor axes. This procedure will calculate the scalar product of the vector of the first data set and each vector of the second; each row, and therefore each category, corresponds to one vector. On each row, the result will be placed in a variable whose name is given by the _NAME_ variable of the first data set. Thus, for the category 'CC >= 200 euros', the scalar product Estimate will be:

$$(2.57107 \times 0.01611) - (1.30669 \times 0.46706) + \ldots = -0.84991.$$

This will be the coefficient of the DISQUAL score function for this category. Thus we obtain the coefficient of each category.

```
PROC SCORE DATA=output (WHERE=(_TYPE_ = "VAR") KEEP=_TYPE_ _VAR_ _
NAME_ Dim1-Dim9 RENAME=(_VAR_=Variable))
SCORE=coeff_score OUT=coeff_disqual ;
VAR Dim2 Dim1 Dim5 Dim3 Dim9 ;
RUN ;
```

Variable	_NAME_	Dim1	Dim2	...	Dim9	Estimate
Accounts	CA >= 200 euros	−0.46706	0.01611	...	1.91183	−0.84991
Accounts	CA < 0 euros	0.16320	0.55458		−0.46835	3.26808
Accounts	CA [0–200 euros]	0.15208	0.42907		0.11839	0.89370
Accounts	No checking account	−0.14265	−0.68119		−0.06083	−2.74699
Credit_history	Credits with non-payments	1.17406	0.63792		0.03760	4.66420
Credit_history	Credits without delay	−0.14471	0.23650		0.09229	0.03387
Credit_history	No credit at any time	−0.05140	−0.69260		−0.20609	−1.48821
Credit_duration	>36 months	1.43311	0.28466		−0.86441	2.46306
...	...	...	...		...	...

We then move from the table of coefficients on the initial variables to the scorecard, as we did previously for logistic regression, and once again we should start by re-evaluating these coefficients over the whole population. This new scorecard uses more variables and is more complex than the previous one.

Because of the greater number of variables, the variables that were present in the first scorecard must now have a lower weight: this is the case for age, the existence of other credit, the current account balance (which is still the most important variable) and the duration of the credit. However, savings have an increased weight, as does the credit history. The latter also now has a weight equal to that of the average current account balance, which seems more reasonable than the limited weight that it had in the first scorecard.

The absence of a guarantor has considerably less weight than in the first scorecard. This is because the two categories of the 'Guarantees' variable are closer to each other, for all the chosen axes, than are the categories of other variables, such as the current account balance. The moderate weight of this variable also appears to be a better reflection of its non-payment rates than the high weight of this variable in the logistic model. The advantage of the DISQUAL method is that we can allow for not only the links with the dependent variable (because of the linear discriminant analysis) but also the links between the independent variables (because of the MCA). In looking for a risk profile, we are interested in the elements of the profile which clearly distinguish the at-risk individuals from the rest, but also distinguish between the individuals. This double aim is reminiscent of PLS regression.

Other variables appear in the scorecard, some of them with a greater weight than the variables that were already present in the previous scorecard: these are seniority in employment, assets owned, marital status and residential status. The purpose of credit also appears, although it was excluded from the logistic regression. In this case it only shows a minor anomaly, in that the 'Education-Business' category has a greater weight than the 'New vehicle' category, although the latter is slightly more risky (38.03% non-payment rate as against 36.54% – see above). However, there is only one point of difference in the scorecard, and furthermore the 'Education-Business' category includes the 'Education' category which is by far the most risky (44% non-payment rate).

In conclusion, this scorecard is more balanced and comprehensive than the previous one. As the areas under the ROC curve are almost identical in learning and testing, we can expect this scorecard to have a good capacity for generalization to other samples of customers. Admittedly, its greater complexity might lead us to expect the opposite, given that a simpler model is usually generalized more satisfactorily, but the rather large number of variables in this case is not achieve at the price of a regression on correlated variables, as in the usual methods, because the regression is carried out on factor axes that are orthogonal by construction. As with PLS regression, we retain more predictors, with weights reduced where necessary, without any problems of collinearity.

Variable	_NAME_	Estimate	nbpoints
Age	>25 years	−0.32343	0
Age	<= 25 years	1.37885	4
Seniority_employment	E >= 4 years	−0.92331	0
Seniority_employment	E [1–4]years	−0.28461	2
Seniority_employment	Unemployed or <1 year	2.09716	8

Other_credits	No credit	−0.22075	0
Other_credits	Other banks or institutions	0.96608	3
Assets	Property	−1.39717	0
Assets	Other assets	0.13477	4
Assets	No known assets	2.06488	9
Accounts	No checking account	−2.74699	0
Accounts	CA >= 200 euros	−0.84991	5
Accounts	CA [0–200 euros]	0.89370	9
Accounts	CA < 0 euros	3.26808	16
Credit_duration	<=15 months	−0.68205	0
Credit_duration	16–36 months	0.16530	2
Credit_duration	>36 months	2.46306	8
Savings	no savings or >500 euros	−3.13341	0
Savings	<500 euros	1.30485	11
Guarantees	Guarantor	−0.71941	0
Guarantees	No guarantor	0.03946	2
Credit_history	No credit at any time	−1.48821	0
Credit_history	Credits without delay	0.03387	4
Credit_history	Credits with non-payments	4.66420	16
Purpose_credit	2nd hand vehicle	−1.33668	0
Purpose_credit	Household	−0.84660	1
Purpose_credit	New vehicle	1.17962	7
Purpose_credit	Education-Business	1.90228	8
Marital_status	Man: unmarried/married/widowed	−0.62477	0
Marital_status	Woman: divorced/separated/married	0.91186	4
Marital_status	Man: divorced/separated	2.34351	8
Status_residence	Owner	−0.72992	0
Status_residence	Not owner	1.81335	7

The scorecard can then be programmed and applied to the whole population, or to another sample. We create a data set which will include the training and validation samples, and also the logistic score whose performance can be tested against that of the DISQUAL score.

```
DATA test_disqual ;
SET scores testout ;
nbpoints_disqual = SUM (
(Age <= 25) * 4, (Seniority_employment = 3) * 2, (Seniority_employment
<= 2) * 8, (Other_credits IN ("A141" "A142")) * 3, (Assets IN
("A122" "A123")) * 4, (Assets = "A124") * 9, (Accounts = 3) * 5,
```

```
(Accounts = 2) * 9, (Accounts = 1) * 16, (Credit_duration >= 16 AND
Credit_duration  <= 36) * 2, (Credit_duration > 36) * 8, (Savings IN (1 2)) *
11, (Guarantees = 'No guarantor' ) * 2, (Credit_history IN ("A32" "A33")) *
4, (Credit_history IN ("A30" "A31")) * 16, (Purpose_credit IN ("A42"
"A43" "A44" "A45")) * 1, (Purpose_credit = "A40") * 7, (Purpose_credit IN
("A46" "A48" "A49")) * 8, (Marital_status = "A92") * 4, (Marital_status =
"A91") * 8, (Status_residence NE "A152") * 7
) ;
 RUN ;
```

Because of the revision of the DISQUAL model and the rounding in the scorecard, the correlation coefficient between the ranks of the DISQUAL score and the corresponding number of points is 'only' 0.99365. This is enough to create a difference in the area under the ROC curve measured over the set of 1000 data sets: the area is 0.793 for the DISQUAL score and 0.791 for the number of points. For the logistic model, this area is 0.790 for both the score and the number of points, but the correlation coefficient of 0.99923 is closer to 1. In order to reduce the differences in the DISQUAL model, we would have to create a scorecard based on 1000 points instead of 100 points. The small loss of simplicity would be offset by the gain in precision.

The number of points in the DISQUAL scorecard is then divided into score bands, as was done previously for logistic regression, using deciles (if the volume were greater, we could use a finer division).

Table of dnbpoints_disqual by Credit

dnbpoints_disqual (Rank for Variable nbpoints_disqual)	Credit		
Frequency Row Pct	Good	Bad	Total
0	100 96.15	4 3.85	104
1	97 91.51	9 8.49	106
2	81 89.01	10 10.99	91
3	85 82.52	18 17.48	103
4	64 76.19	20 23.81	84
5	79 73.15	29 26.85	108
6	66 71.74	26 28.26	92
7	59 55.66	47 44.34	106

8	40 36.70	69 63.30	109
9	29 29.90	68 70.10	97
Total	700	300	1000

Analysis Variable: nbpoints_disqual

Rank for Variable nbpoints_disqual	N Obs	Minimum	Maximum
0	104	1.0000000	16.0000000
1	106	17.0000000	22.0000000
2	91	23.0000000	26.0000000
3	103	27.0000000	30.0000000
4	84	31.0000000	33.0000000
5	108	34.0000000	38.0000000
6	92	39.0000000	42.0000000
7	106	43.0000000	47.0000000
8	109	48.0000000	55.0000000
9	97	56.0000000	80.0000000

As before, score bands appear naturally, and if we cross-tabulate them with the dependent variable (non-payment), we obtain a table that can be compared with the one provided by logistic regression.

```
PROC FORMAT ;
VALUE nbpointsd
0-30 = 'low risk'
31-47 = 'medium risk'
48-high = 'high risk' ;
RUN ;
```

Table of nbpoints_disqual by Credit

nbpoints_disqual	Credit		
Frequency **Percent** **Row Pct**	**Good**	**Bad**	**Total**
low risk	363 36.30 89.85	41 4.10 10.15	404 40.40

(*continued*)

medium risk	268 26.80 68.72	122 12.20 31.28	390 39.00
high risk	69 6.90 33.50	137 13.70 66.50	206 20.60
Total	700 70.00	300 30.00	1000 100.00

The size of the bands is more or less the same as those of the logistic model, but the at-risk files are isolated more clearly, with 20.6% of the population having a non-payment rate of 66.5%, as against the previous model where 19.8% of files had a non-payment rate of 63.6%. On the other hand, the threshold between the fourth and fifth deciles (there is another between the third and fourth) is less clearly marked, so the less risky files are slightly less well identified, but this is less important. This can be seen if we superimpose the ROC curves of the two scorecards on the same graph (Figure 12.8). Neither of the models appears to be superior to the other overall: the DISQUAL model is more discriminating on the most risky files and less so on the others, and therefore its ROC curve is above the other on the left of the graph, but falls below it on the right. This graph is plotted, for the whole sample, using the SAS syntax given in Section 2.24 of my book *Étude de Cas en Statistique Décisionnelle* (see Section 12.7 above); this syntax can also be downloaded from the Éditions Technip website.

In conclusion, for all the reasons given above, namely the more similar performances in training and validation, the more comprehensive scorecard, and the better identification of the more risky files, I tend to prefer the second scorecard obtained by DISQUAL discriminant analysis.

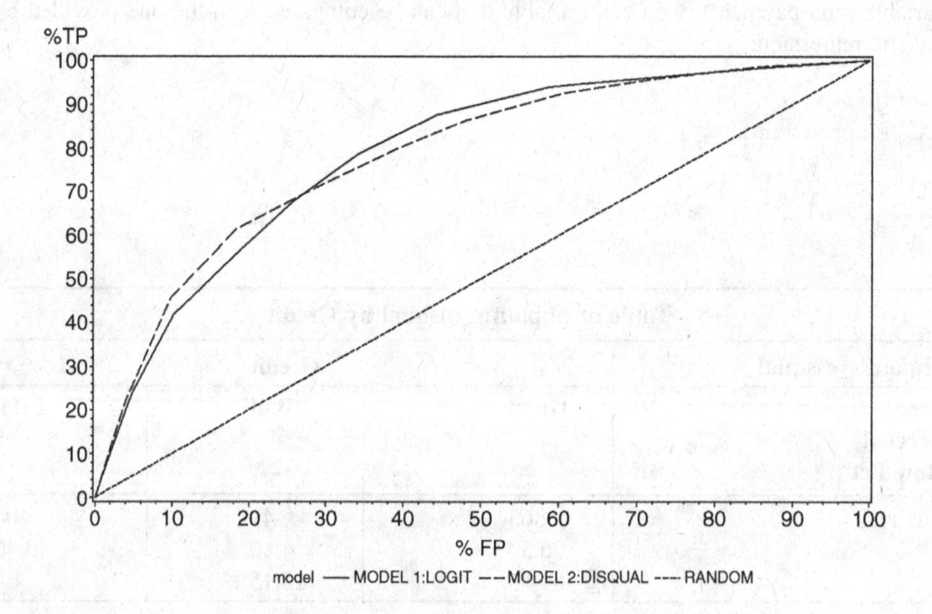

Figure 12.8 ROC curves for the two scorecards.

12.10 A brief history of credit scoring

The origins of credit scoring lie in the work[8] of David Durand, who showed in 1941 that the risk of non-payment of a borrower could be modelled by discriminant analysis on the basis of a number of characteristics such as the age and sex of the individual. At that time, Fisher's discriminant analysis was a mere 5 years old! After the Second World War, this theoretical work was taken up by businesses facing a shortage of credit analysts. With the advent of the first computers, in the late 1950s, credit scoring began to be used on an industrial scale, particularly by consultancy firms, including the pioneers, Fair, Isaac and Company (FICO). These were responsible for the first credit scoring system in 1958. From the 1960s onwards, the development of computing tools enabled the constantly increasing numbers of credit applications from individuals to be processed *en masse*. Credit scoring was subsequently applied to businesses, with the well-known Z-score of Edward I. Altman,[9] which was a discriminant function of five financial ratios, capable of predicting the bankruptcy of a business one year in advance with a reliability of about 94%. This function, invented in 1968, has been updated and refined ever since. The initial Z-function was:

$$Z = 0.012X1 + 0.014X2 + 0.033X3 + 0.006X4 + 0.999X5$$

where

X1 = working capital / total assets
X2 = retained earnings / total assets
X3 = earnings before interest and taxes / total assets
X4 = market value equity / book value of total debt
X5 = sales / total assets.

Numerous organizations then made use of discriminant analysis to construct credit scoring models, defined by L.J. Mester in 1997 as 'a statistical method used to predict the probability that a prospective borrower or an existing debtor will default'.

In its earliest form, credit scoring was simply a matter of classifying credit applicants to identify those who were least and most at risk, for the purpose of a quick approval or refusal of credit. Later, it was used to 'price' credit, in other words to modulate the interest rate of the credit according to the score to incorporate a risk premium into the rate. The most advanced models also incorporate the attractiveness of the rate offered to the applicant.

A major step forward in credit scoring resulted from the establishment of the rules on the bank solvency ratio (the Basel II ratio – see Section 12.2), which led to the wider use of scoring in many banks' portfolios (for lending to private customers, businesses, individual entrepreneurs, etc.). The rules also made it necessary to take into account not only the probability of individual default, but also the correlation between the defaults of different borrowers. It also became necessary to predict a probability of default in 12 months, over a number of years, with allowance for economic cycles. Here again, the aim must be not only to order a set of

[8] Durand, D. (1941) *Risk Elements in Consumer Instalment Financing*. New York: National Bureau of Economic Research.

[9] Altman, E.I. (1968). Financial Ratios Discriminant Analysis and the Prediction of Corporate Bankruptcy. *Journal of Finance*. 23(4), 589–609.

individual credit applications in isolation, but also to take the economic environment into account and provide a medium-term forecast.

Various methods of prediction are used, including the nearest-neighbours algorithm, neural networks, more recently support vector machines, and also genetic algorithms for developing new optimal scorecards from initial scorecards, using the mechanisms of mutation and crossing. Since 1962, Markov chains have been used experimentally for modelling the behaviour of customers on the basis of their previous behaviour. However, logistic regression is by far the most commonly used method in credit scoring, after extensive use of linear discriminant analysis. LDA provides a simple example of a score function. This is the Z function (according to Altman's terminology) of the Banque de France, which was used for a considerable period to score French industrial companies. Eight financial ratios were taken from their balance sheets and used to calculate Z according to the formula

$$100 \times Z = -85.544 - 1.255 \times X1 + 2.003 \times X2 - 0.824 \times X3 + 5.221 \times X4 - 0.689 \times X5 - 1.164 \times X6 + 0.706 \times X7 + 1.408 \times X8,$$

where

- X1 = financial costs/gross earnings (%)
- X2 = liable capital/invested capital (%)
- X3 = cash flows from operations/overall indebtedness (%)
- X4 = gross earnings/turnover before tax (%)
- X5 = trade payables/net purchases (days)
- X6 = rate of change of value added (%)
- X7 = [stocks of work in progress – advance payments from customers + trade debtors]/ production (days)
- X8 = physical investments/value added (%).

If $Z > 0.125$, the business is assumed to be in good health; if $Z < -0.25$, the business is at risk; between these two values is an area of uncertainty.

The interested reader can consult the following articles which are reasonably comprehensive comparative surveys of the data mining techniques used in credit scoring:

References

Galindo, J. and Tamayo, P. (2000) Credit risk assessment using statistical and machine learning: basic methodology and risk modeling applications. *Computational Economics*, 15, 107–143.

Hand, D.J. and Henley, W.E. (1997) Statistical classification methods in consumer credit scoring: a review. *Journal of the Royal Statistical Society, Series A*, 160(3), 523–541.

Hand, D.J. and Henley, W.E. (1993–4) Can reject inference ever work? *IMA Journal of Mathematics Applied in Business and Industry*, 5, 45–55.

13

Factors for success in a data mining project

The aim of this chapter is to present the factors for success in a data mining project in business, particularly where the project is implemented in-house rather than outsourced. It will describe the pitfalls to be avoided and provide an outline of the expected return on investment.

13.1 The subject

The subject of the study must of course be one that requires the use of data mining tools and cannot be dealt with by simple descriptive statistics. Data mining will not help us to find the '20% of customers who generate 80% of the profits', but it is useful for determining their profile or for discovering those who do not form part of this group at present, but who will in future.

The subject, the target population and the objectives must be precisely specified. We must avoid constructing a score on a certain customer segment and then extending it to another for which it is inappropriate. The results must be capable of being measured. We must try to estimate the return on investment.

The objectives must be realistic: if the rate of response to a mailing is 1%, it may perhaps be increased to 3%, but certainly not 10%. Unrealistically ambitious objectives can lead to disappointment which will harm the credibility of data mining and its wider application in the business.

The business must have at least a degree of expertise on the subject.

The subject must be a challenge for the enterprise, and must offer some real benefits. This is particularly true of a first project, which must be convincing and develop loyalty.

The business must be both willing and able to implement the solutions proposed by data mining. For example, it is necessary to check the IT and electronic publishing resources: there is no point in devising customized mailings if they cannot be provided at an acceptable cost.

Data Mining and Statistics for Decision Making, First Edition. Stéphane Tufféry.

13.2 The people

The project must be supported by a business decision. The *decision makers* must be made aware of the project and must back it.

The *specialist staff* of the business must be mobilized:

- before the project, to specify its content, outline the correct underlying concepts, identify useful information sources, and supply the necessary data and definitions for the study;
- during the study, to assess relevance and identify elements for closer examination among the phenomena discovered by the statistician;
- after the study, to use the results and take the appropriate action.

IT specialists are needed to extract the data, construct the database to be supplied to the statistician, and if necessary to program statistical models subsequently in an industrial computing environment.

Statisticians are need to analyse and format the data, detect any anomalies, choose the appropriate modelling techniques, implement them correctly, produce effective models from these, test the models and analyse the results of their application.

A thorough knowledge of statistics, data, and the nature and customers of the business is essential if we are to decide whether a clustering is correct and usable (the clusters must be homogeneous, consistent and readable) and whether a model is correct (the coefficients of the variables must have relevant values and signs, and sufficient reliability), and for the purposes of regrouping or transforming the data and constructing good indicators or excluding redundant data. A model or a classification may be far from self-evident and may even include unexpected elements (which is what makes data mining useful), but it must not be unlikely, incomprehensible or unusable by the people for whom it is designed.

We can see that the business will require a lot of in-house skills. The active cooperation of all these people is essential. Specialist staff and future users must be involved in the progress of the study: the knowledge contained in the data is only one part of the business's general know-how. Data mining on its own will not provide the best models; these will be created by the interplay between the knowledge extracted from the data and the experience of specialist staff. However, it is preferable for a single person, the *data miner*, to be skilled in three areas (knowledge of the business, statistics, and information technology) to ensure the fast and effective deployment of the project.

Finally, we must add the necessary *legal* expertise to decide on the legal and regulatory aspects of using the data and carrying out the planned processing.

13.3 The data

We must have data which are known, reliable and usable, and there must be enough of them (see Sections 2.4 and 3.3–3.5). We need to archive the data that change over time and whose variation is to be analysed, or those which will be used to predict subsequent phenomena and behaviour. We need to keep all information on earlier business operations: who has been contacted, by what channel, who has responded, after what time interval, who has been followed up, how many times, who has accepted or refused, what the cost of the process was, and so on. This will help us avoid the problems mentioned in Section 11.16.2. At the very least, we must preserve the number of business contacts with each customer and their results. This is because data mining does not guess the profile of 'good' customers, but extrapolates from the data provided, mainly the results of earlier operations, which can be used to extract positive and negative profiles relating to risk, propensity, etc. It is therefore absolutely essential to store this information on business operations.

The multiplication of customer contact and distribution channels also offers many sources of information, which tends to be scattered accordingly. To make the best use of this information, we must be able to consolidate them into a coherent form in a synthetic database, in order to obtain a unique and comprehensive view of each customer. This is not easy, and is not always achieved.

13.4 The IT Systems

If a business is implementing data warehousing and data mining projects at the same time, it is preferable to execute them in parallel rather than in sequence. To archive the data first and then carry out the data mining would not be an acceptable procedure, because to some extent it is the data mining itself that determines what must be archived in a reliable form, and how to go about this task. It would also be unfortunate if we waited several years for the completion of the data warehouse and then found, in the early stages of data mining, that we lacked some important data which were not considered when the warehouse was designed. There are at least four good reasons for the early application of data mining:

- it will generate a return on investment which will provide evidence of the value of data mining, especially to managers in charge of budgets;
- it will identify the most important data and indicators for the construction of relevant models;
- it will start the process of archiving business operations, as mentioned in Section 13.3;
- it will support and develop the skills of the participants.

For a first trial, or pilot project, it is possible and may even be preferable to use existing tools, or new tools that are easily implemented, for data collection and output, and to wait for the results of the initial trials before deciding on major changes in the IT systems for full-scale operation.

It will also be necessary to archive numerous files, requiring a large storage capacity. Unlike conventional management information systems, which back up data on media suitable for limited, one-off retrieval, data mining often has to process several years of archives simultaneously. Whereas a simple back-up system only has to retrieve a given stored file on a given day, the data mining system must be able to put a very large number files – up to several tens of terabytes – on-line.

Data mining also uses specialized data models. We cannot directly 'mine' the production data or the tables of a data centre or data warehouse. We must first set up a special *data mart*, known as a *mining mart*, a modelling base (Section 2.3), resulting in an even greater increase in the volume of computer data to be stored. However, we can try to standardize the descriptions of the modelling bases so that they can be used for a number of different studies or applications in the business.

Finally, it is obvious that a successful implementation of data mining is dependent on full integration into the users' workstations, especially those of the customer service representatives, who need to have fast and straightforward access to the data, given the context in which they use them. The output must be user-friendly and allow for direct, simple and unambiguous interpretation. If necessary it must be made simpler than the detailed results supplied by the data mining algorithms. For example, the average human mind will find it difficult to think of more than six or seven customer segments simultaneously and instantly place a customer in the correct segment.

13.5 The business culture

Data mining must form part of the business culture. The business must ensure that it maintains its expertise in data mining and statistics, as well as the quality of the data gathered and stored. Every business operation dependent on data mining must be carefully managed in its implementation and monitoring (recording the results). The iterative nature of data mining must be clearly understood, and the results of an operation dependent on a data mining study must be used automatically to enrich the next study.

Care must be taken in presenting and 'selling' data mining to marketing and sales managers and to field staff, who may think it calls their know-how into question or is designed to replace it. They must be persuaded that data mining only offers an aid to decision making, not the decision itself, which is always up to them. They must also be reminded to keep the marketing databases up to date, especially as regards the results of campaigns, namely the acceptances and refusals. Customer service representatives must be made aware of the gains in productivity and security that they can expect from data mining.

Marketing managers must also be involved in data mining studies, so that they do not feel that they have lost control of the choice of target customers. Data mining will not make their experience obsolete, but should incorporate it, not after the identification of targets for campaigns, in the definition of each target, but rather beforehand, in the design of the data mining models. In some businesses it will be essential to make the change from 'product-orientated' to 'customer-orientated' marketing. Conventional product-orientated marketing starts with a product i, looks for the period of the year P_i which will be best for selling it, looks for the customers C_i who are likely to buy it, and targets customers C_1 in period P_1, customers C_2 in period P_2, and so on. The drawback of this method is that the intersection of the C_i is not necessarily empty: in other words, the same customer may be targeted several times without any consistency in the marketing communications and trading logic. At best, this is useless; at worst, it detracts from the customer's image of the business. Of course, it would be possible to excluded previously targeted customers from each campaign, but there would be no guarantee that the order of targeting would lead to the best results. Conversely, some customers will never be targeted. Moving to customer-orientated marketing means that marketing operations are carried out according to the profile of customers, their requirements or their life events, rather than according to the events in the life of the products. We know that a given customer belongs to a given segment characterized by a certain consumption of products, services and means of access, and that this customer should therefore be offered certain products and services in a certain order of priority and via a certain channel. It will be the strong trends in the customer segment that determine the priorities for this customer, not the sequence of marketing campaigns and the randomness of targeting. This will require a radical review of working habits, or even of whole organizations, where the marketing management is structured according to product lines rather than customer segments.

13.6 Data mining: eight common misconceptions

13.6.1 No *a priori* knowledge is needed

It is true that some descriptive methods such as cluster analysis can be used without knowing what the resulting clusters will be, or even the appropriate number of clusters.

However, it is important to know that the result of the clustering is influenced by the choice of data and their coding at the input of the algorithm (an example is the standardization of continuous variables), and it is therefore impossible to be completely neutral. We could imagine a system in which all the available computer data were fed into the clustering process. But even if this were technically feasible, such a solution would mean that the result of the classification would be dependent on the computer data model, rather than on the business or statistical requirements, which would obviously be unsatisfactory. Additionally, for purely technical reasons, there may be redundant data which could distort the result of the cluster analysis.

As for the predictive methods of data mining, these require some *a priori* input in all cases, because it is necessary to choose a target (dependent) variable whose definition and categories will be carefully weighted.

In any case, someone who knows what he is looking for is more likely to find it!

13.6.2 No specialist staff are needed

The assistance of professional specialists (in production, engineering, risk assessment, marketing, etc.) is indispensable at several stages of a data mining study. First of all, it is required for the definition of the objectives. For example, before drawing up a risk score for a financial establishment, we need to agree on the definition of a risk: is it a delayed payment, a downgrading of debt, or a financial loss for the establishment? This is not a question for the statistician only. It must be answered by the professional specialist, who will consider the regulatory constraints and the policy of the establishment among other matters.

The assistance of specialists is also required in building up the store of useful and legally usable data, including both raw and composite data. It is useful to know which data are considered to be relevant by the specialists, and which may have concealed pitfalls, even if the statistician may subsequently question certain prejudices about the importance of some of the data, such as the debt ratio for the granting of credit.

Finally, such assistance is essential for analysing the results. Given two classifications of equal statistical merit, a marketing analyst may prefer one which he considers to be more suitable for business use. On seeing the initial results of a study, a professional specialist may also say whether he considers them to be predictable, new and worth investigating, or surprising and highly suspect, in which case the validity of the data, the sampling, and the use of the data mining tools will be called into question. The professional specialist may also be consulted to discover if a correlation between the dependent variable and an independent variable is created simply by the definition of the variable, or if it can be considered valid. In some complex problems such as the analysis of the financial health of businesses based on their accounting data, the cooperation of the statistician and the professional specialist is essential if errors of interpretation are to be avoided.

13.6.3 No statisticians are needed ('you can just press a button')

In any data mining study, the most time-consuming and most decisive stage is data processing. It is entirely dependent on the use of statistical analyses for verifying the reliability of the variables, their distribution, their correlations, etc., and for carrying out reliability improvements, transformations, discretizations and groupings on categories and the like, before the data mining algorithms are used. These operations are not performed in the same way for every algorithm. Not all algorithms can accept every type of input variable. Variables with missing values can be retained in some cases, but not in others. Some algorithms also require preliminary sampling (see below). In predictive methods, we must ensure that variables correlated by definition with the dependent variable are not included among the independent variables. We must also be wary of the phenomenon of overfitting. Finally, the setting of the parameters of data mining algorithms can have a considerable effect on the results, and certain seemingly fine adjustments can lead to surprising differences. Simply encoding a qualitative variable as a 'discrete numeric' variable may be enough to distort the results completely, even if it is only one variable among a hundred other correctly coded ones.

Finally, the data processing phase is interleaved with the modelling phase, as the first models produced are hardly ever completely satisfactory, and require further data transformations before the operations are repeated.

On completion of the data processing, the reading of the results may be deceptive; for example, correlation may be confused with causation.

In conclusion, I quote Philippe Besse, from his course on 'Statistical modelling and learning' at the University of Toulouse (France):

> With the tools now available, it is becoming so easy to start the computation process that some people compare a data miner with a driver, saying that you do not need to be a skilled mechanic to drive a car. However, the designer of a modelling, segmentation or discrimination procedure has to make more or less implicit decisions which are far from being neutral and which are far more complex than the simple choice of a fuel by a driver at a service station.

13.6.4 Data mining will reveal unbelievable wonders

The models produced by data mining are rarely marvellous or extraordinary; they normally make use of variables considered to be discriminating by professional specialists, in a common-sense way. So what does data mining offer us? Simply the fact that there are thousands of common-sense combinations of variables known to be discriminating for any given problem area, and that data mining enables us to detect the very best possible combination (or one of the best), together with the precise parameter that should be assigned to each of the variables. Ultimately, a small improvement in each rule among a set of several targeting rules is enough to multiply the response rate by a factor of 3–4.

13.6.5 Data mining is revolutionary

Data mining incorporates conventional data analysis, and only differs from it in the following ways (see also Section A.1.2):

- some of the techniques used, such as decision trees and neural networks, are exclusive to data mining;
- the number of individuals studied is often larger in data mining, where the optimization of the algorithms for data processing may be crucial;
- data mining sometimes prefers a slightly less precise model, if it is much more understandable;
- data mining models are integrated into industrialized data processing procedures, with automatic updates, computation and outputs.

In spite of everything, we cannot claim that data mining is really a radically new approach.

13.6.6 You must use all the available data

We might think that the results of a data mining model will improve as the number of input variables increases. However, this is not the case. Models are degraded by unreliable or incomplete variables and by the presence of outliers; furthermore, redundant variables may affect a cluster analysis, variables with categories having irregular frequencies may affect a factor analysis, poorly discriminating or excessively intercorrelated variables may reduce the predictive power of a discriminant analysis, and an excessive number of variables may swamp a neural network. Quite often, when a good score model has been built, an attempt is made to improve it by incorporating a new variable, but, even though the relevance and reliability of this variable have been ascertained, it actually degrades the quality, and above all the robustness, of the model.

13.6.7 You must always sample

It is always tricky to achieve satisfactory sampling. A thorough knowledge of the population to be sampled is a prerequisite. Since this knowledge is not always available, especially with the kind of unstable populations formed by customers, we must avoid sampling as far as possible. As an example of the problems caused by sampling, if the distribution of a variable in the training sample differs from its distribution across the whole population, this may have a major impact on a method using this variable. It is also best to avoid sampling when we are looking for rare phenomena (e.g. types of fraud) or narrow customer segments.

13.6.8 You must never sample

Predictive methods based on modelling (inductive methods) require sampling, because they work by building a model based on part of the population, and then testing the model on another part of the population. The test phase is essential for selecting the best of the resulting models.

It may also be desirable to work on a sample of the population in order to avoid prohibitive computing time for large volumes of data. Sometimes it is best to sample and perform more in-depth calculations on a sample, rather than more superficial calculations on the total population. In the words of Jerome H. Friedman: 'a powerful computationally intense procedure operating on a subsample of the data may in fact provide superior accuracy than a less sophisticated one using the entire data base'.[1]

13.7 Return on investment

The return on investment (ROI) is generated by an increase in the response rate to marketing campaigns, an increase in the productivity of sales staff, a better distribution of resources, an increase in customer loyalty, a reduction in defaults, etc.

Many figures have been quoted on the subject of this ROI. The truth is that it is often difficult to quantify, because the gains due to the use of data mining are not always distinguished from those due to good communication, effective marketing, and motivated personnel. In some cases, these various factors cannot be separated: one example is that of the bank which, having established a risk score, a propensity score for consumer credit and a monthly repayment capacity for each customer, sent a customized offer of credit to each of its customers having a good score for risk and propensity (the 'core target' group). The amount of credit offered to each customer was not a standard (rather low) amount such as €1000, €2000 or €3000, but an amount corresponding to his capacity to repay, which was itself calculated according to his profile, income, expenditure, commitments and scores. This was a clear example of 'one-to-one' marketing. The results were much better than usual, as demonstrated both quantitatively (in the increased take-up rate) and qualitatively (in the appreciation of the sales personnel and telephone sales staff). How much of this was due to the quality of targeting, and how much was due to the customization of the mailing and the amount offered, which were highly appreciated by the customers? The answer will never be known, and in any case is irrelevant, since the customization would not have been possible without the information provided by data mining. Clearly, the essential factor in the return on investment is not the possession of the best data mining tools (although this certainly cannot be disregarded), but the ability to use them in an integrated database marketing strategy.

Data mining is only one element in database marketing, among others such as:

- the marketing communication style;
- the sales dialogue used;
- the format of the mailings sent to customers (colour or black and white? etc.);
- the provision of a dedicated telephone number;
- a system of telephone follow-ups;
- the training of the sales staff;
- the quality of the data output from data mining;
- the recording and storage of information supplied by customers;

[1] Friedman, J.H. (1997) Data mining and statistics: what's the connection? http://www-stat.stanford.edu/~jhf/ftp/dm-stat.pdf

- the adaptation of the marketing processes (changing from 'product' to 'customer' marketing);
- the adaptation of the sales procedures (including decision-making powers).

However, if we really need to provide accurate information on the quality of a targeting process based on data mining, because there will always be some managers concerned about

Table 13.1 Calculated return on investment.

		Conventional targeting	**Targeting using data mining**
A	number of customers targeted	30 000	15 000
B	cost of each mailing	€1	€1
C	cost of each telephone follow-up	€5	€5
D	total cost (= A × (B + C))	€180 000	€90 000
E	number of new subscriptions	1 000	1 500
F	subscription rate (= E/A)	3.33%	10%
G	cost per subscription (= D/E)	€180	€60
H	annual turnover per subscription	€150	€175 (larger amounts taken up)
I	total annual turnover (= H × E)	€150 000	€262 500
	ROI (= I/D)	83%	292%

Table 13.2 Calculated return on investment due to increased loyalty.

A	cost of acquiring a new customer	€150
B	annual profit from each departing customer	€450
C	customer activation time	0.5 year
D	loss due to a departure (= A + (B × C))	€375
E	cost of increasing loyalty of a detected 'departing' customer	€50
F	total number of customers	1 000 000
G	number of departures per year	80 000
H	attrition rate (= G/F)	8%
I	number of 'departing customers' detected (correctly or incorrectly)	40 000
J	total cost of increasing loyalty (= E × I)	€2 000 000
K	number of actual departing customers retained	8 000
L	losses avoided (= D × K)	€3 000 000
	net total profit (= L – J)	€1 000 000

the soundness of their investments, and also because it is important to measure performance in order to improve it, there is one way of achieving this. This is to add a random 'control' sample of customers from one (or more) conventional target groups, identified by the marketing department, to the marketing target generated by data mining. We must then treat all the customers in the same way (using the same channels, the same media, the same communications, the same follow-ups, etc.) and compare the results at the end of the campaign. They can be presented as in Table 13.1, where the last row shows the ROI, which is greater than 100% if it is achieved in less than one year, and less than 100% otherwise.

In another field, the *development of customer loyalty* is also an important source of profit for a business. We can attempt to estimate this as in Table 13.2. For the sake of completeness, we should deduct the software costs and the salaries of data miners from the profits and ROI. However, these costs are often small compared with the savings they offer.

14

Text mining

I mentioned in Chapter 3 that there was a special class of data, namely text data. The earlier chapters discussed tools for manipulating data consisting of codes and quantities; the aim of this chapter is to complete our survey of data mining by showing how it can be combined with linguistics and lexicometry for the automatic analysis and use of text data.

14.1 Definition of text mining

Text mining is the set of techniques and methods used for the *automatic processing* of *natural language text data* available in reasonably large quantities in the form of computer files, with the aim of *extracting and structuring their contents and themes*, for the purposes of rapid (non-literary) analysis, the discovery of hidden data, or automatic decision making. It is different from *stylometry*, which studies the style of texts in order to identify an author or date the work, but it has much in common with *lexicometry* or *lexical statistics* (also known as 'linguistic statistics' or 'quantitative linguistics'); indeed, it is an extension of the latter science using advanced methods of multidimensional statistics.

We can show this schematically as:

$$\text{Text mining} = \text{Lexicometry} + \text{Data mining}$$

Like data mining, text mining originated partly in response to the huge volume of text data created and diffused in our society (think of the amounts of laws, orders, regulations, contracts, for example), and partly for the purpose of quasi-generalized input and storage of these data in computer systems. It also owes its acceptance to developments in statistical and data processing tools whose power has increased greatly in recent years. Thus, following the work of researchers such as Jean-Baptiste Estoup, George Kingsley Zipf, Benoît Mandelbrot, George Udny Yule, Pierre Guiraud, Charles Muller, Gustav Herdan, Etienne Brunet, Jean-Paul Benzécri, Ludovic Lebart and André Salem, there has been an exponential growth in the use of statistics, probabilities, data analysis, Markov chains and artificial

Data Mining and Statistics for Decision Making, First Edition. Stéphane Tufféry.

intelligence tools based on data mining for the processing of text material, and we have made considerable progress since the early days of simple calculation of percentages. Beginning in 1916 (Estoup[1]) and 1935 (Zipf[2]), the frequency of appearance of a word in a text has been studied by statistical methods, giving rise to Zipf's law, a well-known formula which links the frequency of a word to its rank in the table of frequencies. The example of James Joyce's *Ulysses* is famous: the 10th word appears 2653 times, the 100th word appears 265 times, the 1000th word appears 26 times and the 10 000th word appears twice. We find that the product of the rank r and the frequency f is virtually constant:

$$rf = \text{constant}.$$

This law is not always valid with the same degree of accuracy, but it is truly universal, because it applies to all types of text in all languages. Wentian Li[3] demonstrated in 1992 that it could be applied to a text in which the words were created by drawing letters (and a 'space' character) at random from an alphabet with a uniform distribution.

The formula shown above has now been revised to

$$r^a f = \text{constant},$$

where a is an exponent which depends on the language and the type of speaker. It generally ranges from 1.1 to 1.3, and is close to 1.6 in children's language. As a general rule, it decreases with the richness of the corpus, measured as the ratio of the number of different words V (the vocabulary) to the total number of words (V is generally proportional to the square root of N).

Zipf's law has since been extended to other rank-size problems, such as the rank of cities in a country related to their size, the rank of businesses related to their turnover, the rank of individuals related to their income, etc.

One interesting consequence of Zipf's law for text mining is that a few tens of words are enough to represent a large part of any corpus, enabling the depth and complexity of analyses to be limited.

As in data mining, there are two types of method in text mining. Descriptive methods can be used to search for themes dealt with in a set (corpus) of documents, without knowing these themes in advance. Predictive methods find rules for automatically assigning a document to one of a number of predefined themes. This may be done, for example, for the purpose of automatically forwarding a letter or a CV to the appropriate department. The corpus analysed must meet the following conditions:

- it must be in a data processing format (the automatic reading of handwriting, used in the processing of cheques and mail, is a different problem);
- it must include a minimum number of texts;
- it must be sufficiently comprehensible and coherent;

[1] Estoup, J.-B. (1916) *Gammes Sténographiques*, 4th edn. Paris: Imprimerie Moderne.

[2] Zipf, G.K. (1935) *The Psycho-biology of Language*. Boston: Houghton-Mifflin. The definitive formulation can be found in Zipf, G.K. (1949) *Human Behavior and the Principle of Least Effort*. Cambridge, MA: Addison-Wesley.

[3] Li, W. (1992) Random texts exhibit Zipf's-Law-like word frequency distribution. *IEEE Transactions on Information Theory*, 38(6), 1842–1845.

- there must not be too many different themes in each text;
- it must avoid, as far as possible, the use of innuendo, irony and antiphrasis (saying the opposite of what one thinks, e.g. 'Oh, brilliant!' in response to a particularly stupid blunder).

14.2 Text sources used

The main sources of texts analysed by text mining are opinion polls, customer satisfaction surveys, letters of complaint, telephone interview transcriptions, electronic mail, reports of marketing or medical interviews, press surveys, despatches from news agencies, experts' documentation and reports, technology monitoring, competition monitoring, strategic and economic monitoring, the Internet and on-line databases, and more recently curricula vitae. Users of the information analysed may be financial analysts, economists, marketing professionals, customer relations services, recruiters or decision makers.

14.3 Using text mining

Some periodical analyses in which the presentation is always identical can be automated by using text mining. This generates quick analyses without the need for repetitive and tedious computation. The applications include the automatic generation of satisfaction surveys, reports on a business's image or the state of the competition, and the automatic indexing of documents.

Text mining is also used to discover hidden information ('descriptive method'), for example new research fields (in filed patents), or information to be added to marketing databases on customers' areas of interest and plans. It can even be used by a business wishing to communicate with its customers in the vocabulary that they use, and to adapt its marketing presentations to each customer segment. It can be used in search engines on the web.

Finally, text mining is an aid for decision making ('predictive methods'), for example in automatic mail routeing, email filtering (to identify spam and non-spam, technical and business subject matter, etc.), data filtering and news filtering.

The discovery of hidden information and decision making are mainly classed as forms of information retrieval, while quick analysis is a form of information extraction.

Information retrieval is concerned with documents in their totality and with the themes which they deal with, and is used to compare documents and detect types of documents. It aims to detect all the themes that are present. The analysis is global.

Information extraction is a search for specific information in the documents, without any comparison of the documents, taking the order and proximity of words into account to discriminate between different statements which have identical keywords. It is only concerned with themes related to the 'target' database. Information extraction starts with natural-language data and uses them to build up a structured database. It is a matter of scanning the natural language text to detect words or phrases corresponding to each field of the database. The analysis is local. In one sense, information extraction is a more complex process, because it requires the use of lexical and morpho-syntactic analysis to recognize the constituents of the text (words and phrases), their nature and their relationships.

14.4 Information retrieval

This section will describe the different analyses, first linguistic and then statistical, that are required for the automatic updating of the themes contained in a corpus of documents. These analyses follow the Strasbourg School which, following Charles Muller, does not apply statistical methods directly to the text, but to its underlying lexicon, found by a sequence of operations described below for disambiguation, categorization, lemmatization and combination. These operations consist of identifying units (the *graphic forms* which are sequences of non-separator characters) in the text sequence, grouping them into equivalent classes (up to the level of the *theme*) and performing counts and statistical analyses on these classes. This is not the only approach, and other methods have been proposed by researchers who have pointed out that the text sequence cannot be reduced to a series of unrelated units, and that the meaning of a text is highly dependent on the relative positioning, juxtapositions and co-occurrences of the graphic forms (even before considering the equivalence class). Étienne Brunet expressed this in a humorous way: 'Some people may regret the loss of the raw forms, whose opaque materiality could conceal a degree of mystery. They may be repelled by a pale, bloodless lemma reduced to a set of abstract properties'.[4] Having said that, this method of content analysis is effective because it can be used very successfully in conjunction with data mining tools. It is also implemented in some of the leading text mining tools, such as IBM SPSS® Text Analytics and (under the name of 'text parsing') in the Text Miner add-in of SAS® Enterprise Miner™. Although largely automatic, it still needs to be adapted, sometimes manually, to the needs of the user and the vocabulary of his society: this is done by creating a list of prohibited or obligatory terms and a dictionary of synonyms and compound words.

14.4.1 Linguistic analysis

Language identification

It should be noted that the Web obliges us to deal with multilingualism, even within a single document in some cases. Some lovers of linguistic curiosities know about an extreme case of multilingualism: this is the 'polyglot' phrase that has different meanings in different languages. For example, at the time of Watergate, the English headline 'Nixon put dire comment on tape' is also a French sentence meaning 'Nixon could tell you how to type'. In English-speaking parts of Canada there may be posters for a 'Garage Sale', which to French speakers simply means 'Dirty Garage'!

Identification of grammatical categories (grammatical labelling)

The next step is to identify the nouns, verbs, adjectives and adverbs in the texts of the corpus, which requires a grammatical analysis. This can be complicated by the presence of homographs (e.g. 'in a flood of tears, she tears up the letter').

Disambiguation

There are many sources of ambiguity in a natural language text. They may be due to the polysemy of words (the fact that a word has several meanings), to ellipses (in a 'telegraphic'

[4] Brunet, E. (2002) Le lemme comme on l'aime. In *JADT 2002: 6th International Conference on the Statistical Analysis of Textual Data.*

style), homographs ('lead me to the lead mine!'), or antiphrasis and irony, the last of these being particularly difficult to detect automatically.

Anaphora also gives rise to ambiguities that must be removed. 'Anaphora' is used here in its linguistic sense, meaning avoiding repetition of a word by using another word, most often a pronoun, to refer to it ('he', 'she', 'it', 'this', etc.).

The format of data processing texts is another source of ambiguities, such as the ambiguity between the number '0' and the letter 'O', the ambiguities due to line breaks without hyphens, or those due to poor typography. The word 'hit' can be interpreted as a noun, a verb, an adjective ('a hit song') or a past participle.

In personal notes which may have been recorded on an electronic notepad, the ambiguities are even more numerous, owing to personal abbreviations, incorrect spelling, and the often non-syntactical and non-logical order of entry of the words of a sentence.

This stage is tricky, and some ambiguities can only be removed by analysing the whole text, or even by an arbitrary decision. We can also consider the probability of the appearance of a form: 'sate' is more likely to be a verb meaning 'satisfy completely' than the past tense of 'sit', at least in a modern text.

Recognition of compound words

It is necessary to recognize that expressions such as '2 April 2005', 'the Governor of the Central European Bank' and 'the Proceedings of the Royal Institution' are groups meaning a date, a person and a publication. 'Term' can denote either this kind of sequence of graphic forms, or a graphic form with a length of 1 (a 'word'). Then we must allow for the specialist lexicon of the field of activity concerned. Thus, the banking lexicon includes terms such as Visa card, current account, housing savings scheme, etc. The lexicon of business intelligence will include the terms data mining, text mining, data warehouse, etc. It will be useful to create a specific lexicon for the business, identifying sequences of graphic forms (often two or three forms) which are repeated many times in the corpus, or even to compile such a lexicon 'manually'.

Lemmatization

The steps described above will have improved the understanding of the texts. They must then be simplified, without changing the meaning of course, so that the main themes can be extracted more easily.

We need to start by lemmatizing the texts: this means putting the terms in their canonical form, so that nouns would be put in the singular and the various forms of verbs would be put in the infinitive. This is the form in which the words are set out in an ordinary dictionary, which may contain about 60 000 entries covering 700 000 different forms, such as plurals of nouns and different tenses of verbs. French, Spanish, Russian and German have many inflected forms (conjugations and declensions). German also has the distinctive feature of creating compound words by stringing several nouns together, and we may have to decide whether to divide these units into elementary fragments.

Grouping the variants

A second stage of simplification is to group together the variants of terms found in texts. The graphic variants (realise = realize), syntactic variants (name of a man = a man's name), semantic variants ('X buys Y from Z' = 'Z sells Y to X'), synonyms (US = USA = United States = Uncle Sam), parasynonyms (words with closely related meanings: discontent, anger, dissatisfaction), and full forms of abbreviations (€ = EUR = euro, BBC = B.B.C. = British Broadcasting Corporation) are all recorded. Like the dictionary of words, the dictionary of synonyms can be saved in a file specified by the text mining software.

Expressions and metaphors are identified: for instance, 'Empire of the Rising Sun' is replaced with 'Japan' and 'Threadneedle Street' is replaced with 'Bank of England'.

Figure 14.1 shows an example of the results of a text mining analysis using SAS Text Miner.

Grouping the analogies

We can then group the analogies. We group the terms in families of derivative terms, as in a thesaurus, which may include the following group of terms, for example:

- credit/loan/undertaking/debt/borrow/borrower/debtor.

Intensity markers are also grouped, for example:

- a little/less/very little/–
- much/more/very/ +

Identification of themes

The text analysis is completed by grouping all the terms around level 1 themes, then grouping all the level 1 themes around level 2 themes. The first transition will be of the following type:

- cheque/bank card/draft/currency/... ⇔ means of payment

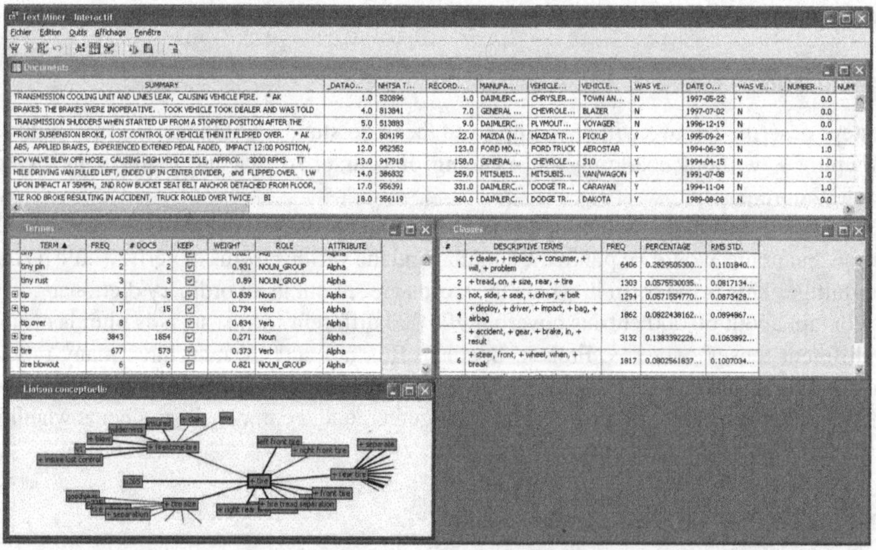

SUMMARY	_DATAO...	NHTSA T...	RECORD...	MANUFA...	VEHICLE...	VEHICLE...	WAS VE...	DATE O...	WAS VE...	NUMBER...	NUM
TRANSMISSION COOLING UNIT AND LINES LEAK, CAUSING VEHICLE FIRE. * AK	1.0	520896	1.0	DAIMLERC...	CHRYSLER...	TOWN AN...	N	1997-05-22	Y	0.0	
BRAKES: THE BRAKES WERE INOPERATIVE. TOOK VEHICLE TOOK DEALER AND WAS TOLD	4.0	813841	7.0	GENERAL ...	CHEVROLE...	BLAZER	N	1997-07-02	N	0.0	
TRANSMISSION SHUDDERS WHEN STARTED UP FROM A STOPPED POSITION AFTER THE	5.0	513883	9.0	DAIMLERC...	PLYMOUT...	VOYAGER	N	1996-12-19	N	0.0	
FRONT SUSPENSION BROKE, LOST CONTROL OF VEHICLE THEN IT FLIPPED OVER. * AK	7.0	804195	22.0	MAZDA (N...	MAZDA TR...	PICKUP	Y	1995-09-24	N	1.0	
ABS, APPLIED BRAKES, EXPERIENCED EXTENED PEDAL FADED, IMPACT 12:00 POSITION,	12.0	952352	123.0	FORD MO...	FORD TRUCK	AEROSTAR	Y	1994-06-30	N	1.0	
PCV VALVE BLEW OFF HOSE, CAUSING HIGH VEHICLE IDLE, APPROX. 3000 RPMS. TT	13.0	947918	158.0	GENERAL ...	CHEVROLE...	S10	Y	1994-04-15	N	1.0	
HILE DRIVING VAN PULLED LEFT, ENDED UP IN CENTER DIVIDER, and FLIPPED OVER. LW	14.0	386832	259.0	MITSUBIS...	MITSUBIS...	VAN/WAGON	Y	1991-07-08	N	1.0	
UPON IMPACT AT 35MPH, 2ND ROW BUCKET SEAT BELT ANCHOR DETACHED FROM FLOOR,	17.0	956391	331.0	DAIMLERC...	DODGE TR...	CARAVAN	Y	1994-11-04	N	1.0	
TIE ROD BROKE RESULTING IN ACCIDENT, TRUCK ROLLED OVER TWICE. BI	18.0	356119	360.0	DAIMLERC...	DODGE TR...	DAKOTA	Y	1989-08-08	N	0.0	

TERM ▲	FREQ	# DOCS	KEEP	WEIGHT	ROLE	ATTRIBUTE
tiny pin	2	2	☑	0.931	NOUN_GROUP	Alpha
tiny rust	3	3	☑	0.89	NOUN_GROUP	Alpha
⊞ tip	5	5	☑	0.839	Noun	Alpha
⊞ tip	17	15	☑	0.734	Verb	Alpha
tip over	8	6	☑	0.834	Verb	Alpha
⊞ tire	3843	1854	☑	0.271	Noun	Alpha
⊞ tire	677	573	☑	0.373	Verb	Alpha
tire blowout	6	6	☑	0.821	NOUN_GROUP	Alpha

#	DESCRIPTIVE TERMS	FREQ	PERCENTAGE	RMS STD.
1	+ dealer, + replace, + consumer, + will, + problem	6406	0.2829505300...	0.1101840...
2	+ tread, on, + size, rear, + tire	1303	0.0575530035...	0.0817134...
3	not, side, + seat, + driver, + belt	1294	0.0571554770...	0.0873428...
4	+ deploy, + driver, + impact, + bag, + airbag	1862	0.0822438162...	0.0894867...
5	+ accident, + gear, + brake, in, + result	3132	0.1383392226...	0.1063892...
6	+ steer, front, + wheel, when, + break	1817	0.0802561837...	0.1007034...

Figure 14.1 Terms disambiguated, labelled and lemmatized.

while the second transition could be:

- means of payment/money/cash/... ⇔ bank

14.4.2 Application of statistics and data mining

When the analysis of texts and their themes is completed, we can filter the themes or terms to be examined. We can use either a statistical criterion (selecting terms and themes by their frequency) or a semantic criterion (centred on a given subject), or a corpus (identifying offensive words to avoid and their derivation, in order to 'clean up' a document). With the statistical criterion, we can use a number of weighting rules, for example preferring terms which appear frequently but in few texts (weight = frequency of the term/number of texts containing it).

These terms, having been disambiguated, labelled, lemmatized, grouped and selected, are then treated with data mining methods, with the individuals (in the statistical sense) being the texts or documents (e.g. emails) and the characters of the individuals (their variables) being the themes or terms in the documents. Thus we can produce lexical tables in which each cell c_{ij} is a number of occurrences of term j (or an indicator of presence/absence) in document i, to which the conventional statistical methods are applied. The cell c_{ij} can also be the number of occurrences of term j in the set of documents relating to customer i (letters, reports of interviews, etc.)

These tables can be processed by correspondence analysis, which simplifies the problem by reducing the initial variables (corresponding to the terms), often present in very large numbers (several thousand) although the preliminary transformations may have decreased their number, to about a hundred factors (which no longer correspond to terms: this is the drawback of the method). At the end of this transformation, continuous variables will have been substituted for the initial discrete variables, and conventional data analysis techniques can be used – classification, clustering, etc. This method is incorporated, under the name of SVD (singular value decomposition), in SAS Text Miner. Some techniques such as regularized regression (see Section 11.7.2) are useful when we need to process a large number of variables compared with the number of individuals.

14.4.3 Suitable methods

Text mining can respond to two types of request. *Open requests* (or *free text requests*) are requests in the form of keywords or free text, used to search for relevant documents in a corpus that changes slowly (such as a yearbook or an electronic library), with the most relevant sections of text highlighted. *Predefined requests* are requests relating to a number of fixed terms, applied to a corpus that changes in a dynamic way with time (e.g. categorization of documents, routeing/filtering of mail or news). They are subject to the same problems as classification.

Like data mining, text mining includes descriptive and predictive methods. In the *predictive* domain, the classification (or *categorization*) of documents is carried out according to predefined themes (nomenclature). It is used for predefined requests (routeing, filtering) and is based on decision trees (CART, C5.0) and supervised learning neural networks.

Markov chains can be used for open requests. A Markov chain can be briefly described as follows. Imagine that we have n boxes, each filled with numbered balls. We draw a ball at random from the first box; its number indicates the box from which the next ball is to be drawn. We continue to draw balls until we reach an empty box. The set of boxes that we have passed through, in sequence, is a Markov chain. The probability of drawing a given ball depends on the box from which it is drawn, and therefore on all the previous drawings. The same applies to a sentence: the probability of the appearance of a word depends on the preceding words, and not all sequences of words have the same probability of occurring. Markov chains are used for speech and handwriting recognition, spelling correction, voice control of automatic systems, and natural language human–machine interfaces in general.

In the *descriptive* domain, a *corpus clustering* is carried out according to non-predefined themes (discovered in the documents) and can be followed by automatic extraction of keywords (terms which are frequent in the cluster and rare in the set of documents). The clustering can be carried out by a Kohonen network or an agglomerative hierarchical algorithm. It is also possible to carry out a multiple correspondence analysis by matching the text data with the other data. For example, we can match the response to a questionnaire with the respondent's socio-occupational category. It is possible to create a *document map* and identify isolated themes, themes forming homogeneous sets, the strength of the links between the themes in a single set (the vocabulary common to the themes) and the number of documents for each theme. Figure 14.2 shows by way of example a graphic representation of the Tropes software from Acetic.

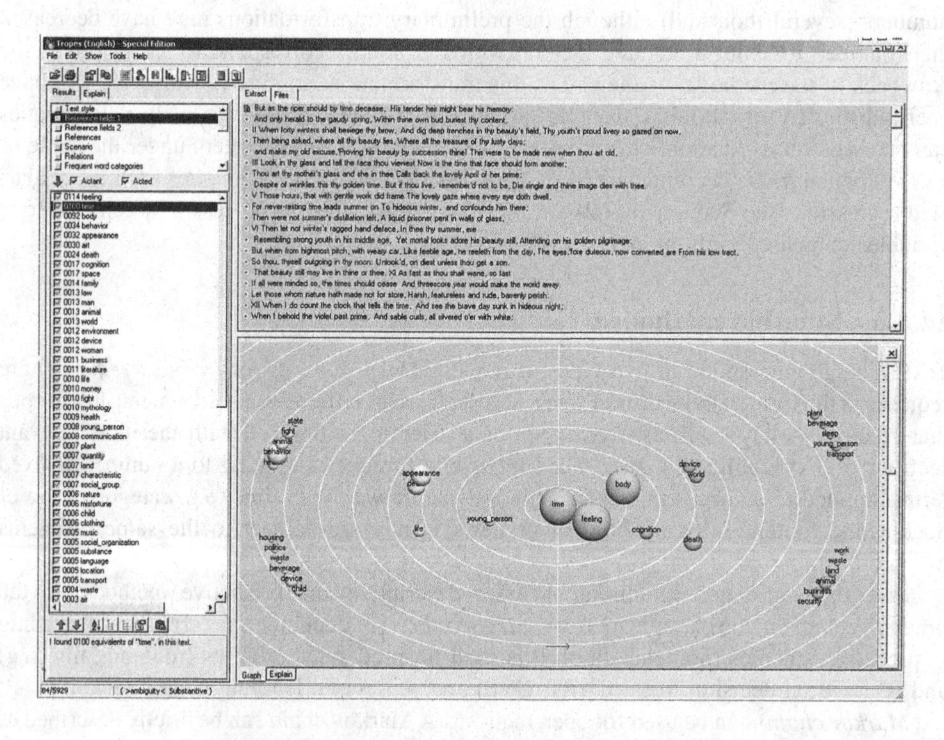

Figure 14.2 Themes in Shakespeare's *Sonnets*.

14.5 Information extraction

14.5.1 Principles of information extraction

Information extraction systems are made up of trigger words (verbs or nouns), linguistic forms, and constraints which limit the application of the trigger. These systems require specific semantic dictionaries for the domain or business, as well as syntactic analysers that can recognize the general linguistic forms (subject, verb, direct object, etc.). Using a target to be extracted from and predefined fields to be filled, information extraction systems detect the relevant sentences and extract the desired information.

The main applications of information extraction are:

- automatic completion of predefined forms from free texts;
- automatic construction of bibliographic databases from research papers (fields to be extracted: title, author, journal, publication date, research establishment, etc.);
- automatic scanning of Reuters despatches on the acquisition of one company by another (fields to be extracted: purchaser, vendor, price, industrial sector, turnover, stock exchange quotation, etc.);
- automatic scanning of the financial press (the 'people' section, on chief executives' moves between companies);
- automatic detection of the plans or requirements of the customers of a business, based on the records of sales staff (fields to be extracted: name of customer, type of product or service offered, type of plan or requirement of the customer, amount, customer's deadline, customer's response (take-up/refusal), reason for customer's response, other suppliers used by the customer, etc.), and the use of the extracted information in a propensity score.

The performance of information extraction is summarized by two indicators. The *accuracy rate* is the number of correctly completed fields divided by the number of completed fields. The *recall rate* is the number of correctly completed fields divided by the number of fields to be completed.

14.5.2 Example of application: transcription of business interviews

If sales staff discover that their customers have plans that can be financed (buying a house, changing their car, etc.), they offer the customer a credit proposal and note their reaction in a report. If the reaction is positive, the completion of the report is less important, because it will be evident that the product has been taken up. If the reaction is negative, the existence of the report is more important, as it indicates that a product has been offered to the customer. It is also useful to be able to analyse these marketing reports automatically in order to determine the reasons for the customers' refusal, and then deduce predictive models and typologies, or even adapt the products offered.

One problem that arises is that these reports, written in natural language, are obviously not standardized, and may contain:

- spelling mistakes;
- personal abbreviations;
- stream-of-consciousness writing;
- ellipsis ('telegraphic' style);
- an illogical order in sentences in some cases (related words may be separated by a certain distance);
- negations which are not always explicit (the sentence 'construction Brighton – finance NatWest' is a negation if the bank where the salesperson works is not the NatWest!).

Faced with the difficulty of automatic normalization of reports, we need powerful text mining tools for information extraction, not just keyword search tools.
Report analysis by text mining can be a highly penetrating method, offering these benefits:

- detection of customers resistant to certain kinds of credit (useful information for building a propensity score);
- automatic detection of certain reasons for refusing to take up a product (customer 'opposed to credit', better offer from the competition, no need for credit, etc.);
- detection of customers having plans for future dates (enabling us to market to them at the right time).

14.6 Multi-type data mining

A very promising method of data mining, known as *multi-type* data mining, can simultaneously examine text data (from text mining processes), paratextual data (such as the date and purpose of a document, the type of document, the recipient of the document in the business, etc.), and contextual data (such as information about the author of the document, his relations with the business, the products he has bought, the services he has used, etc.).

Text data are converted into coded data and then stored with the other data in marketing databases. The matching of all the data (textual and non-textual) makes multi-type data mining a very powerful tool. For example, an attrition study will be more precise if it takes letters of complaint and other exchanges between the business and the customer into account.

15

Web mining

This brief chapter will take a look at *web mining*, which is the recent application of data mining to data obtained from Internet servers on the way in which users browse the websites of businesses and organizations. It can be used for analysing the behaviour of web users and can also be linked to analyses of other data sources.

15.1 The aims of web mining

Just as market baskets provide useful information on the associations of products that have been bought, so that a store can adjust its stocks, the analysis of a web user's movements in a website can supply valuable information to those who know how to use it: this is the aim of web mining, the application of statistics and data mining to web browsing data. Web mining covers a range of methods from the simple counting of visits to a page to the modelling of users' movements in the site.

Using web mining, we can:

- optimize browsing on a site by analysing the behaviour of users, in order to maximize their convenience, increase the number of pages viewed and enhance the impact of links and advertising banners;
- identify the focus of interest, and therefore the expectations, of users visiting the site;
- improve the business's knowledge of the customers who log on under their own names, by matching their browsing data with their personal data held by the site owner.

Each of three areas corresponds to a particular level of analysis, described in one of the sections of this chapter. There is a fourth topic, which is not tackled in this book. This is the whole area of information searching on the web and the 'web crawling' methods used by search engines. It can be classed as 'web content mining', as opposed to the 'web usage mining' which is described below.

Data Mining and Statistics for Decision Making, First Edition. Stéphane Tufféry.

15.2 Global analyses

15.2.1 What can they be used for?

Some browsing information is useful even if it cannot be associated with any specific users. Association rules such as '40% of users visiting page A also visit page B' or that '20% of users visiting page A visit page C immediately afterwards' are interesting in themselves, because they enable browsing in the website to be optimized, or because they are an aid to the effective positioning of advertising banners or links to other pages. The same applies to descriptive statistics which inform us that '70% of users have visited three pages or fewer' or that '40% of the users access the site without going through the home page'. A *transition matrix* between the pages of the site can be drawn up. On the subject of association rules, we should note that they take the order of the items into account, unlike the usual methods of market basket analysis (see Chapter 10).

The analysis of browsing on the site can also be used to construct taxonomies of users, according to the original sites, the entry pages, the number of pages viewed, the time spent on the pages, the files downloaded, the exit pages, etc. The taxonomies are interesting in themselves, although it is even better to match them with the customer databases of the business. A second kind of taxonomy can be created in web mining: this is web page taxonomy, which groups the pages by their content.

15.2.2 The structure of the log file

These global analyses are based on the 'log' file, which is a text file saved on the website server, in which one line is written for each user request (e.g. for a change of page, or for downloading a file). There are several log file formats, including common log format (CLF), extended log format (XLF) and special formats for certain sites, such as secure sites.

The common log format contains the IP address of the client computer, the date and time of the request, the type of request, the requested URL, the HTTP protocol, the server return code (see below) and the size (in bits) of the object returned. For example, the line

```
130.5.48.74 [22/May/2006:12:16:57 -0100] "GET
/content/index.htm HTTP/1.1" 200 1243
```

indicates a successful request (return code = 200) for the download (GET) of an object containing 1243 bits, on 22 May 2006 at 12:16 hours with a time difference of −1 hour (−0100) from GMT.

The extended log format also contains the original page ('referrer') and the 'user agent' (designating the browser, the operating system of the client computer, and any other parameters required). An example is:

```
130.5.48.74 - [22/May/2006:12:16:57 -0100] "GET
/content/news.htm HTTP/1.1" 200 4504 "/content/index.htm"
"Mozilla/4.0 (compatible; MSIE 6.0; Windows NT 5.1; SV1)".
```

The user has come from the page/content/index.htm using Internet Explorer 6.0 installed in Windows XP SP2. The reader will find a list of 'user agents' on Wikipedia at http://en.

wikipedia.org/wiki/User_agent. The user's login is shown between the IP address and the date, if it is known; in most cases it is not known and is replaced by a dash.

A few explanations are required concerning the request type and the return code. The main values of the *request type* are:

- GET: download an object from the server;
- PUT: store an item on the server;
- DELETE: delete an item on the server;
- HEAD: a variant of GET (sometimes used by robots).

The main values of the return code are:

- 200: request completely successful;
- 2xx: request partially successful;
- 3xx: redirection;
- 401: access refused;
- 404: URL not found;
- 4xx: other errors;
- 5xx: server errors.

Be careful about the IP address, which may not be fixed for a user, as it is assigned dynamically by the Internet service provider at the time of connection (there are exceptions in the case of ADSL). The same difficulty is encountered if the user changes his computer or uses his company network.

15.2.3 Using the log file

As with any data mining analysis, we have to begin by formatting the data, in this case the log file. The log files are very large (up to several hundred megabytes per day for a medium-sized web server) and they must be cleaned. We remove any lines recording the following:

- pages visited by less than five IP addresses;
- image files (GIF, JPEG, etc.) or scripts, which contribute nothing to the analysis;
- accesses by robots, agents or link testers;
- anomalous IP addresses.

Some data mining software, such as SAS Web Analytics and IBM SPSS Modeler Web Mining, can clean log files in a largely automatic way (Figure 15.1) and then apply all the data mining techniques (clustering, detection of association rules, etc.) to them.

The log file provides a set of very useful information for analysing visits by users. But how do we define a visit? It is a set of requests from the same IP address, from the same 'user agent'

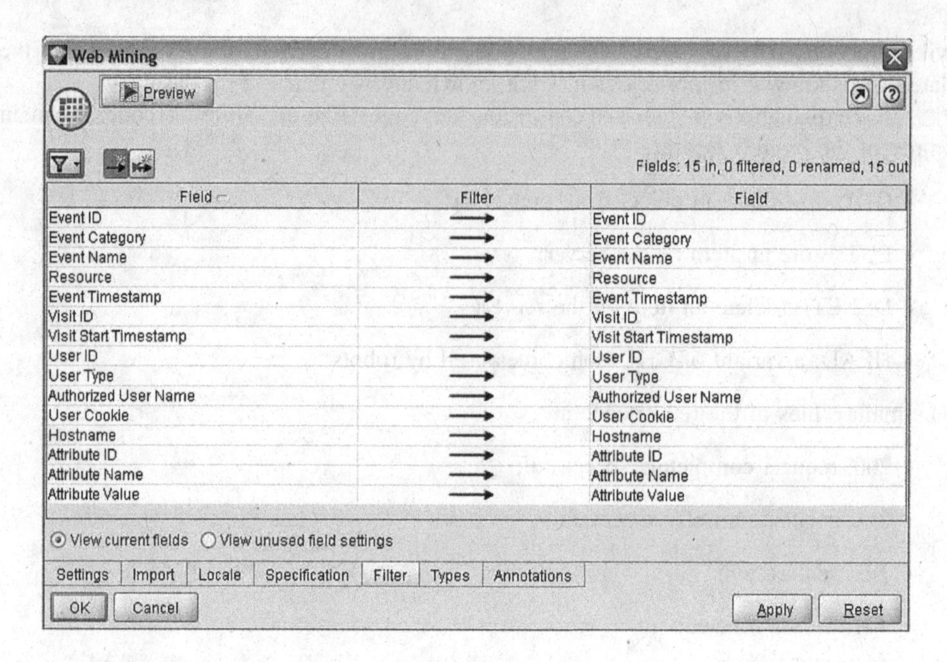

Figure 15.1 Fields supplied by the log file in IBM SPSS Modeler Web Mining.

(and especially the same browser), separated from each other by a maximum time interval. The interval is generally set at 30 minutes, which means that, if a request follows the previous one more than 30 minutes later, it is counted as the start of a new visit; in other words, if a request is not followed by another within a period of 30 minutes, the visit is terminated.

The data extracted from the log file are: the identifier (IP address), the date of the visit, the start and end time of the visit (working hours/evening and night/weekend and public holidays), the browser type (Internet Explorer, Firefox, Netscape, Opera, Safari, Google Chrome, etc.), the visitor's operating system (Windows, Linux, Mac, etc.), the geographic origin, the pages visited, the number of pages visited, the mean time spent on each page and the mean number of clicks.

The mean time spent on each page is one of the indicators for distinguishing human visitors from robots: only robots spend less than a second on each page! Another way of identifying a robot is to look in the 'user agent' for terms such as 'bot', 'crawler', 'libwww-perl' or 'Java/', or an expression such as the following (the example given here refers to Google): "Mozilla/5.0 (compatible; Googlebot/2.1; + http://www.google.com/bot.html)". A third indicator is that a correctly programmed robot will request the file 'robots. txt'.

The IP enables us to identify the visitor's country (Figure 15.2), possibly his city, and his Internet service provider (see, for example, www.ip2location.com and www.dnsstuff.com). The visitor's browsing can be followed line by line in the log file, using the fields containing the requested URL and the original URL.

In addition to commercial log file analysis software, such as the very comprehensive WebTrends used by large companies, there is some free software for reporting and creating charts based on log files: the leading brands are Analog, AWStats and Webalizer. Google

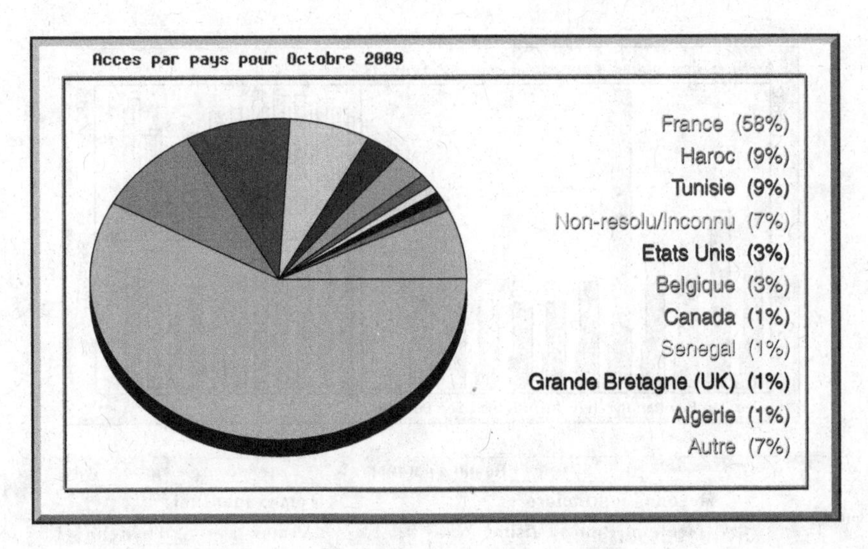

Figure 15.2 Webalizer performance chart.

Analytics is a very useful newcomer. A list of these web analytics packages can be found on Wikipedia.[1]

In the screenshots captured in Figure 15.3, we can see the difference between the numbers of requests (hits), files and pages. The files are the satisfied requests (return code 200), and the pages are the HTML files (excluding images, JavaScript, etc.). We can also find the dates of the first and last visit, the number of visits, and the mean duration of visits for each IP address (and each user agent). By cross-tabulating these data with the others, we can create a taxonomy of visitors.

15.3 Individual analyses

We may wish to establish that '35% of web users who visit the web page for a crime novel by Patricia Highsmith also view the web page for a Hitchcock film in the next two months'. In other words, we would like to go beyond a global analysis and obtain 'one-to-one' analyses. In this case, the log file is not enough. We need to make use of 'cookies', which are small text files stored on the hard disks of users when they connect to certain websites. When a user connects to a site that uses cookies, a text file is created on his computer, containing a specific identifier for the computer together with other information on the user's browsing, the number of pages viewed, the entry and exit pages, the sites from which he has come, the downloaded files, and possibly some personal information requested by the site. When the user (or rather the computer) connects again, the identifier contained in the cookie is transmitted to the website server, which can then update the content of the cookie, and offer customized web pages and links, relevant advertisements, and the like to the user.

[1] http://en.wikipedia.org/wiki/List_of_web_analytics_software

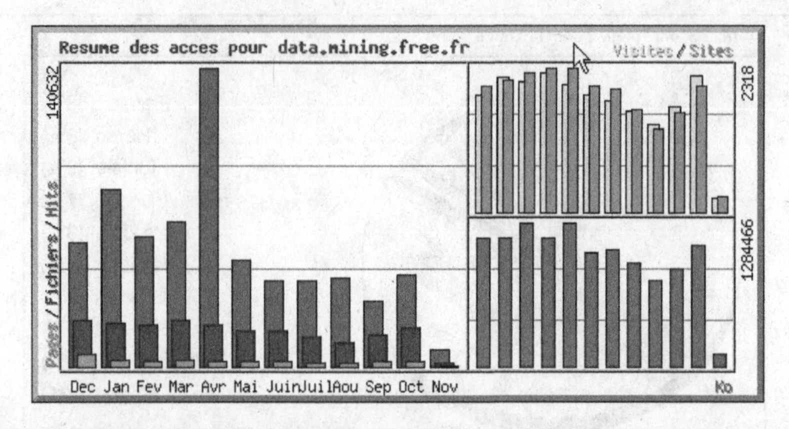

Mois	Moyenne journalière				Totaux mensuels					
	Hits	Fichiers	Pages	Visites	Sites	Ko	Visites	Pages	Fichiers	Hits
Nov 2009	2062	457	65	56	254	112721	225	263	1831	8248
Oct 2009	1401	599	88	70	2026	1076645	2199	2749	18579	43456
Sep 2009	1037	486	76	56	1608	874129	1685	2295	14596	31110
Aou 2009	1355	372	58	45	1347	761733	1413	1799	11548	42018
Juil 2009	1318	455	65	52	1643	923940	1632	2039	14129	40870
Juin 2009	1346	571	74	59	1980	1036395	1787	2224	17157	40399
Mai 2009	1619	557	79	60	2020	1009612	1889	2474	17275	50218
Avr 2009	4687	648	87	68	2302	1284466	2047	2629	19452	140632
Mar 2009	2199	696	94	71	2318	1143287	2225	2933	21585	68193
Fev 2009	2177	714	96	75	2234	1271964	2102	2706	20000	60971
Jan 2009	2688	656	93	69	2122	1140466	2159	2894	20359	83337
Dec 2008	1886	703	196	60	2034	1142611	1880	6092	21821	58467
Totaux						11777969	21243	31097	198332	667919

Table title: Résumé par mois

Figure 15.3 Webalizer performance chart.

Analyses based on the use of cookies have the advantage of speed, reliability, and transparency for the user, who, without be aware of it, provides large amounts of information on his interests.

Their disadvantage is that, in the present state of the law and Internet software technology a user can refuse a cookie or destroy it when he logs off, or even block it with a firewall. Another drawback is that a cookie does not identify an individual but a computer, which is equivalent to a whole household in the case of a family computer.

15.4 Personal analysis

The most detailed analysis of a user's browsing habits is possible when a website requires the personal identification of the customer, as is the case with on-line banking sites for viewing accounts and carrying out banking transactions. In this case, the user is a known customer of the company, and the information on his browsing can be compared with other information held by the establishment on its customer, and can enrich this. In particular, the pages viewed and the requests made for simulation provide detailed indicators of the customer's interest in this or that product and his planned purchases or investments, and this information can be used in propensity scores. The comparison of the user taxonomies with the customer databases is a very useful exercise for the business, which can predict the behaviour of a customer by this means. The Internet browsing data can form a valuable addition to the other customer data in the business's databases.

Appendix A

Elements of statistics

This appendix provides a basic introduction to statistics. It will be useful for an understanding of the methods described in this book and the outputs of statistical and data mining software, enabling the user to set the correct parameters and manipulate the data without erroneous interpretation. The reader should consult the books cited in Section B.1 of Appendix B for further information about statistical methods.

A.1 A brief history

A.1.1 A few dates

Table A.1 provides a summary timeline. Note that, although some older methods (such as Fisher's discriminant analysis) have lost none of their usefulness, some recent techniques such as bagging and boosting have already become widespread. In comparison to other scientific disciplines, statistics is remarkable for the very short time lag between the discovery of a new method and its widespread application. This is a powerful incentive for statisticians working in business (and for other users) to improve their knowledge.

A.1.2 From statistics ... to data mining

Traditional statistics (up to the 1950s):

- a few hundred individuals
- several variables defined with a special protocol (sampling, experimental design, etc.)
- firm assumptions regarding the statistical distributions involved (linearity, normality and homoscedasticity)
- models developed theoretically and compared with the data

Data Mining and Statistics for Decision Making, First Edition. Stéphane Tufféry.

Table A.1 Timeline.

1875	Francis Galton's linear regression
1888	Francis Galton's correlation
1896	Karl Pearson's formula for the correlation coefficient
1900	Karl Pearson's χ^2
1933	Harold Hotelling's factor analysis
1934	Chester Bliss's probit model
1936	Discriminant analysis, developed by Ronald A. Fisher and Prasanta Chandra Mahalanobis
1936	Harold Hotelling's canonical correlation analysis
1941	Guttman's correspondence analysis
1943	Formal neuron invented by the neurophysiologist Warren McCulloch and the logician Walter Pitts
1944	Joseph Berkson's logistic regression
1958	Frank Rosenblatt's perceptron
c.1960	Appearance of the concept of exploratory data analysis in France (Jean-Paul Benzécri) and the USA (John Wilder Tukey)
1962	Jean-Paul Benzécri's correspondence analysis
1964	AID decision tree (precursor of CHAID) invented by J.P. Sonquist and J.-A. Morgan
1965	E. W. Forgy's moving centres method
1967	J. MacQueen's *k*-means method
1970	Ridge regression proposed by Arthur E. Hoerl and Robert W. Kennard
1971	Edwin Diday's dynamic cloud method
1972	Generalized linear model formulated by John A. Nelder and Robert W. Wedderburn
1972	David Cox's proportional hazards regression model
1975	John Holland's genetic algorithms
1975	Gilbert Saporta's DISQUAL classification method
1979	Bootstrap method proposed by Bradley Efron
1980	CHAID decision tree developed by Gordon V. Kass
1982	Teuvo Kohonen's self-organizing maps (Kohonen networks)
1983	Herman and Svante Wold's PLS regression
1984	CART tree proposed by Leo Breiman, Jerome H. Friedman, R.A. Olshen and Charles J. Stone
1986	Multilayer perceptron invented by David E. Rumelhart and James L. McClelland
1990	Generalized additive model proposed by Trevor Hastie and Robert Tibshirani
c.1990	First appearance of the data mining concept
1991	Jerome H. Friedman's multivariate adaptive regression splines (MARS)
1993	J. Ross Quinlan's C4.5 tree
1993	Apriori algorithm proposed by R. Agrawal *et al.* for detecting association rules
1995	Vladimir Vapnik's learning theory and support vector machines

1995	Robert Tibshirani's lasso method of linear regression
1996	DBSCAN clustering algorithm proposed by M. Ester, H.-P. Kriegel, J. Sander and X. Xu
1996	Leo Breiman's bagging method
1996	Yoav Freund's and Robert E. Shapire's boosting method
1998	Leo Breiman's arcing method
2000	PLS logistic regression formulated by Michel Tenenhaus
2001	Leo Breiman's random forests
2005	Elastic net linear regression proposed by Zou and Hastie
2007	Grouped lasso method proposed by Yuan and Lin

- probabilistic and statistical methods
- used in the laboratory.

Data analysis (1960–1980):

- a few thousand individuals
- several tens of variables
- construction of 'individuals × variables' tables
- importance of computing and visual representation.

Data mining (1990s onwards):

- several millions or tens of millions of individuals
- several hundreds or thousands of variables
- numerous non-numeric variables, such as textual variables (or variables containing images)
- weak assumptions regarding the statistical distributions involved
- data collected before the study, and often for other purposes
- constantly changing population (difficulty of sampling)
- presence of 'outliers' (abnormal individuals, at least in terms of the distributions studied)
- imperfect data, with errors of input and coding, and missing values
- fast computing, possibly in real time, is essential
- the aim is not always to find the mathematical optimum, but sometimes the model that is easiest for non-statisticians to understand
- the models are developed from the data, and attempts are sometimes made to draw theoretical conclusions from them

- use of statistical methods, artificial intelligence and machine learning theory
- used in the business world.

A.2 Elements of statistics

A.2.1 Statistical characteristics

The job of statistics, like that of data mining, is often to reduce a large body of data to a small amount of relevant information. The main kinds of information relating to a single variable at a time ('univariate statistics') are classed in three main groups: the characteristics (or parameters) of central tendency, dispersion and shape.

The central tendency (or position) characteristics indicate the order of magnitude of the data and their central value. The dispersion characteristics provide information about the spread of the values of the data about the central value. The shape characteristics indicate the symmetry or asymmetry of the data set, as well as its kurtosis.

The main *central tendency characteristics* are the mode, the means (arithmetic, geometric, harmonic), the median and the other quantiles (percentiles, deciles, quartiles, etc.). In particular, the quartiles q_1, q_2 (the median), q_3 divide the data set into four equal frequencies.

The main *dispersion characteristics* are the range, the interquartile range ($q_3 - q_1$), the variance, the standard deviation and the coefficient of variation. The coefficient of variation CV(X) of a variable X is the standard deviation divided by the mean, expressed as a percentage. This is a dimensionless quantity: multiplying the variable by a constant does not change the coefficient of variation. It is sometimes considered that X is dispersed if CV (X) $\geq 25\%$.

The main *shape characteristics* are Fisher's coefficients of skewness and kurtosis. The *skewness* of a population is $\mathrm{E}[(X-\bar{X})^3]/\sigma^3$, equal to the third-order moment divided by the cube of the standard deviation, and for a series $x = (x_i)$ it is

$$\frac{1}{n}\sum_i \left(\frac{x_i - \bar{x}}{\sigma_x}\right)^3.$$

It is 0 if the data set is symmetrical, positive if it is elongated towards the right (the tail of the distribution towards the right), and negative if it is elongated towards the left (the tail of the distribution towards the left); see Figure A.1.

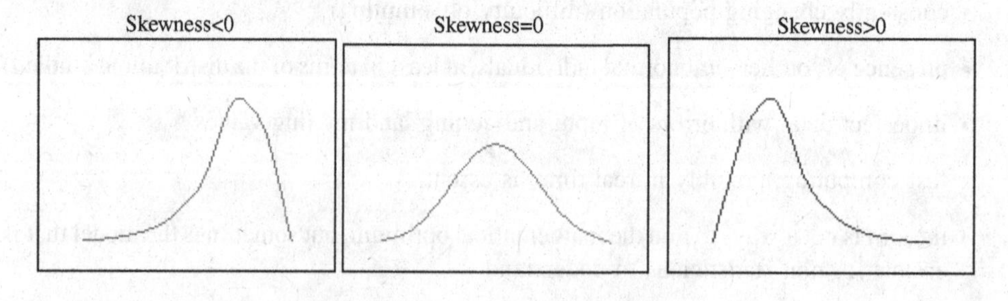

Figure A.1 Coefficient of skewness.

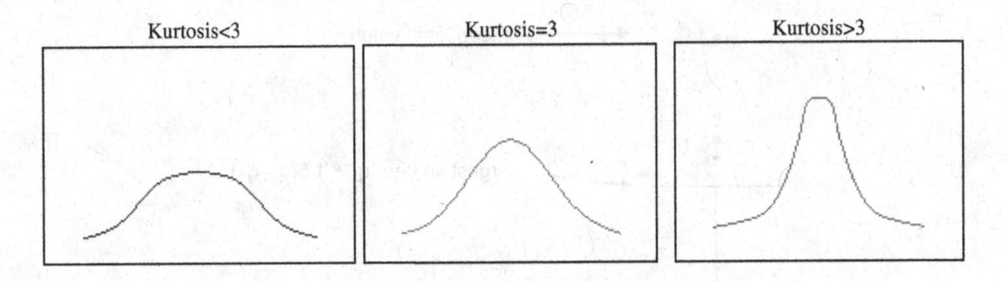

Figure A.2 Coefficient of kurtosis.

The *kurtosis* of a population is $E[(X-\bar{X})^4]/\sigma^4$, equal to the fourth-order moment divided by the fourth power of the standard deviation, and for a series $x=(x_i)$ it is:

$$\frac{1}{n}\sum_i \left(\frac{x_i-\bar{x}}{\sigma_x}\right)^4.$$

It is 3 for a normal distribution, regardless of the variance. The kurtosis is 3 when the series has the same kurtosis (or concentration) as that of a normal distribution having the same variance ('mesokurtic' distribution), more than 3 if it is more concentrated than a normal distribution ('leptokurtic' distribution), and less than 3 if it is more flattened than a normal distribution ('platykurtic' distribution); see Figure A.2. The distributions being compared must, of course, have the same variance, as seen in the example of logistic distribution in Section 11.8.2: the logistic distribution of parameter 1 with a kurtosis of 4.2 is less flattened than the standard normal distribution, but this is because it should be compared with the logistic distribution of parameter $\sqrt{3/\pi^2}$ (with a variance of 1). The coefficient of kurtosis is still greater than 1 and also greater than or equal to 1 + skewness^2. It is 1.8 for the uniform distribution between 0 and 1. Some authors subtract the term '3' from it to normalize it to 0 for the normal distribution; this is done in the R, SAS and IBM SPSS Statistics software.

Note that the skewness and kurtosis of a sample include a correction term, like the sample standard deviation.

A.2.2 Box and whisker plot

The box and whisker plot (or box plot), devised by J.W. Tukey, is a very popular and simple summary representation of the dispersion of a data set. It is constructed by placing the values of certain quantiles on a vertical or horizontal scale (Figure A.3). It provides an instant snapshot of some of the central tendency, dispersion and shape characteristics of the variables. It can also be used to compare two populations, or to detect the individual outliers that must be excluded from the analysis to avoid falsifying the results.

A.2.3 Hypothesis testing

This is done in order to confirm a hypothesis H_1 which may be, for example:

- that a mean measured in a sample is significantly different from the mean in the population ('significantly' means that the result is not simply due to chance);
- that the means measured in a number of samples are significantly different;

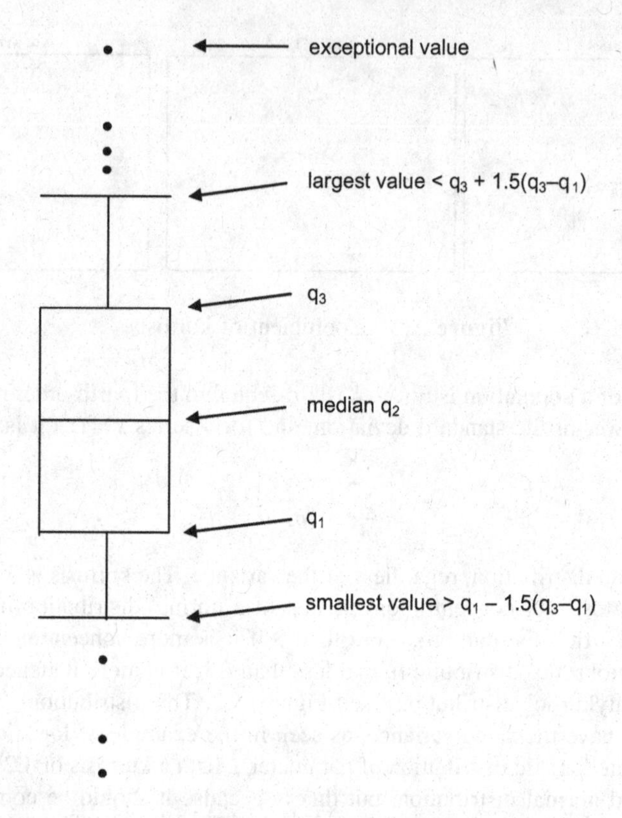

Figure A.3 Box plot.

- that a variable does not follow a given theoretical distribution;
- that two variables are significantly different;
- that a sample is not homogeneous, but is composed of a number of sub-populations.

To accept H_1, we subject the contrary hypothesis H_0 to a test T which must be passed if H_0 is true, and then we show that T is not passed and that we must therefore reject H_0 and accept H_1. H_0 is called the null hypothesis and H_1 is the alternative hypothesis.

To create the test T, we associate the null hypothesis H_0 with a statistic, based on the observations, which follows a known theoretical distribution if H_0 is true. For example, if the null hypothesis is H_0: $\mu = \mu_0$, then $(\bar{x} - \mu_0/(\sigma/\sqrt{n}))$ follows a standard normal distribution, that is, a normal distribution with mean 0 and standard deviation 1.

In this distribution, we choose a rejection region (one- or two-tailed) characterized by a probability α of being in this zone. In many cases, $\alpha = 0.05$ (5%). The complement of this is the acceptance region (if $\alpha = 0.05$, this is the region around the mean where 95% of the values of the statistics are found; see Figure A.4).

The value of the statistic for the sample is measured and compared with the theoretical values of the distribution. If this measured value falls in the rejection region, H_0 is rejected; otherwise, it is accepted. Rejecting H_0 does not mean that we have 'proved' H_1, because the

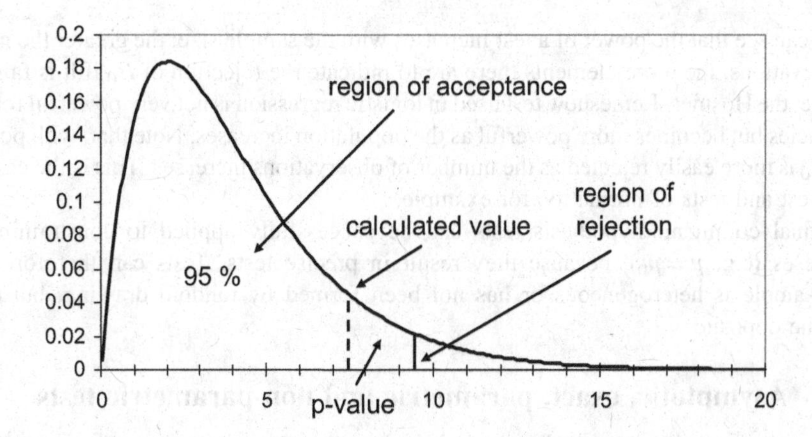

Figure A.4 Acceptance and rejection regions in a statistical test.

calculation of the statistic associated with the test may have been affected by random factors, especially if the number of observations is small.

The *significance level*, the *degree of significance*, or the *p-value* refers to the probability of finding a test statistic which is as extreme ($\geq$ or $\leq$) as the value measured for the sample where H_0 is true.

If the *p*-value is greater than or equal to α (as in Figure A.4), H_0 is not rejected. If the *p*-value is less than α, H_0 is rejected, because we consider that such a low *p*-value is too unlikely to exist if H_0 is true, and therefore we cannot accept that H_0 is true. However, there is always a probability α that H_0 will be rejected even though it is true. This is called the *type I error.* The other possible error in a test is the non-rejection of a false H_0, called the *type II error*, the probability of which is conventionally denoted by β. It is impossible to reduce α and β simultaneously.

		REALITY	
		H_0 true	**H_0 false**
DECISION	**H_0 not rejected**	decision correct ($1-\alpha$)	error β (type II)
	H_0 rejected	error α (type I)	decision correct ($1-\beta$)

By definition, the power of the test is $1-\beta$. This is the probability that H_0 will be rejected if it is false. The error β and the power $1-\beta$ depend on:

- the true value of the population parameter (the farther it is from the tested value, the lower the error β will be);
- the standard deviation σ of the population (the higher it is, the lower β is);
- the chosen significance level α (the higher it is, the lower β is);
- the size n of the sample (as n increases, β decreases).

A powerful test is also described as 'liberal'. The opposite is a 'conservative' test.

We can see that the power of a test increases with the sample size: the greater the number of observations, the more elements there are to indicate the rejection of H_0 if it is false. For example, the Hosmer–Lemeshow test used in logistic regression is not very powerful for small frequencies but becomes more powerful as the population increases. Note that, with powerful tests, H_0 is more easily rejected as the number of observations increases. This is the case with the χ^2 test and tests of normality, for example.

A final comment: hypothesis tests can be successfully applied to constraining null hypotheses (e.g. $\mu = \mu_0$), because they result in precise tests. Tests can therefore prove that a sample is heterogeneous or has not been formed by random drawing, but cannot prove the opposite.

A.2.4 Asymptotic, exact, parametric and non-parametric tests

An *asymptotic test* is an approximation which is valid when the frequencies are relatively large and the data sets are relatively dense. For example, the χ^2 test is not valid if the theoretical frequencies are less than 5.

An *exact test* can be used on sparse data, as it is created by a direct probability calculation taking all possible cases into account. The opposite side of the coin is that this calculation can required a lot of machine time. An alternative to these costly calculations is approximation by the Monte Carlo method, a variant offered by software such as SAS, IBM SPSS Statistics and S-PLUS. Some asymptotic tests have exact equivalents; for example, the counterpart of the χ^2 test is Fisher's exact test.

A *parametric test* assumes that the data come from a given distribution (normality, homoscedasticity, etc.). It may be more powerful than a non-parametric test, but is not often much more powerful. Student's *t* test and ANOVA are parametric tests.

A *non-parametric test*, such as the Wilcoxon or Kruskal–Wallis test, does not assume that the variables follow a given distribution, and is often based on the ranks of the values of the variables, rather than the values themselves. This reduces its sensitivity to outliers. It is the preferable method where sample sizes are small (less than 10). When sample sizes are large (several tens or hundreds of thousands), the computation time may become very long. By definition, distribution matching tests (such as tests of normality) are non-parametric.

A.2.5 Confidence interval for a mean: student's *t* test

Consider a variable X with a mean μ_X and standard deviation σ_X defined in a population from which random samples of size n are taken. For each of these samples, we calculate the empirical mean $\mu_m = n^{-1}\sum_i x_i$, and we examine the distribution $\{\mu_m\}$ of the empirical means. If X follows a normal distribution or if n is large (> 30), then the distribution of the empirical means also follows a normal distribution whose mean is μ_X and whose standard deviation is $\sigma_X/\sqrt{n}$. Given that 95% of the values of a normal distribution (μ,σ) are in the range $[\mu - 1.96\sigma,\ \mu + 1.96\sigma]$, the empirical mean measured on a random sample of size n will have a 95% probability of being in the range $[\mu_X - 1.96\sigma_X/\sqrt{n},\ \mu_X + 1.96\sigma_X/\sqrt{n}]$.

Although the principle of this test is simple it can only be applied if the following two conditions are met:

1. The distribution of X is normal, or $n > 30$.

2. The standard deviation σ_X in the whole population is known.

If the standard deviation σ_X is not known, the more generally applicable Student's t test must be used.

Student's test is used to compare the empirical mean in a random sample with a theoretical value (the value μ_X of the mean for the whole population), and to decide whether the difference observed between these two means is significant or may be due to chance. There are two ways of using the test:

- either the mean μ_X of the whole population is known, and, by providing the range of values in which the sample means must fall, the test enables us to determine whether a sample is 95% representative; or
- the mean μ_X is not known, and, by assuming that the sample is representative and calculating its empirical mean μ_m, we determine a confidence interval which has a 95% probability of containing the mean μ_X of the population.

Student's t test enables us to replace the unknown standard deviation σ_X with an estimator which is the *sample* standard deviation:

$$\sigma_m = \sqrt{\frac{\sum_i (x_i - \mu_m)^2}{n-1}}.$$

The definition of σ_m includes, in the denominator, a factor $n-1$, instead of n as in the *population* standard deviation, because the *sample* standard deviation must allow for not only the dispersion of the observations x_i about their mean μ_m, but also the dispersion of the mean μ_m about the actual mean μ_X.

Student's test is based on the finding that the above statement that the distribution of the empirical means is a normal distribution with a mean of μ_X and a standard deviation of $\sigma_X/\sqrt{n}$ is equivalent to the statement that the quantity

$$\frac{\mu_m - \mu_X}{\sigma_X/\sqrt{n}}$$

follows the standard normal distribution. Student (whose real name was William Gosset) showed that, if σ_X was replaced by the sample standard deviation σ_m, the quantity

$$t = \frac{\mu_m - \mu_X}{\sigma_m/\sqrt{n}}$$

followed a special distribution called Student's t distribution with $n-1$ degrees of freedom.

As in the case of the standard normal distribution, we can find in a table the value t_α (generally $t_{0.025}$) such that t has a probability of 2α of not being in the range $[-t_\alpha, +t_\alpha]$, or, in other words, that the range

$$\left[\mu_m - t_\alpha \frac{\sigma_m}{\sqrt{n}}, \mu_m + t_\alpha \frac{\sigma_m}{\sqrt{n}}\right]$$

has a probability of $1-2\alpha$ (generally 0.95) of containing the mean μ_X. This range is therefore the $100(1-2\alpha)\%$ confidence interval for μ_X.

The test above is a two-tailed (symmetrical) test: the population mean can be above or below the confidence interval, so there are two rejection regions. In a one-tailed test, the mean is on only one side of the confidence interval, and the value t_α to be found in the table is generally $t_{0.05}$.

Finally, if $n > 30$, the test is simpler, because the t distribution is approximated by the standard normal distribution. Instead of finding t_α or $t_{\alpha/2}$ in the table for the t-distribution with $n-1$ degrees of freedom, we find it in the standard normal distribution table; in particular, $t_{0.025} = 1.96$.

A.2.6 Confidence interval of a frequency (or proportion)

Theory

When an event occurs in a large population with a probability p, this probability can be estimated, from a sample of size n from this population, by using the frequency $f = k/n$ of occurrence of the event (or the observed proportion) in the sample. If we assume that the events are independent, the variable k follows a binomial distribution $B(n,p)$ with a mean of $\mu = np$ and a variance of $\sigma^2 = np(1-p)$. We know that, when $n > 30$, $np > 15$ and $np(1-p) > 5$, the binomial distribution $B(n,p)$ tends to a normal distribution of parameters (μ,σ), so that the frequency f follows a normal distribution with a mean of $\mu/n = p$ and a variance of $\sigma^2/n^2 = p(1-p)/n$. Given that 95% of the values of a normal distribution (μ,σ) are found in the range $[\mu - 1.96\sigma, \mu + 1.96\sigma]$, the frequency f has a probability of 0.95 of falling within the confidence interval:

$$\left[p - 1.96\sqrt{\frac{p(1-p)}{n}}, p + 1.96\sqrt{\frac{p(1-p)}{n}}\right]$$

Therefore the range

$$\left[f - 1.96\sqrt{\frac{f(1-f)}{n}}, f + 1.96\sqrt{\frac{f(1-f)}{n}}\right]$$

has a probability of 0.95 of containing the true value of p.

Note that, in reality, the assumption of the independence of events is by no means true in all cases. For example, in a survey, each person is only asked questions on one occasion, and a survey of n persons is not the equivalent of n independent surveys according to a Bernoulli distribution. Strictly speaking, sampling with replacement should be replaced by sampling without replacement, and therefore the binomial distribution should be replaced by the hypergeometric distribution. In practice, the binomial distribution is considered to be sufficiently close to the hypergeometric distribution when the size N of the population is more than $10n$ (where n is the size of the sample), meaning that the variance $\sigma^2 = np(1-p)(N-n)/(N-1)$ of the variable k following a hypergeometric distribution tends towards the variance of a binomial distribution.

The 95% confidence interval is the one most commonly used. The formula for the confidence interval shows that it is widest in the region of $p = 0.5$. For a 99% confidence interval, the constant 1.96 above should be replaced by 2.58.

Example 1

An exit poll of 96 voters at a two-candidate election finds that 52% voted for candidate X and 48% for candidate Y. The formula shown above, with $f = 0.52$ and $n = 96$, shows that the actual proportion of votes for X is $52\% \pm 10\%$ with 95% confidence. Thus the exit poll cannot separate the two candidates. In a poll of 9604 voters, the same results give a 95% confidence interval of $52\% \pm 1\%$. We can therefore predict, with only a 5% risk of error, that X will be elected.

Example 2

If a survey question receives 50% 'yes' responses, we can deduce the confidence interval for 'yes' according to the sample size n, as follows (with 95% confidence, given that all the above values of n would have to be multiplied by 1.727 at the 1% threshold):

n	range	precision
96	[40%, 60%]	10%
119	[41%, 59%]	9%
150	[42%, 58%]	8%
196	[43%, 57%]	7%
267	[44%, 56%]	6%
384	[45%, 55%]	5%
600	[46%, 54%]	4%
1067	[47%, 53%]	3%
2401	[48%, 52%]	2%
9604	[49%, 51%]	1%

Note that, as indicated by the formula above, the sample size n must be multiplied by x^2 in order to divide the width of the confidence interval by x. Precision, then, comes at a high price. It depends on the size of the sample, not the size of the population.

Similarly, if 40% of the responses to a question are 'yes', the sample sizes needed to achieve a certain degree of precision are almost identical to those shown above. If 30% of the responses to a question are 'yes', the necessary sample sizes will be slightly smaller than those above.

n	range	precision
81	[20%, 40%]	10%
...	...	...
8068	[29%, 31%]	1%

If 20% of the responses to a question are 'yes', the necessary sample sizes will again be a little smaller than those above, and so on.

A.2.7 The relationship between two continuous variables: the linear correlation coefficient

The strength of the linear relationship between two continuous variables can be measured by the *linear correlation coefficient* (Pearson's r). The relationship is:

- *zero* if the correlation coefficient is 0 (i.e. the cloud of points is circular or parallel to one of the two axes corresponding to the variables);
- *perfect* if the correlation coefficient is $+1$ or -1 (the cloud of points is a straight line)
- *strong* if the correlation coefficient is greater than $+0.8$ or less than than -0.8 (the cloud is elliptical and elongated).

The linear correlation coefficient is positive when the two variables change in the same direction: that is, both variables increase or decrease together. A negative correlation coefficient indicates the opposite kind of change: one increases when the other decreases.

Note that a non-linear relationship, particularly a non-monotonic relationship, cannot always be measured with Pearson's linear correlation coefficient. This is the case with a (second degree) parabolic relationship such as that shown in Figure A.5 (D). And even a linear relationship may not be detected by the Pearson coefficient if there are extreme values or outliers present.

Thus, in Figure A.6, the decreasing linear relationship is masked by an outlier point, so that Pearson's r becomes positive instead of negative. Another example is that of Anscombe, described in Section 11.7.6 on linear regression. Figure A.7 shows how the shape of the cloud of points representing two variables varies as a function of the linear correlation coefficient of these two variables. Each variable has been constructed five times, by random sampling

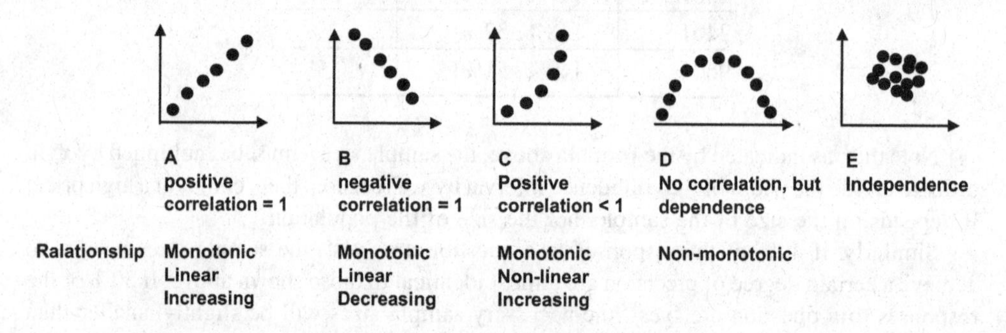

Figure A.5 Correlation and dependence of numeric variables.

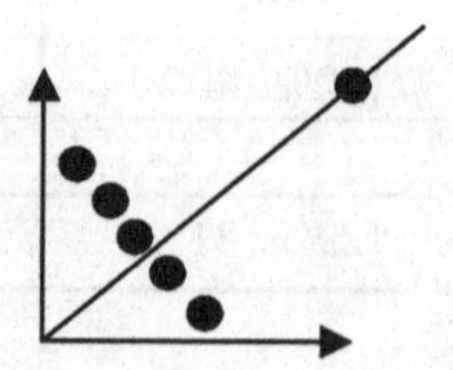

Figure A.6 Correlation masked by the presence of outlier points.

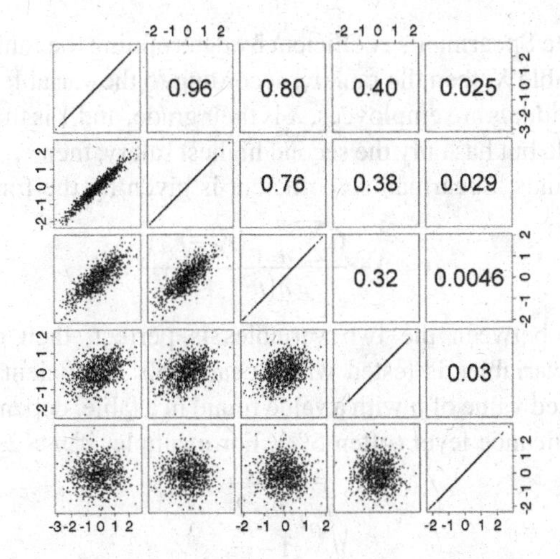

Figure A.7 Clouds of points as a function of the correlation coefficient.

according to the normal distribution, the five sampling procedures yielding ten different correlation coefficients, from 0.0046 to 0.96. The correlation coefficient of each cloud is in the cell which is symmetrical about the diagonal, the clouds of the diagonal corresponding to a correlation coefficient of 1 (correlation of the variable with itself).

A.2.8 The relationship between two numeric or ordinal variables: Spearman's rank correlation coefficient and Kendall's tau

Whereas Pearson's linear correlation coefficient can only be used with continuous variables, Spearman's rank correlation coefficient (or 'rho') can be used to measure the relationship between two variables X and Y which may be continuous, discrete or ordinal. Even for continuous variables, Spearman's ρ is preferable to Pearson's r if the variables have extreme values or do not follow a normal distribution. Spearman's ρ is also good at detecting all monotonic relationships, even if they are non-linear. Thus, in Figure A.5 (C), the non-linearity of the monotonic relationship will give rise to a low value of r, but ρ will remain high. A striking example is provided by the calculation of r and ρ for the variable $X = 0, 1, 2, 3, \ldots$ and the variable $Y = e^0, e^1, e^2, e^3, \ldots$. In this case, $\rho = 1$ as would be expected, but r is less than 1 and becomes smaller and smaller.

It is always useful to compare the Pearson and Spearman coefficients, as the latter can be used in more situations:

- if $r > \rho$, there may be exceptional values present;
- if $r < p$, there may be a non-linear relationship.

In theory, Spearman's correlation coefficient is calculated in the same way as Pearson's coefficient, after replacing the values of the variables with their ranks:

$$\rho = \frac{\text{cov}(r_x, r_y)}{\sigma_{r_x}, \sigma_{r_y}}.$$

In practice, we calculate Spearman's coefficient by determining the rank r_{xi} of each individual i according to the variable X, then the rank r_{yi} according to the variable Y. Assume, by way of example, that the individuals are employees, X is their grade, and Y is their salary. If individual 1 is on the highest grade but has only the second highest salary, then $r_{x1} = 1$ and $r_{y1} = 2$. If n is the number of individuals, Spearman's coefficient is given by the formula

$$\rho = 1 - \frac{6\sum_{i=1}^{n}(r_{xi} - r_{yi})^2}{n(n^2-1)}.$$

If the association between the two variables is perfect, then of course $\rho = 1$. The independence of two variables is tested with Spearman's coefficient in the usual way, by comparing the calculated value of ρ with a value found in a table, depending on the degrees of freedom and the significance level (often 5%). For example, if $n > 25$, the quantity

$$\sqrt{\rho^2 \frac{n-2}{1-\rho^2}}.$$

follows a Student t distribution with $n - 2$ degrees of freedom; we can then conduct a t test.

This test is robust, because Spearman's ρ is calculated on the ranks of the values of the variables, not on the values themselves, enabling us to avoid the constraining assumption of the normality of the variables: it is a non-parametric test.

Spearman's coefficient has the advantage of being easily interpreted, like the linear correlation coefficient. However, it is considered to be imperfect if a variable has numerous cases of tied ranks in the population; in this situation another statistical quantity, called *Kendall's tau* (τ), is preferred. Without going into the complexities of its calculation and interpretation, I will point out that τ is not a correlation coefficient, but is the difference between:

- the probability that, for the observed data, X and Y are in the same order, and
- the probability that, for the observed data, X and Y are in different orders.

It follows that Kendall's tau is equal to 2 times the first probability -1, and it lies between -1 and $+1$.

Kendall's tau τ and Spearman's rho ρ are related by the Siegel–Castellan formula: $-1 \leq 3\tau - 2\rho \leq 1$.

A.2.9 The relationship between *n* sets of several continuous or binary variables: canonical correlation analysis

Canonical correlation analysis is a very broad generalization of the correlation of two continuous variables. It deals with n (≥ 2) sets $\{U_i\}$, $\{V_i\}$, ... of a number of continuous or binary variables, and the aim is to find the linear combinations (called *canonical variables*) which maximize the correlation between $\sum_i a_i U_i$, $\sum_i \mu_i V_i$, ... It can be implemented by the CANCORR procedure in SAS (if $n = 2$) and OVERALS in IBM SPSS Statistics. It is a generalization of all the following methods:

- multiple regression ($n = 2$ and one of the sets of variables contains one variable only);
- linear discriminant analysis ($n = 2$ and one of the sets of variables contains the indicators of the partition to be discriminated);

- correspondence analysis ($n = 2$ and each set of variables contains the indicators of a qualitative variable).

A.2.10 The relationship between two nominal variables: the χ^2 test

The χ^2 test is used to check that the distribution of a variable X follows a given probability distribution, by comparing the observed distribution (of frequencies $\{O_i\}$) with the theoretical distribution (of frequencies $\{T_i\}$). If the null hypothesis, $H_0 = \{O_i = T_i$, for every $i = 1, \ldots, n\}$, is true, then each element $(O_i - T_i)/T_i$ tends towards a standard normal distribution (according to the central limit theorem) if the theoretical frequency T_i is large enough (conventionally ≥ 5). Therefore, if H_0 is true and if $T_i \geq 5$ for every i, the quantity

$$\sum_i \frac{(O_i - T_i)^2}{T_i}$$

follows a distribution which is a sum of p squares of independent $N(0,1)$ distributions: such a distribution is called a χ^2 distribution with p degrees of freedom. Note the absence of any assumption on the theoretical distribution of the variable X: the χ^2 test is non-parametric (Section A.2.4). However, the quantity $\sum_i (O_i - T_i)^2/T_i$ only follows a χ^2 distribution asymptotically: if the theoretical frequencies are small (< 5), an exact probability calculation is required. The exact test which replaces the χ^2 test for a 2×2 contingency table is called Fisher's test and is based on the *hypergeometric distribution*. This distribution is defined for two variables A and B with two categories, for which the 2×2 contingency table has frequencies denoted a, b, c and d, and it states that, if A and B are independent, the probability of having a table (a,b,c,d) with fixed margins $a + c$, $b + d$, $a + b$, $c + d$ is given by

$$P(a,b,c,d) = \frac{(a+c)!(b+d)!(a+b)!(c+d)!}{a!b!c!d!(a+b+c+d)!}.$$

The χ^2 test is often used to test the *independence of two variables* X and Y, given that X and Y are independent (the null hypothesis, H_0). Then, for all i and j,

$$\#\{X = i \text{ and } Y = j\} = \#\{X = i\} \times \#\{Y = j\} \times 1/N,$$

where $\#\{..\}$ is to be read as 'the number of individuals such that' and N is the total number of individuals. If O_{ij} denotes the term on the left-hand side of the equality and T_{ij} denotes the term on the right, the test of independence of X and Y is the χ^2 test applied to the statistic

$$\chi^2 = \sum_i \sum_j \frac{(O_{ij} - T_{ij})^2}{T_{ij}}$$

The number of degrees of freedom is

$$p = (\text{number of rows} - 1) \times (\text{number of columns} - 1).$$

A.2.11 Example of use of the χ^2 test

The χ^2 test is used to decide if the value (A or B) taken by a variable depends on the class (1 or 2) to which the individual belongs.

For a small frequency

Consider the case of a population of 150 individuals.

	Class 1	Class 2	Total
Observed frequencies:			
A	55	45	100
B	20	30	50
Total	75	75	150
Frequencies expected if the variable is independent of class:			
A	50	50	100
B	25	25	50
Total	75	75	150
Probability of the $\chi^2 = 0.0833$			

In the population of 150 individuals, 66.67% of individuals take the value A. In class 1, 73.33% of individuals take the value A. Is this difference in the distribution of the categories between the two classes significant? The χ^2 calculated for the above distribution is

$$\frac{(55\text{–}50)^2}{50} + \frac{(45\text{–}50)^2}{50} + \frac{(20\text{–}25)^2}{25} + \frac{(30\text{–}25)^2}{25} = 3,$$

and the number of degrees of freedom is $(2-1)(2-1) = 1$. Now, for 1 degree of freedom, if the variables $\{A, B\}$ and {class 1, class 2} are independent, the probability that $\chi^2 \geq 3$ is 0.0833. This probability is greater than 0.05, indicating that the null hypothesis of independence can be accepted.

For a large frequency

We shall now conduct the same test on a population of 1500 individuals. All the previous frequencies are multiplied by 10, to demonstrate the effect of the frequency on the result of the χ^2 test.

	Class 1	Class 2	Total
Observed frequencies:			
A	550	450	1000
B	200	300	500
Total	750	750	1500
Frequencies expected if the variable is independent of the class:			
A	500	500	1000
B	250	250	500
Total	750	750	1500
Probability of $\chi^2 = 4.3205 \times 10^{-8}$			

In the population of 1500 individuals we have that 66.67% of individuals take the value A, while in class 1 we have that 73.33% of individuals take the value A, as before. The χ^2 calculated for the above distribution is

$$\frac{(550-500)^2}{500}+\frac{(450-500)^2}{500}+\frac{(200-250)^2}{250}+\frac{(300-250)^2}{250}=30,$$

and the number of degrees of freedom is $(2-1)(2-1)=1$. Now, for 1 degree of freedom, if the variables $\{A, B\}$ and {class 1, class 2} are independent, the probability that $\chi^2 \geq 30$ is 4.3205×10^{-8}. Because this probability is less than 0.05, the hypothesis of independence is rejected in this case.

Conclusion

When the size of the population increases, the least difference becomes significant at the normal significance levels (5% or 1%), even if the proportions are unchanged. It must be remembered that the χ^2 test was devised at a time when statistics dealt with small samples of no more than a few hundred individuals. The significance levels were fixed accordingly, but must be reduced when the frequencies increase.

A.2.12 The relationship between two nominal variables: Cramér's coefficient

Let r be the number of rows and s the number of columns. In the above description, the maximum value of the variable

$$\chi^2=\sum_i\sum_j\frac{(O_{ij}-T_{ij})^2}{T_{ij}}$$

found when the contingency table has a single non-zero cell in each row ($r > s$) or a single non-zero cell in each column ($r < s$), or when it has non-zero values on the diagonal only ($r = s$), is:

$$\chi^2_{max} = \text{frequency} \times [\min(r, s) - 1].$$

From this value χ^2_{max} we can deduce a new quantity, Cramér's coefficient

$$V=\sqrt{\frac{\chi^2}{\chi^2_{max}}},$$

which varies between 0 (zero relationship) and 1 (maximum relationship) and can be used to evaluate the strength of the relationship between two qualitative variables, without using a χ^2 table. This is because this quantity is scaled by definition and incorporates the number of degrees of freedom via χ^2_{max}. The probability can be used in the same way, but it lacks readability when it is very small, and above all it is sensitive to the size of the population, as mentioned in the last section. Cramér's V, on the other hand, is insensitive to this, because it incorporates the population size via χ^2_{max}. It therefore has the major advantage of providing an absolute measure of the strength of the relationship between two nominal variables, independently of the number of their categories and the population size. Thus, for the tables in the last two sections, Cramér's V takes the same value, being the square root of 3/150 or 30/1500.

Other coefficients can be found from χ^2, namely Pearson's contingency coefficient and Chuprov's coefficient; they are located in the range from 0 to 1, but Pearson's coefficient never reaches 1 and Chuprov's coefficient can only reach it if $r = s$. Pearson also used the formula $\Phi^2 = \chi^2/N$ to find an indicator Φ which is not normalized but is independent of the population size N. As their significance level depends on r and s, unlike that of Cramér's V, and is not associated with a statistical test (their distribution is not known), unlike χ^2, these indicators are less useful than Cramér's V.

A.2.13 The relationship between a nominal variable and a numeric variable: the variance test (one-way ANOVA test)

Let X be a nominal variable having k categories $x_1, x_2, \ldots, x_k$, with frequencies $n_1, n_2, \ldots, n_k$. Let Y be a numeric variable, with mean μ. Let n be the total number of individuals. We wish to test the independence of X and Y.

Let μ_i be the mean of Y calculated for all the individuals for which X is equal to x_i. This is the mean of Y for the ith group. Similarly, we define V_i as the variance of Y calculated for all the individuals for which X is x_i, i.e. the sum of $(Y - \mu_i)^2$ calculated for the n_i individuals y_j for which X is x_i, divided by $(n_i - 1)$: this is the variance within the class $\{x_i\}$.

For example, let X be the sex of an individual and Y his salary. We examine μ_1 and V_1, which are the mean and variance of the men's salaries, and μ_2 and V_2, which are the mean and variance of the women's salaries; we then check to see if X and Y are independent, in other words if the means μ_i are equal to each other. This equality of the means is the null hypothesis in the ANOVA test. The ANOVA test can also be used to compare the crop yields of a number of fields, the effectiveness of a number of fertilizers, or the outputs of a number of factories. Thus we are generally comparing more than two groups: ANOVA is a generalization of Student's t test where $k > 2$. In this case, the null hypothesis is that $\mu_1 = \mu_2 = \ldots = \mu_k$, and the alternative hypothesis is that one or more of the means is different. This does not mean that $\mu_1 \neq \mu_2 \neq \ldots \neq \mu_k$, and the ANOVA test does not reveal which means are significantly different. To discover this, we must use the Bonferroni test, or the more conservative Scheffé test (which does not reject the null hypothesis very easily), as these tests are more suitable than the more liberal method in which multiple comparisons are made by Student's t tests. It may seem strange to use the term 'analysis of variance' for what is really a test of equality of the means; this is due to the way in which the test is carried out, by dividing the variance of the variable Y into two parts, as follows:

- the part due to the differences between groups (between-class variance);
- and the part due to random variations (within-class variance, also called 'error').

The strength of the relationship between X and Y is commonly measured by the R^2, defined thus:

$$R^2 = \frac{\text{between-class sum of squares}}{\text{total sum of squares}} = \frac{\sum_{i=1}^{k} n_i(\mu_i - \mu)^2}{\sum_{j=1}^{n} (y_j - \mu)^2}.$$

The relationship is perfect if $R^2 = 1$ and zero if $R^2 = 0$.

The difference between the total sum of squares TSS and the between-class sum of squares ESS (E for 'explained') is the within-class sum of squares RSS (R for 'residual'), and these three sums of squares are related by the equality

$$\text{TSS} = \text{ESS} + \text{RSS},$$

that is,

$$\sum_{j=1}^{n}(y_j-\mu)^2 = \sum_{i=1}^{k} n_i(\mu_i-\mu)^2 + \sum_{i=1}^{k}\sum_{j:X=x_i}(y_j-\mu_i)^2.$$

The variances are found in the usual way by dividing the sum of squares by the number of degrees of freedom, which gives:

$$\text{total variance} = \text{TSS}/(n-1),$$

$$\text{between-class variance} = \text{ESS}/(k-1),$$

$$\text{within-class variance} = \text{RSS}/(n-k).$$

The within-class variance is also equal to the sum of the variances in each class:

$$\sum_i \text{V}_i.$$

Another way of measuring the strength of the relationship between X and Y is to calculate the F-ratio:

$$F = \frac{\text{between-class variance}}{\text{within-class variance}}.$$

The independence of X and Y is tested by formulating the null hypothesis H_0 that all the means μ_i are equal. If H_0 is true, F must follow Fisher's distribution, and to check this we can compare the calculated value of F with the value found in the Fisher–Snedecor distribution table, as a function of the chosen threshold and the degrees of freedom of the between-class variance $(k-1)$ and the within-class variance $(n-k)$: this is the Fisher–Snedecor test. If the F-ratio exceeds the critical value of Fisher's distribution, H_0 must be rejected and the residual variations are weak compared with the effect of the differences between groups.

As a general rule, the Fisher–Snedecor test is used to demonstrate that a ratio of two variances v_1/v_2 $(v_1 > v_2)$ is significantly greater than 1, for a specified significance level and degrees of freedom (those of v_1 and v_2).

However, there are limitations on the conditions for using Fisher's test for ANOVA. These can be demonstrated if we write the general ANOVA model in another form. The value Y_{ij} of observation j in group i is

$$Y_{ij} = \mu + \alpha_i + \varepsilon_{ij},$$

where $\alpha_i = \mu_i - \mu$ and ε_{ij} is a residual value which must meet the following conditions:

- it is normally distributed in all the groups (Fisher's test lacks robustness if normality is not present);
- its mean is 0 in all the groups;
- its variance is equal in all the groups (homoscedasticity);
- the ε_{ij} are idendepent for all i and j, which means that an observation must not be dependent on others in the group, and that the observations in one group must not depend on those in other groups.

Returning to the previous example, the salary must be normally distributed in the group of men and the group of women, it must have the same variance in both groups, and the salary of one individual must be independent of that of another individual.

In conclusion, we can say that ANOVA is generalized in two directions by dealing with:

- m (> 1) independent nominal variables (m-way ANOVA);
- more than one continuous variable to be explained (MANOVA).

If more than one variable is explained simultaneously, this may reveal explanatory factors which would not be detected by a number of ANOVAs carried out separately on each of the variables to be explained.

A.2.14 The Cox semi-parametric survival model

In a survival analysis, the observations for each individual are repeated in time at the instants $t_1, t_2, \ldots, t_N$. We examine the occurrence of an event (e.g. the death of a patient or the departure of a customer) at time t_i, which is modelled by the dependent variable defined as follows:

- $y_k = 0$ if $k < i$;
- $y_i = 1$;
- there is no observation if $k > i$.

The event does not always occur in the period under investigation. We have $y_k = 0$ for every $k \leq N$ if the event does not occur (and if the individual is observed until the end of the period). In this case, we only know the lower bound of the time elapsing before the occurrence of the event, and we say that this data element is censored. The data element is also censored if the individual ceases to be monitored before the end of the period and before the event occurs.

The aim of survival analysis is to explain the variable 'length of survival' in order to reveal the factors promoting survival. Even if the data are censored (because the individual is still alive and still present at the end of the period), they must be taken into account, since the longest lifetimes are, by definition, the most censored. Therefore the aim is to find models that can handle both censored and uncensored data.

The basic approach (the Kaplan–Meier model) is to calculate a non-parametric estimate of the survival function: $S(t) = \text{Prob(length of life} > t)$.

Cox's proportional hazards regression model (1972)[1] is the most widely used method in survival analysis. It enables us to add p independent variables ('covariables') and estimate their coefficients in the survival function, and consequently their effect on the length of survival (such variables may include sex, number of cigarettes smoked per day, etc.) This is a semi-parametric model, with a parametric form for the effects of the covariables, and a non-parametric form of the survival function.

For each individual i with covariables $\{x_{ij}\}_j$, the survival function is expressed as follows:

$$S(t, x_i) = S_0(t)^{\exp\left(\sum_{j=0}^{p} \beta_j x_{ij}\right)},$$

where $S_0(t)$ is the basic survival function ('basic hazard'), and $x_{i0} = 1$ for every i. To find the vector $\{\beta_j\}$ of regression coefficients, which is assumed to be independent of i, we proceed as in logistic regression, by maximizing a likelihood function. There are several methods of selecting the covariables (forward, backward, or stepwise), and the odds ratios can be interpreted. As in regression, the covariables must be present in a larger number than the observations, and they must not be strongly correlated; if they are, we can use the Cox PLS model, an extension of the Cox model which bears the same kind of relation to the basic model as PLS regression does to linear regression.[2]

The censored data are not used in the calculation of $\{\beta_j\}$, but they play a part in the calculation of $S_o(t)$.

The term 'proportional hazards' refers to the basic assumption that two individuals having different configurations of covariables have a constant hazard ratio over time. For example, if two individuals are such that $x_{1j} = 1$ and $x_{2j} = 0$ for the jth covariable, and if they coincide for the other covariables, the ratio $S(t,x_1)/S(t,x_2)$ will be constant and equal to $\exp(\beta_j)$.

A.3 Statistical tables

A.3.1 Table of the standard normal distribution

Table A.2 shows the probability $\Pr(0 \le Z \le z_0)$: the first decimal place of z_0 is read down the leftmost column, and the second decimal place of z_0 is read along the top row. For example, we can see that $\Pr(0 \le Z \le 0.53) = 0.2019$.

A.3.2 Table of Student's *t* distribution

Table A.3 shows the critical values of t, in other words the values t_α such that $\Pr(t \ge t_\alpha) = \alpha$, where α is the probability read along the top row, and the number of degrees of freedom is read down the leftmost column. For example, for 6 degrees of freedom, the probability that $t \ge 1.943180$ is 5%.

---

[1] Cox, D. R. (1972) Regression Models and Life Tables (with Discussion). *Journal of the Royal Statistical Society*, Series B, 34(2), 187–220.

[2] Bastien, P. and Tenenhaus, M. (2001) PLS generalized linear regression: Application to the analysis of life time data. In *PLS and Related Methods, Proceedings of the PLEASE '01 International Symposium*, CISIA-CERESTA, pp. 131–140.

Table A.2 Standard normal distribution.

z_0	0.00	0.01	0.02	0.03	0.04	0.05	0.06	0.07	0.08	0.09
0.0	0.0000	0.0040	0.0080	0.0120	0.0160	0.0199	0.0239	0.0279	0.0319	0.0359
0.1	0.0398	0.0438	0.0478	0.0517	0.0557	0.0596	0.0636	0.0675	0.0714	0.0753
0.2	0.0793	0.0832	0.0871	0.0910	0.0948	0.0987	0.1026	0.1064	0.1103	0.1141
0.3	0.1179	0.1217	0.1255	0.1293	0.1331	0.1368	0.1406	0.1443	0.1480	0.1517
0.4	0.1554	0.1591	0.1628	0.1664	0.1700	0.1736	0.1772	0.1808	0.1844	0.1879
0.5	0.1915	0.1950	0.1985	0.2019	0.2054	0.2088	0.2123	0.2157	0.2190	0.2224
0.6	0.2257	0.2291	0.2324	0.2357	0.2389	0.2422	0.2454	0.2486	0.2517	0.2549
0.7	0.2580	0.2611	0.2642	0.2673	0.2704	0.2734	0.2764	0.2794	0.2823	0.2852
0.8	0.2881	0.2910	0.2939	0.2967	0.2995	0.3023	0.3051	0.3078	0.3106	0.3133
0.9	0.3159	0.3186	0.3212	0.3238	0.3264	0.3289	0.3315	0.3340	0.3365	0.3389
1.0	0.3413	0.3438	0.3461	0.3485	0.3508	0.3531	0.3554	0.3577	0.3599	0.3621
1.1	0.3643	0.3665	0.3686	0.3708	0.3729	0.3749	0.3770	0.3790	0.3810	0.3830
1.2	0.3849	0.3869	0.3888	0.3907	0.3925	0.3944	0.3962	0.3980	0.3997	0.4015
1.3	0.4032	0.4049	0.4066	0.4082	0.4099	0.4115	0.4131	0.4147	0.4162	0.4177
1.4	0.4192	0.4207	0.4222	0.4236	0.4251	0.4265	0.4279	0.4292	0.4306	0.4319
1.5	0.4332	0.4345	0.4357	0.4370	0.4382	0.4394	0.4406	0.4418	0.4429	0.4441
1.6	0.4452	0.4463	0.4474	0.4484	0.4495	0.4505	0.4515	0.4525	0.4535	0.4545
1.7	0.4554	0.4564	0.4573	0.4582	0.4591	0.4599	0.4608	0.4616	0.4625	0.4633
1.8	0.4641	0.4649	0.4656	0.4664	0.4671	0.4678	0.4686	0.4693	0.4699	0.4706
1.9	0.4713	0.4719	0.4726	0.4732	0.4738	0.4744	0.4750	0.4756	0.4761	0.4767
2.0	0.4772	0.4778	0.4783	0.4788	0.4793	0.4798	0.4803	0.4808	0.4812	0.4817
2.1	0.4821	0.4826	0.4830	0.4834	0.4838	0.4842	0.4846	0.4850	0.4854	0.4857
2.2	0.4861	0.4864	0.4868	0.4871	0.4875	0.4878	0.4881	0.4884	0.4887	0.4890
2.3	0.4893	0.4896	0.4898	0.4901	0.4904	0.4906	0.4909	0.4911	0.4913	0.4916
2.4	0.4918	0.4920	0.4922	0.4925	0.4927	0.4929	0.4931	0.4932	0.4934	0.4936
2.5	0.4938	0.4940	0.4941	0.4943	0.4945	0.4946	0.4948	0.4949	0.4951	0.4952
2.6	0.4953	0.4955	0.4956	0.4957	0.4959	0.4960	0.4961	0.4962	0.4963	0.4964
2.7	0.4965	0.4966	0.4967	0.4968	0.4969	0.4970	0.4971	0.4972	0.4973	0.4974
2.8	0.4974	0.4975	0.4976	0.4977	0.4977	0.4978	0.4979	0.4979	0.4980	0.4981
2.9	0.4981	0.4982	0.4982	0.4983	0.4984	0.4984	0.4985	0.4985	0.4986	0.4986
3.0	0.4987	0.4987	0.4987	0.4988	0.4988	0.4989	0.4989	0.4989	0.4990	0.4990

A.3.3 Chi-Square table

Table A.4 shows the critical values of χ^2, in other words the values c_α such that $\Pr(\chi^2 \geq c_\alpha) = \alpha$, where α is the probability read along the top row, and the number of degrees of freedom is read down the leftmost column. For example, for 6 degrees of freedom, the probability that $\chi^2 \geq 12.59159$ is 5%.

Table A.3 Student's t distribution.

t_α	0.25	0.10	0.05	0.025	0.01	0.005	0.0005
1	1.000000	3.077684	6.313752	12.70620	31.82052	63.65674	636.6192
2	0.816497	1.885618	2.919986	4.30265	6.96456	9.92484	31.5991
3	0.764892	1.637744	2.353363	3.18245	4.54070	5.84091	12.9240
4	0.740697	1.533206	2.131847	2.77645	3.74695	4.60409	8.6103
5	0.726687	1.475884	2.015048	2.57058	3.36493	4.03214	6.8688
6	0.717558	1.439756	1.943180	2.44691	3.14267	3.70743	5.9588
7	0.711142	1.414924	1.894579	2.36462	2.99795	3.49948	5.4079
8	0.706387	1.396815	1.859548	2.30600	2.89646	3.35539	5.0413
9	0.702722	1.383029	1.833113	2.26216	2.82144	3.24984	4.7809
10	0.699812	1.372184	1.812461	2.22814	2.76377	3.16927	4.5869
11	0.697445	1.363430	1.795885	2.20099	2.71808	3.10581	4.4370
12	0.695483	1.356217	1.782288	2.17881	2.68100	3.05454	4.3178
13	0.693829	1.350171	1.770933	2.16037	2.65031	3.01228	4.2208
14	0.692417	1.345030	1.761310	2.14479	2.62449	2.97684	4.1405
15	0.691197	1.340606	1.753050	2.13145	2.60248	2.94671	4.0728
16	0.690132	1.336757	1.745884	2.11991	2.58349	2.92078	4.0150
17	0.689195	1.333379	1.739607	2.10982	2.56693	2.89823	3.9651
18	0.688364	1.330391	1.734064	2.10092	2.55238	2.87844	3.9216
19	0.687621	1.327728	1.729133	2.09302	2.53948	2.86093	3.8834
20	0.686954	1.325341	1.724718	2.08596	2.52798	2.84534	3.8495
21	0.686352	1.323188	1.720743	2.07961	2.51765	2.83136	3.8193
22	0.685805	1.321237	1.717144	2.07387	2.50832	2.81876	3.7921
23	0.685306	1.319460	1.713872	2.06866	2.49987	2.80734	3.7676
24	0.684850	1.317836	1.710882	2.06390	2.49216	2.79694	3.7454
25	0.684430	1.316345	1.708141	2.05954	2.48511	2.78744	3.7251
26	0.684043	1.314972	1.705618	2.05553	2.47863	2.77871	3.7066
27	0.683685	1.313703	1.703288	2.05183	2.47266	2.77068	3.6896
28	0.683353	1.312527	1.701131	2.04841	2.46714	2.76326	3.6739
29	0.683044	1.311434	1.699127	2.04523	2.46202	2.75639	3.6594
30	0.682756	1.310415	1.697261	2.04227	2.45726	2.75000	3.6460
∞	0.674490	1.281552	1.644854	1.95996	2.32635	2.57583	3.2905

A.3.4 Table of the Fisher-Snedecor distribution at the 0.05 significance level

Table A.5 shows the critical values of the Fisher–Snedecor distribution $F(\nu_1; \nu_2)$ at the 5% significance level, in other words the values f_α such that $\Pr(F(\nu_1; \nu_2) \geq f_\alpha) = 0.05$. The number

Table A.4 The χ^2 distribution.

c_α	0.995	0.990	0.975	0.950	0.900	0.100	0.050	0.025	0.010	0.005
1	0.00004	0.00016	0.00098	0.00393	0.01579	2.70554	3.84146	5.02389	6.63490	7.87944
2	0.01003	0.02010	0.05064	0.10259	0.21072	4.60517	5.99146	7.37776	9.21034	10.59663
3	0.07172	0.11483	0.21580	0.35185	0.58437	6.25139	7.81473	9.34840	11.34487	12.83816
4	0.20699	0.29711	0.48442	0.71072	1.06362	7.77944	9.48773	11.14329	13.27670	14.86026
5	0.41174	0.55430	0.83121	1.14548	1.61031	9.23636	11.07050	12.83250	15.08627	16.74960
6	0.67573	0.87209	1.23734	1.63538	2.20413	10.64464	12.59159	14.44938	16.81189	18.54758
7	0.98926	1.23904	1.68987	2.16735	2.83311	12.01704	14.06714	16.01276	18.47531	20.27774
8	1.34441	1.64650	2.17973	2.73264	3.48954	13.36157	15.50731	17.53455	20.09024	21.95495
9	1.73493	2.08790	2.70039	3.32511	4.16816	14.68366	16.91898	19.02277	21.66599	23.58935
10	2.15586	2.55821	3.24697	3.94030	4.86518	15.98718	18.30704	20.48318	23.20925	25.18818
11	2.60322	3.05348	3.81575	4.57481	5.57778	17.27501	19.67514	21.92005	24.72497	26.75685
12	3.07382	3.57057	4.40379	5.22603	6.30380	18.54935	21.02607	23.33666	26.21697	28.29952
13	3.56503	4.10692	5.00875	5.89186	7.04150	19.81193	22.36203	24.73560	27.68825	29.81947
14	4.07467	4.66043	5.62873	6.57063	7.78953	21.06414	23.68479	26.11895	29.14124	31.31935
15	4.60092	5.22935	6.26214	7.26094	8.54676	22.30713	24.99579	27.48839	30.57791	32.80132
16	5.14221	5.81221	6.90766	7.96165	9.31224	23.54183	26.29623	28.84535	31.99993	34.26719
17	5.69722	6.40776	7.56419	8.67176	10.08519	24.76904	27.58711	30.19101	33.40866	35.71847
18	6.26480	7.01491	8.23075	9.39046	10.86494	25.98942	28.86930	31.52638	34.80531	37.15645
19	6.84397	7.63273	8.90652	10.11701	11.65091	27.20357	30.14353	32.85233	36.19087	38.58226
20	7.43384	8.26040	9.59078	10.85081	12.44261	28.41198	31.41043	34.16961	37.56623	39.99685
21	8.03365	8.89720	10.28290	11.59131	13.23960	29.61509	32.67057	35.47888	38.93217	41.40106
22	8.64272	9.54249	10.98232	12.33801	14.04149	30.81328	33.92444	36.78071	40.28936	42.79565
23	9.26042	10.19572	11.68855	13.09051	14.84796	32.00690	35.17246	38.07563	41.63840	44.18128
24	9.88623	10.85636	12.40115	13.84843	15.65868	33.19624	36.41503	39.36408	42.97982	45.55851
25	10.51965	11.52398	13.11972	14.61141	16.47341	34.38159	37.65248	40.64647	44.31410	46.92789
26	11.16024	12.19815	13.84390	15.37916	17.29188	35.56317	38.88514	41.92317	45.64168	48.28988
27	11.80759	12.87850	14.57338	16.15140	18.11390	36.74122	40.11327	43.19451	46.96294	49.64492
28	12.46134	13.56471	15.30786	16.92788	18.93924	37.91592	41.33714	44.46079	48.27824	50.99338
29	13.12115	14.25645	16.04707	17.70837	19.76774	39.08747	42.55697	45.72229	49.58788	52.33562
30	13.78672	14.95346	16.79077	18.49266	20.59923	40.25602	43.77297	46.97924	50.89218	53.67196

Table A.5 The Fisher–Snedecor distribution at the 0.05 significance level.

	1	**2**	**3**	**4**	**5**	**6**	**7**	**8**	**9**	**10**
1	161	200	216	225	230	234	237	239	241	242
2	18.5	19.0	19.2	19.2	19.3	19.3	19.4	19.4	19.4	19.4
3	10.1	9.55	9.28	9.12	9.01	8.94	8.89	8.85	8.81	8.79
4	7.71	6.94	6.59	6.39	6.26	6.16	6.09	6.04	6.00	5.96
5	6.61	5.79	5.41	5.19	5.05	4.95	4.88	4.82	4.77	4.74
6	5.99	5.14	4.76	4.53	4.39	4.28	4.21	4.15	4.10	4.06
7	5.59	4.74	4.35	4.12	3.97	3.87	3.79	3.73	3.68	3.64
8	5.32	4.46	4.07	3.84	3.69	3.58	3.50	3.44	3.39	3.35
9	5.12	4.26	3.86	3.63	3.48	3.37	3.29	3.23	3.18	3.14
10	4.96	4.10	3.71	3.48	3.33	3.22	3.14	3.07	3.02	2.98
11	4.84	3.98	3.59	3.36	3.20	3.09	3.01	2.95	2.90	2.85
12	4.75	3.89	3.49	3.26	3.11	3.00	2.91	2.85	2.80	2.75
13	4.67	3.81	3.41	3.18	3.03	2.92	2.83	2.77	2.71	2.67
14	4.60	3.74	3.34	3.11	2.96	2.85	2.76	2.70	2.65	2.60
15	4.54	3.68	3.29	3.06	2.90	2.79	2.71	2.64	2.59	2.54
16	4.49	3.63	3.24	3.01	2.85	2.74	2.66	2.59	2.54	2.49
17	4.45	3.59	3.20	2.96	2.81	2.70	2.61	2.55	2.49	2.45
18	4.41	3.55	3.16	2.93	2.77	2.66	2.58	2.51	2.46	2.41
19	4.38	3.52	3.13	2.90	2.74	2.63	2.54	2.48	2.42	2.38
20	4.35	3.49	3.10	2.87	2.71	2.60	2.51	2.45	2.39	2.35
21	4.32	3.47	3.07	2.84	2.68	2.57	2.49	2.42	2.37	2.32
22	4.30	3.44	3.05	2.82	2.66	2.55	2.46	2.40	2.34	2.30
23	4.28	3.42	3.03	2.80	2.64	2.53	2.44	2.37	2.32	2.27
24	4.26	3.40	3.01	2.78	2.62	2.51	2.42	2.36	2.30	2.25
25	4.24	3.39	2.99	2.76	2.60	2.49	2.40	2.34	2.28	2.24
26	4.23	3.37	2.98	2.74	2.59	2.47	2.39	2.32	2.27	2.22
27	4.21	3.35	2.96	2.73	2.57	2.46	2.37	2.31	2.25	2.20
28	4.20	3.34	2.95	2.71	2.56	2.45	2.36	2.29	2.24	2.19
29	4.18	3.33	2.93	2.70	2.55	2.43	2.35	2.28	2.22	2.18
30	4.17	3.32	2.92	2.69	2.53	2.42	2.33	2.27	2.21	2.16
40	4.08	3.23	2.84	2.61	2.45	2.34	2.25	2.18	2.12	2.08
50	4.03	3.18	2.79	2.56	2.40	2.29	2.20	2.13	2.07	2.03
60	4.00	3.15	2.76	2.53	2.37	2.25	2.17	2.10	2.04	1.99
70	3.98	3.13	2.74	2.50	2.35	2.23	2.14	2.07	2.02	1.97
80	3.96	3.11	2.72	2.49	2.33	2.21	2.13	2.06	2.00	1.95
90	3.95	3.10	2.71	2.47	2.32	2.20	2.11	2.04	1.99	1.94
100	3.94	3.09	2.70	2.46	2.31	2.19	2.10	2.03	1.97	1.93
200	3.89	3.04	2.65	2.42	2.26	2.14	2.06	1.98	1.93	1.88
500	3.86	3.01	2.62	2.39	2.23	2.12	2.03	1.96	1.90	1.85
∞	3.84	3.00	2.60	2.37	2.21	2.10	2.01	1.94	1.88	1.83

(*continued*)

Table A.5 (*Continued*)

	15	**20**	**30**	**40**	**50**	**100**	**200**	**500**	**∞**
1	246	248	250	251	252	253	254	254	254
2	19.4	19.4	19.5	19.5	19.5	19.5	19.5	19.5	19.5
3	8.70	8.66	8.62	8.59	8.58	8.55	8.54	8.53	8.53
4	5.86	5.80	5.75	5.72	5.70	5.66	5.65	5.64	5.63
5	4.62	4.56	4.50	4.46	4.44	4.41	4.39	4.37	4.37
6	3.94	3.87	3.81	3.77	3.75	3.71	3.69	3.68	3.67
7	3.51	3.44	3.38	3.34	3.32	3.27	3.25	3.24	3.23
8	3.22	3.15	3.08	3.04	3.02	2.97	2.95	2.94	2.93
9	3.01	2.94	2.86	2.83	2.80	2.76	2.73	2.72	2.71
10	2.85	2.77	2.70	2.66	2.64	2.59	2.56	2.55	2.54
11	2.72	2.65	2.57	2.53	2.51	2.46	2.43	2.42	2.40
12	2.62	2.54	2.47	2.43	2.40	2.35	2.32	2.31	2.30
13	2.53	2.46	2.38	2.34	2.31	2.26	2.23	2.22	2.21
14	2.46	2.39	2.31	2.27	2.24	2.19	2.16	2.14	2.13
15	2.40	2.33	2.25	2.20	2.18	2.12	2.10	2.08	2.07
16	2.35	2.28	2.19	2.15	2.12	2.07	2.04	2.02	2.01
17	2.31	2.23	2.15	2.10	2.08	2.02	1.99	1.97	1.96
18	2.27	2.19	2.11	2.06	2.04	1.98	1.95	1.93	1.92
19	2.23	2.16	2.07	2.03	2.00	1.94	1.91	1.89	1.88
20	2.20	2.12	2.04	1.99	1.97	1.91	1.88	1.86	1.84
21	2.18	2.10	2.01	1.96	1.94	1.88	1.84	1.83	1.81
22	2.15	2.07	1.98	1.94	1.91	1.85	1.82	1.80	1.78
23	2.13	2.05	1.96	1.91	1.88	1.82	1.79	1.77	1.76
24	2.11	2.03	1.94	1.89	1.86	1.80	1.77	1.75	1.73
25	2.09	2.01	1.92	1.87	1.84	1.78	1.75	1.73	1.71
26	2.07	1.99	1.90	1.85	1.82	1.76	1.73	1.71	1.69
27	2.06	1.97	1.88	1.84	1.81	1.74	1.71	1.69	1.67
28	2.04	1.96	1.87	1.82	1.79	1.73	1.69	1.67	1.65
29	2.03	1.94	1.85	1.81	1.77	1.71	1.67	1.65	1.64
30	2.01	1.93	1.84	1.79	1.76	1.70	1.66	1.64	1.62
40	1.92	1.84	1.74	1.69	1.66	1.59	1.55	1.53	1.52
50	1.87	1.78	1.69	1.63	1.60	1.52	1.48	1.46	1.45
60	1.84	1.75	1.65	1.59	1.56	1.48	1.44	1.41	1.40
70	1.81	1.72	1.62	1.57	1.53	1.45	1.40	1.37	1.36
80	1.79	1.70	1.60	1.54	1.51	1.43	1.38	1.35	1.34
90	1.78	1.69	1.59	1.53	1.49	1.41	1.36	1.33	1.31
100	1.77	1.68	1.57	1.52	1.48	1.39	1.34	1.31	1.30
200	1.72	1.62	1.52	1.46	1.41	1.32	1.26	1.22	1.21
500	1.69	1.59	1.48	1.42	1.38	1.28	1.21	1.16	1.14
∞	1.67	1.57	1.46	1.39	1.35	1.24	1.17	1.11	1.00

Table A.6 The Fisher–Snedecor distribution at the 0.10 significance level.

	1	2	3	4	5	6	7	8	9	10
1	39.9	49.5	53.6	55.8	57.2	58.2	58.9	59.4	59.9	60.2
2	8.53	9.00	9.16	9.24	9.29	9.33	9.35	9.37	9.38	9.39
3	5.54	5.46	5.39	5.34	5.31	5.28	5.27	5.25	5.24	5.23
4	4.54	4.32	4.19	4.11	4.05	4.01	3.98	3.95	3.94	3.92
5	4.06	3.78	3.62	3.52	3.45	3.40	3.37	3.34	3.32	3.30
6	3.78	3.46	3.29	3.18	3.11	3.05	3.01	2.98	2.96	2.94
7	3.59	3.26	3.07	2.96	2.88	2.83	2.78	2.75	2.72	2.70
8	3.46	3.11	2.92	2.81	2.73	2.67	2.62	2.59	2.56	2.54
9	3.36	3.01	2.81	2.69	2.61	2.55	2.51	2.47	2.44	2.42
10	3.29	2.92	2.73	2.61	2.52	2.46	2.41	2.38	2.35	2.32
11	3.23	2.86	2.66	2.54	2.45	2.39	2.34	2.30	2.27	2.25
12	3.18	2.81	2.61	2.48	2.39	2.33	2.28	2.24	2.21	2.19
13	3.14	2.76	2.56	2.43	2.35	2.28	2.23	2.20	2.16	2.14
14	3.10	2.73	2.52	2.39	2.31	2.24	2.19	2.15	2.12	2.10
15	3.07	2.70	2.49	2.36	2.27	2.21	2.16	2.12	2.09	2.06
16	3.05	2.67	2.46	2.33	2.24	2.18	2.13	2.09	2.06	2.03
17	3.03	2.64	2.44	2.31	2.22	2.15	2.10	2.06	2.03	2.00
18	3.01	2.62	2.42	2.29	2.20	2.13	2.08	2.04	2.00	1.98
19	2.99	2.61	2.40	2.27	2.18	2.11	2.06	2.02	1.98	1.96
20	2.97	2.59	2.38	2.25	2.16	2.09	2.04	2.00	1.96	1.94
21	2.96	2.57	2.36	2.23	2.14	2.08	2.02	1.98	1.95	1.92
22	2.95	2.56	2.35	2.22	2.13	2.06	2.01	1.97	1.93	1.90
23	2.94	2.55	2.34	2.21	2.11	2.05	1.99	1.95	1.92	1.89
24	2.93	2.54	2.33	2.19	2.10	2.04	1.98	1.94	1.91	1.88
25	2.92	2.53	2.32	2.18	2.09	2.02	1.97	1.93	1.89	1.87
26	2.91	2.52	2.31	2.17	2.08	2.01	1.96	1.92	1.88	1.86
27	2.90	2.51	2.30	2.17	2.07	2.00	1.95	1.91	1.87	1.85
28	2.89	2.50	2.29	2.16	2.06	2.00	1.94	1.90	1.87	1.84
29	2.89	2.50	2.28	2.15	2.06	1.99	1.93	1.89	1.86	1.83
30	2.88	2.49	2.28	2.14	2.05	1.98	1.93	1.88	1.85	1.82
40	2.84	2.44	2.23	2.09	2.00	1.93	1.87	1.83	1.79	1.76
50	2.81	2.41	2.20	2.06	1.97	1.90	1.84	1.80	1.76	1.73
60	2.79	2.39	2.18	2.04	1.95	1.87	1.82	1.77	1.74	1.71
70	2.78	2.38	2.16	2.03	1.93	1.86	1.80	1.76	1.72	1.69
80	2.77	2.37	2.15	2.02	1.92	1.85	1.79	1.75	1.71	1.68
90	2.76	2.36	2.15	2.01	1.91	1.84	1.78	1.74	1.70	1.67
100	2.76	2.36	2.14	2.00	1.91	1.83	1.78	1.73	1.69	1.66
200	2.73	2.33	2.11	1.97	1.88	1.80	1.75	1.70	1.66	1.63
500	2.72	2.31	2.09	1.96	1.86	1.79	1.73	1.68	1.64	1.61
∞	2.71	2.30	2.08	1.94	1.85	1.77	1.72	1.67	1.63	1.60

(*continued*)

Table A.6 (*Continued*)

	15	**20**	**30**	**40**	**50**	**100**	**200**	**500**	**∞**
1	61.2	61.7	62.3	62.5	62.7	63.0	63.2	63.3	63.3
2	9.42	9.44	9.46	9.47	9.47	9.48	9.49	9.49	9.49
3	5.20	5.18	5.17	5.16	5.15	5.14	5.14	5.14	5.13
4	3.87	3.84	3.82	3.80	3.80	3.78	3.77	3.76	3.76
5	3.24	3.21	3.17	3.16	3.15	3.13	3.12	3.11	3.10
6	2.87	2.84	2.80	2.78	2.77	2.75	2.73	2.73	2.72
7	2.63	2.59	2.56	2.54	2.52	2.50	2.48	2.48	2.47
8	2.46	2.42	2.38	2.36	2.35	2.32	2.31	2.30	2.29
9	2.34	2.30	2.25	2.23	2.22	2.19	2.17	2.17	2.16
10	2.24	2.20	2.16	2.13	2.12	2.09	2.07	2.06	2.06
11	2.17	2.12	2.08	2.05	2.04	2.01	1.99	1.98	1.97
12	2.10	2.06	2.01	1.99	1.97	1.94	1.92	1.91	1.90
13	2.05	2.01	1.96	1.93	1.92	1.88	1.86	1.85	1.85
14	2.01	1.96	1.91	1.89	1.87	1.83	1.82	1.80	1.80
15	1.97	1.92	1.87	1.85	1.83	1.79	1.77	1.76	1.76
16	1.94	1.89	1.84	1.81	1.79	1.76	1.74	1.73	1.72
17	1.91	1.86	1.81	1.78	1.76	1.73	1.71	1.69	1.69
18	1.89	1.84	1.78	1.75	1.74	1.70	1.68	1.67	1.66
19	1.86	1.81	1.76	1.73	1.71	1.67	1.65	1.64	1.63
20	1.84	1.79	1.74	1.71	1.69	1.65	1.63	1.62	1.61
21	1.83	1.78	1.72	1.69	1.67	1.63	1.61	1.60	1.59
22	1.81	1.76	1.70	1.67	1.65	1.61	1.59	1.58	1.57
23	1.80	1.74	1.69	1.66	1.64	1.59	1.57	1.56	1.55
24	1.78	1.73	1.67	1.64	1.62	1.58	1.56	1.54	1.53
25	1.77	1.72	1.66	1.63	1.61	1.56	1.54	1.53	1.52
26	1.76	1.71	1.65	1.61	1.59	1.55	1.53	1.51	1.50
27	1.75	1.70	1.64	1.60	1.58	1.54	1.52	1.50	1.49
28	1.74	1.69	1.63	1.59	1.57	1.53	1.50	1.49	1.48
29	1.73	1.68	1.62	1.58	1.56	1.52	1.49	1.48	1.47
30	1.72	1.67	1.61	1.57	1.55	1.51	1.48	1.47	1.46
40	1.66	1.61	1.54	1.51	1.48	1.43	1.41	1.39	1.38
50	1.63	1.57	1.50	1.46	1.44	1.39	1.36	1.34	1.33
60	1.60	1.54	1.48	1.44	1.41	1.36	1.33	1.31	1.29
70	1.59	1.53	1.46	1.42	1.39	1.34	1.30	1.28	1.27
80	1.57	1.51	1.44	1.40	1.38	1.32	1.28	1.26	1.24
90	1.56	1.50	1.43	1.39	1.36	1.30	1.27	1.25	1.23
100	1.56	1.49	1.42	1.38	1.35	1.29	1.26	1.23	1.21
200	1.52	1.46	1.38	1.34	1.31	1.24	1.20	1.17	1.14
500	1.50	1.44	1.36	1.31	1.28	1.21	1.16	1.12	1.09
∞	1.49	1.42	1.34	1.30	1.26	1.18	1.13	1.08	1.00

of degrees of freedom of the numerator is read along the top row, and the number of degrees of freedom of the denominator is read down the leftmost column. For example, for 10 degrees of freedom in the numerator and 30 in the denominator, the probability that $F(\nu_1; \nu_2) \geq 2.16$ is 0.05.

A.3.5 Table of the Fisher-Snedecor distribution at the 0.10 significance level

Table A.6 shows the critical values of the Fisher–Snedecor distribution $F(\nu_1; \nu_2)$ at the 10% significance level, in other words the values f_α such that $\Pr(F(\nu_1; \nu_2) \geq f_\alpha) = 0.10$. The number of degrees of freedom of the numerator is read along the top row, and the number of degrees of freedom of the denominator is read down the leftmost column. For example, for 10 degrees of freedom in the numerator and 30 in the denominator, the probability that $F(\nu_1; \nu_2) \geq 1.82$ is 0.10.

Appendix B

Further reading

B.1. Statistics and data analysis

Alan Agresti: *An Introduction to Categorical Data Analysis*, Wiley, 2nd edn, 2007.
An excellent book which provides a survey of categorical (i.e. qualitative) data. Two chapters are added in the second edition on the subject of correlated categorical data, which occur in longitudinal studies with repeated measurements.

T.W. Anderson: *An Introduction to Multivariate Statistical Analysis*, Wiley-Interscience, 3rd edn, 2003.
A classic from 1958, updated and still very useful. Not the easiest read, but rigorous and very comprehensive, it covers clustering, factor analysis (PCA and MCA), classification, bootstrapping, etc.

Jean-Paul Benzécri: *Histoire et Préhistoire de l'Analyse des Données*, Dunod, new edn, 1982 (out of print).
A fascinating story, written in a sparkling style by a leading statistician who is also a thinker.

George Casella and Roger L. Berger: *Statistical Inference*, Duxbury Press, 2nd edn, 2001.
An excellent textbook, comprehensive and rigorous, for advanced students.

Christophe Croux, Jean-Jacques Droesbeke, Pierre-Louis Gonzalez, Christian Gourieroux, Gentiane Haesbroeck, Michel Lejeune, Gilbert Saporta, and Michel Tenenhaus: *Modèles Statistiques pour Données Qualitatives*, Éditions Technip, 2005.
The proceedings of a very interesting seminar organized by the Société Française de Statistique, on classification methods, logistic regression, the log-linear model, counting models, generalized linear models, and PLS regression, with applications in medicine and insurance. Worth reading to discover the state of the art on these subjects.

Data Mining and Statistics for Decision Making, First Edition. Stéphane Tufféry.

Bradley Efron and Robert J. Tibshirani: *An Introduction to the Bootstrap*, Chapman & Hall, 1994.

A reference work on the bootstrap method, clear and comprehensive (the first author devised the method).

Brigitte Escofier and Jérôme Pagès: *Analyses Factorielles Simples et Multiples*, Dunod, 4th edn, 2008.

A very comprehensive book, helpful for the reader who wishes to be able to read a factor analysis.

Stanton A. Glantz and Bryan K. Slinker: *Primer of Applied Regression and Analysis of Variance*, McGraw-Hill, 2000.

A very full description of linear regression and analysis of variance, with an introduction to the Cox semi-parametric survival models.

Joseph F. Hair, Bill Black, Barry Babin, Rolph E. Anderson, and Ronald L. Tatham: *Multivariate Data Analysis*, Prentice Hall, 6th edn, 2005.

A real doorstop of a book (more than 900 pages), but full of excellent and accessible practical information, without too many equations. It provides examples of each type of analysis (factor analysis, canonical analysis, MANOVA, multiple regression, discriminant analysis, conjoint analysis, structural equations, etc.), with SAS and SPSS syntax, and an interpretation of the outputs. It also covers data cleaning and missing values.

David J. Hand: *Information Generation: How Data Rule Our World*, Oneworld Publications, 2007.

An excellent survey of statistics and its place in the modern world, written in a clear and attractive style by a leading expert, of interest to statisticians and others.

David J. Hand: *Statistics: A Very Short Introduction*, Oxford University Press, 2008.

A very useful introduction to statistics, detailed and concise, covering the basics of the subject, from data collection to modelling and computing, via probability theory and inference.

David W. Hosmer and Stanley Lemeshow: *Applied Logistic Regression*, Wiley, 1989; 2nd edn, 2000.

A well-known work on logistic regression, with numerous examples from the field of biostatistics. Starting from a basic knowledge of the linear model, this book provides a remarkably clear explanation of the principles and applications of logistic regression.

Ludovic Lebart, Alain Morineau, and Marie Piron: *Statistique Exploratoire Multidimensionnelle: Visualisations et Inférences en Fouille de Données*, Dunod, 4th edn, 2006.

One of the best titles on 'French-style' data analysis (factor analysis), also covering developments in clustering, discriminant analysis, log-linear models, decision trees, neural networks, validation methods, etc.

Kanti V. Mardia, J. T. Kent, and J. M. Bibby: *Multivariate Analysis*, Academic Press, 1980.

This book is less recent and does not cover subjects such as the bootstrap, but it is still a standard work on multivariate analysis, noted for its clear and elegant presentation of the subject.

Peter McCullagh and John A. Nelder: *Generalized Linear Models*, Chapman & Hall, 2nd edn, 1989.

A fundamental treatise on generalized linear models. John Nelder was one of the inventors of this method. Rather a challenging read, but of excellent quality.

R. H. Myers, D. C. Montgomery, and G. C. Vining: *Generalized Linear Models with Applications in Engineering and the Sciences*, Wiley-Interscience, 2001.

A clear introduction to generalized linear models.

Patrick Naïm, Pierre-Henri Wuillemin, Philippe Leray, Olivier Pourret, and Anna Becker: *Réseaux Bayésiens*, Éditions Eyrolles, 3rd edn, 2007.

A reference work on Bayesian networks, written by specialists and practitioners in this field.

Jean-Pierre Nakache and Josiane Confais: *Statistique Explicative Appliquée*, Éditions Technip, 2003.

A recent book on the main predictive methods: linear discriminant analysis, logistic regression and decision trees. The theory is described in a concise and detailed way, followed by a variety of illuminating examples produced with the SAS and SPAD software.

Jean-Pierre Nakache and Josiane Confais: *Approche Pragmatique de la Classification: Arbres Hiérarchiques et Partitionnements*, Éditions Technip, 2004.

A recent and very comprehensive text which is to descriptive clustering methods as the previous book is to predictive methods – an excellent reference work in a style characterized by its thoroughness in the theoretical sections and an educational approach in the examples of application, with many references to the recent literature, Internet sites and the latest versions of software, mainly, but not exclusively, SAS and SPAD.

Olivier Pourret, Patrick Naïm, and Bruce Marcot: *Bayesian Networks: A Practical Guide to Applications*, John Wiley & Sons Ltd, 2008.

A general introduction to Bayesian networks, illustrated with 20 case studies in the fields of medicine, science, engineering, robotics, finance, risk, etc.

Gilbert Saporta: *Probabilités, Analyse des Données et Statistique*, Éditions Technip, 2nd edn, 2006.

This is the standard work (in French), one to be kept handy at all times, offering a precise and comprehensive treatment of the subject. It contains all the essentials of probability calculation, multidimensional data analysis (factor analysis, clustering) and statistics for decision making (tests, estimation, regression and discrimination).

Michel Tenenhaus: *La Régression PLS: Théorie et Pratique*, Éditions Technip, 1998.

Everything you need to know about PLS regression, used ever more widely in industry for manipulating a large number of strongly collinear independent variables.

Sylvie Thiria, Olivier Gascuel, Yves Lechevallier, and Stéphane Canu: *Statistique et Méthodes Neuronales*, Dunod, 1997.

A collection of technical papers providing a very thorough survey of neural networks (several papers on the multilayer perceptron), their application to problems of classification, prediction and clustering, and Vapnik's learning theory.

Stéphane Tufféry: *Étude de Cas en Statistique Décisionnelle*, Éditions Technip, 2009.

Based on a data set from the insurance sector, available on the publisher's website, this book applies the principles of statistics to a case study covering two classic problems, namely the construction of customer segmentation, and the creation of a propensity score for the purchase of a product. The resources of the SAS software, especially SAS/STAT, are used to show that rigour and efficiency can be combined.

Larry Wasserman: *All of Statistics: A Concise Course in Statistical Inference*, Springer, 2004. In just over 400 pages, this book provides an excellent overview of probability and statistics, without assuming a large amount of previous knowledge on the reader's part, and gives a concise and clear introduction to the main principles, including the most recent ones such as the bootstrap, support vector machines, Bayesian inference and Markov chain Monte Carlo methods.

B.2. Data mining and statistical learning

Michael J. A. Berry and Gordon Linoff: *Data Mining Techniques: for Marketing, Sales, and Customer Relationship Management*, John Wiley & Sons, 2nd edn, 2004.

A readable book on data mining, more useful for its examples than its technical content (it does not cover factor analysis, discriminant analysis, logistic regression, or their more recent developments).

Michael J. A. Berry and Gordon Linoff: *Mastering Data Mining: The Art and Science of Customer Relationship Management*, John Wiley & Sons, 2000.

The second book on data mining by the same authors, with 20 case studies.

L. Breiman, J.H. Friedman, R.A. Olshen, and C.J. Stone: *Classification and Regression Trees*, Wadsworth, 1984.

A basic work on decision trees, written by the inventors of CART.

Bertrand Clarke, Ernest Fokoue, and Hao Helen Zhang: *Principles and Theory for Data Mining and Machine Learning*, Springer, 2009.

A recent reference work on data mining, methods of selecting variables, clustering, regression, ensemble methods, etc., with numerous examples for which the R code is provided.

Richard O. Duda, Peter E. Hart, and David G. Stork: *Pattern Classification*, Wiley-Interscience, 2nd edn, 2000.

A new edition of the classic from 1973, also very well illustrated, accompanied by exercises, covering numerous techniques ranging from neural networks to Markov models, taking in mixture models for clustering, with interesting sections on the bias–variance dilemma, overfitting and ensemble methods.

Paolo Giudici and Silvia Figini: *Applied Data Mining: for Business and Industry*, Wiley, 2nd edn, 2009.

This book is aimed at a broad spectrum of readers interested in data mining, applied statistics, databases and econometrics, providing a simple description of data mining in the 150-page introductory section. The second part presents seven case studies, each ten pages long, in the following fields: web mining, market basket analysis, credit risk, lifetime value, etc. The question of software is considered and some of the case studies use R.

David J. Hand, Heikki Mannila, and Padhraic Smyth: *Principles of Data Mining*, MIT Press, 2001.

A very well-written reference work, by experienced teachers of the subject, with rather fewer details of the algorithms than the book by Hastie, Tibshirani and Friedman.

Trevor Hastie, Robert Tibshirani, and Jerome H. Friedman: *The Elements of Statistical Learning: Data Mining, Inference and Prediction*, 2nd edn, Springer, 2009.

A high-level work on the statistical aspects of data mining, written by renowned statisticians who have invented several of the major data mining techniques. Both comprehensive and thorough. A major work of reference. A further advantage of this book is that it can be read while using the R package called *ElemStatLearn* which contains functions and databases described in the text.

Simon Haykin: *Neural Networks and Learning Machines*, Prentice Hall, 3rd edn, 2008.

A very comprehensive work on neural networks, multilayer perceptrons, radial basis function networks, Kohonen maps (SOMs), support vector machines, etc.

Alan Julian Izenman: *Modern Multivariate Statistical Techniques: Regression, Classification, and Manifold Learning*, Springer, 2008.

A recent book covering similar subjects to those discussed by Hastie, Tibshirani and Friedman, but more accessible, suitable for students. It provides numerous examples and uses R, S-PLUS and Matlab.

Olivia Parr Rud: *Data Mining Cookbook*, Wiley, 2000.

A practical guide with useful advice, accompanied with numerous examples of modelling using SAS software. This is not a very recent book, so it does not cover the enhancements in the latest versions of SAS, the outputs are not particularly attractive and the programming is not always very sophisticated, but it is accessible and provides full details of the proposed solutions. There is a version with a CD-ROM of SAS code included.

Brian D. Ripley: *Pattern Recognition and Neural Networks*, Cambridge University Press, 2008.

A very good review of the state of the art and the theoretical bases of neural networks.

Vladimir N. Vapnik: *The Nature of Statistical Learning Theory*, Springer, 2nd edn, 1999.

A 'historic' survey of statistical learning theory by one of its main proponents, with a concise description of his contribution, and an introduction to support vector machines (and to support vector regression in the second edition). For readers interested in theory.

Vladimir N. Vapnik: *Statistical Learning Theory*, Wiley-Interscience, 1998.

A much longer and more demanding work than the previous one, dealing with the theoretical bases of the concepts of learning, VC dimension and structural risk minimization, and detailing the theory and practice of support vector machines.

Christopher Westphal and Teresa Blaxton: *Data Mining Solutions: Methods and Tools for Solving Real-World Problems*, John Wiley & Sons, 1998.

A useful reference work, especially for those interested in visual methods.

Ian H. Witten and Eibe Frank: *Data Mining: Practical Machine Learning Tools and Techniques*, Morgan Kaufmann, 2nd edn, 2005.

Much appreciated for its clarity and simplicity as well as its practical aspects.

Xindong Wu and Vipin Kumar: *The Top Ten Algorithms in Data Mining*, Chapman & Hall/ CRC, 2009.

A survey of the 'top 10' data mining algorithms. Chosen from the most important methods identified in December 2006 by the IEEE International Conference on Data Mining (ICDM),[1] they are as follows: C4.5, CART, *k*-means, support vector machines, Apriori, EM (expectation–maximization), PageRank, AdaBoost, *k* nearest neighbours and the naive Bayesian classifier. Each of these is described and illustrated with practical examples.

B.3. Text mining

Ludovic Lebart and André Salem: *Statistique Textuelle*, Dunod, 1994.
Lebart, L., Salem, A., Berry, L. (1998). *Exploring textual data*, Kluwer, Dordrecht.
An enthralling classic account.

B.4. Web mining

Mylène Bazsalicza and Patrick Naïm: *Data mining pour le Web*, Eyrolles, 2001.
A book with a strongly educational approach, which starts by discussing the basics of the Internet and data mining and goes on to describe the statistical processing of web data.

Michael J.A. Berry and Gordon S. Linoff: *Mining the Web*, John Wiley & Sons, 2002.
The latest title from the well-known consultants of Data Miners Inc., Boston.

B.5. R software

John M. Chambers: *Programming with R*, Springer, 2008.
Written by one of the creators of the S language (on which R language is based), this book is unquestionably the best resource for advanced programming in R. It starts with the basics and ends by giving the reader all the information he needs to create his own packages.

Pierre-André Cornillon: *Statistiques avec R*, Presses Universitaires de Rennes, 2008.
This very well-produced book consists of two parts. The first part is a general course on R, providing the basic principles, showing how data are manipulated and represented, and outlining basic programming in R. The second part gives examples of the main statistical analysis and modelling procedures, in the form of separate sections with several pages each. The book is accompanied by exercises with answers.

Michael J. Crawley: *Statistics: An Introduction Using R*, John Wiley & Sons Ltd, 2005.
A basic work, very comprehensive, with numerous examples.

Michael J. Crawley: *The R Book*, John Wiley & Sons Ltd, 2007.
More than 900 pages on R, covering every aspect, from the basics to the standard statistical tests, and then going on to more advanced models such as time series, survival analyses, generalized linear models and generalized additive models.

---

[1] There is an interesting article about the same 'top 10': Xindong Wu, Vipin Kumar, J. Ross Quinlan *et al.* (2008), 'Top 10 algorithms in data mining', *Knowledge and Information Systems*, 14(1), 1–37.

Robert A. Muenchen: *R for SAS and SPSS Users*, Springer, 2009.
A well-designed book, useful for those familiar with SAS and SPSS, as it compares the languages by creating a 'Rosetta Stone', with numerous programs written in all three languages. So it is also useful for anyone wishing to compare SAS and SPSS with each other. It also features a 'trilingual' glossary and further interesting information on the publisher's site. A new way of presenting R, which I can heartily recommend.

A full list of literature on R is available from: http://www.r-project.org/doc/bib/R-books.html

B.6. SAS software

Art Carpenter : *Carpenter's Complete Guide to the SAS Macro Language*, SAS Publishing, 2nd edition, 2004.
The reference book on the subject, written by a recognized expert.

Ron Cody : *Learning SAS by Example: A Programmer's Guide*, SAS Publishing, 2007.
It is a comprehensive book, clear and concise, whose level is higher than the The Little SAS Book. Some exercises are corrected.

Ron Cody : *SAS Functions by Example*, SAS Publishing, 2nd edition, 2010.
A comprehensive guide to SAS functions, including what is new in 9.2. For each function, it gives a brief description of its purpose, the syntax and clear examples with useful explanations.

Olivier Decourt: *Reporting avec SAS: Mettre en forme et diffuser vos résultats avec SAS 9 et SAS 9 BI*, Dunod, 2008.
A book on the graphic resources of SAS, describing SAS GRAPH, ODS and ODS GRAPHICS. A very useful source, because graphic functionality has not always been the strong point of SAS, and it still suffers from a poor reputation in this field, even though the latest versions can produce excellent results.

Olivier Decourt and Hélène Kontchou Kouomegni: *SAS: Maîtriser SAS Base et SAS Macro, SAS 9.2 et versions antérieures*, Dunod, 2nd edn, 2007.
An excellent presentation, clear and precise, of SAS Base (and its macro language), including the latest functionality.

Geoff Der, Brian S. Everitt : *A Handbook of Statistical Analyses using SAS*, Chapman and Hall/CRC, 3d edition, 2008.
A very good overview of what can be done with SAS in statistics, from descriptive statistics to survival analysis, through regression, analysis of variance, logistic regression, generalized linear model, longitudinal data analysis, factor analysis and cluster analysis. With accompanying exercises and examples of SAS macros.

Lora D. Delwiche and Susan J. Slaughter: *The Little SAS Book: A Primer*, SAS Publishing, 4th edn, 2008.
An accessible, comprehensive book on the resources of SAS BASE, including the recent functionality in Version 9, such as ODS GRAPHICS. Concise and supported with numerous examples, it is a pleasure to read. However, it lacks detail in the areas of the macro language, ODS and graphics (it does not cover GPLOT or GCHART).

Stephen McDaniel and Chris Hemedinger: *SAS for Dummies*, John Wiley & Sons Ltd, 2nd edn, 2010.
A basic reference.

Sébastien Ringuedé: *SAS Version 9.2: Introduction au décisionnel: méthode et maîtrise du langage*, Pearson Education, 2008.
In this serious and exhaustive book, Sébastien Ringuedé, an academic partner of SAS, provides the essential information (and more!) that will be needed to achieve 'SAS Base Programming for SAS®9' certification, with accompanying exercises (the solutions are available on the companion website).

B.7. IBM SPSS software

Arthur Griffith: *SPSS for Dummies*, John Wiley & Sons Ltd, 2010.
A basic reference.
Paul Kinnear and Colin Gray: *IBM SPSS Statistics 18 Made Simple*, Psychology Press, 2010.
A book that is accessible to novices, leading on to more complex matters such as statistical tests, experimental design, regression, discriminant analysis and factor analysis, all illustrated with examples.

Naresh Malhotra, Jean-Marc Décaudin, and Afifa Bouguerra: *Recherche et études Marketing avec SPSS*, Pearson Education, 2004.
A book with an accompanying CD-ROM, including many detailed examples. It provides an introduction to new techniques such as conjoint analysis and multidimensional positioning.

B.8. Websites

Modulad magazine, a mine of practical information (on software, events, etc.) and very interesting articles on statistics: www.modulad.fr/

The website of Philippe Besse, Professor at the University of Toulouse, providing very full coverage of statistics and data mining: http://www.math.univ-toulouse.fr/~besse/

The website of Gilbert Saporta, Professor at CNAM, with a very rich content including many study courses: http://cedric.cnam.fr/~saporta/

StatNotes Online Textbook – David Garson's online resource, with much well-presented information on all aspects of statistics and data analysis, with details of implementation in IBM SPSS Statistics: http://www2.chass.ncsu.edu/garson/pa765/statnote.htm

The StatSoft site for statistics and data mining: www.statsoft.com/textbook/stathome.html
The website for the book *The Elements of Statistical Learning* (Hastie, Tibshirani and Friedman), with further information, data, R packages, errata, etc.: http://www-stat.stanford.edu/~tibs/ElemStatLearn/

The website for the book *Introduction to Data Mining* (Pang-Ning Tan, Michael Steinbach and Vipin Kumar), Addison-Wesley, offering a wide range of resources (extracts, PowerPoint slides, etc.): http://www-users.cs.umn.edu/~kumar/dmbook/index.php

Re-sampling Methods in Statistical Modeling: a very good course devised by Professor Bontempi, of the Free University of Brussels, on predictive models and contribution of jackknife and bootstrap techniques, including ensemble methods such as bagging and boosting: www.ulb.ac.be/di/map/gbonte/Stat104.html

The very comprehensive on-line help for the SAS software: http://support.sas.com/documentation/onlinedoc/

Olivier Decourt's website, with a very interesting FAQ section on SAS, statistics and data mining, including many concise and well-written descriptions of various technical aspects: www.od-datamining.com/index.htm

Numerous resources on SPSS: www.spsstools.net/

The R software site: www.r-project.org/
The website of *Lexicometrica* magazine, where articles on text data mining can be downloaded: www.cavi.univ-paris3.fr/lexicometrica/index.htm

A course on web mining by Gregory Piatetsky-Shapiro: www.kdnuggets.com/web_mining_course/

Real data to illustrate statistical methods, sorted by method (University of Massachusetts): http://www.umass.edu/statdata/statdata/index.html

A very good glossary on statistics: http://dorakmt.tripod.com/mtd/glosstat.html

An eclectic blog written by Arthur Charpentier, Professor at the University of Rennes 1, about statistics, probability, actuarial science, econometrics, R, and more: http://blogperso.univ-rennes1.fr/arthur.charpentier/

Eric Weisstein's *World of Mathematics*, an on-line mathematical encyclopaedia with more than 11 000 entries and 5000 diagrams: http://mathworld.wolfram.com/

Index